Outstanding Contributions to Logic

Volume 13

More information about this series at http://www.springer.com/series/10033

Gerhard Jäger · Wilfried Sieg
Editors

Feferman on Foundations

Logic, Mathematics, Philosophy

Springer

Editors
Gerhard Jäger
Institut für Informatik
Universität Bern
Bern
Switzerland

Wilfried Sieg
Department of Philosophy
Carnegie Mellon University
Pittsburgh, PA
USA

ISSN 2211-2758 ISSN 2211-2766 (electronic)
Outstanding Contributions to Logic
ISBN 978-3-030-09691-5 ISBN 978-3-319-63334-3 (eBook)
DOI 10.1007/978-3-319-63334-3

Printed on acid-free paper

This Springer imprint is published by the registered company Springer International Publishing AG part of Springer Nature
The registered company address is: Gewerbestrasse 11, 6330 Cham, Switzerland

Foreword

Sol Feferman and I have quite similar backgrounds. We were both born in the Bronx in 1928 to East European Jewish parents. As teenagers we had similar interests: we both thought of theoretical physics as a likely professional goal, and we both read the same science fiction authors as well as the same popularizers James Jeans and Arthur Eddington. And we both moved from physics to mathematics and indeed to logic and foundations.

I was at the 5-week Institute for Logic at Cornell during the summer of 1957 that Sol wrote about in his autobiography. I do not remember whether I heard Sol's talk there about his dissertation on the arithmetization of metamathematics, but I certainly read the manuscript and was delighted at the clarity of the exposition, which, in particular, eliminated the penumbra of vagueness about the concept of a formula expressing the consistency of a formal system. Sol was part of what some of us thought of as Tarski's cohort. Tarski had been allocated an afternoon speaker slot every day for the remarkable Berkeley logic group he had developed. His habit of commenting aggressively on talks by non-Berkeley speakers did not always go over well, and there was some tension in the air. But all in all, it was an exciting, stimulating event and an important influential experience for all of us.

As Sol mentioned in his autobiography, during the academic year 1959–1960, he and I were regular attendees at the logic seminar at Princeton that Church led. I particularly enjoyed Sol's talks on progressions of theories. Sol's work was based on and extended Turing's *Ordinal Logic* paper which I had found quite difficult. I admired the clarity and rigor of Sol's exposition with his previous work on the arithmetization of metamathematics in the background.

Although I frequently met Sol at conferences over the years, my next professional interaction occurred in connection with the publication of Gödel's "Collected Works". Sol was the chief editor of this daunting project which resulted in five meticulously produced volumes with introductions by experts for each article. Sol asked me to write the introduction for a manuscript found in Gödel's Nachlass. The third volume consists of such previously unpublished work, following the first two that cover his published contributions. The fourth and fifth volumes are devoted to

Gödel's correspondence. The superb result of this undertaking is a tribute to the care and precision to be found in all of Sol's endeavors.

This same insistence on precision and rigor may not be unrelated to Sol's well-known agnosticism regarding the concepts of set theory. This skeptical attitude is perhaps most directly evident in his characterization of the continuum hypothesis (which after all only asks whether there is an uncountable set of real numbers which cannot be mapped one–one to the set of all real numbers) as "inherently vague".

I was on a panel with Sol on the future of logic in Padua in 1988. As I recall, I spoke of possible connections of Gödel incompleteness with mathematical practice. Sol took a much more practical point of view, speaking of problems one might suggest to a student with some reasonable hope for success. In fact, Sol has been quite openly skeptical about the need for axioms going beyond Zermelo–Fraenkel to decide important mathematical questions. Citing the success with Fermat's Last Theorem, he stressed the need to "try harder".

Sol was instrumental in making Stanford University a world center for proof theory. His autobiography mentions that Georg Kreisel, who had already done important work in this area, was a "second mentor" for him. Sol admired Hermann Weyl's predicative development of classical analysis and worked on extending it. His determining the proof-theoretic ordinal of the resulting system was a major achievement. Not completely satisfied because of the ramified character of the extended system, Sol found an equivalent unramified system. Despite his admiration for Weyl's system, Sol insisted that he was not a predicativist, and indeed he (necessarily) used non-predicative methods for obtaining the proof-theoretic ordinal of such systems.

Grigori Mints was an expert in Hilbert's ε-calculus and the ε-substitution method in proof theory. Educated in Leningrad, he applied to leave the Soviet Union because of the pervasive antisemitism that had become endemic in Russian mathematics. Subsequently denied employment, he moved to Estonia when the Soviet Union dissolved. Sol was determined to have him for Stanford. I was a very small part of this effort, writing a letter in support when Sol asked me to do so. Sol's effort was successful, and "Grisha" became a vibrant part of the Stanford logic community.

Sol's skepticism regarding set theory did not lead him to question the use of Grothendieck universes in Wiles's proof of Fermat's Last Theorem. Indeed, he considered even larger structures to provide a set-theoretic foundation for category theory and has expressed confidence in the consistency of the Zermelo–Fraenkel axioms based on the iterative concept of set. In Koellner's essay in this volume, he presents Sol with a challenge: can he really coherently hold, together with his set-theoretic skepticism, his acceptance of the natural number concept as sufficiently clear that every sentence of arithmetic can be said to have a definite truth value?

Alas Koellner's challenge will remain unanswered, and this volume is published with a huge lacuna because of Sol's death. As set out, the plan for this series would have had Sol provide his own comments on the various essays making up the book. But this is not to be. Instead of a full autobiography, we have but an initial fragment. And we will never know what Sol would have written to explain his set-theoretic views. I myself would love to know what he thought about projective determinacy and the picture of the projective set hierarchy that emerges from assuming it.

The depth and breadth of Sol's work are reflected in the variety of topics found in this volume. His influence over the course of his long career on directions taken in foundational investigations has been immense. He has had a number of outstanding students who have been making their own valuable contributions to the field. Sol's clear precise voice is deeply missed.

Berkeley, USA Martin Davis

Martin Davis was born in 1928. He studied with Emil Post at City College in New York and with Alonzo Church in Princeton. He is known for work on automated deduction and on Hilbert's Tenth Problem. His book "Computability & Unsolvability" has been called "one of the few real classics in computer science." Davis is a Professor Emeritus at New York University. He and his wife of 66 years now live near the campus of the University of California at Berkeley where he is a Visiting Scholar. Davis's book "The Universal Computer: The Road from Leibniz to Turing," intended for a general audience, is about to appear in a third updated edition.

Preface

On January 27, 2014, we received this e-mail from our teacher, mentor, colleague and friend Solomon Feferman:

> Dear Gerhard and Wilfried,
>
> I am forwarding a message below from Sven Ove Hansson proposing a volume devoted to my contributions to logic for a series that he is editing for Springer, along with my (tentatively) positive response. The attraction is obvious but, I must say, I have a couple of reservations about proceeding with this, first because the overall character of my work is somewhat different from that of the others that are already out or lined up for the series. (Though perhaps my addition would signal a shift in that direction to include some other obvious senior choices not mentioned by him.) The second reservation I have is that the series is being published by Springer, and I am afraid of it getting lost in their impersonal sea of volumes. I would be very interested to hear your thoughts about this.
>
> Hansson was fully in agreement with my suggestion to have two editors if we proceed with this, and of course both your names were the first that came to my mind. It would mean a great deal to me if you saw your way to accepting. I realize that that would mean for each of you taking on an extra burden that should not be considered at all lightly. But if you agreed, I would help in any way I could, for example by organizing my work into a number of useful categories and suggesting authors. I myself would be responsible for providing a full bibliography (no problem about that), an autobiography, and–later–responses to individual chapters, as Hansson suggested in a further message. (The whole organization of these volumes is reminiscent of the *Philosopher of the Century* series, the closest being the one for Hintikka.) Also, given my age, the sooner we could work together on this if you are willing to go ahead, the better.
>
> I hope to hear from you soon, but at the same time want you to take your time that it all deserves.
>
> Warmest best wishes,
> Sol

By the very beginning of February both of us "were in"—with great enthusiasm. We turned our attention to "substantive questions"; WS formulated three in a note to Sol on February 3.

> The first question seems to be: Is the volume to be systematically organized (according to major categories of your work)? The second crucial question: Who is going to (be invited to) contribute? A third very practical issue: What is the timeline for our work?

WS had forgotten one prior question, namely, what should the title be for the volume?—That was very important to Sol.

The three of us had quite a bit of e-mail exchange concerning the forgotten question and, in particular, also concerning the systematic structure of the volume that was indeed to be shaped by the major categories of Sol's work. In any event, on March 23 we sent a tentative table of contents to Sven Ove Hansson and suggested with Sol's full approval "Feferman on Foundations" as the title for our volume with the extension "Logic, Mathematics, Philosophy". The table of contents underwent changes in the subsequent weeks until Sol was really happy with it. The title, in contrast, had obviously been fixed forever. By the beginning of July, all the contractual matters were settled with Springer, and we wrote to potential contributors. It was wonderful for us and extremely gratifying for Sol that almost all potential contributors turned themselves very quickly into real ones. The self-imposed deadline for getting all the contributions was the end of March 2015; we hoped that we would complete the volume by the end of 2015. As usual, circumstances were in many different ways extremely challenging. On September 5, 2015, we wrote to Sven Ove answering his inquiry concerning the status of FoF:

> Dear Sven Ove,
>
> Thanks for your note. As to the title of the book, it is simply to be: "Feferman on Foundations". We have made progress (or rather, the contributors to the volume have made progress). We have some papers in hand, have requests by some to extend the deadline to the end of this calendar year, and don't know the status of some. We intend to write a brief note to our colleagues asking them about progress and likely completion date. When we have heard from everyone, we'll write to you again.

Here we are in December of 2016 more than a year later, expecting the final version of all the contributions to arrive within the next couple of days. We hope to submit the volume to Springer by the beginning of next year, i.e., January 2017 almost exactly 3 years after we received Sol's note asking us to serve as editors of a volume on his contributions to logic.[1]

In the PS to his note from January 2014, Sol mentioned that we would find as an attachment "a short autobiographical fragment" he had just finished for a volume in honor of Leon Henkin; that fragment was to be the starting point for the autobiography he had agreed to write for this volume. He had been working on it again at the beginning of 2016 and reached the mid-1980s, with quite a bit of his life's work still to be covered. The much-expanded autobiographical fragment was to remain, however, a fragment. After a difficult trip to New York in April, where he participated in a Workshop at Columbia to honor Charles Parsons, he was diagnosed as

---

[1] This note was obviously written in December of 2016. One year later, the volume is nearing completion and should be published at the very beginning of 2018.

having had a "mild stroke": Sol was hospitalized, underwent some rehabilitation, and finally returned home. His health deteriorated and he died on July 25. We are still mourning his death: the loss of a great logician, a thoughtful scholar, a man of integrity, and a dear friend. This volume is a testimony to him.

Bern, Switzerland Gerhard Jäger
Pittsburgh, USA Wilfried Sieg

Contents

Contributors

Wilfried Buchholz Mathematisches Institut der Universität München, München, Germany

Andrea Cantini Dipartimento di Lettere e Filosofia, Universitá di Firenze, Firenze, Italy

Laura Crosilla School of Philosophy, Religion and History of Science, University of Leeds, Leeds, UK

Martin Davis Berkeley, CA, USA

Ali Enayat Department of Philosophy, Linguistics, and Theory of Science, University of Gothenburg, Göteborg, Sweden

Solomon Feferman Stanford, CA, USA

Fernando Ferreira Faculdade de Ciências, Departamento de Matemática, Universidade de Lisboa, Lisboa, Portugal

Kentaro Fujimoto Department of Philosophy and School of Mathematics, University of Bristol, Bristol, England

Paul Gorbow Department of Philosophy, Linguistics, and Theory of Science, University of Gothenburg, Göteborg, Sweden

Volker Halbach Faculty of Philosophy, University of Oxford and New College, Oxford, England

Geoffrey Hellman Department of Philosophy, University of Minnesota, Minneapolis, MN, USA

Wilfrid Hodges Herons Brook, Sticklepath, Okehampton, England

Gerhard Jäger Institut für Informatik, Universität Bern, Bern, Switzerland

Peter Koellner Department of Philosophy, Harvard University, Cambridge, MA, USA

Zachiri McKenzie Department of Philosophy, Linguistics, and Theory of Science, University of Gothenburg, Göteborg, Sweden

Dag Normann Department of Mathematics, The University of Oslo, Oslo, Norway

Charles Parsons Department of Philosophy, Harvard University, Cambridge, MA, USA

Wolfram Pohlers Institute for Mathematical Logic and Foundational Research Universität Münster, Münster, Germany

Michael Rathjen Department of Pure Mathematics, University of Leeds, Leeds, UK

Helmut Schwichtenberg Mathematisches Institut der Universität München, München, Germany

Stewart Shapiro Department of Philosophy, The Ohio State University, Columbus, OH, USA

Wilfried Sieg Department of Philosophy, Carnegie Mellon University, Pittsburgh, PA, USA

Thomas Strahm Institut für Informatik, Universität Bern, Bern, Switzerland

William Tait Chicago, IL, USA

Albert Visser Philosophy, Faculty of Humanities, Utrecht University, Utrecht, The Netherlands

Stanley S. Wainer School of Mathematics, University of Leeds, Leeds, UK

Dag Westerståhl Department of Philosophy, Stockholm University, Stockholm, Sweden

Jeffery Zucker Department of Computing and Software, McMaster University, Hamilton, ON, Canada

Introduction: Solomon Feferman's Autobiography from 1928 to 1981 and Extensions

In 2014, Solomon Feferman began drafting an autobiography to be included in this volume. The draft built on two earlier biographical essays, namely, *A fortuitous year with Leon Henkin* and *Philosophy of mathematics, 5 questions*. Sol used as a title for the draft, *An Intellectual (mostly) Autobiography*. Indeed, it gives a detailed account of his intellectual development and his professional work, but it covers also key events in his personal life. The draft is quite polished, but Sol never completed it: at the time of his death, he had traced developments through the very early 1980s. This partial autobiography is presented first as *Part A*. The next part, *Part B*, contains his CV with important milestones, accomplishments, and honors; it also has a full list of his Ph.D. students.

The expanding range of topics for his research can be gleaned from his bibliography that he had prepared for this volume early on.[2] The latter indicates also the great number of his collaborators. Sol, of course, interacted with many colleagues in the departments of mathematics, philosophy, and computer science. During the time of Sol's work at Stanford, the university evolved into a world center for mathematical logic and the foundations of mathematics. The list of associated Stanford colleagues and of short and long-term visitors we are aware of is extremely impressive; but we decided not to construct it, as it would most likely be incomplete and, in addition, it would require judgments about the significance of interactions we are not confident in making.

[2]Indeed, he sent the bibliography to Jäger and Sieg on August 12, 2015 with this note: "Dear Gerhard and Wilfried, Attached is the latest version of my list of publications for FoF. Some items have been added and I have uniformized the format of the citations and expanded the publication information in various ways. I do not plan to do any more work on this until we are closer to the volume publication. So I think it could be useful to the authors of the individual chapters and would be happy to have you circulate it to them now." He added the remark: "The work on my autobiography is progressing slowly. I have only 60 years to go."

However, we did try to indicate the projects Sol was actively engaged in at the beginning of 2016; there was, first of all, his work on this very volume: the autobiography was to be completed during the summer of 2016; after that, Sol intended to respond to the individual contributions to *Feferman on Foundations* in the fall of the same year. In *Part C* we list the other projects Sol was pursuing during that period. (The photo below was taken by Sommer around 2004 on the balcony of the Library of the Mathematics Department.)[3]

Pittsbursh, USA Wilfried Sieg
Stanford, USA Rick Sommer

[3]We thank Julie Feferman Perez and Ivano Caponigro for their support and detailed, helpful information. We have not edited the text, except for correcting obvious mis-spellings and completing some references. Thus, it is the document as Sol left it.

Part A: An Intellectual (Mostly) Autobiography

Early years. I was born on December 13, 1928 in the Bronx to working class parents who had emigrated to the United States after WW I and had met and married in New York: my father, Leon Feferman, was from Omsk, Russia, and worked as a housepainter, while my mother, Helen Grand Feferman, came from Warsaw, Poland, and was a dressmaker. Neither had had any advanced education. They identified themselves as Jewish culturally but were not religious. Besides the English that they acquired in the U.S., their languages were Yiddish and Russian. We lived in a brick walkup of four or five storeys not far from the Bronx Zoo. I played in the streets but preferred reading, and in school was good at Arithmetic and Spelling.

When I was 9 years old, in 1938, my family moved to Los Angeles in the hopes of a better life and a (then) more salubrious climate. At the outset, we lived in Boyle Heights in East Los Angeles, at that time a Jewish enclave (that turned into a Latino enclave in later years). I finished sixth grade there at the age of ten (I had been "skipped" a couple of times), and started middle school in Boyle Heights. I was perhaps first attracted to science through serials such as "Buck Rogers in the 21st Century" at the Saturday afternoon movie matinees. I wondered if I would ever live to see the 21st century, since I calculated that I would have to still be alive at age 72 in order to achieve that. The 1939 World's Fair in New York fascinated me from a distance since I never got to go. But at that remove I thrilled to the iconic Trylon and Perisphere structures and the marvels of the Futurama exhibit, the first television set, and the Time Capsule that was to be opened 5000 years hence. On top of all that, Albert Einstein gave a lecture on cosmic rays, and Superman (so they said) made an appearance in person.

In 1940, we moved to Hollywood, not at all in the glamorous part, but modestly comfortable: that was to be our home for many years. On December 7, 1941, our middle school principal called a general assembly to tell us that the Japanese had bombed Pearl Harbor; that meant we would enter WW II. Six days later my parents had a party to celebrate my 13th birthday. I met my wife to be, Anita Burdman, for the first time at that party, but she hardly spoke to me there since she was a year older and I was just a "kid" in her eyes. So she hung around with my sister (4 years my senior) and her friends, while I hid in the back with mine.

In those years, I started reading science fiction, going back to Jules Verne and H. G. Wells but then moving on to the stories in the then current pulps such as *Amazing Science Fiction* and *Astounding Science Fiction* and authors such as Robert Heinlein and Isaac Asimov. I would also take a streetcar downtown to the Main Library where I could venture into the "adult" section for detective fiction and such topical authors as Sinclair Lewis and John Dos Passos. My most ambitious reading was Thomas Mann's *The Magic Mountain*. In general, I was puzzled by the adult relationships in these novels.

At Hollywood High School, which I entered in 1942 at age 13, I excelled in Mathematics, English, History, and Art, and did well in Physics, Chemistry, and

German. My Algebra teacher was primarily a gymnastics coach and did not know what the Calculus was; the most advanced mathematics one could take at Hollywood following Algebra and Plane Geometry and Trigonometry was "Solid Geometry". Somewhere along the line I somehow decided that I was going to be a theoretical physicist and a professor, and I read popular books about relativity theory and quantum mechanics by such authors as James Jeans and A.S. Eddington. I also read philosophy by Bertrand Russell and John Dewey, and for a few years took seriously the so-called General Semantics movement of Alfred Korzybski, whose book consisted of a mish-mash of type theory, non-Aristotelian logic, and colloidal chemistry. Then I somehow discovered Rudolf Carnap's *Der logische Aufbau der Welt* and carried that around with me to show off but could not really penetrate it, though it had some sort of magical hold on me.

College years; CalTech. The high school schedule was accelerated due to the war and the necessity of some of the male students to finish before being drafted or joining up with the military. By taking summer classes, I graduated in the middle of the school year 1944–1945 soon after I had turned 16, and was ready to go to college. The only choice I considered was between UCLA and CalTech (California Institute of Technology); the former was free and co-educational and was where most of my friends were going or had already gone, while the latter was relatively expensive and (then) for men only but was the place to study physics and science more generally. I did well on an entry exam at CalTech, was accepted to begin in early 1945, and was offered a partial fellowship that I could supplement with waiter work in the Athenaeum, the faculty club. But it was still a financial burden for my family, gladly undertaken out of pride for my being a student there.

In my first semester that spring at CalTech, I took Calculus and loved it, and Physics—which was mostly about systems of pulleys that was as far from the romance of relativity and quantum theory as one could get—and did not love it. The students were a mix of naïve youngsters like me and returning veterans who knew the ways of the world and women. The war was still going on though winding down in Europe; in April 1945 we were shocked when Franklin Delano Roosevelt died; he had been our rock through all of the depression years and the war. The Vice President and Missourian, Harry Truman, became President, and we did not know how he could possibly fill Roosevelt's shoes. Within a month, Germany surrendered, and in early August, Truman ordered atomic bombs to be dropped on Hiroshima and Nagasaki, thus horrifically ending the war with Japan. Within a year, as the Communists took over Eastern Europe, the Cold War was underway.

Coming out of a bookstore near Hollywood and Vine late in the summer of 1945 I ran into Anita Burdman. I had just finished the summer session at CalTech, thus completing my first year there. Anita and I had not had any contact in high school, except to say "Hi" in passing on the way to classes. I thought she was beautiful and smart and so appealing, and would have liked to date her, but she was just leaving for studies at U.C. Berkeley, having already finished a year at UCLA. But we were able to see each other when she returned home for Christmas vacation; we started dating then and in the later summer vacations, the relationship began to become serious.

Meanwhile, at CalTech I continued to enjoy doing mathematics in such applicable courses as differential equations and vector analysis, but physics itself was turning out to be a disappointment. I found that I did not have the requisite physical intuitions, and the mathematics involved in the physics courses was make do, not treated as a subject for its own interest. A high point was a course on general relativity by Linus Pauling that attracted a large audience of both undergraduates and graduate students, and was pitched way over my head. Meanwhile, I had decided by default to switch my major to mathematics. But the upper division courses in that turned out to be somewhat of a culture shock; it was no longer techniques to master and problems to solve, but now abstract concepts ("group", "linear space", "topology") to understand and proofs to follow. My place in that subject did not begin to open up until I took a course on logic by Eric Temple Bell, the number theorist and popular historian of mathematics (and author of science fiction novels under the pseudonym, John Taine). The course was a hodgepodge because Bell did not really understand the modern logic (I learned later that he was a fan of Lukasiewicz' three-valued logic). While the material was of great appeal, I did not then see where it would take me. I thought my greatest personal achievement while in college was reading James Joyce's *Ulysses*. My greatest impact may have been through a job I had one summer collecting air samples on top of the roof of a seven-storey downtown Los Angeles building for a study of smog. That had already begun to be a serious air pollution problem and the chemical analysis of the samples (carried out by others) showed that it was primarily due to the unfiltered contents of automobile exhausts.

Having decided on an academic career, at the end of my undergraduate studies at CalTech in 1948, I applied for graduate work in mathematics at the University of Chicago and U.C. Berkeley. Accepted at both, it did not take much for me to decide which to choose, since I was offered a teaching assistantship at Berkeley and since Anita Burdman was still there. Within 4 months of my arrival, we were married, 4 days shy of my 20th birthday. Having already finished at UC, she had a job at a psychiatric institute in San Francisco as an assistant teacher on the pediatric ward. After a year of that demanding work she decided to return to school and get a credential as an elementary school teacher.

Graduate studies, pursuit, and interruption. I spent my first year at Berkeley, 1948–1949, taking the required mathematics courses for the Ph.D. program in real and complex variables, and in modern algebra. Teaching assistant duties consisted in holding office hours and helping grade papers and exams for undergraduate courses offered by faculty in the department. But in the summers I received extra income teaching basic algebra and calculus; I found that I was good at it and enjoyed teaching a lot. The Math graduate students, irrespective of level, were housed in the wooden "T" buildings in the North end of campus; those structures had been put up during the war and were supposedly temporary, but lasted for decades after. One of the students I became friendly with there was Frederick B. Thompson, who was working on a Ph.D. thesis with Alfred Tarski. He raved about

Tarski and encouraged me to take his year-long course on metamathematics, which I proceeded to do in my second year, 1949–1950.

As I have related elsewhere, when I did do so I knew immediately that this was to be my subject and Tarski would be my professor. He explained everything with such passion and, at the same time, with such amazing precision and clarity, spelling out the details with obvious pleasure and excitement as if they were as new to him as they were to us. He wrote on the blackboard with so much force that the chalk literally exploded in his hand, but step by step a coherent picture emerged. Methodically yet magically, he conveyed a feeling of suspense, a drama that managed somehow to leave us with a question hanging in the air at the end of the hours.[4]

That same year I started taking part in Tarski's logic seminar that was attended by novices like me along with doctoral students in various stages of progress; besides Fred Thompson, these included Julia Robinson, who was close to finishing, Wanda Szmielew, also well advanced, and Anne Davis Morel. Within a year we were joined by Robert Vaught and Chen-Chung Chang. And not long after that, coming up rapidly from behind, were the bright as can be youngsters, Richard Montague and Dana Scott. In that first seminar, I impressed by making a good presentation of some simple results about Boolean algebras. Such presentations were a trial by fire, especially for those who could never state things with the clarity, exactitude, and adherence to his notation that Tarski demanded; they would be endlessly interrupted by him and forced to go through things until they got them right, if they could manage that at all. For some reason, I never had difficulties of that sort and could see that Tarski looked favorably on me as a result. But it took me most of the year to work up the courage to ask him to be my dissertation advisor; to my relief, he readily agreed.

However, in order to be advanced to candidacy for a Ph.D. in mathematics, I had first to pass an unusual and demanding qualifying exam. The setup was that the chair of the department, Griffith C. Evans, would assign a topic in the research literature far distant from one's own expertise and direction of interests; the material in question was to be studied on one's own and then presented to a committee in an oral examination, for which no time limit was set. The topic Evans chose for me was "Asymptotic eigenvalues of vibrating membranes". It took me much of a year to master to my satisfaction the substantial underlying material in partial differential and integral equations and Tauberian theorems and their specific applications to the given problem; all the while, Tarski kept urging me to get done with that work and move on to logic. In the event, in the spring of 1951 I impressed the examining committee of mathematical analysts with my command of the material and clarity of presentation and was duly advanced to candidacy.

[4] Anita B. Feferman and Solomon Feferman, *Alfred Tarski. Life and Logic*, Cambridge University Press 2004, p. 171. In recent years I found by looking at the detailed notes that I took for the course in question that it was quite slow going. A painful amount of time—close to a full term—was spent on developing an elementary theory of concatenation as a basis for syntax. Nowadays, if one gave a nod to that material at all, it would be done in a week.

In the meantime, I was attending Tarski's graduate course in Set Theory, another essential branch of mathematical logic, and continued to take part in his seminars. Much time that year and next was spent on his then primary interests in algebraic logic via so-called cylindric algebras and in model theory. From these, he proposed two possible thesis problems for me: the first was to obtain a representation theorem for locally finite cylindric algebras, and the second was to obtain a decision procedure for the first-order theory of ordinals under addition. The former would provide the completeness of the axioms for cylindric algebras as an algebraic analogue of the completeness of first-order logic, while the latter would be an extension of a decision procedure that Tarski and his former student Andrzej Mostowski had established some years earlier for the first-order theory of the ordering of the ordinals.

In the years 1951–1953, I acted both as Tarski's Course Assistant for his graduate courses in Metamathematics and in Set Theory, and as his Research Assistant. The former relieved me of routine teaching assistant responsibilities and at the same time allowed me to gain a greater command of the material, which were their subjects. The latter often involved working with him—as was his wont—into the wee hours in his smoke-filled study at his home, with no concern for one's stamina or one's personal life outside of his demands. One frequent task was to help put his articles in final form in preparation for the typist and thence for publication, going over all the details and advising as to the choice of words in English, since Polish was his native tongue. But his own English was excellent (spoken accent aside) though rather "correct", and most often, after considerable discussion, he would choose his own word in preference to my suggestions.

A more substantial long-term task assigned to me was to reformulate in terms of Tarski's theory of arithmetical classes the work of Wanda Szmielew's thesis providing a decision procedure for the first-order theory of Abelian groups that had been obtained via the syntactic method of elimination of quantifiers. As I engaged in that I began to come to the conclusion that the aim in question was a greatly misguided attempt to put Szmielew's procedure in so-called ordinary mathematical terms. The background to that goal was Tarski's constant efforts to try to interest mainline mathematicians in the work of logic, and especially in that of his school. He thought (perhaps rightly so) that an obstacle to their appreciation of such was in the constant use by logicians of the notions of formal symbolic languages and in particular in the notions of formulas and sentences for such languages. But what the theory of arithmetical classes did was to provide model-theoretic surrogates for those notions while kicking away the traces of the formal languages that dictated their choice. In principle, translating Szmielew's work into the language of arithmetical classes should have been routine, painful as that might be. What it would not do, and what puzzled me in the whole enterprise, was that it would provide no illumination for why her procedure worked in the first place. I thought there should be some underlying explanation for that from the algebraic facts about finite and infinite Abelian groups as looked at in model-theoretic terms, and spent a lot of time trying to see what that would look like without really doing what I was assigned to do. In the end, Tarski was extremely annoyed with me (justifiably so) and had

Szmielew herself carry that through. But as later work showed, my instincts (the first where I was thinking for myself) were sound: Abraham Robinson found simple model-theoretic necessary and sufficient conditions for the eliminability of quantifiers in general, and then Paul Eklof and Edward Fischer established those conditions for the theory of Abelian groups by making substantial use of the known mathematical facts about the structure of countable Abelian groups.

Meanwhile, I was supposed to be doing my own research toward a dissertation and the first thing I tackled was a representation theorem for locally finite cylindric algebras. (The pursuit of cylindric algebras was another attempt to recast logical notions in "ordinary" mathematical terms.) I turned first to reexamination of proofs of the completeness theorem for first-order logic of which the representation theorem would be an analogue. The simplest such proof was that provided by Leon Henkin in a modified form due to Gisbert Hasenjaeger; I found that the ideas for that proof could be converted into prima facie algebraic terms and lead to the desired result. However, Tarski thought that my proof was not algebraic enough and pushed me to improve on it. I did not see how that could be done, and left it at that. Years later, I learned that what I had found (but never published) was the eventual "standard" proof of the representation theorem in question provided by Henkin himself.

The second thesis problem on which I worked and that Tarski had proposed was to provide a decision procedure for the theory of ordinals under addition. In that case, building on Mostowski's work on powers of theories as a means to reduce the decidability of the theory of natural numbers under multiplication to that of the numbers under addition ("Presburger arithmetic"), I introduced a notion of generalized powers of theories that could be applied to my problem. What I ended up showing by those means was that the decision problem in question could be reduced to the decision problem for the weak second-order theory of ordinals under the less-than relation, but I did not succeed in establishing the latter itself. Still I thought that the combination of that with the representation theorem would be satisfactory for a thesis, but Tarski refused to accept it.

Meanwhile, the Los Angeles draft board was breathing down my neck and wanting to know why I was taking so long with my graduate studies. The draft had continued from WW II through the Korean War and I had received regular deferments all along as a graduate student. The board thought that 5 years should be enough for a Ph.D. and could not be persuaded that I should be deferred any longer. Thus it was that I was drafted into the US Army beginning in September 1953 and my graduate work was suspended.

The army years and completion of graduate studies. While up to then I had always been the youngest in my group, in basic training at Fort Ord, California, I found myself surrounded by 18- and 19-year olds who regarded me at age 24 as an old man. Physically, too, years of sitting at a desk had exacted their toll, and it took a while to toughen up and manage the long marches and runs with rifles. On the firing range, I was lucky to hit the target. Fortunately, the fighting in Korea had ended with an armistice in July 1953, and the prospects of being sent into battle while in the army were considerably lessened. At the end of the 3 months in basic training, I was assigned to the Signal Corps in Fort Monmouth, New Jersey and my wife and I drove there in December 1953. On top of everything else, we had learned that she was pregnant on the eve of my being drafted. At Fort Monmouth, married soldiers could live off base, and we found a small house that was just right for us and our child to be; our first daughter was born there in May 1954.

Thanks to my mathematical background, I was assigned to a research group where we mainly spent the time calculating "kill" probabilities of Nike missile batteries around New York and Washington against possible incoming missile attacks. The fact that these were never even close to 100% had to give one pause. My fellow workers in the research office were mostly draftees like me who were closer to me in age, having been student deferees too, but we had a civilian boss. There was not much real work to do and there was a lot of time for casual conversation or to read William Feller's book on probability theory. But any sort of political discussion was highly discouraged, since Senator Joe McCarthy and his group had come through Ft. Monmouth earlier in the fall as part of his witch hunt to unearth communists under every possible rock, and that had left a residue of fear.

My research responsibilities did not exclude me from being assigned KP ("Kitchen Police") or night guard duty from time to time. And at home, finances were more than tight and there was much to do to help out with our new baby. Still I managed to keep my logical studies alive (when sleep deprivation and breathing space allowed) by reading Kleene's *Introduction to Metamathematics* (1952) in order to get a better understanding of recursion theory and Gödel's theorems than I had obtained in my Berkeley courses. As it happened, out of the blue one day when I was well advanced in those studies, I received a postcard from Alonzo Church asking if I would review for *The Journal of Symbolic Logic* an article by Hao Wang (1951) on the arithmetization of the completeness theorem for the classical first-order predicate calculus. I do not know what led Church to me, since we had had no previous contact, and I was not known for expertise in that area; perhaps my name had been recommended to him by Dana Scott who had left Berkeley in order to study with Church in Princeton, after a breakdown of his relations with Tarski due to the dereliction of his duties as Research Assistant. Quite fortuitously, my work on that review led me directly down the path to my dissertation.

The completeness theorem for the first-order predicate calculus is a simple consequence of the statement that if a sentence of that language is logically consistent then it has a model, and in fact a countable one. Actually, Gödel had shown that this holds for any set of sentences T. A theorem due to Paul Bernays in Hilbert and Bernays (1939) tells us that any first-order sentence can be formally modeled in the natural numbers if one adjoins the statement of its consistency to PA, the Peano Axioms; Wang generalized this to the statement that if T is any recursive set of sentences, then T is interpretable in PA augmented by a sentence Con_T that expresses in arithmetic the consistency of T. Wang's somewhat sketchy proof more or less followed the lines of Gödel's original proof of the completeness theorem. In my review, I noted that his argument could be simplified considerably by following the Henkin–Hasenjaeger proof instead, by then much preferred in expositions. But in addition I criticized Wang's statement on the grounds that it contained an essential ambiguity. Namely, there is no canonical number-theoretical statement Con_T expressing the consistency of an infinite recursive set of sentences T, since there are infinitely many ways in which membership in T (or more precisely, the set of Gödel numbers of sentences in T) can be defined in arithmetic, and the associated statements of consistency of T need not be equivalent. So that led me to ask what conditions should be placed on the way that a formula of PA defines membership in T in order to obtain a precise version of Wang's theorem. Moreover, the same question could be raised about formulations of Gödel's second incompleteness theorem for arbitrary recursive theories T.

By the time I was released from the army in September 1955 and returned to Berkeley to complete my doctoral studies, I had decided to devote myself to the precise study of formal consistency statements and the arithmetization of metamathematics in some generality, including both the completeness theorem and the incompleteness theorems. It happened that Tarski was on sabbatical in Europe that year, and he asked Leon Henkin to take over as acting advisor in his absence. Though Henkin's own major interests were in model theory and algebraic logic, he offered me a willing ear and a great deal of encouragement, and with the prod of weekly meetings, I soon made significant progress. It was thus that when Tarski returned from his sabbatical in May 1956, I had a body of work that I was sure was thesis worthy, and presented it to him as such. This included generalizations of both Gödel's incompleteness theorems and the Bernays–Wang completeness theorems for which I showed that there was an essential distinction between the two in terms of the conditions to be imposed on the formula used to express membership and thence provability in a system. In addition, it opened up in novel ways the study of the interpretability relation between theories, a relation that had been of particular interest to Tarski. But to my dismay, instead of pronouncing it "excellent!" he hemmed and hawed. Perhaps, he was irked that the subject was not the original one *he* had suggested and not in any of his own main directions of research; instead, it sharpened and extended considerably the method of arithmetization that Gödel had introduced in 1931 to prove his incompleteness theorems. Perhaps, the old rivalry Tarski felt with Gödel over those theorems was awakened. In any event, he decided not to decide on his own whether the work was sufficiently important and instead asked me to send a summary of the results to Andrzej Mostowski in Poland. This took more time and created more tension. To my relief, Mostowski found the results new and interesting and strongly encouraged Tarski's approval. Mostowski's intervention was decisive, and so Tarski agreed at last to accept the results of my research for the dissertation. But dotting all the i's and crossing all the t's took another year before conferral of the Ph.D., by which time I was installed as an instructor at Stanford University.

Stanford; the early years. To cap off my fortuitous year at Berkeley with Leon Henkin, I learned from him of an opening for an instructorship at Stanford to teach logic and mathematics; the information came from Patrick Suppes of the Philosophy Department at Stanford. The subject of logic was based there in that department, since mathematics was a bastion of classical analysis in those days.[5] After a personal visit to meet Suppes, an appointment with a joint position in mathematics and philosophy was made, and I came to Stanford in 1956. Except for leaves of absence of one sort or another since then, that was to become my permanent academic home. Our first personal home there was a tract house of modern open design that we purchased in South Palo Alto. Along with all our possessions, we arrived with our two daughters in tow; the second one had been born in July 1956.

Activity at Stanford in the area of logic was initially greatly spurred by the efforts of Suppes, who had come to the Philosophy Department in 1951; he was joined at that time by J.C.C. McKinsey with whom he collaborated on the axiomatic foundations of physics until the latter's tragic death (a probable suicide) in 1953. McKinsey had earlier done important joint work with Alfred Tarski, and it was Tarski who suggested to Suppes that he be invited to teach logic at Stanford. After his death, McKinsey was succeeded in the Philosophy Department by Robert McNaughton, who later became famous for his contributions to automata theory and computer science. But then Suppes, no doubt inspired by the example and influence of Tarski, aimed to establish logic as an active interdepartmental subject at Stanford. Through his great powers of persuasion and with the help of Halsey Royden in mathematics, Suppes worked to bring about faculty appointments in logic at the junior and senior levels many of them jointly in the philosophy and mathematics departments. Thus a couple of years after I came to Stanford, Georg Kreisel began spending part of each year as Visiting Professor; his appointment was made permanent in 1964. Other additions to the faculty were those of John Myhill in the early 60s, then Dana Scott, and later Harvey Friedman, while Bill Tait and Jaako Hintikka were brought into the Philosophy Departments. The period of the 60s saw, too, the beginning of a steady stream of visitors and the production of first-class Ph.D. students.

Of particular significance to me in this stimulating group of colleagues was Georg Kreisel, who was to become my second and more lasting mentor in logic. I had first met him during the period in early 1956 when I was well into the research for my hoped-for dissertation; Kreisel happened to be visiting Berkeley for a month or so at that time and Dana Scott had told him to look me up. Our initial personal contact clicked wonderfully for me: I had hardly to begin explaining what I had done and what I was in the process of working on to see that Kreisel understood immediately and that it related to things he had thought about and to a whole body of literature in which he was completely at home. His positive reception of my ideas confirmed my views of the significance of what I was up to, and added to, my determination to make this work my thesis, despite Tarski's reservations. In addition to the active encouragement and regular monitoring of the work given to me by Henkin, the boost provided by Kreisel's quick appreciation was psychologically crucial at that agonizing time. Furthermore, Kreisel opened up a new world to me

[5] Its most distinguished faculty members were George Pólya and Gabor Szegö.

through his interests in constructivity, predicativity, and proof theory, interests to which I was naturally attracted and that would come to dominate my own subsequent work.

It was not only in subject matter that Kreisel differed from Tarski. In personality, he was courtly and charming with a quick wit, sometimes sly and sometimes devastating. In his technical work and expositions, he was much more concerned to explain, at length, the significance of the work than to set it out in an organized step-by-step fashion. His attitude seemed to be that if one has the right ideas the details would look after themselves. And they did amazingly often; details bored him, and if necessary, he could rely on more disciplined collaborators to supply them or, if necessary, patch things up. He was also very quick to take in others' ideas and proofs, as well as to anticipate trouble spots. Under Kreisel's influence and thorough critiques, I learned to write papers with the main aims up front instead of plunging into the details of formalism, but I never gave up the Tarskian concern for clarity and precision in the statement of results and the spelling out of proofs.

My teaching at Stanford was at first largely in mathematics—mostly in the lower division calculus sequences—while my logic teaching in Philosophy was of some elementary introductions to logic. I greatly enjoyed lecturing and being entirely responsible for my own courses. In later years, I expanded the material I taught in mathematics to various upper division undergraduate courses, including differential equations, linear algebra, algebra, number theory, history of mathematics, and foundations of analysis. My notes for this last led me to my first book, *The Number Systems* (1964). And in Philosophy, I progressed to teaching upper division and graduate courses in logic and set theory. In 1961–1962, I gave a graduate course in metamathematics that covered model theory, recursion theory, and proof theory over three-quarters;[6] my notes for that were bound in the Mathematics Library as *Lecture Notes in Metamathematics* (1962), but never published. All of this teaching was invaluable in deepening my understanding of various areas of mathematics and logic. In subsequent years, we replaced the year-long metamathematics course by courses of two- or three-quarters devoted separately to model theory, recursion theory, proof theory, and set theory, in alternating years. In still later years, I divided my time equally between mathematics and philosophy, and added both Philosophy of Mathematics and Theories of Truth to my regular offerings.

In my first year 1956–1957 at Stanford, besides finishing up my thesis for the Ph.D. at Berkeley, I began thinking about the question of what one could obtain by transfinitely iterating the process of adjunction of consistency statements to a theory in order to overcome Gödelian incompleteness. I learned that this had been considered by Alan Turing in his dissertation with Alonzo Church at Princeton University in 1939 under the rubric of "ordinal logics," and I set out to study what he had accomplished. This was slow going for me as the presentation was couched

[6] Stanford was then and is to this date on the quarter system rather than the semester system. Quarters consisted of 10 weeks (compared to 15-week semesters), and one could teach courses lasting one, two- or three- quarters.

in the language of Church's lambda calculus, with which I had had no experience and initially found very obscure.

Another thing that was begun that year was work on a monograph in collaboration with Richard Montague on the method of arithmetization and some of its applications. The idea was, essentially, to provide a combined presentation of the results of both our doctoral theses. Montague's thesis work concerned non-finitizability results in axiomatic set theory via relative consistency proofs, making use to some extent of my precise treatment of proof predicates and consistency statements. Montague left Berkeley in early 1955 for a position at UCLA, where, like me, he continued working on the exposition of his results to meet what he took to be Tarski's exacting standards (in the end, I think they out-Tarskied Tarski). Also like me, his Ph.D. was not awarded until 1957. Over the following years, we invested considerable effort in the preparation of the joint monograph, but there were many partial drafts, and because of the technology of those days (handwritten MSS turned into typescript by secretaries) and because of the distance between us, progress was slow. At a certain point we both realized that we should publish the results of our respective theses as separate articles; mine appeared in 1960 (cf. [4][7]) and Montague's in 1961. (We had already each presented our main results to a broad audience of logicians at the 1957 Cornell Summer Institute to be described below.) But in subsequent years our paths steadily diverged and our thoughts and energies became largely directed elsewhere. Even so, since an agreement had been made early on with the North-Holland Publishing Co. for publication of the monograph, we continued to work on it sporadically, frequently prodding each other to take the next step. But even before Montague's awful murder in 1971, I had ceased to have any heart for the project. Moreover, research by others in the meantime had overtaken us and would have had to be incorporated in some way in order to remain up to date. In particular, Montague's dissertation work was pretty much superseded by a paper of Kreisel and Lévy in 1968 on applications of partial truth definitions to non-axiomatizability of various systems of arithmetic and set theory by statements of bounded complexity.

Starting in this same period I also worked with Bob Vaught on a paper on generalized products of structures. We had first begun talking of a collaboration in 1955, when I explained to him my work on generalized powers of structures that I had introduced in my (only partially successful) attack on the decision problem for ordinal addition, reducing that to the decision problem for the Boolean algebra of sets of ordinals with the less-than relation. Vaught's thesis had provided a proof of Los' conjecture that if a theory is closed under arbitrary finite ordinary (Cartesian) products of its models, then it is closed under arbitrary products of its models. Vaught recognized that my work and his could be combined to reduce properties of generalized products of structures to properties of the factors on the one hand and a certain Boolean algebra of subsets of the index set for the product on the other hand.

[7] This and all the other references in square brackets refer to Feferman's bibliography in this volume, starting on p. lxix. (WS and RS).

But preparation of a joint paper was slow going because he was then at the University of Washington, and producing and exchanging typewritten drafts took considerable time as with Montague; also both Vaught and I accepted the demands of Tarski-style precision, though he was even more meticulous than I. We did not complete the paper until 1958 and it was published in 1959 (cf. [2]).

Topping off that intense first year at Stanford, a 5-week-long AMS Summer Institute in Symbolic Logic was held at Cornell University beginning on July 1, 1957. The 1950s had witnessed a great increase in activity in mathematical logic and there had been some meetings on special topics but this was the first devoted to the broad spectrum of the field. As we described it in our biography, *Alfred Tarski. Life and Logic*:

> The conference was unusual for its length and breadth; the speakers represented all branches of mathematical logic and, for the first time, a large number of computer scientists took part. Most of the participants lived in the college dormitories and ate in the communal Cornell Hotel School dining room. It was like being at a summer camp… For weeks the green, hilly campus buzzed with talk about model theory, recursion theory, set theory, proof theory, many-valued logics, and significantly, the logical aspects of computation; there was also the usual discussion about whose work was the most important. There was novelty, rivalry, conviviality, and [even] scandal. (Feferman and Feferman 2004, p.220)

The inspiration for the conference came from Paul Halmos, a younger self-described "brash" mathematician at the University of Chicago, who worked in a number of mainline fields and had taken an interest in algebraic logic, thus connecting him with Tarski and his students and collaborators' ongoing work in the subject. With the assistance of Tarski and the other American leaders in the field—Alonzo Church, Stephen Kleene, Willard van Orman Quine, and Barkley Rosser—Halmos succeeded in getting the sponsorship of the American Mathematical Society (AMS) and financial support from the National Science Foundation (NSF). Tarski of course pushed to have the institute take place in Berkeley, while Rosser was adamant that it should be in Cornell; one argument was that the majority of participants would be coming from the East Coast; reluctantly, Tarski acceded.

> The issue of who would be invited to speak also created heat. There was quick agreement about the most prominent senior scholars, but discussion—mostly by correspondence—about who to choose among the up-and-coming younger crowd went on for many months, with each of the main organizers giving preference to his own disciples. On this score, Tarski did very well; about one fourth of the speakers were under his influence in one way or another, and many of them gave two or three talks. In this way, he succeeded in positioning himself as the leading man of the occasion. Because the reclusive Gödel, whose name was first on the invitation list, had declined to attend, there was no direct challenge to Tarski's assumption of that role. (ibid., p. 221)

There were also two "wild cards", Abraham Robinson (unrelated to Berkeley's Raphael M. Robinson) and Georg Kreisel:

> [They] came to the Cornell Institute unfettered by a link to a mentor. Both would soon have enormous influence on their younger colleagues… Both men were European, like Tarski,

> but much younger than he and more recently arrived in the United States. Robinson, born in Germany, had lived in Israel, France, England, and Canada, and he would return to Israel once more for a few years before settling in the United States. Kreisel—Austrian born, educated in England, and a frequent visitor to France—had a position in Reading, England. At the invitation of Kurt Gödel, he had spent the preceding two years at the Institute for Advanced Study [IAS] in Princeton. Largely self-taught in logic and less bound by tradition, neither Robinson nor Kreisel owed allegiance to a single methodology. Both had worked in applied mathematics in England during World War II, and as a result their style was much more experimental and free-wheeling than Tarski's. Also, each of these men in his own way exuded intellectual self-confidence; neither was afraid to lock horns with Tarski. (ibid., p. 223)

Robinson worked on problems concerning the applications of model theory to algebra, which of course interested Tarski very much. He introduced novel general approaches to the subject, among which (later on) were model-theoretic methods to establish decidability of various theories without having to make use of the method of elimination of quantifiers for which Tarski had been the standard bearer. In 1960, Robinson became famous for the creation of "non-standard analysis," a model-theoretic foundation for the systematic use of infinitesimal (and infinitely large) quantities in mathematical analysis.

Participation in the Cornell Institute was extremely important for my career, first for widening my understanding of the field, and then for the intellectual and personally valuable contacts I made with both senior and junior logicians, and finally for the opportunity to present myself and my work to that group. In consultation with Tarski, I gave two talks there. The first was on the results of my dissertation on the arithmetization of metamathematics [6]. For the second, I had proposed to him to speak about my joint work with Vaught on generalized products of models. But Tarski insisted that I speak instead about some work of Andrzej Ehrenfeucht and Roland Fraïssé [7]. The background was that subsequent to my efforts, Ehrenfeucht (a student of Mostowski's in Poland) had succeeded in establishing the decidability of the theory of ordinals under addition by means of the so-called "back-and-forth" methods that had been introduced by Fraïssé (then working independently in French Algeria and later in France). I have never gotten over Tarski's insistence that should be my second talk, but there seemed to be no way that I could get around him for that. And Tarski never seemed to realize the significance of my work with Vaught, though after its publication in 1959 it became a much-cited landmark in the field of model theory. That, too, made me much more aware of his blind spots.

Back at Stanford, at the end of my first 2 years there, I was promoted from the rank of instructor to the tenure-track position of Assistant Professor of Mathematics and Philosophy. The challenge in the following years would be to make tenure and, for that, more substantial work would have to be produced and recognized. In particular, in 1958–1959, after mastering Turing's work on ordinal logics, I reframed it as the study of transfinite recursive progressions of first-order axiomatic theories with standard formalization. The theories in such progressions are indexed along paths in the Church–Kleene system O of recursion-theoretic notations for all "constructive" ordinals. I re-proved Turing's completeness result for Pi-0-1 number-theoretical sentences in a progression based on the iteration of consistency statements along certain paths, and showed that even iterating the local reflection principle did not give completeness for Pi-0-2 sentences along any path in or through O. My main result—only obtained after a considerable struggle—was that transfinite iteration of the global (or uniform) reflection principle is complete for all sentences of arithmetic along suitable paths in the set O of all constructive ordinal notations [8].

A year at the Princeton IAS. Due to Kurt Gödel's presence at the Princeton Institute for Advanced Study, that increasingly became a mecca for logicians at all levels, and so I applied for and succeeded in obtaining a National Science Foundation (NSF) Post-doctoral Fellowship to visit the IAS in the academic year 1959–1960, for which I was granted a leave of absence from Stanford. Gödel himself had visited the Institute from Vienna in 1933–1934 and again in 1939. To escape being drafted into the Nazi army, he fled Austria in 1939 and joined the Institute as an "Ordinary" Member in 1940; he became a Permanent Member in 1946. What Gödel accomplished in the decade of the 1930s before joining the Institute changed the face of mathematical logic and continues to influence its development to this day. Since the subject of my 1957 dissertation was directly concerned with the method of arithmetization that Gödel had used to prove his famous incompleteness theorems of 1931, and since my main concern after that was to study systematic ways of overcoming incompleteness, the prospect of meeting with Gödel and drawing on him for guidance and further inspiration was particularly exciting. (I did not know at the time what it took to get invited. Hassler Whitney commented for an obituary notice in 1978 that "it was hard to appoint a new member in logic at the Institute because Gödel could not prove to himself that a number of candidates shouldn't be members, with the evidence at hand." That makes it sound like the problem for Gödel was deciding who *not* to invite. Anyhow, I ended up being one of the lucky few.)

Housing for Institute Fellows and their families was provided on Einstein Drive in two-storey apartment units that had been designed in the efficient and modernist esthetic Bauhaus style by Marcel Breuer. Our daughters were then aged five and three, respectively, and our family was given a two-bedroom ground floor apartment. In addition to housing, a day care center was provided on campus for preschool children, so our 3 years old went there while our 5 years old started kindergarten. The atmosphere for adults was very social and we soon made a number of new friends.

Once I got settled, my meetings with Gödel were strictly regulated affairs. There were few colleagues with whom he had extensive contact—in earlier years, Albert Einstein, as everyone knows, and Oskar Morgenstern among them—as well as a few of the more senior visiting logicians. There were some younger logicians who managed to see him more often than I did, but I was too intimidated to take full advantage of him, something I regret to this day. Gödel's office was directly above the one that I shared with another visitor, the Japanese logician, Gaisi Takeuti. We used to think we heard him pacing the floor above us. When I wanted to meet Gödel and figured he was in his office, I would phone him for an appointment and would hear the phone ring and hear him answer. When it worked out, I would walk upstairs to his office. There he would be seated at his desk, and I would sit down across from him; there was no work at the blackboard as is common among mathematicians. It was clear to me that Gödel had read my papers and knew about the work in progress, and the latter would be the focus of our conversations. He would raise some questions and make a few suggestions, and what he had to say would be very much to the point and fruit for further thought. After precisely half an hour the alarm on his watch would go off, and he would say, "I have to take my pills." I took that as my cue to leave. It was only in later years that I learned of his general hypochondria and neurasthenia and of his earlier bouts of mental illness in the 1930s in Vienna, where his position at the University had simply been that of a Dozent following the Ph.D. and Habilitation degrees.

1959–1960 was an exciting year for logicians at the Institute. Among the distinguished senior visitors were Paul Bernays, from the ETH in Zürich, and Kurt Schütte, from the University of Kiel. Bernays had been Hilbert's leading assistant in Göttingen in the 1920–1930s in the development of Hilbert's consistency program and its principal tool of proof theory (*Beweistheorie*) for the foundation of mathematics; it was Bernays who was principally responsible for the preparation of their joint work, *Grundlagen der Mathematik* (in two volumes, 1934–1939). In 1934, Bernays had had to return to Switzerland when the Nazis took over Germany, since he was Jewish. Schütte was Hilbert's last student, in 1933; he subsequently had to spend the years up to 1945 as a meteorologist, and it took him another 5 years after that to reestablish himself in the academic world. He was awarded the Habilitation degree in 1950 at the age of 43.

Schütte was at the Institute without his family, and Bernays was unmarried. Both gentle and friendly, they would often visit us in our Institute flat on Einstein Drive, and Bernays would play the piano for our young daughters. Other visitors in logic that year were Gaisi Takeuti (with whom, as I have already mentioned, I shared an office), Roger Lyndon from Michigan, and Anne Davis Morel, who had studied with Alfred Tarski a few years before me. Lurking in the wings was the brilliant mathematician, Paul Cohen, from the University of Chicago, who was going around asking everybody—and the logicians in particular—what the most important problem would be to solve in their field (the outcome of which to be described anon). Gödel did not conduct a seminar or offer any courses, so, instead, we visitors in logic at the Institute had a regular seminar with the logicians at Princeton University, led by Alonzo Church; Hilary Putnam was there as a young faculty member, and Martin Davis would come down regularly from New York to join us. (Putnam and Davis were then working on Hilbert's 10th problem on the question of decidability of Diophantine equations, eventually solved in the negative by Yuri Matiyasevich after essential steps made by Julia Robinson.) Finally, Raymond Smullyan—who was finishing his Ph.D. under Church—was also an active participant; in addition, at parties he entertained with amazing (to us) card tricks.

At the time I was immersed in my paper on transfinite progressions of axiomatic systems, which reworked and extended Turing's work on ordinal logics that I described above. As it happened, Turing's work had been carried out under Church's direction at Princeton as a doctoral thesis in 1939; curiously, Church had no comment to make about the historical connections when I presented my results in our joint seminar. Though the idea of autonomous progressions was already in the air, I did not start working on them until 1961. I did not know in 1959-1960 how important Schütte's work in his papers of the 1950s and in his book *Beweistheorie* then in progress would turn out to be to me for the treatment of predicative analysis via autonomous ramified progressions. Moreover, I did not yet consider myself to be one of the proof theorists in any usual sense of the word, and paid only modest attention at that time to what *they* were concentrating on. That year, at the Institute, it was Takeuti's Fundamental Conjecture as to the eliminability of cuts in the simple theory of types that held the center of their attention. Takeuti himself had obtained various special cases by syntactic arguments. One of Schütte's main results in that period was a reformulation of the Fundamental Conjecture in the semantic terms of the extendibility of suitable partial valuations to total ones. The Fundamental Conjecture was solved in the latter part of the 1960s in terms of Schütte's semantical reformulation, first by Bill Tait for second-order type theory and later in full and independently by Dag Prawitz and M. Takahashi.

Predicativity, proof theory, and tenure. After returning to Stanford in the summer of 1960, I concentrated on completing for publication my paper on transfinite progressions of theories. As a foil to my completeness results, during a visit to Stanford early that summer, Kleene's brilliant student, Clifford Spector, suggested that one could have incompleteness even for Pi-0-1 statements along inductively defined (Pi-1-1) paths, and in fact that proved to be the case. Spector's idea got incorporated in the definition of a set O* of O-like possibly nonstandard notations that are well-founded with respect to hyperarithmetic descending sequences (or "pseudo-well-ordered"). Progressions of theories T_a can be extended to a in O*; in particular those based on the iteration of the global reflection principle are shown consistent by induction on an arithmetic property. Finally the Pi-1-1 paths P through O turn out to be exactly those that are the restrictions to O of the predecessors of some a in O*–O; hence the consistency of T_a for such a is true but not provable along P. Both my full work on completeness of transfinite progressions and a short paper with Spector on incompleteness along Pi-1-1 paths were eventually published in the *JSL* in 1962 [8, 9]. In another publication of that year [10], I was able to apply on the one hand the methods of my paper to obtain complete hierarchies of recursive functions along relatively short paths in O, and on the other hand the methods of our joint paper to obtain incomplete hierarchies along nonstandard paths in O*. Sadly, Spector did not live to see our joint work and its consequences, since he died unexpectedly of leukemia in 1961.

My next major project was to investigate *predicative analysis*, its potentialities, and its limits. In one respect, this followed directly on from my work on progressions. Namely, the leading open question remaining from that was what are natural conditions to impose on the choice of ordinal notations used to index theories in a progression. One answer was provided for theories whose language contains a formula I(x) for which provability of I(a) implies *a* in O; informally, I(x) would express well-foundedness of the initial segment determined by x in O or a part of O. Given an initial system T_0 that is correct for such statements and a progression $\{T_a\}$ based on iteration of a successor principle that preserves correctness, define the set Aut($\{T_a\}$) of *autonomous* notations for the progression to be the smallest set A containing 0 that is closed under successors and predecessors and is such that whenever b is in A and T_b proves I(a) then a is in A. This notion of autonomy (aka the "boot-strap" condition) had been introduced by Kreisel in 1958 in a proposal to characterize *finitism* and *predicativity* by suitable autonomous progressions of theories, $\{F_a\}$ and $\{R_a\}$, resp. The latter are systems of ramified analysis, i.e., systems of second-order number theory in which the set variables are ranked by ordinal notations or the corresponding ordinals, and the comprehension axiom takes the form that one only asserts the existence of sets of rank α by definitions of the form {x: A(x)} (with 'x' ranging over the natural numbers) when each bound set variable in the formula A has rank less than a.

The idea of ramification—but only for finite ranks—goes back to the work of Bertrand Russell in the early twentieth century in his monumental effort (with Whitehead in the *Principia Mathematica*) to rescue Frege's fatally flawed program to reduce all of mathematics to logic. Russell called properties P(x) *predicative* if they lead only to consistent definitions of sets {x: P(x)}. Henri Poincaré had identified vicious circles as being the underlying source for the paradoxes (Russell's being the one that had undermined Frege's system). The ramification condition was Russell's means to avoid *prima facie* vicious circles and thus meet the predicativity requirement. But Russell could not obtain a satisfactory definition of the natural numbers in his system of ramified type theory without adding two *ad hoc* principles, the Axiom of Infinity and the Axiom of Reducibility; the latter in effect vitiated the distinctions by ramification. Meanwhile, Poincaré had argued that the concept of the natural number system is an irreducible minimum of abstract mathematical thought and any effort to reduce it to logical notions was a misbegotten enterprise.

Kreisel's proposal combined aspects of Poincaré's and Russell's ideas with the natural numbers taken as given at the lowest rank in the language of the progression of ramified theories; moreover, he argued that there is no reason to restrict the ranks to finite levels, as long as the transfinite levels used are only those accessed by the autonomy condition. On that proposal, *predicative analysis* is identified with the provable sentences of the union of the systems in the autonomous progression $\{R_a\}$, and the *predicatively provable ordinals* are just the levels α arrived at in this autonomous manner. Given that the first question would be to determine what the limit is of the predicatively provable ordinals—or, in other terms, what is the least ordinal that is not predicatively provable—the second question would be to see how much of actual classical analysis can be accounted for in terms of the predicative systems R_a. An answer to the first question for an upper bound was already practically at hand in the work of Schütte on ramified systems without restriction on the ranks. That made use of the Veblen hierarchy φ_α of critical functions on the ordinals to obtain the ordinal strength of the ramified systems at arbitrary levels.

The ordinal Γ_0 is defined to be the least ordinal closed under the Veblen hierarchy considered as a function $\varphi(\alpha, \xi)$ of two variables. Using Schütte's work on ramified systems, he and I independently established Γ_0 as an upper bound to the predicatively provable ordinals (cf. [11]). The more difficult part for each of us was showing that it is the least upper bound; we accomplished that by different methods. Since then, Γ_0 has come to be referred to as the ordinal of predicative analysis as well as the Feferman–Schütte ordinal.

The second question to be dealt with in regard to predicative analysis was how much of actual classical analysis could be accounted for in predicatively justified terms. This was a difficult project that I undertook to deal with in the following years (a question that Schütte did not pursue). Since real numbers may be represented by sets of natural numbers via Dedekind sections in the rational numbers, real numbers would be transfinitely ranked in the ramified progression, and that is unsuitable for the actual development of analysis. Thus my first aim was to find unramified second-order theories proof-theoretically equivalent to the autonomous progression of the R_a's. The first such system that I obtained for this purpose was an autonomous progression $\{H_a\}$ of unramified systems based on iteration of the Delta-1-1 Comprehension Rule (also known as the Hyperarithmetic Comprehension Rule (HCR)). Then I obtained a single system with the same theorems as the union of the H_a's for $|a| < \Gamma_0$, based on the rule HCR together with a version of the so-called Bar Rule (also in [11]). As for the actual development of analysis on predicative grounds, that had been initiated by Hilbert's great student and colleague, Hermann Weyl, in a little book called *Das Kontinuum* published in 1918. That accounted for essentially all of nineteenth-century classical analysis; my extension of that to considerable parts of twentieth-century analysis would have to wait a number of years, but the theoretical foundation was now secured and the basic techniques were already in place.

It was through this work that I came to be known as a proof theorist. Meanwhile, proof theory itself was very much the topic *du jour* at Stanford especially due to the deep interest of Kreisel, Tait, and others in Spector's consistency proof of analysis (unramified second-order number theory), obtained by a double-negation interpretation in a formally intuitionistic system followed by an extension of Gödel's functional interpretation, using the so-called bar recursive functionals in place of the primitive recursive functionals. Spector had worked out the main arguments for his result in the year 1960–1961 that he spent at the IAS up to the time of his death, though it was left to Kreisel to organize the details in a form suitable for posthumous publication.[8]

Meanwhile, my body of work to date had made a sufficiently strong impression to lead the Departments of Mathematics and Philosophy at Stanford to promote me to the tenured rank of Associate Professor, commencing with the academic year

[8] In the summer of 1963, Kreisel led a seminar whose main purpose was to examine the extent to which Spector's interpretation could be considered constructive; his conclusion was "not by a long shot"; I attended the seminar along with Bill Tait, Bill Howard, Verena Dyson, and my student Joseph Harrison.

1962–1963. Psychologically that meant, on the one hand, great relief from the uncertainties and pressures that I had experienced as an Instructor and Assistant Professor during the preceding 6 years and, on the other hand, assurance that I could think of my research projects continuing in longer terms in the years ahead, wherever they might lead me. It also meant some major changes for my family. In particular, we could now think of building a home in a new subdivision for faculty and staff residences that Stanford had just opened up behind the campus. After a long period that involved choosing a lot, an architect, a plan and a contractor, our house was completed and ready to move in April 1963.

An excursion into set theory. Coincidentally, in the months leading up to April 1963, Paul Cohen worked out before my eyes his novel method of forcing and generic sets to obtain his solutions to the long outstanding problems concerning the independence of the Axiom of Choice (AC) and the Continuum Hypothesis (CH) in axiomatic set theory. As mentioned above, I had first met Cohen when we spent the year 1959–1960 together at the IAS, but he went on to spend an additional year there. He had received his Ph.D. at the University of Chicago in 1958 with a dissertation in analysis, but was already noted for his ambition to solve the most important problems in mathematics no matter what the field.[9] Cohen seems to have had an interest in logic from early years but never took a course in it, though he had a number of logician friends while a graduate student at Chicago from whom he could absorb a certain amount about the subject including what were the most important problems in that field. At the IAS, though, he kept asking me the same question, and of course the independence of the Axiom of Choice from the axioms of Zermelo–Fraenkel set theory (ZF) kept coming up. In fact, though I was not a set-theorist, I had toyed with that problem myself for a while at the IAS.

After leaving the IAS, Cohen came to Stanford in 1961 and soon after he was a frequent visitor to my office, pumping me about problems in logic. Before long he settled on establishing the consistency of the system of second-order number theory (also called "analysis"). It was not his style to consult the literature, and he did not hesitate to reinvent the wheel as he came to understand things for himself. To begin with, he found a form of Gentzen's consistency proof for number theory. After that, he floated to me various ideas for a consistency proof of the system of analysis, despite obstacles to those approaches that I kept pointing out. Nevertheless, in the spring of 1962, he decided to give a special series of lectures (well attended both by logicians and non-logicians) in which his aim was to present such a consistency proof. Somewhere along the way, he realized that there was a fundamental difficulty and the seminar fell apart with no evidence of embarrassment on his part.

In late 1962, Cohen shifted his attention and began to work in earnest on establishing the independence of AC from ZF; again, he used me as a sounding

[9] Cohen's big result in analysis was the solution of a conjecture of Littlewood's, for which he was to receive a prestigious prize in 1964. It is known that at various times throughout his life, Cohen also tried—without success—to settle the famous Riemann Hypothesis, one of the seven problems on the $1,000,000 Millennium Prize list.

board to try out various approaches, despite the fact that I was not an expert in set theory. But when he came up with the method of forcing and generic sets in the early spring of 1963, it seemed to me that his new methods really did the trick. In the event, not only did he obtain the independence of AC but also he showed that the Continuum Hypothesis CH is independent of ZFC; more precisely, it is consistent with ZFC that the cardinality of the continuum is Aleph-2. So far as I could see, the proofs were correct but the truly novel arguments involved required checking some delicate points, and only the experts could be depended on for final confirmation. Indeed, after he circulated a draft of his work, verifications of most of the arguments came in, but still some questions were raised about delicate points. In desperation, Cohen wrote Gödel seeking his imprimatur, and that was not long in coming, along with a letter lauding Cohen as having made the most important advance in set theory since its axiomatization. Thus assured, Cohen gave a large dramatic public lecture on his work at Stanford in April 1963; after that, he was to lecture on it in Princeton in June 1963 and—most notably—on Independence day July 4, 1963 for the Symposium on the Theory of Models being held in Berkeley that month.

Having acted as some sort of midwife to the birth of Cohen's method, I was the first to really understand and apply it, using some simplifications in the definition of forcing that Dana Scott had suggested to me. The main application I obtained is that it is consistent with ZFC together with the Generalized Continuum Hypothesis that no formula of set theory can serve to define a well-ordering of the continuum; this settled in the negative a conjecture that Hilbert had made in 1900 in the first of his famous list of open mathematical problems. And I also showed that the method of forcing and generic sets could be applied as well in both first-order and second-order number theories. I spoke about this work at the July 1963 Berkeley Symposium and among the listeners was the set-theorist, Azriel Lévy, who was eager to learn more. He spent part of the summer of 1963 with me at Stanford in order to absorb Cohen's method and then to make new applications. In particular, in joint work, we obtained existence of a model of ZF in which there is a countable union of countable sets that is not countable, thus contradicting the Countable Axiom of Choice, a weak form of AC.

Incidentally, toward the end of my 1965 paper on the method of forcing and generic sets [13], I pointed out that a sentence A is true in all generic extensions M [G] of the ground model M iff $\neg\neg A$ is forced by the empty set.[10] The proof immediately generalizes to the statement that A is true in all generic extensions M [G] for which G extends q iff $q \Vdash \neg\neg A$, or as is also written, $q \Vdash^* A$; the latter relation is called *weak forcing*. Joseph Shoenfield showed in 1967 how you could simplify the methodology of independence proofs by turning Cohen's approach on its head. Namely, given a partial ordering P whose members are the forcing conditions q, instead of starting with the forcing relation and defining what it means to be a generic set in terms of that, one defines a generic set G to be one that meets every dense set of forcing conditions, and then defines weak forcing in terms of

[10] This was realized independently by Lévy.

truth of A in all M[G] for which G extends a given condition q. The work then goes into showing how that relation $\Vdash^*$ can be defined in M, so that the properties of M [G] may be reduced to those of M. This methodological inversion became standard and has several advantages among which it is simpler to begin with and much more versatile.[11]

Though I was first out of the gate after Cohen, a stampede soon followed. Among others, Robert Solovay had moved into the subject after attending Cohen's lecture in Princeton in June 1963. His first result was announced in an abstract for the Berkeley Model Theory Symposium with the title, the cardinal number of the continuum "can be anything it ought to be", i.e., that it is consistent with ZFC that its cardinal number is Aleph-α for any α not excluded by König's theorem. Solovay was then still a graduate student at the University of Chicago, working with Saunders MacLane on category theory and higher geometry; he obtained his Ph.D. the following year and then joined the Berkeley Mathematics Department. As it turned out, Solovay was to become a leader in the onslaught of work on set theory that had been unleashed by Cohen. The subject of set theory took off in a dazzling way in the post-Cohen years, I think much to Cohen's surprise; I believe he had wished that his work could be seen as a stand-alone achievement, like Gödel's 25 years before, rather than one that quickly opened the flood gates to one result after another. Nevertheless, he could be well satisfied with receiving the Fields Medal, the preeminent award in mathematics, at the International Congress of Mathematicians that took place in Moscow in 1966 and with the award of the National Medal of Science by President Lyndon B. Johnson a year later.

A summary of my work in set theory was published in the proceedings of the 1963 Berkeley Model Theory Symposium [12] and it later appeared in full in the journal *Fundamenta Mathematicae* for 1965 [13]. After that, I made only one more serious effort to apply the method of forcing in set theory, namely to the measure problem. Since the axiom of choice had been used in an essential way in the proof of the existence of Lebesgue nonmeasurable sets of reals, the question was whether it is consistent to assume all sets are measurable if one drops AC. Solovay's main result (announced in 1964) was that it is consistent with ZF plus the Countable Axiom of Choice that there is a translation invariant extension of Lebesgue measure to arbitrary sets of reals. Following that he showed that, assuming the existence of an inaccessible cardinal, it is in fact consistent that all sets are Lebesgue measurable. Cohen was long disturbed by that assumption until it was shown by Saharon Shelah that Solovay's result cannot be obtained without it.

Paris and Amsterdam. The year 1963–1964 would be the seventh that I spent at Stanford (except for the year at the Institute), and so I was due for a sabbatical leave at half salary in 1964–1965. I thus applied for and was awarded a Senior Post-doctoral NSF Fellowship to make up the balance of my salary; the proposed plan was to spend the first half of the year at the University of Paris and the second

[11] See Shoenfield [1967] p. 364.

half at the University of Amsterdam. The choice of Paris was made on the grounds that Kreisel would be there; indeed, he had been alternating his time on a regular basis between Stanford and Paris, and had helped foster an incipient group in logic in Paris (though as I recall he ended up not being in residence in the autumn of 1964). The choice of Amsterdam was reasonable in view of the fact that since some time it had a strong group in logic that had been built up by Arend Heyting in mathematics and Evert Beth in philosophy. Heyting had been a student of L. E. J. Brouwer, the great topologist and originator and ideological promoter of the brand of constructive mathematics called *intuitionism*, and Heyting had been the first to develop intuitionism on formal grounds. The philosopher and logician Beth had a broad range of interests in other areas of logic, especially model theory after he came under the influence of Alfred Tarski, with whom he became very close. Sadly, Beth died in the spring of 1964, about a year before the planned stay in Amsterdam but after I had made our overall arrangements.

A few weeks after arrival in Paris we found an apartment in the relatively bourgeois 17th Arrondissement, a metro stop away from the Arc de Triomphe. We had originally planned to enroll our daughters in the École Bilingue in Auteuil somewhat to the west of the 17th. But practically facing our new apartment was a public elementary school, sectioned between a school for girls and a school for boys. We asked our daughters if they would be willing to enter the École des Filles, and they said yes, they were game for that, but the total immersion proved to be a great challenge in the first few months that they were finally able to meet. Meanwhile, Anita and I studied French at the Alliance Française on the Left Bank.

In Paris I attended weekly seminars at the Institut Henri Poincaré (IHP) on rue Pierre et Marie Curie near the Sorbonne; the IHP had been established in the 1920s for the use by mathematicians and physicists at an advanced level. (There was no office space for me there and in any case it was my habit to work at home.) Besides Kreisel when he was in residence, there were few logicians at the IHP in those days since the subject had not been at all encouraged in France despite its explosive development in the US and elsewhere in Europe.[12] One of the logicians at the IHP when I visited was Daniel Lacombe, who was by then well established in the field; there were a few others like Jean-Louis Krivine who were beginning to make their mark. That picture was to change radically when I visited 8 years later.

In Amsterdam we were fortunate to get an apartment belonging to the linguist and philosopher Frits Staal and his wife; Staal was later to become a Professor of Philosophy at UC in Berkeley. Though my principal host in the Mathematics Department was Heyting, the person with whom I had most contact was Anne S. Troelstra, an advanced doctoral student of Heyting's. Troelstra was to become one of the most important contributors to the metamathematics of intuitionism and

[12] The last French logician of world note had been Jacques Herbrand, who died in 1931 at the age of 23 in a mountain-climbing accident. The French mathematicians had valued Herbrand for his contributions to algebraic number theory but were ignorant of his fundamental contribution to proof theory; that was recognized instead by the Germans and Austrians such as Hilbert, von Neumann and Gödel, and later Gentzen.

constructivity, to begin with by developing variants of the method of realizability that had been introduced by Kleene. He later collaborated with Kreisel to give a consistency proof for a formal system of Brouwer's theory of choice sequences, done for a different formalism in a different way by Kleene and his student R.E. Vesley.

Proof Theory I: Many-sorted interpolation theorems. I made a number of direct and indirect contributions to proof theory in the period 1966–1971. The first of these was a form of the interpolation theorem for many-sorted languages that gives information about the location of quantifiers over variables of common sort in a way that could not be obtained as a consequence of Craig's form of the theorem. One of the applications of this theorem is a generalization of the Łoš-Tarski theorem characterizing the form of sentences preserved under substructures, and its dual for extensions. Namely, suppose given an indexing I of sorts and a proper subset J of I. Then a formula is preserved under passage to substructures (extensions) in which the domains of J sort are fixed just in case it is provably equivalent to a formula that is essentially universal (existential) outside of J. Another application of the general theorem for formulas invariant under extensions mod J is an eliminability of quantifiers theorem, which when combined with some simple model-theoretic criteria implies various known eliminabilities of quantifiers results in algebra, for example, for real and algebraically closed fields. Later, using many-sorted interpolation arguments I obtained a similar general theorem for formulas invariant under *end-extensions* in the language of set theory [18].

At the time I was doing this work, my student Jon Barwise was working on the model theory and proof theory of the infinitary languages L_A for A an admissible subset of the Hereditarily Countable sets HC.[13] The *admissible sets* in general are the transitive models of the system KP of Kripke–Platek set theory, a weak subsystem of ZF that had been developed as a setting for the generalization of recursion theory to set theory. The analogues of the recursively enumerable sets are just those that are Sigma-1 relative to A, also called the A-r.e. sets; the analogues of the finite sets called the A-finite sets are just the members of A. The formulas of L_A are those elements of A built up from atomic formulas using negation, countable conjunction, countable disjunction and ordinary universal and existential quantification. Barwise showed that the straightforward generalized forms of the completeness theorem, Gentzen's cut-elimination theorem and Craig's interpolation theorem hold for L_A. His main new result was that the compactness theorem holds for L_A with A countable admissible in the following form: if S is an A-r.e. set of L_A formulas and each A-finite subset of A has a model, then A has a model. These results generalize the classical ones by taking A = HF.

---

[13] Barwise was my second Ph.D. student; he received his Ph.D. in 1967. My first student was Joseph Harrison, who extended my work on recursion-theoretic pseudo-hierarchies in interesting and unexpected ways; he finished in 1966. Unlike Barwise, Harrison decided not to pursue a research career but chose instead to devote himself to the cause of social justice. He went on to a teaching position at Emory University, a noted "black" college.

Using the completeness of the cut-free calculus for L_A where A is an admissible subset of HC, I could show that the many-sorted interpolation theorem described above and its applications to preservation and invariance theorems with stationary sorts—as well as the quantifier eliminability theorems—carry over directly to these logics. I explained all of my work described above in a series of lectures on proof theory at the Summer School in Logic held in Leeds, England in August 1967 [19]. I also included useful results on ordinal bounds that had been obtained by Tait for infinitely long derivations: suppose d is a derivation whose is bounded by α and whose cut rank is bounded by ν, then we can convert d to a cut-free derivation d* of the same end formula whose length is bounded by $\varphi(\nu, \alpha)$. Handled constructively, this is related to the proof theory of predicativity, since if $\nu, \alpha < \Gamma_0$ then $\varphi(\nu, \alpha) < \Gamma_0$.[14]

Following Leeds, I attended the Third International Congress of Logic, Methodology and Philosophy of Science at the University of Amsterdam as an invited speaker.[15] My topic was various versions of autonomous systems in connection with predicative mathematics, including some formulated in terms of infinitary languages. The ordinal Γ_0 was shown to be the least upper bound of the provable ordinals in each of them, thus adding some robustness to my characterization of predicativity (cf. [17]).

The cast of characters in logic at Stanford: entrances and exits in the 60s and early 70s. In order to understand my relations with other logicians at Stanford in the 1960s and early 1970s, here is a brief summary of their various entrances and exits. John Myhill had come to Stanford from Berkeley in 1960 but for personal reasons left Stanford for SUNY at Buffalo in 1964. Kreisel was then appointed in his place; he had been coming to Stanford on a visiting basis off and on since 1958, though after 1964 he still divided his time between Stanford and Paris.

When Dana Scott finished his Ph.D. with Church in Princeton in 1958, he moved to the University of Chicago as an Instructor for a couple of years, and then Tarski brought him back to a position in Berkeley, where he was tenured 2 years later. Though there was a non-raiding agreement of some sort between Berkeley and Stanford, in 1963 we managed to attract him away to a joint position in mathematics and philosophy. After a year visiting Amsterdam in 1968–1969, Dana was slated to return to Stanford, but he was attracted back to Princeton for a couple of

---

[14] Tait's bounds were actually somewhat sharper: with each derivation d of length bounded by α and cut rank bounded by $\gamma + \omega^\nu$ is associated a derivation d* of the same end formula whose length is bounded by $\varphi(\nu, \alpha)$ and cut rank is bounded by γ. In particular, for $\nu = 0$, lowering the cut rank by 1 may be achieved by passing to a derivation of length ω^α. This may be applied to the infinitary translations of derivations in PA to obtain ε_0 as a bound for the resulting cut-free derivations.

[15] As described in Chapter 10 of Feferman and Feferman (2004), Tarski had been the principal creator—with the help especially of Evert Beth—of the Division of Logic, Methodology and Philosophy of Science within the International Union of History and Philosophy of Science. Its first congress was held at Stanford in 1960, where Suppes was the highly harassed point man. I gave a contributed talk there on my work on ordinal logics, i.e., non-autonomous transfinite progressions of theories.

years before moving on to Oxford where he finally settled in for a good 10-year run in 1972.

Bill Tait had come to Stanford in the Philosophy Department in 1959 following his doctoral work at Yale with Frederick Fitch and Alan Ross Anderson. He soon actively tackled a number of questions in proof theory and constructive foundations raised by Kreisel. Tait left Stanford in 1965, first for the University of Illinois at Chicago Circle and later at the University of Chicago.[16]

Finally, the prodigy Harvey Friedman came to Stanford in 1967 at age 18 as an Assistant Professor of Philosophy. Having skipped much of high school and undergraduate work, he had just received his Ph.D. at MIT as a student of Gerald Sacks, but his work was completely independent of Sacks. In his thesis, Friedman established a surprising theorem that was to feed into my work in a significant way that will be described later. At Stanford, he immediately allied himself with Kreisel and, taking the latter's ideas as a cue, set out to produce quite a varied series of results. Friedman was tenured 2 years later and stayed at Stanford until 1973 when his relations with Kreisel broke down; he went on to Wisconsin, then Buffalo, and finally Ohio State for his permanent position.

A year at MIT 1967–1968. In the spring of 1967, I had expected to be promoted to Full Professor at Stanford for the following academic year and when I was not, was very annoyed, so much so that I sought out (through Gerald Sacks)—and received—an invitation to visit the MIT Mathematics Department for the year 1967–1968. I guess the idea was that that would show that I should not be taken for granted and that I could be attractive to other institutions and might even receive outside offers. The strategy worked to the extent that I was promoted the following year, but meanwhile it required another disruption for our family, especially as to schools and friendships for our daughters.

In addition to Gerald Sacks and Hartley Rogers at MIT, during that year I had rewarding contacts with the logicians in the Philosophy Department at Harvard—Willard van Orman Quine, Burton Dreben, and Hilary Putnam—all of whom I had first met at the Cornell conference. Anita and I also had contact with Jean van Heijenoort, whom we had first met at Cornell too, and who was now teaching at Brandeis University not far from Cambridge. Van, as he was called, had had a fascinating life in the 1930s in close association with Leon Trotsky and was much later to be the subject of Anita's first biography. He left the Trotskyite movement in the 1940s and, after obtaining a Ph.D. at NYU in mathematical analysis, became an autodidact in logic while teaching mathematics. One of his major contributions to our field was the production, as editor, of *From Frege to Gödel. A Source Book in Mathematical Logic*; that appeared the very year we came to Cambridge. The *Source Book* provided English translations done by Van and others of a series of

[16] Following Tait's departure, Mostowski's former student Andrzej Ehrenfeucht was brought as a visitor in Philosophy for 2 years. Meanwhile, over in mathematics, Rohit Parikh served as an instructor in mathematics from 1962 to 1964, and then we had a succession of visitors: Joe Shoenfield (1964–65), Azriel Lévy (1965–66), and Robin Gandy (1966–67).

fundamental papers spanning the crucial period 1879–1931 that was formative for modern logic. Each paper was also accompanied by an informative introductory note, many of them written by van Heijenoort as well as by Quine, Dreben, and Charles Parsons. In the 1980s Van was to become one of my coeditors of the *Collected Works of Kurt Gödel*, for which the *Source Book* was a model in a number of respects.

Proof Theory II: Proof-theoretic ordinals and theories of iterated inductive definitions. After returning to Stanford in the summer of 1968, I headed back east in August (this time alone) for a banner meeting at SUNY Buffalo on intuitionism and proof theory. "Everybody" was there. I gave two talks, one on systems of ordinal functions and functionals arising from proof-theoretical work, and the second on systems of iterated inductive definitions and subsystems of analysis.

The first of these was on what I called *hereditarily replete functionals* over the ordinals, began with a description of work I had completed the year before on systems $\underline{f} = (f_1,\ldots,f_n)$ of functions of one or more arguments on the set Ω of countable ordinals and their allied systems of representation. The initial purpose of that was to explain what was needed for the system of representation of ordinals up to Γ_0 to carry out the details of the arguments in my work on predicativity, but the leading properties turned out to be applicable to much larger systems of ordinals. With each system $\underline{f}$ and set X of ordinals is associated the closure Cl(X) of $X \cup \{0\}$ under $\underline{f}$. The system $\underline{f}$ is called *complete* if Cl(0) is an ordinal, and is called *replete* if for every α, Cl(α) is an ordinal. The *inaccessibles under* $\underline{f}$ are those α for which Cl (α) $\subseteq \alpha$; the *critical function* $\underline{f}'$ associated with $\underline{f}$ enumerates its inaccessibles. Let Cr ($\underline{f}$) be the system obtained by adjoining $\underline{f}'$ to $\underline{f}$. This system needs not be complete if $\underline{f}$ is, but a basic result in my 1968 paper [16] on systems of ordinal representation was that the property of repleteness is preserved under the critical process Cr, and so also is its transfinite iteration as in the Veblen process.

Moving on from that, in my first Buffalo talk I generalized the notion of repleteness to that of *hereditarily replete functional of finite type* over Ω (cf. [21]). Since the basic Veblen process for constructing larger and larger systems of ordinal functions was obtained by iteration of the critical process Cr at type 2, I considered that as given by a functional $\mathrm{It}^{(3)}$ at type 3. There is a natural generalization of this to $\mathrm{It}^{(n)}$ for each $n \geq 3$, and I showed that each of these is hereditarily replete. If one starts with the simplest nontrivial replete function $f_0(\alpha) = 1 + \alpha$, its successive iterations under the critical process give the functions $f_1(\alpha) = \omega + \alpha$, $f_2(\alpha) = \omega^\alpha$ and $f_3(\alpha) = \varepsilon_\alpha$. I was thus led to consider the ordinal κ generated from 0 and f_0 together with Cr and all the iteration functionals $\mathrm{It}^{(n)}$ and conjectured that κ is the Howard ordinal $\varphi(\varepsilon_{\Omega+1}, 0)$, i.e., the ordinal of the intuitionistic version of the system ID_1 of one arithmetical inductive definition. This was proved 4 years later by my student Richard Weyrauch (1975). The Howard ordinal makes use of the extension of the Veblen process to uncountable ordinals due to Bachmann; I will return to my substantial simplification of that later. My suggestion to consider iteration functionals of transfinite type was taken up by Aczel (1972) who obtained another Bachmann ordinal for what ordinals can be generated thereby.

My second talk at the 1968 Buffalo conference concerned the reduction of theories of iterated Pi-1-1 comprehension schemes to theories of iterated arithmetical inductive definitions. The classical theory ID_1 has for each arithmetical formula A (P, x) in which P has only positive occurrences (and thus is monotonic in P) a symbol P_A with axioms expressing that it is closed under A and that any formula B (x) closed under A (substituting B for P in A) contains P_A. The classical theories ID_α are formulated similarly, except that the formulas used at level α may contain earlier introduced predicate symbols both positively and negatively. Of special interest are accessibility inductive definitions, of which the paradigms are those of the constructive ordinal number classes or constructive tree classes. One of the main theorems I obtained in [22] is that for $\lambda = \omega^\gamma$ where γ is a limit ordinal, (Pi-1-1 CA)$_{<\lambda}$ (i.e., the Pi-1-1 Comprehension Axiom iterated up to λ) is proof-theoretically equivalent to $ID_{<\lambda}$, with conservation for statements arithmetical in O. In particular, for $\lambda = \varepsilon_0$, this work connected directly with the contributions of Friedman and Tait to the Buffalo conference, that are briefly as follows.

In his 1967 dissertation at MIT, Friedman showed by means of a novel model-theoretic argument that, again for $\lambda = \varepsilon_0$, the system (Sigma-1-1 AC) is a conservative extension of (Pi-1-0 CA)$_{<\lambda}$ for Pi-1-2 statements. In his Buffalo talk, Friedman generalized this to the result that (Sigma-1-(n+1) AC) is a conservative extension of (Pi-1-n CA)$_{<\lambda}$ for suitable classes of statements. In particular, by my result above, the system (Sigma-1-2 AC) is a conservative extension of $ID_{<\lambda}$ for statements arithmetical in O.

In Tait's presentation at Buffalo, he gave a consistency proof of (Sigma-1-2 AC) via a constructive cut-elimination theorem in uncountable propositional logic allowing conjunctions and disjunctions over higher tree classes up to ε_0. In principle, it seemed that this could be expressed as a reduction of (Sigma-1-2 AC) to an intuitionistic theory of iterated inductive definitions up to ε_0.

My view of all this was as potential material for a more perspicuous relativized Hilbert program. The greatest advance that had been made in that program prior to 1968 was Takeuti's constructive proof of the consistency of (Pi-1-1 CA) + BI by means of induction on certain accessibility systems that he called ordinal diagrams. However, these did not have a clear interpretation in terms of natural systems of representation for ordinals. Ideally, besides the constructive reduction, one would like to attach an ordinal such as Gentzen's for PA, the Feferman–Schütte ordinal for predicativity and the Howard ordinal for ID_1. My proposal for (Sigma-1-2 AC) was, first, to replace Friedman's model-theoretic proof of its conservation over (Pi-1-1 CA)$_{<\lambda}$ for $\lambda = \varepsilon_0$ by a constructive reduction to that system; thus, by my result (ii) above, one would have the proof-theoretic equivalence of (Sigma-1-2 AC) and classical $ID_{<\lambda}$. The next step would be to reduce the latter proof-theoretically to intuitionistic $ID_{<\lambda}$; one would hope more generally to reduce classical systems of iterated inductive definitions ID_α and $ID_{<\lambda}$ for limit λ in general to the corresponding intuitionistic systems, preferably of accessibility inductive definitions. Finally, one would want to determine the proof-theoretic ordinals of the latter in terms of systems of ordinal representation such as provided by the Bachmann hierarchy. I tackled the first part of this three-pronged attack in my proof-theoretical work of the year after Buffalo that I shall describe next. The second and third parts would turn out to be an achievement by Schütte's students Wolfram Pohlers and Wilfried Buchholz and my student Wilfried Sieg in the latter part of the 1970s. That will be described later.

Proof Theory III: functional interpretation relative to non-constructive functionals and representations of ordinals using higher number classes; Nice and London 1970. The first International Congress of Mathematicians (ICM) that I attended was held in Moscow in August 1966. I gave a contributed talk there on the independence of (Sigma-1-1 AC) from a weakened form of (Delta-1-1 CA).[17] The Moscow Congress was otherwise memorable for a number of non-mathematical reasons.

The second ICM that I attended was held in the city of Nice, September 1–10, 1970, and I was invited to give a half hour talk there. The ICM meetings are usually held in August, but that was excluded at Nice because the Riviera is usually full up during that month of the year. My wife and I rented a villa in St. Paul de Vence in the hills above the coast for the month before the Nice Congress. We were accompanied by our daughters and a friend of theirs; they wasted no time in leaving us to explore other parts of Europe on their own.

For my talk [23] at the ICM, I presented a way to obtain a proof-theoretical reduction for n = 0, 1 of (Sigma-1-n AC) to (Pi-1-n CA) iterated up to ε_0, that implied and thus strengthened Friedman's conservation results, via an adaptation of Gödel's functional interpretation. These results are achieved by the formal adjunction of non-constructive type 2 functionals F in order to transform instances of the Axiom of Choice scheme expressed in the analytic hierarchy into instances that are quantifier-free relative to those functionals. In the case n = 0, that is accomplished by adjunction of the non-constructive minimum operator μ, and in the case n = 1 by adjunction of the Suslin–Kleene operator E_1 that tests for well-foundedness of a relation. A decade later, in joint work with Wilfried Sieg, we found more straightforward reductions using Gentzen and Herbrand style proof theory, to be described below. But the use of the functionals μ and E_1 would make its reappearance in other parts of my work for quite different purposes.

It happened that I was also invited to talk at a logic meeting in London in the last week of August a few days prior to the beginning of the ICM Congress in Nice. The point of departure for my lecture [24] at the 1970 London conference was the notion of *relative categoricity* of systems $\underline{f}$ of ordinal functions that I had introduced in the article on systems of ordinal representation where I had introduced the notion of repleteness described above. One considers terms t built up from variables by the functions in $\underline{f}$. The system $\underline{f}$ is said to be relatively categorical (r.c.) if the ordering of terms obtained by substituting for the variables members of the set In($\underline{f}$) of $\underline{f}$-inaccessibles depends only on the ordering of those inaccessibles. It was shown

[17] That result was never published; the independence from full (Delta-1-1 CA) was obtained later by John Steel.

that if f is r.c. then so also is the system f′ obtained by adjoining the critical function Cr(f) to f, and the same holds for the adjunction of the transfinite iteration of the critical process as in the Veblen hierarchy.[18] In my talk for the London meeting, I was led to consider the naturally associated functors F on the category of subsets A of In(f) whose values are F(A) = (Cl(A), $\leq$, f), where Cl(A) is the closure of A under f. These preserve inclusions and direct limits, and I introduced a more general class of functors that I called *κ-local* for every infinite cardinal κ and a rather general class of concrete categories. The ω-local functors are just those that preserve inclusions and direct limits. The main result was that κ-local functors preserve elementary equivalence and elementary substructure in the languages $L_{\infty,\kappa}$; this had a number of algebraic applications. Later, Paul Eklof () used a 1964 modification of the notion of κ-local functor to obtain a precise formulation of Lefschetz' Principle in algebraic geometry.

What is more important is a conversation that I had with Peter Aczel at the London conference in which I suggested a substantial simplification of the means to represent large countable ordinal numbers for proof theory. Up to then, that had been done by Helmut Pfeiffer and David Isles following the method of Heinz Bachmann by using hierarchies of functions of ordinals in higher ordinal number classes. Writing Ω_α for the αth initial ordinal ($\Omega_0 = \omega$, $\Omega_1 = \Omega$ = the least uncountable ordinal, etc.), Bachmann had extended the Veblen hierarchy by first using a hierarchy of normal functions on Ω_2, from which he could define a system of representation up to a segment of the ordinals $< \Omega_2$, such as $\varepsilon_{\Omega+1}$. At successor ordinals or limit ordinals of cofinality less than Ω, one proceeds as in the Veblen hierarchy, but for α represented as $\lim_{\xi<\Omega}\alpha_\xi$, one diagonalizes, i.e., takes $\varphi(\alpha, \xi) = \varphi(\alpha_\xi, 0)$. (That is how one is led to the Howard ordinal $\varphi(\varepsilon_{\Omega+1}, 0)$.) Pfeiffer (1964) then lifted Bachmann's procedure to higher finite number classes by more and more complicated systems of partial ordinal representation and assignment of fundamental sequences; at each stage one step down successively from segments of higher number classes Ω_{n+1} to Ω_n. At the Buffalo Conference, Isles (1970) showed how to extend this to all number classes up to the least inaccessible ordinal.

My suggestion to Aczel was that one could avoid the complications of the Bachmann–Pfeiffer–Isles procedures without the successive assignment of fundamental sequences in partial systems of representation in the higher number classes by defining a single "long" sequence of functions θ_α defined for each α on arbitrary ordinals which is such that when restricted to any given Ω_ν maps Ω_ν into itself. Namely, suppose we have defined θ_ξ for each $\xi < \alpha$; let $\psi(\xi, \eta) = \theta_\xi(\eta)$ for $\xi < \alpha$ and η arbitrary. The crucial point in defining θ_α as the critical function w.r.t the preceding functions is to apply an autonomy condition, i.e., when defining the closure of an ordinal γ under ψ, one only uses those $\xi < \alpha$ which arise in the closure process; one also throws in Ω_ν whenever the closure contains ν. A simple argument shows that if $\gamma < \Omega_{\mu+1}$ then the cardinality of $Cl_\psi(\gamma)$ is $\leq \aleph_\mu$. Then the collection of

[18] Moreover, one has effective versions of these results, using a natural notion of effective relative categoricity.

ψ-*inaccessibles* is defined to be the class of all γ such that $Cl_\psi(\gamma) \cap \Omega_{\mu+1} = \gamma$ when μ is least with γ less than $\Omega_{\mu+1}$.

The hard work on this idea then went into showing that one had matchups of the θ functions with the functions defined by the Bachmann–Pfeiffer–Isles procedures. That was carried out in part by Aczel in unpublished notes and to a full extent by Jane Bridge in her doctoral dissertation at Oxford (1972). Bridge also showed that the countable ordinals generated by the θ functions are recursive. Subsequent extensions and simplifications were made by Schütte and members of his school beginning in the early 1980s, most efficiently by Wilfried Buchholz (1986), and that has become the current standard method for naming large ordinals in proof theory.

Branching out: Foundations of category theory I. Besides using these categorical notions, in the late 1960s, I had become interested in the question of the foundations of category theory. From the very beginnings of the subject as introduced by Eilenberg and Mac Lane in 1945, it was recognized that the notion of category lends itself naturally to possibly problematic instances of self-application. Not only does it appear reasonable to speak of the category **Grp** of all groups, the category **Top** of all topological spaces, etc., we are also led to consider the category **Cat** of all categories, whose morphisms are just all the functors F: A → B, where A and B are arbitrary categories. Even more, given any two categories A and B, no matter how large, it seems one can form the category B^A of all functors from A to B, whose morphisms from given F to G are all the natural transformations η: F → G. Thus one may contemplate as apparently reasonable mathematical objects such categories as $\mathbf{Grp}^{\mathbf{Top}}$ and $\mathbf{Cat}^{\mathbf{Cat}}$, with each counting as an object of **Cat**.

From the beginnings, too, of the subject it was suggested that some sort of set-theoretical foundation is needed for category theory since such naïve or unrestricted readings having to do with "large" and "super-large" categories appeared to border on the familiar paradoxes. Eventually, Mac Lane (1961, 1969, and elsewhere) pushed for finding a suitable set-theoretical framework to deal with these problems. The "one universe" solution that he settled on as presented in the text Mac Lane (1971) is one of the approaches that is widely accepted. A universe U in a set-theoretical framework is a nonempty transitive set that contains ω, is closed under the operations of pairing, union and power set, and is closed under strong replacement, i.e., (f: a → U) implies $f[a] \in U$ for $a \in U$ and f an arbitrary function. The existence of such a universe U is equivalent to the existence of a strongly inaccessible cardinal. A related approach is that ascribed to Grothendieck; that invokes the assumption of arbitrarily large universes, thus whose existence is equivalent to the existence of infinitely many strongly inaccessible cardinals. Relative to any such universe U, a set is called *small* if it belongs to U and *large* if it is a subset of U but not a member of U; similarly for categories. Special attention is given to *locally small categories*, i.e., those whose Hom sets are all small. Thus, for example, in place of **Grp**, **Top**, and **Cat** one deals in such a set-theoretical reduction with the categories of all small groups, small topological spaces, and small categories, respectively; each is a large, locally small category. In these terms what takes the place of $\mathbf{Grp}^{\mathbf{Top}}$ is now a "super-large" or "meta" category, i.e., one lying beyond U though one existing in a perfectly reasonable way in the class of all sets. It seems to be universally accepted in practice that such distinctions as those of being small, locally small, and large are essential to many of the fundamental theorems of category theory, the prime example being Freyd's Adjoint Functor Theorem (AFT). As Freyd (1964) pp. 85–86 said of his theorem, the crucial "solution set" condition for it "is not baroque" since one has counterexamples to AFT when that condition is dropped. Moreover, such set-theoretical conditions can be verified for the various categories that arise in the applications to such areas as combinatorial topology, homological algebra, and algebraic geometry.

In my first venture into the foundations of category theory [20], I introduced refinements of the Mac Lane and Grothendieck approaches in which the condition on a transitive set U to be a universe is weakened to the requirement that U with the membership relation restricted to it forms an elementary substructure of the class of all sets; in other words, it is a kind of surrogate for the universe of all sets. The resulting theories are conservative over ZFC, so do not require the existence of inaccessible cardinals, but it was not clear if this would take care of all the cases handled by the stronger notion of universe. Later, in a talk for a logic meeting at Orléans in 1972, I began to think instead of ways in which one might obtain a direct foundation of "naïve" or "unlimited" category theory. My first effort in that direction was never published, though it did eventually get posted on my home page when I revisited the subject years later.

Abstract model theory and the Tarski Symposium. Another subject that I turned to in the early 1970s was abstract model theory. The idea of a "logic" considered in model-theoretic terms had developed along three lines, first that initiated by Mostowski on cardinality quantifiers in the late 1950s, then the work of Tarski and his school on infinitary languages in the mid-1960s, and finally the work of Per Lindström on generalized quantifiers and abstract characterizations of first-order logic (1966, 1969). In the latter, Lindström showed among other things that first-order logic is the largest logic satisfying the compactness theorem and the Löwenheim–Skolem theorem. Abstract model theory blossomed as a means to provide a uniform framework in which to organize, compare, and seek out the properties of the many stronger logics that had then come to be recognized. My interest in the subject was drawn to it through the discussion in our logic seminar at Stanford of Lindström's work, as well as a characterization by Barwise of the language $L_{\infty,\omega}$ that allows arbitrarily long conjunctions and disjunctions but only ordinary quantification, as the largest language that is absolute relative to the Kripke–Platek axioms KP.

At the 1971 symposium in Berkeley in honor of Tarski's 70th birthday, I gave a rather ambitious lecture in which I sketched my work in three directions: local functors, functional interpretation with non-constructive functionals, and many-sorted interpolation theorems, all of which I have described above. But in my write up for the symposium volume of these, I restricted myself to the last, but I added relevant new work from abstract model theory. Namely, I introduced notions for model-theoretic languages L and L* of L being *adequate to truth* in L*, and of L being *truth maximal*. The latter means that L is adequate to truth in itself and whenever it is adequate to truth in L*, then L* is contained in L. By the *Δ-interpolation* (or Suslin-Kleene) *property for L* is meant the statement that if K_1 and K_2 are complementary projective classes for L, then K_1 (and hence K_2) is an elementary class for L; when it holds, this implies the Beth-definability property for L, but is stronger. The main new results of my Tarski symposium paper are that L is truth maximal if and only if it has the Δ-interpolation property, and that the logic L_A for A admissible is truth maximal if and only if A is contained in the hereditarily countable sets. Examples of logics that fail to have the Δ-interpolation property are the extensions of first-order logic by the cardinality quantifiers Q_κ for κ an infinite cardinal, and second-order and higher order logics.

I published this work in full in [31] along with another note on abstract model theory in which I dealt with properties φ that are invariant on the range of definable relations **R** between structures for a model-theoretic language L [32]. The main result for L that satisfies the many-sorted interpolation theorem is that if **R** is an L-elementary class (or even L-projective class) of pairs of L-structures and if φ is invariant on the range of **R**, then there is an L-sentence ψ such that φ holds in N if and only if ψ holds in M whenever **R**(M, N). This has as an immediate consequence theorems of Beth, Robinson, Gaifman, Barwise, and Rosenthal.

I did no further direct work in abstract model theory, but contributed to the subject with Jon Barwise by organizing in the early 1980s workshops for and editing the volume *Model-Theoretic Logics* that appeared in 1985. My work influenced various contributions to that volume, including those of Makowsky (1985) and Väänänen (1985) in particular.

On my own; 1972 and anon. The Berkeley symposium in 1971 had been an important opportunity for me to pay my homage to Tarski; as I wrote in my article for the occasion, I was always aware of my great debt to him for his teaching as much in work far distant from his own as in that paper. We had friendly relations and met on a number of academic and social occasions from then until his death in 1983, but he never asked and we never did speak of my interests or work in the field that he had crucially opened up for me.

As to my second mentor, a rupture took place in my relations with Kreisel in 1972 for reasons that I never fully understood. As I related above, I had first met Kreisel in my last year as a graduate student 17 years before, and once he came to Stanford we were constantly and intensely engaged with each other in seminars and with innumerable personal discussions in his office or mine, and often for hours on the telephone. He read and critiqued all my papers and I did the same for him. Not only were we the closest of colleagues but he was friends with Anita and me and we often socialized. So the rupture was a very painful one but in retrospect not a surprising one. I had seen Kreisel became enthusiastic and closely engaged with a number of other logicians and then suddenly break off relations after a few years with no explanation. In contrast, I could console myself that our relationship had worked so well for so long. But nevertheless it was strange to be in the same department and see each other in one way or another in the dozen or so years that followed and never—except for one occasion— say one word to each other.[19] And what was more important—not that I had not already carved out a distinct and distinctive career for myself—I was now fully on my own. It was time.

Theories of finite type and mathematical practice. Up to now, it has been possible to explain the development of my work in fairly chronological sequence. However, beginning in the early 1970s, that turns into various strands that run side by side and weave in and out, and it will not always be so easy to relate that to the progress of my academic career, but here it goes. Moreover, I shall generally take the topics in bigger blocks.

The year 1972–1973 was the second in which I took a sabbatical leave of absence. In the fall of that year, the arrangement was to visit the Mathematical Institute in Oxford; my host at the Institute in those days was Robin Gandy, a fine logician with an inimitable personality. Robin had been a student and friend of Alan Turing and mainly worked in recursion theory, and it was partly through him that I began to take an interest in recursion on functionals of finite type that had been inaugurated by Kleene's groundbreaking work of 1957. Dana Scott came to Oxford that year from Princeton as Professor of Mathematical Logic (a position he held until 1981) and it provided a fortuitous opportunity to renew our old friendship.

From January 1973 on, my visit was to the University of Paris VII, where I gave lectures in French on proof theory, since the English of those to whom I lectured was not then as good as it would become—not that my French was much better. There were many more logicians then in Paris than had been in my previous visit 1964–1965, and they were pursuing all the main branches of mathematical logic. Among those attending my lectures was the young proof theorist Jean-Yves Girard, who had come out of the blue in 1970 with his proof of normalization of terms for a system of analysis using the novel impredicative notion of "candidat de réductibilité". (That method was then applied independently by Per Martin-Löf and Dag Prawitz to proofs of cut-elimination for analysis and type theory, thus rounding out the work on Takeuti's conjecture.) Another logician attending my lectures was the model-theorist Gabriel Sabbagh, with whom I was to become a life-long friend.

My work during that year was primarily devoted to writing up the work described in the preceding section, in particular, the results that I had reported to the Nice conference in 1970. Within a couple of years that was to mesh with an invitation to contribute a chapter to the *Handbook of Mathematical Logic* being organized under the editorship of Jon Barwise. What I proposed for that became "Theories of finite type related to mathematical practice" [34], though by the time I finished it in 1975–1976, I had begun to explore various type-free theories as will be described below. The aim of the *Handbook* chapter (which is rather long) was to give recursion-theoretic and proof-theoretic information about various classical theories of finite type over the natural numbers that account for (i) Bishop-style constructive analysis without restriction on logic, (ii) predicative analysis, and

[19] Kreisel retired from Stanford in 1985.

(iii) some descriptive set theory. The theories themselves include Gödel's primitive recursive functionals both in full and as a restriction of the recursion functional due to Kleene. For (ii), they are augmented by an operator for quantification over N, more specifically via the unbounded least number operator, and for (iii) by the operator that tests for well-foundedness of an ordering relation in N, more specifically the operator that selects the left-most branch through the Brouwer–Kleene ordering. Furthermore, one can consider induction on the natural numbers as given in full or as restricted to quantifier-free formulas. The recursion-theoretic models are both of hereditary effective operations and of recursion in higher types à la Kleene, while the proof theory makes use of my adaptations of Gödel's functional interpretation in [23] to the case of non-constructive operators combined with the double-negation interpretation and relativized forms of Tait's normalization procedures for infinitely long terms. The metatheorems then give various conservation theorems over second-order theories and bounds for the provably recursive ordinals of the systems involved. It was stated in the article that a development in full of this material would be given in a book with the title *Explicit Content of Actual Mathematical Analysis* that was supposed to be in progress, but was never carried out in that form.[20]

In fact, in order to formalize mathematical practice of the indicated kinds in the most direct way in the work on my intended book, I decided that one would have to make use instead of a system of variable finite types (VFT) that I introduced in [56]. I devoted much work in the years 1977–1981 writing up a draft of my book based on the VFT systems. In particular, I outlined there a succession of steps to reduce the VFT (aka VT) systems to the theories of constant finite types that had been described in my *Handbook* chapter; that could then be used to determine the strength of the various theories of variable finite type under consideration. But there were obstacles to the plan, and I never published the book draft based on these systems. Instead, as was shown much later in work with Gerhard Jäger [85, 96] it turned out to be simpler to translate the VFT systems into systems of Explicit Mathematics (cf. next) for which the proof theory could be developed directly without going through the systems of finite type as before.

Systems of Explicit Mathematics I. In early 1974, a meeting was held at Monash University in Melbourne, Australia in which I introduced the first systems of what I call Explicit Mathematics; these were to influence much of my further research and that of my collaborators. I have recently described the motivation for this work and related systems in Sec. 2 of an article called "The operational perspective: three routes" [164], and the following is drawn from that.

While I was thinking about the smoothest way in which to develop mathematics on a predicative basis, Errett Bishop's novel informal approach to constructive analysis (Bishop 1967) had made a big impression on me and I was interested in seeing what kind of more or less direct axiomatic foundation could be given for it

[20] And what an awkward title!

that would explain how it managed to look so much like classical analysis in practice while admitting a constructive interpretation. Closer inspection showed that this depended on dealing with all kinds of objects (numbers, functions, sets, etc.) needed for analysis as if they are given by explicit presentations, each kind with an appropriate "equality" relation, and that operations on them are conceived to lead from and to such presentations preserving the given equality relations. In other words, the objects are conceived of as given *intensionally*, while a classical reading is obtained by working *extensionally* instead with the equivalence classes with respect to the given equality relations. Another aspect of Bishop's work that was more specific to its success was his systematic use of witnessing data as part of what constitutes a given object, such as modulus of convergence for a real number and modulus of (uniform) continuity for a function of real numbers. Finally, his development did not require restriction to intuitionistic logic (though Bishop himself abjured the Law of Excluded Middle).

Stripped to its core, the ontology of Bishop's work is given by a universe of objects, each conceived to be given explicitly, among which are operations and classes (*qua* classifications). This led to my initial formulation of a system T_0 of Explicit Mathematics in [30] in which that approach to constructive mathematics could be directly formalized. In addition, I introduced a second system T_1, obtained by the adjunction of the unbounded minimum operator so as to include a similar foundation of predicative mathematics. The theory T_0 was formulated in a single-sorted language with basic relations =, App, Cl, and η. App(x, y, z) expresses that x is an operation which when applied to y has the value z, while Cl(x) expresses that x is a class (ification) and $y\eta x$ expresses that y has the property given by x when Cl(x) holds. Variables $A, B, C, \ldots X, Y, Z$, are introduced to range over the objects satisfying Cl, and $y \in X$ is also written for $y\eta x$ where $x = X$. The basic logic of T_0 is the classical first-order predicate calculus.[21] The axioms of T_0 include basic operational axioms as a partial combinatory structure with pairing, projections, and definition by cases, while the remaining axioms are operationally given class existence axioms. For example, we have an operation **prod** which takes any pair X, Y of classes to produce their cartesian product, $X \times Y$ and another operation **exp** which takes X, Y to the cartesian power Y^X, also written $X \to Y$. The formation of such classes is governed by an *Elementary Comprehension Axiom* scheme (ECA) that tells which properties determine classes in a uniform way from given classes. These are given by formulas φ in which classes may be used as parameters to the right of the membership relation and in which we do not quantify over classes, and the uniformity is provided by operations $\mathbf{c}_\varphi$ applied to the parameters of φ.[22] But to form general products, we need further notions and an additional axiom. Given a class I, by an I-termed sequence of classes is meant an operation f with domain I such that for each $i \in I$ the value of $f(i)$ is a class X_i; one

[21] In (F 1979) I also examined T_0 within intuitionistic logic.

[22] The scheme ECA can be finitely axiomatized by adding constants for the identity relation, the first-order logical operations for negation, conjunction, existential quantification, and inverse image of a class under an operation.

wishes to use this to define $\Pi X_i[i \in I]$. It turns out that in combination with ECA a more basic operation is that of forming the join (or disjoint sum) $\sum X_i[i \in I]$ whose members are all pairs (i, y) such that $y \in X_i$; an additional *Join axiom* (J) is introduced in T_0 to assure existence of the join as given by an operation $\mathbf{j}(I, f)$. Finally, T_0 contains an operation $\mathbf{i}(A, R)$ and associated axiom (IG) for *Inductive Generation* which produces the class of objects accessible under the relation R (a class of ordered pairs) hereditarily within the class A. In particular, IG may be used to produce the class **N** of natural numbers, then the class **O** of countable tree ordinals, and so on. The system in which **i**(A, R) is simply required to be the least *class* satisfying the closure condition for A and R is called restricted T_0, resp. T_1.

It was shown in [30] how to construct a model of T_0 in which the universe is the set of natural numbers and App(x, y, z) is interpreted as $\{x\}(y) \simeq z$, so that the extensions of the operations range over the partial recursive functions. More generally, one can build a model for T_0 on any domain M containing a copy of N on which is given a class F of partial functions from N to N whose cardinality is not greater than that of M. In particular, one can obtain a model of T_0 in which the extensions of the operations from N to N are arbitrary partial numerical functions. Similarly, one can construct a model of T_1 on the natural numbers in which the extensions of the operations range over the partial Pi-1-1 functions, hence those of the total operations range over the hyperarithmetic functions. And one can construct models of T_1 in which, like T_0, the extensions of the operations from N to N range over arbitrary partial numerical operations.

For the summer European Logic Colloquium that was held in 1978 at the University of Mons in Belgium,[23] I was invited to give a series of lectures and chose to give four on various aspects of T_0, considered in both intuitionistic and classical logics; these were published in [42]. Among other things, I discussed there various informal and formal approaches to constructive mathematics, and sketched the direct formalization of Bishop-style constructive mathematics in T_0. Using forms of realizability interpretation, I established the expected disjunction and existential quantification properties for certain variants of T_0 in intuitionistic logic. By the model constructions just described, the theorems of T_0 serve to generalize results from recursive, constructive, and classical mathematics. In the final section of [42], a number of results were stated about the proof-theoretical strength of various natural subsystems of T_0 in classical logic as measured by familiar systems of second-order arithmetic. The main results were that restricted-T_0 is proof-theoretically equivalent to (Delta-1-1 CA) and in T_0 in full is bounded in strength by (Delta-1-1 CA) + BI. Proofs of these were sketched in the Mons article, with full proofs later given in my chapter with Sieg [53] in the 1981 volume on iterated inductive definitions to be discussed below.

Actually, at a couple of meetings before that of Mons, I had shown how one could make use of extensions of T_0 and T_1 to generalize other parts of mathematics.

[23] The 1978 Mons meeting as a whole was dedicated to the memory of Paul Bernays, who had died the year before, and Kurt Gödel, who had died that year.

The paper [40] was for a meeting on generalized recursion theory held in Oslo in June 1977. There I considered systems $T_0(S)$ and $T_1(S)$ in which S acts like the class of all sets. The system $T_0(S)$ could be used to develop a form of measure theory that generalizes Bishop's first form of constructive redevelopment of that subject (subsequently superseded by his work with Cheng). And both systems could be used to develop theories of accessible ordinal numbers and higher number classes generalizing the classical set theory of these notions.

Still before that, in the article [44] for a 1976 meeting in Jyväskylä, Finland, I had dealt with a theory T_Ω that is interpretable in T_1 in which Ω acts like the class of countable ordinals. In that, one can express a form of the Continuum Hypothesis, CH, by saying that one has an Ω-enumeration of the total functions from N to N. T_Ω has a model in which the order type of Ω is the least admissible ordinal greater than ω, i.e., Church-Kleene ω_1. The point of introducing this theory was to provide a way of accounting for the success of the novel work of Cutland (1972, 1973) that provided an analogue over suitable admissible sets A to various parts of classical model theory in which $\aleph_1$ and CH play a special role. Cutland dealt with models M $\subseteq$ A whose satisfaction relation is Σ_1 over A; he took the elements of A to be the analogues of the countable sets (not the finite sets!), and the Σ_1 subsets of A to be the analogues of sets of cardinality $\leq \aleph_1$. Among Cutland's results were analogues of existence and uniqueness of $\aleph_1$ saturated structures, the Ehrenfeucht–Mostowski theorem, the Vaught two-cardinal theorem, and results connected with categoricity in uncountable powers. What I showed in [44] is that T_Ω serves to generalize Cutland's work and the relevant parts of classical model theory.

Systems of Explicit Mathematics were to be the subject of considerable investigation since then, especially through the work of Gerhard Jäger and his colleagues and students at the University of Bern. In addition, I showed that the operational perspective was adaptable to a wider variety of contexts, including, as we will see later on, systems of operational set theory and the unfolding of schematic systems.

1979–1980. The urge to live. Oxford. After my year in Oxford and Paris in 1972–1973, my next sabbatical was slated for 1979–1980. Michael Dummett—who was then a Fellow at All Souls College in Oxford—urged me to apply for a Visiting Fellowship there, which I did, and succeeded in being invited for the autumn and winter terms. Through Robin Gandy I then arranged to visit Wolfson College for the spring term. In the past, when we left Stanford for any period of time, we always rented out our house to visiting academics, and this year was no different. But this time it was to someone we knew personally and were very happy to have in the house, Paul Benacerraf. All plans were thus in place when on April 13, 1979, I was mugged and shot on a street in San Francisco, and came near to dying.

The circumstances were these. From the mid-1970s on, Anita and I shared a pied-à-terre with a small group of friends in San Francisco. It was a top floor flat in a three-storey apartment building on Jones Street in Russian Hill, one of the nicer but not very affluent areas of the city (since then much more so), rather hilly with great views over the Golden Gate while convenient to many of the city's main attractions. We rotated stays with the other couples, with whom we would meet at the apartment on a Friday once a quarter to map out our schedule over the coming period over drinks and goodies, followed by a dinner out together. It happened that on this occasion, the apartment was ours and instead of returning to it directly from dinner, asked one of the couples to drop us a few blocks from there, since it was better for their return route to Stanford. We then decided to walk up a different way than usual, not one we had tried before, and it turned out to be darker and steeper than expected.

At one point, Anita stopped to look at something on the side, when a man darted out from between two cars and pointed a small silver pistol at me. I stared at the man, thinking, strangely, he could be my brother in looks and size, and said, "I don't have any money on me"—it was true, I had accidentally left my wallet in the apartment before dinner, and our friends had covered us for our part of the bill—and turned slightly to open my jacket to show that the billfold pocket was empty. At that moment his pistol went off, I was wounded in the right side of my chest, and fell to the ground screaming in excruciating pain and agony, "I don't want to die. I don't want to die." I was 51 years old, not young anymore, but felt I had so much more to live for.

As soon as the gun went off the man raced back down the street; all this was in a matter of minutes. Anita was unharmed and aghast and did not know what to do (there were no cell phones in those days), but neighbors heard me screaming, and called an ambulance and came out and covered me with a blanket. The ambulance came in 15 minutes and the nurses assured me I would not die and anesthetized me. I was taken to S.F. General, which happens to have one of the best trauma centers in the world. Exploratory surgery was done; it turns out the bullet had grazed my lung, which was partially collapsed, and it had lodged in the right side of my chest, and was not removed at the time. In a few days, I was transferred to the Stanford Hospital where I could be under the care of my physician and close to family and friends. The bullet was not removed until much later when I was well on the way to recovery, since it was doing no harm where it was lodged.

When I was well enough, I was asked to come to the S.F. Police Department to see if I could identify my shooter from their collections of photos of criminals. I was sure I would be able to spot him if he was there: I kept looking for my "brother", but never found him. We discussed the shooting with the police. Their guess was that it was not intentional, but that the gun had a hair trigger and that is why my mugger ran off.

April 13 happened to be my mother's birthday, a day on which I always called her in Los Angeles to wish her well, but had somehow forgotten to do so that day. Never mind that it was Friday the 13th, it was part of the series of happenstances that made—for me—that day different from all other days.

Iterated inductive definitions and subsystems of analysis. In 1981, the volume of Lecture Notes in Mathematics 897 entitled *Iterated Inductive Definitions and Subsystems of Analysis: Recent Proof-Theoretical Studies* was published under the joint authorship of Wilfried Buchholz, Solomon Feferman, Wolfram Pohlers, and Wilfried Sieg; I shall refer to it as LNM 897 below. As reported in my preface to that volume, it represented the remarkable progress that had been made since 1968 on the problem that I had raised at the Buffalo Conference on proof theory and constructive mathematics. In my view then and in the following years, the central problem to which this work responded was the need for an ordinally informative, conceptually clear, proof-theoretic reduction of classical theories of iterated arithmetical inductive definitions to corresponding constructive systems. The first breakthrough in this area was made by Pohlers (1975) that was to become the core of his Habilitationsschrift with Schütte (Pohlers 1977).[24] That breakthrough was the start of a 5-year sustained effort in developing a variety of approaches to the above problem by my coauthors.

The proof-theoretical work on systems of single and (finitely or transfinitely) iterated arithmetical inductive definitions were the first challenges to obtaining perspicuous ordinal analyses and constructive reductions of impredicative theories. The general problem was both to obtain exact bounds on the provably recursive ordinals and to reduce inductive definitions described "from above" as the least sets satisfying certain arithmetical closure conditions to those constructively generated "from below". In the event, the work on these systems took us only a certain way into the impredicative realm, but the method of local predicativity for semi-formal systems with uncountably infinitary rules of inference that Pohlers developed to deal with them turned out to be of wider application. What I want to emphasize in the following is, first of all, that ordinal analysis and constructive reduction are separable goals and that in various cases, each can be done without the other, and, second, that the aim to carry these out in ever more perspicuous ways has led to recurrent methodological innovations.

The consideration of formal systems of "generalized" inductive definitions originated with Georg Kreisel (1963) in a seminar that he led on the foundations of analysis held at Stanford in the summer of 1963.[25] Kreisel's aim there was to assess the constructivity of Spector's consistency proof of full second-order analysis (Spector 1962) by means of a functional interpretation in the class of so-called bar recursive functionals. The only candidate for a constructive foundation of those functionals would be the hereditarily continuous functionals given by computable representing functions in the sense of (Kleene 1959) or (Kreisel 1959). So Kreisel asked whether the intuitionistic theory of inductive definitions given by monotonic arithmetical closure conditions, denoted $ID_1(mon)^i$ below, serves to generate the class of (indices of) representing functions of the bar recursive functionals. Roughly speaking, $ID_1(mon)$, whether classical or intuitionistic, has a predicate P_A for each arithmetic $A(P, x)$ (with a place-holder predicate symbol P) which has been proved to be monotonic in P, together with axioms expressing that P_A is the least predicate

[24] I first met Pohlers at a workshop on proof theory that was held in Tübingen in April 1973. I lectured there about my work in [22] and possible proof-theoretic attacks on the reduction of classical systems of iterated inductive definitions to corresponding constructive systems. That stimulated Pohlers to make his first attack on that problem.

[25] The notes for that seminar are assembled in the unpublished volume *Seminar on the Foundations of Analysis, Stanford University 1963. Reports*, of which only a few mimeographed copies were made; one copy is available in the Mathematical Sciences Library of Stanford University.

definable in the system that satisfies the closure condition $\forall x(A(P, x) \rightarrow P(x))$. In the event, Kreisel showed that the representing functions for bar recursive functionals of types ≤ 2 can be generated in an $ID_1(mon)^i$ but not in general those of type ≥ 3.

Because of this negative result, Kreisel did not personally pursue the study of theories of arithmetical inductive definitions any further, but he did suggest consideration of theories of finitely and transfinitely iterated such definitions as well as special cases involving restrictions on the form of the closure conditions A(P, x). For example, those A in which the predicate symbol P has only positive occurrences are readily established to be monotonic in P. And of special interest among such A are those that correspond to the accessible (i.e., well-founded part) of an arithmetical relation. And, finally, paradigmatic for those are the classes of recursive ordinal number classes O_α introduced in Church and Kleene (1936) and continued in Kleene (1938). The corresponding formal systems for α times iterated inductive definitions (α an ordinal) are denoted as (in order of decreasing generality) $ID_\alpha(mon)$, $ID_\alpha(pos)$, $ID_\alpha(acc)$, and $ID_\alpha(O)$ in both classical and intuitionistic logics, where the restriction to the latter is signaled with a superscript 'i'.[26] For limit ordinals λ, we shall also be dealing with $ID_{<\lambda}(-)$, the union of the $ID_\alpha(-)$ for $\alpha < \lambda$, of each of these kinds, whether classical or intuitionistic. Finally, when no qualification of ID_α or $ID_{<\lambda}$ is given, it is meant that we are dealing with the corresponding $ID_\alpha(pos)$ or $ID_{<\lambda}(pos)$, since there is a relatively easy reduction of the monotonic case to the positive case. The $ID_\alpha(O)$ theories, or similar ones for constructive tree classes, are of particular interest, because the elements of those classes wear their build-up on their sleeves, i.e., can be retraced constructively; some of the $ID_\alpha(acc)$ classes considered below share that significant feature.

Kreisel's initiative led one to study the relationship between such theories to subsystems of classical analysis considered independently of Spector's approach and as the subject of proof-theoretical investigation in their own right. The first such result was obtained by William Howard sometime around 1965, though it was not published until 1972. He showed in Howard (1972) that the proof-theoretic ordinal of $ID_1(acc)^i$ is $\varphi\varepsilon_{(\Omega+1)}0$, as measured in the hierarchy of normal functions introduced in Bachmann (1950). Howard's method of proof proceeded via an extension of Gödel's functional interpretation. This was the first ordinally informative characterization of an impredicative system using a system of ordinal notation based on a natural system of ordinal functions. What was left open by Howard's work was whether one could obtain a reduction of the general classical ID_1 to $ID_1(acc)^i$ (and even better to $ID_1(O)^i$) and thus show that the proof-theoretic ordinal is the same, and similarly for the systems of iterated inductive definitions more generally.[27]

These results and the prior work of Takeuti (1967) containing constructive proofs of consistency of (Pi-1-1 CA) and (Pi-1-1 CA) + BI, together with the results of my Buffalo conference article [22] reducing these to ID_ω, gave hope that one could obtain

[26] The positivity requirement has to be modified in the case of intuitionistic systems.

[27] As will be explained below, Zucker (1971, 1973) showed the ordinals to be the same without a reduction argument and by a method that did not evidently extend to the iterated case.

a constructive reduction of some of the above second-order systems via a reduction of classical theories of iterated inductive definitions to their intuitionistic counterparts.[28] What Takeuti had done was to carry out his consistency proofs by an extension of Gentzen's methods with cut-reduction steps measured in certain partially ordered systems that Takeuti called ordinal diagrams; these are not based on natural systems of ordinal functions such as those in the Bachmann hierarchy. Takeuti proved the well-foundedness of the ordering of ordinal diagrams by constructive arguments that could be formulated in suitable intuitionistic iterated accessible IDs. These methods were later extended to (Delta-1-2 CA) + BI in Takeuti and Yasugi (1973).

The first successful results on ordinal analysis for theories of iterated inductive definitions were obtained only on the intuitionistic side by Per Martin-Löf (1971) via normalization theorems for the $ID_n(acc)^i$ systems as formulated in calculi of natural deduction. He conjectured the bounds $\varphi\varepsilon(\Omega_n + 1)0$ in the Bachmann–Pfeiffer hierarchies for these and proved that their supremum is the ordinal of $ID_{<\omega}(acc)^i$ by the use of Takeuti (1967).

The first breakthrough on the problems of ordinal analysis for the classical systems was made by Pohlers (1975) to give ordinal upper bounds for the finite ID_n also by an adaptation of the methods of Takeuti (1967); this was extended later in his Habilitationsschrift, Pohlers (1977), to arbitrary α, with the result that

$$|ID_\alpha| \leq \theta\varepsilon(\Omega_\alpha + 1)0$$

as measured in the modified (Feferman–Aczel) hierarchies described above. In addition, Buchholz and Pohlers (1978) showed this to be best possible by verification of

$$\theta\varepsilon(\Omega_\alpha+1)0 \leq |ID_\alpha(acc)^i|$$

using a constructive well-ordering proof of each proper initial segment of a natural recursive ordering of order type $\varphi\varepsilon(\Omega_\alpha + 1)0$. These results lent further hope to the solution of the reductive problem posed above. Independently of their work, in his Stanford dissertation, Sieg (1977) adapted and extended the method of Tait (1970) followed by a formalization of the cut-elimination argument to reduce ID_α to $ID_{\alpha+1}(O)^i$, and thence $ID_{<\lambda}$ to $ID_{<\lambda}(O)^i$, for limit λ, without requiring any involvement of ordinal bounds.

The volume LNM 897. In particular, my preface (Feferman 1981) to the volume fills out the historical picture to that point. Then the first chapter, Feferman and Sieg (1981a), goes over reductive relationships between various subsystems of (Sigma-1-2 AC), systems of iterated inductive definitions, and subsystems of the system T_0 of explicit mathematics from Feferman (1975). The second chapter, Feferman and Sieg (1981b), showed how to obtain the reductions of (Sigma-1-(n+1) AC) to (Pi -1-n CA)$<\varepsilon_0$ by proof-theoretic arguments (based on a method called *Herbrand analysis* by Sieg), in place of the model-theoretic arguments that had been used by Friedman. Following that, Sieg (1981) presented the work of his thesis in providing the reductions of ID_α to $ID_{\alpha+1}(O)^i$ and of $ID_{<\lambda}$ to $ID_{<\lambda}(O)^i$ for limit λ, without the intervention of ordinal analysis. In the next two chapters, Buchholz (1981a, 1981b)

[28] BI is the scheme of Bar Induction, i.e., the implication from well-foundedness to transfinite induction.

introduced uncountably infinitary semi-formal systems making use of a special new $\Omega_{\alpha+1}$-rule in order, in the first of these to obtain the proof-theoretical reduction of the ID_α to suitable $ID_\alpha(acc)^i$ and in the second to reestablish the ordinal bounds previously obtained by Pohlers. Finally, in the last chapter, Pohlers (1981) presented a new approach called the *method of local predicativity*, to accomplish the very same results in a different way. This dispensed with the earlier dependence on the methods of Takeuti's (1967); the more perspicuous method of local predicativity, in its place, utilizes a kind of extension to uncountably branching proof trees of the methods of predicative proof theory. But both Buchholz' and Pohlers' works in the Buchholz et al. (1981) volume required the use of certain syntactically defined collapsing functions, in order to reduce prima facie uncountable derivations to countable ones in a way that allows one to obtain the recursive ordinal bounds. As will be described below, this was superseded a decade later by the work of Buchholz (1992) showing how to obtain the same bounds without the use of such collapsing functions.

- 1979–1980, shooting and the year at Oxford, visits to Munich
- Presidency of the ASL; the Prague ASL "meet"
- Iterated Inductive Definitions next
- Essays, surveys
- The Gödel project
- Why a little bit goes a long way.

Foundations of category theory II. In the discussion of the foundations of category theory above, I said that it seemed natural to have self-membership for categories, such as **Cat** ∈ **Cat**, and that for any two categories A and B, even "large" ones like **Grp**, **Top**, and **Cat**, we should be able to construct the category B^A of all functors from A to B and show that it is a member of **Cat**. In contrast to the various set-theoretical foundations described above, I have called such a potentially consistent framework an unlimited theory of functors and categories. Moreover, I introduced just such a theory at a meeting of the Congrès de Logique d'Orléans, held in Orléans, France in September 1972. The typewritten write up of that work was informally circulated but never published, for reasons to be explained below. Eventually, though, I had that scanned as a pdf file and posted on my home page as [25].

The photos on the previous page are from 1971, 1983, and 2006. The topmost photo was taken at the Tarski Symposium at Berkeley, celebrating Tarski's 70th Birthday. Sol's perspective on that event and the talk he gave is discussed in the section entitled "Abstract model theory and the Tarski Symposium". This photo was taken by Steven Giranti, he gave us permission to publish it here. The larger middle photo captures some of the participants of the Gödel Symposium of 1983 in Salzburg; they play also a significant part in the autobiography: from left to right, Stephen Kleene, John Dawson, Solomon Feferman, Michael Beeson, Georg Kreisel, Azriel Lévy, Dana Scott, and Charles Parsons. The remaining photo shows Feferman lecturing on truth at the Center for Philosophy of Science, University of Pittsburgh. John Norton took that photo and we have his permission to publish it here.

Pittsbursh, USA Wilfried Sieg
Stanford, USA Rick Sommer

Part B: Solomon Feferman's CV

Born: December 13, 1928, New York, New York
Died: July 26, 2016, Stanford, California

Educational Record:

B.S. California Institute of Technology, 1948 (Mathematics)
Ph.D. University of California, Berkeley, 1957 (Mathematics), Advisor: Alfred Tarski
Military Service: U.S. Army, 1953–55

Professional Record:

1956–58 Instructor of Mathematics and Philosophy, Stanford University
1958–62 Assistant Professor of Mathematics and Philosophy, Stanford University
1959–60 NSF Post-doctoral Fellow at the Institute for Advanced Study, Princeton
1958–63 Consultant, Stanford Research Institute
1962–68 Associate Professor of Mathematics and Philosophy, Stanford University
1964–65 NSF Senior Post-doctoral Fellow, University of Paris and University of Amsterdam
1967–68 Visiting Associate Professor, Massachusetts Institute of Technology
1968–03 Professor of Mathematics and Philosophy, Stanford University (Emeritus, 2004)
1972–73 Guggenheim Fellow, University of Oxford and University of Paris
1979–80 Visiting Fellow at All Souls and Wolfson Colleges, University of Oxford

1985–92 Chairman, Department of Mathematics, Stanford (on leave 1986–87 and 1989–90)
1986–87 Guggenheim Fellow, Stanford, ETH Zürich, University of Rome
1989–90 Fellow, Stanford Humanities Center
1993–2003 Patrick Suppes Family Professor, Stanford University (Emeritus, 2004)
1995–96 Fellow, Center for Advanced Study in the Behavioral Sciences, Stanford
2001 Fellow (April and May 2001), Mittag-Leffler Institute, Djursholm, Sweden
2003 Visiting Professor of Philosophy, UC Berkeley, Spring Semester

Honors:

1990 Elected Fellow, American Academy of Arts and Sciences
1999 University of California at Irvine Chancellor's Distinguished Fellow
2003 Rolf Schock Prize in Logic and Philosophy

Professional Activities:

1964–67 Member, Executive Committee and Council, Association for Symbolic Logic
1976–79 Editor, Transactions and Memoirs, American Mathematical Society
1986–2003 Editor, Ergebnisse der Mathematik (Springer-Verlag)
1986–2003 Editor, Perspectives in Mathematical Logic (ASL/Springer-Verlag)
1980–82 Member, AMS Committee on Translations
1980–82 President, Association for Symbolic Logic
1983–90 Advisory Editor, Studies in Proof Theory (Bibliopolis),
1983–86 Member, Steering Committee, 1986 International Congress of Mathematicians
1982–2003 Editor-in-Chief, Kurt Gödel Collected Works

Named Lectures:

1983 Association for Symbolic Logic Retiring Presidential Address, "Reflecting on incompleteness", 30 December, Boston University.
1989 Thoralf Skolem Lectures, "New life for Skolem's finitist program", 4 September, "Foundational ways", 5 Sept 1989, University of Oslo.
1990 E.W. Beth Lectures, "Logics for termination and correctness of functional programs", 20 March, Delft University of Technology, and "Infinity in mathematics: Where is it necessary?", 21 March, University of Amsterdam.
1997 Association for Symbolic Logic Annual Gödel Lecture, "Occupations and preoccupations with Gödel: His Works and the work", 22 March, MIT.
1997 Spinoza Lecture at ESSLLI, "What is a logical operation? (According to Tarski, McGee, and me.)", 13 August, Université de Provence.
2006 Alfred Tarski Lectures, "Truth unbound", 3 April, "The 'logic' question", 5 April, "Real computation", 7 April, University of California at Berkeley.
2007 Ernest Nagel Lecture, "Gödel, Nagel, minds and machines", 27 Sept, Columbia University.
2008 Martin Löb Lectures (inaugural), "Operational set theory and 'small' large cardinals", 12 May, "Gödel, Nagel, minds and machines", 13 May, University of Leeds.

2012 Paul Bernays Lectures (inaugural), "Bernays, Gödel and Hilbert's consistency program", 11 September, "Is the continuum hypothesis a definite mathematical problem?", 12 September, "Foundations of unlimited category theory", 12 September, ETH Zürich.

Ph.D. Students:

1966 Harrison, Joseph
1967 Barwise, K. Jon
1969 Larson, Alan
1970 Nebres, Bienvenido
1971 Zucker, Jeffrey
1976 Weyhrauch, Richard
1977 Sieg, Wilfried
1979 Lindström, Ingrid
1985 Talcott, Carolyn
1986 Mason, Ian
1986 Takahashi, Shuzo
1986 Ungar, Anthony
1989 Mancosu, Paolo
1990 Bellin, Gianluigi
1991 Fernando, Timothy
1998 Pezzoli, Elena
1999 Hofweber, Thomas
2013 Buchholtz, Ulrik

Part C: Active Projects of 2016

Apart from the work on this very volume (as we outlined in the Introduction to this chapter), Feferman was engaged in at least four significant projects that are described in sections **C1** though **C4** below; they remain incomplete. Let us first mention a fifth project he did complete: At the beginning of the year, he wrote a talk for a Symposium in honor of Charles Parsons at Columbia University; he delivered the talk on April 19. The text, quite polished, was posthumously published as *Parsons and I: Sympathies and Differences* in the Journal of Philosophy, volume CXIII, no. 5/6, pp. 234–246.

In **C1**, we present the Proposal Feferman submitted in March 2014 to Oxford University Press for the second volume of essays under the title *Logic, Mathematics and Conceptual Structuralism*. That proposal was approved quickly, but work on it was set aside; the engagement with this volume and other projects took precedence. In **C2**, the reader finds a brief abstract of the book project Feferman was pursuing with Gerhard Jäger and Thomas Strahm. **C3** contains the abstract of Feferman's keynote address to the special AMS/ASL session on applications of Logic, Model Theory, and Theoretical Computer Science to Systems Biology at Seattle, WA; it was delivered on January 9, 2016. On Feferman's website one finds a quite polished paper that was the source of the PowerPoint presentation he gave in Seattle; the PowerPoint slides are also found on his website. Finally, another abstract is presented in **C4**; Feferman sent it to Michael Rathjen on April 16, 2016 under the "subject" *Draft title and abstract for Brouwer volume*. He wrote in the body of the message:

Dear Michael,

Let me know what you think of the draft title and abstract below. I welcome any modifications.

Best,

Sol

The envisioned joint paper was to be published in *Indagationes Mathematicae* on the occasion of the 50th anniversary of Brouwer's death.

Pittsbursh, USA — Wilfried Sieg
Stanford, USA — Rick Sommer

C1. *Logic, Mathematics and Conceptual Structuralism*

Logic, Mathematics and Conceptual Structuralism

Proposal to OUP

Solomon Feferman

March 17, 2014

This is a proposal for OUP to publish a follow-up volume to my collection of essays *In the Light of Logic* that appeared in 1998 in the series Logic and Computation in Philosophy. Since then I have published over thirty more essays of the same character that partially expand on the earlier collection but more importantly take up several new topics. What is proposed now is publication of a selection of twelve of those essays that form a thematically coherent progression of ideas concerning the philosophy of logic and mathematics and that concentrate on these new avenues of thought. A number of these essays appeared in out of the way places so this publication would serve to make them available to a wider audience, while bringing them together will allow the presentation of extended lines of thought. The proposed title for the new volume is *Logic, Mathematics and Conceptual Structuralism.*

Below is a draft of Table of Contents laid out in five parts; the #s refers to the articles as they have been posted on my home page at http://math.stanford.edu/~feferman/papers.html.

Of course, print copies can also be made available if desired. For the moment, the final item, #95, is in draft form only (and is the only such); a final version is in preparation.[29]

[29] This is Feferman's original formulation; the #s in this version refers to the items in the bibliography included in the volume. In the *Proposed Table of Contents* we left out the bibliographical details, as they are contained in the references.

Part I consists of a single essay that appeared in a volume entitled Philosophy of Mathematics: 5 questions, in which each contributor was invited to respond to five stimulating questions, first about how one was drawn to the foundations and/or philosophy of mathematics, followed by examples from one's work to illustrate the use of mathematics for philosophy; then one was asked about the proper role of philosophy of mathematics for various areas and what one considers the most neglected problems and important open problems in that field. My response to the first question includes a fair amount of intellectual autobiography, and the responses to the remaining questions will serve to orient the reader to a number of my concerns that are addressed in the remaining parts of this volume.

Several ways of dealing with mathematical intuition and the mathematical mind are dealt with in the three essays of Part II. The third of these is one of several in which I have critically analyzed the arguments of Gödel, Penrose, Lucas, and others that mind is not mechanical on the basis of the incompleteness theorems; it was written for the volume *Free Will and Modern Science*.

Part III, in contrast, is devoted to questions concerning the nature and limits of classical logic that underlies almost all of mathematical practice. The very influential semantical tradition inaugurated by Tarski is traced historically in the first of these essays. That is followed by two essays in which I critique the claims of Tarski, Sher, and others to explain which classical operations are logical entirely in terms of certain set-theoretical semantical criteria that lead one far beyond what is usually accepted. In the final essay in that part I show, following work of Zucker, that combined semantical and inferential criteria yield exactly the operations that are usually taken for systems of classical logic.

That leads directly to Part IV whose general heading is Conceptual Structuralism, an approach to the philosophy of mathematics that I have been elaborating in recent years. In the first essay, I explain how both classical and constructive logics are to be considered in that framework as well as in their relationship to mathematical practice. The second essay in that part examines five essentially different conceptions of the continuum (from the Euclidean to the set-theoretical) in the light of conceptual structuralism.

In the final part, I deal in two essays with the controversial question whether the continuum hypothesis (CH) is a definite mathematical problem. The current approaches to settling CH assume new axioms for set theory that assert the existence of extremely large cardinal numbers. My argument against the definiteness of CH viewed either as an ordinary mathematical problem or as a logical problem proceeds on the one hand by directly questioning the case for such axioms and on the other hand by appeal to conceptual structuralism.

Most of these essays were written for a general audience from advanced undergraduates to professionals interested in questions of the philosophy of mathematics. For the most part, it assumes some background in logic and mathematics at the advanced undergraduate level but each essay has parts that can be rewarding without presuming much of that. The printed volume is estimated to come out under 250 pages in length.

Logic, Mathematics, and Conceptual Structuralism
(Proposed Table of Contents)
I. My Route
Philosophy of mathematics: 5 questions. (#137)
II. The Mathematical Mind
Mathematical intuition vs. mathematical monsters. (#114)
And so on… Reasoning with infinite diagrams. (#150)
Gödel's incompleteness theorems, free will, and mathematical thought. (#147)
III. What is Classical Logicality?
Tarski's conceptual analysis of semantical notions. (#121)
Logic, logics, and logicism. (#109)
Set-theoretical invariance criteria for logicality. (#144)
Which quantifiers are logical? A combined semantical and inferential criterion. (#161)
IV. Conceptual Structuralism
Logic, mathematics, and conceptual structuralism. (#158)
Conceptions of the continuum. (#142)
V. New Axioms for Mathematics?
Does mathematics need new axioms? (#106)
Is the Continuum Hypothesis a definite mathematical problem? (#148)

C2. *Foundations of Explicit Mathematics*

The aim is to produce a research-level book systematically expositing some central results on Explicit Mathematics. The plan of the book is to begin with an introduction explaining the genesis of Explicit Mathematics. The treatment of Explicit Mathematics itself is divided into the treatment of the

- First-order part of Explicit Mathematics, so-called Applicative Theories and
- Theories of Classes and Names.

The central theories of Explicit Mathematics will be analyzed from a proof-theoretic perspective, comparing them to well-known subsystems of second-order arithmetic and set theory. Typical ontological properties of Explicit Mathematics and the development of "ordinary mathematics" within this framework will be discussed.

The final part of this book is about alternative operational approaches like Feasible Operational Systems, the Unfolding Program, Universes, and Operational Set Theory.

C3. *Many-sorted First-order Model Theory as a Conceptual Framework for Biological and Other Complex Dynamical Systems*

Abstract

When complex biological systems (among others) are conceived reductively, they are modeled in set-theoretical hierarchical terms from the bottom up. But the point of view of Systems Biology (SB) is to deal with such systems from the top down. So in this talk I will suggest the use of many-sorted first-order structures with downward nested sorts as an alternative conceptual framework for modeling them. In particular, the notion of a nested substructure allows one to study parts of a structure in isolation from the rest, while the notion of restriction allows one to study a structure relative to some of its parts treated as black boxes. The temporal dimension can be incorporated both as an additional sort and in the indexing of sorts, allowing for both static and dynamic views of a system. Furthermore, one may make use of a quite general theory of recursion on many-sorted first-order structures which includes both discrete and continuous computations. Some possible applications of this model-theoretic approach to SB include excision or substitution of a part as operations on structures, similarity of biological systems via similarity notions for structures, and homeostasis via least fixed point recursion.

C4. *Semi-intuitionistic Theories of Sets*

Abstract

Brouwer argued that limitation to constructive reasoning is necessary when dealing with "unfinished" totalities such as the natural numbers. As a complement to that, the predicativists such as Poincaré and Weyl (of *Das Kontiuum*) accepted the natural numbers as a "finished" or definite totality, but nothing beyond that. On the other hand, the "semi-intuitionistic" school of descriptive set theory (DST) of Borel et al. in the 1920s took both the natural numbers and the real numbers as definite totalities and explored what could be obtained on that basis alone. From a metamathematical point of view, these and other different levels of definiteness can be treated in the single setting of semi-intuitonistic theories of sets, whose basic logic is intuitionistic, but for which the law of excluded middle (LEM) is accepted for bounded formulas. One may then add the assumption that the set of natural numbers exists, corresponding to the predicative point of view, or that both it and its power set exists, corresponding to the point of view of DST, and so on. One then investigates which propositions are definite, i.e., satisfy LEM, relative to such additional set existence assumptions. This will be a report on the work that has been done so far on that question, for which novel techniques have had to be employed. Finally, some further open problems will be raised in that setting.

Solomon Feferman Publications

Books

Formal Consistency Proofs and Interpretability of Theories, PhD dissertation, University of California, Berkeley, 1957.

The Number Systems. Foundations of Algebra and Analysis. Addison-Wesley, Reading, 1957; 2nd edn. Chelsea Publishing Co., New York; distributed by the American Mathematical Society, Providence, Rhode Island, 1989.

Notes for Lectures on Metamathematics, Stanford University Library, 1962.

In the Light of Logic, Oxford University Press, New York, 1998.

Alfred Tarski. Life and Logic (with A. B. Feferman), Cambridge University Press, New York, 2004.

Edited Books

Iterated Inductive Definitions and Subsystems of Analysis: Recent Proof-theoretical Studies (with W. Buchholz, W. Pohlers, and W. Sieg), Lecture Notes in Mathematics 897, Springer Verlag, Berlin, 1981.

Model-theoretic Logics (with J. Barwise), Springer Verlag, Berlin, 1985.

Kurt Gödel. Collected Works, Vol. I. Publications 1929–1936 (as editor-in-chief, with J.W. Dawson, Jr., S. C. Kleene, G. H. Moore, R. M. Solovay, and J. van Heijenoort), Oxford University Press, New York, 1986.

Kurt Gödel. Collected Works, Vol. II, Publications 1938–1974 (as editor-in-chief, with J.W. Dawson, Jr., S. C. Kleene, G. H. Moore, R. M. Solovay, and J. van Heijenoort), Oxford University Press, New York, 1990.

Kurt Gödel. Collected Works, Vol. III, Unpublished essays and lectures (as editor-in-chief, with J.W. Dawson, Jr., W. Goldfarb, C. Parsons, and R.M. Solovay), Oxford Univ. Press, New York, 1995.

The Collected Works of Julia Robinson, American Mathematical Society, Providence, Rhode Island, 1996.

Kurt Gödel. Collected Works. Vol. IV. Correspondence A-G Oxford University Press, (as editor-in-chief, with J.W. Dawson, Jr. as co-editor-in-chief, and with W. Goldfarb, C. Parsons, W. Sieg and R.M. Solovay), Oxford University Press, Oxford, 2003.

Kurt Gödel. Collected Works. Vol. V. Correspondence H-Z (as editor-in-chief, with J. W. Dawson, Jr. as co-editor-in-chief, and with Warren Goldfarb, Charles Parsons and Wilfried Sieg, co-editors), Oxford University Press, Oxford, 2003.

Proofs, Categories and Computations. Essays in Honor of Grigory Mints (with W. Sieg), College Publications, London, 2010.

Kurt Gödel. Essays for his Centennial (with C. Parsons and S. G. Simpson), Lecture Notes in Logic 33, Association for Symbolic Logic, Cambridge University Press, New York, 2010.

For Feferman

Reflections on the Foundations of Mathematics. Essays in Honor of Solomon Feferman (W. Sieg, R. Sommer, and C. Talcott, eds.), Lecture Notes in Logic 15, Association for Symbolic Logic, A. K. Peters, Ltd., Natick, 2002

Articles

1957

1. Degrees of unsolvability associated with classes of formalized theories, *J. Symbolic Logic* 22, 161–175.

1959

2. The first order properties of products of algebraic systems (with R. L. Vaught), *Fundamenta Mathematicae* 47, 57–103.
3. Representability of recursively enumerable sets in formal theories (with A. Ehrenfeucht), *Archiv für Mathematische Logik und Grundlagenforschung* 5, 37–41.

1960

4. Arithmetization of metamathematics in a general setting, *Fundamenta Mathematicae* 49, 35–92.
5. 1-consistency and faithful interpretations (with G. Kreisel and S. Orey), *Archiv für Mathematische Logik und Grundlagenforschung* 5, pp. 52–63.
6. Formal consistency proofs and interpretability of theories, *Summaries of talks presented at the Summer Institute for Symbolic Logic,* Cornell University 1957, 2nd edn., Institute for Defense Analyses, 71–77.
7. Some recent work of Ehrenfeucht and Fraïssé, *ibid.*, 201–209.

1962

8. Transfinite recursive progressions of axiomatic theories, *J. Symbolic Logic* 27, 259–316.
9. Incompleteness along paths in progressions of theories (with C. Spector), *J. Symbolic Logic* 27, 383–390.
10. Classifications of recursive functions by means of hierarchies, *Transactions American Mathematical Society* 104, 101–122.

1964

11. Systems of predicative analysis, *J. Symbolic Logic* 29, 1–30.

1965

12. Some applications of the notions of forcing and generic sets (Summary), in *The Theory of Models,* Proceedings of the 1963 International Symposium, Berkeley (J. Addison, L. Henkin, and A. Tarski, eds.), North-Holland, Amsterdam, 89–95.
13. Some applications of the notions of forcing and generic sets, *Fundamenta Mathematicae* 56, 325–345.

1966

14. Persistent and invariant formulas relative to theories of higher order (with G. Kreisel), (Research Announcement) *Bulletin American Mathematical Society* 72, 480–485.
15. Predicative provability in set theory, (Research Announcement) *Bulletin American Mathematical Society* 72, 486–489.

1968

16. Systems of predicative analysis, II. Representations of ordinals, *J. Symbolic Logic* 33, 193–220.
17. Autonomous transfinite progressions and the extent of predicative mathematics, in *Logic, Methodology, and Philosophy of Science III,* Proceedings 3rd International Congress, Amsterdam, 1967, (B. van Rootselaar and F. Staal, eds.), North-Holland, Amsterdam, 121–135.
18. Persistent and invariant formulas for outer extensions, *Compositio Mathematica* 20, 29–52.
19. Lectures on proof theory, in *Proceedings of the Summer School in Logic,* Leeds, 1967, Lecture Notes in Mathematics vol. 70, 1–107.

1969

20. Set-theoretical foundations of category theory (with an Appendix by G. Kreisel), in *Reports of the Midwest Category Seminar, III* (S. MacLane, ed.), Lecture Notes in Mathematics vol. 106, 201–247.

1970

21. Hereditarily replete functionals over the ordinals, in *Intuitionism and Proof Theory,* Proceedings of the Summer Conference at Buffalo, 1968 (A. Kino, J. Myhill and R. E. Vesley, eds.), North-Holland, Amsterdam, 289–301.
22. Formal theories for transfinite iterations of generalized inductive definitions and some subsystems of analysis, in *Intuitionism and Proof Theory,* Proceedings of the Summer Conference at Buffalo, 1968 (A. Kino, J. Myhill and R. E. Vesley, eds.), North-Holland, Amsterdam, 303–326.

1971

23. Ordinals and functionals in proof theory, in *Actes du Congrès International des Mathématiciens (Nice) 1970,* vol. 1, Gauthier-Villars, Paris, 229–233.

1972

24. Infinitary properties, local functors, and systems of ordinal functions, in *Conference in Mathematical Logic - London '70* (W. Hodges, ed.) Lecture Notes in Mathematics vol. 255, 63–97.

1973

25. Some formal systems for the unlimited theory of functors and categories, http://math.stanford.edu/~feferman/papers/Unlimited.pdf, unpublished MS.

1974

26. Applications of many-sorted interpolation theorems, in *Proceedings of the Tarski Symposium,* Proceedings of Symposia in Pure Mathematics vol. XXV (L. Henkin, et al., eds.), American Mathematical Society, Providence, RI, 205–223.
27. Predicatively reducible systems of set theory, in *Axiomatic Set Theory,* Proceedings of Symposia in Pure Mathematics vol. XIII Part 2 (T. J. Jech, ed.), American Mathematical Society, Providence, RI, 11–32.
28. Intuitionism (part of an article on "Mathematics, foundations of"), in *Encyclopedia Britannica*, 15th ed. pp. 633–635 and p. 639.
29. Two notes on abstract model theory. I. Properties invariant on the range of definable relations between structures, *Fundamenta Mathematicae* 82, 153–165.

1975

30. A language and axioms for explicit mathematics, in *Algebra and Logic* (J. N. Crossley, ed.), Lecture Notes in Mathematics, vol. 450 87–139.
31. Two notes on abstract model theory. II. Languages for which the set of valid sentences is semi-invariantly implicitly definable, *Fundamenta Mathematicae* 89, 111–130.
32. Impredicativity of the existence of the largest divisible subgroup of an Abelian p-group, in *Model Theory and Algebra. A Memorial Tribute to A. Robinson* (D. H. Saracino and V. B. Weispfenning, eds.), Lecture Notes in Mathematics vol. 498, 117–130.
33. Non-extensional type-free theories of partial operations and classifications, I, in *Logic Conference, Kiel, 1974* (G. H. Müller, et al., eds.) Lecture Notes in Mathematics vol. 500, 73–118.

1977

34. Theories of finite type related to mathematical practice, in *Handbook of Mathematical Logic* (J. Barwise, ed.), North-Holland, Amsterdam, 913–971.

35. Categorical foundations and foundations of category theory, in *Logic, Foundations of Mathematics and Computability Theory,* vol. 1, Proceedings 5th International Congress of Logic, Methodology, and Philosophy of Science, London, Ontario, 1975 (R. E. Butts and J. Hintikka, eds.), Reidel, Dordrecht, 149–169.
36. Review of *Proof Theory* by G. Takeuti, *Bulletin American Mathematical Society* 83, 351–361.
37. Recursion in total functionals of finite type, *Compositio Mathematica* 35, 3–22.

38. Generating schemes for partial recursively continuous functionals (summary), in *Colloque International de Logique,* (Clermont-Ferrand, 1975), pp. 191–198, Èditions du C.N.R.S., Paris.
39. Inductive schemata and recursively continuous functionals, in *Logic Colloquium '76,* Proceedings of the Oxford Conference (R. O. Gandy and M. Hyland, eds.) North-Holland, Amsterdam, 373–392.

1978

40. Recursion theory and set theory: a marriage of convenience, in *Generalized Recursion Theory II* (J. E. Fenstad, et al., eds.), North-Holland, Amsterdam, 55–98.
41. The logic of mathematical discovery vs. the logical structure of mathematics, in *PSA 1978,* Philosophy of Science Association, East Lansing, MI, 309–327; reprinted as Ch. 3 in *In the Light of Logic.*

1979

42. Constructive theories of functions and classes, in *Logic Colloquium '78,* Proceedings of the Mons Colloquium (M. Boffa, et al., eds.), North-Holland, Amsterdam, 159–224.
43. A more perspicuous formal system for predicativity, in *Konstruktionen versus Positionen* I (K. Lorenz, ed.), Walter de Gruyter, Berlin, 68–93.
44. Generalizing set-theoretical model theory and an analogue theory on admissible sets, in *Essays on Mathematical and Philosophical Logic* (J. Hintikka, et al., eds.) Synthese Library, vol. 22, Reidel, Dordrecht, 171–195.
45. Review of *Proof Theory* by K. Schütte, *Bulletin American Mathematical Society (New Series)* 1, 224–228.
46. What does logic have to tell us about mathematical proofs?, *The Mathematical Intelligencer* 2, pp. 20–24; reprinted as Ch. 9 in *In the Light of Logic.*
47. Progressões transfinitas recursivas de teorias axiomáticas (translation of "Transfinite recursive progressions of axiomatic theories"), in *O Teorema de Gödel e a Hipótese do Contínuo,* Fundacão C. Gulbenkian, Lisbon, 573–754.
48. As progressões tranfinitas autónomas e a extensão da matemática predicativa (translation of "Autonomous transfinite progressions and the extent of predicative mathematics"), in *O Teorema de Gödel e a Hipótese do Contínuo,* Fundacão C. Gulbenkian, Lisbon, 755–790.
49. Sistemas de análise predicativa (translation of "Systems of predicative analysis"), in *O Teorema de Gödel e a Hipótese do Contínuo,* Fundacão C. Gulbenkian, Lisbon, 790–866.

1980

50. Consistency of the unrestricted abstraction principle using an intensional equivalence operator (with P. Aczel), in *To H. B. Curry: Essays on Combinatory Logic, Lambda Calculus and Formalism* (J. P. Seldin and J. R. Hindley, eds.) Academic Press, New York, 67–98.

1981

51. How we got from there to here, in *Iterated Inductive Definitions and Subsystems of Analysis: Recent Proof-theoretical Studies,* Lecture Notes in Mathematics, vol. 897 (W. Buchholz, S. Feferman, W. Pohlers, W. Sieg, eds.), 1–15.
52. Iterated inductive definitions and subsystems of analysis (with W. Sieg), *ibid.* 16–77.
53. Proof-theoretic equivalences between classical and constructive theories for analysis (with W. Sieg), *ibid.*, 78–142.

1982

54. Iterated inductive fixed-point theories: application to Hancock's conjecture, in *Patras Logic Symposion* (G. Metakides, ed.), North-Holland, Amsterdam, 171–196.
55. Inductively presented systems and the formalization of meta-mathematics, in *Logic Colloquium '80* (D. van Dalen, et al., eds.), North-Holland, Amsterdam, 95–128.
56. A theory of variable types, *Revista Colombiana de Matématicas* 19, Proceedings 5th Latin American Logic Symposium, Bogotá 1981 (X. Caicedo, et al., eds.), 95–105.
57. Monotone inductive definitions, in *The L. E. J. Brouwer Centenary Symposium* (A. S. Troelstra and D. van Dalen, eds.), North-Holland, Amsterdam, 77–89.

1983

58. Choice principles, the bar rule and autonomously iterated comprehension schemes in analysis (with G. Jäger), *J. Symbolic Logic* 48, 63–70.

1984

59. Toward useful type-free theories, I, *J. Symbolic Logic* 49, 75–111.
60. Foundational ways, in *Perspectives in Mathematics* (W. Jäger, et al., eds.), Birkhäuser, Basel, 147–158; reprinted as Ch. 4 in *In the Light of Logic.*
61. Kurt Gödel: conviction and caution, *Philosophia Naturalis* 21, 546–562; reprinted as Ch. 7 in *In the Light of Logic.*
62. Between constructive and classical mathematics, in *Computation and Proof Theory* (M. M. Richter, et al., eds.), Lecture Notes in Mathematics, vol. 1104 143–162.

1985

63. Working foundations, *Synthese* 62, 229–254.
64. Intensionality in mathematics, *J. Philosophical Logic* 14, 41–55.

1986

65. Gödel's life and work, in *Kurt Gödel. Collected Works, Vol. I. Publications 1929–1936* (S. Feferman, et al., eds.), Oxford Univ. Press, New York, 1–36; reprinted as Ch. 6 in *In the Light of Logic.*

1987

66. Proof theory: a personal report, Appendix to *Proof Theory,* 2nd edn., by G. Takeuti, North-Holland, Amsterdam, 447–485.
67. Infinity in mathematics: is Cantor necessary?, in *L'infinito nella scienza (Infinity in Science),* (G. Toraldo Francia, ed.), Istituto della Enciclopedia Italiana, Rome, 151–209; reprinted between Chs. 2 and 12 in *In the Light of Logic.*

1988

68. Hilbert's program relativized: proof-theoretical and foundational reductions, *J. Symbolic Logic* 53, 364–384.
69. Turing in the land of O(z), in *The Universal Turing Machine. A Half-century Survey* (R. Herken, ed.) Oxford University Press, Oxford, 113–147.

70. Weyl vindicated: *Das Kontinuum* 70 years later, in *Temi e prospettive della logica e della filosofia della scienza contemporanee, vol. I,* CLUEB, Bologna, 59–93; reprinted as Ch. 13 in *In the Light of Logic.*
71. Kurt Gödel: conviction and caution, reprinting in *Gödel's Theorem in Focus* (S. G. Shanker, ed.), Croom Helm, London, 96–114.

1989

72. Finitary inductively presented logics, in *Logic Colloquium '88* (R. Ferro, et al., eds.), North Holland, Amsterdam, 191–220; reprinted in *What is a Logical System?* (D. S. Gabbay, ed.), Clarendon Press, Oxford (1994), 297–328.

1990

73. Polymorphic typed lambda-calculi in a type-free axiomatic framework, in *Logic and Computation,* Contemporary Mathematics, vol. 106, American Mathematical Society, Providence, RI, 101–136.

1991

74. Reflecting on incompleteness, *J. Symbolic Logic*, 56, 1–49.
75. Proofs of termination and the "91" function, in *Artificial Intelligence and Mathematical Theory of Computation. Papers in honor of John McCarthy* (V. Lifschitz, ed.), Academic Press, Boston, 47–63.

1992

76. A new approach to abstract data types, I. Informal development, *Mathematical Structures in Computer Science* 2, 193–229.
77. A new approach to abstract data types, II. Computability on ADTs as ordinary computation, in *Computer Science Logic,* Lecture Notes in Computer Science vol. 626, Springer-Verlag, Berlin, 79–95.
78. Turing's 'Oracle': From absolute to relative computability - and back, in *The Space of Mathematics* (J. Echeverria, et al., eds.), Walter de Gruyter, Berlin, 314–348.
79. Logics for termination and correctness of functional programs, in *Logic from Computer Science* (Y. N. Moschovakis, ed.), MSRI Publications vol. 21, Springer-Verlag, New York, 95–127.
80. Logics for termination and correctness of functional programs, II. Logics of strength PRA, in *Proof Theory* (P. Aczel, et al. eds.), Leeds Proof Theory Programme 1990, Cambridge University Press, Cambridge, 197–225.

1993

81. Working foundations - '91, in *Bridging the Gap: Philosophy, Mathematics and Physics,* Boston Studies in the Philosophy of Science vol. 140 (G. Corsi, et al., eds.) Kluwer, Dordrecht, 99–124; reprinted as Ch. 5 in *In the Light of Logic.*
82. Jean van Heijenoort's scholarly work, Appendix to *Politics, Logic and Love. The Life of Jean van Heijenoort,* by A.B. Feferman, A. K. Peters Ltd., Wellesley, MA, 371–390.

83. The development of programs for the foundations of mathematics in the first third of the 20th century. Appeared in translation as, Le scuole de filosofia della matematica, in *Storia della scienza* (S. Petruccioli, ed.) Istituto della Enciclopedia Italiana, 10 v., 2001–2004, v. VIII (2004) 112–121.
84. Why a little bit goes a long way: Logical foundations of scientifically applicable mathematics, in *PSA 1992*, Vol. II, 442–455; reprinted as Ch. 14 in *In the Light of Logic.*
85. Systems of explicit mathematics with non-constructive μ-operator. Part I (with G. Jäger), *Annals of Pure and Applied Logic* 65, 243–263.
86. Gödel's *Dialectica* interpretation and its two-way stretch, in *Computational Logic and Proof Theory* (G. Gottlob, et al., eds.), Lecture Notes in Computer Science 713, 23–4; reprinted as Ch. 11 in *In the Light of Logic.*
87. What rests on what? The proof-theoretic analysis of mathematics, in *Philosophy of Mathematics*, Part I, Proceedings of the 15th International Wittgenstein Symposium (J. Czermak, ed.), Verlag Hölder-Pichler-Tempsky, Vienna, 141–171; reprinted as Ch. 10 in *In the Light of Logic.*

1994

88. Julia Bowman Robinson, December 8, 1919 - July 30, 1985, *Biographical Memoirs of the National Academy of Sciences* 63, 453–478.
89. Deciding the undecidable: Wrestling with Hilbert's problems, Inaugural lecture, Stanford University, May 13, 1994, published as Ch. 1 in *In the Light of Logic*, 3–27.

1995

90. Definedness, *Erkenntnis* 43, 295–320.
91. Predicative foundations of arithmetic (with G. Hellman), *J. Philosophical Logic* 24, 1–17.
92. Ordinal logics, *The Cambridge Dictionary of Philosophy*, 550–551.
93. Reflection principles, *The Cambridge Dictionary of Philosophy*, 682.
94. Ordinal logics (entry in the *Routledge Encyclopedia of Philosophy*).
95. Introductory note to Gödel's 1933 lecture, in *Kurt Gödel. Collected Works Vol. III. Unpublished Essays and Lectures* (S. Feferman, et al., eds.); reprinted as Ch. 8 in *In the Light of Logic.*

1996

96. Systems of explicit mathematics with non-constructive μ-operator. Part II (with G. Jäger), *Annals of Pure and Applied Logic* 79, 37–52.
97. Kreisel's "unwinding" program, in *Kreiseliana* (P. Odifreddi, ed.), A. K. Peters Ltd., 247–273.
98. Penrose's Gödelian argument, PSYCHE 2, 21–32.
99. Computation on abstract data types. The extensional approach, with an application to streams, *Annals of Pure and Applied Logic* 81, 75–113.
100. Proof theory, in *Encyclopedia of Philosophy Supplement*, MacMillan.
101. Gödel's program for new axioms: Why, where, how and what?, in *Gödel '96* (P. Hájek, ed.), Lecture Notes in Logic 6, 3–22.
102. Three conceptual problems that bug me, http://math.stanford.edu/~feferman/papers/conceptualprobs.pdf, unpublished lecture text for 7th Scandinavian Logic Symposium, Uppsala, 18 August 1996.

1997

103. My route to arithmetization, *Theoria* 63, 168–181.

1998

104. Gödel's functional ("Dialectica") interpretation (with J. Avigad), in *The Handbook of Proof Theory* (S. Buss, ed.), North-Holland, Amsterdam, 337–405.
105. Tarski and Gödel between the lines, in *Tarski and the Vienna Circle* (J. Wolenski and E. Köhler, eds.), Kluwer Academic Publishers, 53–63.

1999

106. Does mathematics need new axioms?, *American Mathematical Monthly* 106, 99–111.
107. Challenges to predicative foundations of arithmetic (with G. Hellman), in *Between Logic and Intuition: Essays in Honor of Charles Parsons* (G. Sher and R. Tieszen, eds.), Kluwer Academic Publishers, 317–339.
108. Ah, Chu. In *JFAK. Essays Dedicated to Johan van Benthem on the Occasion of his Fiftieth Birthday*, Amsterdam University Press, CD-ROM only.
109. Logic, logics, and logicism, *Notre Dame J. of Formal Logic* 40, 31–54.

2000

110. The unfolding of non-finitist arithmetic (with T. Strahm), *Annals of Pure and Applied Logic* 104, 75–96.
111. Highlights in proof theory, in *Proof Theory* (V. F. Hendricks et al., eds.), Kluwer Academic Publishers, 11–31.
112. The significance of Hermann Weyl's *Das Kontinuum, ibid.*, 179–194.
113. Relationships between constructive, predicative and classical systems of analysis, *ibid.*, 221–236.
114. Mathematical intuition vs. mathematical monsters, *Synthese* 125, 317–332.
115. Does reductive proof theory have a viable rationale?, *Erkenntnis* 53, 63–96.
116. Does mathematics need new axioms? (Proceedings of a symposium with H. M. Friedman, P. Maddy, and J. R. Steel), *Bulletin of Symbolic Logic* 6 (2000) 401–446.
117. Why the programs for new axioms need to be questioned, *ibid.* 401–413.
118. In memoriam: Kenneth Jon Barwise, 1942–2000. *Bulletin Symbolic Logic* 6, 505–508.

2003

119. Alfred Tarski and a watershed meeting in logic: Cornell 1957, in (J. Hintikka, et al., eds.) *Philosophy and Logic. In search of the Polish tradition,* Synthese Library vol. 323, Kluwer Acad. Pubs., 151–162.

2004

120. Tarski's conception of logic. *Annals of Pure and Applied Logic* 126 (2004) 5–13.
121. Tarski's conceptual analysis of semantical notions, *Sémantique et épistémologie* (A. Benmakhlouf, ed.) Editions Le Fennec, Casablanca, distributed by J. Vrin, Paris, 79–108. Reprinted in (D. Patterson, ed.) *New Essays on Tarski and Philosophy*, Oxford University Press (2008), 72–93.

122. Typical ambiguity: trying to have your cake and eat it too, in *One Hundred Years of Russell's Paradox* (G. Link, ed.), Walter de Gruyter, Berlin, 131–151.
123. Comments on "Predicativity as a philosophical position" by G. Hellman, *Review Internationale de Philosophie* (special issue, Russell en héritage. Le centenaire des *Principles*, Ph. de Rouilhan, ed.) 229 (no. 3), 313–323.

2005

124. Predicativity, in *The Oxford Handbook of the Philosophy of Mathematics and Logic* (S. Shapiro ed.), Oxford University Press, Oxford, 590–624.
125. The Gödel editorial project: A synopsis, *Bulletin Symbolic Logic* 11, 132–149; reprinted in *Kurt Gödel. Essays for his Centennial* (S. Feferman, C. Parsons, and S. G. Simpson, eds.), Lecture Notes in Logic 33, Association for Symbolic Logic, 3–20.

2006

126. What kind of logic is "Independence Friendly" logic? in *The Philosophy of Jaakko Hintikka* (Randall E. Auxier and Lewis E. Hahn, eds.), Library of Living Philosophers, Open Court, 453–469
127. Enriched stratified systems for the foundations of category theory, in *What is Category Theory?* (G. Sica, ed.) Polimetrica, Milano, 185–203; reprinted in (G. Sommaruga, ed.), *Foundational Theories of Classical and Constructive Mathematics*, Springer, Dordrecht (2011), 127–143.
128. Are there absolutely unsolvable problems? Gödel's dichotomy, *Philosophia Mathematica* 14(2), 134–152.
129. Review of *Incompleteness. The proof and paradox of Kurt Gödel,* by Rebecca Goldstein, *London Review of Books* vol. 28 no. 3 (9 Feb. 2006).
130. The impact of the incompleteness theorems on mathematics, *Notices of the American Mathematical Society* 53, no. 4, 434–439.
131. Turing's thesis, *Notices American Mathematical Society,* 53 no. 10, 1200–1205; reprinted in (A. W. Appel, ed.) *Alan Turing's Systems of Logic*, Princeton University Press, Princeton (2012), 13–26.
132. Tarski's influence on computer science, *Logical Methods in Computer Science* 2, issue 3, paper 6, http://arxiv.org/pdf/cs/0608062.pdf.

2008

133. Lieber Herr Bernays! Lieber Herr Gödel! Gödel on finitism, constructivity and Hilbert's program, *Dialectica* 62 (2008), 179–203. (This is a preprint of the following; the differences are of an editorial nature.)
134. Lieber Herr Bernays! Lieber Herr Gödel! Gödel on finitism, constructivity and Hilbert's program, in (M. Baaz et al., eds.) *Kurt Gödel and the Foundations of Mathematics. Horizons of Truth,* Cambridge University Press, Cambridge, 111–133.
135. Harmonious logic: Craig's interpolation theorem and its descendants, *Synthese* 164, no. 3, 341–357.
136. Axioms for determinateness and truth, *The Review of Symbolic Logic* 1, no. 2, 204–217.
137. Philosophy of Mathematics: 5 questions, in (V. F. Hendricks and H. Leitgeb, eds.) *Philosophy of Mathematics: 5 Questions*, Automatic Press/VIP, 115–135.
138. Andrzej Mostowski: An appreciation, in *Andrzej Mostowski and Foundational Studies* (A. Ehrenfeucht et al. eds.), IOS Press, Amsterdam, 388–392.

2009

139. Gödel, Nagel, minds and machines, *J. of Philosophy* CVI no. 4, 201–219.
140. Operational set theory and small large cardinals, *Information and Computation* 207, 971–979.
141. Modernism in mathematics, review of *Plato's Ghost* by Jeremy Gray, *American Scientist* 97 no. 5, 417.
142. Conceptions of the continuum, *Intellectica* 51, 169–189.

2010

143. The unfolding of finitist arithmetic (with T. Strahm), *The Review of Symbolic Logic* 3, 665–689.
144. Set-theoretical invariance criteria for logicality, *Notre Dame J. of Formal Logic* 51, 3–20.
145. On the strength of some semi-constructive theories, in (S. Feferman and W. Sieg, eds.) *Proof, Categories and Computation: Essays in Honor of Grigori Mints*, College Publications, London (2010); reprinted in (U. Berger, et al., eds.) *Logic, Construction, Computation*, Ontos Verlag, Frankfurt (2012), 201–22.
146. The proof theory of classical and constructive inductive definitions. A 40 year saga, 1968–2008, in (R. Schindler, ed.) *Ways of Proof Theory*, Ontos Verlag, Frankfurt, 7–30.

2011

147. Gödel's incompleteness theorems, free will and mathematical thought, in (R. Swinburne, ed.) *Free Will and Modern Science*, Oxford Univ. Press for the British Academy, Oxford (2011), 102–122.
148. Is the Continuum Hypothesis a definite mathematical problem? lecture text for the *Exploring the Frontiers of Incompleteness Project* (EFI), Harvard, http://math.stanford.edu/~feferman/papers/IsCHdefinite.pdf2012.
149. Axiomatizing truth: Why and how, in (U. Berger, et al., eds.) *Logic, Construction, Computation*, Ontos Verlag, Frankfurt (2012), 185–200.
150. And so on … Reasoning with infinite diagrams, *Synthese* 186, 371–386.
151. Review of Curtis Franks, *The Autonomy of Mathematical Knowledge. Hilbert's program revisited,* in *Philosophia Mathematica. Series III*, 20 no. 3, 387–400.
152. On rereading van Heijenoort's *Selected Essays*, in *Logica Universalis* 6 no. 3–4, 535–552.

2013

153. Foundations of unlimited category theory: What remains to be done, *The Review of Symbolic Logic* 6, 6–15.
154. How a little bit goes a long way: Predicative foundations of analysis, http://math.stanford.edu/~feferman/papers/pfa(1).pdf, unpublished notes dating from 1977–1981, with a new introduction.
155. About and around computing over the reals, in (B. J. Copeland, C. J. Posy and O. Shagrir, eds.) *Computability. Turing, Gödel, Church, and Beyond*, MIT Press, 55–76.
156. Turing's Thesis: Ordinal logics and oracle computability, in *Alan Turing: His Work and Impact* (S. B. Cooper and J. van Leeuwen, eds.) Elsevier, Amsterdam, 145–150.
157. Indescribable cardinals and admissible analogues, lecture text for meeting in celebration of Gerhard Jäger's 60th birthday, http://math.stanford.edu/~feferman/papers/Indes%20Cards%20&%20Admiss.pdf.

2014

158. Logic, mathematics and conceptual structuralism, in (P. Rush, ed.) *The Metaphysics of Logic*, Cambridge University Press, Cambridge, 72–92.
159. A fortuitous year with Leon Henkin, in (M. Manzano, I. Sain and E. Alonso, eds.) *The Life and Work of Leon Henkin*, Studies in Universal Logic, Birkhäuser, Cham, 35–40.
160. The Continuum Hypothesis is neither a definite mathematical problem nor a definite logical problem, revised version of "Is the Contiuum Hypothesis a definite mathematical problem?", http://math.stanford.edu/~feferman/papers/CH_is_Indefinite.pdf.

2015

161. Which quantifiers are logical? A combined semantical and inferential criterion, in (A. Torza, ed.) *Quantifiers, Quantifiers, Quantifiers: Themes in Logic, Metaphysics, and Language*, Synthese Library 373, Springer Verlag, Berlin, 19–30.
162. (with V. Lifschitz), In Memoriam: Grigori E. Mints, 1939–2014, *The Review of Symbolic Logic* 21, 31–33.
163. Theses for computation and recursion on concrete and abstract structures, to appear in a Turing Centennial volume edited by G. Sommaruga and T. Strahm; http://math.stanford.edu/*feferman/papers/CompRecThesesFinal.pdf.[30]
164. The operational perspective: Three routes, to appear in a volume in honor of Gerhard Jäger's 60th birthday; http://math.stanford.edu/*feferman/papers/Op_Persp_3Routes.pdf.[31]

[30] The article was published in this volume, re-titled *Turing's Revolution - The Impact of His Ideas about Computability*, Birkhäuser (2015), 105–126.

[31] The paper was published in (R. Kahle, T. Strahm, T. Studer eds.), *Advances in Proof Theory*, volume 28 of Progress in Computer Science and Applied Logic, Birkhäuser (2016), 269–291.

Part I
Mathematical Logic

From Choosing Elements to Choosing Concepts: The Evolution of Feferman's Work in Model Theory

Wilfrid Hodges

Abstract When Solomon Feferman began his research with Alfred Tarski in the early 1950s, model theory was still in process of becoming a distinct part of mathematical logic. Although Feferman's doctoral thesis was not in model theory, his interests included model theory from the start, and he published a paper in the field roughly once every six years throughout his career. His earliest work in model theory is recognised in the name 'Feferman-Vaught theorem', which stems from some very detailed bare-hands work on sums and products of structures. During the 1960s and 1970s he worked on applications of many-sorted interpolation theorems, in particular to derive results relating implicit and explicit definability in various contexts. In the 1980s he edited with Jon Barwise a monumental collection of essays on 'Model-theoretic logics'. In more recent papers he reflected on the conceptual basis of model theory from a historical point of view.

Keywords Solomon Feferman · Model theory · Feferman-Vaught theorem · Interpolation theorem

2010 Subject Classification 03C99

Solomon Feferman engaged with model theory in three different ways. In the 1950s he did fundamental work on the first-order theories of generalised sums and products, for which the name 'Feferman-Vaught theorem' will ensure that he never goes forgotten. Between the 1960s and the 1980s he followed a trajectory that began with applications of interpolation theorems in model theory, and led to work on set-theoretic generalisations of model theory. In more recent years he reshaped himself as a historian of ideas, recalling and explaining aspects of the development of mathematical logic through the twentieth century; model theory takes a significant place in this work, largely thanks to Feferman's connections with Tarski and the editing of Gödel's collected works.

W. Hodges (✉)
Herons Brook, Sticklepath, Okehampton EX20 2PY, England
e-mail: wilfrid.hodges@btinternet.com

G. Jäger and W. Sieg (eds.), *Feferman on Foundations*, Outstanding Contributions to Logic 13, DOI 10.1007/978-3-319-63334-3_1

These three themes seem rather distinct, though they probably had various interesting links in Feferman's own mind. So we will treat them separately.

My thanks to Jamshid Derakhshan, Janos Makowsky and Jeffery Zucker for various kinds of information and advice, and (posthumously, alas) to Solomon Feferman himself for a number of preprints of his work that he sent me over the years.

1 Logic When Feferman Entered the Field

In his valuable book with Anita Burdman Feferman, on the life and works of Alfred Tarski, Solomon Feferman recalls ([11] p. 171) that he went to work at Berkeley in 1948 as 'an eager new graduate student in mathematics'. He attended Tarski's lectures in metamathematics at Berkeley in 1949/50 ([26] footnote p. 80). By 1953 he was working on his doctoral thesis under Tarski's supervision, and simultaneously acting as Tarski's course assistant ([28] p. 388).

In the early 1950s the boundaries of logic were in a state of flux. Was the kind of thing that Tarski was doing at that date a branch of philosophy or of mathematics, or in the intersection of the two? Tarski himself had distinguished metamathematics from mathematics back in the 1930s. In 1950 he spoke of working 'on the borderline of algebra and metamathematics' [55], and he continued to put some of his work on the metamathematical side of the mathematics/metamathematics divide right up to 1978, as witness the title of [7]. By the time I began work in model theory in 1965, I think most model theorists regarded themselves as straight mathematicians—as I did—and the notion of metamathematics never entered their heads.

One memory I have is that in 1966 my supervisor John Crossley arranged for me to meet Georg Kreisel with a view to getting some advice on possible research topics. Kreisel's suggestions struck me as aimed more at foundations than model theory, and I didn't pursue his advice. This story is relevant to Feferman, because it appears that Kreisel had greater success drawing Feferman towards similar problems a year or so earlier; a background paper of Kreisel had the title 'Set theoretic problems suggested by the notion of potential totality' ([42], written in 1959 and cited in [32]).

Not only was the boundary between mathematics and philosophy unclear; the divisions within purely mathematical logic were unclear too. The now-standard division into Set Theory, Recursion Theory, Proof Theory and Model Theory seems to go back to the early 1960s. It reached the AMS Subject Classification in 1973. Around 1971 I remember Robin Gandy complaining that Jack Silver had been invited to give some talks in model theory and had spoken on set theory (or maybe it was the other way round); Silver's topic had been the use of forcing to prove the consistency of some propositions about model-theoretic two-cardinal theorems. Soon afterwards Saharon Shelah was inventing proper forcing to prove results in his model-theoretic classification theory. The problem that Tarski first proposed to Feferman was about decidability, but Feferman responded to it by proving results in model theory. A few years later Feferman was proving results in model theory by means of proof theory.

There will always be crossovers, but the boundaries between model theory and the other main divisions of mathematical logic have been reasonably robust for the last thirty years or so. The fact that they were not robust when Feferman came into the subject shows up in Feferman's work in two ways. The first is that nearly all of his so-called model-theoretic work contains at least a taste of some other logical discipline. Today this work of his seems to us strikingly interdisciplinary, though it may not have seemed so at the time. The second is that Feferman, being a reflective kind of person, drew on his experience to write a number of very interesting papers about the interactions between different parts of logic in the formative years of model theory.

In his paper for the Proceedings of the Tarski Symposium 1974 ([17] footnote p. 205), Feferman tells us that the paper which he presented to the Symposium itself 'was entitled *Model theory and foundations*. It dealt with three areas of mutually fruitful interaction which I had found between model theory and work in proof theory and constructivity'. One of these involved the application of many-sorted interpolation theorems, and we review it in Sect. 3 below. The second was about model-theoretic functors; we discuss it briefly at the end of Sect. 2 below. The third was on'ordinals and functionals in proof theory', as reported in [15]. Looking at that paper [15], I don't see anything in it that would be regarded as model theory today. But this fact in itself points to questions about the boundaries of model theory, which we come back to in Sect. 4.

2 The Feferman-Vaught Theorem

The Feferman-Vaught theorem is as much a technique as a theorem. A typical case of the theorem is that when the structure A is a cartesian product, say $A = \prod_{i \in I} B_i$ with $I \neq \emptyset$, we can reduce the question whether a first-order sentence ϕ is true in A to a question whether a certain set-theoretic formula Φ_ϕ is satisfied in the power set $\mathcal{P}(I)$ of the index set I by a certain finite sequence of sets $(\|\theta_0\|, \ldots, \|\theta_{n-1}\|)$, where $\|\theta\|$ is the set $\{i \in I : B_i \models \theta\}$. The reduction is effective in the sense that there is an algorithm which computes Φ_ϕ and the sentences $\theta_0, \ldots, \theta_{n-1}$ uniformly from ϕ. The technique that gives this special case applies to a wide range of other constructions.

Some results provoke the question 'How on earth did anybody think of doing that?'. The Feferman-Vaught theorem surely provokes that question; but in this case the history is well recorded, so we can answer the question. The answer runs as follows.

Starting in 1927, Tarski ran a seminar at Warsaw University, in which he and his students developed the method of 'elimination of quantifiers', which had been created a few years earlier by Leopold Löwenheim and Thoralf Skolem and pursued also by Cooper H. Langford. As practised by Skolem and Langford, the method was to take a structure A and show that for each first-order sentence ϕ in the language of A, we can effectively find a first-order sentence ϕ' which is true in A if and only if ϕ is true in A, and which is a boolean combination of sentences of some simple form; we can call these simple sentences the 'elimination set'. In some cases the elimination set would contain only quantifier-free sentences, so that passing from ϕ

to ϕ' eliminates quantifiers completely. But in general these sentences might contain quantifiers, though they would typically be all universal or all existential. (Cf. [11] pp. 72–75.)

Early successes of this programme were Presburger's quantifier elimination for the structure of integers with addition, and Tarski's own quantifier elimination for the structure of the real numbers with addition and multiplication. Tarski also devoted some of his early efforts to extending Langford's work on dense and discrete linear orderings. In 1938 Tarski and Andrzej Mostowski (a loyal student of Tarski, though officially his doctoral thesis supervisor was Kuratowski) began working on quantifier elimination for ordinals, in languages involving one or more of $<$, ordinal addition and ordinal multiplication. Mostowski's notes were lost during the war (they were in the notebook that he famously had to abandon in favour of a loaf of bread, [41] pp. 6f), and he never reconstructed them, though some of the main results were published in an abstract in 1949 [51]. Tarski later published a reconstruction of this work in a joint paper with his student John Doner [7]. One of the results was that the simply ordered structure of the ordinals has a decidable first-order theory. The obvious next result to try to prove was that the ordered structure of the ordinals with ordinal addition has decidable first-order theory, and this was one of the two tasks that Tarski set Feferman for his doctoral thesis ([31] p. 37).

Feferman approached the problem from the point of view of elimination of quantifiers, as Tarski no doubt expected him to do. Tarski may have given him some information from the joint work with Mostowski. But also a very pertinent paper of Mostowski had just appeared [50]. In this paper Mostowski proved quantifier elimination results for structures A that allow a decomposition as a cartesian power of another structure, say $A = B^I$ for some index set I. His elimination set was not in the first-order language of A; instead it also allowed one to refer to $\mathcal{P}(I)$ and sets $\|\theta_i\|$ as above. He showed that his results adapt also to 'weak powers' which are substructures of cardinal powers whose domains consist of all the elements that differ from e in at most finitely many factors, where e is a designated element of B (for example the identity element if B is a group, or 0 if B is an ordinal). From Mostowski's own account it seems that he thought he was not so much striking out in a new direction as finding a common generalisation of several results already proved by quantifier elimination. One application that he mentions is of particular interest to us here. He considers the ordinal α and the weak product of α copies of ω, with sums taken pointwise. The resulting structure is isomorphic to ω^α with natural addition; by Mostowski's results this structure has a decidable first-order theory. This is less interesting than ω^α with ordinal addition, but it suggests ways of approaching ordinal addition. In fact by late 1953 Feferman was able to follow this route and show that for every positive integer n, ω^n with ordinal addition has a decidable first-order theory. His main adjustment to Mostowski's scheme was that he allowed the elimination formulas to refer to other features of the construction of A from B, for example a linear ordering of the index set I.

In 1956 Robert Vaught (also a student of Tarski) noticed that Feferman's methods generalised to the case where A is a product of a family $(B_i : i \in I)$ of distinct structures; there was no need to assume that the factors were all isomorphic. Almost at once this gave a common generalisation both of Feferman's earlier work, and of Vaught's own work on sentences true in cartesian products. They wrote up the results as their joint paper 'The first order properties of products of algebraic systems' [33].

To explain how the technique works, we begin with the case of cartesian products, so that A has the form $\prod_{i \in I} B_i$ with $I \neq \emptyset$. We write L for the first-order language of A and the B_i. If $\bar{a}$ is a tuple of elements of A and $i \in I$, then we write $\pi_i \bar{a}$ for the tuple of projections to B_i of the terms of $\bar{a}$. If $\theta(\bar{x})$ is a formula of L, we write $\|\theta(\bar{a})\|$ for the set

$$\{i \in I : B_i \models \theta(\pi_i \bar{a})\}. \tag{1}$$

Feferman and Vaught prove a theorem stating that for each formula $\phi(\bar{x})$ of L there are a boolean formula Φ_ϕ and formulas $\theta_0(\bar{x}), \ldots, \theta_{k-1}(\bar{x})$ of L such that for any cartesian product $A = \prod_{i \in I} B_i$ of structures for the language L, and every tuple $\bar{a}$ of elements of A,

$$\begin{aligned} &A \models \phi(\bar{a}) \text{ if and only if the sequence } (\|\theta_0(\bar{a})\|, \ldots, \|\theta_{k-1}(\bar{a})\|) \\ &\text{satisfies } \Phi_\phi \text{ in the power set algebra } \mathcal{P}(I) \end{aligned} \tag{2}$$

Moreover Φ_ϕ and the formulas θ_i can be found effectively from ϕ, and the formulas θ_i can be chosen so that independently of A and I, the sets $\|\theta_0(\bar{a})\|, \ldots, \|\theta_{k-1}(\bar{a})\|$ form a partition of I (though some partition sets may sometimes be empty).

We state one corollary about decidability. Suppose the factor structures B_i are required to be models of a decidable theory T (i.e. the set of consequences of T is recursive). Given any sentence ϕ of L, we can determine as follows whether A can be chosen to be a model of ϕ. Find the formula Φ_ϕ and the sentences θ_j $(j < k)$ as in (2). Skolem's quantifier elimination reduces the condition on the righthand side of (2) to a boolean combination of conditions of the form

$$\mathcal{P}(I) \models \text{For at least } p\ i\text{'s}, i \in \bigcap_{j \in Y} \|\theta_j\| \setminus \bigcup_{j \notin Y} \|\theta_j\|, \tag{3}$$

where p is a positive integer and $Y \subseteq k$. Since the sets $\|\theta_j\|$ form a partition of I and I is not empty, (3) is false unless Y is a singleton $\{j_0\}$, and in this case (3) will be true if and only if there are at least p factors B_i that are models of θ_{j_0}. Such factors can be found if and only if $T \cup \{\theta_{j_0}\}$ is consistent, which can be checked since T is decidable. The rest is book-keeping.

The Feferman-Vaught theorem, in the form stated above, is proved by induction on the complexity of ϕ. The most interesting case is where $\phi(\bar{x})$ is $\exists y \phi'(\bar{x}, y)$, and for this reason Feferman and Vaught say that the proof is by the 'method of eliminating quantifiers' ([33] p. 66). I am not entirely convinced that this is the best description. It seems to me more illuminating to compare with Jaakko Hintikka's 'distributive normal forms' [39]. Let me digress briefly on these.

For simplicity suppose the language L has finitely many relation symbols and no function symbols. We will find, for all $n, r < \omega$, a finite set $\Theta_{n,r}$ of formulas $\theta(x_0, \dots, x_{n-1})$ of L, such that

$$\begin{array}{l}\text{For each fixed } n \text{ and } r, \text{ if } \Theta_{n,r} \text{ consists of the distinct formulas} \\ \theta_0, \dots, \theta_{k-1} \text{ then } \vdash \forall \bar{x} \text{ (Exactly one of } \theta_0(\bar{x}), \dots, \theta_{k-1}(\bar{x}) \text{ holds).} \end{array} \tag{4}$$

We will assume that the set $\Theta_{n,r}$ is given a fixed ordering, so that we can regard it as a sequence. (Feferman and Vaught [33] p. 64 call a sequence satisfying the condition in (4) a 'partitioning sequence'.)

When $r = 0$ we choose $\Theta_{n,0}$ so that the formulas describe the complete quantifier-free types of n-tuples.

For $r > 0$ we proceed by induction on r. Suppose $\Theta_{n+1,r}$ has been defined and consists of the formulas $\theta_0(x_0, \dots, x_n), \dots, \theta_{k-1}(x_0, \dots, x_n)$. List as $(s(j') : j' < k' = 2^k)$ all the subsets of k. Then we define $\Theta_{n,r+1}$ to consist of all the formulas $\psi_{j'}(x_0, \dots, x_{n-1})$ $(j' < k')$ where $\psi_{j'}$ is

$$\bigwedge_{j \in s(j')} \exists x_n \theta_j(x_0, \dots, x_n) \wedge \forall x_n \bigvee_{j \in s(j')} \theta_j(x_0, \dots, x_n). \tag{5}$$

We will need the following two consequences of this definition. First,

$$\vdash \exists x_n \theta_j(x_0, \dots, x_n) \leftrightarrow \bigvee \{\psi_{j'}(\bar{x}) : j \in s(j')\} \tag{6}$$

for each $j < k$. And second,

$$\begin{array}{l}\text{We can effectively find, for each formula } \phi(x_0, \dots, x_{n-1}) \text{ of } L, \text{ an} \\ r_0 < \omega, \text{ and for each } r \geqslant r_0 \text{ a subset } \Theta \text{ of } \Theta_{n,r}, \text{ such that } \phi \text{ is} \\ \text{logically equivalent to } \bigvee \Theta. \end{array} \tag{7}$$

See [39] for further information. The families of formulas $\Theta_{n,r}$, or close relatives of them, appear also in the theory of Ehrenfeucht-Fraïssé games, and in Scott sentences.

We return to the Feferman-Vaught theorem, as stated above for cartesian products. It will be helpful to add a clause to the theorem, namely that the formulas $\theta_0, \dots, \theta_{k-1}$ are the formulas of $\Theta_{n,r}$, where n is the length of the tuple $\bar{a}$ and r is found from ϕ. (If ϕ is written so as to remove any nesting of function symbols, then r can be taken to be the quantifier rank of ϕ.) This guarantees that the θ_i can be found effectively from ϕ, and that the sets $\|\theta_i(\bar{a})\|$ always form a partition of I. So the proof of the theorem reduces to finding an appropriate formula Φ_ϕ for (1). We concentrate on the most significant case, which is where $\phi(\bar{x})$ is $\exists x_n \phi'(\bar{x}, x_n)$.

By induction hypothesis there is a formula $\Phi_{\phi'}$ such that for any $A = \prod_{i \in I} B_i$ and any $\bar{a}, b$ in A we have

$$A \models \phi'(\bar{a}, b) \text{ if and only if } \mathcal{P}(I) \models \Phi_{\phi'}(\|\theta_0(\bar{a}, b)\|, \ldots, \|\theta_{k-1}(\bar{a}, b)\|) \quad (8)$$

where $\theta_0, \ldots, \theta_{k-1}$ lists the formulas of $\Theta_{n+1,r-1}$. If $\psi_0(\bar{x}), \ldots, \psi_{k'-1}(\bar{x})$ are the formulas of $\Theta_{n,k}$, we want to find Φ_ϕ so that for all A and $\bar{a}$ as above,

$$A \models \exists y \phi'(\bar{a}, y) \text{ if and only if } \mathcal{P}(I) \models \Phi_\phi(\|\psi_0(\bar{a})\|, \ldots, \|\psi_{k'-1}(\bar{a})\|). \quad (9)$$

Now we know that if the lefthand side of (9) holds then

$$\text{There is a partition } I = I_0 \cup \ldots \cup I_{k-1} \text{ such that } \mathcal{P}(I) \models \Phi_{\phi'}(I_0, \ldots, I_{k-1}) \text{ and for each } j < k, I_j \subseteq \|\exists y \theta_j(\bar{a}, y)\| . \quad (10)$$

Namely, choose b so that $A \models \phi'(\bar{a}, b)$ and put $I_j = \|\theta_j(\bar{a}, b)\|$, recalling that the θ_j are a partitioning sequence. Conversely if (10) holds, then we can use the definition of cartesian product to choose elements from the B_i to form an element b of A with $I_j = \|\theta_j(\bar{a}, b)\|$ for each $j < k$, and then it follows by (8) that $A \models \exists y \phi'(\bar{a}, y)$. So what we need is to find a formula Φ_ϕ so that the righthand side of (9) says the same as (10). This reduces to translating '$I_j \subseteq \|\exists y \theta_j(\bar{a}, y)\|$' into a condition on I_j and the sets $\|\psi_{j'}(\bar{a})\|$, for each $j < k$. But by (6),

$$\|\exists y \theta_j(\bar{a}, y)\| = \bigcup \{\|\psi_{j'}(\bar{a})\| \colon j \in s(j')\}, \quad (11)$$

and from this the definition of Φ_ϕ is straightforward.

Cartesian products are only a particular case for the Feferman-Vaught theorem; but they already contain the worst headaches. Feferman and Vaught show how to extend the technique to other kinds of product or sum by adding extra relations to the power set $\mathcal{P}(I)$; for example a linear ordering on I can be coded up as a relation between singleton subsets of I. Then the required structure can be taken to be a definable substructure of the cartesian product. The extra relations are incorporated into the formulas by taking the formulas in $\Theta_{n,0}$ to be formed not from quantifier-free formulas, but from formulas—added to the language L if necessary—that define sets of the form

$$\{\bar{a} \text{ in } A : \mathcal{P}(I) \models \Phi(\|\eta_0(\bar{a})\|, \ldots, \|\eta_{k-1}(\bar{a})\|)\} \quad (12)$$

for any η_j in L and any suitable Φ. In this way the added complexity is fed in at the bottom level of the induction; the induction steps proceed exactly as before, since they are still done in the cartesian product. This approach has great generality, but for particular applications of the theorem there are often easier routes.

The Feferman-Vaught theorem, distributive normal forms, the Fraïssé back-and-forth method and Ehrenfeucht games have close relationships, and it is not at all surprising that they can be used to prove the same or similar results. In fact Feferman soon became aware that Roland Fraïssé [35] had a back-and-forth proof of results overlapping his own on elementary equivalence of ordinals as ordered sets, and that

Andrzej Ehrenfeucht [8] found a back-and-forth proof of the decidability conjecture that Tarski had given Feferman for his thesis. Feferman reported on these results to the Summer Institute at Cornell in 1957 [12]. His report is clear, but it doesn't go into the details of how back-and-forth equivalences imply degrees of elementary equivalence. This is the part of the back-and-forth theory where the Hintikka distributive normal forms appear. He may not have appreciated at once the common underlying themes of Feferman-Vaught and back-and-forth methods.

Nor did he know Hintikka's ideas. In fact Hintikka had no contacts with the Berkeley group until later. He had visited Williams College in Massachusetts in 1948–9, and he later recalled unsuccessfully trying to explain distributive normal forms to Quine during that visit ([38] p. 11). But in any case Hintikka had only the theory, and nothing like the concrete applications that the Berkeley group could claim. (Fraïssé should count as a member of the Berkeley group. In 1995 I asked Fraïssé about his contacts with Tarski. My recollection is that he told me he attended Tarski's Berkeley seminar for most of a year sometime in the early 1950s. The brief reference in Feferman and Feferman [11] p. 315 is consistent with this.)

It seems that after publishing his joint paper with Vaught, Feferman did no further work in the area. But the Feferman-Vaught theorem took on a life of its own. Model theorists used it to prove results about elementary equivalence and elementary embeddings, and computer scientists used it to prove a wide range of algorithmic results. Both groups extended the scope of the theorem. The generalisation from first-order logic to monadic second-order logic (which has quantifiers ranging over the class of sets of individuals) proved particularly useful. Yuri Gurevich [37] p. 479 comments '... monadic [second-order] logic definitely appears to be the proper framework for examining generalized products'.

One strikingly fruitful area of application in computer science was to the complexity of model-checking of properties of various structures, for example graphs and transition systems. The survey by Janos Makowsky [47] both illustrates and extends the applications to graph properties definable in monadic second-order logic. Ingo Felscher [34] reviews the use of the Feferman-Vaught technique in connection with composition theorems for transition systems. These two papers make clear both the wide range of applications, and the curious status of the Feferman-Vaught theorem on the borderline between theory of models and theory of algorithms. On the model-theoretic side one should note an incisive application of the Feferman-Vaught theorem to the model theory of adeles in number theory, by Jamshid Derakhshan and Angus Macintyre [6].

The work that Feferman did on back-and-forth methods led him to one further development. In [16] he defined a class of functors F between classes of structures, and showed for example that if F is such a functor defined on a class $\mathbf{K}$ of structures, and A, B in $\mathbf{K}$ are equivalent in the infinitary language $L_{\infty,\kappa}$ for a suitable cardinal κ, then $F(A)$ and $F(B)$ are also equivalent in $L_{\infty,\kappa}$. The functor F also preserves elementary embedding in the sense of $L_{\infty,\kappa}$. Feferman notes ([16] footnote p. 83) that he first found these results for the special case of ordinal functions. So proof theory hovers in the background. In fact this paper is one of several where Feferman raises the question whether there is a good notion of 'natural' well-ordering for use in proof theory.

Paul Eklof [9] describes some developments from Feferman's [16]. Not least is Eklof's own theorem that for any two universal domains K_1, K_2 (in the sense of Weil) and any functor F preserving direct limits, if K_1 and K_2 have the same characteristic then $F(K_1)$ is $L_{\infty,\omega}$-equivalent to $F(K_2)$. This is a strong form of Lefschetz's Principle ([9] p. 430).

3 Applications of Interpolation Theorems

In 1953 Evert W. Beth [2] published a remarkable theorem about definability. Suppose T is a first-order theory and R a relation symbol that occurs in sentences of T. We say that R is *implicitly defined by* T if whenever A and B are models of T that have the same domain and agree in the interpretations of all nonlogical symbols except perhaps R, the structures A and B do in fact give R the same interpretation. This is a model-theoretic condition. We say that R is *explicitly defined by* T if there is a formula $\phi(\bar{x})$ in the first-order language of T, not containing any occurrences of R, such that

$$T \vdash \forall\bar{x}\ (\phi(\bar{x}) \leftrightarrow R\bar{x}). \tag{13}$$

This is a logical condition on R and T. But it is not specifically a syntactic or a model-theoretic one, because the consequence relation $\vdash$ can be defined equally well by a proof calculus or model-theoretically; the completeness theorem says we get the same consequence relation either way.

Beth proved that these two conditions on R and T, namely implicit definability and explicit definability, are equivalent. It's trivial that if R is explicitly defined by T then R is implicitly defined by T too. The interest lies in the other direction. By proving that in this first-order context every implicitly definable relation is explicitly definable, Beth turned an old heuristic idea of Padoa [53] into a theorem.

To prove his theorem, Beth had to extract a formula ϕ somehow from the information that R is implicitly defined by T. Seeing no other way, he converted implicit definability into a statement that a certain proof exists, and then used Gentzen's analysis of cut-free proofs to carve ϕ out of the proof. More precisely, he reasoned: let the theory U be the same as T except that in U every occurrence of R is replaced by an occurrence of the new relation symbol S. The completeness theorem tells us that if R is implicitly defined by T, then there is a proof of

$$\forall\bar{x}\ (R\bar{x} \leftrightarrow S\bar{x}) \tag{14}$$

from $T \cup U$.

Beth's argument for extracting ϕ from this proof was not very clean. William Craig, who reviewed Beth's paper for the *Journal of Symbolic Logic* in 1956 [3], thought about it and saw a much neater way to express the relevant proof-theoretic information. This became *Craig's Interpolation Theorem*, which he published in 1957 [4]. Feferman [27] recalls that he first heard Craig's theorem at the July 1957 Cornell Summer Institute, where Craig gave a talk on it. Craig showed, by using his own variant of Gentzen's proof theory, that if ψ and χ are first-order sentences such that

$$\psi \vdash \chi \tag{15}$$

and some relation symbol occurs in both ψ and χ, then there is a first-order sentence θ (the 'interpolant'), which contains no relation symbols except those occurring both in ψ and in χ, such that

$$\psi \vdash \theta \quad \text{and} \quad \theta \vdash \chi. \tag{16}$$

(This is Theorem 5 of [4].)

Beth's result is an easy consequence, as follows (Craig [5]). Use compactness to shrink T down to a single sentence τ, and U to a corresponding sentence σ, and add new constants $\bar{c}$ to the language. Then implicit definability gives that

$$\tau \wedge \sigma \vdash R\bar{c} \rightarrow S\bar{c}. \tag{17}$$

or after rearrangement

$$\tau \wedge R\bar{c} \vdash \sigma \rightarrow S\bar{c} \tag{18}$$

where R occurs only on the left of $\vdash$ and S only on the right. By Craig's theorem there is an interpolant $\theta(\bar{c})$ such that

$$\tau \wedge R\bar{c} \vdash \theta(\bar{c}) \quad \text{and} \quad \theta(\bar{c}) \vdash \sigma \rightarrow S\bar{c}. \tag{19}$$

From this, elementary manipulations give

$$\tau \vdash \forall\bar{x}(R\bar{x} \rightarrow \theta(\bar{x})) \quad \text{and} \quad \sigma \vdash \forall\bar{x}(\theta(\bar{x}) \rightarrow S\bar{x}). \tag{20}$$

Changing S back to R, and noting that R and S occur nowhere in θ, we reach the explicit definition

$$\tau \vdash \forall\bar{x}(R\bar{x} \leftrightarrow \theta(\bar{x})), \tag{21}$$

which proves Beth's theorem.

Note that this derivation ignores Craig's condition that some relation symbol occurs in both ψ and χ. One can justify this by arguing that the condition can always be met by ensuring that the language includes at least one of $=$ and $\bot$, and then making trivial adjustments to ψ and/or χ. Craig's theorem and its various descendents all tend to have this feature, that a correct statement involves some nuisance conditions that make only a marginal difference to applications. The discussion below will ignore these conditions.

Soon after Craig published his theorem, Roger Lyndon [44] showed, using cut elimination, that we can put more conditions on the interpolant. In particular we can require that every relation symbol occurring positively in θ occurs positively in both ψ and χ, and likewise with 'negatively' for 'positively'. He deduced from this [45] that every formula preserved in surjective homomorphisms between models of a first-order theory T is equivalent modulo T to a positive formula. Lyndon's result suggested a programme: find other conditions that can be put on the interpolant formula, and use them to give similar characterisations of other model-theoretic notions. It was above all Feferman who carried this programme through, though as he remarks in [27], he began work on this only some ten years after he heard Craig's talk in 1957.

In his course on proof theory at the 1967 European JSL meeting in Leeds [13], Feferman showed in detail how to adapt cut elimination so as to impose further conditions on the interpolant. His most striking innovation was that the language is many-sorted. Thus for example besides obeying Craig's or Lyndon's condition on relation symbols, the interpolant contains no variables of a given sort unless such variables occur in both ψ and χ; and for every sort s the interpolant contains no universal (resp. existential) quantifiers of sort s unless ψ (resp. χ) contains a universal (resp. existential) quantifier of sort s. Caution: for this result to be correct, we need to assume that the only truth functions in the language are $\wedge$, $\vee$ and $\neg$, and that $\neg$ never occurs in the formulas involved except immediately in front of atomic formulas. (Or equivalently, an occurrence of a quantifier is 'universal' if either the occurrence is positive and the quantifier is $\forall$, or the occurrence is negative and the quantifier is $\exists$; and analogously with 'existential'.) In the title of his [17] Feferman speaks of 'many-sorted interpolation theorems' in the plural. The reason is that his interpolation theorem of [13], like the Feferman-Vaught theorem, is as much a method as a theorem. The method can be adapted to prove various interpolation theorems for various applications.

Speaking loosely, the kinds of application of interpolation theorems that Feferman harvested generally involve two features: a property that is in some generalised sense implicitly definable is shown to be explicitly definable by a certain type of formula, and some relationship between structures is involved. The two features are connected by the fact that the implicit definition is given in terms of the relationship between structures.

Take for example the Łoś-Tarski theorem [43], which is easily derived by Feferman's method, though it was originally found in another way. We have a first-order theory T and a property P of structures. For this theorem the relevant 'implicit definability' of P says that if A and B are both models of T and B is an extension of A, and B has property P, then A also has property P; and that the property P is expressible by a first-order sentence ϕ. The 'explicit definability' says that there is a prenex first-order sentence θ with only universal quantifiers, such that P is expressible by θ in all models of T.

The significance of the many sorts in Feferman's interpolation is that it allows him to talk about two or more structures by assigning different sorts to the structures. Anatoliĭ Mal'tsev [48] had explored a similar use of many-sorted logic.

Applying this idea to the Łoś-Tarski theorem, suppose we take two sorts, 1 and 2, and assign expressions to the sorts by using 1 and 2 as superscripts. Then T can be written in sort 1, say as T^1, and it has an exact copy T^2 in sort 2. Wherever T^1 has a relation symbol R^1, T^2 has a corresponding relation symbol R^2. Feferman allows equality $=$ to run across all sorts. So a pair of structures A, B that are models of T, with B an extension of A, can be written as a model of

$$T^1, T^2, \text{Ext} \tag{22}$$

where Ext says

$$\forall x^1 \exists x^2 (x^1 = x^2) \wedge \forall \bar{x}^1 (R^1 \bar{x}^1 \leftrightarrow R^2 \bar{x}^1) \wedge \ldots \tag{23}$$

and the missing part at '...' repeats for all relation symbols S etc. the statement made for R. The statement that ϕ is implicitly defined in the relevant sense can be written

$$T^1, T^2, \text{Ext}, \phi^2 \vdash \phi^1. \tag{24}$$

Similar adjustments to those we made for Beth's theorem transform this entailment into

$$\tau^1 \wedge \text{Ext} \wedge \neg\phi^1 \ \vdash \ \tau^2 \rightarrow \neg\phi^2. \tag{25}$$

We find an interpolant θ, and without loss of generality we can take it to be prenex. Since no variables or relation symbols of sort 1 occur on the righthand side of (25), θ lies in sort 2. But then no universal quantifiers can occur in θ, since in that case a universal quantifier of sort 2 would have to occur on the lefthand side of (25), and inspection of Ext shows that none do.

So θ is a prenex existential sentence in sort 2, and we have

$$T^2 \ \vdash \ \theta \rightarrow \neg\phi^2. \tag{26}$$

We also have

$$T^1, \text{Ext} \ \vdash \neg\phi^1 \rightarrow \theta \tag{27}$$

which holds in particular when the two sorts are identified. Under this identification Ext becomes logically true, and there remains that

$$T \ \vdash \ \phi \leftrightarrow \neg\theta \tag{28}$$

where $\neg\theta$ is logically equivalent to a prenex sentence with only universal quantifiers. Using constants as in the proof of Beth's theorem allows us to generalise the result from sentences ϕ to formulas $\phi(\bar{x})$.

By varying the sentence Ext, together with other suitable adjustments, we can capture various other relations between structures. One of Feferman's first applications of this approach was a result parallel to the Łoś-Tarski theorem, but for end-extensions between transitive models of set theory. The natural conjecture is that in this case the formula ϕ should be equivalent, modulo set theory, to a Π_1 formula, and in [14] Feferman showed that this conjecture is true. The question arose at least partly from Feferman's discussions with Kreisel about related results in higher order logic ([32], as mentioned in Sect. 1 above).

Another result of a similar ilk appears in [23].

Let me mention two more applications. They have very different characters: the first of them feeds directly into nontrivial questions of algebra, while the second is about logical properties of quantifiers.

The first application [18] can be thought of as a generalisation of Beth's theorem to situations where we add not just a relation symbol but also new elements. We consider structures B that are models of a theory T in a language L, such that B has a relativised reduct A (a substructure in a language L^- where some relation or function symbols of L may be dropped), and A is picked out by a relation symbol R in the language L. The 'implicit definability' condition on T says that if B and B' are two models of T, with corresponding relativised reducts A and A', then every isomorphism from A to A' extends to an isomorphism from B to B'. The 'explicit definability' condition on T (which Feferman calls the 'uniform reduction' property) says that if $\phi(\bar{x})$ is a formula of L, then there is a formula $\theta(\bar{x})$ of L^- such that if B is any model of L with corresponding relativised reduct A, and $\bar{a}$ is a tuple of elements of A, then

$$B \models \phi(\bar{a}) \quad \Leftrightarrow \quad A \models \theta(\bar{a}). \tag{29}$$

The theorem says that when the languages L, L^- are first-order and T satisfies the implicit definability condition, then T has the uniform reduction property. (In fact the theorem in this form is due to Haim Gaifman [36] p. 31, who indicates that Feferman's interpolation results for many-sorted logic can be used to prove it. In [18] Feferman cites Gaifman [36] but gives the theorem in a broader and more abstract form.)

A simple example is where L^- is the language of rings, L is L^- with an added 1-ary relation symbol R, and T says of any model B that B is a field which is the field of fractions of the subring A picked out by R. The theorem tells us that in any model B of T, every first-order property of a tuple $\bar{a}$ of elements of A in B can be translated uniformly into a first-order property of $\bar{a}$ in A. This is not hard to check directly, using the fact that every element of B can be written as a ratio of elements of A. But the theorem applies also in cases where the elements of B are not explicitly definable in terms of elements of A.

In fact it's a natural question whether the Gaifman-Feferman result can be extended to say that if T satisfies the implicit definability condition, then in every model B of T with relativised reduct A, the elements of B can be uniformly defined in terms of elements of A. But group cohomology prevents any such theorem. If it was true, then the automorphism group of A would lift homomorphically to a group of automorphisms of B. The paper [10] presents many counterexamples in finite abelian groups.

In the second application [30], Feferman encodes second-order logic by letting the second sort range over relations on the domain of the first sort. Then a generalised quantifier is encoded as a relation which has one or more arguments of the second sort. For example existential quantification is the relation $\exists(p)$ which is satisfied by a 1-ary relation p of the first sort if and only if p is nonempty. In this setting a Beth-type theorem says that if a quantifier is implicitly definable by a theory (which can be a first-order theory but with its relations expressed as constants of the second sort), then the quantifier is explicitly definable by the theory. If one spells out the details, this ingeniously tightens up a proof sketched by Jeffery Zucker in [58], stating that if a generalised quantifier is completely determined by its quantifier rules (which have to be first-order expressible in the appropriate sense), then it is first-order definable.

By the early 1960s it was known that there are several ways of proving Craig's interpolation theorem without going through proof theory. Abraham Robinson found a proof by building an isomorphism of structures through unions of chains. Similar insights led to proofs using saturated structures, or structures with some variant of saturation. Another cluster of proofs use techniques for building models from atomic sentences upwards; these techniques are closely related to cut-free proofs. They include the consistency properties of Smullyan [46] and the similar construction of models by Hintikka sets. Starting in the 1960s, a number of people adapted these various techniques to give model-theoretic proofs of any and all interpolation theorems (e.g. Stern [54], Otto [52]). Also some results that were first proved by many-sorted interpolation were given proofs not using any kind of interpolation (e.g. Marker [49]). But Feferman's interpolation theorem itself is still impressive for its uniform treatment of a wide range of cases. His use of sorts for encoding is a valuable tool; and he was the first person to make many of the applications, in particular those that combine model theory with ideas from other areas of logic—such as definability of truth, or end-extensions of models of set theory.

As we saw earlier, in the 1950 and 1960s there was no general understanding that model theory had any intrinsic connection with first-order languages. When a result was proved for first-order logic, people would soon ask whether it holds for other languages too. Feferman's interpolation theorem was an important contributor to this work, because it lifts smoothly to any countable admissible language, and hence to $L_{\omega_1\omega}$ which is the union of the countable admissible languages. Feferman's student Jon Barwise wrote his doctoral thesis on the proof theory and model theory of admissible languages.

In 1985 Barwise and Feferman together edited the massive volume *Model-Theoretic Logics* [1], which brought together a vast amount of work on model-theoretic properties of various logics. In his Preface Feferman notes that he and

Barwise had 'turned to other interests in the latter part of the 1970s' (p. vii), and the initiative for the book came not from them but from the editors of 'Perspectives in Mathematical Logic'. In fact Feferman wrote none of the chapters. Notoriously the book has no index, but if it had one there would certainly be many entries under 'Feferman, S.'. Thus Jouko Väänänen in his chapter on 'Set-theoretic definability of logics' [57] remarks (p. 599) that an important tool throughout his chapter is Feferman's notion of *adequacy to truth*. This notion is a formalisation of the notion of implicit definability of truth within a logic, and it comes from a paper [19] where Feferman applies the interpolation theorem machinery to relate implicit and explicit definability of truth in a logic.

We should add too that Feferman contributed to this area of research not only through specific results that he proved, but also through helping to establish it as a substantive part of logic. For example Janos Makowsky kindly tells me that a preprint of Feferman's paper [17] was in circulation in 1971 and led him to the topic of his PhD thesis. He names Daniele Mundici, Jouko Väänänen, Jonathan Stavi and Saharon Shelah as other logicians whose work around that time took inspiration from this preprint.

A paper that touches on model-theoretic logics, but is rather orthogonal to Feferman's other work in model theory, is his study [20] of some results of Nigel Cutland on effective model theory. Unlike most writers in this vein, Feferman looks not for an effective analogue of model theory, but for a common abstract framework that includes both the classical set-theoretic version and effective analogues. This leads us neatly into our final section.

4 The Concept of Model Theory

In his paper for the Proceedings of the 1974 Tarski Conference [17] p. 206, Feferman refers to

> … the growing subject of "abstract logic" or, more precisely, the theory of *model-theoretic languages* , initiated by Lindström … and carried on by Barwise …. The ground for such work was especially prepared by Tarski's efforts over the years to isolate appropriate basic notions for a systematic development of model theory, beginning with the fundamental ones of *truth* and *satisfaction* … (30)

It's clear that one reason why Feferman mentions Tarski's aims in this direction is that they resonate with Feferman's own thinking. Feferman writes about Tarski's conceptual analyses in several places (e.g. [24, 26]). In one unpublished paper he notes several conceptual problems that drove much of his own work [21]—though none of his listed problems are directly model-theoretic.

That said, I am not entirely clear how Feferman in 1974 saw Tarski's conceptual work as related to the 'model-theoretic languages' that he refers to. Did he think that identifying the class of 'model-theoretic languages' was itself a task parallel to defining truth and satisfaction? Or did he merely mean that Tarski by giving a precise definition of satisfaction, rather than leaving it to intuition, had made it easier to develop model theory in an abstract setting?

A third possibility is that Feferman regarded it as a conceptual advance, similar in some sense to Tarski's conceptual advances, to think of *assessing* logics by formal properties that they have. This is after all what Lindström contributed, with his proof that first-order logic can be picked out from other logics by some very general and abstract properties that it has (for example compactness and downward Löwenheim-Skolem). This possibility is worth mentioning because Feferman himself quite often used this kind of assessment of logics.

One central example of this is Feferman's interest in the question of what terms (i.e. meaningful expressions) in a language are 'logical'. Two of his published papers [22, 29] discuss criteria for terms to be counted as 'logical', for example Tarski's criterion in terms of permutations of the universe. He also has a paper [25] in which he says his 'main purpose' is to examine in what sense Hintikka's Independence-Friendly Logic 'deserves to be called a logic' (p. 454). In all these papers, Feferman discusses and assesses criteria for a term to be regarded as 'logical'. In fact his work on Zucker's paper, discussed in the previous section, was intended as an application of just such a criterion. The criterion was that a 'logical' quantifier should be determined uniquely, up to logical equivalence in an obvious sense, by its proof rules. Feferman's result, following Zucker, shows that any such quantifier is explicitly definable in first-order logic, and this supports Feferman's own intuition that 'logical' terms should be limited to first-order ones.

Here I am reporting what Feferman says, though I think I am not the only logician who finds it hard to see what question is being answered. Expressions like 'logic' and 'logical' get their meaning from their usage among people called logicians. This usage is likely to vary from year to year as the subject develops. When the community of logicians fragments, as it inevitably does, the notion of 'logic' comes to vary between the fragments. So Feferman's appeal to 'the traditional conception of logic' ([29] p. 17) could be an appeal to a set of concepts that nobody still uses. (Compare 'the traditional conception of probability', bearing in mind that probability used to be treated as a branch of logic.) But let me hold my sceptical tongue.

In [27] Feferman recalls the correspondence between Beth, Tarski and himself in 1953, about Beth's definability theorem and its proof. He says:

A draft of Beth's proof was sent to Tarski in May 1953, and Tarski discussed it with me at that time (I was in Berkeley then, working with him as a student). From Tarski's point of view, since the statement of Beth's definability theorem is model-theoretic, there ought to be a model-theoretic proof, and there was correspondence (via me) with Beth about how that might be, accomplished (cf. Van Ulsen 2000, pp. 136ff). (31)

This interests me very much. To my eye the correspondence reported by Van Ulsen [56] p. 138 can more easily be read in the opposite sense. On 22 June 1953 Feferman wrote to Beth

Your solution of the problem is really a solution of a problem in proof theory and only incidentally an application of [the completeness theorem]. Indeed, it seems to me that your main result has its proper phrasing as follows: If A, B are formulas symmetric in a, b (which are consistent), and if $\forall x_1 \ldots \forall x_k (a(x_1, \ldots, x_k) \leftrightarrow b(x_1, \ldots, x_k))$ is derivable from $A \wedge B$ by elementary logic, then [and here follow statements about derivability in first-order logic]. From this theorem it is, of course, a quick step to the solution of your problem via the [completeness theorem]. I believe it is worthwhile putting the problem in this form, since then the difference between your problem and the problem of exhibiting models for independence of axioms is quite sharply pointed up. (32)

In short, Feferman in 1953 argued that Beth's theorem was really a proof-theoretic theorem and ought to be stated and proved in a form which makes this clear.

Feferman's statement in 1953 seems to me very much in line with what we know Tarski was saying about Padoa's method in the 1920 and 1930s. Namely, the method should be made rigorous by removing the model-theoretic content and rewriting it as a statement about deducibility in the appropriate logic. (See for example [40].) In the 1930s the appropriate logic was higher-order; for Beth's theorem it is first-order. Interestingly Van Ulsen ([56] p. 137) does include an item that closely fits Feferman's later account in [27]:

… in particular, Tarski has emphasized the desirability of establishing the Interpolation Theorem by methods independent of the theory of proof. (33)

But this is a quotation from Lyndon [44] in 1959!

The point to note is that Van Ulsen's quotation shows Feferman in 1953 describing Beth's theorem as a solution of a problem in proof theory, in line with Tarski's earlier perceptions. Today I think everybody regards Beth's theorem as relating an overtly model-theoretic notion (implicit definability) to a more basic logical notion (explicit definability). How far is this a different perception of the theorem itself? How far is it a reflection of how the boundaries between the branches of logic were moving in the mid 1950s?

So here we have a question not about what should count as logic, but about what did actually count as proof theory, or as semantics, or as model theory, in the early to mid 1950s. It was a time of paradigm change, and Solomon Feferman was right at the heart of the action, with a mental ear well tuned to uses and adjustments of concepts. Everything that he has told us about the understanding of those notions in Berkeley at that time is gold dust for present and future historians of logic.

References

1. Barwise, J., Feferman, S. (eds.): Model-Theoretic Logics. Perspectives in Mathematical Logic. Springer, New York (1985)
2. Beth, Evert W.: On Padoa's method in the theory of definition. Nederl. Akad. Wetensch. Proc. Ser. A (= *Indagationes Mathematicae 15*) **56**, 330–339 (1953)
3. Craig, W.: Review of [2]. J. Symb. Log. **21**(2), 194–195 (1956)
4. Craig, W.: Linear reasoning: a new form of the Herbrand-Gentzen theorem. J. Symb. Log. **22**(3), 250–268 (1957)
5. Craig, W.: Three uses of the Herbrand-Gentzen theorem in relating model theory and proof theory. J. Symb. Log. **22**(3), 269–285 (1957)
6. Derakhshan, J., Macintyre, A.: Some supplements to Feferman-Vaught related to the model theory of adeles. Ann. Pure Appl. Log. **165**(11), 1639–1679 (2014)
7. Doner, J.E., Mostowski, A., Tarski, A.: The elementary theory of well-ordering – a metamathematical study. In: Macintyre, A., Pacholski, L., Paris, J. (eds.) Logic Colloquium 77, pp. 1–54. North-Holland, Amsterdam (1978)
8. Ehrenfeucht, A.: Application of games to some problems of mathematical logic. Bull. de l'Acad. Pol. Sci. Cl **3**(5), 35–37 (1957)
9. Eklof, P.C.: Applications to algebra, in [1] pp. 423–441
10. Evans, David M., Hodges,W., Hodkinson, Ian M.: Automorphisms of bounded abelian groups. Forum Math. **3**, 523–542 (1991)
11. Feferman, A.B., Feferman, S.: Alfred Tarski: Life and Logic. Cambridge University Press, Cambridge (2004)
12. Feferman, S.: Some recent work of Ehrenfeucht and Fraisse. In: Proceedings of the Summer Institute of Symbolic Logic, Cornell Ithaca (1957), Communications Research Division, pp. 201–209. Institute for Defense Analyses, Princeton NJ, (1960)
13. Feferman, S.: Lectures on proof theory. In: Proceedings of the Summer School in Logic, Leeds 1967, Lecture Notes in Mathematics 70. pp. 1–107. Springer, Berlin (1968)
14. Feferman, S.: Persistent and invariant formulas for outer extensions. Compos. Math. **20**, 29–52 (1968)
15. Feferman, S.: Ordinals and functionals in proof theory. Actes du Congrès Internationale des Mathématiciens **1**, 229–233. Nice (1970)
16. Feferman, S.: Infinitary properties, local functors, and systems of ordinal functions. In: Hodges, w. (ed.) Conference in Mathematical Logic: London 1970, Lecture Notes in Mathematics 255, pp. 63–97. Springer, Berlin (1972)

17. Feferman, S.: Applications of many-sorted interpolation theorems. In: Henkin, L. et al., (eds.) Proceedings of the Tarski Symposium. pp. 205–223. American Mathematical Society, Providence RI (1974)
18. Feferman, S.: 'Two notes on abstract model theory. I. Properties invariant on the range of definable relations between structures. Fundam. Math. **82**, 153–165 (1974)
19. Feferman, S.: Two notes on abstract model theory. II. Languages for which the set of valid sentences is semi-invariantly implicitly definable. Fundam. Math. **89**, 111–130 (1975)
20. Feferman, S.: Generalizing set-theoretical model theory and an analogue theory on admissible sets. In: Hintikka, J., Niiniluoto, I., Saarinen, E. (eds.) Essays on Mathematical and Philosophical Logic: Proceedings of the Fourth Scandinavian Logic Symposium and of the First Soviet-Finnish Logic Conference, Jyväskylä, Finland, June 29–July 6, 1976, pp. 171–195. Reidel, Dordrecht (1979)
21. Feferman, S.: Three conceptual problems that bug me. Draft lecture notes for 7th Scandinavian Logic Symposium, Uppsala, August 18–20. (1996). math.stanford.edu/~feferman/papers/conceptualprobs.pdf
22. Feferman, S.: Logic, logics, and logicism. Notre Dame J. Form. Log. **40**(1), 31–54 (1999)
23. Feferman, S.: Ah, Chu!. In: JFAK, Essays Dedicated to Johan Van Benthem on the Occasion of his 50th Birthday (1999) (compact disc; the paper is available at math.stanford.edu/~feferman/papers)
24. Feferman, S.: Tarski's conception of logic. Ann. Pure Appl. Log. **126**(1), 5–13 (2004)
25. Feferman, S.: What kind of logic is Independence-Friendly logic?. In: Randall E. Auxier., Hahn, L.E. (eds.) The Philosophy of Jaakko Hintikka, vol. XXX, pp. 453–469. Library of Living Philosophers, Open Court, Chicago IL (2006)
26. Feferman, S.: Tarski's conceptual analysis of semantical notions. In: Patterson, D. (ed.) New Essays on Tarski and Philosophy, pp. 72–93. Oxford University Press, Oxford (2008)
27. Feferman, S.: Harmonious logic: Craig's interpolation theorem and its descendants. Synthese **164**(3), 341–357 (2008)
28. Feferman, S.: Andrzej Mostowski: An Appreciation. In: Ehrenfeucht, A., Marek, V.W., Srebrny, M. (eds.) Andrzej Mostowski and Foundational Studies, pp. 388–392. IOS Press, Amsterdam (2008)
29. Feferman, S.: Set-theoretical invariance criteria for logicality. Notre Dame J. Form. Log. **51**(1), 3–20 (2010)
30. Feferman, S.: Which quantifiers are logical? A combined semantical and inferential criterion. Preprint 2013, math.stanford.edu/~feferman/papers/ WhichQsLogical(text).pdf
31. Feferman, S.: A fortuitous year with Leon Henkin. In: Manzano, M., Sain, I., Alonso, E. (eds.) The Life and Work of Leon Henkin, pp. 35–40. Birkhäuser, Heidelberg (2014)
32. Feferman, S., Kreisel, G.: Persistent and invariant formulas relative to theories of higher order. Bull. Am. Math. Soc. **72**(3), 480–485 (1966)
33. Feferman, S., Vaught, R.: The first order properties of products of algebraic systems. Fundam. Math. **47**, 57–103 (1959)
34. Felscher, I.: Model composition in model-checking, Doctoral Dissertation, Hochschulbibliothek der Rheinisch-Westfälischen Technischen Hochschule Aachen, (2014)
35. Fraïssé, R.: Sur quelques classifications des relations, basées sur des isomorphismes restreints. II. Applications aux relations d'ordre, et construction d'exemples montrant que ces classifications sont distinctes. In: Publ. Sci. de l'Université d'Alger, Série A (Mathematiques), vol. 2, pp. 273–295. (1957)
36. Gaifman, H.: Operations on relational structures, functors and classes. I. In: Henkin, L. et al., (eds.) Proceedings of the Tarski Symposium. pp. 21–39. American Mathematical Society, Providence RI (1974)
37. Gurevich, Y.: Monadic Second-Order Theories, in [1] pp. 479–506
38. Hintikka, J.: Intellectual autobiography. In: Randall E. Auxier., Hahn, L.E. (eds.) The Philosophy of Jaakko Hintikka. Library of Living Philosophers vol. XXX, pp. 3–84. Open Court, Chicago IL (2006)

39. Hintikka, J.: Distributive Normal Forms in the Calculus of Predicates, Acta Philosophica Fennica 6. Philosophical Society of Finland, Helsinki (1953)
40. Hodges, W.: Tarski on Padoa's Method: a test case for understanding logicians of other traditions. In: Chakraborti, Mihir K. et al., (eds.) Logic, Navya-Nyāya and Applications: Homage to Bimal Krishna Matilal, pp. 155–169. College Publications, London (2008)
41. Krajewski, S., Srebrny, M.: On the life and work of Andrzej Mostowski (1913–1975). In: Ehrenfeucht, A., Marek, V.W., Srebrny, M. (eds.) Andrzej Mostowski and Foundational Studies, pp. 3–14. IOS Press, Amsterdam (2008)
42. Kreisel, G.: Set theoretic problems suggested by the notion of potential totality, in Infinitistic Methods. Pergamon Press, Warsaw (1959)
43. Łoś, Jerzy: On the extending of models (I). Fundam. Math. **42**, 38–54 (1955)
44. Lyndon, Roger C.: An interpolation theorem in the predicate calculus. Pac. J. Math. **9**, 129–142 (1959)
45. Lyndon, Roger C.: Properties preserved under homomorphism. Pac. J. Math. **9**, 143–154 (1959)
46. Makkai, M.: An application of a method of Smullyan to logics on admissible sets. Bull. Pol. Acad. Sci. **17**, 341–346 (1969)
47. Makowsky, J.A.: Algorithmic uses of the Feferman-Vaught Theorem. Ann. Pure Appl. Log. **126**(1), 159–213 (2004)
48. Mal'tsev, Anatoliĭ I.: Model'niye sootvestvija, Izvestiya Rossiiskoi Akademii Nauk Seriya Matematicheskaya 23(3), pp. 313–336; trans. as 'Model correspondences.: In Mal'cev, A.I. (ed.) The Metamathematics of Algebraic Systems: Collected Papers 1936–1967, ed. and trans. Benjamin Franklin Wells III, pp. 66–94. North-Holland, Amsterdam (1971)
49. Marker, D.: A model theoretic proof of Feferman's preservation theorem. Notre Dame J. Form. Log. **25**(3), 213–216 (1984)
50. Mostowski, A.: On direct products of theories. J. Symb. Log. **17**(1), 1–31 (1952)
51. Mostowski, A., Tarski, A.: Arithmetical classes and types of well ordered systems. Preliminary report. Bull. Am. Math. Soc. **55**, 65 (1949)
52. Otto, M.: An interpolation theorem. Bull. Symb. Log. **6**(4), 447–462 (2000)
53. Padoa, A.: Essai d'une théorie algébrique des nombres entiers, précédé d'une introduction logique à une théorie déductive quelconque. In: Bibliothèque du Congrès international de philosophie, Paris, 1900, vol. 3, Armand Colin, Paris 1902, pp. 309–365; translated in part as 'Logical introduction to any deductive theory' in *From Frege to Gödel*, ed. Jean van Heijenoort, pp. 118–123. Harvard University Press, Cambridge Mass. (1967)
54. Stern, J.: A new look at the interpolation problem. J. Symb. Log. **40**(1), 1–13 (1975)
55. Tarski, A.: Some notions and methods on the borderline of algebra and metamathematics. In: Proceedings of the International Congress of Mathematicians: Cambridge, Massachusetts. vol. 1, pp. 705–720. American Mathematical Society, Providence RI (1952)
56. Ulsen, Paul van.:*E. W. Beth als logicus*, Doctoral Dissertation, Institute for Logic, Language and Computation, Amsterdam (2000); ILLC Dissertation Series DS–2000–04
57. Väänänen, J.: Set-Theoretic Definability of Logics, in [1] pp. 599–643
58. Zucker, J.I.: The adequacy problem for classical logic. J. Philos. Log. **7**(1), 517–535 (1978)

Author Biography

Wilfrid Hodges FBA is Emeritus Professor of Mathematics at Queen Mary, University of London. Apart from one year at UCLA and one at Boulder, Colorado, his career was spent in London University, at Bedford College and then Queen Mary. He is the author of five textbooks of logic or model theory at various levels (one joint with Ian Chiswell), and a sixth (historical) text is in preparation.

Feferman on Computability

Jeffery Zucker

To the memory of Sol Feferman, who, with his unfailing kindness, patience and good humour, was a constant source of inspiration to me, ever since his supervision of my graduate studies at Stanford

Abstract Solomon Feferman has left his mark on computability theory, as on many other areas of foundational studies. The purpose of this paper is, by means of reviewing a selected few of his many papers in this area, to give an idea of his impressive insights and developments in this field.

Keywords Feferman · Generalized computability · Abstract data types
Computing on streams · Computing on reals

AMS 2010 Classification 01-02 · 03-02 · 03-03 · 03D65 · 03D70 · 03D75 · 03D78

Introduction

Classical computability theory (or recursion theory) investigates the computability of functions on the domain of natural numbers $\mathbb{N}$, or (equivalently) strings over a finite alphabet. This study can be generalized in two directions: investigating (i) functionals

J. Zucker—Thanks to Wilfried Sieg and an anonymous referee for very helpful comments on an earlier draft, and to Mark Armstrong for technical assistance with the typesetting. This research was supported by a grant from the Natural Sciences and Engineering Research Council (Canada).

J. Zucker (✉)
Department of Computing and Software, McMaster University, Hamilton,
ON L8S 4K1, Canada
e-mail: zucker@mcmaster.ca

G. Jäger and W. Sieg (eds.), *Feferman on Foundations*, Outstanding
Contributions to Logic 13, DOI 10.1007/978-3-319-63334-3_2

of higher types over $\mathbb{N}$, and/or (ii) function(al)s over more general domains, such as the reals, with their distinctive topological and other properties. Over a period of about four decades, Sol Feferman (hereinafter SF) carried out highly significant investigations in both directions. The aim of this paper is not to give a complete survey of his work in this area, which would be very difficult because of the profusion of his writings, but to examine a few of his more noteworthy papers spanning this period, so as to get a taste of his research here.

We will investigate four of SF's papers, spanning this period:

1. "Inductive Schemata and Recursively Continuous Functionals" (1977) [15],
2. "A New Approach to Abstract Data Types", Parts I and II (1992) [17, 18],
3. "Computation on Abstract Data Types: The Extensional Approach, with an Application to Streams" (1996) [19], and
4. "About and Around Computing over the Reals" (2013) [20].

A section is devoted to each of these papers below.

It is hoped that the summaries given here will encourage researchers, students and historians to read the original papers. In such cases, one will be impressed not only by the contents, but by the clarity and elegance of SF's scientific literary style. For the same reason, as will be seen, I have frequently taken to quoting directly from these papers, since I often found that to be the best way to convey SF's ideas.[1]

1 Inductive Schemata and Recursively Continuous Functionals

The first paper to be reviewed here is [15]. Its topic is a particular approach to generalized recursion theory (g.r.t.), based on *monotone inductive schemata over arbitrary structures*. It has 9 sections, and is moreover divided into two main parts.

Part I is an informal introduction to this theory. To quote SF: "the ideas ... are due independently to Moschovakis and the author, but the detailed development is due almost entirely to Moschovakis. A principal source for these ideas is in Platek's work on fixed-point schemata". This refers to Richard Platek's PhD thesis [48] written under SF's supervision at Stanford.[2]

Section 2 (still in Part I) gives a review of Platek's work, and how it was modified and expanded for the present purpose. As SF puts it, Platek's main idea was that the central feature of recursion theory which makes sense for *arbitrary structures* $\mathcal{A} = \langle A, \ldots \rangle$ is the process of recursion itself, i.e., the definition of a function φ as the *least fixed point* of an equation $\varphi = \Phi(\varphi)$; or, equivalently, $\mathbf{FP}(\Phi)$ is the least φ "closed under" Φ: $\Phi(\varphi) \subseteq \varphi$. For this to make sense, we must assume that the

[1]To avoid confusion, sections in SF's papers are referred to as Sect. 1, etc., and sections in the present article as §1 and §2, etc.

[2]And regrettably never published, but with a far-ranging influence, as we will see in some of the other papers investigated here.

functional Φ is a *monotonic* operator on partial functions on A. Platek introduced a structure of *hereditarily monotonic functionals*[3] of finite type over A, denoted $\tilde{\mathcal{A}} = \langle \tilde{A}_\tau \rangle_\tau$, where τ ranges over all finite types.

Next, for any class $\mathcal{F}$ of functionals in $\tilde{\mathcal{A}}$, let **Ind**($\mathcal{F}$) be the class of functionals inductively defined from $\mathcal{F}$, obtained by closing under *explicit definition* and **FP** at all types.

Two of the main results of [48] are:

- If every member of $\mathcal{F}$ has type level $\leq n+1$ and $\varphi \in$ **Ind**($\mathcal{F}$) has type level $\leq n$ then we can get φ from $\mathcal{F}$ by explicit definition and the **FP** scheme at type levels $\leq n$ only. In particular, with the structure $\mathcal{A} = (A, \mathcal{F})$, if $\mathcal{F}$ has only total functions of level 1, we need only use Φ of type level 2.
- The system of functionals given by Kleene's schemata S1–S9 [33][4] on the *maximal type structure* over $\mathbb{N}$, i.e. $\mathcal{N} = (\mathbb{N}_n)_{n \in \mathbb{N}}$ (where $\mathbb{N}_0 = \mathbb{N}$ and $\mathbb{N}_{n+1} = \mathbb{N}_n^{\mathbb{N}}$), is equivalent to **Ind**(0, ***Sc***),[5] in the sense that there is an embedding of $\mathcal{N}$ into $\tilde{\mathcal{N}}$ such that Φ on $\mathcal{N}$ is Kleene partial recursive (i.e., derivable from S1–S9) iff its image in $\tilde{\mathcal{N}}$ is in **Ind**(0, ***Sc***).

We should note that Platek makes some special assumptions on the structure $\langle A, \mathcal{F} \rangle$, namely that $\mathcal{F}$ (or **Ind**($\mathcal{F}$)) contains pairing and projection functions, and distinct elements 0,1 of A. With these special assumptions on a level 1 structure $\langle A, \mathcal{F} \rangle$, the functions in **Ind**($\mathcal{F}$) turn out to be the same as the *prime computable* functions of Moschovakis [37] on $\langle \mathcal{F}, =_A \rangle$.

SF then notes certain limitations on Platek's theory:

(1) These special assumptions effectively introduce $\mathcal{N}$ as a substructure of $\mathcal{A}$, and hence ordinary recursion theory as part of **Ind**($\mathcal{F}$). SF argues against this (in Sect. 4, as we will see).
(2) It does not generalize recursion theories based on *relations*, such as Post–Smullyan systems [50] or search computability [37].
(3) Recursion on $\mathcal{N}$ is derived indirectly, from embedding into $\tilde{\mathcal{N}}$, which is "messy". It would be preferable, conceptually, to *identify* the S1–S9 functionals over $\mathcal{N}$ with **Ind**($\mathcal{F}$) over the ground domain system $\langle \mathbb{N}_n \rangle_n$ for suitable $\mathcal{F}$, i.e., more generally, associate a suitable recursion theory with any ground structure $\langle \langle A_i \rangle_{i \in I}, \mathcal{F} \rangle$, for which recursion on $\langle \mathbb{N}_n \rangle_n$ would be one example.

Section 3 of this paper presents SF's general theory of monotone inductive definitions which overcomes these three limitations. SF found that there was a large overlap of his theory with that of Moschovakis, which had been developed independently (and also included a theory of non-monotone inductive definitions) [34, 40].

Briefly:Suppose given a *domain system* $\langle A_i \rangle_{i \in I}$ and a collection $\mathcal{X}$ of relations on these, closed under unions of chains. Assume each $X \in \mathcal{X}$ has an *arity* $\nu =$

---

[3]This terminology is SF's.

[4]Kleene's notion of *partial recursive functional*, given by his schemata S1–S9 [33] or (equivalently for type levels ≤ 2) as in [31, p. 326], will feature in every one of the four articles discussed here.

[5]Where '***Sc***' is the successor function on $\mathbb{N}$.

$(\nu(1), \ldots, \nu(m))$, with $X \subseteq A_{\nu(1)} \times \cdots \times A_{\nu(m)}$. A *monotone schema* is a functional $\Phi\colon \mathcal{X}_{\nu(1)} \times \cdots \times \mathcal{X}_{\nu(m)} \to \mathcal{X}_\mu$ which is monotonic, in the sense that

$$X_i \subseteq X_i' \; (1 \leq i \leq m) \implies \Phi(X_1, \ldots, X_m) \subseteq \Phi(X_1', \ldots, X_m').$$

With each such Φ we can associate a least fixed point $\mathbf{FP}(\Phi)$. Then, given a collection $\mathcal{F}$ of schemata, we take $\mathbf{Ind}(\mathcal{F})$ to be the smallest collection of schemata containing $\mathcal{F}$ and closed under explicit definition and **FP**.

Given a domain system $\langle A_i \rangle_{i \in I}$, there are two principal choices for $\mathcal{X}$: (i) all relations on $\langle A_i \rangle_{i \in I}$, and (ii) all partial functions on $\langle A_i \rangle_{i \in I}$. These give rise to (respectively) *relational* and *functional* inductive theories. From these Moschovakis's *prime computability* and *search computability* [37] can easily be constructed on an arbitrary A (or rather the closure of $A \cup \{0\}$ under pairing).

Another example is recursion on the *maximal type structure* $\mathcal{N} = \langle \mathbb{N}_n \rangle_n$ over $\mathbb{N}$. Here we take the class $\mathbf{Ind}(\mathcal{F})$ where $\mathcal{F}$ contains 0, successor, some basic functions and two schemata for application and abstraction at all types. Now we can derive, in $\mathbf{Ind}(\mathcal{F})$, Kleene's [33] schema S8 (higher order function abstraction, combined with application) and an "enumeration" schema S9:

$$\varphi(\vec{\alpha}, z) \simeq \{z\}(\vec{\alpha}).$$

where z is a number variable, and '$\simeq$' means that the l.h.s. is defined iff the r.h.s. is. In fact $\mathbf{Ind}(\mathcal{F})$ is equivalent to Kleene's S1–S9, since the **FP** schema can be obtained from the functional form of Kleene's recursion theorem.

SF ends this section by posing the question: Is there an interesting *relational* inductive theory over $\mathcal{N}$ – or rather over $\mathcal{S} = \langle S_n \rangle_n$, where $S_0 = \mathbb{N}$ and S_{n+1} is the power set of S_n?

In Sect. 4, SF briefly considers what he calls *axiomatic enumerative g.r.t.*, characterized by an axiomatic approach to Kleene's S9, and developed by Wagner and Strong [51], Moschovakis [38], Fenstad [22], Hyland [30] and others. It is shown in [34] that under quite general hypotheses a theory $\mathbf{Ind}(\mathcal{F})$ is an enumerative g.r.t. SF remarks on what he considers two defects of such enumerative approaches: (1) the *ad hoc* character of such codings, and (2) the necessity to incorporate $\mathcal{N}$ as part of the structure.

Section 5 gives some ideas ("which remain to be developed") for more restricted kinds of inductive schemata given by syntactic closure conditions or inference rules, where the course of the induction can be represented by a *derivation* or *computation* tree.

To take an example, consider *syntactic closure conditions*, and the *relational case*. We take a formula Γ in a language $\mathcal{L}$ over a structure $\mathcal{A} = \langle \langle A_i \rangle_{i \in I}, \ldots \rangle$ augmented by relation parameters which occur only positively in Γ. This gives a monotone schema Φ_Γ with a (least) fixed point, defining a function from tuples of relations (of the correct arity) to relations. We can then identify $\mathcal{F}$ with the class of such formulas Φ, and so $\mathbf{Ind}(\mathcal{F})$ is the set of functions so defined. SF gives two examples.

(1) Let $\mathcal{A} = (\mathbb{N}, 0, \mathbf{\mathit{Sc}})$, and let $\mathcal{F}$ be the class of *existential formulas* (positive in their relational parameters). Then $\mathbf{Ind}(\mathcal{F})$ is the class of all r.e. relations; cf. the Post–Smullyan approach [50].
(2) Let $\mathcal{A}$ be any structure, $\mathcal{L}$ the corresponding full first order language, and $\mathcal{F}$ all formulas positive in their relation parameters. Then $\mathbf{Ind}(\mathcal{F})$ is the set of relations inductively definable over $\mathcal{A}$ in the sense of Moschovakis [39].

As another example, we can associate, with certain inductive definitions, *rules of inference* and *derivation trees*. The inductive definition has the form: X is the least solution of $\Phi(X) \subseteq X$, i.e. the least set such that $\forall\vec{x}\big[\Gamma(\vec{x}, X) \rightarrow \vec{x} \in X\big]$, where elements that are put into X at any stage are related by elementary conditions (given by Γ) to elements already in X. We can write such a *closure condition* as an *inference rule* $\Gamma(\vec{x}, X)/X(\vec{x})$. Such inference rules give rise to a (possibly infinite) *derivation tree* for generating membership of X.

We turn to Part II (and Sect. 6), which deals with recursion on structures of *continuous functionals* over $\mathcal{N}$. SF sketches two similar constructions by Kleene [32] and Kreisel [35] of a finite type structure $\mathcal{C} = \langle C_n \rangle_n$ of *hereditarily* (*total*) *continuous functionals* over $C_0 = \mathbb{N}$. (Another approach: the structure $\mathcal{C}^{\smile}$ of *hereditarily partial continuous functionals*, developed by Ershov, will be discussed below shortly.)

The principal question now is: are the resulting theories equivalent to those given by inductive schemata?

Considering first, *total continuous* (or "countable", to use Kleene's terminology) functionals: we consider functionals α^n of type n, where the value of α^{n+1} at any β^n is given by a finite amount of information U^n about β, given by "formal neighborhoods" which can be coded as natural numbers. The details are available in SF's paper, and, of course, in [32, 35]. Hence for all n, any such α^n can be *represented* by a type-one "associate" denoted by $\alpha^{(n+1)}$ or (here) by $\ulcorner\alpha\urcorner$.

The main difference between Kleene's and Kreisel's approach is that $\alpha^{n+1}(\beta^n)$ makes sense for all $\beta^n \in \mathbb{N}_n$ in Kleene's approach, but only for $\beta^n \in C_n$ in Kreisel's.

A functional $\varphi \in \mathcal{C}$ is said to be *recursively continuous* (or *recursively countable*) if it has a (total) recursive associate $\ulcorner\varphi\urcorner$. With Kleene's identification of C_n with part of $\mathbb{N}_n$, a central problem was to find the relationship between recursively continuous functionals and those generated by Kleene's schemata S1–S9.

In one direction: Kleene showed [32] that if φ is generated by his schemata, and *if* φ *is total on* $\mathcal{C}$, then φ (restricted to arguments in $\mathcal{C}$) is recursively continuous.

The question was raised by Kreisel [35] if the converse holds, in the sense that every recursively continuous function on $\mathcal{C}$ is the restriction of a function generated by S1–S9. A counterexample was found by Tait (unpublished), namely a modulus of uniform continuity functional at type level 3.

The situation with K-K (Kleene–Kreisel) recursiveness is thus not satisfactory as in stands. In Sect. 7, SF points to a possible way forward, by turning to a more general theory of higher-order *partial recursion*, which would reduce to the theory of K-K recursiveness in the special case of total functions.

The problem here is that even a *definition* for such a concept is problematic. He presents one proposed by Robert Winternitz, a former student of his at Stanford.

(We omit details, but urge the reader to consult this paper.) This definition satisfies (among other good properties) an *enumeration* or *universality* property at all types.

Also important in the further development of the theory is the concept of *potential partial recursiveness*, which is satisfied by a function φ if there is some partial recursive continuous $\psi \supseteq \varphi$. With this definition, Kleene's partial positive result above can be can be re-cast in the form: Each φ which is (S1–S9) partial recursive on $\mathcal{N}$ has its restriction to $\mathcal{C}$ potentially partial recursive. (But see below.)

SF writes here: "I consider the main defect of this work on schemata to be taking $\mathcal{N}$ as the point of departure, rather than working entirely in the context of $\mathcal{C}$". This makes a difference in interpreting Kleene's schema S8:

$$\varphi(\alpha^{n+2}, \vec{\gamma}) \;\simeq\; \alpha(\lambda\beta^n \cdot \psi(\alpha, \beta, \vec{\gamma}))$$

in which even though α^{n+2} and $\vec{\gamma}$ range over $\mathcal{C}$, the "abstracted" variable β^n is taken to range over all of $\mathcal{N}$ (of the appropriate type). It would be more appropriate to have β^n ranging over $\mathcal{C}$. This would make the r.h.s. (and hence the l.h.s.) of this equation definable in more cases. (The other schemata, S1–S7 and S9, are not affected by such re-interpretation.) Then, as before, every functional generated by S1–S9 in this new interpretation on $\mathcal{C}$ is potentially partially recursive, and we can again inquire about the converse. However the uniform continuity functional at type level 3 still provides a counterexample.[6]

Section 8 deals with the other approach to computation on higher types noted above, based on the structure $\mathcal{C}^{\smile} = (C_n^{\smile})_n$ of hereditarily partial continuous functionals developed by Ershov [11]. Here $C_0^{\smile} = \mathbb{N}$, and $C_{n+1}^{\smile}$ is the set of continuous partial functions from $C_n^{\smile}$ to $\mathbb{N}$ (suitably defined). Then C_n can be successively mapped into $C_n^{\smile}$ [12]. To quote SF again: "Now there is also a natural definition of *partial recursive functional* on $\mathcal{C}^{\smile}$. I studied the schematic generation of these functionals in [14], centering attention on so-called "search" operators introduced in Moschovakis [37], namely $\nu x[\psi(x, \vec{\alpha}) \simeq 0]$ which is interpreted as 'an x such that $\psi(x, \vec{\alpha}) \simeq 0$'. By S10 we mean the scheme

$$\varphi(\vec{\alpha}) \;\simeq\; \nu x[\psi(x, \vec{\alpha}) \simeq 0]$$

which, without further restriction, must lead to multivalued functions ..." Single-valuedness can be recovered by a suitable restriction in the use of this scheme, resulting in a scheme (S10 !).

The main results of [14] were that the multi-valued partial recursively continuous functionals over $\mathcal{C}^{\smile}$ are exactly those generated by (S1–S8) + (S10), and the single-valued ones are those generated by (S1–S8) + (S10 !). However this is not completely satisfactory, since (S10 !) is not a monotone scheme.

[6] Gandy to SF, personal communication.

This result was improved by Winternitz, by incorporating the "strong or" operator introduced in [48][7]:

$$\mathsf{OR}^{+}(\varphi_1, \varphi_2) \simeq 0 \iff \varphi_1(0) = 0 \ \vee \ \varphi_2(0) = 0. \tag{1.1}$$

Winternitz showed that

at type level 2, the functionals generated by (S1–S9) +OR^+ *are exactly the partial recursive functionals over* $\mathcal{C}^{\smile}$.

However the result does not hold at higher type levels. The solution turns out to come from work of Sazonov [49] who introduced the functional $\exists^3$, defined by

$$\exists^3(\alpha^2) \simeq \begin{cases} 0 & \text{if } \alpha(\Lambda^1) \simeq 0 \\ 1 & \text{if } \alpha(\delta_n) \simeq 1 \text{ for some } n \end{cases}$$

where Λ^1 is the completely undefined function at level 1, and $\delta_n(x) \simeq z \iff x = 0 \wedge z = n$. Sazonov's result can then be formulated (in this framework) as:

A (partial) functional is partial recursively continuous over $\mathcal{C}^{\smile}$ iff
it is generated by the schemata (S1–S9) + OR^+ + $\exists^3$.

This shows that *partial recursion over* $\mathcal{C}^{\smile}$ is equivalent to a *monotonic inductive schematic* theory.

Ershov's work is discussed again below in Sect. 3, in connection with SF's providing recursion-theoretic interpretations for abstract computational procedures over his stream algebras.

In Sect. 9 SF makes some concluding remarks:

> By [12] the recursively continuous functions over $\mathcal{C}$ are restrictions to $\mathcal{C}$ of those over $\mathcal{C}^{\smile}$, and hence just those functionals total on $\mathcal{C}$ generated by the above schemata. However there remains the question of whether we can generate these functionals directly over $\mathcal{C}$. What we are really after are monotone schemata …

Now it might at first be thought that a non-monotone theory could be found, for example (S1–S8) + (S10 !). However, as Winternitz showed, although the partial recursively continuous functionals on $\mathcal{C}$ *are* closed under (S1–S7) and (S10 !), they are *not* closed under (S8), even when abstracting over types 0 and 1 only. SF continues:

> With this we can complete our remark in Section 7 about Kleene's partial positive result … [for] the statement of closure under (S1–S9) given in Sect. 7 above involves *potential* partial recursiveness in an essential way. … The main question with which we are left is the following.

Question. Is there a natural monotone collection $\mathcal{F}$ over $\mathcal{C}$ such that the partial recursively continuous functionals over $\mathcal{C}$ are exactly those generated by **Ind**($\mathcal{F}$)?

---

[7] The notation and presentation have been changed here to match that in [19], discussed in §3 below.

SF concludes this fascinating paper with the remark:

> As long as this question remains unsettled, the inductive schematic approach to g.r.t. is not completely vindicated. But I hope that the considerations in this paper combined with the detailed work of [34, 40], which demonstrate its scope otherwise, will lead one to give serious attention to this program.

2 A New Approach to Abstract Data Types, Parts I and II

In this pair of papers [17, 18], SF switches gears, and focuses on computation not on the classical structures (such as naturals or ordered reals, and higher types on these) but on *abstract data types*, which can be taken (for now) as classes of algebras of a given signature Σ, closed under Σ-isomorphism. SF motivates the topic by saying: "The concept of abstract data types (ADTs) has emerged in the last fifteen years or so as one of the major programming design tools, with the emphasis on modular construction of large-scale programs."

The first paper (Part I) is an informal introduction to ADTs, while the second gives a more formal development, with the emphasis again on computation over these. We thus give a very brief overview of Part I, as an introduction to Part II, which is our main concern.

SF begins Part I by stating:

> [J]ust as in mathematics generally, one is concerned in computational practice with general algebraic notions such as orderings, rings, fields, polynomials, etc. There is additional concern in computer science with other kinds of ADTs such as lists, stacks, trees, records, arrays, streams, etc. A coherent account of how these are all to be treated for computational purposes requires answers to such questions as:
>
> Q1. What are ADTs and how may they be specified?
>
> Q2. What does it mean to implement an ADT?
>
> Q3. How can we construct new ADTs from old ones?
>
> Q4. What does it mean to compute with ADTs?

SF's aim is to answer the above four questions, especially Q4, in the course of this and the following paper. Since this paper gives only a semi-formal development, we discuss it very briefly, before turning to Part II for a more detailed account.

SF refers to previous foundational approaches: algebraic [27], computational or recursion-theoretic [8, 52] and type-theoretic [43]. He argues that all previous foundational approaches fail in certain cases.

> A new, constructively based, approach is proposed here to provide a sufficiently general account ... within the conceptual framework of the school of constructive mathematics associated with Bishop ... [5]. The formal foundations will be provided by theories of operations and classes in which that style of constructive mathematics can be formalized ... [T]hese will be taken up in Part II.

SF gives some examples of ADTs, noting that "there is a basic division into those structures whose objects are described or generated in a finitary way, and those whose description is infinitary." Examples of finitary ADTs dicussed are: lists over arbitrary types; lists over preordered types; binary and finitely branching trees; and finite sets. Examples of infinitary ADTs are *infinite streams* and *infinite precision reals* – the latter following Bishop's implementation [1, pp. 18–19].

We turn to Part II. As SF says in the opening sentence, the main purpose of this paper is to give a precise definition of *abstract computational procedures*[8] F on ADTs, with interpretations $\mathsf{F}^{\mathcal{A}}$ over structures

$$\mathcal{A} = (A_0, \ldots, A_n, =_{A_0}, \ldots, =_{A_n}, F_0, \ldots, F_m) \tag{2.1}$$

of signature Σ, say, where $=_{A_i}$ is an "equality" relation[9] on A_i ($i = 0, \ldots, n$), A_0 is the boolean type $\mathbb{B} = \{\mathrm{tt}, \mathrm{ff}\}$, $=_{A_0}$ is the identity on $\mathbb{B}$, and F_j is either a constant (0-ary function), partial (level 1) function or partial (level 2) functional over $\mathcal{A}$ of a specified arity. The constants of $\mathcal{A}$ include tt and ff, and the functions and functionals among F_j *preserve equality* and are *monotonic*.

The aim here is to clarify the concept of *abstract computational procedure* F satisfying, for any $\mathcal{A}$ as above, the following criteria:

C1. F associates with $\mathcal{A}$ an object $\mathsf{F}^{\mathcal{A}}$ of specified arity over $\mathcal{A}$.
C2. $\mathsf{F}^{\mathcal{A}}$ is determined by the (individual, function and functional) constants of $\mathcal{A}$.
C3. The map $\mathcal{A} \mapsto \mathsf{F}^{\mathcal{A}}$ preserves Σ-isomorphism.
C4. $\mathsf{F}^{\mathcal{A}}$ preserves the equality relations on $\mathcal{A}$.
C5. For domains A_i of $\mathcal{A}$ contained in $\mathbb{N}$, $\mathsf{F}^{\mathcal{A}}$ reduces to an ordinary computational procedure (see Remark 2 below).

SF explains:

> These requirements are met here by a *generalized recursion theory*[10] (g.r.t.) which provides a notion of computability over arbitrary structures of the kind described above. In order to satisfy C3 we must insure that whenever an object is defined by recursion it is uniquely specified. For (partial) functions this will be as a least fixed point (LFP) of a suitable monotonic functional. There are two forms of g.r.t. available in the literature which feature LFP as a central scheme … namely those of Moschovakis [41, 42] and the earlier Platek [48].[11]

In both of these definition by recursion is implemented as the least fixed point of a suitable monotonic functional. SF adapts Moschovakis's version, but with schemata like Kleene's S1–S9 [33], with S9 replaced by a simple LFP recursion.

SF continues:

> The g.r.t. developed here … applies to a wide variety of data universes V, with weak closure conditions on the classes of partial functions and functionals over V. There are two extremes of interpretation: (i) V is the full cumulative hierarchy and "all" functions and functionals are

---

[8] Written 'π' in [17]. Notation changed here to match [19]; cf. §3 below.

[9] Actually a congruence relation w.r.t. the F_k's.

[10] Emphasis added. The meaning of this phrase is discussed below.

[11] As noted in §1 above.

admitted; (ii) $V = \omega$[12] and we only admit partial recursive functions and functionals. The setting (i) is the usual one for g.r.t., while the setting (ii) serves for the precise formulation of the criterion C5; the statement of that is the main new contribution of this paper.

Remarks.

(1) We need only consider structures as in (2.1) with function(al)s up to type level 2, because of Platek's result given in §1 above.

(2) By "ordinary computation procedure", SF means the structure constructed by Kleene over $\mathbb{N}$ in [33]. Also recall Footnote 4 for the concept of partial recursive functional.

A structure $\mathcal{A}$ has a "data universe": an underlying universe $\boldsymbol{V}$ containing the data objects $a, b, \ldots, x, y, \ldots$, a collection $\boldsymbol{DT}$ of subsets $A, B, \ldots, X, Y, \ldots$ of $\boldsymbol{V}$, called *data types*, containing $\mathbb{B}$, $\mathbb{N}$ and $\boldsymbol{V}$, and closed under Cartesian product.

There is a collection $\boldsymbol{PFn}$ of *partial functions* $\varphi, \psi, \ldots$ on $\boldsymbol{V}$, and more specifically, collections of partial functions[13] $\varphi\colon A \overset{\sim}{\to} B$ between the various data types.

$\mathcal{A}$ has (as in (2.1) above) *basic domains* $A_0, \ldots, A_n$, with $A_0 = \mathbb{B}$. We let i, j, k range over $\{0, \ldots, n\}$ and $\bar{\imath}, \ldots,$ over finite (possibly empty) sequences of these. For $\bar{\imath} = (i_1, \ldots, i_\nu)$, put $A_{\bar{\imath}} = A_{i_1} \times \cdots \times A_{i_\nu}$. Then a partial function φ on $\mathcal{A}$ has *arity* $\bar{\imath} \to j$ if $\varphi\colon A_{\bar{\imath}} \overset{\sim}{\to} A_j$. We let $\sigma, \tau, \ldots$ range over arities of partial functions on $\mathcal{A}$. For $\bar{\sigma} = \sigma_1, \ldots, \sigma_\mu$, put $A_{\bar{\sigma}} = A_{\sigma_1} \times \cdots \times A_{\sigma_\mu}$. Then a partial functional F on $\mathcal{A}$ has *arity* $(\bar{\sigma}, \bar{\imath}) \to j$ if $F\colon A_{\bar{\sigma}} \times A_{\bar{\imath}} \overset{\sim}{\to} A_j$. When $\varphi = (\varphi_1, \ldots, \varphi_\mu)$ and $x = (x_1, \ldots, x_\nu)$, we write $F(\varphi, x)$ for $F(\varphi_1, \ldots, \varphi_\mu, x_1, \ldots, x_\nu)$. Generally F has type level 2, but when $\mu = 0$, $\nu > 0$ then F is a partial function on $\mathcal{A}$ of level 1, and when $\mu = \nu = 0$ then F is a constant of sort j of level 0.

Suppose $\mathcal{A}$ (as in (2.1) above) has the basic functionals

$$F_k\colon A_{\bar{\sigma}_k} \times A_{\bar{\imath}_k} \to A_{j_k} \ (k = 1, \ldots, n).$$

Then $\mathcal{A}$ has the *signature* $\Sigma(\mathcal{A}) = \big(n,\ \langle \bar{\sigma}_k, \bar{\imath}_k, j_k \rangle_{1 \le k \le m}\big)$.

In Sect. 4, four interpretations of the structure are presented, which we describe here, (mainly in SF's own words). For convenience I refer to these as four "models".

Model 1: The full set-theoretic interpretation.
"Here V is the class of all sets in the cumulative hierarchy. The types range over all sets in V ... Functionals are just those partial functions in V of the form $F(\varphi, x)$ where the φ_k's are partial functions in V."

Model 2: The set-theoretic interpretation on computational data.
"For computational purposes, all data should be represented in in finite symbolic form; without loss of generality, we can take the universe to be $V = \mathbb{N}$. ... The partial functions here are arbitrary $\varphi\colon \mathbb{N} \overset{\sim}{\to} \mathbb{N}$ ... Partial functionals $F(\varphi, x)$ in this

[12]Below I use '$\mathbb{N}$' for 'ω'.

[13]This notation '$\overset{\sim}{\to}$' for partial functions is not the same as SF's here, but is used for consistency with his notation in the paper [19] discussed in §3 below.

interpretation take *arbitrary partial function arguments*[14] φ on $\mathbb{N}$. Special interest attaches below to those F which are *partial recursive* (p.r.) or have a p.r. extension. Note that p.r. functionals are *not closed under abstraction* when the remaining function arguments are not p.r."

Model 3: The recursion-theoretic interpretation, extensional form.
"Here again we take $V = \mathbb{N}$ .... The types ... range over arbitrary subsets of $\mathbb{N}$ (or, more generally, over any collection of subsets closed under arithmetical definability). The collection ***PFn*** is taken to be the p.r. functions on $\mathbb{N}$, and ***PFnl*** the p.r. functionals of *p.r. function arguments*[15] Thus we have *closure under abstraction* in this case."

Model 4: The recursion-theoretic interpretation, intensional form.
This is like model 3, except that ***PFn*** consists of *Gödel numbers* or indices e of p.r. functions, with application given by $e\,x = \{e\}x$. Now ***PFnl*** coincides with ***PFn***, but with the emphasis on indices which are "extensional" or "effective" in the recursion-theoretic sense.

Returning to a general structure $\mathcal{A}$ (as in (2.1)), SF defines a *partial order* on ***PFn***: for φ, ψ of arity $\bar{\imath} \to j$ on $\mathcal{A}$[16]:

$$\varphi \subseteq_A \psi \iff \forall x \in A_{\bar{\imath}}\left[\varphi(x)\downarrow \implies \psi(x)\downarrow =_{A_i} \varphi(x)\right]. \tag{2.2}$$

A partial functional F of arity $(\bar{\sigma}, \bar{\imath}) \to j$ is $\mathcal{A}$*-monotonic* if

$$\forall \varphi, \psi \in A_{\bar{\sigma}}, \forall x \in A_{\bar{\imath}}\left[F(\varphi, x)\downarrow \wedge\, \varphi \subseteq_A \psi \implies F(\psi, x)\downarrow =_{A_i} F(\varphi, x)\right]. \tag{2.3}$$

Next, for partial functionals F_1, F_2 of arity $(\bar{\sigma}, \bar{\imath}) \to j$ on $\mathcal{A}$, we define

$$F_1 \subseteq_A F_2 \iff \forall \varphi \in A_{\bar{\sigma}} \forall x \in A_{\bar{\imath}}\left[F_1(\varphi, x)\downarrow \implies F_2(\varphi, x)\downarrow =_{A_i} F_1(\varphi, x)\right]. \tag{2.4}$$

Next, in order to determine the *least fixed point* of a functional $G\colon A_\sigma \times A_{\bar{\imath}} \to \mathcal{A}_j$, where $\sigma = (\bar{\imath} \to j)$, we define $\widehat{G}\colon A_\sigma \to A_\sigma$ by $(\widehat{G}\varphi)x = G(\varphi, x)$ (i.e. a "curried" version of G). Then, supposing G is $\mathcal{A}$-monotonic, so is $\widehat{G}$, and further, a (unique) *least fixed point* $\mathbf{L}G$ of G can be found from $\widehat{G}$, as follows.

First, $\mathbf{L}$ is called an LFP operator on $\mathcal{A}$ if for any $\sigma = \bar{\imath} \to j$ and $\mathcal{A}$-monotonic $G\colon A_\sigma \times A_{\bar{\imath}} \xrightarrow{\sim} A_j$, we have:

(i) $\mathbf{L}G \in A_\sigma$ and $\widehat{G}(\mathbf{L}G) \subseteq_A \mathbf{L}G$;
(ii) whenever $\psi \in A_\sigma$ and $\widehat{G}(\psi) \subseteq_A \psi$ then $\mathbf{L}G \subseteq_A \psi$;
(iii) For $\mathcal{A}$-monotonic G_1, G_2 of the same arity, $G_1 \subseteq_A G_2 \implies \mathbf{L}(G_1) \subseteq_A \mathbf{L}(G_2)$.

---

[14] Emphasis added. This gives the essential difference between models 2 and 3.
[15] Emphasis added. This gives the essential difference between models 2 and 3.
[16] Recall (2.1) the equality relations '$=_{A_i}$' on $\mathcal{A}$.

The question of the *existence* of LFP operators in the various interpretations must still be discussed (see below).

Recall (2.1) we are assuming each domain A_i has an "equality" relation. Next SF defines the relation of $\mathcal{A}$-*equality* between partial functions:

$$\varphi =_A \psi \iff \varphi \subseteq_A \psi \ \wedge \ \psi \subseteq_A \varphi$$

and hence the concept of a functional F *preserving* $\mathcal{A}$-*equalities*. Then F is said to be *strongly* $\mathcal{A}$-*monotonic* if it is $\mathcal{A}$-monotonic and also preserves $\mathcal{A}$-equalities.

We come to ***schemata*** for ***abstract computational procedures*** (***ACPs***) over Σ. Suppose given a Σ-structure $\mathcal{A}$, with basic functionals $F_1, \ldots, F_m$, where each F_k has a specified arity $\bar{\sigma}_k \times \bar{\imath}_k \to j_k$. We make some *general assumptions*:

(i) $A_0 = \mathbb{B} = \{\mathtt{tt}, \mathtt{ff}\}$, where $=_{A_0}$ is the identity relation, and
(ii) $\mathtt{tt}$ and $\mathtt{ff}$ are themselves in Σ.

A formal language of computational procedures over Σ is defined, with variables $a, \ldots, x, \ldots$ of all Σ-sorts, partial function varables $\varphi, \psi, \ldots$ of each arity, and functional symbols $\mathsf{F}, \mathsf{G}, \mathsf{H}, \ldots$ of each appropriate arity. It is further assumed:

(iii) For each $\mathcal{A}$, $\mathsf{L}^{\mathcal{A}}$ is an operator from $\mathcal{A}$-functionals of arity $\sigma \times \bar{\imath} \to j$, where $\sigma = \bar{\imath} \to j$, to A_σ-partial functions (for each $\bar{\imath}$, j).

Note that at this stage $\mathsf{L}^{\mathcal{A}}$ is not yet assumed to be an LFP operator on $\mathcal{A}$.

There follows a list of *schemata* (for Σ):

I (Initial fns) $\mathsf{F}(\varphi, x) \simeq \mathsf{F}_k(\varphi, x)$ $(k = 0.\ldots, m)$
II (Identity) $\mathsf{F}(x) = x$
III (Application) $\mathsf{F}(\theta, x) \simeq \theta(x)$
IV (Conditional) $\mathsf{F}(\varphi, x, v) \simeq [\text{if } v = \mathtt{tt} \text{ then } \mathsf{G}(\varphi, x) \text{ else } \mathsf{H}(\varphi, x)]$
V (Structural) $\mathsf{F}(\varphi, x) \simeq \mathsf{G}(\varphi_f, x_g)$
VI (Indiv. subst.) $\mathsf{F}(\varphi, x) \simeq \mathsf{G}\big(\varphi, x, \mathsf{H}(\varphi, x)\big)$
VII (Func. subst.) $\mathsf{F}(\varphi, x) \simeq \mathsf{G}\big(\varphi, \lambda u.\mathsf{H}(\varphi, x, u), x\big)$
VIII (LFP) $\mathsf{F}(\varphi, x, u) \simeq \big[\mathsf{L}(\lambda\theta, w.G(\varphi, \theta, x, w))\big](u)$

In the above schemata, φ and x are (respectively) tuples of function and individual variables; in schema IV v is boolean variable; in schema V, f and g are (respectively) mappings of the indices of the variable tuples φ and x, and φ_f and x_g are the corresponding mappings of these tuples.

An *ACP* for Σ is then a partial functional F generated by the above schemata.

Note the resemblance of the above schemata (other than VIII) to Kleene's schemata [33] apart from S5 (primitive recursion on the integers) and S9 (enumeration).

So for each Σ-structure $\mathcal{A}$ and each ACP F generated by the schemata, there is an associated partial functional $\mathsf{F}^{\mathcal{A}}$, where the intended semantics is clear. Here we must make one more general assumption:

(iv) $\mathsf{L}^{\mathcal{A}}$ is an LFP operator on $\mathcal{A}$.

SF then states and proves his two main theorems:

Theorem 1 (Preservation of strong monotonicity) *If each* F_k $(k = 1, \ldots, m)$ *is strongly monotonic, then so is* $\mathsf{F}^{\mathcal{A}}$ *for every ACP* F *generated by the schemata I–VIII.*

The proof amounts to showing that the property of strong monotonicity is preserved by an application of any of the schemata, including notably the LFP schema VIII.

Theorem 2 (Invariance under isomorphism) *For any* Σ*-ACP* F*, an isomorphism between two* Σ*-structures* $\mathcal{A}$ *and* $\mathcal{A}'$ *induces an isomorphism between* $\mathsf{F}^{\mathcal{A}}$ *and* $\mathsf{F}^{\mathcal{A}'}$*. Note that Theorems 1 and 2 imply (respectively) criteria C4 and C3 of the five criteria listed above.*

Next (in Sect. 10) we suppose given an $\mathcal{A}$*-monotonic functional* $G\colon A_\sigma \times A_{\bar{\imath}} \overset{\sim}{\to} A_j$*, with* $\sigma = (\bar{\imath} \to j)$*. The construction of the LFP operator* $\mathsf{L}^{\mathcal{A}}(G)$ *is shown for each of the four types of interpretation* $\mathcal{A}$ *considered in Sect. 4 (see above).*

Briefly: in the case of all four models, $\mathsf{L}^{\mathcal{A}}(G)$ is constructed as the union of a transfinite sequence of approximations from below. In the case of model 4, the Myhill–Shepherdson Theorem [44] is needed.

We come now to the final important result of the paper (in Sect. 11), which shows how computation on ADTs can reduce (under certain conditions) to ordinary computation.

SF restricts consideration to models with $V = \mathbb{N}$, and the extensional version of the recursion-theoretic interpretation (model 3). So we assume that all the domains of $\mathcal{A}$ are contained in $\mathbb{N}$, and all the F_k are strongly $\mathcal{A}$-monotonic. SF then shows (**Theorem 3**) that for each ACP F, $\mathsf{F}^{\mathcal{A}}$ can be taken as the p.r. functional $\mathsf{F}^{\mathcal{V}}$ on $\mathcal{A}$. Hence $\mathsf{F}^{\mathcal{A}}$ satisfies criterion C5, and thus all the desired criteria C1–C5 for ACPs.

Another version of this theorem will be encountered as Theorem 6 in [19], discussed in §3 below.

In the concluding Sect. 12, SF presents some applications of this theory, revisiting some of his examples in Part 1. As he says:

> [A]ll the standard finitary examples of ADTs, such as lists, trees, sets, etc., as described in Part I of this paper ... have implementations whose F_k are simply partial recursive functions.

"Infinitary" data types, such as infinite streams and infinite precision reals, are discussed briefly. In the case of infinite streams over a structure $(A, =_A)$, SF uses the following structure, first defined in Part 1 above[17]:

$$\textit{\textbf{Stream}}(A) \;=\; (S,\, A,\, \mathbb{N},\, =_S,\, =_A,\, =_{\mathbb{N}},\, \ldots,\, \textit{\textbf{Cons}},\, \textit{\textbf{Hd}},\, \textit{\textbf{Tl}},\, \textit{\textbf{Sim}}) \qquad (2.5)$$

This has three domains, A, S and $\mathbb{N}$ for data, streams of data and naturals respectively; the standard operations on $\mathbb{N}$, shown here as '...', and the stream operations $\textit{\textbf{Cons}}\colon A \times S \to S$, $\textit{\textbf{Hd}}\colon S \to A$ and $\textit{\textbf{Tl}}\colon S \to S$, where (informally)

[17] The notation here has been modified to conform to that in [19], cf. §3 below.

$$\begin{aligned} \textbf{\textit{Cons}}\,(b,\ \langle a_0, a_1, a_2 \ldots \rangle) &= \langle b, a_0, a_1, a_2, \ldots \rangle \\ \textbf{\textit{Hd}}\,\langle a_0, a_1, a_2, \ldots \rangle &= a_0, \\ \textbf{\textit{Tl}}\,\langle a_0, a_1, a_2, \ldots \rangle &= \langle a_1, a_2, \ldots \rangle. \end{aligned}$$

In order to characterize infinite stream structures up to isomorphism, we also need a second order "simulating" functional $\textbf{\textit{Sim}}: (\mathbb{N} \to A) \to S$ which is a bijection from the set of "all" functions $f: \mathbb{N} \to A$ onto S. (Without this, we can only ensure the existence of eventually constant streams.)

SF comments: "Here, it seems, only the intensional recursion-theoretic interpretation[18] is appropriate." This is accomplished by interpreting S as the set of all indices of total recursive functions from $\mathbb{N}$ to $\mathbb{N}$, and $\textbf{\textit{Sim}}$ as the identity on $\mathbb{N}$.[19]

The ADT of infinite precision reals is likewise given an intensional recursion-theoretic interpretation, in which the reals are interpreted as indices of effective Cauchy sequences.

This perspective will shift quite dramatically in the following paper [19], with the investigation of higher order ("extensional") models of streams, and a second order '***Sim***' operator.

3 Computation on ADTs: The Extensional Approach

This paper[20] carries the dedication "With profound gratitude to Stephen C. Kleene" (then recently deceased) with the footnote "Kleene was not my mentor, official or otherwise, but through his exceptional development of our subject I learned as much from him as if he had been."

I quote from the introduction:

> This paper is a continuation of the work of Feferman [17, 18] which initiated an approach through a form of *generalized recursion theory* (g.r.t.) to computation on *abstract data types* (*ADT*s), including intensionally presented types ...
>
> [W]e separate out the extensional part of the theory and show how it may be applied to computation on streams as an ADT. One of the main new contributions here is an explanation of how this is to be done for finite "nonterminating" streams as well as infinite streams, and even more general partial ("gappy") streams.

Logically (as stated in the previous section) an ADT is just a class of structures closed under isomorphism. "[O]ne is mainly interested in structures determined by categorical or relatively categorical conditions." Paradigmatic examples considered are, for any data set A: A-lists and A-streams.

---

[18] That is, model 4 described above.

[19] There is an error here, corrected by SF, with an improved presentation, in Sect. 11 of [19] (§3 below). This and a few other corrigenda for the present paper are listed in App. C of that paper.

[20] Previously reviewed by me in [60], which lists some (minor) slips in this paper.

Continuing with the framework of the previous paper (§2 above). the general theory in this paper applies, for a given signature Σ, to many-sorted Σ-structures

$$\mathcal{A} = (A_0, \ldots, A_n, F_1, \ldots, F_m) \tag{3.1}$$

where $A_0 = \mathbb{B} = \{\mathrm{tt}, \mathrm{ff}\}$, and each F_j is a partial functional (or function, or constant, including tt and ff) of type level ≤ 2. It will also be assumed that each F_j is $\mathcal{A}$-*monotonic*.

As SF says, "computation on streams is subsumed under a general theory of computation for arbitrary structures.[21]"

Remark (*Treatment of equality*) Note the difference between the structures displayed here (3.1) and in (2.1): here an ("intensional") equality predicate is *not* automatically assumed at each sort. In fact, infinite stream equality is generally non-computable. SF turns to this issue later in the paper.

In Sect. 3 a system of formal schemata for a given signature Σ is presented. This is the same as the system of schemata (I–VIII) given in [18] and shown above (§2); and, as before, with each Σ-structure $\mathcal{A}$ and each ACP F generated by the schemata, is associated a partial functional (or function, or constant) $\mathsf{F}^{\mathcal{A}}$ of type level ≤ 2.

As shown in [18] (cf. §2 above), an $\mathcal{A}$-monotonic functional has an associated *least fixed point* (LFP), which is also monotonic, by the **LFP Monotonicity Lemma**[22]. Hence for schema VIII to make sense on $\mathcal{A}$, it must be verified that $\mathsf{F}^{\mathcal{A}}$ is $\mathcal{A}$-monotonic. More generally, it is shown (**Monotonicity Lemma**[23]) that the ACPs generated by all the schemata are $\mathcal{A}$-monotonic, assuming that the initial functionals $F_1, \ldots, F_m$ are.

Remarks (*Extensionality*) The assumption of extensionality in this paper leads to two interesting divergences from the theory developed in the previous paper [18] (§2), with intensional equality:

(1) In [18] the significant property of ACPs was *strong monotonicity*, i.e., monotonicity plus equality-preservation. Here it is simply monotonicity.
(2) In [18] the existence of the LFP operator had to be explicitly assumed ("General assumption (iv)" at the beginning of Sect. 8) and proved for the four types of interpretations ("Models 1–4"). Here the construction of the LFP operator (the LFP Monotonicity Lemma) can be shown quite generally as a consequence of the more general Monotonicity Lemma.[24] Again, it is constructed as the union of a transfinite sequence of approximations from below.

Back to Sect. 3: It is next shown that the schemata are *invariant under isomorphism*, i.e., if $\mathcal{A}$ and $\mathcal{A}'$ are Σ-isomorphic, then so are $\mathsf{F}^{\mathcal{A}}$ and $\mathsf{F}^{\mathcal{A}'}$ for all F generated by the schemata. This justifies the terminology "*abstract* computation procedures",

[21] As in (3.1).

[22] My terminology.

[23] My terminology.

[24] Or more accurately, simultaneously with this lemma.

and the notation **ACP(Σ)** for the collection of ACPs generated by the schemata over Σ, and **ACP($\mathcal{A}$)** for the collection of all $\mathsf{F}^{\mathcal{A}}$ for any Σ-structure $\mathcal{A}$.

Next (Sect. 4) this paper deals with an important *substructure property* of Σ-structures.

First, some definitions. Suppose given a Σ-structure $\mathcal{A}$ as in (3.1) above, and subsets $B_i \subseteq A_i$ for $i = 1, \ldots, n$. Write $A = (A_0, \ldots, A_n)$ and $B = (B_0, \ldots, B_n)$.

For a level-1 partial function φ over A, the *restriction* $\varphi \restriction B$ and the concept "B is *closed under* φ" are defined in the standard way.

For a level-2 partial functional $F\colon A_{\bar{\sigma}} \times A_{\bar{\imath}} \to A_j$, the *restriction* $F \restriction B$ means the function $\lambda\varphi \in B_{\bar{\sigma}} \cdot \lambda x \in B_{\bar{\imath}} \cdot F(\varphi, x)$, and B is said to be *closed under* F if[25]

$$\forall\varphi \in A_{\bar{\sigma}}, \forall x \in B_{\bar{\imath}} \big[B \text{ closed under } \varphi \;\wedge\; F(\varphi, x)\downarrow \Longrightarrow \\ F(\varphi, x) \in B_j \;\wedge\; F(\varphi \restriction B, x) = F(\varphi, x)\big].$$

We then say that B *determines a substructure of* A if $B_0 = A_0 = \mathbb{B}$ and B is closed under F_k for $k = 1, \ldots, m$.

SF then states and proves his

Substructure Theorem (Theorem 1 + Corollary).
Suppose B determines a substructure of $\mathcal{A}$. Then putting

$$\mathcal{B} = (B_0, \ldots, B_n, F_1 \restriction B, \ldots, F_m \restriction B),$$

B is closed under $\mathsf{F}^{\mathcal{A}}$ for each Σ-ACP F, and

$$\mathsf{F}^{\mathcal{B}} = \mathsf{F}^{\mathcal{A}} \restriction B.$$

This result turns out to be very useful for the rest of the paper, as we will see.

Next (Sect. 5) the paper deals with *continuity* of functionals. A functional F on $A = (A_0, \ldots, A_n)$ of type level 2 is said to be *continuous* if for any φ, x, y,

$$F(\varphi, x) \simeq y \implies \exists \text{ finite } \tilde{\varphi} \subseteq \phi : F(\tilde{\varphi}) \simeq y.$$

SF next states and proves the

Continuity Theorem (Theorem 2). If each basic functional F_k of $\mathcal{A}$ is continuous, then for each ACP F, $\mathsf{F}^{\mathcal{A}}$ is continuous.

The proofs of the Substructure and Continuity Theorems are (as one would expect) by induction on the generation of F by the schemata. However they are far from routine.

Next there is a small section (Sect. 6) on computation on *first-order structures*.

[25] As SF points out, this implies that $F \restriction B\colon B_{\bar{\sigma}} \times B_{\bar{\imath}} \stackrel{\sim}{\to} B_j$, but not conversely.

A signature $\Sigma = \big(n,\ \langle \bar{\sigma}_k, \bar{\imath}_k, j_k\rangle_{1\le k\le m}\big)$ is said to be *first-order* if $\bar{\sigma}_k$ is empty for $k = 1, \ldots, m$. In that case Σ-structures $\mathcal{A} = (A_0, \ldots, A_n,\ F_1, \ldots, F_m)$ are *first-order* in the sense that each F_k has type level 1 or 0.

In that case also, all the F_k are vacuously *monotonic* and *continuous*, and so (by the Monotonicity Lemma and Continuity Theorem) all ACPs over $\mathcal{A}$ are monotonic and continuous.

Further, for computation on *first-order* structures, schema VII (for function substitution) can be omitted, since (as SF shows) if Σ is a first-order signature, then the system $\mathbf{ACP}_0(\Sigma)$ of ACPs over Σ obtained by omitting schema VII is closed under that schema.

As an example, consider the "*ur*-structure for recursion theory"

$$\mathcal{N} \;=\; (\mathbb{N},\ \boldsymbol{Sc},\ \boldsymbol{Pd},\ \mathbf{0},\ \boldsymbol{Eq_0}). \tag{3.2}$$

where the booleans are coded as $\{0, 1\}$, $\boldsymbol{Pd}$ is the predecessor function with $\boldsymbol{Pd}(0) = 0$, and $\boldsymbol{Eq_0}$ tests for equality with 0.

SF then states and proves the following interesting Z. (For the concept of "partial recursive (p.r.) functional", see Footnote 4.)

Theorem 5 (Characterizing **ACP**$(\mathcal{N})$).

(i) The ACPs of type level 1 over $\mathcal{N}$ are exactly the p.r. functions.
(ii) The ACPs of type level 2 over $\mathcal{N}$ are exactly the p.r. functionals when restricted to total function arguments.
(iii) Every ACP of type level 2 over $\mathcal{N}$ is p.r. on partial function arguments, but not conversely.

A counterexample for the failure of the converse for (iii) is given by the "strong or" functional OR^+ (cf. (1.1)) as shown by Platek [48]. In fact, the equivalence

$$\mathbf{ACP}(\mathcal{N}) + \mathsf{OR}^+ \iff \text{p.r. on } \boldsymbol{PFn}(\mathcal{N})$$

for level 2 functionals was proved by Winternitz.[26]

Next, SF defines the concept of ***partial recursive structure in*** $\mathbb{N}$. This is a structure of the form

(i) $\mathcal{A} = (A_0, \ldots, A_n,\ F_1, \ldots, F_m)$, where
(ii) each $A_i \subseteq \mathbb{N}$, $A_0 = \mathbb{B} = \{0, 1\}$.
(iii) each F_k is the restriction to $A = (A_0, \ldots, A_n)$ of a p.r. functional F_k^* on $\mathbb{N}$, under which A is closed.

Using the Substructure Theorem, SF then derives:

Theorem 6 (ACPs of p.r. structures in $\mathbb{N}$). *Suppose $\mathcal{A}$ is a p.r. structure in $\mathbb{N}$ of signature Σ. Then for each F in* **ACP**(Σ), *$\mathsf{F}^{\mathcal{A}}$ is the restriction to $\mathcal{A}$ of a p.r. functional F^*, and is (therefore) continuous.*

[26] This has already been discussed in [15] (cf. §1 above).

This provides another version of Theorem 3 in [18, Sect. 11] (cf. §2 above). To convey the significance of this result (and of the Substructure Theorem), let me quote SF here:

> Most examples of abstract data types ... which contain partial recursive structures are those whose domains are generated by finitely many finitary operations, or are obtained from such by restriction, such as lists, finites sets, finite trees, records, etc. ...
> [I]f $A = \{a_0, a_1, \ldots, a_n, \ldots\}$ is any countable set, we can realize lists-of-A's as a partial recursive structure, no matter how A is identified as a subset of $\mathbb{N}$ ... For example, ... A might be a nonrecursive subset of $\mathbb{N}$, such as the set of Gödel numbers of total recursive functions ... That is why no restriction was made on the A_i's in the definition [of p.r. structures in $\mathbb{N}$] other than that they be subsets of $\mathbb{N}$.

The next section (8) illustrates the theory of Sect. 7 with the ADT of lists over a structure A. To quote SF: "The case of abstract computational procedures on (relativized) list structures is paradigmatic for finitary data types in many respects, and is useful for comparison with computation on infintary data types, of which streams form the main example in this paper."

SF shows how all the standard list operations can be defined as ACPs. He presents a number of formulations of definition by *list recursion*, and demonstrates the use of the Substructure Theorem in the case of p.r. list structures.

We will not describe this development in detail, moving rather on to the next section (9) dealing with the infinitary data type of *streams*.

This (together with the following sections) is the most interesting part of the paper. It deals with A-streams, or (potentially) infinite sequences of members of A. Keeping to the framework of computation on ADTs, SF develops and investigates the structure $\mathcal{S}$ of A-streams with basic sets A and S.

To quote SF: "[T]hough a standard interpretation of S consists of *second-order* objects, in the present approach they are to be treated as *first-order* objects in $\mathcal{S}$."

However, treating streams as first-order objects, like lists, leads to trouble. For consider an axiomatization of a first-order structure of streams[27]:

$$\mathcal{S}^{(1)} \;=\; (A,\ S,\ \boldsymbol{Cons},\ \boldsymbol{Hd},\ \boldsymbol{T\ell}) \tag{3.3}$$

(the superscript "1" indicating a *first-order* structure) where

(i) $A \neq \emptyset$
(ii) $\boldsymbol{Cons}\colon A \times S \to S, \quad \boldsymbol{Hd}\colon S \to A, \quad \boldsymbol{T\ell}\colon S \to S,$
(iii) $\forall a \in A\, \forall s \in S\big[\boldsymbol{Hd}(\boldsymbol{Cons}(a,s)) = a \;\wedge\; \boldsymbol{T\ell}(\boldsymbol{Cons}(a.s)) = s\big].$

As SF writes: "The main point against this is that these (and similar) conditions do not uniquely determine $\mathcal{S}^{(1)}$ up to isomorphism, given A. Two nonisomorphic structures are obtained by interpreting S in the first instance to be the set $(\mathbb{N} \to A)$ of *all* functions from $\mathbb{N}$ to A, and in the second instance to be the subset $(\mathbb{N} \overset{\text{fin}}{\to} A)$ ... of eventually constant functions."

[27] Note that this is the stream structure (2.5) without the equalities.

These two structures will be denoted here, respectively, by $\mathcal{S}^{(1)}[\mathbb{N} \to A]$ and $\mathcal{S}^{(1)}[\mathbb{N} \overset{\text{fin}}{\to} A]$, the latter clearly a substructure of the former.

The second problem with this approach is that these conditions do not guarantee closure under some standard computation procedures, such as recursion on streams. One can easily find examples of recursively defined functions from $\mathbb{N}$ to A which are not in $\mathcal{S}^{(1)}[\mathbb{N} \overset{\text{fin}}{\to} A]$, and hence (by the Substructure Theorem) also not ACPs in $\mathcal{S}^{(1)}[\mathbb{N} \to A]$.

The answer is to work with *second-order* stream structures. It will be shown how to obtain functionals for (e.g.) recursion schemes for streams as ACPs on *second order* stream structures $\mathcal{S}^{(2)}$. It turns out that the simplest effective way to construct such a structure is to adjoin to the stream signature a level 2 functional $\boldsymbol{Sim} : (\mathbb{N} \to A) \to S$ which *simulates* every function $\varphi : \mathbb{N} \to A$ as a level 0 object $\boldsymbol{Sim}(\varphi) \in S$.

The next step is to extend the domain of $\boldsymbol{Sim}$ to include *potentially infinite* streams. As SF says: "These arise naturally both from mathematical computations and physical phenomena." An example of the first kind is obtained by *filtering* an infinite stream according a suitable condition on the items, where we may not know in advance whether the condition applies to finitely or infinitely many items in the stream. An example of the second kind is provided by irregularly received signals from an extraterrestial source, where we do not know at any point whether there will be any further signals.

Actually, the simplest and most elegant theory is obtained by allowing the domain of $\boldsymbol{Sim}$ to include all *partial functions* on $\mathbb{N}$, giving rise to "gappy" streams, from which the ACPs for potentially infinite streams can be obtained as a special case. This leads to second-order structures of the form

$$\mathcal{S} = (A,\ S,\ \boldsymbol{Cons},\ \boldsymbol{Hd},\ \boldsymbol{T\ell},\ \boldsymbol{Sim},\ \mathcal{N}) \tag{3.4}$$

where $\mathcal{N}$ is as in (3.2) and

(i) $A \neq \emptyset$,
(ii) $\boldsymbol{Cons} : A \times S \to S, \quad \boldsymbol{Hd} : S \overset{\sim}{\to} A, \quad \boldsymbol{T\ell} : S \to S, \quad \boldsymbol{Sim} : (\mathbb{N} \overset{\sim}{\to} A) \to S,$
(iii) $\forall a \in A \forall s \in S\big[\boldsymbol{Hd}(\boldsymbol{Cons}(a, s)) = a \ \wedge\ \boldsymbol{T\ell}(\boldsymbol{Cons}(a.s)) = s\big]$,
(iv) $\forall \varphi \in (\mathbb{N} \overset{\sim}{\to} A)\, \forall n \in \mathbb{N}\big[\boldsymbol{Hd}(\boldsymbol{T\ell}^n(\boldsymbol{Sim}(\varphi))) \simeq \varphi(n)\big]$, and
(v) $s, s' \in S \ \wedge\ \forall n\big[\boldsymbol{Hd}(\boldsymbol{T\ell}^n(s)) \simeq \boldsymbol{Hd}(\boldsymbol{T\ell}^n(s'))\big] \implies s = s'$.

Let ***P-STREAM*** ("P" for "partial") be the ADT of all such structures.

This is the starting point for the analysis of computation on streams in Sect. 10. Since we are working with partial streams, a more refined concept of monotonicity is required than that given earlier in this paper (Sect. 2), namely *hereditary monotonicity*,[28] requiring *chain-completeness*[29] of the partial orderings $\sqsubseteq_i$ on all basic domains A_i.

[28] As in Platek's finite type structures [15] (cf. §1 above and Footnote 3).

[29] I.e., any linearly ordered subset of A_i has a l.u.b in A_i.

Notation. For $\mathcal{S}$ as in (3.4), and $s \in S,\ n \in \mathbb{N},\ a \in A$, we write

- $(s)_n$ for $\boldsymbol{Hd}(\boldsymbol{T\ell}^n(s))$,
- $\langle a; s\rangle$ for $\boldsymbol{Cons}(a, s)$, and
- $s^{\rightarrow}$ for $\boldsymbol{T\ell}(s)$.

Then the ***partial ordering*** on the basic domains of $\mathcal{S}$ is defined by:

(i) $s \subseteq_S s'$ for $\forall n[(s)_n \downarrow \implies (s')_n \downarrow = (s)_n]$.
(ii) $\subseteq_A$ and $\subseteq_{\mathbb{N}}$ are equality on A and $\mathbb{N}$.

This makes all three basic orderings in $\mathcal{S}$ chain-complete.

Now consider the general situation of a structure $\mathcal{A}$ as in (3.1) where all the basic domains A_i $(i = 1, \ldots, n)$ have chain-complete partial orderings $\subseteq_i$. We define, for certain function types σ, the domains, orderings, and concepts of monotonicity for that type.

First, taking $\sigma = (\bar{\imath} \overset{\sim}{\rightarrow} j)$, A_σ is the set of all $\varphi\colon A_{\bar{\imath}} \overset{\sim}{\rightarrow} A_j$ which are *monotonic*, in the sense that

$$\forall x, y \in A_{\bar{\imath}}, \left[\varphi(x) \downarrow \wedge\ x \subseteq_{\bar{\imath}} y \implies \varphi(y) \downarrow \wedge\ \varphi(x) \subseteq_j \varphi(y)\right]$$

(where the orderings $\subseteq_{i_k}$ on A_{i_k} are extended termwise to orderings $\subseteq_{\bar{\imath}}$ on $A_{\bar{\imath}}$ in the obvious way). Then the ordering $\subseteq_\sigma$ on A_σ is defined by: for all $\varphi, \psi \in A_\sigma$:

$$\varphi \subseteq_\sigma \psi \iff \forall x \in A_{\bar{\imath}}\left[\varphi(x) \downarrow \implies \psi(x) \downarrow \wedge\ \varphi(x) \subseteq_j \psi(x)\right]. \tag{3.5}$$

Next, *monotonicity* of level 2 functionals is defined by: $F\colon A_{\bar{\sigma}} \times A_{\bar{\imath}} \overset{\sim}{\rightarrow} A_j$ is monotonic if

$$\begin{aligned}\forall \varphi, \psi \in A_{\bar{\sigma}}\, \forall x, y \in A_{\bar{\imath}}\big[\, F(\varphi, x) \downarrow \wedge\ \varphi \subseteq_{\bar{\sigma}} \psi\ \wedge\ x \subseteq_{\bar{\imath}} y \implies \\ F(\psi, y) \downarrow \wedge\ F(\varphi, x) \subseteq_j F(\psi, y)\big].\end{aligned} \tag{3.6}$$

Note that the basic function(al)s of $\mathcal{S}$ are all monotonic.

Then for a level 2 type $\tau = \sigma \times \bar{\imath} \overset{\sim}{\rightarrow} j$, we can define the domain A_τ of monotonic (partial) functionals of type τ, with the ordering[30]

$$\begin{aligned}&F_1 \subseteq_\tau F_2 \iff \\ &\quad \forall \varphi \in A_{\bar{\sigma}}\, \forall x \in A_{\bar{\imath}}\big[\, F_1(\varphi, x) \downarrow \implies F_2(\varphi, x) \downarrow \wedge\ F_1(\varphi, x) \subseteq_\tau F_2(\varphi, x)\big].\end{aligned} \tag{3.7}$$

Note the similarity – and difference! – between the definitions given here ((3.5)–(3.7)) and those in [18] (§2 above): ((2.2)–(2.4)), where (essentially) it was assumed that the partial order on each basic domain is the identity.

[30] Not given explicitly in [18].

Note also that the orderings defined in this way on these higher level domains are chain-complete (assuming the basic orderings are). So the technique used in §2 to construct LFPs and (hence) ACPs can be adapted to structures of the form (3.1) with chain-complete basic orderings and monotonic basic functionals F_i. This forms the basis of a *least fixed point* semantics for partial stream structures, using a version of the LFP Monotonicity Lemma of Sect. 3.

In particular, this theory can be applied to ***P-STREAM***, providing a justification for the *recursive schemes* for defining ACPs on partial stream structures given in Sect. 10, to which we turn below.

Discussion: Why a partial structure on streams?
There are two points here.

(1) We noted above that the theory of LFPs is simpler when each basic domain has the identity as (trivial) partial order, as was done in §2. But that would make the basic functional ***Sim*** not monotonic.
(2) We *could* recover monotonicity for ***Sim*** by having *total streams* only in the stream domain S. However, this would complicate the theory of recursion schemes on streams. As SF says: "[T]he recursion schemata for partial streams (such as needed for the *Filter* operation) come out much more simply than they do for total streams."[31]

First, we note that the '***Sim***' functional characterizes stream structures up to *isomorphism*:

Categoricity Theorem for *P-STREAM* (Theorem 8). Suppose structures $\mathcal{S} = (A, \ldots)$ and $\mathcal{S}' = (A', \ldots)$ both satisfy conditions (i)–(v). Then an isomorphism $A \cong A'$ can be extended to an isomorphism $\mathcal{S} \cong \mathcal{S}'$.

Now SF turns to the problem of a general formulation of stream recursion. He begins by stating: "By stream recursion we mean any general computational scheme for producing streams as values."

He does not attempt a single "most general" form (assuming that is even possible), but presents a number of schemes which have good practical applications, of which I'll give one example.

Note first that *partialness* (or "gappiness") of streams is sometimes a looser condition than we want. A more useful concept may be *potential infiniteness*, where a stream $s \in S$ is said to be potentially infinite (or "non-gappy") if

$$\forall n, m\big[(s)_n \downarrow \wedge\ m < n \implies (s)_m \downarrow\big].$$

We denote by S_{potinf} the subset of S consisting of these.

Let $\mathcal{S}^+ = (\mathcal{S}, \ldots)$ be an expanded structure with $\mathcal{S} \in$ ***P-STREAM***.

[31] As we will see with the recursion scheme shown below.

Recursion Scheme (Theorem 10). Let C be a subset of one of the basic domains in $\mathcal{S}^+$. Given ACPs $G\colon C \to A$, $H_0, H_1\colon C \to C$ and $D\colon C \to \mathbb{B}$ over $\mathcal{S}^+$, we can find an ACP F over $\mathcal{S}^+$ satisfying

(i) $F\colon C \to S_{\mathsf{potinf}}$,
(ii) $F(c) = \big[\text{if}\, D(c) = \mathtt{tt}\, \text{then} \langle G(c);\ F(H_0 c)\rangle\ \text{else}\ F(H_1 c)\big]$ for all $c \in C$, and
(iii) if $F'\colon C \to S_{\mathsf{potinf}}$ is any function satisfying (ii) then $F(c) \subseteq F'(c)$ for all $c \in C$.

As SF points out[32]: "While F solves a fixed-point equation (ii), it cannot be described as its LFP, since that is the completely undefined function. Here, in contrast, F is total and is characterized by (iii) among all total solutions of (ii) as the one which is least pointwise in C."

An interesting application of this scheme is the *filtering operation* with respect to a predicate $\varphi\colon A \to \mathbb{B}$, where

$$\textbf{\textit{Filter}}\colon (A \to \mathbb{B}) \times S_{\mathsf{inf}} \ \to\ S_{\mathsf{potinf}}$$

is defined by

$$\textbf{\textit{Filter}}(\varphi, s) \ = \ \begin{cases} \langle (s)_0;\ \textbf{\textit{Filter}}(\varphi, s^{\to})\rangle & \text{if } \varphi((s)_0) = \mathtt{tt} \\ \textbf{\textit{Filter}}(\varphi, s^{\to}) & \text{otherwise.} \end{cases}$$

This produces, from an infinite stream, a potentially infinite stream, which may or may not actually be finite ("extensionally" speaking).

We turn to Sect. 11, dealing with the recursion-theoretic interpretation of computation on stream structures over $\mathbb{N}$ – more precisely, structures for A-streams, where $A \subseteq \mathbb{N}$.[33]

The *standard realization* for these takes $\mathcal{S}(A) = (\mathbb{N} \overset{\sim}{\to} A)$. In particular, the standard realization for $A = \mathbb{N}$ is

$$\mathcal{S}(\mathbb{N}) \ = \ (\mathbb{N},\ S(\mathbb{N}),\ \ldots)$$

as in (3.4) with A replaced by $\mathbb{N}$, and ***Sim*** being the identity on $(\mathbb{N} \overset{\sim}{\to} \mathbb{N})$.

The substructure of $\mathcal{S}(\mathbb{N})$ induced by $A \subseteq \mathbb{N}$ is then

$$\mathcal{S}(A) \ = \ (A,\ S(A),\ \ldots)$$

with ***Sim*** the identity on $(\mathbb{N} \overset{\sim}{\to} A)$.

By the Categoricity Theorem for ***P-STREAM***, every member $\mathcal{S}$ of ***P-STREAM*** on A has $\mathcal{S} \cong \mathcal{S}(A)$. Further, applying the Substructure Theorem in Sect. 4 to $\mathcal{S}(A)$, we have:

---

[32]In connection with another recursion scheme, but it is still appropriate here.

[33]This supersedes the presentation of this topic in Sect. 12 of [18] (§2 above). See Footnote 19.

Substructure Theorem for $\mathcal{S}(A)$.
For each ACP F in Σ(***P-STREAM***), $\mathcal{S}(A)$ is closed under $\mathsf{F}^{\mathcal{S}(\mathbb{N})}$, and

$$\mathsf{F}^{\mathcal{S}(\mathbb{N})} \restriction \mathcal{S}(A) \;=\; \mathsf{F}^{\mathcal{S}(A)}.$$

Thus, for any $A \subseteq \mathbb{N}$, a recursion-theoretic description of any ACP over $\mathcal{S}(A)$ is obtainable simply as the *restriction* of that ACP over $\mathcal{S}(\mathbb{N})$. To clarify this: replace $\mathcal{S}(\mathbb{N})$ by the structure

$$\mathcal{E}(\mathbb{N}) \;=\; (\mathbb{N},\ S(\mathbb{N}),\ \boldsymbol{Eval},\ \boldsymbol{Sim},\ \boldsymbol{Sc},\ \boldsymbol{Pd},\ \mathbf{0},\ \boldsymbol{Eq}_0).$$

where $\boldsymbol{Eval}$: $S(\mathbb{N}) \times \mathbb{N} \ \tilde{\to}\ \mathbb{N}$ is given by $\boldsymbol{Eval}(s, n) \simeq s(n)$. This structure is easily seen to be equivalent to $\mathcal{S}(\mathbb{N})$, in the sense that every ACP over $\mathcal{S}(\mathbb{N})$ is obtainable as one over $\mathcal{E}(\mathbb{N})$, and conversely.

Note that $\mathcal{E}(\mathbb{N})$ has two basic domains: $A_0 = \mathbb{N}$ and $A_1 = S(\mathbb{N})$, of type levels 0 and 1 respectively. This suggests a straightforward interpretation of $\mathcal{E}(\mathbb{N})$ into the finite type structure over $\mathcal{N}$ (cf. (3.2)), which contains only one basic domain, $A_0 = \mathbb{N}$ of level 0. By this interpretation, monotonic partial functionals over $\mathcal{S}(\mathbb{N})$ (or $\mathcal{E}(\mathbb{N})$) of type level 2 (or less) are identified with monotonic partial functionals of type level 3 (or less) over $\mathcal{N}$.

Hence, for a recursion-theoretic interpretation of ACPs over $\mathcal{S}(\mathbb{N})$, we need an extension of the notion of *partial recursiveness* to functionals of type level 3 over $\mathcal{N}$. This was provided by Ershov [11] (as outlined in §1 above) in a structure $C^{\smile} = (C_n^{\smile})_n$ with a notion of *partial recursiveness* for functionals of arbitrary finite type, with hereditarily partial continuous arguments over $C_0^{\smile} = \mathbb{N}$, based on an abstract theory of f-spaces (a special kind of topological space).

This theory was then simplified by SF [14, 15] to a "concrete" theory $\boldsymbol{PR}/C^{\smile}$ of *partial recursively continuous functionals* of finite type, analogous to functionals of hereditarily *total* continuous arguments [32, 35], using (again) a system of *formal neighbourhoods*. The precise definitions are given in the paper. The following result can then be obtained via Ershov's theory of f-spaces [11]. A simpler, direct proof[34] is possible via SF's version of the theory indicated above.

Theorem (Closure of $\boldsymbol{PR}/C^{\smile}$ under ACP).
$\boldsymbol{PR}/C^{\smile}$ is closed under the extension of ACP schemata to arbitrary finite types.

It follows that ACPs over $\mathcal{S}(\mathbb{N})$ of level ≤ 2 can be re-interpreted as *partial recursively continuous functionals* over $\mathcal{N}$ of level ≤ 3.

The paper has three appendices. I shall only remark here on *Appendix B: Comparison with the work of Tucker and Zucker*. This concerns research that John Tucker and I have done in a series of publications[35] on many-sorted models of computation $\mathcal{A}$. One model over $\mathcal{A}$ that we have investigated is $\boldsymbol{\mu PR}^*(\mathcal{A})$, consisting of schemes for

[34]Unfortunately never published (personal communication by SF).

[35]See e.g. [56] and the references therein.

primitive recursion over $\mathcal{A}^*$ plus μ (the "constructive least number operator"), where $\mathcal{A}^*$ is formed by adding, to each carrier set A of $\mathcal{A}$, a set A^* of all finite sequences from A (with associated basic operations). Note that to formulate this model, we must assume that $\mathcal{A}$ includes, as a subalgebra, the algebra $\mathcal{N}$ of naturals — or add it on. This model also forms a basis for a *generalized Church-Turing Thesis*.[36]

SF noted that for first-order functions on $\mathcal{A}$, $\boldsymbol{\mu PR}^*(\mathcal{A}) \subseteq \mathbf{ACP}(\mathcal{A}^*)$. He conjectured the reverse inclusion, leading to the equality

$$\boldsymbol{\mu PR}^*(\mathcal{A}) \;=\; \mathbf{ACP}(\mathcal{A}^*).$$

This was proved in [58], assuming a modification of SF's LFP schemes (cf. §2 above) formed by replacing the simple LFP scheme (VIII) by a *simultaneous* LFP. This is discussed further in the last of our four papers by SF (§4 below), to which we now turn.

4 About and Around Computing over the Reals

An overview of this paper is provided by the first section, in which SF sets the stage by referring to a "very interesting and readable" article by Lenore Blum [6], which explains the so-called BSS model of computation over the reals due to Blum, Shub and Smale [7], expounded also in the well-known book by Blum, Cucker, Shub and Smale [3].

Blum claimed that the BSS model of computation on reals is the appropriate foundation for *scientific computing*.

Braverman and Cook [2] argued rather for a *bit computation* model, *prima facie* incompatible with the BSS model. This goes back to ideas of Banach and Mazur in the 1930s, improved by Grzegorczyk and Lacombe (independently, in the 1950s). SF proposes rather to name these "*effective approximation*" models of computation. Later in this paper he discusses such models further (see below).

We should note that there are functions computable in each of these two models which are not computable in the other.

We note also that the bit-computable model only computes *continuous* functions. I consider this a positive rather than a negative property of the model, in keeping with the *continuity principle*:

$$\textit{computability} \implies \textit{continuity}. \tag{4.1}$$

This is related to Hadamard's principle [28] which, as (re-)formulated by Courant and Hilbert ([9, p. 227ff.], [29]) states that for a scientific problem to be well posed,

[36]Discussed further in §4 below.

the solution must (apart from existing and being unique) depend continuously on the data.[37]

On the topic of comparing models, I now quote SF extensively. He asks:

> Despite their incompatibility, is there any way that these can both be considered to be reasonable candidates for computation on the real numbers?

– and gives an "obvious answer":

> [T]he BSS model may be considered to be given in terms of computation over the reals as an *algebraic structure*, while … the effective approximation model can be given … as a *topological structure* of a particular kind, or alternatively as a *second order structure over* the rationals. But all such explanations presume a general theory of *computation over an arbitrary structure*.

He continues:

> After explaining the BSS and effective [approximation] models respectively in Sects. 2 and 3 below, my main purpose here is to describe three theories of computation over (more or less) arbitrary strutures in Sects. 4 and 5, the first due to Harvey Friedman, the second due to John Tucker and Jeffery Zucker, and the third due to the author, adapting to earlier work of Richard Platek and Yiannis Moschovakis. Finally, and in part as an aside, I shall relate the effective approximation approach to the foundations of constructive analysis in its groundbreaking form due to Errett Bishop.

SF concludes the Introduction by touching on the relevance of these structures to *scientific computation* (or "*numerical analysis*"):

> The justification for particular techniques varies with the areas of application but there are common themes that have to do with identifying the source and control of errors and with efficiency of computation. However, there is no concern in the literature on scientific computation with the underlying nature of computing with the reals as exact objects. For, in practice, those computations are made in "floating point arithmetic" using finite decimals with relatiely few significant digits, for which computation *per se* reduces to computation with rational numbers.

He also notes:

> Besides offering a theory of computation on the real numbers, the main emphasis in the articles [2, 6] and the book [3] is on the relevance to the subject of scientific computation in terms of *measures of complexity* … While complexity issues must certainly be taken into account in choosing between the various theories of computation over the reals on offer as a foundation for scientific computation, I take no position as to which of these is most appropriate for that purpose. Rather, my main aim here is to compare them on *purely conceptual grounds*.[38]

Section 2 provides a quick survey of ***BSS-type models***. SF refers to [6], "a brief but informative description", and to the more detailed description in [3].

The BSS definition makes sense for any ring or field, including $\mathbb{R}$, $\mathbb{C}$, $\mathbb{R}^n$, etc., making it an "*algebraic* conception of computability":

---

[37]These issues are discussed in [56, §7],[55, §4.2.14].

[38]Emphasis added. Following SF, I shall focus on the *real-computability* aspect of the various models, rather than their complexity-theoretical or scientific-computational aspects.

> This is reflected in the fact that inputs to a machine for computing a given algorithm are *unanalyzed entities* in the algebra A, and that a basic admitted step in a computation procedure is to test *whether two machine contents* x and y *are equal or not* ... [and] in case A is ordered, ...*whether* $x < y$ *or not.*[39]

There are finite and infinite dimensional versions. An example of an algorithm in this formalism in the finite-dimenional case is the Newton algorithmn for $\mathbb{R}$ or $\mathbb{C}$. An example for the infinite-dimensional case is testing whether a finite set of polynomials over $\mathbb{C}$ have a common zero; this is related to the Hilbert Nullstellensatz.

It is pointed out in [3] that in the finite-dimensional case, a BSS algorithm can be implemented as a form of register machine, and in the infinite-dimensional case, as a form of Turing machine with 2-way infinite tapes. In the case of rings and fields, only piecewise polynomial and rational functions (respectively) are computable.

> In the opposite direction, one may ask what BSS algorithms can actually be carried out on a computer. Here Tarski's decision procedure for the algebra of real numbers is relevant, as it reduces a question of the Hilbert Nullstellensatz type, concerning common roots of a set of polynomials, to a quantifier-free condition on the their coefficients. On the face of it, Tarski's procedure runs in time complexity as a tower of exponentials.

This can be improved to doubly exponential upper bounds by the method of cylindrical algebra decomposition [10].

In ***Section 3***, SF turns to ***effective approximation*** (***EA***) ***models***. We consider functions $f: \mathbb{I} \to \mathbb{R}$, where $\mathbb{I}$ is a (finite or infinite) real interval.

There are two main approaches here:

(1) S-effective approximation: working with sequencess of approximating arguments and values;
(2) P-effective approximation: approximating functions by polynomials.

To illustrate (1): Suppose $f(x) = y$. We work with a sequential representation of x, i.e., a Cauchy sequence of rationals $\langle q_n \rangle$ with limit x, to effectively determine a Cauchy sequence $\langle r_n \rangle$ of rationals with limit y. We may also assume the Cauchy sequences are "fast", i.e., $|q_n - x| \leq 2^{-n}$. With the sequences $\langle q_n \rangle$ and $\langle r_n \rangle$ coded as functions $\varphi, \psi\colon \mathbb{N} \to \mathbb{N}$ in a standard way, the S-effective approximation computability of f reduces to finding an *effectively computable* functional $F: \mathbb{N}^{\mathbb{N}} \to \mathbb{N}^{\mathbb{N}}$ of type level 2 over $\mathbb{N}$.

Writing $\mathcal{P}$ and $\mathcal{T}$ for the classes of partial and total functions (respectively) from $\mathbb{N}$ to $\mathbb{N}$: for "*effective computability*" of functionals $F: \mathcal{P} \to \mathcal{P}$ we can use Kleene's notion of *partial recursiveness* of functionals – as has already been done above in all three previous papers.[40] This definition ensures that F is monotonic and continuous on $\mathcal{P}$, and hence, as SF writes, "computable functionals $F: \mathcal{T} \to \mathcal{T}$ may be defined as those partial recursive functionals whose value for each total function φ is a total function $F(\varphi)$."

---

[39] Emphasis added.

[40] See footnote 4.

SF continues: "[T]here are several other ways of defining which are the computable functionals $F: \mathcal{T} \to \mathcal{T}$ without appealing to the notion of partial recursive functionals." For example, Grzegorczyk [25] used a generalization of Kleene's schemata [31] for general recursive functions with both primitive recursion and the least number operator.

In fact this notion of computable real function was shown by Grzegorczyk [26] to be equivalent to one formulated by him and Lacombe [36] independently. In the simple version stated in [47], this says (roughly) that a function from the reals to the reals is computable if (i) it maps computable sequences of points to computable sequences of points; and (ii) it satisfies an effective locally uniform continuity condition. We call this *Grzegorczyk/Lacombe* computability. This turns out to be equivalent to Weihrauch's Type-2 Theory of Effectivity (TTE) [57]. To quote SF:

> In my view, Kleene's notion of partial recursive functional is the fundamental one, in that it specializes to Grzegorczyk's (or Weihrauch's) in the following sense: if F is a partial recursive functional $F: \mathcal{P} \to \mathcal{P}$, and $F|\mathcal{T}$, the restriction of F to $\mathcal{T}$, maps $\mathcal{T}$ to $\mathcal{T}$, then $F|\mathcal{T}$ is definable by the Grzegorczyk schemata, as may easily be shown.

SF continues:

> It is a consequence of the continuity of partial recursive functionals that if F effectively represents a real-valued function f on its domain $\mathbb{I}$, then f is continuous at each point of $\mathbb{I}$. ...Thus, unlike the BSS model, the order relation on $\mathbb{R}$ is not computable.

To formulate essentially the same phenomenon differently: In all versions of the S-effective and P-effective computation theories on the reals considered here (Grzegorczyk/Lacombe, Weihrauch, etc., and see the next section), and in contrast to the BSS model, order and equality on the reals is *not* computable – specifically, equality on $\mathbb{R}$ is co-semicomputable.

In ***Section 4***: ***The view from generalized recursion theory*** (***g.r.t.***) SF considers two generalizations of recursion theory to arbitrary structures.

First, there is Harvey Friedman's adaptation of the register machine approach [24]. He dealt with structures of the form

$$\mathcal{A} = (A, c_1, \ldots, c_j, f_1, \ldots, f_k, R_1, \ldots, R_m)$$

Comparing this to the stucture in (3.1) above, we note that (unlike the latter) there is here only one domain A (to be changed later); and further (as in (3.1)) equality is not necessarily assumed as a basic operation on A.

A *finite algorithmic procedure* (*fap*) π on $\mathcal{A}$ is given by a finite list of instructions $I_1, \ldots, I_t$. There are also register names $r_0, r_1, r_2, \ldots$ (functioning as variables), with r_0 reserved for output. The instructions include assignments to the r_i and branching on a conditional given by one of the R_i. The class of fap computable functions is denoted by **FAP**($\mathcal{A}$).

For the structure $\mathcal{N}$ of naturals (as in (3.2)) **FAP**($\mathcal{N}$) is equal to the partial recursive functions.

This notion can be generalized to many-sorted structures $\mathcal{A}$ as in (3.1).

Friedman also introduced the class **FAPC**($\mathcal{A}$) (*faps with counting over* $\mathcal{A}$), corresponding to **FAP**($\mathcal{A},\mathcal{N}$), where ($\mathcal{A},\mathcal{N}$) denotes $\mathcal{A}$ augmented by $\mathcal{N}$.

A further extension was made by Moldestad, Stoltenberg-Hansen and Tucker [45, 46] to incorporate *stack registers* in the model, producing the structure **FAPS**($\mathcal{A}$) of *faps with stacks over* $\mathcal{A}$, and **FAPCS**($\mathcal{A}$) of *faps with counting and stacks over* $\mathcal{A}$.

To bring this into the realm of the main concern of this article, consider the structure of the reals

$$\mathcal{R} = (\mathbb{R}, 0, 1, +, -, \times, {}^{-1}, =, <)$$

(with decidable equality and order). Friedman and Mansfield [23] showed the equivalence of **FAPS**($\mathcal{R}$) to the BSS model.

SF also discusses my collaborative work with John Tucker (see, e.g., the lengthy survey paper [53] or the more recent [56]) dealing with a high level 'while' programming language over abstract many-sorted algebras.[41] It is assumed that such an algebra $\mathcal{A}$, of sort Σ, is *standard* in the sense of containing the sort of booleans, with the standard boolean operations.

We also consider expansions of $\mathcal{A}$: $\mathcal{A}^N$, which includes the algebra $\mathcal{N}$ of naturals, and $\mathcal{A}^*$, which includes (further) for each basic domain A_i of $\mathcal{A}$, also a domain A_i^* of finite sequences of elements of A, with associated basic operations, having signatures Σ^N and Σ^* respectively. For a signature Σ of such an algebra, we consider the class of ***While***(Σ) program statements generated by:

$$S ::= \mathsf{skip} \mid \mathsf{x} := t \mid S_1 \,;\, S_2 \mid \mathsf{if}\ b\ \mathsf{then}\ S_1\ \mathsf{else}\ S_2\ \mathsf{fi} \mid \mathsf{while}\ b\ \mathsf{do}\ S_0\ \mathsf{od}$$

where the variable x and term t have the same Σ-sort. Then ***While***($\mathcal{A}$), ***While***^{N}($\mathcal{A}$) and ***While****($\mathcal{A}$) are the classes of (partial) functions on $\mathcal{A}$ definable (respectively) by ***While***(Σ), ***While***(Σ^N) and ***While***(Σ^*) procedures. It follows from [45, 53] that for any standard algebra $\mathcal{A}$,

$$\begin{aligned} \textbf{\textit{While}}(\mathcal{A}) &= \textbf{FAP}(\mathcal{A}) \\ \textbf{\textit{While}}^N(\mathcal{A}) &= \textbf{FAPC}(\mathcal{A}) \\ \textbf{\textit{While}}^*(\mathcal{A}) &= \textbf{FAPCS}(\mathcal{A}). \end{aligned}$$

On the basis of this and other results, John Tucker and I have presented a *generalized Church-Turing thesis for algebraic computability* on standard many-sorted algebras $\mathcal{A}$ involving ***While**** computability on $\mathcal{A}$ [56].

Remark (Compatibility between algebraic and EA models). Recall the two approaches to computability on the reals signalled by SF at the beginning of this

[41] We use "algebra" rather than "structure" to indicate that their signatures contain function (and constant) but not relation symbols.

article: *algebraic* (as exemplified by the BSS model) and *effective approximation* (EA) (e.g., Grzegorczyk-Lacombe or Weierstrass). Our ***While**** model would fall into the "algebraic" class. Let us call this approach to computability on the reals "abstract", and the EA approach "concrete". There would seem to be an *incompatibility* between the abstract and concrete approaches, since in e.g. the BSS model, but *not* in the topological models, equality and order on the reals are total and (hence) not continuous, but nonetheless computable. In the EA (but not the BSS) models, equality and order are given as *partial* (boolean-valued) operations, which are continuous, and also computable. This is in accordance with the continuity principle discussed above (cf. (4.1)).

The point here is that by considering *topological partial algebras* on the reals, in which the basic operations may be partial and are all continuous, we can recover *compatibility between abstract and concrete models*.

In fact, in [54] we proved the equivalence of *abstract* (*computable*) *approximability* by ***While**** programs augmented by a "countable choice" operator, and *concrete computability*, on *metric partial algebras*, under restrictions of effective locally uniform continuity. This result applies, for example, to algebras such as a partial version $\mathcal{R}_\mathsf{p}$ of $\mathcal{R}$.

In ***Sect. 5: The higher type approach***, SF applies his theory **ACP**($\mathcal{A}$) developed in [18, 19] (§§2 and 3 above) of *abstract computational procedures* over many-sorted higher order algebras $\mathcal{A}$ to the special case $\mathcal{A} = \mathcal{N}$. Recall that these ACPs are monotonic, continuous, partial functionals generated by schemata, including (notably) a schema for the least fixed point functional. Further, the sets $\mathbf{ACP}^1(\mathcal{N})$ and $\mathbf{ACP}^2(\mathcal{N})$ of ACPs of levels 1 and 2 over $\mathcal{N}$ correspond to Kleene's partial recursive functions and functionals at these levels.[42]

The interesting point here is that the reals are taken, not as elements of one of the basic domains A_i, but as functions of type level 1, $f : \mathbb{N} \to \mathbb{N}$, representing effective Cauchy sequences of rationals (under suitable coding).

Let us write ***Rep*** for the class of such functions $f, g, \ldots$. Under this representation, the computable functions over the reals can be identified with those functionals in $\mathbf{ACP}^2(\mathcal{N})$ which map ***Rep*** to ***Rep***, preserving the '$\equiv$' relation on ***Rep***, where $f \equiv g$ means that their corresponding Cauchy sequences have the same limit. To quote SF again:

> So now the S-approximation theory of effective computability of functions of real numbers is explained essentially as in Section 3 above in terms of total recursive functionals in $\mathbf{ACP}^2(\mathcal{N})$. ... Thus abstract computational procedures provide another way of subsuming the two approaches to computation over the real numbers at a basic conceptual level. Of course, this in no way adjudicates the dispute over the proper way to found scientific computation on the real numbers or to deal with the relevent questions of complexity.

SF makes one more important point in this section: how the above illustrates the difference between *extensional* and *intensional* aspects of computation.

---

[42] See footnote 4.

On the face of it, the BSS approach is extensional, while that of S-effective approximation theory is intensional in its essential use of ***Rep*** and $\equiv$ on ***Rep***. But there is an even more basic difference ... Namely, functions $f, g, h, \ldots$ there are tacitly understood in the usual set-theoretic sense for which the extensionality principle ... holds, i.e., if $f(n) = g(n)$ for all n in $\mathbb{N}$, then $f = g$. By the *intensional recursion-theoretic interpretation* of **ACP**($\mathcal{N}$) I mean what one gets by taking the function variables $f, g, h, \ldots$ to range instead over *indices* of partial recursive functions .... Now one proves inductively for this interpretation that each F in $\mathbf{ACP}^2(\mathcal{N})$ preserves extensional equality and hence is an effective operator in the sense of Myhill and Shepherdson (1955), i.e., if $f \equiv g$ then $F(f) \equiv F(g)$. .... In the end, when speaking about actual computation, we have intensionality throughout, since computers only work with finite symbolic representations of the objects being manipulated.

Finally, ***Sect. 6: the Bishop approach to constructive analysis*** is interesting, in that it is the only approach to computing on the reals discussed in this paper which is based on an *informal* notion of computation. This comes from an investigation into the concept of *constructive analysis* carried out by Errett Bishop in his book [5] and his book with Douglas Bridges [1]. SF gives a good summary of the philosophical and technical aspects of this program, the details of which I omit here.

Briefly, the point of Bishop's constructive mathematics (hereinafter BCM) is that existential assertions must produce witnesses of an existential claim. These witnesses then provide a constructive function of the other parameters of the problem. The problem, however, is (to quote SF again *in extenso*):

What is not clear from Bishop's [5] or that of Bishop and Bridges [1] is how the computational content of the results obtained is to be accounted for in recursion-theoretic terms, in the sense of ordinary or generalized recursion theory as discussed in Sects. 3–5 above. From the logical point of view, this may be accomplished by formalizing the work of [1] (and BCM more generally) in a formal system T that has recursive interpretations. A variety of such systems were proposed in the 1970s, first by Bishop himself and then by Nicolas Goodman, Per Martin-Löf, John Myhill, Harvey Friedman and me, and surveyed in [16] ... (cf. also [4]).

About these formal systems, SF continues:

Roughly speaking, those account for the computational content of [5] in two different ways: the first treats witnessing information implicitly and depends for its extraction on the fact that the systems are formalized in intuitionistic logic, while the second kind treats witnessing information explicitly as part of the package explaining each notion and does not require the logic to be intuitionistc. For the first kind of system, the method of extraction is by ...the method of recursive realizability introduced by Kleene or by the use of (recursive) functional interpretations originated by Gödel. Only the system T_0 of Explicit Mathematics introduced in [13], and applied to BCM in [16] is of the second kind ...

I omit a description of T_0 except to say that it has variables of two kinds, for individuals and classes, and the basic relation between individuals, besides identity, is the 3-place relation $\boldsymbol{App}(x, y, z)$, with the meaning $\{x\}y \simeq z$ in ordinary recursion theory.

Then case studies of typical arguments in BCM show that it can be formalized in a subsystem of T_0 of the same strength as Peano Arithmetic [16]; in fact, work of Feng Ye [59] suggests that this can already be done in a subsystem of the strength of Primitive Recursive Arithmetic.

Turning finally to the issue of *feasibility*, let SF have the last word on BCM:

> [T]he practice of Bishop style constructive analysis needs to be examined directly for turning its results that predict computability in principle to ones that demonstrate computability in practice. Presumably all of the specific methods of scientific computation are subsumed under Bishop style constructive mathematics. Assuming that is the case, here is where a genuine connection might be made between constructive mathematics, the theory of computation, and scientific computation, which puts questions of complexity up front.

5 Conclusion

This brings to a close my view of SF's groundbreaking work in generalized computability theory, by means of a close look at four of his papers through the years.

Lack of time and space have prevented discussion of more papers. Let me at least recommend one of his last papers [21], in which he examined various proposals for generalizing the Church-Turing Thesis to concrete and abstract structures.

References

1. Bishop, E., Bridges, D.: Constructive Analysis. Springer (1985)
2. Braverman, M., Cook, S.: Computing over the reals: foundations for scientific computation. Not. Am. Math. Soc. **51**, 318–329 (2006)
3. Blum, L., Cucker, F., Shub, M., Smale, S.: Complexity and Real Computation. Springer (1998)
4. Beeson, M.: Foundations of Constructive Mathematics. Springer (1985)
5. Bishop, E.: Foundations of Constructive Analysis. McGraw-Hill (1967)
6. Blum, L.: Computability over the reals: where turing meets Newton. Not. Am. Math. Soc. **51**, 1024–1034 (2004)
7. Blum, L., Shub, M., Smale, S.: On a theory of computation and complexity over the real numbers: NP-completeness, recursive functions and universal machines. Bull. Am. Math. Soc. **21**, 1–46 (1989)
8. Bergstra, J.A., Tucker, J.V.: Initial and final algebra semantics for data type specifications: two characterization theorems. SIAM J. Comput. **12**, 366–387 (1983)
9. Courant, R., Hilbert, D.: Methods of Mathematical Physics, vol. II. Interscience (1953). Translated and revised from the German edition (1937)
10. Caviness, B.F., Johnson, J.R. (eds.) Quantifier Elimination and Cylindrical Algebraic Decomposition. Springer (1998)
11. Ershov, Y.L.: Computable functions of finite types. Algebra and Logic, **11**, 203–242 (1972). Translated from Algebra i Logika, **11**, 367–437 (1972)
12. Ershov, Y.L.: Maximal and everywhere-defined funtionals. Algebra and Logic, **13**, 210–225 (1974). Translated from Algebra i Logika, **13**, 374–397 (1974)
13. Feferman, S.: A language and axioms for explicit mathematics. In: Crossley, J.N. (ed.) Algebra and Logic: Papers from the 1974 Summer Research Institute of the Australasian Mathematical Society, vol. 450. Lecture Notes in Mathematics, pp. 87–139. Springer (1975)
14. Feferman, S.: Generating schemes for partial recursively continuous functionals (Summary). In: Colloque International de Logique, Clermont-Ferrand, 1975, vol. 249. Colloq. Int. du Centre National de la Recherche Scientifique, Paris, pp. 191–198 (1977)

15. Feferman, S.: Inductive schemata and recursively continuous functionals. In: Gandy, R.O., Hyland, J.M.E. (eds.) Logic Colloquium 76: Proceedings of a Conference held in Oxford in July 1976, pp. 191–198. North Holland (1977)
16. Feferman, S.: Constructive theories of functions and classes. In: Boffa, M., van Dalen, D., McAloon, K. (eds.) Logic Colloquium '78: Proceedings of a Colloquium held in Mons, August 1978, pp. 159–224. North Holland (1979)
17. Feferman, S.: A new approach to abstract data types, I: informal development. Math. Struct. Comput. Sci. **2**, 193–229 (1992)
18. Feferman, S.: A new approach to abstract data types, II: computability on adts as ordinary computation. In: Börger, E. et al. (ed.) Computer Science Logic: Proceedings 5th Workshop, Berne, Switzerland, October 1991, vol. 626. Lecture Notes in Computer Science, pp. 79–95. Springer (1992)
19. Feferman, S.: Computation on abstract data types: the extensional approach, with an application to streams. Ann. Pure Appl. Log. **81**, 75–113 (1996)
20. Feferman, S.: About and around computing over the reals. In: Copeland, B.J., Posy, C.J., Shagrir, O. (eds.) Computability: Turing, Gödel, Church, and Beyond, pp. 55–76. MIT Press (2013)
21. Feferman, S.: Theses for computation and recursion on abstract structure. In: Sommaruga, G., Strahm, T. (eds.) Turing's Revolution: The Impact of his Ideas about Computability. Birkhäuser/Springer, Basel (2015)
22. Fenstad, J.E.: On axiomatizing recursion theory. In: Fenstad, J.E., Hinman, P.G. (eds.) Generalized Recursion Theory: Proceedings 1972 Oslo Symposium, pp. 385–404. North Holland (1974)
23. Friedman, H., Mansfield, R.: Algebraic procedures. Trans. Am. Math. Soc. **332**, 297–312 (1992)
24. Friedman, H.: Algebraic procedures, generalized Turing algorithms, and elementary recursion theory. In: Gandy, R.O., Yates, C.M.E. (eds.) Logic Colloquium '69, pp. 361–389. North Holland (1971)
25. Grzegorczyk, A.: Computable functions. Fundamenta Mathematicae **42**, 168–202 (1955)
26. Grzegorczyk, A.: On the defintions of computable real continuous functions. Fundamenta Mathematicae **44**, 61–71 (1957)
27. Goguen, J.A., Thatcher, J.W., Wagner, E.G., Wright, J.B.: Initial algebra semantics and continuous algebras. J. Assoc. Comput. Mach. **24**, 68–95 (1977)
28. Hadamard, J.: Lectures on Cauchy's Problem in Linear Partial Differential Equations. Dover (1952). Translated from the French edition (1922)
29. Hadamard, J.: La Théorie des Équations aux Dérivées Partielles. Éditions Scientifiques (1964)
30. Hyland, J.M.E.: Recursion Theory on the Conntable Functionals. D.Phil. thesis, Oxford University (1975)
31. Kleene, S.C.: Introduction to Metamathematics. North Holland (1952)
32. Kleene, S.C.: Countable functionals. In: Heyting, A. (ed.) Constructivity in Mathematics: Proceedings of the Colloquium held at Amsterdam, 1957, pp. 81–100. North Holland (1959)
33. Kleene, S.C.: Recursive functionals and quantifiers of finite types i. Trans. Am. Math. Soc. **91**, 1–52 (1959)
34. Kechris, A.S., Moschovakis, Y.N.: Recursion in higher types. In: Barwise, J. (ed.) Handbook of Mathematical Logic, pp. 681–737. North Holland (1977)
35. Kreisel, G.: Interpretation of analysis by means of constructive functionals of finite type. In: Heyting, A. (ed.) Constructivity in Mathematics: Proceedings of the Colloquium held at Amsterdam, 1957, pp. 101–128. North Holland (1959)
36. Lacombe, D.: Extension de la notion de fonction récursive aux fonctions d'une ou plusieurs variables réelles, I, II, III. C.R. Acad. Sci. Paris (1955), **240**, 2470–2480, **241**, 13–14, 151–153
37. Moschovakis, Y.N.: Abstract first order computability I. Trans. Am. Math. Soc. **138**, 427–464 (1969)
38. Moschovakis, Y.N.: Axioms for computation theories — first draft. In: Gandy, R.O., Yates, C.E.M. (eds.) Logic Colloquium '69: Proceedings Summer School Colloquium in Mathematical Logic, Manchester, August 1969, pp. 199–255. North Holland (1971)

39. Moschovakis, Y.N.: Elementary Induction on Abstract Structures. North Holland (1974)
40. Moschovakis, Y.N.: On the basic notions in the theory of induction. In: Proceedings of the 5th International Congress in Logic, Methodology and Philosophy of Science, London, Ontario, 1975, pp. 207–236 (1976)
41. Moschovakis, Y.N.: Abstract recursion as a foundation for the theory of recursive algorithms. In: Richter, M.M. et al. (ed.) Computation and Proof Theory: Proceedings Logic Colloquium Aachen 1983, Part II, vol. 1104. Lecture Notes in Mathematics, pp. 289–364. Springer (1984)
42. Moschovakis, Y.N.: The formal language of recursion. J. Symb. Log. **54**, 1216–1252 (1989)
43. Mitchell, J.C., Plotkin, G.D.: Abstract types have extensional type. ACM Trans. Progr. Lang. Syst. **10**, 470–502 (1985)
44. Myhill, J., Shepherdson, J.: Effective operations on partial recursive functions. Zeitschr. Msth. Logik u. Grundlagen. Math. **1**, 310–317 (1955)
45. Moldestat, J., Stoltenberg-Hansen, V., Tucker, J.V.: Finite algorithmic procedures and computation theories. Mathematica Scandinavica **46**, 77–94 (1980)
46. Moldestat, J., Stoltenberg-Hansen, V., Tucker, J.V.: Finite algorithmic procedures and inductive definability. Mathematica Scandinavica **46**, 62–76 (1980)
47. Pour-El, M.B., Richards, J.I.: Computability in Analysis and Physics. Springer (1989)
48. Platek, R.A.: Foundations of Recursion Theory. Ph.D. thesis, Department of Mathematics, Stanford University (1966)
49. Sazonov, V.Y.: Degrees of parallelism in computation. In: Mazurkiewicz, A. (ed.) Proceedings of the 5th Symposium Mathematical Foundations of Computer Science (MFCS'76), Gdansk, September 1976
50. Smullyan, R.M.: Theory of Formal Systems, vol. 47. Annals of Mathematical Studies. Princeton University Press (1961)
51. Strong Jr., H.R.: Translating recursion equations into flowcharts. J. Comput. Syst. Sci. **5**, 254–285 (1971)
52. Tucker, J.V., Zucker, J.I.: Program Correctness over Abstract Data Types, with Error-State Semantics, vol. 6. CWI Monographs. North Holland (1988)
53. Tucker, J.V., Zucker, J.I.: Computable functions and semicomputable sets on many-sorted algebras. In: Abramsky, S., Gabbay, D., Maibaum, T. (eds.) Handbook of Logic in Computer Science, vol. 5, pp. 317–523. Oxford University Press (2000)
54. Tucker, J.V., Zucker, J.I.: Abstract versus concrete computation on metric partial algebras. ACM Trans. Comput. Log. **5**, 611–668 (2004)
55. Tucker, J.V., Zucker, J.I.: Continuity of operators on continuous and discrete time streams. Theoret. Comput. Sci. **412**, 3378–3403 (2011)
56. Tucker, J.V., Zucker, J.I.: Generalizing computability to abstract algebras. In: Sommaruga, G., Strahm, T. (eds.) Turing's Revolution: The Impact of his Ideas about Computability. Birkhauser/Springer, Basel (2015)
57. Weihrauch, K.: Computable Analysis: An Introduction. Springer (2000)
58. Jian, Xu, Zucker, Jeffery: First and second order recursion on abstract data types. Fundamenta Informaticae **21**, 1–43 (2004)
59. Ye, Feng: Toward a constructive theory of unbounded linear operators. J. Symb. Log. **65**, 357–370 (2000)
60. Zucker, J.I.: Review of [18]. Bull. Symb. Log. **8**, 538–542 (2002)

Author Biography

Jeffery Zucker was born and raised in South Africa. He received his undergraduate education at the University of the Witwatersrand in Johannesburg. He completed his graduate studies at Stanford University, obtaining his PhD in Mathematics under Sol Feferman's supervision in 1971. He has held academic appointments at the University of Illinois in Chicago, the Universities of Amsterdam, Eindhoven and Utrecht in the Netherlands, the Centrum Wiskunde & Informatica in Amsterdam, Bar Ilan University in Israel, the University at Buffalo, and (since 1989) McMaster University in Hamilton, Canada. His research is mainly in the area of generalized computation theory, and especially, for the last few years, in computation on topological partial algebras based on the reals, and analog computing.

On the Computability of the Fan Functional

Dag Normann and William Tait

Abstract We give a self-contained presentation of the original argument for why the fan functional is countably recursive, but not Kleene computable. This proof, from 1958, is due to the second author and has not been published before. A lemma concerning the Kleene computable functionals used in the proof is unnecessarily strong, which is fortunate since there is a counterexample, due to the first author, which we include.

Keywords Fan functional · Recursively countable · Kleene computable

Mathematics subject classification: 03D65

Solomon Feferman took an early interest in my work, and whenever we met, he was a source of inspiration. We met at conferences, but also in more relaxed places like a beach in Patras or a cottage in Wales. I was honored to be invited to write this paper in his memory, and to do so jointly with Bill Tait. DAG NORMANN

I was very pleased by the original invitation to contribute to a volume of papers in honor of Solomon Feferman; but of course that pleasure is now saturated with the sadness of his sudden death. I began my career at Stanford a year after Sol. We were colleagues there for several years and had been friends for fifty-eight years. I am especially pleased to share with Dag Normann in offering this paper in his memory. BILL TAIT

D. Normann (✉)
Department of Mathematics, The University of Oslo, P.O. Box 1053, Blindern, 0316 Oslo, Norway
e-mail: dnormann@math.uio.no

W. Tait
5522 S. Everett Ave, Chicago, IL 60637, USA
e-mail: williamtait@mac.com

G. Jäger and W. Sieg (eds.), *Feferman on Foundations*, Outstanding Contributions to Logic 13, DOI 10.1007/978-3-319-63334-3_3

1 Introduction

This paper may be read partly as an historical note. In 1959, the second author, Tait, introduced the *fan functional* and proved that it is not Kleene computable, though recursively countable. An announcement of the result appears in the abstract [16]. The proof was never published and exists only in a copy of a letter to S.C. Kleene, dated May 25, 1959, and in Lecture 71 of [11], a series of lectures by Kreisel on Constructive Mathematics in 1958–59. The argument is that a Kleene computable function Φ of the relevant type satisfies a certain continuity property, which the fan functional does not possess.

As the second author, Normann, noted, this continuity property can be expressed as possession of a winning strategy in a certain two-person game. In these terms, the original argument (both in the abstract, the letter to Kleene and in Kreisel's lecture notes) mistakenly contained the unnecessary assertion that there is a uniform bound on the length of the games associated with Φ, depending only on Φ. A letter to Tait from Robin Gandy shortly thereafter (the exact date is unknown) contains a copy of part of a letter to Kreisel in which Gandy exposes the fallacy in the argument that the bound is uniform—and offers an alternative measure-theoretic proof that the fan functional is not Kleene computable. A third proof was ultimately published by Gandy and Hyland in [3]. Section 5.3 of the present paper contains a counterexample, due to Normann, to the uniformity of the bound.

The main purpose of this paper is to give the original proof of the non-computability of the fan functional in game-theoretic terms. This will be given in Sect. 4. In Sect. 2, we will put the problem into a historical context and in Sect. 3 we will introduce the concepts of recursively countable functions and Kleene computable functions, at least for functions of the relevant types. In Sect. 5 we will discuss the error of the uniform bound in the abstract [11, 16].

Normann has published some general expositions [13–15], the latter jointly with John Longley, and several research papers, where it has been assumed that the proof in Gandy and Hyland [3] more or less was as the original one. His motivation for co-authoring this paper is partly to have the original ideas of Tait presented in todays terminology in the form of a complete proof, and partly to introduce a counterexample to the claim of uniformity. It may be of interest that the source of this counterexample is work by Martín Escardó [1] published as late as in 2013, so the counterexample could not have been easily seen in 1959.

2 Background

The finite types are given by the inductive definition

0 is a type
If $t_0, \ldots t_k$ are types, then $\langle t_0, \ldots, t_k \rangle$ is a type.

If t is a type, we define the type level $|t|$ by

$$|0| = 0 \text{ and} |\langle t_0, \ldots, t_k \rangle| = \max_i \{|t_i| + 1\}.$$

A hierarchy $\mathcal{T}$ of functions of finite type, called a *type structure*, attaches to each finite type t a set $\mathcal{T}_t$ of objects such that

$$\mathcal{T}_0 = \mathbb{N}$$

and

$$\mathcal{T}_{\langle t_0, \ldots, t_k \rangle} \subseteq (\mathcal{T}_{t_0} \times \cdots \times \mathcal{T}_{t_k}) \to \mathbb{N}.$$

We will mainly be concerned with the *pure* types denoted by integers 0 and $n + 1$ for $\langle n \rangle$, and with types of level ≤ 3. Moreover, we are interested in two type structures, the *maximal* type structure, where each type is interpreted as the full set of functions in question, and the one in which each type is interpreted as the *countable* objects of that type (to be defined below for the relevant cases).

The *countable* or *continuous* functionals were independently introduced by Kleene [10] and Kreisel [12], both published in 1959. Actually, the definitions were not identical, the Kreisel functionals are formed as the extensional collapse of the Kleene functionals which are elements of the maximal type structure. We will follow Kreisel, and let the domain of a countable functional consist of countable objects only. In parallel, Kleene [9] introduced the partially computable functionals of higher types, via the schemes S1–S9 (see below).

In the late 1950's and early '60's, inspired by the example of Gödel's *Dialectica interpretation* [5], the idea of employing functions of higher types in the extended Hilbert program of finding constructive interpretations of theorems of classical mathematics seemed promising. (It turns out that Gödel already had this idea by 1933, see Gödel [4].) There seemed to be three *prima facie* candidates for a type structure $\mathcal{T}$, three different ways to constructively conceive of higher type functions: the hereditarily (extensional) effective operations, the (recursively) countable functionals (Kleene [10] and Kreisel [12]) and the Kleene computable functionals (Kleene [9]. An obvious question about them concerns the relationships among them. Kleene raised this question in [9], p. 11, footnote 9, with respect to the computable functions of finite type and the countable functions, as did Kreisel in [12], page 118. Working with a typescript of Kleene [9] in the winter of 1958–59, Tait introduced the fan functional (to be defined below) and showed that it is a recursively countable function of type $\langle 1, 2 \rangle$, but not the restriction of a Kleene-computable function to arguments (g, ϕ), where ϕ is a continuous function of type 2.

A function g of type 1, i.e. $g : \mathbb{N} \to \mathbb{N}$, determines a fan

$$Fan_g = \{f : \mathbb{N} \to \mathbb{N} \mid \forall x[f(x) \leq g(x)]\}.$$

The value of the fan functional for the argument (g, ϕ) is the least modulus of uniform continuity with respect to ϕ of all $f \in Fan_g$. The term "fan functional" is inspired of course by the fan theorem of Brouwer.

Relatively soon after, Joseph Harrison [8] proved that the (extensional) effective operations of finite type are equivalent in a natural sense to the hereditarily recursive continuous functionals, completing the answer to the question of relationships. That work was never published, either, but also in this case a proof appears in Gandy and Hyland [3].

Actually, there were two separate questions: one, just mentioned, concerns the appropriate *constructive* notion of function of higher type, the other concerns the appropriate extension of the concept of a recursive (as was the standard terminology of the time) function to higher types. It would seem that the effective operations would provide the answer to the first question from the point of view of the Russian school of constructive mathematics, according to which all higher type objects are computable. The 'point free' approach to functions of finite type provided by the countable functions would seem more appropriate to Brouwer's intuitionistic concept of mathematics, but perhaps less so from the point of view of Bishop's constructive mathematics. But whatever the case may be for the recursive functions of finite type being the right answer to the first question, they seem a more convincing answer to the second. We mention this especially here because of the evidence for this answer afforded by Feferman's "Recursion in total functionals of finite type" [2], in which he recaptures the Kleene hierarchy as the special case of the single ground $\mathbb{N}$ of a different analysis of the idea of computing with total objects of finite type involving some finite number of other arbitrary ground types.

For an updated exposition of the relations between various type structures of interest, see [13].

3 Preliminaries

In this section we will introduce the countable (or continuous) functionals, the fan functional and Kleene computability to the extent needed for the proof. We will consider a special case of the fan functional, where we fix the type 1 argument to the constant 1 and consider it as a functional of pure type 3. It is a matter of taste if we consider Φ defined below as a partial function from the set of all functions of type 2, or as a total function on the set of continuous ones:

Definition 3.1 Let Φ be the type 3 object defined by

$$\Phi(\phi) = \mu n.\forall f, g \in \{0, 1\}^{\mathbb{N}}[\bar{f}(n) = \bar{g}(n) \rightarrow \phi(f) = \phi(g)],$$

where μ denotes "the least" and $\bar{f}(n) = (f(0), \ldots, f(n-1))$ considered as a function defined on $\{0, \ldots, n-1\}$.

From now on, we will refer to Φ as *the fan functional*. The advantage of considering Φ and not the general fan functional is mainly notational: the extension to the impure case is routine. As for conventions, Ψ will denote a partial function of type 3, ϕ and ψ will denote (continuous) functions of type 2, f, g will denote total functions of type $1, a, b, d, e$ will denote integers when used as arguments for, or values of, algorithms or functions (including as Kleene indices), n, k, i, j, p, r will be used for integers used for other purposes as indices in enumerations etc.

Definition 3.2 (*The countable objects $Ct(k)$ of pure type k*)

(a) $Ct(0) = \mathbb{N}$. If $a \in \mathbb{N}$, $\{a\}$ is at the same time a *formal neighborhood* such that $\{a\} \prec_0 a$ and the one and only *associate* for a. For technical reasons, we also let $\emptyset$ be a formal neighborhood with $\emptyset \prec_0 a$ for all a. Two formal neighborhoods of type 0 are *consistent* if one is included in the other.

(b) Assume that we have defined the formal neighborhoods of type k, the countable objects of type k with their associates, the consistency relation on the set of formal neighborhoods and the approximation relation $\prec_k$ between formal neighborhoods and countable objects of type k. We then let

(i) A *basic neighborhood* of type $k+1$ is a pair (σ, a) where σ is a formal neighborhood of type k and $a \in \mathbb{N}$

(ii) A *formal neighborhood* of type $k+1$ is a finite set

$$\{(\sigma_1, a_1), \ldots, (\sigma_n, a_n)\}$$

of basic neighborhoods of type k such that $a_i = a_j$ whenever σ_i and σ_j are consistent.

Two formal neighborhoods of type $k+1$ are *consistent* if the union is a formal neighborhood.

(iii) If $\xi : Ct(k) \to \mathbb{N}$ and $\tau = \{(\sigma_1, a_1), \ldots, (\sigma_n, a_n)\}$ is a formal neighborhood of type $k+1$, we let $\tau \prec_{k+1} \xi$ if for all $x \in Ct(k)$ and i we have that

$$\sigma_i \prec_k x \to \xi(x) = a_i.$$

(iv) If $\xi : Ct(k) \to \mathbb{N}$ and α is a set of basic neighborhoods of type k, we say that α is *an associate* for ξ if $\tau \prec_{k+1} \xi$ for all finite $\tau \subset \alpha$, and moreover, for all $x \in Ct(k)$ and all associates β for x there is a finite subset $\sigma \subset \beta$ such that $(\sigma, \xi(x)) \in \alpha$.

(v) $Ct(k+1)$ will be the set of functions $\xi : Ct(k) \to \mathbb{N}$ with associates.

When $\sigma \prec_k \xi$ we say that σ *approximates* ξ.

It is easy to see that the countable objects of type 1 will be all functions $f : \mathbb{N} \to \mathbb{N}$, and that if we consider the standard product topology on this set, the countable objects of type 2 will be exactly the continuous functions $\phi : \mathbb{N}^{\mathbb{N}} \to \mathbb{N}$.

By standard methods, the formal neighborhoods of a fixed type can be represented by numbers. Using classical terminology, we say that an associate is *recursive* if the set of number codes for the elements is so, and a function is *recursively countable* if it has a recursive associate.

Definition 3.3 Let σ be a formal neighborhood of type 1

We let

$$U_\sigma = \{f : \mathbb{N} \to \mathbb{N} \mid \sigma \prec_1 f\}.$$

U_σ will be an open set (and a closed one) in the topology of $\mathbb{N}^{\mathbb{N}}$.

Lemma 3.4 *The fan functional* Φ *is a recursively countable object.*

Proof Let ϕ be a continuous function of type 2, let β be an associate for ϕ and let $f : \mathbb{N} \to \{0, 1\}$. Then there is a finite formal neighborhood $\sigma_f \prec_1 f$ such that $(\sigma_f, \phi(f)) \in \beta$.

$$\{U_{\sigma_f} \mid f : \mathbb{N} \to \{0, 1\}\}$$

is an open covering of the compact set $\{0, 1\}^N$, so there will be a finite subcovering. This means that there is a formal neighborhood

$$\{(\sigma_1, b_1), \ldots, (\sigma_n, b_n)\} \subset \beta$$

such that

$$\{U_{\sigma_1}, \ldots, U_{\sigma_n}\}$$

is an open covering of $\{0, 1\}^{\mathbb{N}}$.

Disregarding β, whenever $\tau = \{(\sigma_1, b_1), \ldots, (\sigma_n, b_n)\}$ is a formal neighborhood such that $\{U_{\sigma_1}, \ldots, U_{\sigma_n}\}$ is an open covering of $\{0, 1\}^{\mathbb{N}}$, we can effectively extract the constant value of Φ on any type 2 ϕ being approximated by τ. This is used to construct a recursive associate for Φ. We omit further details.

We will now give a brief introduction to Kleene computability. Kleene defined a relation

$$\{e\}(x_1, \ldots, x_n) = a$$

where e is an integer acting as a *Kleene index* or a *Gödel number* for an algorithm, $x_1, \ldots, x_n$ are total functionals of pure type and $a \in \mathbb{N}$. Since we are only interested in the computability of the fan functional Φ, we will give an alternative definition that is equivalent to the original one from Kleene [9] when x_1 is an object of pure type 2 and the rest of the inputs are of type 0. We may leave out scheme S5 for primitive recursion, as this can be reduced to the other schemes. We may also leave out S7, since we do not consider arguments of type 1.

Definition 3.5 The relation

$$\{e\}(\phi, a_1, \ldots, a_n) = a$$

is defined by positive induction as follows:

S1 If $e = \langle 1 \rangle$ then $\{e\}(\phi, a_1, \ldots, a_n) = a_1 + 1$.
S2 If $e = \langle 2, b \rangle$ then $\{e\}(\phi, a_1, \ldots, a_n) = b$.
S3 If $= \langle 3 \rangle$ then $\{e\}(\phi, a_1, \ldots, a_n) = a_1$.
S4 If $e = \langle 4, e_1, e_2 \rangle$, then $\{e\}(\phi, a_1, \ldots, a_n) = a$ if there is some $b \in \mathbb{N}$ such that $\{e_1\}(\phi, a_1, \ldots, a_n) = b$ and $\{e_2\}(\phi, b, a_1, \ldots, a_n) = a$.
S6 If $e = \langle 5, e_1, \pi \rangle$ where π is a permutation of the set $\{1, \ldots, n\}$, then $\{e\}(\phi, a_1, \ldots, a_n) = a$ if $\{e_1\}(\phi, a_{\pi(1)}, \ldots, a_{\pi(n)}) = a$.
S8 If $e = \langle 8, d \rangle$ and there is some $f : \mathbb{N} \to \mathbb{N}$ such that

$$\{d\}(\phi, a, a_1, \ldots, a_n) = f(a)$$

for all $a \in \mathbb{N}$, then $\{e\}(\phi, a_1, \ldots, a_n) = \phi(f)$.
S9 If $e = \langle 9, m \rangle$ where $m \leq n$, then

$$\{e\}(\phi, d, a_1, \ldots, a_n) = a$$

if $\{d\}(\phi, a_1, \ldots, a_m) = a$.

Since this relation is defined by induction, there will be an ordinal rank assigned to each identity

$$\{e\}(\phi, a_1, \ldots, a_n) = a,$$

obtained in this way and it is easy to prove by induction on this rank that a will be unique when existing. So, what we have done is to define, for each $n \geq 0$, a partial function $\{e\}$ on the set of tuples $(\phi, a_1, \ldots, a_n)$. Kleene [10] proved that if $\{e\}$ is total over $Ct(2) \times \mathbb{N}^n$ when defined this way, it will be recursively countable. We will not give that argument. We may consider this as defining a generalization of classical recursion theory, now also accepting an input of type 2, and we do, for instance, have a version of the recursion theorem and versions of the $s_{n,m}$-theorems for Kleene computations. In Sect. 4 we will establish a continuity property of total Kleene-computable objects of type 3 and prove that the fan functional Φ does not satisfy this property. In Sect. 5 we will show that there are Kleene computable objects not satisfying a stronger continuity property claimed in [11, 16] to hold for all Kleene computable objects.

4 The Modified Proof

4.1 The Games

Let Ψ be a partial functional of type 3. We will consider Ψ only for continuous arguments ϕ. For each Ψ and ϕ we will consider a game $\Gamma_{\Psi,\phi}$ played as follows:

1. I selects a finite set of functions $f_{1,1}, \ldots, f_{1,n_1}$
2. II chooses numbers $r_{1,1}, \ldots, r_{1,n_1}$ such that ϕ is constant on the set $U_{1,j}$ of extensions of $\bar{f}_{1,j}(r_{1,j})$ for each $j \leq n_1$. Each set $U_{1,j}$ will be both closed and open (clopen).
3. I selects a finite set of functions $f_{2,1}, \ldots, f_{2,n_2}$
4. II selects numbers $r_{2,1}, \ldots, r_{2,n_2}$ in analogy with II's previous move, defining the clopen sets $U_{2,j}$.
5. and so on
6. and so on $\cdots\cdots\cdots$

After p pairs of moves the players will have produced a finite set of clopen neighborhoods $U_{1,1}, \ldots, U_{p.n_p}$, and for each i, j in question, ϕ will take a constant value $a_{i,j}$ on $U_{i,j}$. So the game will produce, for each p, a formal neighborhood

$$\{(\bar{f}_{1,1}(r_{1,1}), a_{1,1}), \ldots, (\bar{f}_{p,n_p}(r_{p,n_p}), a_{p,n_p})\}$$

that approximates ϕ.

I wins the game after each player has made p moves if for all ψ for which $\Psi(\psi)$ has a value, and for which

$$\{(\bar{f}_{1,1}(r_{1,1}), a_{1,1}), \ldots, (\bar{f}_{p,n_p}(r_{p,n_p}), a_{p,n_p})\}$$

approximates ψ, we have that $\Psi(\phi) = \Psi(\psi)$.

II wins if the game goes on forever without I winning.

In the original proof, the existence of a winning strategy in p moves for player I was expressed by a single formula. Writing $U_{f,r}$ for the set of extensions of $\bar{f}(r)$, the formula was

$$\exists p \exists n_1 \exists f_{1,1} \cdots \exists f_{1,n_1} \forall r_{1,1} \cdots \forall r_{1,n_1} \cdots \exists n_p \exists f_{p,1} \cdots \exists f_{p,n_p} \forall r_{p,1} \cdots \forall r_{p,n_p}$$

$$\forall \psi \left[\bigwedge_{j \leq p, i \leq n_j} \forall f \in U_{f_{i,j}, r_{i,j}} (\phi(f) = \psi(f)) \rightarrow \Psi(\psi) = \Psi(\phi) \right]$$

where ψ ranges over continuous functions of type 2. The only change needed is to view the main body of the prefix as a *game quantifier* and move $\exists p$ to the end of the, now infinite, quantifier prefix:

$$\exists n_1 \exists f_{1,1} \cdots \exists f_{1,n_1} \forall r_{1,1} \cdots \forall r_{1,n_1} \exists n_2 \exists f_{2,1} \cdots \exists f_{2,n_2} \forall r_{2,1} \cdots \forall r_{2,n_2} \cdots$$

$$\exists p \forall \psi \left[\bigwedge_{j \leq p, i \leq n_j} \forall f \in U_{f_{i,j}, r_{i,j}} (\phi(f) = \psi(f)) \rightarrow \Psi(\psi) = \Psi(\phi) \right].$$

For this kind of game, where at least one of the players only needs finitely many moves to win, one of the players will have a winning strategy. The relevant continuity property for Ψ will be that I has a winning strategy for all arguments ϕ for which $\Psi(\phi)$ has a value.

In this game, player I may merge finitely many strategies into one, simply by playing the union of the finite sets of functions required in each move. The merge of finitely many winning strategies for games $\Gamma_1, \ldots, \Gamma_k$ will be a strategy for winning all the games simultaneously.

4.2 *Player I Wins for Kleene Computable Functions*

Theorem 4.1 *If Ψ is the partial function*

$$\Psi(\phi) = \{e\}(\phi, a_1, \ldots, a_n)$$

for some fixed e and $a_1, \ldots, a_n$, and $\Psi(\phi)$ is defined, then I has a winning strategy for the game $\Gamma_{\Psi,\phi}$

Proof The proof is by induction on the ordinal rank of the computation of

$$\{e\}(\phi, a_1, \ldots, a_n) = a.$$

We describe the strategy of I for each of the Kleene-schemes.

1. In the case of S1, S2 or S3, I has won the game before they start playing, since the output of the algorithm is independent of ϕ.
2. In the case S4, I merges the strategies for

   $$\{e_1\}(\phi, a_1, \ldots, a_n) = b$$

   and

   $$\{e_2\}(\phi, b, a_1, \ldots, a_n) = a.$$

3. The cases S5 and S9 are even simpler, I will employ the strategy for the immediate subcomputation.
4. S8 is the only interesting case:

   $$\{e\}(\phi, a_1, \ldots, a_n) = \phi(f) = a$$

   where

   $$f(b) = \{d\}(\phi, b, a_1, \ldots, a_n).$$

Then I plays $\{f\}$ as the first move. II must respond with an r such that ϕ is constant a on $U_{\bar{f}(r)}$.
By the induction hypothesis, for each $b < r$, I has a winning strategy for the game securing that

$$\{d\}(\psi, b, a_1, \ldots, a_n) = \{d\}(\phi, b, a_1, \ldots, a_n)$$

whenever the former has a value and ϕ and ψ agree on the closed open neighborhoods produced. In the case of S8, I will continue with the merge of those strategies, and is bound to win.

This ends the proof.

4.3 Player II Wins the Fan Functional Games

Theorem 4.2 *Let Φ be the fan functional and let ϕ be any continuous object of type 2. Then II has a winning strategy in the game $\Gamma_{\Phi,\phi}$.*

Proof The nature of the game is as follows: In turns, I selects finite sets of functions while II selects clopen coverings of these finite sets. II is free to choose these coverings arbitrarily small. In particular, II may, at each move, choose a clopen covering of the new finite set of functions sufficiently small for the play never to produce a covering of $\{0, 1\}^{\mathbb{N}}$.
Let II follow a strategy to this effect (at move no. p for II, the uniform r_p will only depend on p and the number n_p of finite functions just played by I). Then, for any p there will be a non-empty clopen subset of $\{0, 1\}^{\mathbb{N}}$ disjoint from all the neighborhoods $U_{1,1}, \ldots, U_{p,n_p}$. We may then find a continuous ψ_p agreeing with ϕ on $U_{1,1}, \ldots, U_{p,n_p}$ but with a modulus of uniform continuity larger than $\Phi(\phi)$, so I is not in a winning position. Thus, the described strategy is a winning strategy for II. This ends the proof.

Corollary 4.3 *The fan functional is not Kleene computable.*

Remark 4.4 This way of proving Corollary 4.3 was, until early 2016, unknown to the first author. The argument given here captures in essence the original proof given by the second author in 1958. To our knowledge, the proof has not previously been published in this form, and we find it of historical interest to present it here.

5 The False Claim

In this section, we will discuss the claim in [16], repeated in [11], that the length p of the game $\Gamma_{\Psi,\phi}$ in Theorem 4.1 depends only on Ψ, and produce a counterexample. The counterexample is based on ideas of Escardó [1]. We make the example simpler by working within S1–S9, while the results in [1] actually show that there are counterexamples expressible in Gödel's T.

5.1 The Claim

Tait [16] essentially characterizes the Kleene computable functions in terms of the game we described above in Sect. 4. But he further claimed that if $\{e\}(\phi, a_1, \ldots, a_n)$ is defined for all $(\phi, a_1, \ldots, a_n)$, then there will be a p and a strategy for I winning the game in p moves. The argument is that p is the depth of the nesting of occurrences of S8 in computing $\Psi(\phi, a_1, \ldots, a_n)$; but as Gandy pointed out in the letter mentioned above, the depth depends upon the arguments $\phi, a_1, \ldots, a_n$ as well. As we will see below, there is a counter example to the stronger property.

5.2 Sets Allowing Kleene-Computable Search

Let $\omega \leq \alpha < \omega_1^{CK}$ be a computable ordinal, and let $\prec$ be a computable (non reflexive) well ordering of $\mathbb{N}$ of length α, with $\preceq$ as its reflexive companion. For each $n \in \mathbb{N}$, let f_n be the characteristic function of $\{m \mid m \prec n\}$, and let f be the constant 1. Let $X_\alpha = \{f, f_n \mid n \in \mathbb{N}\}$. Then X_α is a closed subset of $\{0, 1\}^{\mathbb{N}}$, and order isomorphic to $\alpha + 1$ when ordered by the lexicographical ordering.

Theorem 5.1 *Let α and $X = X_\alpha$ be as above. Then the following functional of pure type 3 is Kleene-computable:*

$$\Phi_X(\phi) = \begin{cases} 0 \text{ if } \forall f \in X(\phi(f) = 0) \\ 1 \text{ if } \exists f \in X(\phi(f) > 0) \end{cases}$$

Proof Let ϕ be any total functional of type 2, not necessarily continuous. Using the recursion theorem for Kleene's S1-S9 we define g_n and g uniformly computable in ϕ:

- $g_n(m) = 0$ if $n \preceq m$.
- $g_n(m) = 1$ if $m \prec n$ and $\phi(g_m) = 0$.
- $g_n(m) = 0$ if $m \prec n$ and $\phi(g_m) > 0$.
- $g(m) = 1$ if $\phi(g_m) = 0$.
- $g(m) = 0$ if $\phi(g_m) > 0$.

By the recursion theorem, g_n and g are well defined as partial functions. By induction on the rank of n in $\prec$ we see that each g_n is total, and then that g is total.

If there is some n such that $\phi(f_n) > 0$, let n_0 be the $\prec$-least such n. Then

- $g_n = f_n$ for $n \preceq n_0$.
- $g_n = f_{n_0}$ for $n_0 \prec n$.
- $g = f_{n_0}$.

If there is no such n, then $g_n = f_n$ for all n, and $g = f$. All these claims are easily proved by $\prec$-induction. Then

$$\exists h \in X(\phi(h) > 0) \Leftrightarrow \phi(g) > 0,$$

and it follows that the functional Φ_X is Kleene computable.

Remark

This proof is a direct adjustment of an argument due to Escardó [1] to S1-S9-recursion.

5.3 The Conterexample

We will now let X be as in the previous section with order type $\omega^\omega + 1$. We will prove that there is no p such that I has a strategy for winning the game $\Gamma_{\Phi_X,\phi}$ in p moves when ϕ is the constant zero function of type 2.

Each $f \in X$ will have an ordinal rank $\alpha_f \leq \omega^\omega$. We say that the level of $f \in X$ is infinite if $\alpha_f = \omega^\omega$ and the level is n if the Cantor Normal Form of α_f ends with ω^n.

We may now follow the proof of the non Kleene-computability of the fan functional. When p is fixed, we can describe a strategy for II that keeps the game going for more than p moves. After II has made k moves, they have essentially produced a clopen set Y_k. As long as this does not contain the set X as a subset, I is not in winning position, as we may find a ψ agreeing with ϕ on Y_k while not being constant zero on X. The p-strategy for II is to ensure that after move k for II, there is an $f \in X \setminus Y_k$ with level $> p - k$. Since the functions with levels $> (p - k) + 1$ are cluster points of functions with levels $> p - k$, and I only selects finitely many functions at the time, this can be achieved.

If II follows this strategy, X is not a subset of Y_p, so I cannot be certain of winning this game in p moves for any prefixed p. Φ_X is not what Kreisel [11] called *continuous in the Tait topology*, but nonetheless Kleene computable.

References

1. Escardó, M.H.: Infinite sets that satisfy the principle of omniscience in any variety of constructive mathematics. J. Symb. Log. **78**(3), 764–784 (2013)
2. Feferman, S.: Recursion in total functionals of finite type. Compos. Math. **35**, 3–22 (1977)
3. Gandy, R.O., Hyland, J.M.E.: Computable and recursively countable functions of higher type. In: Gandy, R.O., Hyland, J.M.E. (eds.) Logic Colloquium '76, pp. 407–438. North Holland, Amsterdam (1977)
4. Gödel, K.: The present situation in the foundations of mathematics (1933). Collected Works, vol. III, pp. 45–53. Oxford University Press, Oxford (1995)

5. Gödel, K.: Über eine bisher noch nicht benützte Erweiterung des finiten Standpunktes. Dialectica **12**, 280–287 (1958). Reprinted with an English translation in Gödel [7]. Gödel [6] is a revised version
6. Gödel, K.: On an extension of finitary mathematics which has not yet been used (1972) in [7], pp. 271–289 (1990)
7. Gödel, K.: Collected Works, vol. II. Oxford University Press, Oxford (1990)
8. Harrison, J.: Equivalence of the effective operations and the hereditarily continuous functionals. In: Stanford Seminar Report, pp. 204–206 (1963). Appendix C to Tait [17]
9. Kleene, S.C.: Recursive functionals and quantifiers of finite type I. Trans. Am. Math. Soc. **91**, 1–52 (1959)
10. Kleene, S.C.: Countable functionals. In: Heyting, A. (ed.) Constructivity in Mathematics: Proceedings of the Colloquium held at Amsterdam 1957, pp. 81–100. North-Holland, Amsterdam (1959)
11. Kreisel, G.: Lecture notes from a course on Constructive Mathematics, Stanford University (1958, 1959)
12. Kreisel, G.: Interpretation of analysis by means of constructive functionals of finite types. In: Heyting, A. (ed.) Constructivity in Mathematics: Proceedings of the Colloquium held at Amsterdam 1957, pp. 101–128. North-Holland, Amsterdam (1959)
13. Longley, J.R., Normann, D.: Higher Order Computability. Springer, Berlin (2015)
14. Normann, D.: Recursion on the Countable Functionals. Springer, Berlin (1980)
15. Normann, D.: The continuous functionals. In: Griffor, E.R. (ed.) Handbook of Computability Theory, pp. 251–275. Elsevier, Amsterdam (1999)
16. Tait, W.W.: Continuity Properties of Partial Recursive Functionals of Finite Type, International Congress for Logic, Methodology and Philosophy of Science: Abstracts of contributed papers. Stanford University, Stanford, California (1960)
17. Tait, W.W.: A second order theory of functionals of higher type, with two appendices. Appendix A: Intensional functionals. Appendix B: An interpretation of functionals by convertible terms. Stanford Seminar Report, pp. 171–206 (1963). Unpublished. A published version is Tait [18]
18. Tait, W.W.: Intensional interpretations of functionals of finite type I. J. Symb. Log. **32**, 198–212 (1967)

Author Biographies

Dag Normann has published papers and monographs in mathematical logic, mainly in Higher Computability Theory. He introduced the concept of Set Recursion, but has mainly been concerned with continuity aspects of computability theory, partly its links to theoretical computer science. He is the author of Springer Lecture Notes in Mathematics *Recursion on the countable functionals* from 1980 and he recently co-authored the volume *Higher-Order Computability* with John Longley, Springer Verlag 2015.

William Tait has published papers in logic, foundations of mathematics and on the history of foundations of mathematics. On the latter two topics, he has published a collection of his essays in *The Provenance of Pure Reason: Essays on the Philosophy of Mathematics and Its History* (Oxford University Press, 2005). He has also edited two collections of essays: A Special issue of Synthese on Philosophy of Mathematics (Volume 84, No. 2–3: 1990) and *Early Analytic Philosophy; Frege, Russell, Wittgenstein: Essays in Honor of Leonard Linsky* (Open Court, 1997).

A Survey on Ordinal Notations Around the Bachmann–Howard Ordinal

Wilfried Buchholz

Abstract Various ordinal functions which in the past have been used to describe ordinals not much larger than the Bachmann–Howard ordinal are set into relation. Special efforts are made to reveal the intrinsic connections between Feferman's θ-functions and the Bachmann hierarchy.

Keywords Bachmann–Howard ordinal · Bachmann hierarchy · Fundamental sequence · Klammersymbol · Normal function

2010 Mathematics Subject Classification 03F15

1 Introduction

In the early seventies one of the central problems in ordinally informative proof theory was to obtain a Gentzen-style ordinal analysis for theories of iterated arithmetical inductive definitions. For the intuitionistic theory ID_1 of noniterated inductive definitions this problem was already solved by Howard [15] who had shown that the proof-theoretic ordinal of ID_1 is bounded by the ordinal $\phi_{\varepsilon_{\Omega+1}}(0)$ which since then is called the Bachmann–Howard ordinal. Bachmann [3] had introduced a new method for generating 'long' sequences of normal functions $\phi_\alpha : \Omega \to \Omega$ where Ω is the first uncountable ordinal and α ranges over an uncountable segment of ordinals including $\varepsilon_{\Omega+1}$ (the first ε-number greater than Ω). The essential feature of Bachmann's method was the assignment of a fundamental sequence $(\alpha[\xi])_{\xi<\tau_\alpha}$ to each limit index α. The function ϕ_α was then defined by referring to the sequence of previously defined normal functions $\phi_{\alpha[\xi]}$ $(\xi < \tau_\alpha)$. Later Pfeiffer (in [18]) extended Bachmann's method by considering also normal functions $f^n_\alpha : \Omega_{n+1} \to \Omega_{n+1}$ on the finite higher number

---

This is a revised and slightly extended version of [8].

W. Buchholz (✉)
Mathematisches Institut der Universität München, Theresienstr. 39, 80333 München, Germany
e-mail: buchholz@math.lmu.de

G. Jäger and W. Sieg (eds.), *Feferman on Foundations*, Outstanding Contributions to Logic 13, DOI 10.1007/978-3-319-63334-3_4

classes and a simultaneous assignment of fundamental sequences to their indices α. Isles (in [16]) went even further and made use also of transfinite number classes. The Bachmann–Pfeiffer–Isles approach in the end became so complex that it was practically unfeasable for proof-theoretic applications. On the other side it seemed most likely that functions of this kind would be needed for the ordinal analysis of (transfinitely) iterated inductive definitions. Therefore Feferman, in unpublished work around 1970, proposed an entirely different and much simpler method for generating hierarchies of normal functions θ_α ($\alpha \in On$), which was intended to replace the extremely complicated Bachmann–Pfeiffer–Isles definition procedure (see e.g. [12, p.67ff], [13, 462]). This proposal brought a new impetus into ordinally informative proof theory and turned out as the starting point of a rather successful development.

Definition (Feferman 1970) By transfinite recursion on α one defines functions $\theta_\alpha : On \to On$ as follows:

$\theta_\alpha :=$ ordering function of $\mathrm{In}_\alpha := \{\beta \in On : C(\alpha, \beta) \cap \beta^{\oplus} \subseteq \beta\}$, where
$C(\alpha, \beta) :=$closure of $\beta \cup \{0\}$ under the functions $+$, $\lambda\xi.\Omega_{1+\xi}$ and

$$\theta|\alpha : \alpha \times On \to On, (\xi, \eta) \mapsto \theta_\xi(\eta),$$

$$\beta^{\oplus} := \min\{\Omega_\sigma : 0 < \sigma \ \& \ \beta \leq \Omega_\sigma\} \text{ and } \Omega_\sigma := \begin{cases} 0 & \text{if } \sigma = 0 \\ \aleph_\sigma & \text{if } \sigma > 0. \end{cases}$$

A simple cardinal argument shows that $C(\alpha, \beta)$ has cardinality $\aleph_0 \cup card(\beta)$, no matter how large α is. This yields $\forall\beta < \Omega_{\sigma+1}(\theta_\alpha(\beta) < \Omega_{\sigma+1})$, for any α. Hence the function $\lambda\alpha.\theta_\alpha(\Omega_\sigma)$ maps On into $\Omega_{\sigma+1}$, i.e., it is a so-called *collapsing function*. First important contributions to the investigation of the θ-functions were made by Weyhrauch, Aczel and J. Bridge. Aczel (in [1]) showed how θ_α ($\alpha < \Gamma_{\Omega+1}$) corresponds to Bachmann's ϕ_α. Independently, Weyhrauch [28] established the same results for $\alpha < \varepsilon_{\Omega+1}$. In addition, Aczel generalized Feferman's definition and conjectured that the generalized hierarchy (θ_α) matches up with the Isles functions. This conjecture was proved by J. Bridge in her dissertation (Oxford 1972) the results of which were published in [4]. She also obtained partial results on the recursiveness of the notation system associated with (θ_α) [4]. Starting from Bridge's thesis, recursiveness of the full θ-notation system was established by the author in his dissertation (München 1974), parts of which were published in [6]. There also a variant $\overline{\theta}_\alpha$ of θ_α was introduced which has the advantage that the $<$-relation between $\overline{\theta}$-terms can be characterized in a particularly simple way. Later it turned out that in proof-theoretic applications actually only the values of $\overline{\theta}_\alpha$ (or θ_α) at initial ordinals Ω_σ are used, which led to the idea to define directly functions ψ_σ corresponding to $\lambda\alpha.\theta_\alpha(\Omega_\sigma)$.

Definition (Buchholz 1981) $\psi_\sigma(\alpha) := \min\{\beta \geq \Omega_\sigma : C(\alpha, \beta) \cap \Omega_{\sigma+1} \subseteq \beta\}$, where $C(\alpha, \beta) :=$ closure of $\beta \cup \{0\}$ under $+$ and $\psi|\alpha : On \times \alpha \to On$, $(\rho, \xi) \mapsto \psi_\rho(\xi)$. Note that $\psi_\sigma(0) = \Omega_\sigma$ (for $\sigma > 0$) and therefore $C(\alpha, \beta)$ is also closed under $\lambda\sigma.\Omega_\sigma$.

Of course this definition can be generalized in the same way as Aczel [1] generalized Feferman's definition, namely by incorporating closure under countably many cardinal valued functions into the definition of $C(\alpha, \beta)$. More substantial extensions of the θ/ψ-approach were developed by Schütte (unpublished), Pohlers [19] and, most important, Jäger [17]. There in addition to the ψ_σ's which are collapsing functions for successor regulars, also collapsing functions for limit regulars, i.e. weakly inaccessible cardinals, are introduced. In order to be able to treat both kinds of collapsing functions uniformly Jäger denoted the ψ-function for a regular cardinal κ, i.e. one with values below κ, by ψ_κ. In this notation the former ψ_σ becomes $\psi_{\Omega_{\sigma+1}}$.

Definition (Jäger 1984): $\psi_\kappa(\alpha) := \min\{\beta \geq \kappa^- : C(\alpha,\beta) \cap \kappa \subseteq \beta\}$, where
$C(\alpha,\beta) :=$ closure of β under $+$, $\lambda xy.I_x(y)$ and

$$\psi|\alpha : \mathfrak{R} \times \alpha \to On,\ (\pi,\xi) \mapsto \psi_\pi(\xi)$$

$\mathfrak{R} :=$ class of all uncountable regular cardinals.
$I_\rho :=$ ordering function of the topological closure of $\{\kappa \in \mathfrak{R} : \forall \xi < \rho(I_\xi(\kappa) = \kappa)\}$,
$\kappa^- := I_\rho(\sigma)$, if $\kappa = I_\rho(\sigma+1)$ with $\rho, \sigma < \kappa$.
Later Rathjen developed further extensions up to the use of large cardinals [20–22]. In the first of these extensions Rathjen assumed the existence of a weakly Mahlo cardinal M and utilized the fact that the regular cardinals are stationary in M.

Definition (Rathjen 1990): Actually this is a variant of Rathjen's definition.

$$\psi_\kappa(\alpha) := \begin{cases} \min\{\beta \in \mathfrak{R} : \alpha \in C(\alpha,\beta)\ \&\ C(\alpha,\beta) \cap \kappa \subseteq \beta\} & \text{if } \kappa = M \\ \min\{\beta \in On : \kappa \in C(\alpha,\beta)\ \&\ C(\alpha,\beta) \cap \kappa \subseteq \beta\} & \text{if } \kappa < M \end{cases}$$

$C(\alpha,\beta) :=$ closure of β under $+$, $\lambda x.\omega^{M+x}$, $\psi|\alpha$

The Mahlo property is used to prove that $\psi_M(\alpha) < M$ for all $\alpha \in On$.

While the above sketched development aims at generating larger and larger segments of recursive ordinals, in recent years a renewed interest in ordinal notations around the (comparatively small) Bachmann–Howard ordinal $\phi_{\varepsilon_{\Omega+1}}(0)$ has evolved, mainly caused by Solomon Feferman's unwinding program and Gerhard Jäger's metapredicativity program. Therefore it seems worthwile to review some important results of this area and to present detailed and streamlined proofs for them. The results in question are mainly comparisons of various functions which in the past have been used for describing ordinals not much larger than the Bachmann–Howard ordinal. We start with a treatment of the Bachmann hierarchy $(\phi_\alpha)_{\alpha \leq \Gamma_{\Omega+1}}$ from [3], and a presentation of Schütte's Klammersymbols [24] within that hierarchy. Then, in Sect. 3, we give an alternative characterization of the Bachmann hierarchy which instead of fundamental sequences $(\alpha[\xi])_{\xi<\tau_\alpha}$ uses finite sets $K\alpha \subseteq \Omega$ of *coefficients* ("Koeffizienten"). For $\alpha < \varepsilon_{\Omega+1}$, $K\alpha$ is almost identical with the set $C(\alpha)$ of *constituents* in [14]. In Sect. 4 we show how Feferman's functions θ_α $(\alpha < \Gamma_{\Omega+1})$ can be defined using the $K\alpha$'s (instead of the closure sets $C(\alpha,\beta)$). This leads to an easy comparison of the hierarchies $(\phi_\alpha)_{\alpha<\Gamma_{\Omega+1}}$ and $(\theta_\alpha)_{\alpha<\Gamma_{\Omega+1}}$ which becomes particularly simple if one switches to the fixed-point-free versions: $\overline{\phi}_\alpha(\beta) = \overline{\theta}_\alpha(\beta)$ for all $\alpha < \Gamma_{\Omega+1}, \beta < \Omega$. In Sects. 5, 6 we deal with the unary functions $\vartheta : \varepsilon_{\Omega+1} \to \Omega$ and $\psi : \varepsilon_{\Omega+1} \to \Omega$ which play an important rôle in [23]. We show that $\overline{\theta}_{1+\alpha}(\beta) = \vartheta(\Omega\alpha + \beta)$ (for $\alpha < \varepsilon_{\Omega+1}$, $\beta < \Omega$) and refine a result from [23] on the relationship between ϑ and ψ. In Sect. 7, largely following [28], we show how the Bachmann hierarchy below $\varepsilon_{\Omega+1}$ can be defined by means of functionals of finite higher types.

A nice survey on the history of the subject can be found in [12].

Preliminaries. The letters $\alpha, \beta, \gamma, \delta, \xi, \eta, \zeta$ always denote ordinals. On denotes the class of all ordinals and Lim the class of all limit ordinals. We are working in ZFC. So, every ordinal α is identical to the set $\{\xi \in On : \xi < \alpha\}$, and we have $\beta < \alpha \Leftrightarrow \beta \in \alpha$ and $\beta \leq \alpha \Leftrightarrow \beta \subseteq \alpha$. By $\mathbb{H}$ we denote the class $\{\gamma > 0 : \forall \alpha, \beta < \gamma(\alpha + \beta < \gamma)\} = \{\omega^\alpha : \alpha \in On\}$ of all *additive principal numbers* (Hauptzahlen), and by $\mathbb{E}$ the class $\{\alpha \in On : \omega^\alpha = \alpha\} = \{\varepsilon_\alpha : \alpha \in On\}$ of all *epsilon-numbers*. A *normal function* is a strictly increasing continuous function $F : On \to On$. The normal functions $\varphi_\alpha : On \to On$ ($\alpha \in On$) are defined by: $\varphi_0(\beta) := \omega^\beta$, and $\varphi_\alpha :=$ ordering (or enumerating) function of $\{\beta : \forall \xi < \alpha(\varphi_\xi(\beta) = \beta)\}$, if $\alpha > 0$. The family $(\varphi_\alpha)_{\alpha \in On}$ is called *the Veblen hierarchy over* $\lambda\xi.\omega^\xi$. An ordinal α is called *strongly critical* iff $\varphi_\alpha(0) = \alpha$. The class of all strongly critical ordinals is denoted by SC, and its enumerating function by $\lambda\alpha.\Gamma_\alpha$. It is well-known that $\lambda\alpha.\Gamma_\alpha$ is again a normal function, and that $\Gamma_{\Omega_\sigma} = \Omega_\sigma$ for each $\sigma > 0$.

2 Fundamental Sequences and the Bachmann Hierarchy

The following stems from Bachmann's seminal paper [3], but in some minor details we deviate from that paper. We start by assigning to each limit number $\alpha \leq \Gamma_{\Omega+1}$ a fundamental sequence $(\alpha[\xi])_{\xi < \tau_\alpha}$ with $\tau_\alpha \leq \Omega$. The definition of $\alpha[\xi]$ is based on the normal form representation of α in terms of $0, +, \cdot, F$, where $(F_\alpha)_{\alpha \in On}$ is the Veblen hierarchy over $\lambda x.\Omega^x$, i.e., $F_0(\beta) := \Omega^\beta$, and $F_\alpha :=$ ordering function of $\{\beta : \forall \xi < \alpha(F_\xi(\beta) = \beta)\}$, if $\alpha > 0$. The relationship between F_α and φ_α for $\alpha > 0$ is given by

$$F_\alpha(\beta) = \varphi_\alpha(\widetilde{\alpha} + \beta) \text{ with } \widetilde{\alpha} := \begin{cases} \Omega{+}1 & \text{if } 0 < \alpha < \Omega \\ 1 & \text{if } \alpha = \Omega \\ 0 & \text{if } \Omega < \alpha \end{cases} .$$

From this it follows that $\Gamma_{\Omega+1}$ is the least fixed point of $\lambda\alpha.F_\alpha(0)$.

For completeness note, that $F_0(\beta) = \varphi_0(\Omega\beta)$.

Abbreviations.

1. $\Lambda := \Gamma_{\Omega+1} = \min\{\alpha : F_\alpha(0) = \alpha\}$.
2. $\alpha|\gamma \;:\Leftrightarrow\; \exists\xi(\gamma = \alpha{\cdot}\xi)$
3. $\alpha =_{\mathrm{NF}} \gamma + \Omega^\beta\eta \;:\Leftrightarrow\; \alpha = \gamma + \Omega^\beta\eta \;\&\; 0 < \eta < \Omega \;\&\; \Omega^{\beta+1}|\gamma$.
4. $\gamma =_{\mathrm{NF}} F_\alpha(\beta) \;:\Leftrightarrow\; \alpha, \beta < \gamma = F_\alpha(\beta)$.

Propositions.

(a) For each $0 < \delta < \Lambda$ there are unique γ, β, η such that $\delta =_{\mathrm{NF}} \gamma + \Omega^\beta\eta$.

(b) For each $\delta \in \mathrm{ran}(F_0) \cap \Lambda$ there are unique α, β such that $\delta =_{\mathrm{NF}} F_\alpha(\beta)$.

(c) $\delta < \Lambda \;\Rightarrow\; (\delta =_{\mathrm{NF}} F_\alpha(\beta) \Leftrightarrow \beta < \delta = F_\alpha(\beta))$.

Definition of a fundamental sequence $(\lambda[\xi])_{\xi<\tau_\lambda}$ for each limit number $\lambda \leq \Lambda$

1. $\lambda =_{\mathrm{NF}} \gamma + \Omega^\beta \eta \notin \mathrm{ran}(F_0)$:
   1.1. $\eta \in \mathit{Lim}$: $\tau_\lambda := \eta$ and $\lambda[\xi] := \gamma + \Omega^\beta \cdot (1+\xi)$.
   1.2. $\eta = \eta_0+1$: $\tau_\lambda := \tau_{\Omega^\beta}$ and $\lambda[\xi] := \gamma + \Omega^\beta \eta_0 + \Omega^\beta[\xi]$.
2. $\lambda =_{\mathrm{NF}} F_\alpha(\beta)$:
   2.1. $\beta \in \mathit{Lim}$: $\tau_\lambda := \tau_\beta$ and $\lambda[\xi] := F_\alpha(\beta[\xi])$
   2.2. $\beta \notin \mathit{Lim}$: Let $\lambda^- := \begin{cases} 0 & \text{if } \beta = 0 \\ F_\alpha(\beta_0)+1 & \text{if } \beta = \beta_0+1 \end{cases}$
   2.2.0. $\alpha = 0$: Then $\beta = \beta_0+1$. $\tau_\lambda := \Omega$ and $\lambda[\xi] := \Omega^{\beta_0} \cdot (1+\xi)$.
   2.2.1. $\alpha = \alpha_0+1$: $\tau_\lambda := \omega$ and $\lambda[n] := F_{\alpha_0}^{(n+1)}(\lambda^-)$.
   2.2.2. $\alpha \in \mathit{Lim}$: $\tau_\lambda := \tau_\alpha$ and $\lambda[\xi] := F_{\alpha[\xi]}(\lambda^-)$.
3. $\tau_\Lambda := \omega$ and $\Lambda[0] := 1$, $\Lambda[n+1] := F_{\Lambda[n]}(0)$.

Definition. For each limit $\lambda \leq \Lambda$ we set $\lambda[\tau_\lambda] := \lambda$. Further $\tau_0 := 0$, $0[\xi] := 0$ and $\tau_{\alpha+1} := 1$, $(\alpha+1)[\xi] := \alpha$ for all ξ.

Remark (i) $\lambda \in \mathit{Lim} \cap \Omega \;\Rightarrow\; \tau_\lambda = \lambda$ and $\lambda[\xi] = 1+\xi$.
(ii) $\lambda =_{\mathrm{NF}} \gamma + \Omega^{\beta+1}(\eta+1) \;\Rightarrow\; \tau_\alpha = \Omega$ and $\lambda[\xi] = \gamma + \Omega^{\beta+1}\eta + \Omega^\beta \cdot (1+\xi)$.

Lemma 2.1 $\lambda =_{\mathrm{NF}} F_\alpha(\beta) < \Lambda \;\&\; \beta \in \mathit{Lim} \;\&\; 1 \leq \xi < \tau_\beta \;\Rightarrow$
$\lambda[\xi] =_{\mathrm{NF}} F_\alpha(\beta[\xi])$.

Proof Cf. Appendix.

Lemma 2.2 *Let $\lambda \in \mathit{Lim} \cap (\Lambda+1)$.*

(a) $\xi < \eta \leq \tau_\lambda \;\Rightarrow\; \lambda[\xi] < \lambda[\eta]$.
(b) $\lambda = \sup_{\xi<\tau_\lambda} \lambda[\xi]$.
(c) $\eta \in \mathit{Lim} \cap (\tau_\lambda + 1) \;\Rightarrow\; \lambda[\eta] \in \mathit{Lim} \;\&\; \tau_{\lambda[\eta]} = \eta \;\&\; \forall \xi < \eta(\lambda[\eta][\xi] = \lambda[\xi])$.
(d) $\xi < \tau_\lambda \;\&\; \lambda[\xi] < \delta \leq \lambda[\xi+1] \;\Longrightarrow\; \lambda[\xi] \leq \delta[1]$.

The proof of (a),(b),(c) is left to the reader. The proof of (d) will be given in the Appendix.

Definition. An Ω-*normal function* is a strictly increasing continuous function $f : \Omega \to \Omega$. A set $M \subseteq \Omega$ is Ω-*club* (*closed and unbounded in* Ω) iff $\forall X \subseteq M(X \neq \emptyset \;\&\; \sup(X) < \Omega \Rightarrow \sup(X) \in M)$ and $\forall \alpha < \Omega \exists \beta \in M(\alpha < \beta)$.
It is well-known that $M \subseteq \Omega$ is Ω-club if, and only if, M is the range of some Ω-normal function. Hence the ordering function of any Ω-club set is Ω-normal.

The collection of Ω-club sets has the following closure properties:

1. If f is Ω-normal then $\{\beta \in \Omega : f(\beta) = \beta\}$ is Ω-club.
2. If $(M_\xi)_{\xi<\alpha}$ is a sequence of Ω-club sets with $0 < \alpha < \Omega$ then $\bigcap_{\xi<\alpha} M_\xi$ is Ω-club.
3. If $(M_\xi)_{\xi<\Omega}$ is a sequence of Ω-club sets then also $\{\alpha \in \Omega : \alpha \in \bigcap_{\xi<\alpha} M_\xi\}$ is Ω-club.

Drawing upon 1.–3. and upon the above assignment of fundamental sequences we now define Bachmann's hierarchy of Ω-normal functions ϕ_α $(\alpha \le \Lambda)$.

Definition. $\phi_\alpha : \Omega \to \Omega$ is the ordering function of the Ω-club set R_α, where R_α is defined by recursion on α as follows:

$R_0 := \mathbb{H} \cap \Omega$,

$R_{\alpha+1} := \{\beta \in \Omega : \phi_\alpha(\beta) = \beta\}$,

$$R_\alpha := \begin{cases} \bigcap_{\xi<\tau_\alpha} R_{\alpha[\xi]} & \text{if } \tau_\alpha \in \Omega \cap Lim \\ \{\beta \in \Omega \cap Lim : \beta \in \bigcap_{\xi<\beta} R_{\alpha[\xi]}\} & \text{if } \tau_\alpha = \Omega \end{cases}$$

Notes. 1. In Lemma 2.3g we will show that $R_\alpha = \{\beta \in \Omega : \phi_{\alpha[\beta]}(0) = \beta\}$ if $\tau_\alpha = \Omega$.
2. As mentioned above, our definition of the Bachmann hierarchy (and of F_α) diverges in some minor points from [Ba50]. As a consequence of this, Bachmann's ordinals $H(1) = \varphi_{F_\Omega(1)+1}(1)$ and $\varphi_{F_{\omega_2+1}(1)}(1)$ are $\phi_{F_\Omega(0)}(0)$ and $\phi_\Lambda(0)$, respectively, in the present paper. For more details cf. [2, Note on p.35].

Remark (i) $R_{\alpha+1} \subseteq R_\alpha$; (ii) $0 < \alpha \Rightarrow R_\alpha \subseteq R_1 \subseteq Lim$.

Lemma 2.3 *For $\alpha \le \Lambda$ the following holds:*

(a) $R_\alpha \subseteq \bigcap_{n<\omega} R_{\alpha[n]}$.
(b) $\phi_\alpha(0) < \phi_{\alpha+1}(0)$
(c) $\xi+1 < \tau_\alpha \Rightarrow R_{\alpha[\xi+1]} \subseteq R_{\alpha[\xi]+1}$.
(d) $\eta < \xi < (\tau_\alpha+1) \cap \Omega \Rightarrow \phi_{\alpha[\eta]}(0) < \phi_{\alpha[\xi]}(0)$
(e) $0 < \alpha \;\&\; n < \omega \Rightarrow \phi_{\alpha[n]}(0) < \phi_\alpha(0)$.
(f) $\xi \le \tau_\alpha < \Omega \Rightarrow \xi \le \phi_{\alpha[\xi]}(0)$.
(g) $\tau_\alpha = \Omega \Rightarrow R_\alpha = \{\beta \in \Omega : \phi_{\alpha[\beta]}(0) = \beta\}$.

Proof (a) For $\tau_\alpha < \Omega$ the claim is trivial. For $\tau_\alpha = \Omega$ we have $R_\alpha = \{\beta \in \Omega : \beta \in Lim \cap \bigcap_{\xi<\beta} R_{\alpha[\xi]}\} \subseteq \bigcap_{n<\omega} R_{\alpha[n]}$.
(b) $0 < \phi_{\alpha+1}(0) \Rightarrow \phi_\alpha(0) < \phi_\alpha(\phi_{\alpha+1}(0)) = \phi_{\alpha+1}(0)$.
(c) By induction on δ we prove: $\alpha[\xi] < \delta \le \alpha[\xi+1] \Rightarrow R_\delta \subseteq R_{\alpha[\xi]+1}$.

1. $\delta = \delta_0+1$ with $\alpha[\xi] \le \delta_0$: Then either $\alpha[\xi]+1 = \delta$ or $R_\delta \subseteq R_{\delta_0} \overset{\text{IH}}{\subseteq} R_{\alpha[\xi]+1}$.
2. $\delta \in Lim$: By Lemma 2.2a, d, $\alpha[\xi] < \delta[2] < \alpha[\xi+1]$.

   Hence $R_\delta \overset{\text{(a)}}{\subseteq} R_{\delta[2]} \overset{\text{IH}}{\subseteq} R_{\alpha[\xi]+1}$.

(d) Induction on ξ:

1. $\xi \in Lim$: $R_{\alpha[\xi]} \overset{\text{L.2.2c}}{\subseteq} R_{\alpha[\eta+1]} \overset{\text{(c)}}{\subseteq} R_{\alpha[\eta]+1} \Rightarrow \phi_{\alpha[\eta]}(0) \overset{\text{(b)}}{<} \phi_{\alpha[\eta]+1}(0) \le \phi_{\alpha[\xi]}(0)$.
2. $\xi = \xi_0+1$: $\phi_{\alpha[\eta]}(0) \overset{\text{IH}}{\le} \phi_{\alpha[\xi_0]}(0) \overset{\text{(b)}}{<} \phi_{\alpha[\xi_0]+1}(0) \overset{\text{(c)}}{\le} \phi_{\alpha[\xi]}(0)$.

(e) $R_\alpha \overset{\text{(a)}}{\subseteq} R_{\alpha[n+1]} \overset{\text{(c)}}{\subseteq} R_{\alpha[n]+1}$ and thus $\phi_{\alpha[n]}(0) \overset{\text{(b)}}{<} \phi_{\alpha[n]+1}(0) \le \phi_\alpha(0)$.
(f) follows from (d) by induction on ξ.
(g) $R_\alpha = \{\beta \in \Omega \cap Lim : \beta \in \bigcap_{\xi<\beta} R_{\alpha[\xi]}\} \overset{\text{L.2.2c}}{=} \{\beta \in \Omega : \beta \in R_{\alpha[\beta]}\} \overset{\text{(f)}}{=} \{\beta \in \Omega : \phi_{\alpha[\beta]}(0) = \beta\}$.

The following theorem provides a more compact characterization of ϕ_α.

Definition.
$\tau_\alpha^\beta := \begin{cases} \beta & \text{if } \tau_\alpha = \Omega \\ \tau_\alpha & \text{otherwise} \end{cases}.$

$\delta <_\beta^1 \alpha \; :\Leftrightarrow \; \exists \xi < \tau_\alpha^\beta (\delta = \alpha[\xi])$.

$<_\beta$:= transitive closure of $<_\beta^1$.

Theorem 2.4 *(a)* $\beta \in R_\alpha \;\Leftrightarrow\; \beta \in R_0 \;\&\; \forall \xi < \tau_\alpha^\beta (\phi_{\alpha[\xi]}(\beta) = \beta)$.
(b) $\beta \in R_\alpha \;\Leftrightarrow\; \beta \in R_0 \;\&\; \forall \delta <_\beta \alpha (\phi_\delta(\beta) = \beta)$.

Proof (a) 1. $\tau_\alpha \in \{0, 1\}$: trivial.

2. $\tau_\alpha \in Lim \cap \Omega$: $\beta \in R_\alpha \;\Leftrightarrow\; \forall \xi < \tau_\alpha (\beta \in R_{\alpha[\xi]}) \overset{\text{L.2.3c}}{\Leftrightarrow} \forall \xi < \tau_\alpha (\beta \in R_{\alpha[\xi]+1})$.

3. $\tau_\alpha = \Omega$: $\beta \in R_\alpha \;\Leftrightarrow\; \beta \in Lim \;\&\; \forall \xi < \beta (\beta \in R_{\alpha[\xi]}) \overset{\text{L.2.3c}}{\Leftrightarrow} \beta \in R_0 \;\&\; \forall \xi < \beta$ $(\beta \in R_{\alpha[\xi]+1})$

(b) "$\Leftarrow$": trivial consequence of (a).

"$\Rightarrow$" Induction on α. Assume $\beta \in R_\alpha$ and $\delta <_\beta \alpha$. Then, for some $\xi < \tau_\alpha^\beta$, $\delta \le_\beta \alpha[\xi]$. By (a) we get $\phi_{\alpha[\xi]}(\beta) = \beta$. If $\delta = \alpha[\xi]$, we are done. Otherwise $\delta <_\beta \alpha[\xi] \;\&\; \beta \in R_{\alpha[\xi]}$ from which $\phi_\delta(\beta) = \beta$ follows by IH.

Corollary. $0 < \alpha < \Omega \Rightarrow R_\alpha = \{\beta \in \Omega : \forall \xi < \alpha(\phi_\xi(\beta) = \beta)\}$.

Schütte's Klammersymbols

In [24], building on [27], Schütte introduced a system of ordinal notations based on so-called 'Klammersymbols'. A Klammersymbol is a matrix $\binom{\xi_0 \; \dots \; \xi_n}{\alpha_0 \; \dots \; \alpha_n}$ with $\alpha_0 < \alpha_1 < \cdots < \alpha_n < \Omega$ and $\xi_0, \dots, \xi_n < \Omega$. Two Klammersymbols A and B are defined to be equal (in symbols $A \equiv B$) if they are identical after deleting all columns of the form $\binom{0}{\alpha_i}$. Therefore one can identify the Klammersymbol $\binom{\xi_0 \; \dots \; \xi_n}{\alpha_0 \; \dots \; \alpha_n}$ with the ordinal $\Omega^{\alpha_n}\xi_n + \cdots + \Omega^{\alpha_0}\xi_0$. Under this identification the $<$-relation between ordinals induces a well-ordering $\prec$ on the Klammersymbols, which coincides exactly with the "lexikographische Anordnung der Klammersymbole" as defined in [24, p.17]. By recursion on this well-ordering, to each Ω-normal function f and each Klammersymbol A an ordinal $fA < \Omega$ is assigned in such a way that the following holds:

(K1) $f\binom{\xi}{0} = f(\xi)$;
(K2) if $A \equiv B$ then $fA = fB$;
(K3) if $0 < \xi_1$ and $0 < \alpha_1 < \cdots < \alpha_n < \Omega$ then $\lambda x.f\binom{x \; \xi_1 \; \dots \; \xi_n}{0 \; \alpha_1 \; \dots \; \alpha_n}$ enumerates the set $\{\beta \in \Omega : \forall \xi < \xi_1 \forall \alpha_0 < \alpha_1 [f\binom{\beta \; \xi \; \dots \; \xi_n}{\alpha_0 \; \alpha_1 \; \dots \; \alpha_n} = \beta]\}$.

By $\prec$-induction one easily sees the following:
(U) For fixed f, the values fA are uniquely determined by (K1)–(K3).

In this subsection we will locate the values $\phi_0 A$ within the Bachmann hierarchy, i.e., we will prove ($\diamond$) $\phi_0\binom{\beta\ \ \xi_1\ \ \dots\ \ \xi_n}{0\ \ 1+\alpha_1\ \dots\ 1+\alpha_n} = \phi_{\Omega^{\alpha_n}\xi_n+\dots+\Omega^{\alpha_1}\xi_1}(\beta)$.

To make the reader a little bit familiar with the subject we first consider two special cases of ($\diamond$).

Proposition. (a) $\phi_0\binom{\beta\ \alpha}{0\ 1} = \phi_\alpha(\beta)$, if $\alpha, \beta < \Omega$.
(b) $\phi_0\binom{\beta\ 1}{0\ 2} = \phi_\Omega(\beta)$, if $\beta < \Omega$.

Proof (a) Induction on α: 1. $\alpha = 0$: $\phi_0\binom{\beta\ 0}{0\ 1} \overset{(K2)}{=} \phi_0\binom{\beta}{0} \overset{(K1)}{=} \phi_0(\beta)$.

2. $0 < \alpha < \Omega$: By (K3), $\lambda x.\phi_0\binom{x\ \alpha}{0\ 1}$ enumerates $\{\beta \in \Omega : \forall\xi < \alpha(\phi_0\binom{\beta\ \xi}{0\ 1} = \beta)\}$. Since $R_\alpha = \{\beta \in \Omega : \forall\xi < \alpha(\phi_\xi(\beta) = \beta)\}$, the claim follows by IH.
(b) By (K3), $\lambda x.\phi_0\binom{x\ 1}{0\ 2}$ enumerates $S := \{\beta \in \Omega : \phi_0\binom{\beta\ 0}{0\ 2} = \beta\ \&\ \phi_0\binom{\beta\ 0}{1\ 2} = \beta\}$. Further we have $S \overset{(K2)}{=} \{\beta \in \Omega : \phi_0\binom{\beta\ 0}{0\ 1} = \beta\ \&\ \phi_0\binom{0\ \beta}{0\ 1} = \beta\} \overset{(a)}{=} \{\beta \in \Omega : \phi_0(\beta) = \beta\ \&\ \phi_\beta(0) = \beta\} = \{\beta \in \Omega : \phi_\beta(0) = \beta\} \overset{L.2.3g}{=} R_\Omega$

We now turn to the proof of ($\diamond$). For this purpose the following definition is particularly useful.

Definition. Due to the fact that every ordinal can be uniquely represented in the form $\Omega\alpha + \beta$ with $\beta < \Omega$ it is possible to code the binary function $(\alpha, \beta) \mapsto \phi_\alpha(\beta)$ ($\alpha \leq \Lambda, \beta < \Omega$) into a unary one by
$\phi\langle\Omega\alpha + \beta\rangle := \phi_\alpha(\beta)$ ($\alpha \leq \Lambda, \beta < \Omega$).

Below, in Lemma 2.8b, we will prove:
(K3*) If $0 < \xi_1$ and $0 < \alpha_1 < \dots < \alpha_n < \Omega$ then $\lambda x.\phi\langle\Omega^{\alpha_n}\xi_n + \dots + \Omega^{\alpha_1}\xi_1 + x\rangle$ enumerates the set
$\{\beta \in \Omega : \forall\xi < \xi_1 \forall\alpha_0 < \alpha_1[\phi\langle\Omega^{\alpha_n}\xi_n + \dots + \Omega^{\alpha_1}\xi + \Omega^{\alpha_0}\beta\rangle = \beta]\}$.

Together with (U) this yields the announced result ($\diamond$).

Theorem 2.5 *For $\alpha_0 < \dots < \alpha_n < \Omega$ and $\beta, \xi_0, \dots, \xi_n < \Omega$:*
(a) $\phi_0\binom{\xi_0\ \dots\ \xi_n}{\alpha_0\ \dots\ \alpha_n} = \phi\langle\Omega^{\alpha_n}\xi_n + \dots + \Omega^{\alpha_0}\xi_0\rangle$.
(b) $\phi_0\binom{\beta\ \ \xi_0\ \ \dots\ \ \xi_n}{0\ \ 1+\alpha_0\ \dots\ 1+\alpha_n} = \phi_{\Omega^{\alpha_n}\xi_n+\dots+\Omega^{\alpha_0}\xi_0}(\beta)$.

Proof (a) For each Klammersymbol $A = \binom{\xi_0\ \dots\ \xi_n}{\alpha_0\ \dots\ \alpha_n}$ let $\phi_0{\cdot}A := \phi\langle\Omega^{\alpha_n}\xi_n + \dots + \Omega^{\alpha_0}\xi_0\rangle$. Then (K1), (K2) are satisfied by $\phi_0{\cdot}A$ in place of fA with $f = \phi_0$. Further, (K3*) says that also (K3) is satisfied by $\phi_0{\cdot}A$. Therefore (U) yields $\phi_0 A = \phi_0{\cdot}A$.

(b) $\phi_0\binom{\beta\ \ \xi_0\ \ \dots\ \ \xi_n}{0\ \ 1+\alpha_0\ \dots\ 1+\alpha_n} \overset{(a)}{=} \phi\langle\Omega^{1+\alpha_n}\xi_n + \dots + \Omega^{1+\alpha_0}\xi_0 + \Omega^0\beta\rangle = \phi\langle\Omega{\cdot}(\Omega^{\alpha_n}\xi_n + \dots + \Omega^{\alpha_0}\xi_0) + \beta\rangle$.

For the proof of (K3*) we need the following two technical lemmata on the relation $<_\omega$ defined above.

Lemma 2.6 *(a)* $\alpha \in Lim \ \& \ \xi{+}1 < \tau_\alpha \ \Rightarrow \ \alpha[\xi]{+}1 \le_\omega \alpha[\xi+1]$.
(b) $\alpha \in Lim \ \& \ \eta < \xi < (\tau_\alpha + 1) \cap \Omega \ \Rightarrow \ \alpha[\eta] <_\omega \alpha[\xi]$.
(c) $\tilde{\alpha} \le_\omega \alpha \ \Rightarrow \ R_\alpha \subseteq R_{\tilde{\alpha}}$.

Proof (a) By induction on δ one proves: $\alpha[\xi] < \delta \le \alpha[\xi{+}1] \Rightarrow \alpha[\xi] + 1 \le_\omega \delta$. Let us assume that $\delta \in Lim$. Then by Lemma 2.2a, d, $\alpha[\xi] < \delta[2] < \alpha[\xi{+}1]$, and therefore, by IH, $\alpha[\xi] + 1 \le_\omega \delta[2] <^1_\omega \delta$.

(b) Proof by induction on ξ, using (a) and Lemma 2.2c. Cf. proof of Lemma 2.3d.

(c) One easily shows that $\tilde{\alpha} <^1_\omega \alpha$ implies $R_\alpha \subseteq R_{\tilde{\alpha}}$ (using Lemma 2.3a in case $\tau_\alpha = \Omega$).

Lemma 2.7 *Assume* $\alpha =_{\mathrm{NF}} \gamma + \Omega^{\alpha_1}\xi_1$ *with* $\alpha_1 < \Omega$.

(a) $\xi < \xi_1 \ \Rightarrow \ \gamma + \Omega^{\alpha_1}(\xi{+}1) \le_\omega \alpha$.
(b) $\xi < \xi_1 \ \& \ \alpha_0 < \alpha_1 \ \Rightarrow \ \gamma + \Omega^{\alpha_1}\xi + \Omega^{\alpha_0+1} \ \le_\omega \ \alpha$.

Proof (a) Let $\hat{\alpha} := \gamma + \Omega^{\alpha_1+1}, \eta := -1 + (\xi + 1)$, and $\eta_1 := -1 + \xi_1$. Then $\hat{\alpha}[\eta] = \gamma + \Omega^{\alpha_1}(\xi{+}1)$, $\hat{\alpha}[\eta_1] = \gamma + \Omega^{\alpha_1}\xi_1 = \alpha$, and $\eta \le \eta_1 < \tau_{\hat{\alpha}}$. Hence $\gamma + \Omega^{\alpha_1}(\xi{+}1) = \hat{\alpha}[\eta] \le_\omega \hat{\alpha}[\eta_1] = \alpha$ by Lemma 2.6b.

(b) Let $\gamma_1 := \gamma + \Omega^{\alpha_1}\xi$. Since $\alpha_0 < \alpha_1$, we have $\alpha_1 = \delta + n$ with ($\alpha_0 < \delta \in Lim$ or $\delta = \alpha_0 + 1$). Then, $\gamma_1 + \Omega^{\alpha_0+1} \le_\omega \gamma_1 + \Omega^\delta <_\omega \gamma_1 + \Omega^{\delta+1} <_\omega \cdots <_\omega \gamma_1 + \Omega^{\delta+n} = \gamma + \Omega^{\alpha_1}(\xi{+}1) \overset{\text{(a)}}{\le_\omega} \alpha$.

Now we are ready to prove (K3*), i.e. Lemma 2.8b.

Lemma 2.8 *Assume* $\alpha =_{\mathrm{NF}} \gamma + \Omega^{\alpha_1}\xi_1$ *with* $\alpha_1 < \Omega$.

(a) $\beta \in R_\alpha \ \Leftrightarrow \ \forall \xi < \xi_1 [\, \beta \in R_{\gamma+\Omega^{\alpha_1}\xi+1} \ \& \ \forall \alpha_0 < \alpha_1 (\beta \in R_{\gamma+\Omega^{\alpha_1}\xi+\Omega^{\alpha_0+1}}) \,]$.
(b) *If* $0 < \alpha_1$ *then* $\lambda x.\phi\langle \gamma + \Omega^{\alpha_1}\xi_1 + x\rangle$ *enumerates the set*
$\{\beta \in \Omega : \forall \xi < \xi_1 \forall \alpha_0 < \alpha_1 (\phi\langle \gamma + \Omega^{\alpha_1}\xi + \Omega^{\alpha_0}\beta\rangle = \beta)\}$.
(c) *If* $\alpha_1 = \alpha_0{+}1$ *then* $\lambda x.\phi\langle \gamma + \Omega^{\alpha_1}\xi_1 + x\rangle$ *enumerates the set*
$\{\beta \in \Omega : \forall \xi < \xi_1 (\phi\langle \gamma + \Omega^{\alpha_1}\xi + \Omega^{\alpha_0}\beta\rangle = \beta)\}$.

Proof (a) "$\Leftarrow$": We distinguish the following cases:
1. $\xi_1 \in Lim$: Then $\beta \in \bigcap_{\xi<\xi_1} R_{\gamma+\Omega^{\alpha_1}(1+\xi)} = R_\alpha$.
2. $\xi_1 = \xi_0{+}1$:
   2.1. $\alpha_1 = 0$: Then $\beta \in R_{\gamma+\Omega^{\alpha_1}\xi_0+1} = R_\alpha$.
   2.2. $\alpha_1 = \alpha_0{+}1$: Then $\beta \in R_{\gamma+\Omega^{\alpha_1}\xi_0+\Omega^{\alpha_0+1}} = R_\alpha$.
   2.3. $\alpha_1 \in Lim$: Since $\alpha_1 < \Omega$, we have $\tau_\alpha = \alpha_1$ and $\alpha[\xi] = \gamma + \Omega^{\alpha_1}\xi_0 + \Omega^{1+\xi}$.
   From $\forall \alpha_0 < \alpha_1 (\beta \in R_{\gamma+\Omega^{\alpha_1}\xi_0+\Omega^{\alpha_0+1}})$ we get
   $\beta \in \bigcap_{\eta<\tau_\alpha} R_{\alpha[\eta+1]} \overset{\text{L.2.3c}}{\subseteq} \bigcap_{\eta<\tau_\alpha} R_{\alpha[\eta]} = R_\alpha$.
   "$\Rightarrow$": By Lemma 2.7 we have
   $\forall \xi < \xi_1 [\, \gamma + \Omega^{\alpha_1}\xi + 1 \le_\omega \alpha \ \& \ \forall \alpha_0 < \alpha_1 (\gamma + \Omega^{\alpha_1}\xi + \Omega^{\alpha_0+1} \le_\omega \alpha) \,]$.
   Together with Lemma 2.6c this yields the claim.

(b) Take $\overline{\gamma}$ and $\overline{\alpha}_1$ such that $\gamma = \Omega{\cdot}\overline{\gamma}$ and $\alpha_1 = 1 + \overline{\alpha}_1$. Then $\alpha = \Omega{\cdot}\overline{\alpha}$, with $\overline{\alpha} := \overline{\gamma} + \Omega^{\overline{\alpha}_1}\xi_1$. By definition, $\phi_{\overline{\alpha}}(x) = \phi\langle\gamma + \Omega^{\alpha_1}\xi_1 + x\rangle$, and therefore, $\lambda x.\phi\langle\gamma + \Omega^{\alpha_1}\xi_1 + x\rangle$ enumerates $R_{\overline{\alpha}}$.

On the other side: $\beta \in R_{\overline{\alpha}} \overset{\text{(a), L.2.3g}}{\Leftrightarrow}$
$\forall\xi < \xi_1[\phi_{\overline{\gamma}+\Omega^{\overline{\alpha}_1}\xi}(\beta) = \beta \;\&\; \forall\alpha_0 < \overline{\alpha}_1(\phi_{\overline{\gamma}+\Omega^{\overline{\alpha}_1}\xi+\Omega^{\alpha_0}\beta}(0) = \beta)] \Leftrightarrow$
$\forall\xi < \xi_1[\phi\langle\gamma + \Omega^{\alpha_1}\xi + \beta\rangle = \beta \;\&\; \forall 1 \le \alpha_0 < \alpha_1(\phi\langle\gamma + \Omega^{\alpha_1}\xi + \Omega^{\alpha_0}\beta\rangle = \beta)] \Leftrightarrow$
$\forall\xi < \xi_1\forall\alpha_0 < \alpha_1(\phi\langle\gamma + \Omega^{\alpha_1}\xi + \Omega^{\alpha_0}\beta\rangle = \beta)$
(c) This can be seen by a simple modification of the proofs of (a) and (b).

Lemma 2.9 *(n+1-ary Veblen function). For $\xi_0, \ldots, \xi_n < \Omega$ let $\varphi^{n+1}(\xi_n, \ldots, \xi_0) := \phi\langle\Omega^n\xi_n + \cdots + \Omega^0\xi_0\rangle$. Then the following holds:*

(i) $\varphi^{n+1}(0, \ldots, 0, \beta) = \phi_0(\beta)$.
(ii) If $0 < k \le n$ *and* $\xi_k > 0$, *then* $\lambda x.\varphi^{n+1}(\xi_n, \ldots, \xi_k, 0, \ldots, 0, x)$ *enumerates* $\{\beta \in \Omega : \forall\xi < \xi_k(\varphi^{n+1}(\xi_n, \ldots, \xi_{k+1}, \xi, \beta, 0, \ldots, 0) = \beta)\}$.

Proof of (ii): Let $\gamma := \Omega^n\xi_n + \cdots + \Omega^{k+1}\xi_{k+1}$. Then $\varphi^{n+1}(\xi_n, \ldots, \xi_k, \vec{0}, x) = \phi\langle\gamma + \Omega^k\xi_k + \Omega^0 x\rangle$, and, by Lemma 2.8c, $\lambda x.\phi\langle\gamma + \Omega^k\xi_k + \Omega^0 x\rangle$ enumerates $\{\beta \in \Omega : \forall\xi < \xi_k(\phi\langle\gamma + \Omega^k\xi + \Omega^{k-1}\beta\rangle = \beta)\}$.

Note. φ^{n+1} $(n \ge 1)$ is known as the n+1-ary Veblen function.
Usually it is *defined* by (i), (ii).

3 Characterization of ϕ_α via $K\alpha$

In [14] the Bachmann hierarchy (ϕ_α) restricted to $\alpha < \varepsilon_{\Omega+1}$ is studied, and thereby, as a technical tool, the sets $C(\alpha)$ and $ND(\alpha)$ (of *constituents* and *nondistinguished constituents* of α) are defined. From [14, Lemmata 3.1, 3.2] and [14, Theorems 3.1, 3.3] one can derive the following interesting result which provides an alternative definition of the Bachmann hierarchy not referring to fundamental sequences:

$$(G)\; R_\alpha = \{\beta \in R_0 : C(\alpha) \subseteq \beta{+}1 \;\&\; ND(\alpha) \subseteq \beta \;\& \\ \forall\delta < \alpha(C(\delta) \subseteq \gamma \to \phi_\delta(\beta) = \beta)\}\; (\alpha < \varepsilon_{\Omega+1}).$$

In the following we will directly prove an analogue of (G), namely Theorem 3.6, and then exemplarily derive Gerber's Theorems 4.1, 4.3 (our 3.9, 3.10) from that.

We start by defining for each $\alpha \le \Lambda$ a finite set $K\alpha$ (corresponding to $C(\alpha)$ in [14]) of ordinals $< \Omega$.

Definition of $K\alpha$ for $\alpha \le \Lambda$

$$1.\; \alpha \le \Omega\colon\; K\alpha := \begin{cases} \emptyset & \text{if } \alpha \in \{0, \Omega\} \\ \{\alpha\} & \text{if } \alpha \in Lim \cap \Omega \\ K\alpha_0 & \text{if } \alpha = \alpha_0{+}1 < \Omega \end{cases}$$

2. $\Omega < \alpha =_{\mathrm{NF}} \gamma + \Omega^{\beta}\eta \notin \mathrm{ran}(F_0)$: $K\alpha := K\gamma \cup K\beta \cup K\eta$.
3. $\Omega < \alpha =_{\mathrm{NF}} F_{\xi}(\eta) < \Lambda$: $K\alpha := K'\xi \cup K\eta$ with $K'\xi := \begin{cases} \emptyset & \text{if } \xi = 0 \\ \{\omega\} \cup K\xi & \text{if } \xi > 0 \end{cases}$
4. $K\Lambda := \{\omega\}$.

Remark $K(\alpha_0{+}1) = K\alpha_0$.

Lemma 3.1 *(a)* $K\alpha = K\alpha[1] \cup K\tau_{\alpha}$.
(b) $K\xi \subseteq K\alpha[\xi] \subseteq K\alpha[1] \cup K\xi$, *for all* $\xi < \tau_{\alpha}$.

Proof For $\alpha \notin Lim$ the statements are trivial. The limit case is covered by the following proposition.

Proposition. $\lambda \in Lim\ \&\ 1 \leq \xi \leq \tau_{\lambda} \Rightarrow K\lambda[\xi] = K\lambda[1] \cup K\xi$.

Proof by induction on λ:

1. $\lambda =_{\mathrm{NF}} \gamma + \Omega^{\beta}\eta \notin \mathrm{ran}(F_0)$:

1.1. $\eta \in Lim$: $\tau_{\lambda} = \eta$ and $\lambda[\xi] = \gamma + \Omega^{\beta}(1{+}\xi)$.

$\xi \leq \eta \Rightarrow K\lambda[\xi] = K\gamma \cup K\beta \cup K\xi$.

1.2. $\eta = \eta_0{+}1$: $\tau_{\lambda} = \tau_{\Omega^{\beta}}$ and $\lambda[\xi] = \gamma + \Omega^{\beta}\eta_0 + \Omega^{\beta}[\xi]$.

$K\lambda[\xi] = K\gamma \cup K(\Omega^{\beta}\eta_0) \cup K(\Omega^{\beta}[\xi]) \overset{\mathrm{IH}}{=} K\gamma \cup K(\Omega^{\beta}\eta_0) \cup K(\Omega^{\beta}[1]) \cup K\xi$.

2. $\lambda =_{\mathrm{NF}} F_{\alpha}(\beta)$:

2.1. $\beta \in Lim$: Then by Lemma 2.1, $\lambda[\xi] =_{\mathrm{NF}} F_{\alpha}(\beta[\xi])$ and thus $K\lambda[\xi] = K'\alpha \cup K(\beta[\xi]) \overset{\mathrm{IH}}{=} K'\alpha \cup K\beta[1] \cup K\xi = K\lambda[1] \cup K\xi$.

2.2. $\beta \notin Lim$: In Sect. 2 we defined $\lambda^- := \begin{cases} 0 & \text{if } \beta = 0 \\ F_{\alpha}(\beta_0){+}1 & \text{if } \beta = \beta_0{+}1 \end{cases}$. Hence

$$K\lambda^- = \begin{cases} K'\alpha \cup K\beta & \text{if } \beta = \beta_0{+}1\ \&\ \beta_0 < F_{\alpha}(\beta_0) \\ K\beta & \text{otherwise} \end{cases}$$

and $K\lambda = K'\alpha \cup K\beta = K'\alpha \cup K\lambda^-$.

2.2.0. $\alpha = 0$: Then $\lambda = \Omega^{\beta_0+1}$, $\tau_{\lambda} = \Omega$ and $\lambda[\xi] = \Omega^{\beta_0}(1{+}\xi)$. Hence $K\lambda[\xi] = K\beta_0 \cup K\xi$.

2.2.1. $\alpha = \alpha_0{+}1$: Then $\tau_{\lambda} = \omega$ and, for $\xi < \omega$, $K\lambda[\xi] = K(F_{\alpha_0}^{(\xi+1)}(\lambda^-)) = K'\alpha \cup K\lambda^-$ and $K\xi = \emptyset$.

Further $K\lambda[\omega] = K\lambda = K'\alpha \cup K\lambda^- = K'\alpha \cup K\lambda^- \cup K\omega$.

2.2.2. $\alpha \in Lim$: For $1 \leq \xi < \tau_{\lambda} = \tau_{\alpha}$ we have $K\lambda[\xi] = KF_{\alpha[\xi]}(\lambda^-) = K\alpha[\xi] \cup \{\omega\} \cup K\lambda^- \overset{\mathrm{IH}}{=} K\alpha[1] \cup \{\omega\} \cup K\lambda^- \cup K\xi$.

Further $K\lambda = K\alpha \cup \{\omega\} \cup K\lambda^- \overset{\mathrm{IH}}{=} K\alpha[1] \cup \{\omega\} \cup K\lambda^- \cup K\tau_{\alpha}$.

3. $\lambda = \Lambda$: For $1 \leq \xi \leq \omega$ we have $K\Lambda[\xi] = \{\omega\}$, whence $K\Lambda[\xi] = K\Lambda[1] \cup K\xi$.

Lemma 3.2 *(a)* $\alpha \in Lim\ \&\ \alpha[\xi] \leq \delta \leq \alpha[\xi{+}1] \Rightarrow K\alpha[\xi] \subseteq K\delta$.
(b) $\tau_{\alpha} = \Omega\ \&\ \delta < \alpha\ \&\ K\delta \subseteq \beta \in Lim \Rightarrow \exists\xi < \beta(\delta < \alpha[\xi])$.

Proof (a) Induction on δ:

1. $\alpha[\xi] = \delta$: trivial.

2. $\alpha[\xi] < \delta$:

Then, by Lemma 2.2d, $\alpha[\xi] \leq \delta[1] < \alpha[\xi{+}1]$. Hence $K\alpha[\xi] \overset{\mathrm{IH}}{\subseteq} K\delta[1] \overset{\mathrm{L.3.1a}}{\subseteq} K\delta$.

(b) Assume $\alpha[0] \leq \delta$. Then by Lemma 2.2a, b, c there exists $\zeta < \tau_\alpha$ such that $\alpha[\zeta] \leq \delta < \alpha[\zeta{+}1]$. By (a) and Lemma 3.1b we get $K\zeta \subseteq K\alpha[\zeta] \subseteq K\delta \subseteq \beta$. Hence $\delta < \alpha[\zeta{+}1]$ with $\zeta{+}1 < \beta$, since $\beta \in Lim$.

Definition. $\mathsf{k}(\alpha) := \max(\{0\} \cup K\alpha)$. $\mathsf{k}^+(\alpha) := \max\{\mathsf{k}(\alpha[1]){+}1, \mathsf{k}(\alpha)\}$.

Remark. As mentioned above, $K\alpha$ corresponds to $C(\alpha)$ of [Ge67]. Moreover, $K\alpha[1]$ corresponds to $ND(\alpha)$. Therefore, the condition $\mathsf{k}^+(\alpha) \leq \beta$ corresponds to $C(\alpha) \subseteq \beta{+}1$ & $ND(\alpha) \subseteq \beta$ in (G).

Lemma 3.3 *(a)* $\mathsf{k}(\alpha) = \max\{\mathsf{k}(\alpha[1]), \mathsf{k}(\tau_\alpha)\}$ *and*
$\mathsf{k}^+(\alpha) = \max\{\mathsf{k}(\alpha[1]) + 1, \mathsf{k}(\tau_\alpha)\}$.

(b) $\mathsf{k}^+(\alpha) \leq \mathsf{k}(\alpha) + 1 = \mathsf{k}^+(\alpha{+}1)$.

(c) $\tau_\alpha^\beta \leq \max\{\mathsf{k}(\alpha), \beta\}$.

(d) $\mathsf{k}^+(\alpha) \leq \phi_\alpha(0)$.

Proof (a) The first equation follows from Lemma 3.1a. The second follows from the first.

(b) By (a), $\mathsf{k}(\alpha[1]) \leq \mathsf{k}(\alpha)$, whence $\mathsf{k}^+(\alpha) \leq \mathsf{k}(\alpha) + 1 = \max\{\mathsf{k}(\alpha) + 1, \mathsf{k}(\alpha)\} = \mathsf{k}^+(\alpha{+}1)$.

(c) If $\tau_\alpha < \Omega$ then $\tau_\alpha^\beta = \tau_\alpha \overset{(a)}{\leq} \mathsf{k}(\alpha)$. Otherwise $\tau_\alpha^\beta = \beta$.

(d) Induction on α:

1. $\mathsf{k}^+(0) = 1 \leq \phi_0(0)$.

2. $\alpha > 0$: $\mathsf{k}(\alpha[1]) \leq \mathsf{k}^+(\alpha[1]) \overset{\text{IH}}{\leq} \phi_{\alpha[1]}(0) \overset{\text{L.2.3e}}{<} \phi_\alpha(0)$ and $\mathsf{k}(\tau_\alpha) \overset{\text{L.2.3f}}{\leq} \phi_\alpha(0)$.

Lemma 3.4 $\mathsf{k}^+(\alpha) \leq \beta$ & $\xi < \tau_\alpha^\beta \Rightarrow \mathsf{k}^+(\alpha[\xi]) \leq \beta$.

Proof $\mathsf{k}^+(\alpha[\xi]) = \max\{\mathsf{k}(\alpha[\xi][1]) + 1, \mathsf{k}(\alpha[\xi])\} \overset{\text{L.3.3a}}{\leq} \mathsf{k}(\alpha[\xi]) + 1 \overset{\text{L.3.1b}}{\leq}$

$\leq \max\{\mathsf{k}(\alpha[1]), \mathsf{k}(\xi)\} + 1 \leq \max\{\mathsf{k}^+(\alpha), \tau_\alpha^\beta\} \overset{\text{L.3.3c}}{\leq} \beta$.

Lemma 3.5 $\mathsf{k}^+(\alpha) \leq \beta \in Lim \Rightarrow (\delta <_\beta \alpha \Leftrightarrow \delta < \alpha$ & $K\delta \subseteq \beta)$.

Proof by induction on α: Assume $\mathsf{k}^+(\alpha) \leq \beta \in Lim$. Then by Lemma 3.4, $\forall \xi < \tau_\alpha^\beta(\mathsf{k}^+(\alpha[\xi]) \leq \beta)$ $(*)$.

"$\Rightarrow$": From $\delta <_\beta \alpha$ we get $\delta \leq_\beta \alpha[\xi]$ for some $\xi < \tau_\alpha^\beta$.

Case $\delta = \alpha[\xi]$: $\mathsf{k}(\alpha[1]) < \mathsf{k}^+(\alpha) \leq \beta$ & $\xi < \tau_\alpha^\beta \Rightarrow$
$\mathsf{k}(\delta) \overset{\text{L.3.1b}}{\leq} \max\{\mathsf{k}(\alpha[1]), \mathsf{k}(\xi)\} \overset{\text{L.3.3c}}{<} \beta$.

Case $\delta <_\beta \alpha[\xi]$: Then the claim follows by IH and $(*)$.

"$\Leftarrow$": Assume $\delta < \alpha$ & $K\delta \subseteq \beta$.

(1) There exists $\xi < \tau_\alpha^\beta$ such that $\delta \leq \alpha[\xi]$.

Proof of (1): If $\tau_\alpha < \Omega$, then $\tau_\alpha^\beta = \tau_\alpha$ and $\exists \xi < \tau_\alpha(\delta \leq \alpha[\xi])$. If $\tau_\alpha = \Omega$, then $\tau_\alpha^\beta = \beta$ and from $\delta < \alpha$ & $K\delta \subseteq \beta \in Lim$ we obtain $\exists \xi < \beta(\delta < \alpha[\xi])$ by Lemma 3.2b.

From $\delta \leq \alpha[\xi]$ we obtain $\delta = \alpha[\xi]$ or, by IH, $\delta <_\beta \alpha[\xi]$. Together with $\alpha[\xi] <_\beta^1 \alpha$ this yields $\delta <_\beta \alpha$.

Theorem 3.6 $\beta \in R_\alpha \Leftrightarrow \mathsf{k}^+(\alpha) \le \beta \in R_0 \;\&\; \forall \delta < \alpha(K\delta \subseteq \beta \to \phi_\delta(\beta) = \beta)$

Proof By Theorem 2.4b and Lemma 3.3d we have

$\beta \in R_\alpha \Leftrightarrow \mathsf{k}^+(\alpha) \le \beta \in R_0 \;\&\; \forall \delta <_\beta \alpha(\phi_\delta(\beta) = \beta)$.

From this the claim follows by Lemma 3.5.

The Fixed-Point-Free Functions $\overline{\phi}_\alpha$

Definition.

$\overline{\phi}\alpha\beta := \phi\alpha(\beta + \tilde{\iota}\alpha\beta)$ where $\tilde{\iota}\alpha\beta := \begin{cases} 1 \text{ if } \beta = \beta_0 + n \text{ with } \phi\alpha\beta_0 \in K\alpha \cup \{\beta_0\} \\ 0 \text{ otherwise} \end{cases}$

$\overline{R}_\alpha := \mathrm{ran}(\overline{\phi}_\alpha)$

Notation. From now on we mostly write $\phi\alpha\beta, \overline{\phi}\alpha\beta$ for $\phi_\alpha(\beta), \overline{\phi}_\alpha(\beta)$.

Theorem 3.7 *(a)* $\overline{\phi}_\alpha$ *is order preserving.*
(b) $\overline{R}_\alpha = \{\phi\alpha\beta : K\alpha \cup \{\beta\} \subseteq \phi\alpha\beta\} = \{\gamma \in R_\alpha : K\alpha \subseteq \gamma < \phi\alpha\gamma\}$.
(c) $\overline{\phi}\alpha\beta = \min\{\gamma \in R_\alpha : \forall \eta < \beta(\overline{\phi}\alpha\eta < \gamma) \;\&\; K\alpha \cup \{\beta\} \subseteq \gamma\}$

Proof (a) If $\beta_1 < \beta_2$ then $\beta_1 + \tilde{\iota}\alpha\beta_1 < \beta_2$ or $\beta_1 + \tilde{\iota}\alpha\beta_1 = \beta_2$.

In the latter case $\tilde{\iota}\alpha\beta_2 = \tilde{\iota}\alpha\beta_1 = 1$.

(b) Due to Lemma 3.3d, $\mathsf{k}(\alpha) \le \phi\alpha 0$. Hence (1) $K\alpha \cup \{\beta\} \subseteq \phi\alpha\beta$ and (2) $\phi\alpha\beta \in K\alpha \to \beta = 0$. Further (3) $\phi\alpha\beta = \beta \to \beta \in Lim$. By (2), (3) and the definition of $\overline{\phi}_\alpha$ we get $\overline{R}_\alpha = R_\alpha \setminus \{\phi\alpha\beta : \phi\alpha\beta \in K\alpha \cup \{\beta\}\}$. Together with (1) this yields the first equation. The second equation is an immediate consequence of the first, since $\gamma = \phi\alpha\beta$ implies $(\gamma < \phi\alpha\gamma \Leftrightarrow \beta < \phi\alpha\beta)$.

(c) Let $X := \{\gamma \in R_\alpha : \forall \eta < \beta(\overline{\phi}\alpha\eta < \gamma) \;\&\; K\alpha \cup \{\beta\} \subseteq \gamma\}$.

From (a), (b) and the definition of $\overline{\phi}_\alpha$ we get $\overline{\phi}\alpha\beta \in X$.

It remains to prove $\forall \gamma \in X(\overline{\phi}\alpha\beta \le \gamma)$. Let $\gamma \in X$.

Case $\gamma < \phi\alpha\gamma$: Then $\gamma \in \overline{R}_\alpha$ and thus $\overline{\phi}\alpha\beta \le \gamma$, since $\forall \eta < \beta(\overline{\phi}\alpha\eta < \gamma)$.

Case $\gamma = \phi\alpha\gamma$: Then $\gamma \in Lim$ and $\beta < \gamma$, since $\gamma \in X$.

Hence $\overline{\phi}\alpha\beta \le \phi\alpha(\beta{+}1) < \phi\alpha\gamma = \gamma$.

Lemma 3.8 *(a)* $\xi < \alpha \;\&\; K\xi \cup \{\eta\} \subseteq \overline{\phi}\alpha\beta \Rightarrow \overline{\phi}\xi\eta < \overline{\phi}\alpha\beta$.
(b) $K\alpha \cup \{\beta\} \subseteq \overline{\phi}\alpha\beta$.

Proof (a) $\xi < \alpha \;\&\; K\xi \cup \{\eta\} \subseteq \overline{\phi}\alpha\beta \in R_\alpha \Rightarrow \overline{\phi}\xi\eta \le \phi\xi(\eta{+}1) < \phi\xi\overline{\phi}\alpha\beta \overset{\text{Th.3.6}}{=} \overline{\phi}\alpha\beta$.

(b) follows immediately from Theorem 3.7c.

Lemma 3.9 *Let* $\gamma_i = \overline{\phi}\alpha_i\beta_i$ *(*$i = 1, 2$*).*

(a) $\gamma_1 < \gamma_2$ *if, and only if, one of the following holds:*

(i) $\alpha_1 < \alpha_2 \;\&\; K\alpha_1 \cup \{\beta_1\} \subseteq \gamma_2$*;*

(ii) $\alpha_1 = \alpha_2 \;\&\; \beta_1 < \beta_2$*;*

(iii) $\alpha_2 < \alpha_1$ & $K\alpha_2 \cup \{\beta_2\} \not\subseteq \gamma_1$.

(b) $\gamma_1 = \gamma_2 \Rightarrow \alpha_1 = \alpha_2$ & $\beta_1 = \beta_2$.

Proof (a) Let $Q(\alpha_1, \beta_1, \alpha_2, \beta_2) :\equiv (i) \vee (ii) \vee (iii)$.
To prove: $\gamma_1 < \gamma_2 \Leftrightarrow Q(\alpha_1, \beta_1, \alpha_2, \beta_2)$.
From Theorem 3.7a and Lemma 3.8 we get the implications
(1) $Q(\alpha_1, \beta_1, \alpha_2, \beta_2) \Rightarrow \gamma_1 < \gamma_2$ and (2) $Q(\alpha_2, \beta_2, \alpha_1, \beta_1) \Rightarrow \gamma_2 < \gamma_1$.
Obviously, (3) $\neg Q(\alpha_1, \beta_1, \alpha_2, \beta_2) \Rightarrow Q(\alpha_2, \beta_2, \alpha_1, \beta_1) \vee (\alpha_1 = \alpha_2$ & $\beta_1 = \beta_2)$.
From (2) and (3) we get: $\neg Q(\alpha_1, \beta_1, \alpha_2, \beta_2) \Rightarrow \neg(\gamma_1 < \gamma_2)$.
(b) It suffices to prove: $\gamma_1 = \gamma_2 \Rightarrow \alpha_1 = \alpha_2$. So we assume $\gamma_1 = \gamma_2$ and $\alpha_1 \neq \alpha_2$. W.l.o.g. we may assume $\alpha_1 < \alpha_2$. Then by Lemma 3.8b we have $\alpha_1 < \alpha_2$ & $K\alpha_1 \cup \{\beta_1\} \subseteq \gamma_2$. Hence $\gamma_1 < \gamma_2$ by Lemma 3.8a. Contradiction.

Lemma 3.10 *For each* $\gamma \in R_0 \cap \phi_\Lambda(0)$ *there exists* $\alpha < \Lambda$ *such that* $\gamma \in \overline{R}_\alpha$.

Proof Assume $\omega < \gamma$. Then $K\Lambda \subseteq \gamma \notin R_\Lambda$. Let α_1 be the least ordinal such that $K\alpha_1 \subseteq \gamma \notin R_{\alpha_1}$. Then by Theorem 3.6 and Lemma 3.3b there exists $\alpha < \alpha_1$ such that $K\alpha \subseteq \gamma < \phi\alpha\gamma$. By minimality of α_1 we get $\gamma \in R_\alpha$. Hence $\gamma \in \overline{R}_\alpha$ by Theorem 3.7b.

The following will prove useful in Sect. 5.

Theorem 3.11 *Let* $\overline{\phi}\langle\Omega\alpha + \beta\rangle := \overline{\phi}\alpha\beta$ $(\alpha \leq \Lambda, \beta < \Omega)$. *Then for all* $\alpha < \Lambda + \Omega$,

$$\overline{\phi}\langle\alpha\rangle = \min\{\gamma \in R_0 : \forall\xi < \alpha(K\xi \subseteq \gamma \rightarrow \overline{\phi}\langle\xi\rangle < \gamma) \;\&\; K\alpha \subseteq \gamma\}$$

Proof $\overline{\phi}\langle\Omega\alpha + \beta\rangle = \overline{\phi}\alpha\beta \overset{\text{Th.3.7c}}{=} \min\{\gamma \in R_\alpha : \forall\eta < \beta(\overline{\phi}\alpha\eta < \gamma) \;\&\; K\alpha \cup \{\beta\} \subseteq \gamma\} \overset{\text{Th.3.6}}{=} \min\{\gamma \in R_0 : \forall\xi < \alpha(K\xi \subseteq \gamma \rightarrow \forall\eta < \gamma(\overline{\phi}\xi\eta < \gamma)) \;\&$
$\forall\eta < \beta(\overline{\phi}\alpha\eta < \gamma) \;\&\; K\alpha \cup \{\beta\} \subseteq \gamma\} \overset{(*)}{=}$
$\min\{\gamma \in R_0 : \forall\xi < \alpha\forall\eta(K\xi \cup K\eta \subseteq \gamma \rightarrow \overline{\phi}\langle\Omega\xi + \eta\rangle < \gamma) \;\&$
$\forall\eta < \beta(K\alpha \cup K\eta \subseteq \gamma \rightarrow \overline{\phi}\langle\Omega\alpha + \eta\rangle < \gamma) \;\&\; K\alpha \cup K\beta \subseteq \gamma\} =$
$\min\{\gamma \in R_0 : \forall\zeta < \Omega\alpha{+}\beta(K\zeta \subseteq \gamma \rightarrow \overline{\phi}\langle\zeta\rangle < \gamma) \;\&\; K(\Omega\alpha + \beta) \subseteq \gamma\}$.

$(*)$ For $\alpha = \beta = 0$ the equation is trivial. Otherwise it follows from the fact that for $1 < \gamma \in R_0$ we have $\forall\eta < \Omega(K\eta \subseteq \gamma \Leftrightarrow \eta < \gamma)$.

4 Comparison of $\phi_\alpha, \overline{\phi}_\alpha$ with $\theta_\alpha, \overline{\theta}_\alpha$

In this section we will compare the Bachmann functions ϕ_α with Feferman's θ_α. We will prove that $\phi\alpha\beta = \theta\alpha(\widehat{\alpha} + \beta)$ for all $\alpha \leq \Lambda$, $\beta < \Omega$, where $\widehat{\alpha} := \min\{\eta : \mathsf{k}^+(\alpha) \leq \theta\alpha\eta\}$. This result is already stated in [1, Theorem 3][1] and, for $\alpha < \varepsilon_{\Omega+1}$, proved in [28].

[1]Actually Aczel's Theorem 3 looks somewhat different, but it implies the above formulated result. A proof of Theorem 3 can be extracted from the proof of Theorem 3.5 in [4].

Notation: $\theta\alpha\beta := \theta_\alpha(\beta)$.

Basic Properties of the Functions θ_α
($\theta 1$) $\theta_\alpha : On \to On$ is a normal function with $\mathrm{ran}(\theta_\alpha) = \mathrm{In}_\alpha$.
($\theta 2$) (i) $\mathrm{In}_0 = \mathbb{H}$;
(ii) $\mathrm{In}_{\alpha+1} = \{\beta \in \mathrm{In}_\alpha : \alpha \in C(\alpha,\beta) \to \theta\alpha\beta = \beta\}$;
(iii) $\mathrm{In}_\alpha = \bigcap_{\xi<\alpha} \mathrm{In}_\xi$ if $\alpha \in Lim$.
($\theta 3$) $\mathrm{In}_\alpha \cap \Omega = \{\beta \in \Omega : C(\alpha,\beta) \cap \Omega \subseteq \beta\}$
($\theta 4$) $\theta_\alpha(\Omega) = \Omega \in C(\alpha,\beta)$.
(For proofs cf. [4, p. 174], [6, Sect. 1], [25, Sect. 24]).

Recall from the introduction that $C(\alpha,\beta)$ is the closure of $\beta \cup \{0\}$ under $+$, $\lambda\xi.\Omega_{1+\xi}$ and $\theta|\alpha$.

Our first goal is to prove that '$\alpha \in C(\alpha,\beta)$' in ($\theta 2$)(ii) can be replaced by '$K\alpha \subseteq \beta$' (cf. Lemma 4.3c). Once this is established we can easily show that $\mathrm{In}_\alpha = \{\beta \in \mathrm{In}_0 : \forall\delta < \alpha(K\delta \subseteq \beta \to \theta\delta\beta = \beta)\}$ (Theorem 4.4). Together with Theorem 3.6 this yields an exact comparison of ϕ_α and θ_α (Theorem 4.6).

Lemma 4.1 *(a)* $\alpha < \theta\alpha(\Omega{+}1)\ \&\ \Omega \le \beta \Rightarrow (\beta \in \mathrm{In}_{\alpha+1} \Leftrightarrow \theta\alpha\beta = \beta)$.
(b) $0 < \alpha \le \Lambda \Rightarrow F_\alpha(\beta) = \theta\alpha(\Omega + 1 + \beta)$.

Proof (a) "$\Leftarrow$": immediate consequence of ($\theta 2ii$).
"$\Rightarrow$": Assume $\beta \in \mathrm{In}_\alpha$ and $(\alpha \in C(\alpha,\beta) \to \theta\alpha\beta = \beta)$. For $\beta = \Omega$ the claim follows directly from ($\theta 4$). Otherwise: $\theta\alpha\Omega \overset{(\theta 4)}{=} \Omega < \beta \in \mathrm{In}_\alpha \Rightarrow \theta\alpha(\Omega{+}1) \le \beta \Rightarrow \alpha < \beta \Rightarrow \alpha \in C(\alpha,\beta) \Rightarrow \theta\alpha\beta = \beta$.
(b) Let $J := \{\beta : \Omega < \beta\}$. We prove $\mathrm{ran}(F_\alpha) = \mathrm{In}_\alpha \cap J$ which is equivalent to the claim $\forall\beta(F_\alpha(\beta) = \theta\alpha(\Omega + 1 + \beta))$.
The proof proceeds by induction on α.
1. $\alpha = 1$: $\mathrm{ran}(F_1) = \{\beta : \Omega^\beta = \beta\} = \{\beta : \Omega < \omega^\beta = \beta\} \overset{(\theta 2)}{=} \mathrm{In}_1 \cap J$.
2. $\alpha = \alpha_0{+}1 > 1$: $\mathrm{ran}(F_\alpha) = \{\beta : F_{\alpha_0}(\beta) = \beta\} \overset{\mathrm{IH}}{=}$
$= \{\beta : \theta\alpha_0(\Omega{+}1{+}\beta) = \beta\} = \{\beta : \Omega < \theta\alpha_0\beta = \beta\} \overset{(*)}{=} \mathrm{In}_\alpha \cap J$.
$(*)$ $\alpha_0 < \Lambda \Rightarrow \alpha_0 < F_{\alpha_0}(0) \overset{\mathrm{IH}}{=} \theta_{\alpha_0}(\Omega{+}1) \overset{(a)}{\Rightarrow} \forall\beta > \Omega(\theta\alpha_0\beta = \beta \Leftrightarrow \beta \in \mathrm{In}_\alpha)$.
3. $\alpha \in Lim$: $\mathrm{ran}(F_\alpha) = \bigcap_{\xi<\alpha} \mathrm{ran}(F_\xi) \overset{\mathrm{IH}}{=} \bigcap_{\xi<\alpha} \mathrm{In}_\xi \cap J \overset{(\theta 2iii)}{=} \mathrm{In}_\alpha \cap J$.

Definition $E_\Omega(\alpha) = \begin{cases} \emptyset & \text{if } \alpha \in \{0,\Omega\} \\ \{\alpha\} & \text{if } \alpha \in \mathbb{E} \setminus \{\Omega\} \\ \bigcup_{i\le n} E_\Omega(\alpha_i) & \text{if } \alpha = \omega^{\alpha_0}\#\ldots\#\omega^{\alpha_n} \notin \mathbb{E} \end{cases}$

Recall that $\mathbb{E} = \{\alpha : \omega^\alpha = \alpha\}$.

Lemma 4.2 *(a)* $E_\Omega(\Omega + \alpha) = E_\Omega(\Omega{\cdot}\alpha) = E_\Omega(\Omega^\alpha) = E_\Omega(\alpha)$.
(b) $\alpha =_{\mathrm{NF}} \gamma + \Omega^\beta\eta \Rightarrow E_\Omega(\alpha) = E_\Omega(\gamma) \cup E_\Omega(\beta) \cup E_\Omega(\eta)$.
(c) *If* $0 < \alpha$, *then* $\forall\gamma(\gamma \in C(\alpha,\beta) \Leftrightarrow E_\Omega(\gamma) \subseteq C(\alpha,\beta))$.
(d) $\alpha < \varepsilon_{\Omega+1}\ \&\ \delta \in \mathbb{E} \Rightarrow (E_\Omega(\alpha) \subseteq \delta \Leftrightarrow K\alpha \subseteq \delta)$.

Proof (a) Let $\alpha = \omega^{\alpha_0} + \cdots + \omega^{\alpha_n}$ with $\alpha_0 \geq \cdots \geq \alpha_n$.

1. $E_\Omega(\Omega + \alpha) = \begin{cases} E_\Omega(\alpha) & \text{if } \Omega < \alpha_0 \\ E_\Omega(\Omega) \cup E_\Omega(\alpha) & \text{if } \Omega \geq \alpha_0 \end{cases}$

2. $E_\Omega(\Omega{\cdot}\alpha) = E_\Omega(\omega^{\Omega+\alpha_0} + \cdots + \omega^{\Omega+\alpha_n}) = \bigcup_{i\leq n} E_\Omega(\Omega + \alpha_i) \overset{1.}{=} \bigcup_{i\leq n} E_\Omega(\alpha_i)$ $= E_\Omega(\alpha)$.

3. $E_\Omega(\Omega^\alpha) = E_\Omega(\omega^{\Omega\cdot\alpha}) = E_\Omega(\Omega{\cdot}\alpha) \overset{2.}{=} E_\Omega(\alpha)$.

(b) Let $\eta = \omega^{\eta_0} + \cdots + \omega^{\eta_m}$ with $\eta_0 \geq \cdots \geq \eta_m$. Then

$\Omega^\beta\eta = \omega^{\Omega\cdot\beta} \cdot (\omega^{\eta_0} + \cdots + \omega^{\eta_m}) = \omega^{\Omega\beta+\eta_0} + \cdots + \omega^{\Omega\beta+\eta_m}$. Hence $E_\Omega(\Omega^\beta\eta) = \bigcup_{i\leq m} E_\Omega(\Omega\beta + \eta_i) = \bigcup_{i\leq m}(E_\Omega(\beta) \cup E_\Omega(\eta_i)) = E_\Omega(\beta) \cup \bigcup_{i\leq m} E_\Omega(\eta_i) = E_\Omega(\beta)$ $\cup\, E_\Omega(\eta)$ and thus $E_\Omega(\alpha) = E_\Omega(\gamma) \cup E_\Omega(\beta) \cup E_\Omega(\eta)$.

(c) "$\Rightarrow$" is shown by induction on the inductive generation of $C(\alpha, \beta)$.

"$\Leftarrow$" is shown by transfinite induction on γ.

For the proof observe that $\theta 0\eta = \omega^\eta$ and $\theta\xi\eta \in \mathbb{E}$ if $\xi > 0$.

(d) Induction on α:

1. $\alpha \in \{0, \Omega\}$: $E_\Omega(\alpha) = \emptyset = K\alpha$.

2. $\alpha < \Omega$: $E_\Omega(\alpha) \subseteq \delta \Leftrightarrow \alpha < \delta \Leftrightarrow K\alpha \subseteq \delta$.

3. $\Omega < \alpha =_{\mathrm{NF}} \gamma + \Omega^\beta\eta$: $E_\Omega(\alpha) \subseteq \delta \overset{(b)}{\Leftrightarrow} E_\Omega(\gamma) \cup E_\Omega(\beta) \cup E_\Omega(\eta) \subseteq \delta \overset{\mathrm{IH}}{\Leftrightarrow} K\gamma \cup K\beta \cup K\eta \subseteq \delta \Leftrightarrow K\alpha \subseteq \delta$.

Lemma 4.3 *For $\alpha < \Lambda$ we have:*

(a) $\xi < \alpha \;\&\; \eta < F_\xi(\eta) < \Lambda \;\Rightarrow\; (\xi, \eta \in C(\alpha, \beta) \;\Leftrightarrow\; F_\xi(\eta) \in C(\alpha, \beta))$.

(b) $\forall\delta \leq \alpha(\delta \in C(\alpha, \beta) \;\Leftrightarrow\; K\delta \subseteq C(\alpha, \beta))$.

(c) $\mathrm{In}_{\alpha+1} = \{\beta \in \mathrm{In}_\alpha : K\alpha \subseteq \beta \to \theta\alpha\beta = \beta\}$

Proof (a) For $\xi = 0$ the claim follows from Lemma 4.2a, c. Assume now $\xi > 0$ and let $\gamma := F_\xi(\eta)$. Then, by Lemma 4.1b, we have $\xi, \eta_1 < \gamma = \theta\xi\eta_1$ with $\eta_1 := \Omega{+}1{+}\eta$. By Lemma 4.2a,c we also have (†) $(\eta \in C(\alpha, \beta) \Leftrightarrow \eta_1 \in C(\alpha, \beta))$.

"$\Rightarrow$": From $\xi < \alpha \;\&\; \xi, \eta \in C(\alpha, \beta)$ and (†) it follows that $\gamma = \theta\xi\eta_1 \in C(\alpha, \beta)$.

"$\Leftarrow$" Assume $\gamma \in C(\alpha, \beta)$. If $\gamma < \beta$ then also $\xi, \eta < \beta$. Assume now $\beta \leq \gamma$. Then there exist $\xi_0, \eta_0 \in C(\alpha, \beta)$ with $\eta_0 < \gamma = \theta\xi_0\eta_0$ and $\xi_0 < \alpha$. Since $\gamma > \Omega$, there exists ζ such that $\eta_0 = \Omega + 1 + \zeta$. By Lemma 4.1b, $\theta\xi_0\eta_0 = F_{\xi_0}(\zeta)$. Hence $\gamma = F_\xi(\eta) = F_{\xi_0}(\zeta)$ with $\eta, \zeta < \gamma$. From this we conclude $\xi = \xi_0 \in C(\alpha, \beta)$ and $\eta = \zeta$, whence $\Omega{+}1{+}\eta = \eta_0 \in C(\alpha, \beta)$. The latter yields $\eta \in C(\alpha, \beta)$.

(b) Induction on δ: Assume $\delta \le \alpha$, and let $C := C(\alpha, \beta)$.

1. $\delta \in \{0, \Omega\}$: $\delta \subset C$ & $K\delta = \emptyset$.

2. $\delta = \delta_0{+}1$: $\delta \in C \Leftrightarrow \delta_0 \in C$, and $K\delta = K\delta_0$.

3. $\delta \in Lim \cap \Omega$: $K\delta = \{\delta\}$.

4. $\delta =_{\mathrm{NF}} \gamma + \Omega^\beta\eta \notin \mathrm{ran}(F_0)$:

$\delta \in C \overset{\mathrm{L.4.2c}}{\Leftrightarrow} E_\Omega(\delta) \subseteq C \overset{\mathrm{L.4.2b}}{\Leftrightarrow} E_\Omega(\gamma) \cup E_\Omega(\beta) \cup E_\Omega(\eta) \subseteq C \overset{\mathrm{L.4.2c}}{\Leftrightarrow} \gamma, \beta, \eta \in C \overset{\mathrm{IH}}{\Leftrightarrow} K\gamma \cup K\beta \cup K\eta \subseteq C \Leftrightarrow K\delta \subseteq C$.

5. $\delta =_{\mathrm{NF}} F\xi\eta$: $\delta \in C \overset{(a)}{\Leftrightarrow} \xi, \eta \in C \overset{\mathrm{IH}}{\Leftrightarrow} K\xi \cup K\eta \subseteq C \overset{(*)}{\Leftrightarrow} K\delta \subseteq C$.

$(*)$ $\omega = \theta 01 \in C$.

(c) follows from ($\theta 2ii$), ($\theta 3$), (b) and the fact that $K\alpha \subseteq \Omega$.

Theorem 4.4 $\alpha \le \Lambda \Rightarrow \mathrm{In}_\alpha = \{\beta \in \mathbb{H} : \forall\xi < \alpha(K\xi \subseteq \beta \rightarrow \theta\xi\beta = \beta)\}$.

Proof by induction on α: 1. $\alpha = 0$: By ($\theta 2i$) we have $\mathrm{In}_0 = \mathbb{H}$.

2. $\alpha = \alpha_0{+}1$: $\mathrm{In}_\alpha \overset{\mathrm{L.4.3c}}{=} \{\beta \in \mathrm{In}_{\alpha_0} : K\alpha_0 \subseteq \beta \rightarrow \theta\alpha_0\beta = \beta\} \overset{\mathrm{IH}}{=} \{\beta \in \mathbb{H} : \forall\xi < \alpha_0(K\xi \subseteq \beta \rightarrow \theta\xi\beta = \beta)$ & $(K\alpha_0 \subseteq \beta \rightarrow \theta\alpha_0\beta = \beta)\}$.

3. $\alpha \in Lim$: Then, by ($\theta 2iii$), $\mathrm{In}_\alpha = \bigcap_{\xi<\alpha} \mathrm{In}_\xi$ and the assertion follows immediately from the IH.

Definition. $\widehat{\alpha} := \min\{\eta : \mathsf{k}^+(\alpha) \le \theta\alpha\eta\}$.

Lemma 4.5 $\alpha \le \Lambda$ & $K\alpha \subseteq \theta\alpha\beta \Rightarrow (\theta\alpha(\widehat{\alpha} + \beta) = \beta \Leftrightarrow \theta\alpha\beta = \beta)$.

Proof "$\Rightarrow$": This follows from $\beta \le \theta\alpha\beta \le \theta\alpha(\widehat{\alpha} + \beta)$.

"$\Leftarrow$": If $K\alpha \subseteq \beta = \theta\alpha\beta$ then $\widehat{\alpha} \le \mathsf{k}^+(\alpha) \le \mathsf{k}(\alpha){+}1 < \beta \in \mathbb{H}$ and thus $\widehat{\alpha} + \beta = \beta$.

Theorem 4.6 *If* $\alpha \le \Lambda$, *then* $R_\alpha = \{\gamma \in \Omega : \mathsf{k}^+(\alpha) \le \gamma \in \mathrm{In}_\alpha\}$, *and thus* $\forall\beta < \Omega(\phi\alpha\beta = \theta\alpha(\widehat{\alpha} + \beta))$.

Proof by induction on α*: For* $\beta < \Omega$ *we have:*

$\beta \in R_\alpha \overset{\mathrm{Th.3.6}}{\Leftrightarrow} \mathsf{k}^+(\alpha) \le \beta \in \mathbb{H}$ & $\forall\xi < \alpha(K\xi \subseteq \beta \rightarrow \phi\xi\beta = \beta) \overset{\mathrm{IH+L.4.5}}{\Leftrightarrow} \mathsf{k}^+(\alpha) \le \beta \in \mathbb{H}$ & $\forall\xi < \alpha(K\xi \subseteq \beta \rightarrow \theta\xi\beta = \beta) \overset{\mathrm{Th.4.4}}{\Leftrightarrow} \mathsf{k}^+(\alpha) \le \beta \in \mathrm{In}_\alpha$.

The functions $\overline{\theta}_\alpha$. In [6] the fixed-point-free functions $\overline{\theta}_\alpha$ are introduced, which are more suitable for proof-theoretic applications than the θ_α's. By definition, $\overline{\theta}_\alpha$ is the $<$-isomorphism from $\{\eta \in On : S\mu(\alpha) \le \eta\}$ onto $\overline{\mathrm{In}}_\alpha$ where $\overline{\mathrm{In}}_\alpha := \mathrm{In}_\alpha \setminus \mathrm{In}_{\alpha+1}$, $\mu(\alpha) := \min\{\eta : \theta\alpha\eta \in \overline{\mathrm{In}}_\alpha\}$, $S\mu(\alpha) := \min\{\Omega_\xi : \mu(\alpha) < \Omega_{\xi+1}\}$. As we will show in a moment, $S\mu(\alpha) = 0$ for all $\alpha < \Lambda$, and therefore, $\overline{\theta}_\alpha$ is the ordering function of $\overline{\mathrm{In}}_\alpha$ for all $\alpha < \Lambda$.

Theorem 4.7 $\overline{\phi}\alpha\beta = \overline{\theta}\alpha\beta$ *for all* $\alpha < \Lambda$, $\beta < \Omega$.

Proof I. From $\forall\beta(\beta{+}1 < \theta\alpha(\beta{+}1))$, Lemma 4.3c and ($\theta 4$) we obtain $\forall\beta \in \Omega$ $(\mathsf{k}(\alpha) \leq \beta \rightarrow \theta\alpha(\beta{+}1) \in \overline{\mathrm{In}}_\alpha \cap \Omega)$. Since $\mathsf{k}(\alpha) < \Omega$, it follows that $S\mu(\alpha) = \Omega_0 = 0$ and $\overline{\mathrm{In}}_\alpha \cap \Omega$ is unbounded in Ω. This implies that $\overline{\theta}_\alpha \restriction \Omega$ is the ordering function of $\overline{\mathrm{In}}_\alpha \cap \Omega$.
II. By Theorem 3.7a, $\overline{\phi}_\alpha$ is the ordering function of $\overline{R}_\alpha$. So it remains to prove that $\overline{R}_\alpha = \overline{\mathrm{In}}_\alpha \cap \Omega$: $\overline{R}_\alpha \overset{\text{Th.3.7b}}{=} \{\gamma \in R_\alpha : K\alpha \subseteq \gamma < \phi\alpha\gamma\} \overset{\text{Th.4.6}}{=} \{\gamma \in \mathrm{In}_\alpha \cap \Omega : K\alpha \subseteq \gamma < \theta\alpha(\widehat{\alpha} + \gamma)\} \overset{\text{L.4.5}}{=} \{\gamma \in \mathrm{In}_\alpha \cap \Omega : K\alpha \subseteq \gamma < \theta\alpha\gamma)\} \overset{\text{L.4.3c}}{=} \overline{\mathrm{In}}_\alpha \cap \Omega$.

5 The Unary Functions $\vartheta^{\mathbb{X}}$ and $\psi^{\mathbb{X}}$

As we have seen above, $\overline{\theta}_\alpha$ is the ordering function of $\overline{\mathrm{In}}_\alpha = \mathrm{In}_\alpha \setminus \mathrm{In}_{\alpha+1}$ (if $\alpha < \Lambda$). From this together with ($\theta 2ii$) and ($\theta 3$) one easily derives the following equation
(1) $\overline{\theta}\alpha 0 = \min\{\beta : C(\alpha, \beta) \cap \Omega \subseteq \beta \;\&\; \alpha \in C(\alpha, \beta)\}$
which motivates the definition of $\vartheta\alpha$ in [23]:
(2) $\vartheta\alpha := \min\{\beta : \tilde{C}(\alpha, \beta) \cap \Omega \subseteq \beta \;\&\; \alpha \in \tilde{C}(\alpha, \beta)\}$ $(\alpha < \varepsilon_{\Omega+1})$
where $\tilde{C}(\alpha, \beta)$ is the closure of $\{0, \Omega\} \cup \beta$ under $+$, $\lambda\xi.\omega^\xi$ and $\vartheta\restriction\alpha$.
On the other side, by Theorems 4.7, 3.11 we have:
(3) $\overline{\theta}\alpha 0 = \overline{\phi}\langle\Omega\alpha\rangle = \min\{\beta \in \mathbb{H} : \forall\xi < \Omega\alpha(K\xi \subseteq \beta \rightarrow \overline{\phi}\langle\xi\rangle < \beta) \;\&\; K\alpha \subseteq \beta\}$.

In the light of (1)–(3) the following theorem suggests itself.

Theorem 5.1 $\alpha < \varepsilon_{\Omega+1} \Rightarrow$
$\vartheta\alpha = \min\{\beta \in \mathbb{E} : \forall\xi < \alpha(K\xi \subseteq \beta \rightarrow \vartheta\xi < \beta) \;\&\; K\alpha \subseteq \beta\}$.

Proof I. From [23, 1.1 and 1.2(1)–(4)] we obtain
$\vartheta\alpha \in \mathbb{E} \;\&\; \forall\xi < \alpha(E_\Omega(\xi) \subseteq \vartheta\alpha \rightarrow \vartheta\xi < \vartheta\alpha) \;\&\; E_\Omega(\alpha) \subseteq \vartheta\alpha$.
II. Assume $\beta \in \mathbb{E} \;\&\; \forall\xi < \alpha(E_\Omega(\xi) \subseteq \beta \rightarrow \vartheta\xi < \beta) \;\&\; E_\Omega(\alpha) \subseteq \beta$.
We will prove that $\vartheta\alpha \leq \beta$.
For this let $Q := \{\gamma : E_\Omega(\gamma) \subseteq \beta\}$. Since $\beta \in \mathbb{E}$, we have $Q \subseteq \beta$. Moreover, as one easily sees, $\{0, \Omega\} \subseteq Q$ and Q is closed under $+$, $\lambda\xi.\omega^\xi$ and $\vartheta\restriction\alpha$. Hence $\tilde{C}(\alpha, \beta) \subseteq Q$ and thus $\tilde{C}(\alpha, \beta) \cap \Omega \subseteq Q \cap \Omega \subseteq \beta$. It remains to show that $\alpha \in \tilde{C}(\alpha, \beta)$. But this follows immediately from $E_\Omega(\alpha) \subseteq \beta \subseteq \tilde{C}(\alpha, \beta)$ and [23, 1.2(4)].
From I. and II. we get
$\vartheta\alpha = \min\{\beta \in \mathbb{E} : \forall\xi < \alpha(E_\Omega(\xi) \subseteq \beta \rightarrow \vartheta\xi < \beta) \;\&\; E_\Omega(\alpha) \subseteq \beta\}$,
which together with Lemma 4.2d yields the claim.

Relativization

Comparing the recursion equations for $\vartheta\alpha$ and $\overline{\phi}\langle\alpha\rangle$ in Theorems 5.1, 3.11 one notices that these equations are almost identical. The only difference is that in the equation for $\vartheta\alpha$ there appears $\mathbb{E}$ where in the equation for $\overline{\phi}\langle\alpha\rangle$ we have R_0 (i.e. $\mathbb{H}$). In order to establish the exact relationship between ϑ and $\overline{\phi}$ we go back to the definition of the Bachmann hierarchy in Sect. 2 and replace the initial clause

"$R_0 := \mathbb{H} \cap \Omega$" of this definition by "$R_0 := \mathbb{X} \cap \Omega$" where here and in the sequel $\mathbb{X}$ *always denotes a subclass of* $\{1\} \cup Lim$ *such that* $\mathbb{X} \cap \Omega$ *is* Ω*-club*. Then the whole of Sects. 2, 3 remains valid as it stands. To make the dependency on $\mathbb{X}$ visible we write $R^{\mathbb{X}}_\alpha$, $\overline{R}^{\mathbb{X}}_\alpha$, $\phi^{\mathbb{X}}_\alpha$, $\overline{\phi}^{\mathbb{X}}_\alpha$, $\phi^{\mathbb{X}}\langle\alpha\rangle$, $\overline{\phi}^{\mathbb{X}}\langle\alpha\rangle$ instead of R_α, $\overline{R}_\alpha$, ...

Remark Theorems 5.1, 3.11 yield $\vartheta\alpha = \overline{\phi}^{\mathbb{E}}\langle\alpha\rangle$ and $\vartheta(\Omega\alpha + \beta) = \overline{\phi}^{\mathbb{E}}_\alpha(\beta)$ ($\alpha < \varepsilon_{\Omega+1}$, $\beta < \Omega$).

The previous explanations motivate the following definition.

Definition.
$\vartheta^{\mathbb{X}}\alpha := \min\{\beta \in \mathbb{X} : \forall\xi < \alpha(K\xi \subseteq \beta \to \vartheta^{\mathbb{X}}\xi < \beta)\ \&\ K\alpha \subseteq \beta\}$ ($\alpha \leq \Lambda$).

Theorem 5.1 now reads: $\vartheta\alpha = \vartheta^{\mathbb{E}}\alpha$ for $\alpha < \varepsilon_{\Omega+1}$. Further, by Theorem 3.11 we have

($\vartheta 0$) $\vartheta^{\mathbb{X}}(\Omega\alpha + \beta) = \overline{\phi}^{\mathbb{X}}_\alpha(\beta)$, if $\beta < \Omega$.

Therefore, properties of $\vartheta^{\mathbb{X}}$ can be proved by deriving them from corresponding properties of $\overline{\phi}$. But for various reasons it is also advisable to work directly from the above definition.

Let us first mention that for $\beta < \Omega$ the set $\{\xi < \alpha : K\xi \subseteq \beta\}$ is countable too, and therefore $\vartheta^{\mathbb{X}}\alpha < \Omega$. Moreover, directly from the definition of $\vartheta^{\mathbb{X}}$ we obtain:

($\vartheta 1$) $K\alpha \subseteq \vartheta^{\mathbb{X}}\alpha \in \mathbb{X}$,

($\vartheta 2$) $\alpha_0 < \alpha\ \&\ K\alpha_0 \subseteq \vartheta^{\mathbb{X}}\alpha \Rightarrow \vartheta^{\mathbb{X}}\alpha_0 < \vartheta^{\mathbb{X}}\alpha$,

($\vartheta 3$) $\beta \in \mathbb{X}\ \&\ K\alpha \subseteq \beta < \vartheta^{\mathbb{X}}\alpha \Rightarrow \exists\xi < \alpha(K\xi \subseteq \beta \subseteq \vartheta^{\mathbb{X}}\xi)$,

and then

($\vartheta 4$) $\vartheta^{\mathbb{X}}\alpha_0 = \vartheta^{\mathbb{X}}\alpha_1 \Rightarrow \alpha_0 = \alpha_1$ [from ($\vartheta 1$), ($\vartheta 2$)],

($\vartheta 5$) $\beta \in \mathbb{X}\ \&\ \beta < \vartheta^{\mathbb{X}}\Lambda \Rightarrow \exists\xi < \Lambda(\beta = \vartheta^{\mathbb{X}}\xi)$.

Proof of ($\vartheta 5$): If $\beta \leq \omega$ then $\beta \in \{\vartheta 0, \vartheta 1\}$. Otherwise we have $K\Lambda \subseteq \beta < \vartheta^{\mathbb{X}}\Lambda$, and the assertion follows by transfinite induction from ($\vartheta 3$).

Note on Klammersymbols. As we mentioned above, Sects. 2, 3 remain valid if ϕ is replaced by $\phi^{\mathbb{X}}$. So by Theorem 2.5, for $A = \begin{pmatrix}\xi_0 & \cdots & \xi_n \\ \alpha_0 & \cdots & \alpha_n\end{pmatrix}$ and $\alpha = \Omega^{\alpha_n}\xi_n + \cdots + \Omega^{\alpha_0}\xi_0$ we have $\phi^{\mathbb{X}}_0 A = \phi^{\mathbb{X}}\langle\alpha\rangle$ from which one easily derives $\overline{\phi^{\mathbb{X}}_0} A = \overline{\phi}^{\mathbb{X}}\langle\alpha\rangle$,[2] whence (by Theorem 3.11) $\overline{\phi^{\mathbb{X}}_0} A = \vartheta^{\mathbb{X}}\alpha$. Via Theorem 5.1 this fits with Schütte's result $\overline{\phi^{\mathbb{E}}_0} A = \vartheta\alpha$ in [26].

The Function $\psi^{\mathbb{X}}$

In [8] (actually already in [7]) the author introduced the functions $\psi_\sigma : On \to \Omega_{\sigma+1}$ and proved, via an ordinal analysis of ID_ν, that $\psi_0\varepsilon_{\Omega_\nu+1} = \theta_{\varepsilon_{\Omega_\nu+1}}(0)$. In [11] ordinal analyses of several impredicative subsystems of 2nd order arithmetic are carried out by means of the ψ_σ's. The definition of ψ_σ in [11] differs in some minor respects from that in [8]; for example, $\lambda\xi.\omega^\xi$ is a basic function in [11] but not in [8]. In [23] Rathjen and Weiermann compare their ϑ with $\psi_0{\restriction}\varepsilon_{\Omega+1}$ from [11] which they abbreviate by ψ. In Sect. 6 we will present a refinement of this comparison which

[2] $\overline{\varphi}A$ is the 'fixed-point-free version' of φA defined in [24, Sect. 3].

is based on Schütte's definition of the Veblen function φ (below Γ_0) in terms of ψ, given in Sect. 7 of [11].

Similarly as Theorem 5.1 one can prove
$\psi\alpha = \min\{\beta \in \mathbb{E} : \forall\xi < \alpha(K\xi \subseteq \beta \to \psi\xi < \beta)\}$, for $\alpha < \varepsilon_{\Omega+1}$.
This motivates the following

Definition of $\psi^{\mathbb{X}}\alpha$ for $\alpha \leq \Lambda+1$
$\psi^{\mathbb{X}}\alpha := \min\{\beta \in \mathbb{X} : \forall\xi < \alpha(K\xi \subseteq \beta \to \psi^{\mathbb{X}}\xi < \beta)\}$.
For the rest of this section we assume $\mathbb{X}$ *to be fixed, and write* ϑ, ψ *for* $\vartheta^{\mathbb{X}}, \psi^{\mathbb{X}}$.

Remark Immediately from the definitions it follows that $\psi\alpha \leq \vartheta\alpha$.

Before turning to the announced exact comparison of ϑ and ψ, we prove a somewhat weaker (but still very useful) result which can be obtained with much less effort. This corresponds to [23, p. 64] which in turn stems from [9, 10].

Lemma 5.2 *For* $\alpha \leq \Lambda$.

(a) $\alpha_0 \leq \alpha \;\Rightarrow\; \psi\alpha_0 \leq \psi\alpha$.

(b) $\alpha_0 < \alpha \;\&\; K\alpha_0 \subseteq \psi\alpha \;\Rightarrow\; \psi\alpha_0 < \psi\alpha$.

(c) $\psi\alpha < \psi(\alpha+1) \;\Leftrightarrow\; K\alpha \subseteq \psi\alpha$.

(d) $\alpha \in Lim \;\Rightarrow\; \psi\alpha = \sup_{\xi<\alpha}\psi\xi$.

(e) $\psi\alpha = \min\{\gamma \in \mathbb{X} : \forall\xi < \alpha(K\xi \subseteq \psi\xi \to \psi\xi < \gamma)\}$.

Proof (a), (b) follow directly from the definition.

(c) "$\Rightarrow$": Assume $\neg(K\alpha \subseteq \psi\alpha)$. Then from $\psi\alpha \in \mathbb{X} \;\&\; \forall\xi < \alpha(K\xi \subseteq \psi\alpha \to \psi\xi < \psi\alpha)$ we conclude $\psi\alpha \in \mathbb{X} \;\&\; \forall\xi < \alpha+1(K\xi \subseteq \psi\alpha \to \psi\xi < \psi\alpha)$, and thus $\psi(\alpha+1) \leq \psi\alpha$.

"$\Leftarrow$": From $\alpha < \alpha+1 \;\&\; K\alpha \subseteq \psi\alpha \subseteq \psi(\alpha+1)$ we conclude $\psi\alpha < \psi(\alpha+1)$ by (b).

(d) By (a) we have $\gamma := \sup_{\xi<\alpha}\psi\xi \leq \psi\alpha$. Assume $\gamma < \psi\alpha$. Then $\gamma \in \mathbb{X} \cap \psi\alpha$, and therefore by definition of $\psi\alpha$ there exists $\xi < \alpha$ with $K\xi \subseteq \gamma \subseteq \psi\xi$. Hence by (c), $\exists\xi < \alpha(\gamma < \psi(\xi+1))$. Contradiction

(e) 1. We have $\psi\alpha \in \mathbb{X}$ and, by (a), (b), $\forall\xi < \alpha(K\xi \subseteq \psi\xi \to \psi\xi < \psi\alpha)$.

2. Notice that $(K\xi \subseteq \psi\xi \to \psi\xi < \gamma)$ implies $(K\xi \subseteq \gamma \to \psi\xi < \gamma)$. Therefore, if $\gamma \in \mathbb{X} \;\&\; \forall\xi < \alpha(K\xi \subseteq \psi\xi \to \psi\xi < \gamma)$ then $\gamma \in \mathbb{X} \;\&\; \forall\xi < \alpha(K\xi \subseteq \gamma \to \psi\xi < \gamma)$ which yields $\psi\alpha \leq \gamma$.

Definition. Let $\alpha \leq \Lambda$ with $K\alpha \subseteq \psi\Lambda$. Then by Lemma 5.2d there exists $\xi < \Lambda$ such that $K\alpha \subseteq \psi\xi$, and we can define

$\tilde{\mathsf{g}}(\alpha) := \min\{\xi < \Lambda : K\alpha \subseteq \psi\xi\}$,

$\mathsf{g}(\alpha) := \tilde{\mathsf{g}}(\alpha)\dot{-}1$, where $\beta\dot{-}1 := \begin{cases} \beta_0 \text{ if } \beta = \beta_0+1 \\ \beta \text{ otherwise} \end{cases}$,

$\mathsf{h}(\alpha) := \mathsf{g}(\alpha) + \Omega^{\alpha}$. (Note that $\mathsf{h}(\alpha) \leq \Lambda$.)

Lemma 5.3 *Assume* $\alpha \leq \Lambda \;\&\; K\alpha \subseteq \psi\Lambda$.

(a) $\psi 0 \leq \mathsf{k}(\alpha) \Rightarrow \psi \mathsf{g}(\alpha) \leq \mathsf{k}(\alpha) < \psi(\mathsf{g}(\alpha)+1)$.

(b) $K\mathsf{g}(\alpha) \subseteq \psi \mathsf{g}(\alpha)$.

(c) $K\mathsf{h}(\alpha) \subseteq \psi \mathsf{h}(\alpha)$.

(d) $\alpha_0 < \alpha \;\&\; K\alpha_0 \subseteq \psi \mathsf{h}(\alpha) \Rightarrow \psi \mathsf{h}(\alpha_0) < \psi \mathsf{h}(\alpha)$.

Proof (a) From $\psi 0 \leq \mathsf{k}(\alpha)$ and Lemma 5.2d it follows that $0 < \tilde{\mathsf{g}}(\alpha) \notin Lim$. Therefore $\tilde{\mathsf{g}}(\alpha) = \mathsf{g}(\alpha)+1$, which yields the assertion.
(b) Follows from (a) and Lemma 5.2c.
(c) $K(\mathsf{g}(\alpha) + \Omega^\alpha) \subseteq K\mathsf{g}(\alpha) \cup K\alpha \overset{(b),(a)}{\subseteq} \psi(\mathsf{g}(\alpha) + 1) \subseteq \psi(\mathsf{g}(\alpha) + \Omega^\alpha)$.
(d) From $\alpha_0 < \alpha \;\&\; K\alpha_0 \subseteq \psi\mathsf{h}(\alpha)$ by (a) we obtain $\alpha_0 < \alpha \;\&\; \mathsf{g}(\alpha_0) < \mathsf{h}(\alpha) = \mathsf{g}(\alpha) + \Omega^\alpha$ and then $\mathsf{h}(\alpha_0) = \mathsf{g}(\alpha_0) + \Omega^{\alpha_0} < \mathsf{h}(\alpha)$. This together with $K\mathsf{h}(\alpha_0) \subseteq \psi\mathsf{h}(\alpha_0)$ (cf. (c)) yields $\psi\mathsf{h}(\alpha_0) < \psi\mathsf{h}(\alpha)$ by Lemma 5.2a, b.

Theorem 5.4 $\alpha \leq \Lambda \;\&\; K\alpha \subseteq \psi\Lambda \Rightarrow \vartheta\alpha \leq \psi\mathsf{h}(\alpha)$.

Proof by induction on α: By Lemma 5.3a, d, $K\alpha \subseteq \psi\mathsf{h}(\alpha) \in \mathbb{X} \;\&\; \forall\xi < \alpha(K\xi \subseteq \psi\mathsf{h}(\alpha) \to \psi\mathsf{h}(\xi) < \psi\mathsf{h}(\alpha))$. Hence by IH, $K\alpha \subseteq \psi\mathsf{h}(\alpha) \in \mathbb{X} \;\&\; \forall\xi < \alpha(K\xi \subseteq \psi\, \mathsf{h}(\alpha) \to \vartheta\xi < \psi\mathsf{h}(\alpha))$ which yields $\vartheta\alpha \leq \psi\mathsf{h}(\alpha)$.

Corollary 5.5 *(a)* $\alpha = \Omega^\alpha \leq \Lambda \;\&\; K\alpha \subseteq \psi\alpha \Rightarrow \vartheta\alpha = \psi\alpha$.
(b) $\vartheta\varepsilon_{\Omega+1} = \psi\varepsilon_{\Omega+1} \;\&\; \vartheta\Lambda = \psi\Lambda$.

Proof (a) $K\alpha \subseteq \psi\alpha \;\&\; \alpha = \Omega^\alpha \Rightarrow \mathsf{g}(\alpha) < \alpha = \Omega^\alpha \Rightarrow \mathsf{h}(\alpha) = \mathsf{g}(\alpha) + \Omega^\alpha = \alpha \overset{\text{Th.5.4}}{\Rightarrow} \vartheta\alpha \leq \psi\alpha \leq \vartheta\alpha$.
(b) are instances of (a).

Note. In the appendix of [5] it is shown that $\psi^{\mathrm{SC}}\Lambda$ equals Bachmann's $\varphi_{F_{\omega_2+1}(1)}(1)$. In the present context this equation can be derived as follows

$$\psi^{\mathrm{SC}}\Lambda \overset{\text{Cor.5.5}}{=} \vartheta^{\mathrm{SC}}\Lambda \overset{(\vartheta 0)}{=} \overline{\phi}^{\mathrm{SC}}_\Lambda(0) = \phi^{\mathrm{SC}}_\Lambda(0) \overset{\text{L.5.6}}{=} \phi^{\mathbb{H}}_\Lambda(0) = \text{`}\varphi_{F_{\omega_2+1}(1)}(1)\text{'}.$$

Lemma 5.6 *(a)* $K\gamma = \emptyset \;\&\; \mathbb{Y} \cap \Omega = R^{\mathbb{X}}_\gamma \Rightarrow \phi^{\mathbb{Y}}_\alpha = \phi^{\mathbb{X}}_{\gamma+\alpha}$.
(b) $\mathrm{SC} \cap \Omega = R^{\mathbb{H}}_\Omega$.

Proof (a) Induction on α using Theorem 3.6 and the fact that $K\gamma = \emptyset$ implies $K(\gamma + \alpha) = K\alpha$ and $\mathsf{k}^+(\gamma + \alpha) = \mathsf{k}^+(\alpha)$ for all α.

(b) By definition we have $\forall\alpha < \Omega(\phi^{\mathbb{H}}_\alpha = \varphi_\alpha)$, which together with Lemma 2.3g yields $\mathrm{SC} \cap \Omega = \{\alpha \in \Omega : \phi^{\mathbb{H}}_\alpha(0) = \alpha\} = R^{\mathbb{H}}_\Omega$.

Corollary 5.7 *(i)* $K\gamma = \emptyset \;\&\; \mathbb{Y} \cap \Omega = R^{\mathbb{X}}_\gamma \Rightarrow$
$\overline{\phi}^{\mathbb{Y}}_\alpha = \overline{\phi}^{\mathbb{X}}_{\gamma+\alpha}$ *and* $\vartheta^{\mathbb{Y}}\alpha = \vartheta^{\mathbb{X}}(\Omega\gamma + \alpha)$.
(ii) $\phi^{\mathbb{E}}_\alpha = \phi^{\mathbb{H}}_{1+\alpha}$, $\phi^{\mathrm{SC}}_\alpha = \phi^{\mathbb{H}}_{\Omega+\alpha}$, $\vartheta^{\mathbb{E}}\alpha = \vartheta^{\mathbb{H}}(\Omega + \alpha)$, *and*
$\vartheta^{\mathrm{SC}}\alpha = \vartheta^{\mathbb{H}}(\Omega^2 + \alpha) = \vartheta^{\mathbb{E}}(\Omega^2 + \alpha)$.

Proof (i) follows from Lemma 5.6a and $(\vartheta 0)$.
(ii) follows from Lemma 5.6, (i), and $\mathbb{E} \cap \Omega = R^{\mathbb{H}}_1$.

6 Exact Comparison of ϑ and ψ

Let $\mathbb{X} \subseteq \mathbb{H}$ be fixed such that $\mathbb{X} \cap \Omega$ is Ω-club. As before we write ϑ, ψ for $\vartheta^{\mathbb{X}}$, $\psi^{\mathbb{X}}$. In this section we always assume $\alpha < \Lambda$ and $K\alpha \cup \{\beta\} \subseteq \psi\Lambda$.

Lemma 6.1 *(a)* $\alpha_0 < \alpha \;\&\; \forall\xi(\alpha_0 \le \xi < \alpha \to \psi\xi = \psi(\xi{+}1)) \Rightarrow \psi\alpha_0 = \psi\alpha$.
(b) $\psi\alpha_0 < \psi\alpha \;\Rightarrow\; \exists\alpha_1(\alpha_0 \le \alpha_1 < \alpha \;\&\; K\alpha_1 \subseteq \psi\alpha_1 = \psi\alpha_0)$.

Proof (a) follows from Lemma 5.2a, d by induction on α.

(b) From $\psi\alpha_0 < \psi\alpha$ by Lemmata 5.2a, 6.1a we obtain $\exists\xi(\alpha_0 \le \xi < \alpha \;\&\; \psi\xi < \psi(\xi{+}1))$. Let $\alpha_1 := \min\{\xi \ge \alpha_0 : \psi\xi < \psi(\xi{+}1)\}$. Then $\alpha_0 \le \alpha_1 < \alpha$ and, by (a) and Lemma 5.2c, $K\alpha_1 \subseteq \psi\alpha_1 = \psi\alpha_0$.

Lemma 6.2 *(a)* $\psi\alpha < \gamma \in \mathbb{X} \;\Rightarrow\; \psi(\alpha{+}1) \le \gamma$.
(b) $\gamma \in \mathbb{X} \cap \psi\Lambda \;\Rightarrow\; \exists\alpha(K\alpha \subseteq \psi\alpha = \gamma)$.
(c) $\Omega^\alpha|\gamma \;\&\; \delta < \Omega^\alpha \;\&\; K(\gamma+\delta) \subseteq \psi(\gamma+\delta) \;\Rightarrow\; K\gamma \subseteq \psi\gamma$.
(d) $\Omega^\alpha|\gamma \;\&\; \psi\gamma < \psi(\gamma+\Omega^\alpha) \;\Rightarrow\; K\gamma \subseteq \psi\gamma$.

Proof (a) $\mathbb{X} \ni \gamma < \psi(\alpha+1) \;\Rightarrow\; \exists\xi < \alpha{+}1(\gamma \le \psi\xi) \;\Rightarrow\; \gamma \le \psi\alpha$.

(b) By Lemma 5.2a, d it follows that $\psi\alpha \le \gamma < \psi(\alpha{+}1)$ for some $\alpha < \Lambda$.

By (a) it follows that $\psi\alpha = \gamma$.

(c) Induction on δ: Since $\Omega^\alpha|\gamma \;\&\; \delta < \Omega^\alpha$ we have $K\gamma \subseteq K(\gamma+\delta)$. Therefore, if $\psi\gamma = \psi(\gamma+\delta)$ then $K\gamma \subseteq \psi\gamma$. If $\psi\gamma < \psi(\gamma+\delta)$, then by Lemma 6.1b there exists $\delta_0 < \delta$ such that $K(\gamma+\delta_0) \subseteq \psi(\gamma+\delta_0)$; thence, by IH, $K\gamma \subseteq \psi\gamma$.

(d) By Lemma 6.1b there exists $\delta < \Omega^\alpha$ such that $K(\gamma+\delta) \subseteq \psi(\gamma+\delta)$.

Hence $K\gamma \subseteq \psi\gamma$ by (c).

Lemma 6.3 $\delta =_{\mathrm{NF}} \gamma + \Omega^\alpha\xi \;\;\&\;\; K(\Omega^\alpha\xi) \subseteq \psi(\gamma+\Omega^{\alpha+1}) \;\;\Rightarrow\;\; K(\Omega^\alpha\xi) \subseteq \psi\delta$.

Proof For $\psi\delta = \psi(\gamma+\Omega^{\alpha+1})$ the claim is trivial. Otherwise, by Lemma 6.1b there exists δ_1 with $\delta \le \delta_1 < \gamma+\Omega^{\alpha+1}$ and $K\delta_1 \subseteq \psi\delta_1 = \psi\delta$. Then $\delta_1 = \gamma + \Omega^\alpha\beta + \delta_2$ with $\xi \le \beta < \Omega$ and $\delta_2 < \Omega^\alpha$. Hence $K(\Omega^\alpha\beta) \subseteq K\delta_1 \subseteq \psi\delta$. Now assume $\beta > 0$. Then $K\alpha \cup K\beta = K(\Omega^\alpha\beta) \subseteq \psi\delta$ which together with $\xi \le \beta < \Omega$ yields $K(\Omega^\alpha\xi) \subseteq K\alpha \cup K\xi \subseteq \psi\delta$.

Definitions. 1. $\dot{\psi}\alpha := \begin{cases} 0 & \text{if } \alpha = 0 \\ \psi\alpha & \text{if } \alpha > 0 \end{cases}$

2. If $\alpha \le \beta$ then $-\alpha+\beta$ denotes the unique γ such that $\alpha+\gamma = \beta$.

The following definition is an extension and modification of the corresponding definition on p. 26 of [11].

Definition of $[\alpha, \beta] < \Lambda$

Abbreviation. $\mathsf{k}(\alpha, \beta) := \max(K\alpha \cup \{\beta\})$.

By Lemma 5.2a, d there exists $\eta < \Lambda$ such that

$\dot{\psi}(\Omega^{\alpha+1}\eta) \le \mathsf{k}(\alpha, \beta) < \psi(\Omega^{\alpha+1}(\eta{+}1))$.

Let $\gamma := \Omega^{\alpha+1}\eta$. Then $\dot{\psi}\gamma \le \mathsf{k}(\alpha,\beta) < \psi(\gamma+\Omega^{\alpha+1})$.
If $\Omega\alpha+\beta<\omega$ then $[\alpha,\beta] := \beta$, else
$[\alpha,\beta] := \gamma+\Omega^{\alpha}(1+\xi)$ with $\xi := \begin{cases} -\dot{\psi}\gamma+\beta & \text{if } K\alpha \subseteq \dot{\psi}\gamma \\ \beta & \text{otherwise} \end{cases}$

Remark $\omega \le \Omega\alpha+\beta \Rightarrow \omega \le [\alpha,\beta]$.

Lemma 6.4 *(a)* $K[\alpha,\beta] \subseteq \psi[\alpha,\beta]$*; (b)* $K(\Omega\alpha+\beta) \subseteq \psi[\alpha,\beta]$.

Proof Assume $\omega \le \Omega\alpha+\beta$ (otherwise $K[\alpha,\beta]=\emptyset$ and $K(\Omega\alpha+\beta)=\emptyset$). Then $[\alpha,\beta] =_{\mathrm{NF}} \gamma+\Omega^{\alpha}(1+\xi)$ with $\dot{\psi}\gamma \le \mathsf{k}(\alpha,\beta) < \psi(\gamma+\Omega^{\alpha+1})$ and $\xi \le \beta \le \dot{\psi}\gamma+\xi$.

(a) By Lemmata 6.2d and 5.2a, b we obtain $K\gamma \subseteq \dot{\psi}\gamma < \psi[\alpha,\beta]$.

$K(\Omega^{\alpha}(1+\xi)) = K\alpha \cup K\xi \;\&\; \xi \le \beta < \Omega \;\&\; K\alpha\cup\{\beta\} \subseteq \psi(\gamma+\Omega^{\alpha+1}) \Rightarrow K(\Omega^{\alpha}(1+\xi)) \subseteq \psi(\gamma+\Omega^{\alpha+1})$.

$[\alpha,\beta] =_{\mathrm{NF}} \gamma+\Omega^{\alpha}(1+\xi) \;\&\; K(\Omega^{\alpha}(1+\xi)) \subseteq \psi(\gamma+\Omega^{\alpha+1}) \overset{\text{L.6.3}}{\Rightarrow} K(\Omega^{\alpha}(1+\xi)) \subseteq \psi[\alpha,\beta]$.

(b) By (proof of) (a) we have $K\alpha\cup\{\xi\} \subseteq K[\alpha,\beta] \subseteq \psi[\alpha,\beta]$ and $\dot{\psi}\gamma < \psi[\alpha,\beta]$. From this together with $\beta \le \dot{\psi}\gamma+\xi$ and $\psi[\alpha,\beta] \in \mathbb{X} \subseteq \mathbb{H}$, we obtain $K(\Omega\alpha+\beta) = K\alpha\cup\{\beta\} \subseteq \psi[\alpha,\beta]$.

Lemma 6.5 $\Omega\alpha_0+\beta_0 < \Omega\alpha_1+\beta_1 \;\&\; K(\Omega\alpha_0+\beta_0) \subseteq \psi[\alpha_1,\beta_1] \Rightarrow [\alpha_0,\beta_0] < [\alpha_1,\beta_1]$.

Proof 1. $\Omega\alpha_1+\beta_1<\omega$: Then $[\alpha_0,\beta_0]=\beta_0<\beta_1=[\alpha_1,\beta_1]$.

2. $\Omega\alpha_0+\beta_0<\omega\le\Omega\alpha_1+\beta_1$: Then $[\alpha_0,\beta_0]=\beta_0<\omega\le[\alpha_1,\beta_1]$.

3. $\omega\le\Omega\alpha_0+\beta_0$: Then $[\alpha_i,\beta_i] =_{\mathrm{NF}} \gamma_i+\Omega^{\alpha_i}(1+\xi_i)$ $(i=0,1)$, and $\dot{\psi}\gamma_0 \le \mathsf{k}(\alpha_0,\beta_0) < \psi[\alpha_1,\beta_1]$.

3.1. $\alpha := \alpha_0=\alpha_1 \;\&\; \beta_0<\beta_1$:

3.1.1. $\gamma_0<\gamma_1$: Then $[\alpha_0,\beta_0]=\gamma_0+\Omega^{\alpha}(1+\xi_0) < \gamma_0+\Omega^{\alpha+1} \le \gamma_1 \le [\alpha_1,\beta_1]$.

3.1.2. $\gamma:=\gamma_0=\gamma_1$: To prove $\xi_0<\xi_1$. We have $\xi_i = \begin{cases} -\dot{\psi}\gamma+\beta_i & \text{if } K\alpha \subseteq \dot{\psi}\gamma \\ \beta_i & \text{otherwise} \end{cases}$.

Hence $\xi_0<\xi_1$ follows from $\beta_0<\beta_1$.

3.2. $\alpha_0<\alpha_1$: From $\dot{\psi}\gamma_0 < \psi[\alpha_1,\beta_1]$ and $0<\alpha_1$ we get $\gamma_0 < [\alpha_1,\beta_1] = \gamma_1+\Omega^{\alpha_1}(1+\xi_1)$, and then $\gamma_0+\Omega^{\alpha_1} \le [\alpha_1,\beta_1]$.
Further we have $[\alpha_0,\beta_0] = \gamma_0+\Omega^{\alpha_0}(1+\xi_0) < \gamma_0+\Omega^{\alpha_0+1} \le \gamma_0+\Omega^{\alpha_1}$.

Lemma 6.6 $\vartheta(\Omega\alpha+\beta) \le \psi([\alpha,\beta]) < \psi\Lambda$.

Proof by induction on $\Omega\alpha+\beta$: Let $\gamma_0 := \psi[\alpha,\beta]$.

To prove: $\gamma_0 \in \mathbb{X} \;\&\; K(\Omega\alpha+\beta) \subseteq \gamma_0 \;\&\; \forall\zeta<\Omega\alpha+\beta(K\zeta\subseteq\gamma_0 \to \vartheta\zeta<\gamma_0)$.
1. By definition of ψ and Lemma 6.4b we have $\gamma_0\in\mathbb{X} \;\&\; K(\Omega\alpha+\beta)\subseteq\gamma_0$.
2. Assume $\Omega\xi+\eta<\Omega\alpha+\beta \;\&\; K(\Omega\xi+\eta)\subseteq\gamma_0$. Then, by Lemma 6.5, $[\xi,\eta]<[\alpha,\beta]$. From this by Lemmata 6.4a, 5.2a, b and the IH we obtain $\vartheta(\Omega\xi+\eta) \le \psi[\xi,\eta] < \psi[\alpha,\beta] = \gamma_0$.

Definition of $\overline{\delta} < \Lambda$ for $\delta < \Lambda$.[3]

1. If $\delta < \omega$ then $\overline{\delta} := \delta$.

2. If $\omega \le \delta =_{\rm NF} \gamma + \Omega^{\alpha}(1+\xi)$ then $\overline{\delta} := \Omega\alpha + \beta$ with $\beta := \begin{cases} \dot{\psi}\gamma + \xi & \text{if } K\alpha \subseteq \dot{\psi}\gamma \\ \xi & \text{otherwise} \end{cases}$

Remark $\omega \le \delta \Rightarrow \omega \le \overline{\delta}$.

Lemma 6.7 $\overline{[\alpha,\beta]} = \Omega\alpha + \beta$.

Proof $[\alpha,\beta] =_{\rm NF} \gamma + \Omega^{\alpha}(1+\xi)$ with $\xi = \begin{cases} -\dot{\psi}\gamma + \beta & \text{if } K\alpha \subseteq \dot{\psi}\gamma \\ \beta & \text{otherwise} \end{cases}$

Hence $\overline{[\alpha,\beta]} = \Omega\alpha + \tilde{\beta}$ with $\tilde{\beta} := \begin{cases} \dot{\psi}\gamma + \xi & \text{if } K\alpha \subseteq \dot{\psi}\gamma \\ \xi & \text{otherwise} \end{cases}$. Obviously $\tilde{\beta} = \beta$.

Lemma 6.8 *Let* $\delta, \delta' < \Lambda$.

(a) $K\delta \subseteq \psi\delta \;\&\; \overline{\delta} = \Omega\alpha + \beta \;\Rightarrow\; \delta = [\alpha,\beta]$.

(b) $K\delta \subseteq \psi\delta \;\Rightarrow\; \vartheta\overline{\delta} \le \psi\delta$.

(c) $K\delta \subseteq \psi\delta \;\&\; K\delta' \subseteq \psi\delta' \;\&\; \overline{\delta} = \overline{\delta'} \;\Rightarrow\; \delta = \delta'$.

Proof (a) 1. $\delta < \omega$: Then $\Omega\alpha + \beta = \overline{\delta} = \delta < \omega$ and thus $[\alpha,\beta] = \beta = \delta$.

2. Otherwise: Then $\omega \le \delta =_{\rm NF} \gamma + \Omega^{\alpha}(1+\xi)$ with $\beta = \begin{cases} \dot{\psi}\gamma + \xi & \text{if } K\alpha \subseteq \dot{\psi}\gamma \\ \xi & \text{otherwise} \end{cases}$.

The latter yields $\dot{\psi}\gamma \le \mathsf{k}(\alpha,\beta)$. From $K\delta \subseteq \psi\delta$ by Lemma 6.2c we get $K\gamma \subseteq \psi\gamma$ and then $\psi\gamma < \psi\delta$. Now we have $K\alpha \cup K\xi \subseteq K\delta \subseteq \psi\delta \in \mathbb{H} \;\&\; \psi\gamma < \psi\delta$ which implies $K\alpha \cup \{\beta\} \subseteq \psi\delta \subseteq \psi(\gamma + \Omega^{\alpha+1})$.

It follows that $[\alpha,\beta] = \gamma + \Omega^{\alpha}(1+\tilde{\xi})$ where $\tilde{\xi} := \begin{cases} -\dot{\psi}\gamma + \beta & \text{if } K\alpha \subseteq \dot{\psi}\gamma \\ \beta & \text{otherwise} \end{cases}$.

Obviously $\tilde{\xi} = \xi$ and therefore $[\alpha,\beta] = \delta$.

(b) Take α, β such that $\overline{\delta} = \Omega\alpha + \beta$. Then by Lemma 6.6 and (a) we obtain $\vartheta\overline{\delta} = \vartheta(\Omega\alpha + \beta) \le \psi[\alpha,\beta] = \psi\delta$.

(c) By (a) there are $\alpha, \beta, \alpha', \beta'$ such that

$\overline{\delta} = \Omega\alpha + \beta \;\&\; \delta = [\alpha,\beta]$ and $\overline{\delta}' = \Omega\alpha' + \beta' \;\&\; \delta' = [\alpha',\beta']$.

Therefore from $\overline{\delta} = \overline{\delta'}$ one concludes $\alpha = \alpha' \;\&\; \beta = \beta'$ and then $\delta = \delta'$.

Theorem 6.9 $\delta < \Lambda \;\&\; K\delta \subseteq \psi\delta \;\Rightarrow\; \vartheta\overline{\delta} = \psi\delta$.

Proof by induction on δ: By Lemma 6.8b we have $\vartheta\overline{\delta} \le \psi\delta$. Assumption: $\vartheta\overline{\delta} < \psi\delta$. Then by Lemma 6.2b there exists γ s.t. $K\gamma \subseteq \psi\gamma = \vartheta\overline{\delta} < \psi\delta$. Hence $\gamma < \delta$ and therefore, by IH, $\psi\gamma = \vartheta\overline{\gamma}$. From $\vartheta\overline{\delta} = \psi\gamma = \vartheta\overline{\gamma} \;\&\; K\delta \subseteq \psi\delta \;\&\; K\gamma \subseteq \psi\gamma$ by (ϑ4) and Lemma 6.8c we obtain $\delta = \gamma$. Contradiction.

Corollary 6.10 *(a)* $\vartheta(\Omega\alpha + \beta) = \psi[\alpha,\beta]$.

[3]This definition is closely related to clause 5 in Definition 3.6 of [23] But be aware that there $\overline{\delta}$ has a different meaning.

(b) $K\alpha \subseteq \psi\Omega^{\alpha+1} \Rightarrow \vartheta(\Omega\alpha) = \psi\Omega^{\alpha}$.

Proof (a) Let $\delta := [\alpha, \beta]$. Then by Lemma 6.4a $K\delta \subseteq \psi\delta$, and therefore $\vartheta(\Omega\alpha + \beta) \overset{\text{L.6.7}}{=} \vartheta\overline{\delta} = \psi\delta = \psi[\alpha, \beta]$.
(b) $\alpha < \Lambda \;\&\; K\alpha \subseteq \psi\Omega^{\alpha+1} \Rightarrow \vartheta(\Omega\alpha) = \psi[\alpha, 0] = \psi(\Omega^{\alpha}(1+0)) = \psi\Omega^{\alpha}$.

7 Defining the Bachmann Hierarchy by Functionals of Higher Type

This section is based on [28, (3.2.9)–(3.2.11), (3.2.15)].

Convention. n ranges over natural numbers ≥ 1.

Definition. Let M be an arbitrary nonempty set.
1. $\mathsf{M}^1 := \mathsf{M}$. 2. $\mathsf{M}^{n+1} :=$ set of all functions $F : \mathsf{M}^n \to \mathsf{M}^n$.

Notation. If $1 \leq m < n$ and $F_i \in \mathsf{M}^i$ for $m \leq i \leq n$,
then $F_n F_{n-1} \ldots F_m := F_n(F_{n-1}) \ldots (F_m)$.

Abbreviation: $\mathrm{Id}^{n+1} := \mathrm{Id}_{\mathsf{M}^n} \in \mathsf{M}^{n+1}$.

Assumption. ∇ is an operation such that for every family $(X_\xi)_{\xi<\alpha}$ with $0 < \alpha \leq \Omega$ the following holds:
$\forall\xi < \alpha(X_\xi \in \mathsf{M}^1) \Rightarrow \nabla_{\xi<\alpha} X_\xi \in \mathsf{M}^1$.

Definition. If $n > 1$ and $\forall\xi < \alpha(F_\xi \in \mathsf{M}^{n+1})$ then
$\nabla_{\xi<\alpha} F_\xi \in \mathsf{M}^{n+1}$ is defined by $(\nabla_{\xi<\alpha} F_\xi)G := \nabla_{\xi<\alpha}(F_\xi G)$.

Lemma 7.1 *If* $0 < \alpha \leq \Omega \;\&\; \forall\xi < \alpha(F_\xi \in \mathsf{M}^{n+1}) \;\&\; H \in \mathsf{M}^{n+1}$, *then*
$(\nabla_{\xi<\alpha} F_\xi) \circ H = \nabla_{\xi<\alpha}(F_\xi \circ H)$.

Proof For each $G \in \mathsf{M}^n$ we have
$((\nabla_{\xi<\alpha} F_\xi) \circ H)G = (\nabla_{\xi<\alpha} F_\xi)(HG) = \nabla_{\xi<\alpha}(F_\xi(HG)) = \nabla_{\xi<\alpha}((F_\xi \circ H)G) = (\nabla_{\xi<\alpha}(F_\xi \circ H))G$.

Definition. For $F \in \mathsf{M}^{n+1}$ and $\alpha \leq \Omega$ we define $F^{(\alpha)} \in \mathsf{M}^{n+1}$ by
(i) $F^{(0)} := \mathrm{Id}^{n+1}$; (ii) $F^{(\alpha+1)} := F \circ F^{(\alpha)}$; (iii) $F^{(\alpha)} := \nabla_{\xi<\alpha} F^{(1+\xi)}$ if $\alpha \in Lim$.

Definition.

(i) Let $\mathrm{I}_2 \in \mathsf{M}^2$ be given;
(ii) For $m \geq 2$ we define $\mathrm{I}_{m+1} \in \mathsf{M}^{m+1}$ by $\mathrm{I}_{m+1} F := F^{(\Omega)}$.

Definition of $[\![\alpha]\!]_m$

For $m \geq 2$ and $\alpha < \varepsilon_{\Omega+1}$ we define $[\![\alpha]\!]_m \in \mathsf{M}^m$ by recursion on α:
(i) $[\![0]\!]_m := \mathrm{Id}^m$; (ii) If $\alpha =_{\mathrm{NF}} \gamma + \Omega^{\beta}\eta$, then $[\![\alpha]\!]_m := ([\![\beta]\!]_{m+1}\mathrm{I}_m)^{(\eta)} \circ [\![\gamma]\!]_m$.

Lemma 7.2 *For $m \geq 2$ and $\alpha < \varepsilon_{\Omega+1}$:*

(a) $[\![\alpha+1]\!]_m = \mathrm{I}_m \circ [\![\alpha]\!]_m$;

(b) $\alpha \in Lim \Rightarrow [\![\alpha]\!]_m = \nabla_{\xi<\tau(\alpha)}[\![\alpha[\xi]]\!]_m$.

Proof (a) $[\![\gamma + \Omega^0(\eta+1)]\!]_m = ([\![0]\!]_{m+1}\mathrm{I}_m)^{(\eta+1)} \circ [\![\gamma]\!]_m = \mathrm{I}_m^{(\eta+1)} \circ [\![\gamma]\!]_m = \mathrm{I}_m \circ (\mathrm{I}_m^{(\eta)} \circ [\![\gamma]\!]_m) = \mathrm{I}_m \circ [\![\gamma + \Omega^0{\cdot}\eta]\!]_m$.

(b) Induction on α:

1. $\alpha =_{\mathrm{NF}} \gamma + \Omega^\beta\eta$ with $\eta \in Lim$: Then $\tau(\alpha) = \eta$ and $\alpha[\xi] = \gamma + \Omega^\beta(1+\xi)$.
$[\![\alpha]\!]_m = ([\![\beta]\!]_{m+1}\mathrm{I}_m)^{(\eta)} \circ [\![\gamma]\!]_m = (\nabla_{\xi<\eta}([\![\beta]\!]_{m+1}\mathrm{I}_m)^{(1+\xi)}) \circ [\![\gamma]\!]_m = \nabla_{\xi<\eta}(([\![\beta]\!]_{m+1}\mathrm{I}_m)^{(1+\xi)} \circ [\![\gamma]\!]_m) = \nabla_{\xi<\eta}[\![\alpha[\xi]]\!]_m$.

2. $\alpha =_{\mathrm{NF}} \gamma + \Omega^\beta(\eta+1)$ with $\beta = \beta_0+1$:
Then $\tau(\alpha) = \Omega$ and $\alpha[\xi] = \gamma + \Omega^\beta\eta + \Omega^{\beta_0}(1+\xi)$.
$[\![\alpha]\!]_m = ([\![\beta]\!]_{m+1}\mathrm{I}_m)^{(\eta+1)} \circ [\![\gamma]\!]_m = ([\![\beta_0+1]\!]_{m+1}\mathrm{I}_m) \circ ([\![\beta]\!]_{m+1}\mathrm{I}_m)^{(\eta)} \circ [\![\gamma]\!]_m \overset{(a)}{=} (\mathrm{I}_{m+1}([\![\beta_0]\!]_{m+1}\mathrm{I}_m)) \circ [\![\gamma + \Omega^\beta\eta]\!]_m = (\nabla_{\xi<\Omega}([\![\beta_0]\!]_{m+1}\mathrm{I}_m)^{(1+\xi)}) \circ [\![\gamma + \Omega^\beta\eta]\!]_m = \nabla_{\xi<\Omega}(([\![\beta_0]\!]_{m+1}\mathrm{I}_m)^{(1+\xi)} \circ [\![\gamma + \Omega^\beta\eta]\!]_m) = \nabla_{\xi<\Omega}[\![\alpha[\xi]]\!]_m$.

3. $\alpha =_{\mathrm{NF}} \gamma + \Omega^\beta(\eta+1)$ with $\beta \in Lim$:
Then $\tau(\alpha) = \tau(\beta)$ and $\alpha[\xi] = \gamma + \Omega^\beta\eta + \Omega^{\beta[\xi]}$.
$[\![\alpha]\!]_m = ([\![\beta]\!]_{m+1}\mathrm{I}_m)^{(\eta+1)} \circ [\![\gamma]\!]_m = ([\![\beta]\!]_{m+1}\mathrm{I}_m) \circ ([\![\beta]\!]_{m+1}\mathrm{I}_m)^{(\eta)} \circ [\![\gamma]\!]_m = ([\![\beta]\!]_{m+1}\mathrm{I}_m) \circ [\![\gamma + \Omega^\beta\eta]\!]_m \overset{\mathrm{IH}}{=} (\nabla_{\xi<\tau(\beta)}([\![\beta[\xi]]\!]_{m+1}\mathrm{I}_m)) \circ [\![\gamma + \Omega^\beta\eta]\!]_m = \nabla_{\xi<\tau(\beta)}(([\![\beta[\xi]]\!]_{m+1}\mathrm{I}_m) \circ [\![\gamma + \Omega^\beta\eta]\!]_m) = \nabla_{\xi<\tau(\alpha)}[\![\alpha[\xi]]\!]_m$.

Corollary 7.3 *For $X \in \mathsf{M}^1$ and $\alpha < \varepsilon_{\Omega+1}$ the following holds:*

(i) $[\![0]\!]_2X = X$;

(ii) $[\![\alpha+1]\!]_2X = \mathrm{I}_2([\![\alpha]\!]_2X)$;

(iii) $[\![\alpha]\!]_2X = \nabla_{\xi<\tau(\alpha)}([\![\alpha[\xi]]\!]_2X)$ *if* $\alpha \in Lim$.

Now we fix M, I_2 and ∇ as follows:

1. $\mathsf{M} :=$ set of all Ω-club subsets of Ω.
2. $\mathrm{I}_2 : \mathsf{M} \to \mathsf{M}$, $I_2(X) := \{\beta \in \Omega : \mathrm{en}_X(\beta) = \beta\}$, where en_X is the enumerating function of X.
3. If $\forall\xi < \alpha(X_\xi \in \mathsf{M})$ then

$$\nabla_{\xi<\alpha}X_\xi := \begin{cases} \bigcap_{\xi<\alpha} X_\xi & \text{if } \alpha < \Omega \\ \{\beta \in \Omega \cap Lim : \beta \in \bigcap_{\xi<\beta} X_\xi\} & \text{if } \alpha = \Omega \end{cases}$$

Then by transfinite induction on α from the above Corollary and the definition of R_α^X we conclude

Theorem 7.4 $R_\alpha^X = [\![\alpha]\!]_2X$, *for all $\alpha < \varepsilon_{\Omega+1}$ and $X \in \mathsf{M}$.*

Appendix

This appendix is devoted to the proof of Lemmata 2.1, 2.2d.

Lemma A.1 *(a)* $\lambda \in Lim \Rightarrow 0 < \lambda[0]$.
(b) $\gamma + \Omega^\beta < \Omega^\alpha \ \& \ \eta < \Omega \Rightarrow \gamma + \Omega^\beta\eta < \Omega^\alpha$.
(c) $\lambda =_{\mathrm{NF}} \gamma + \Omega^\beta\eta \notin \mathrm{ran}(F_0) \ \& \ \Omega^\alpha < \lambda \Rightarrow \Omega^\alpha \leq \lambda[0]$.

Proof of (c): From $\Omega^\alpha < \lambda = \gamma + \Omega^\beta\eta$ by (b) we get $\Omega^\alpha \leq \gamma + \Omega^\beta$. If $\eta \in Lim$ then $\lambda[0] = \gamma + \Omega^\beta$. If $1 < \eta = \eta_0{+}1$ then $\gamma + \Omega^\beta \leq \gamma + \Omega^\beta\eta_0 \leq \lambda[0]$. If $\eta = 1$ then $0 < \gamma$ (since $\lambda \notin \mathrm{ran}(F_0)$) and therefore $\Omega^{\beta+1} \leq \gamma$ which together with $\Omega^\alpha < \lambda = \gamma + \Omega^\beta$ yields $\Omega^\alpha \leq \gamma \leq \lambda[0]$.

Lemma A.2 $\lambda =_{\mathrm{NF}} F_\alpha(\beta) \ \& \ 0 < \beta \Rightarrow F_\alpha(\beta[n]) \leq \lambda[n]$.

Proof 1. $\beta \in Lim$: $F_\alpha(\beta[n]) = \lambda[n]$.
2. $\beta = \beta_0{+}1$:
  2.1. $\alpha = 0$: $F_\alpha(\beta[n]) = \Omega^{\beta_0} \leq \Omega^{\beta_0}{\cdot}(1{+}n) = \lambda[n]$.
  2.2. $\alpha > 0$: $F_\alpha(\beta[n]) = F_\alpha(\beta_0) < \lambda^- \leq \lambda[n]$.

Lemma A.3 $F_\zeta(\mu) < \lambda \leq F_\zeta(\mu{+}1) \Rightarrow F_\zeta(\mu) \leq \lambda[0]$.

Proof 0. $\lambda = F_\zeta(\mu + 1)$:
  0.1. $\zeta = 0$: $\lambda = \Omega^{\mu+1}$, $\lambda[\xi] = \Omega^\mu(1{+}\xi)$, $\lambda[0] = F_0(\mu)$.
  0.2. $\zeta > 0$: $F_\zeta(\mu) < \lambda^- < F_{\zeta[0]}(\lambda^-) = \lambda[0]$.
1. $\lambda < F_\zeta(\mu + 1)$:
  1.1. $\lambda =_{\mathrm{NF}} \gamma + \Omega^\beta\eta \notin \mathrm{ran}(F_0)$:
  $F_\zeta(\mu) \in \mathrm{ran}(F_0) \ \& \ F_\zeta(\mu) < \lambda \overset{\text{L.A1c}}{\Rightarrow} F_\zeta(\mu) \leq \lambda[0]$.
  1.2. $\lambda =_{\mathrm{NF}} F_\alpha(\beta)$: Then $\alpha < \zeta$ and thus $F_\alpha(F_\zeta(\mu)) = F_\zeta(\mu) < F_\alpha(\beta)$. Hence $F_\zeta(\mu) < \beta$ and therefore $F_\zeta(\mu) \overset{\text{IH}}{\leq} \beta[0] \leq F_\alpha(\beta[0]) \overset{\text{A2}}{\leq} \lambda[0]$.

Definition. $r(\gamma) := \begin{cases} -1 & \text{if } \gamma \notin \mathrm{ran}(F_0) \\ \alpha & \text{if } \gamma =_{\mathrm{NF}} F_\alpha(\beta) \\ \gamma & \text{if } \gamma = \Lambda \end{cases}$

Lemma A.4 *(a)* $r(F_\alpha(\beta)) = \max\{\alpha, r(\beta)\}$.
(b) $\lambda[0] < \delta < \lambda \Rightarrow r(\delta) \leq r(\lambda)$.
(c) $\lambda =_{\mathrm{NF}} F_\alpha(\beta) \ \& \ \beta \notin Lim \ \& \ \lambda^- < \eta < \lambda \Rightarrow \lambda^- \leq \eta[1]$.

Proof (a) 1. $\beta < F_\alpha(\beta)$:
  Then $r(F_\alpha(\beta)) = \alpha$ and ($r(\beta) = -1$ or $\beta =_{\mathrm{NF}} F_{\beta_0}(\beta_1)$ with $\beta_0 \leq \alpha$).
  2. $\beta = F_\alpha(\beta)$: Then $\beta =_{\mathrm{NF}} F_{\beta_0}(\beta_1)$ with $\alpha < \beta_0 = r(\beta) = r(F_\alpha(\beta))$.
(b) 1. $\lambda =_{\mathrm{NF}} \gamma + \Omega^\beta\eta \notin \mathrm{ran}(F_0)$:

1.1. $\eta \in Lim$: $\gamma + \Omega^\beta = \lambda[0] < \delta < \gamma + \Omega^\beta \eta \overset{\text{A1b}}{\Rightarrow} \delta \notin \mathrm{ran}(F_0)$.

1.2. $\eta = \eta_0{+}1$:

$\gamma + \Omega^\beta \eta_0 < \lambda[0] < \delta < \gamma + \Omega^\beta(\eta_0{+}1) \notin \mathrm{ran}(F_0)\ \&\ \Omega^{\beta+1}|\gamma \Rightarrow \delta \notin \mathrm{ran}(F_0)$.

2. $\lambda =_{\mathrm{NF}} F_\alpha(\beta)$: If $\lambda < F_{\alpha+1}(0)$ then also $\delta < F_{\alpha+1}(0)$ and thus $r(\delta) \le \alpha = r(\lambda)$. Otherwise there exists μ such that $F_{\alpha+1}(\mu) < \lambda < F_{\alpha+1}(\mu{+}1)$. Then by Lemma A3 we get $F_{\alpha+1}(\mu) \le \lambda[0] < \delta < F_{\alpha+1}(\mu{+}1)$ and thus $\delta \notin \mathrm{ran}(F_{\alpha+1})$, i.e. $r(\delta) \le \alpha = r(\lambda)$.

3. $\lambda = \Lambda$: $r(\delta) < \Lambda = r(\Lambda)$.

(c) For $\beta = 0 \vee \eta = \eta_0{+}1$ the claim is trivial. Assume now $\beta = \beta_0{+}1\ \&\ \eta \in Lim$.

$$F_\alpha(\beta_0) < \eta < F_\alpha(\beta_0{+}1) \overset{\text{L.A3}}{\Rightarrow} \lambda^- = F_\alpha(\beta_0){+}1 \le \eta[0] + 1 \le \eta[1].$$

Lemma 2.1 $\lambda =_{\mathrm{NF}} F_\alpha(\beta)\ \&\ \beta \in Lim\ \&\ 1 \le \xi < \tau_\beta \Rightarrow \lambda[\xi] =_{\mathrm{NF}} F_\alpha(\beta[\xi])$.

Proof We have $\lambda[\xi] = F_\alpha(\beta[\xi])\ \&\ \beta[0] < \beta[\xi] < \beta$. By Lemma A4b this yields $\lambda[\xi] = F_\alpha(\beta[\xi])\ \&\ r(\beta[\xi]) \le r(\beta) \le \alpha$, whence $\lambda[\xi] =_{\mathrm{NF}} F_\alpha(\beta[\xi])$.

Lemma 2.2d $\xi{+}1 < \tau_\lambda\ \&\ \lambda[\xi] < \delta \le \lambda[\xi{+}1] \Rightarrow \lambda[\xi] \le \delta[1]$.

Proof by induction on $\delta\#\lambda$: If $r(\delta) < r(\lambda[\xi])$ then, by Lemma A4b, $\lambda[\xi] \le \delta[0]$.

Assume now that $r(\lambda[\xi]) \le r(\delta)$ (†).

1. $\lambda =_{\mathrm{NF}} \gamma + \Omega^\beta \eta \notin \mathrm{ran}(F_0)$.

1.1. $\eta \in Lim$:

$\gamma + \Omega^\beta(1{+}\xi) = \lambda[\xi] < \delta < \lambda[\xi{+}1] = \gamma + \Omega^\beta(1{+}\xi) + \Omega^\beta \Rightarrow \lambda[\xi] \le \delta[0]$.

1.2. $\eta = \eta_0{+}1$: $\gamma + \Omega^\beta \eta_0 + \Omega^\beta[\xi] = \lambda[\xi] < \delta \le \lambda[\xi{+}1] = \gamma + \Omega^\beta \eta_0 + \Omega^\beta [\xi{+}1] \Rightarrow \delta = (\gamma + \Omega^\beta \eta_0) + \delta_0$ with $\Omega^\beta[\xi] < \delta_0 \le \Omega^\beta[\xi{+}1] \Rightarrow \delta[0] = \gamma + \Omega^\beta \eta_0 + \delta_0[0]$ with $\Omega^\beta[\xi] \overset{\text{IH}}{\le} \delta_0[0] \Rightarrow \lambda[\xi] \le \delta[0]$.

2. $\lambda =_{\mathrm{NF}} F_\alpha(\beta)\ \&\ \beta \in Lim$: Then (1) $\lambda[\xi] = F_\alpha(\beta[\xi])$, and (2) $\lambda[\xi] < \delta < \lambda$.

From $\alpha \overset{(1)}{\le} r(\lambda[\xi]) \overset{(\dagger)}{\le} r(\delta) \overset{(2),\text{L.A4b}}{\le} r(\lambda) = \alpha$ we get $r(\delta) = \alpha$, i.e. $\delta =_{\mathrm{NF}} F_\alpha(\eta)$ for some η. Now from $\lambda[\xi] < \delta \le \lambda[\xi{+}1]$ we conclude $\beta[\xi] < \eta \le \beta[\xi{+}1]$ and then, by IH, $\beta[\xi] \le \eta[0]$. Hence $\lambda[\xi] \le F_\alpha(\eta[0]) \overset{\text{L.A2}}{\le} \delta[0]$.

3. $\lambda =_{\mathrm{NF}} F_\alpha(\beta)\ \&\ \beta \notin Lim$:

3.1. $\alpha = 0$: Then $\beta = \beta_0{+}1$, and $\lambda[\xi] = \Omega^{\beta_0}(1{+}\xi) < \delta \le \Omega^{\beta_0}(1{+}\xi) + \Omega^{\beta_0}$ implies $\lambda[\xi] \le \delta[0]$.

3.2. $\alpha = \alpha_0{+}1$: Then $\lambda[\xi] = F_{\alpha_0}^{\xi+1}(\lambda^-)$.

Hence, by (†), $\delta =_{\mathrm{NF}} F_\zeta(\eta)$ with $\alpha_0 \le \zeta$.

3.2.1. $\alpha_0 < \zeta$: $\lambda^- < F_\zeta(\eta) \Rightarrow \lambda[\xi{+}1] = F_{\alpha_0}^{\xi+2}(\lambda^-) < F_\zeta(\eta)$. Contradiction.

3.2.2. $\zeta = \alpha_0$: Then from $F_{\alpha_0}^{\xi+1}(\lambda^-) = \lambda[\xi] < \delta = F_{\alpha_0}(\eta) \le \lambda[\xi{+}1]$ we conclude $F_{\alpha_0}^{\xi}(\lambda^-) < \eta \le \lambda[\xi]$. As we will show, this implies $F_{\alpha_0}^{\xi}(\lambda^-) \le \eta[1]$, thence $F_{\alpha_0}^{\xi+1}(\lambda^-) \le F_\zeta(\eta[1]) \overset{\text{L.A2}}{\le} \delta[1]$.

Proof of $F_{\alpha_0}^{\xi}(\lambda^-) \le \eta[1]$:

(i) $\xi = n{+}1$: Then the claim follows by IH from $\lambda[n] = F_{\alpha_0}^{\xi}(\lambda^-) < \eta \le \lambda[n{+}1]$.

(ii) $\xi = 0$: $\lambda^- < \eta < \lambda \overset{\text{L.A4c}}{\Rightarrow} \lambda^- \le \eta[1]$.

3.3. $\alpha \in Lim$: $\lambda[\xi] = F_{\alpha[\xi]}(\lambda^-)$, and by (†) we have $\delta =_{\rm NF} F_\zeta(\eta)$ with $\alpha[\xi] \le \zeta$.
3.3.1. $\alpha[\xi{+}1] < \zeta$: $\lambda^- < F_\zeta(\eta) \Rightarrow F_{\alpha[\xi+1]}(\lambda^-) < F_\zeta(\eta) = \delta$. Contradiction.
3.3.2. $\alpha[\xi] < \zeta \le \alpha[\xi{+}1]$:
(i) $\eta \in Lim$: Then $\lambda^- < \delta[1] = F_\zeta(\eta[1])$ (for $\beta = 0$, $\lambda^- = 0$. If $\beta = \beta_0{+}1$, then $F_\alpha(\beta_0) < \delta < F_\alpha(\beta_0{+}1)$ and thus, by Lemma A3, $F_\alpha(\beta_0) \le \delta[0]$).
$\alpha[\xi] < \zeta$ & $\lambda^- < \delta[1] \Rightarrow \lambda[\xi] = F_{\alpha[\xi]}(\lambda^-) < \delta[1]$.
(ii) $\eta \notin Lim$: By IH $\alpha[\xi] \le \zeta[1]$. Further $\lambda^- \le \delta^-$.
Proof of $\lambda^- \le \delta^-$: Assume $\beta = \beta_0{+}1$.
$F_\alpha(\beta_0) < \delta = F_\zeta(\eta)$ & $\zeta < \alpha \Rightarrow 0 < \eta \Rightarrow \eta = \eta_0{+}1$.
$F_\alpha(\beta_0) < F_\zeta(\eta_0{+}1)$ & $\zeta < \alpha \Rightarrow F_\alpha(\beta_0) \le F_\zeta(\eta_0)$.
From $\alpha[\xi] \le \zeta[1]$ and $\lambda^- \le \delta^-$ we conclude $\lambda[\xi] = F_{\alpha[\xi]}(\lambda^-) \le F_{\zeta[1]}(\delta^-) \le \delta[1]$.
3.3.3. $\zeta = \alpha[\xi]$: This case is similar to 3.2.2(ii):

$$\lambda[\xi] = F_\zeta(\lambda^-) < F_\zeta(\eta) < F_\alpha(\beta) \Rightarrow \lambda^- < \eta < F_\alpha(\beta) \Rightarrow \lambda[\xi] = F_\zeta(\lambda^-) \overset{\text{L.A4c}}{\le} F_\zeta(\eta[1]) \overset{\text{L.A2}}{\le} \delta[1].$$

4. $\lambda = \Lambda$: This case is very similar to 3.3, but considerably simpler.

References

1. Aczel, P.: A New Approach to the Bachmann Method for Describing Large Countable Ordinals (Preliminary Summary). unpublished
2. Aczel, P.: Describing ordinals using functionals of transfinite type. J. Symbol. Logic **37**(1), 35–47 (1972)
3. Bachmann, H.: Die Normalfunktionen und das Problem der ausgezeichneten Folgen von Ordnungszahlen. *Vierteljschr. Naturforsch.Ges. Zürich*, pp. 115–147 (1950)
4. Bridge, J.: A simplification of the Bachmann method for generating large countable ordinals. J. Symbol. Logic **40**, 171–185 (1975)
5. Buchholtz, U.T.: Unfolding of systems of inductive definitions. PhD thesis, Stanford University (2013)
6. Buchholz, W.: Normalfunktionen und konstruktive Systeme von Ordinalzahlen. In Diller, J., Müller, G.H. (eds.), Proof Theory Symposium, Kiel 1974. Lecture Notes in Mathematics, vol. 500, pp. 4–25. Springer, Berlin (1975)
7. Buchholz, W.: Collapsingfunktionen (1981). http://www.mathematik.uni-muenchen.de/~buchholz/Collapsing.pdf
8. Buchholz, W.: A survey on ordinal notations around the Bachmann-Howard ordinal. In: Kahle, R., Strahm, T., Studer, T. (eds.) Advances in Proof Theory. Progress in Computer Science and Applied Logic, vol. 28, pp. 1–29. Birkhäuser, Basel (2016)
9. Buchholz, W., Schütte, K.: Die Beziehungen zwischen den Ordinalzahlsystemen Σ und $\overline{\theta}(\omega)$. Arch. Math. Logik und Grundl. **17**, 179–189 (1976)
10. Buchholz, W., Schütte, K.: Ein Ordinalzahlensystem für die beweistheoretische Abgrenzung der Π^1_1-Separation und Bar-Induktion. Bayr. Akad. Wiss. Math.-Naturw. Kl., pp. 99–13 (1983)
11. Buchholz, W., Schütte, K.: Proof Theory of Impredicative Subsystems of Analysis. Studies in Proof Theory Monographs, vol. 2. Bibliopolis, Napoli (1988)
12. Crossley, J.N., Bridge-Kister, J.: Natural well-orderings. Arch. Math. Logik und Grundl. **26**, 57–76 (1986/87)
13. Feferman, S.: Proof theory: a personal report. In: Takeuti, G. (ed.) Proof Theory. Studies in Logic and the Foundations of Mathematics, 2nd edn. pp. 447–485. North-Holland (1987)

14. Gerber, H.: An extension of Schütte's Klammersymbols. Math. Ann. **174**, 203–216 (1967)
15. Howard, W.: A system of abstract constructive ordinals. J. Symbol. Logic **37**, 355–374 (1972)
16. Isles, D.: Regular ordinals and normal forms. In: Kino, A., Myhill, J., Vesley, R.E. (eds.), Intuitionism and Proof Theory (Proceedings Conference, Buffalo N.Y., 1968), pp. 339–362. North-Holland, 1970
17. Jäger, G.: ρ-inaccessible ordinals, collapsing functions and a recursive notation system. Archiv für mathematische Logik **24**, 49–62 (1984)
18. Pfeiffer, H.: Ausgezeichnete Folgen für gewisse Abschnitte der zweiten und weiterer Zahlenklassen. PhD thesis, Hannover (1964)
19. Pohlers, W.: Ordinal notations based on a hierarchy of inaccessible cardinals. Ann. Pure Appl. Logic **33**, 157–179 (1987)
20. Rathjen, M.: Ordinal notations based on a weakly Mahlo cardinal. Arch. Math. Logic **29**, 249–263 (1990)
21. Rathjen, M.: Proof theory of reflection. Ann. Pure Appl. Logic **68**, 181–224 (1994)
22. Rathjen, M.: Recent advances in ordinal analysis: Π^1_2-CA and related systems. Bull. Symbol. Logic **1**(4), 468–485 (1995)
23. Rathjen, M., Weiermann, A.: Proof-theoretic investigations on Kruskal's theorem. Ann. Pure Appl. Logic **60**(1), 49–88 (1993)
24. Schütte, K.: Kennzeichnung von Ordnungszahlen durch rekursiv erklärte Funktionen. Math. Ann. **127**, 15–32 (1954)
25. Schütte, K.: Proof Theory, Grundlehren der Mathematischen Wissenschaften, vol. 225. Springer, Berlin (1977)
26. Schütte, K.: Beziehungen des Ordinalzahlensystems OT(ϑ) zur Veblen-Hierarchie. unpublished (1992)
27. Veblen, O.: Continous increasing functions of finite and transfinite ordinals. Trans. Am. Math. Soc. **9**, 280–292 (1908)
28. Weyhrauch, R.: Relations between some hierarchies of ordinal functions and functionals. PhD thesis, Stanford University (1976). (Completed and circulated in 1972)

Author Biography

Wilfried Buchholz (born 1948) studied mathematics at the Ludwig-Maximilians-Universität München (LMU), where he received his Ph.D. in 1974. From 1978 to 2013 he taught mathematics, especially proof theory, at LMU. In the academic year 1990/91 he was a visiting scholar at the Philosophy Department of Carnegie Mellon University. Since 2013 he is Emeritus Professor at LMU.

The Interpretation Existence Lemma

Albert Visser

Abstract The present paper contains a fairly detailed verification of a reasonably general form of the Interpretation Existence Lemma. In first approximation, this lemma tells us that if a theory U proves the consistency of V then U interprets V. What are theories here? What is *proving the consistency of* V in this context? The paper will explain how the lemma works, providing rather general answers to these questions. We apply the Interpretation Existence Lemma to verify well-known characterization theorems for interpretability: the Friedman Characterization and the Orey–Hájek Characterization. Finally, we provide three randomly chosen examples of application of the Interpretation Existence Lemma.

Keywords Interpretations · Degrees · Sequential theories

2000 Mathematics Subject Classification: 03A05 · 03B25 · 03F45

This paper is dedicated to the memory of Solomon Feferman. His ideas have been a constant inspiration.

1 Introduction

The Interpretation Existence Lemma tells us, in first approximation, that, if we have a theory U with a modicum of arithmetic plus a consistency statement for a given theory V, then we can build an interpretation of V in U. This interpretation gives us a

---

I am grateful to Ali Enayat, Paula Henk and Volodya Shavrukov for helpful comments, references and scholarly support. I thank the anonymous referee for his/her suggestions.

A. Visser (✉)
Philosophy, Faculty of Humanities, Utrecht University, Janskerkhof 13,
3512BL Utrecht, The Netherlands
e-mail: a.visser@uu.nl

G. Jäger and W. Sieg (eds.), *Feferman on Foundations*, Outstanding Contributions to Logic 13, DOI 10.1007/978-3-319-63334-3_5

uniform construction of internal models of V in models of U. The construction yields all kinds of further insights concerning interpretability and non-standard models.

The aim of the present paper is to give a detailed presentation of the proof of a fairly general version of the Interpretation Existence Lemma.[1] After this presentation, I will connect the result with various characterization theorems for interpretability. Finally, I will supply three more or less randomly selected examples of applications of the result.

1.1 Historical Remarks

In this subsection, I give an, admittedly sketchy, overview of the development of the Interpretation Existence Lemma. The reader is referred to [4] for a more detailed discussion of a part of the development.

The story of the Interpretation Existence Lemma starts with Löwenheim's Theorem [23].

Theorem 1.1 ([23]) *If a sentence of predicate logic is satisfiable, then it is satisfiable in the natural numbers.*

The next step is the Model Existence Lemma, which is just the contrapositive of Gödel's Completeness Theorem. This theorem was first published in Gödel's doctoral dissertation *Über die Vollständigkeit des Logikkalküls*. The journal publication of the result is [10].

Theorem 1.2 ([13]) *If a sentence of predicate logic is (syntactically) consistent, then it is satisfiable.*

The Model Existence Lemma was given a sharper form by David Hilbert and Paul Bernays in [13], pp. 234–253. The progress here is the insight that the model that fulfills the consistent sentence can be built entirely inside arithmetic. Said differently, we can find arithmetical predicates such that substitution of these predicates for the predicate letters of the sentence yields an arithmetical truth.

Theorem 1.3 ([13]) *If a sentence of predicate logic is consistent, then it has a true arithmetical interpretation.*

Stephen Cole Kleene in [19] determines more precisely the complexity of the arithmetical substitution one needs. Moreover, we move from a single sentence to a recursively enumerable set of sentences.

Theorem 1.4 ([19]) *If a recursively enumerable set of predicate logical sentences is consistent then it has a true arithmetical interpretation in the Δ^0_2-formulas.*

[1] My most extensive earlier attempt to give a presentation of the proof was in [37]. However, the present one gives more detail and isolates the special features of the construction in a better way.

Hilary Putnam in [29] improves Kleene's estimate by showing:

Theorem 1.5 ([29]) *If a predicate logical sentence is consistent then it has a true arithmetical interpretation in the Boolean combinations of Σ^0_1-formulas.*

I do not know whether Putnam's result can be improved to the case where the interpretation is identity preserving. Putnam's proof is designed to yield the result for a single sentence, but, with some minor modifications, one can extend his proof to the case where we have a recursively enumerable set of sentences in finite signature.

Putnam's improvement is off our main track, so we take a step back in time and consider Hao Wang's papers [50, 51]. In these papers, Wang shifts the focus from the construction of models in the real word to the construction of interpretations inside theories.

Theorem 1.6 ([50, 51]) *Consider any recursively enumerable theory U. We have: Peano Arithmetic plus the arithmetization $\mathsf{con}(U)$ of the consistency statement for U interprets U. In our notation:* $(\mathsf{PA} + \mathsf{con}(U)) \rhd U$.

In Wang's result interpretability is taken to be one-dimensional and without domain relativization.[2] This is not terribly important since we are working over PA, where any interpretation has a definably isomorphic counterpart that is one-dimensional and unrelativized. In case the domain of the interpretation is infinite, we can also make the interpretation identity-preserving. In [51], Wang follows Gödel's original proof and, in [50], he adapts Henkin's proof of the Model Existence Lemma [14] rather than Gödel's.

Sol Feferman set out to extend and clarify Wang's ideas. He describes how his ideas developed around that time in [7]. An important ingredient of his work was the realization of the intentionality of consistency statements. As a consequence he could prove Wang's result in a more precise setting and with much larger scope. Also, completely new consequences follow. For example, we can have a notion of consistency such that (i) Peano Arithmetic proves its own consistency and (ii) the Interpretation Existence Lemma still works for this notion of consistency. These insights immediately give us the existence of a non-trivial internal model for every model of Peano Arithmetic. Also, we obtain the existence of an Orey-sentence for Peano Arithmetic, to wit a sentence O such that Peano Arithmetic PA interprets both $\mathsf{PA} + O$ and $\mathsf{PA} + \neg O$.

Feferman's insights concerning the Interpretation Existence Lemma were part of his Ph.D. thesis and were, for a large part, published in [6]. Steven Orey employed Feferman's ideas to give the well-known Orey–Hájek characterization. Moreover, he pointed out the existence of Orey-sentences. See [26]. The characterization was rediscovered by Petr Hájek. See [11, 12, 15]. The name 'Orey–Hájek Characterization' seems to go back to [22].

---

[2] *One-dimensional* means that the formula representing the object domain of the interpretation just has one free variable, apart from variables serving as parameters. See Sect. 2.2.

1.2 Interpretation Existence Without Induction

In this paper, we will prove the Interpretation Existence Lemma in a more general setting than Feferman and Orey did. The main difference is (the lack of) the use of induction. Where Feferman and Orey worked with extensions of Peano Arithmetic, we will work in a setting that is essentially induction-free. Our basic tool will be Buss's theory S^1_2. This theory still contains a modicum of induction, but in principle that can be eliminated too, since S^1_2 is interpretable on a definable cut in Robinson's Arithmetic Q which does not contain any induction.[3]

Since induction seems to be at the heart of Henkin's proof of the Model Existence Lemma, it is somewhat puzzling that we can get rid of it. The solution to the puzzle requires a new way of looking at numbers in a theory. The solution is, in a nutshell: *don't be faithful to your numbers*. As soon as your numbers are progressive with respect to a certain property P, but fail to to deliver the desired conclusion that every number satisfies P, then we simply switch to a definable cut of the given numbers on which we do have the desired conclusion that every number satisfies P.

The methodology of induction avoidance employs *Solovay's method of shortening cuts*. It was invented by Robert Solovay (see: [34]) and further developed by Pavel Pudlák and the Prague group (see e.g. [28]), by Harvey Friedman (see e.g. [33]) and by Alex Wilkie and Jeff Paris (see e.g. [53]). An impressive foundational program was based on Solovay's idea by Eduard Nelson in [25]. These developments take place roughly between 1975 and 1990.

It is fair to say that something like the version of the Interpretation Existence Lemma that we develop in the present paper was known to researchers like Pudlák and Friedman in the period between 1975 and 1990, certainly in the sense that the statement of the present result would not be any surprise for them and in the sense that, if needed, the relevant researchers could easily have produced it. All the ingredients were present. However, I know of no full statement of the result in that period.

Why would we want a proof of the Interpretation Existence Lemma in an induction-free context? This is not just to satisfy an indefinite craving for generality or based on an idiosyncratic preference for weak theories. The important point is that many applications use the more general form. For example, the Friedman Characterization of interpretability between finitely axiomatized sequential theories works between, say, a finite extension of ACA_0 and a finite extension of GB. This characterization involves weak theories, even if the theories to which the characterization is applied are reasonably strong.

1.3 What Is in the Paper?

In Sect. 2, we introduce the basic notions and notations needed in the paper. Section 3 is the heart of the paper. It contains a detailed verfication of the Interpretation Exis-

[3] Alternatively, we could start from the theory PA^-.

tence Lemma in a fairly general form. In Sect. 4 we consider the world of $\mho$: a group of special theories associated with given theories that form a basic tool for proving Interpretation Existence. Specifically, various characterization-theorems of interpretability are based on theories of the $\mho$-kind. The section contains proofs of the Friedman Characterization and the Orey–Hájek Characterization. Finally, in Sect. 5, we provide three more or less randomly chosen examples of application of Interpretation Existence.

1.4 What Is Not in the Paper?

In the present paper, we do not consider the refinement of the Interpretation Existence Lemma without the methodology of cuts, where one connects the Henkin construction to the Low Basis Theorem and the Low Low Basis Theorem. The reader is referred to the text books [16, 21] for more information about this line of research.

1.5 Prerequisites

Some knowledge of the materials from [16, 18] would certainly help to read the paper.

2 Basic Notions and Facts

In the present section, we provide the basics needed for the rest of the paper. The reader who wants to get on quickly to more exciting stuff could quickly look over the relevant subsections and, if needed, return to them later.

2.1 Theories

Signatures are officially finite relational signatures in this paper. Unofficially, we will often use terms. Terms can be efficiently eliminated by the term elimination algorithm. We will diverge from the finiteness of the signature and the exclusive use of relations only in the case of the use of Henkin constants.

In this paper we will study theories with finite relational signature. Theories are given by a signature plus a set of axioms. There is, in this paper, no a priori complexity constraint on the set of axioms.

If a theory is finitely axiomatizable, *par abus de langage*, we use the variables like A and B for it, making the letters do double work: they stand both for the theory and for a single axiom.

When we diverge from our general format this will always be explicitly mentioned. If we speak of a theory qua set of theorems we will stress this by writing $\mathfrak{T}, \mathfrak{U}$, etcetera. The mapping that sends a theory to the set of theorems is $T \mapsto \overline{T}$. We write $=_{\mathsf{ext}}$ for 'having the same theorems'. So, $T =_{\mathsf{ext}} U$ iff $\overline{T} = \overline{U}$.

An important special theory that we will employ is Buss' theory S^1_2. This theory is finitely axiomatizable. We will employ it in its finitely axiomatized form. The reader is referred to [2, 16] for details about S^1_2. In fact the precise features of S^1_2 are not really relevant for our purposes. We could as well work with a sufficiently large finitely axiomatized fragment of $\mathrm{I}\Delta_0 + \Omega_1$. Here Ω_1 is the axiom that states that the function ω_1 with $\omega_1(x) = 2^{|x|^2}$ is total. What is most relevant is that (i) the theory we employ is (i) finitely axiomatized, (ii) that it is mutually interpretable with other weak theories like Q, PA^- and adjunctive set theory AS, (iii) that it is sequential, i.e. it has a good theory of sequences and (iv) that arithmetization of syntax is easy in the theory and can be done without special tricks (in contrast to e.g. $\mathrm{I}\Delta_0$).

2.2 *Translations and Interpretations*

We present the notion of *m-dimensional interpretation without parameters*. There are two extensions of this notion: we can consider piecewise interpretations and we can add parameters. We refer to our paper [47] for some explanation of these extra features.

2.2.1 Translations

Consider two signatures Σ and Θ. An m-dimensional translation $\tau : \Sigma \to \Theta$ is a quadruple $\langle \Sigma, \delta, \mathcal{F}, \Theta \rangle$, where $\delta(v_0, \ldots, v_{m-1})$ is a Θ-formula and where, for any n-ary predicate P of Σ, $\mathcal{F}(P)$ is a formula $A(\vec{v}_0, \ldots, \vec{v}_{n-1})$ in the language of signature Θ, where $\vec{v}_i = v_{i,0}, \ldots, v_{i,(m-1)}$. Both in the case of δ and A, all free variables are among the variables shown. Moreover, if $i \neq j$ or $k \neq \ell$, then $v_{i,k}$ is syntactically different from $v_{j,\ell}$.

We demand that we have $\vdash \mathcal{F}(P)(\vec{v}_0, \ldots, \vec{v}_{n-1}) \to \bigwedge_{i<n} \delta(\vec{v}_i)$. Here $\vdash$ is provability in predicate logic. This demand is inessential, but it is convenient to have.

We define B^τ as follows:

- $(P(x_0, \ldots, x_{n-1}))^\tau := \mathcal{F}(P)(\vec{x}_0, \ldots, \vec{x}_{n-1})$.
- $(\cdot)^\tau$ commutes with the propositional connectives.
- $(\forall x\, A)^\tau := \forall \vec{x}\, (\delta(\vec{x}\,) \to A^\tau)$.
- $(\exists x\, A)^\tau := \exists \vec{x}\, (\delta(\vec{x}\,) \wedge A^\tau)$.

There are two worries about this definition. First, what variables $\vec{x}_i$ on the side of the translation A^τ correspond with x_i in the original formula A? The second worry is that substitution of variables in δ and $\mathcal{F}(P)$ may cause variable-clashes. These worries are never important in practice. For example, we choose 'suitable' sequences $\vec{x}$ to correspond to variables x, and we avoid clashes by α-conversion. However, if we want to give precise definitions of translations and, for example, of composition of translations, these problems come into play. There are a number of solution strategies for the first worry. The first is to design an *ad hoc* regime of translating variables. E.g., if our dimension is m, we could always send the variable x_i to the sequence $x_{m\cdot i}, x_{m\cdot i+1}, \dots, x_{m\cdot i+m-1}$. The second is to make the choice of the translating variables part of the data of the translation. The third is to work with a variable-free formalism. Let us say, for definiteness that we opt for the first strategy. Similarly, for the α-conversions we will stipulate some fixed regimen.

We allow the identity predicate to be translated to a formula that is not identity. An m-dimensional translation τ is *unrelativized*, if $\delta_\tau(\vec{x}\,) := \top$. A relation is *identity preserving* if $\vec{x} =_\tau \vec{y}$ iff $x_i = y_i$, for all $i < m$. A translation is *direct* if it is both unrelativized and identity-preserving.

There are several important operations on translations.

- $\mathsf{id}_\Sigma : \Sigma \to \Sigma$ is the identity translation. We take $\delta_{\mathsf{id}_\Sigma}(v) := v = v$ and $\mathcal{F}(P) := P(\vec{v}\,)$.
- We can compose translations. Suppose $\tau : \Sigma \to \Theta$ and $\nu : \Theta \to \Lambda$. Then $\nu \circ \tau$ or $\tau\nu$ is a translation from Σ to Λ. We define:
  - $\delta_{\tau\nu}(\vec{v}_0, \dots, \vec{v}_{m_\tau-1}) := \bigwedge_{i<m_\tau} \delta_\nu(\vec{v}_i) \wedge (\delta_\tau(v_0, \dots, v_{m_\tau-1}))^\nu$.
  - $P_{\tau\nu}(\vec{v}_{0,0}, \dots, \vec{v}_{0,m_\tau-1}, \dots \vec{v}_{n-1,0}, \dots, \vec{v}_{n-1,m_\tau-1}) :=$ $\bigwedge_{i<n, j<m_\tau} \delta_\nu(\vec{v}_{i,j}) \wedge (P(v_0, \dots, v_{n-1})^\tau)^\nu$.
- Let $\tau, \nu : \Sigma \to \Theta$ and let A be a sentence of signature Θ. We define the disjunctive translation $\sigma := \tau\langle A\rangle\nu : \Sigma \to \Theta$ as follows. We take $m_\sigma := \mathsf{max}(m_\tau, m_\nu)$. We write $\vec{v} \restriction n$, for the restriction of $\vec{v}$ to the first n variables, where $n \leq \mathsf{length}(\vec{v})$.
  - $\delta_\sigma(\vec{v}) := (A \wedge \delta_\tau(\vec{v} \restriction m_\tau)) \vee (\neg A \wedge \delta_\nu(\vec{v} \restriction m_\nu))$.
  - $P_\sigma(\vec{v}_0, \dots, \vec{v}_{n-1}) := (A \wedge P_\tau(\vec{v}_0 \restriction m_\tau, \dots, \vec{v}_{n-1} \restriction m_\tau))\vee$ $(\neg A \wedge P_\nu(\vec{v}_0 \restriction m_\nu, \dots, \vec{v}_{n-1} \restriction m_\nu))$

Note that in the definition of $\tau\langle A\rangle\nu$ we used a padding mechanism. In case, for example, $m_\tau < m_\nu$, the variables $v_{m_\tau}, \dots, v_{m_\nu-1}$ are used 'vacuously' when we have A. If we had piecewise interpretations, where domains are built up from pieces with possibly different dimensions, we could avoid padding by building the domain directly of disjoint pieces with different dimensions.

A translation $\tau : \Sigma \to \Theta$ maps a model $\mathcal{M}$ of signature Θ to an internal model $\widetilde{\tau}(\mathcal{M})$ of signature Σ, provided that $\mathcal{M}$ satisfies $\exists x\, \delta_\tau(\vec{x}\,)$ and the τ-translations of the identity axioms for $=_\tau$.

2.2.2 Interpretations

A translation relates signatures; an interpretation relates theories. An interpretation $K : U \to V$ is a triple $\langle U, \tau, V\rangle$, where U and V are theories and $\tau : \Sigma_U \to \Sigma_V$. We demand: for all theorems A of U, we have $V \vdash A^\tau$. Equivalently, we could demand that, for all axioms U-axioms B, we have $V \vdash B^\tau$. In this case we should include the identity axioms among the axioms of U, since we allow identity to be translated to a formula different from identity.[4] Here are some further definitions.

- $\mathsf{ID}_U : U \to U$ is the interpretation $\langle U, \mathsf{id}_{\Sigma_U}, U\rangle$.
- Suppose $K : U \to V$ and $M : V \to W$. Then, $KM := M \circ K : U \to W$ is $\langle U, \tau_M \circ \tau_K, W\rangle$.
- Suppose $K : U \to (V + A)$ and $M : U \to (V + \neg A)$. Then $K\langle A\rangle M : U \to V$ is the interpretation $\langle U, \tau_K\langle A\rangle\tau_M, V\rangle$. In an appropriate category $K\langle A\rangle M$ is a special case of a product.

An interpretation is unrelativized if its underlying translation is unrelativized. It is direct if its underlying translation is direct. Etcetera.

An interpretation $K : U \to V$ gives us a mapping $\widetilde{K} := \widetilde{\tau}_K$ from $\mathsf{MOD}(V)$, the class of models of V, to $\mathsf{MOD}(U)$, the class of models of U. If we build a category of theories and interpretations, usually MOD with $\mathsf{MOD}(K) := \widetilde{K}$ will be a contravariant functor.

2.2.3 Reduction Relations

We use $K : U \lhd V$ or $K : V \rhd U$ as alternative notations for $K : U \to V$. The alternative notations $\lhd$ and $\rhd$ are used in a context where we are interested in interpretability as a preorder or as a provability analogue. This way of looking is the primary interest in this paper.

We write: $U \lhd V$ and $V \rhd U$, for: there is an interpretation $K : U \lhd V$. We use $U \equiv V$, for: $U \lhd V$ and $U \rhd V$.

We write $K : U \lhd_{\mathsf{faith}} V$ or $K : V \rhd_{\mathsf{faith}} U$ for: K is *a faithful interpretation* of U in V. This means that: for all U-sentences A, we have: $U \vdash A$ iff $V \vdash A^{\tau_K}$. We use $U \lhd_{\mathsf{faith}} V$ or $V \rhd_{\mathsf{faith}} U$, for: there is a faithful interpretation $K : U \lhd_{\mathsf{faith}} V$. We use $\equiv_{\mathsf{faith}}$ for the induced equivalence relation of $\lhd_{\mathsf{faith}}$.

We write $U \lhd_{\mathsf{mod}} V$ or $V \rhd_{\mathsf{mod}} U$ for: every model $\mathcal{M}$ of V has an internal model $\mathcal{N}$ of U, in other words, for every model $\mathcal{M}$ of V there is a translation τ such that $\widetilde{\tau}(\mathcal{M}) \models U$.[5]

---

[4]The verification of the equivalence between axioms-interpretability and theorems-interpretability demands Σ_1-collection. See [37] for more details.

[5]Here we restrict ourselves to the case without parameters. We can add parameters to the definition in different ways, either by employing a parameter domain or by stipulating parameters locally for each model. These ways are more closely connected than one would think thanks to the Omitting Types Theorem. One can find examples of theories U and V, where V model-interprets U with parameters but does not do so without parameters.

We write $U \lhd_{\mathsf{loc}} V$ or $V \rhd_{\mathsf{loc}} U$ for: for all finite subtheories U_0 of U, $U_0 \lhd V$. We pronounce this as: U is locally interpretable in V or V locally interprets U. We use $\equiv_{\mathsf{loc}}$ for the induced equivalence relation of $\lhd_{\mathsf{loc}}$.

We have: $U \lhd V$ implies $U \lhd_{\mathsf{mod}} V$ and $U \lhd_{\mathsf{mod}} V$ implies $U \lhd_{\mathsf{loc}} V$. The second implication is by a simple compactness argument. None of the two arrows is reversible, even when we restrict ourselves to recursively enumerable sequential theories.

2.3 *Provability, Arithmetization, Complexity*

In this paper we follow Feferman's example in [6] by fixing a proof system and an arithmetization in the background.

2.3.1 Proof System

In this paper we will employ a minor variant of the standard Genzen system $\mathsf{G1c}$ given in [35], Sect. 3.1. It is presented in Fig. 1. In the formulation of the system, we suppose that some axiom set α of a theory is given. The quantifier rules are subject to the usual clauses. Specifically, we demand that in $\mathsf{L}\exists$ and $\mathsf{R}\forall$, we have $y \notin \mathsf{FV}(\Gamma, \Delta)$, and y is either syntactically identical to x or $y \notin \mathsf{FV}(A)$.

$$\frac{}{A \Rightarrow A}\ \mathsf{Share} \qquad \frac{}{\bot \Rightarrow}\ \mathsf{L}\bot \qquad \frac{}{\Rightarrow \top}\ \mathsf{R}\top$$

$$\frac{}{\Rightarrow A}\ A \in \alpha \qquad \frac{\Gamma \Rightarrow \Delta}{\Gamma', \Gamma \Rightarrow \Delta, \Delta'}\ \mathsf{W} \qquad \frac{\Gamma \Rightarrow \Delta, A \quad A, \Gamma' \Rightarrow \Delta'}{\Gamma, \Gamma' \Rightarrow \Delta, \Delta'}\ \mathsf{Cut}$$

$$\frac{A, \Gamma \Rightarrow \Delta}{A \wedge B, \Gamma \Rightarrow \Delta}\ \mathsf{L}\wedge \qquad \frac{B, \Gamma \Rightarrow \Delta}{A \wedge B, \Gamma \Rightarrow \Delta}\ \mathsf{L}\wedge \qquad \frac{\Gamma \Rightarrow \Delta, A \quad \Gamma \Rightarrow \Delta, B}{\Gamma \Rightarrow \Delta, A \wedge B}\ \mathsf{R}\wedge$$

$$\frac{A, \Gamma \Rightarrow \Delta \quad B, \Gamma \Rightarrow \Delta}{A \vee B, \Gamma \Rightarrow \Delta}\ \mathsf{L}\vee \qquad \frac{\Gamma \Rightarrow \Delta, A}{\Gamma \Rightarrow \Delta, A \vee B}\ \mathsf{R}\vee \qquad \frac{\Gamma \Rightarrow \Delta, B}{\Gamma \Rightarrow \Delta, A \vee B}\ \mathsf{R}\vee$$

$$\frac{\Gamma \Rightarrow \Delta, A \quad B, \Gamma \Rightarrow \Delta}{A \rightarrow B, \Gamma \Rightarrow \Delta}\ \mathsf{L}\!\rightarrow \qquad \frac{A, \Gamma \Rightarrow \Delta, B}{\Gamma \Rightarrow \Delta, A \rightarrow B}\ \mathsf{R}\!\rightarrow \qquad \frac{A[x/t], \Gamma \Rightarrow \Delta, A}{\forall x\, A, \Gamma \Rightarrow \Delta}\ \mathsf{L}\forall$$

$$\frac{\Gamma \Rightarrow \Delta, A[x/y]}{\Gamma \Rightarrow \Delta, \forall x\, A}\ \mathsf{R}\forall \qquad \frac{A[x/y], \Gamma \Rightarrow \Delta}{\exists x\, A, \Gamma \Rightarrow \Delta}\ \mathsf{L}\exists \qquad \frac{\Gamma \Rightarrow \Delta, A[x/t]}{\Gamma \Rightarrow \Delta, \exists x\, A}\ \mathsf{R}\exists$$

Fig. 1 A variant of the Genzen system $\mathsf{G1c}$

$$\frac{}{t_0 = u_0, \cdots, t_{n-1} = u_{n-1}, P(t_0, \cdots, t_{n-1}) \Rightarrow P(u_0, \cdots, u_{n-1})}\ \mathsf{leib} \qquad \frac{}{\Rightarrow t = t}\ \mathsf{id}$$

Fig. 2 Identity rules

We will in our system use a finite set of identity rules. See Fig. 2. It is clear that these rules can be replaced by the obvious axioms. When in translating a theory we demand that the translations of the identity axioms are provable, we mean these axioms.

2.3.2 Arithmetization

We work with a finite alphabet. Thus, for example, we treat names of variables as consisting of more than one symbol.

We employ a system of arithmetization that is based on the length-first ordering of strings. We first count all strings of our given alphabet of length 0 in alphabetical order, then of length 1, etcetera. The number associated to a string of length n will be of order 2^{cn}, for a fixed standard c. As a consequence, concatenation (and, thus syntactical operations like conjunction) will have the growth behavior of multiplication and substitution will be roughly Buss' smash function.

An important property of our coding is *monotonicity*: a strict substring is always coded by a strictly smaller number than the original string.

Remark 2.1 The fact that we do our coding in arithmetic and not directly in, say, a weak set theory or a weak theory of syntax is for a large part a matter of *legacy*. There is one further argument for the arithmetic route: the presence of a linear ordering on the codes makes Rosser-style arguments possible.

We will use modal notations like $\Box_U A$ for $\mathsf{prov}_U(\underline{\ulcorner A \urcorner})$. (Here $\ulcorner A \urcorner$ is the Gödel number of A and $\underline{n}$ is the (efficient) numeral of n.) Similarly, $\Diamond_U A$ will stand for $\mathsf{con}(U + A)$. In case we want to stress the formula α representing the axioms, we write $\Box_\alpha A$. We will write $\Box_{\alpha,\Gamma}$ for provability where the formulas allowed in the proof from α have to be from the formula class Γ. We write $\Box_{\alpha,m,n}$ for provability only using α-axioms $\leq m$ and formulas of complexity $\leq n$, for a complexity measure. We also write e.g. $\Box_{\alpha,m,\infty}$, when we have no bound on the complexity of the formulas in the proof. Etcetera.

2.3.3 Depth of Quantifier Alternations

We will employ *depth of quantifier alternations* as our main complexity measure. This measure has the following desirable properties:

$$
\begin{array}{lcl}
\Sigma_0^* & := & \emptyset \\
\Pi_0^* & := & \emptyset \\
\Sigma_{n+1}^* & ::= & \mathsf{AT} \mid \neg\,\Pi_{n+1}^* \mid (\Sigma_{n+1}^* \wedge \Sigma_{n+1}^*) \mid (\Sigma_{n+1}^* \vee \Sigma_{n+1}^*) \mid (\Pi_{n+1}^* \to \Sigma_{n+1}^*) \mid \\
 & & \exists v\, \Sigma_{n+1}^* \mid \forall v\, \Pi_n^* \\
\Pi_{n+1}^* & ::= & \mathsf{AT} \mid \neg\,\Sigma_{n+1}^* \mid (\Pi_{n+1}^* \wedge \Pi_{n+1}^*) \mid (\Pi_{n+1}^* \vee \Pi_{n+1}^*) \mid (\Sigma_{n+1}^* \to \Pi_{n+1}^*) \mid \\
 & & \forall v\, \Pi_{n+1}^* \mid \exists v\, \Sigma_n^*
\end{array}
$$

Fig. 3 Complexity classes for depth of quantifier alternations

i. Eliminating terms in favour of a relational formulation raises the complexity only by a fixed standard number.
ii. Translation of a formula via the translation τ raises the complexity of the formula by a fixed standard number depending only on τ.
iii. The tower of exponents involved in cut-elimination should be of height linear in the complexity of the formulas involved in the proof.

Both Philipp Gerhardy (see [8, 9]) and Sam Buss (see [3]) study such complexity measures. We will follow Buss' approach. Buss shows that the bound on the height of the tower of exponents in cut-elimination is $d + O(1)$ for d alternations.

We work over a signature Θ. The formula-classes we define are officially called $\Sigma_n^*(\Theta)$ and $\Pi_n^*(\Theta)$. However, we will suppress the Θ when it is clear from the context. Let AT be the class of atomic formulas for Θ, extended with $\top$ and $\bot$. The definition of the complexity classes is given in Fig. 3.

Buss uses Σ_{n+1} and Π_{n+1} where we use Σ_{n+1}^* and Π_{n+1}^*. We employ the asterix to avoid confusion with the usual complexity classes in the arithmetical hierarchy where bounded quantifiers also play a role. Secondly, we modified Buss' inductive definition a bit in order to get unique generation histories. For example, Buss adds Π_n^* to Σ_{n+1}^* in stead of $\forall v\, \Pi_n^*$. In addition our Σ_0^* and Π_0^* are empty, where Buss' corresponding classes consist of the quantifier-free formulas.

Here is the parse tree of $\forall x\, (\forall y\, \exists z\, Pxyz \to \exists u\, \exists v\, Qxuv)$ as an element of Σ_4^*.

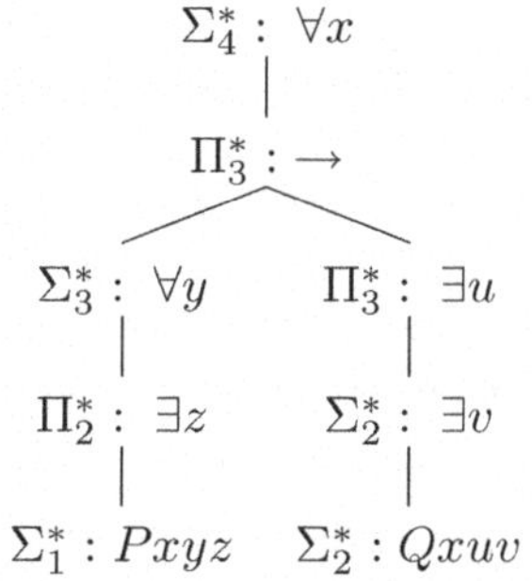

We give the complexity measure $\rho(A)$ such that $\rho(A)$ is the smallest n such that A is in Σ_n^*. This measure is very close to the measure that was employed in [38]. We recursively define this measure by taking $\rho := \rho_\exists$, where $\rho_\exists$ is defined as follows:

- $\rho_\exists(A) := \rho_\forall(A) = 1$, if A is atomic.

- $\rho_\exists(\neg B) := \rho_\forall(B)$, $\rho_\forall(\neg B) := \rho_\exists(B)$.
- $\rho_\exists(B \wedge C) := \mathsf{max}(\rho_\exists(B), \rho_\exists(C))$, $\rho_\forall(B \wedge C) := \mathsf{max}(\rho_\forall(B), \rho_\forall(C))$.
- $\rho_\exists(B \vee C) := \mathsf{max}(\rho_\exists(B), \rho_\exists(C))$, $\rho_\forall(B \vee C) := \mathsf{max}(\rho_\forall(B), \rho_\forall(C))$.
- $\rho_\exists(B \to C) := \mathsf{max}(\rho_\forall(B), \rho_\exists(C))$, $\rho_\forall(B \to C) := \mathsf{max}(\rho_\exists(B), \rho_\forall(C))$.
- $\rho_\exists(\exists v\, B) := \rho_\exists(B)$, $\rho_\forall(\exists v\, B) := \rho_\exists(B) + 1$.
- $\rho_\exists(\forall v\, B) := \rho_\forall(B) + 1$, $\rho_\forall(\forall v\, B) := \rho_\forall(B)$.

Let $\tau : \Sigma \to \Theta$ be a translation. We define $\rho(\tau)$ to be the maximum of $\rho_\exists(\delta_\tau)$, $\rho_\forall(\delta_\tau)$, the $\rho_\exists(P_\tau)$ and the $\rho_\forall(P_\tau)$, for P in Σ. If K is an interpretation, then $\rho(K) := \rho(\tau_K)$. We have:

Theorem 2.2 $\rho(A^\tau) \leq \rho(A) + \rho(\tau)$.

The proof is by a simple induction on A.[6]

2.4 Sequential Theories

The notion of sequentiality is due to Pavel Pudlák. See, e.g., [16, 24, 27, 28]. For a detailed treatment of sequential theories, see [44].

A theory is m-sequential if it directly interprets Adjunctive Set Theory AS via an m-dimensional interpretation.

$$\mathsf{AS1} \vdash \exists y\, \forall x\; x \notin y,$$
$$\mathsf{AS2} \vdash \forall x\, \forall y\, \exists z\, \forall u\; (u \in z \leftrightarrow (u \in y \vee u = x)).$$

We say that a theory *polysequential* if it is m-sequential for some m. A theory is *sequential* if it is 1-sequential. See [44] for discussion. We follow the tradition by stating most results for the sequential case. However, the results usually also hold more generally for the polysequential case. (We only have a rather artificial example of a theory that is polysequential but not sequential.)

Examples of sequential theories are PA^-, S^1_2, EA, $\mathrm{I}\Sigma_1$, PA, ACA_0, ZF.[7] The theory Q is not sequential (not even polysequential), but it still is mutually interpretable with a sequential theory like PA^- or S^1_2. For a proof of the sequentiality of PA^-, see [17]. For proofs of the non-sequentiality of Q, see [17, 41] or [48].

Via bootstrapping, we interpret S^1_2 in AS and develop a good theory of sequences of all objects in the domain. For details concerning the bootstrap see e.g. the textbook [16] and also [24, 42–44]. Using the sequences, we can build satisfaction predicates for the formula classes Σ^*_n. The precise claim is as follows.

Theorem 2.3 *Suppose U is a sequential theory of finite signature. Let the interpretation that witnesses the sequentiality of U be τ. Then we can find an N :*

[6] In my forthcoming paper *Restricted Theories*, I will treat the various properties of the complexity measure in more detail.

[7] We note that in the case of many sorted theories, we need to use a standard translation into a one-sorted theory to apply the definition.

$\mathsf{S}^1_2 \lhd U$ *and a satisfaction predicate* sat_n *for* Σ^*_n*-formulas with complexity of order* $\rho(\tau) + k_0 n + k_1$, *for fixed constants* k_0 *and* k_1. *Here the variables that range over codes of formulas are relativized to* N.

($\rho(N)$ *has a similar bound that is linear in* n.)

I will provide the construction of the satisfaction predicates in a lot of detail in my forthcoming paper *Restricted Theories*. The main idea is that $\mathsf{sat}_{n+1}(x, y)$ is constructed as $\Phi(x, y, \mathsf{sat}_n)$, for a fixed second order formula $\Phi(x, y, S)$. Thus, the transition from sat_n to sat_{n+1} only adds a fixed standard $n = \rho(\Phi)$ to the complexity. The interpretation N is a cut of an given interpretation $N_0 : \mathsf{S}^1_2 \lhd U$: we shorten our interpretation of S^1_2 to compensate for the lack of the inductions that we need to verify the elementary properties of our satisfaction predicate, like commutation and good behaviour for substitution.[8]

A consequence of the presence of satisfaction predicates is the local reflexivity of sequential theories. A theory U is *locally reflexive* iff, for every n, we have $U \rhd (\mathsf{S}^1_2 + \Diamond_{U,n,n}\top)$. This insight follows directly from the following theorem.

Theorem 2.4 *Suppose* A *is finitely axiomatized and sequential. We have, for any* n, *that* $A \rhd (\mathsf{S}^1_2 + \Diamond_{A,A,n}\top)$. *Here the* ρ*-complexity of the witnessing interpretation is linear in* n.

The proof just follows the usual track of proving consistency using satisfaction. We have to choose a 'short' interpretation N to compensate for the lack of induction.

A fundamental result is Pudlák's Theorem [28] that any two interpretations of S^1_2 in a sequential theory have a common cut.

For any interpretation $N : \mathsf{S}^1_2 \lhd U$, *an* ω_1*-cut* (w.r.t. N) is a definable cut of N that is downwards closed w.r.t. $<_N$, and is closed under addition, multiplication and ω_1 in N. Here $\omega_1(x) = 2^{|x|^2}$.

Theorem 2.5 ([29]) *Suppose* U *is sequential. Let* $N, N' : \mathsf{S}^1_2 \lhd U$. *Then, there is a definable* N*-*ω_1*-cut* I *and a definable* N'*-*ω_1*-cut* I' *in* U *such that there is a formula* $F(\vec{x}, \vec{y})$ *that is* U*-verifiably an isomorphism between* I *and* I'. *(Here* $\vec{x}$ *has the dimension of* N *and* $\vec{y}$ *has the dimension of* N'.)

An alternative formulation is as follows. Suppose U *is sequential. Let* $N, N' : \mathsf{S}^1_2 \lhd U$. *Then there is an* $N^* : \mathsf{S}^1_2 \lhd U$ *and formulas* $F(\vec{x}, \vec{y})$ *and* $F'(\vec{x}, \vec{y})$ *such that* U*-verifiably* F *is an initial embedding of* N^* *in* N *and* F' *is an initial embedding of* N^* *in* N'. *(Here the dimension of* N^* *can be taken to be* m *if* U *is* m*-sequential.)*

2.5 Shortening Cuts

Consider any theory U and suppose $N : \mathsf{S}^1_2 \lhd U$. Suppose X is a U-definable subclass of δ_N that is U-verifiably progressive. This means that:

[8] Our Sect. 4.2 contains a bit more detail on the construction of satisfaction predicates.

- $U \vdash \exists \vec{x} \in \delta_N\, (\mathsf{Z}_N(\vec{x}\,) \wedge X(\vec{x}\,))$,
- $U \vdash \forall \vec{x} \in \delta_N\, (X(\vec{x}\,) \to \exists \vec{y} \in \delta_N\, (\mathsf{S}_N(\vec{x}, \vec{y}\,) \wedge X(\vec{y}\,)))$.

In that case we can shorten X to an ω_1-cut. The basic methodology for shortening cuts is due to Robert Solovay [34].

We treat as an example closure under addition. We define the definable class $Y := \{y \in X \mid \forall x \in X\ (x + y) \in X\}$. (We write $+$ for $+_N$.) Clearly, $0 \in Y$ and $1 \in Y$. We show that Y is closed under addition (and hence also under successor). Suppose y_0 and y_1 are in Y and $x \in X$. Then, $(x + y_0) \in X$ and, hence, $(x + (y_0 + y_1)) = ((x + y_0) + y_1) \in X$.

The reader is referred to [16] for more details.

3 The Interpretation Existence Lemma

Usually the Interpretation Existence Lemma is stated in the form: if a theory U satisfies such-and-such conditions with respect to V, then U interprets V (where the interpretation has such-and-such further desirable properties). I will employ a slightly different style. The idea is that we find a minimal theory $\mathsf{sf}(V)$ that satisfies the desired properties with respect to V. The theory U will then satisfy the desired properties if $U \rhd \mathsf{sf}(V)$. This way of stating the result has the immediate advantage of making fully clear that only the resources of $\mathsf{sf}(V)$ were used in proving the theorem. Moreover, we can now ask interesting questions about the new theory $\mathsf{sf}(V)$.

3.1 *Auxiliary Theories*

Consider any finite signature Σ. We define the theory SF_Σ as follows. We expand the signature of arithmetic by two unary predicates α and Γ. Say the resulting signature is Λ. The theory SF_Σ is a Λ-theory with the following axioms:

$\mathsf{Sf}_\Sigma 1$ $\vdash \mathsf{S}^1_2$.[9]
$\mathsf{Sf}_\Sigma 2$ $\vdash \Gamma \subseteq \mathsf{form}_\Sigma$.
$\mathsf{Sf}_\Sigma 3$ $\vdash \alpha \subseteq \Gamma \cap \mathsf{sent}_\Sigma$.
$\mathsf{Sf}_\Sigma 4$ $\vdash \forall v_0 \in \mathsf{var} \ldots \forall v_{n-1} \in \mathsf{var}\ \ P(v_0, \cdots, v_{n-1}) \in \Gamma$,
for any Σ-predicate P, where $n = \mathsf{ar}(P)$. Note that $P(v_0, \cdots, v_{n-1})$ is a sloppy notation for $\ulcorner \underline{P} \urcorner * \ulcorner \underline{(} \urcorner * v_0 * \ulcorner \underline{,} \urcorner * \cdots * \ulcorner \underline{,} \urcorner * v_{n-1} * \ulcorner \underline{)} \urcorner$.
$\mathsf{Sf}_\Sigma 5$ $\vdash \ulcorner \underline{\bot} \urcorner \in \Gamma, \vdash \ulcorner \underline{\top} \urcorner \in \Gamma$.
$\mathsf{Sf}_\Sigma 6$ $\vdash \forall A \in \Gamma\, \forall B \in \mathsf{sub}(A)\ B \in \Gamma$.
Here $\mathsf{sub}(A)$ is the set of subformulas of A.

---

[9]The induction axioms are only for the original language of arithmetic.

$\mathsf{Sf}_{\Sigma}7$ $\vdash \forall A \in \Gamma\, \forall \sigma \in \mathsf{admis}(A)\ \sigma(A) \in \Gamma$.
Here admis codifies that σ is an admissible substitution of variables for the variables in A.

$\mathsf{Sf}_{\Sigma}8$ $\vdash \Diamond_{\alpha,\Gamma}\top$.
Here α, Γ-provability is provability from α using only Γ-formulas in the proof.

We could have opted to demand only closure of Γ under *direct* subformulas working with a weaker theory, say, SF^{-}_{Σ}. However, it is easy to see that we can interpret SF_{Σ} in SF^{-}_{Σ} on a definable cut. We note that we only defined a unique theory when we have a coding of syntax fixed in the background. However, various versions of SF_{Σ} are synonymous as long as we choose reasonable codings.[10]

Suppose X is a set of sentences of signature Σ. We define:

$\mathsf{Sf}_{\Sigma,X}$ $\vdash \underline{\ulcorner A \urcorner} \in \alpha$, for each A in X.

We define $\mathsf{sf}_{\Sigma}(X) := \mathsf{SF}_{\Sigma} + \mathsf{Sf}_{\Sigma,X}$. Since we consider a theory V as being equipped with both an axiom set ax_V and a signature Σ_V, we will simply write $\mathsf{sf}(V)$ for $\mathsf{sf}_{\Sigma_V}(\mathsf{ax}_V)$.

Let Pred_{Σ} be predicate logic in signature Σ. We can view this as the Σ-theory with the empty axiom set. Thus, $\mathsf{SF}_{\Sigma} = \mathsf{sf}_{\Sigma}(\emptyset) = \mathsf{sf}(\mathsf{Pred}_{\Sigma})$.

3.2 The Theorem and Its Proof

This subsection is the heart of the paper. We present the proof of the Interpretation Existence Lemma.

Theorem 3.1 *We have:* $\mathsf{sf}(V) \rhd V$.

Proof Let $\Sigma := \Sigma_V$. We work in SF_{Σ}. We construct a translation η that supports an interpretation of predicate logic for Σ in SF_{Σ}, such that $\mathsf{SF}_{\Sigma} \vdash \underline{\ulcorner A \urcorner} \in \alpha \to A^{\eta}$.

We are going to implement a version of the standard Henkin construction. However, we have to be careful since some relevant steps have to be feasible in S^1_2. Where the resources we have do not suffice for a desired induction, we have to compensate for this by going to a definable cut.

We extend the language of V inductively as follows:

- $t ::= v \mid c[B]$,
- $A ::= \bot \mid \top \mid P(t, \cdots, t) \mid t = t \mid \neg A \mid (A \wedge A) \mid (A \vee A) \mid (A \to A) \mid (A \leftrightarrow A) \mid \forall v\, A \mid \exists v\, A$.
- $B ::= \exists v\, A \mid \forall v\, A$.

Here the atomic predicates P are from the signature of V and the length of the sequence of terms reflects correctly the arity of the relevant predicate. We remind the reader that all our languages are officially of finite relational signature. The Henkin

[10] For the notion of *synonymity of theories*, see e.g. [40].

constants are the only exception we allow. Note that we never interpret a Henkin theory. The Henkin theory is just a tool for building an interpretation.

Let the extended class of formulas be $\mathsf{form}^{\mathsf{h}}_{\Sigma}$. The definition of $\mathsf{form}^{\mathsf{h}}_{\Sigma}$ can be executed in S^1_2. An *extended subformula* is a subformula in the sense of this inductive definition. So $P(c[\exists x\, Qx])$ has e.g. Qx as extended subformula. An element of $\mathsf{form}^{\mathsf{h}}_{\Sigma}$ is in Γ^{h} if all its extended subformulas are substitution instances of Γ-formulas.

Since our coding is monotonic, we cannot have 'loops' of Henkin constants. It is not possible to have a constant $c[A]$ in an extended subformula of A, otherwise A would occur as a strict substring of itself.

We define $\Phi(A) := \{A, B(c[A])\}$ if A is of the form $\exists x\, B(x)$ and $\Phi(A) = \{A\}$, otherwise. Similarly, we define $\Phi^{\perp}(A) := \{A, B(c[A])\}$ if A is of the form $\forall x\, B(x)$ and $\Phi^{\perp}(A) = \{A\}$, otherwise.

We describe first the stage construction of the Henkin theory for α in the usual terms and then discuss how this definition can be implemented in the context of S^1_2. It is pleasant to think of the $\mathsf{form}^{\mathsf{h}}_{\Sigma}$-sentences as enumerated by A_i. Since, we confuse sentences and their Gödel numbers anyway, we can take $A_i := \top$ if i is not the Gödel number of a $\mathsf{form}^{\mathsf{h}}_{\Sigma}$-sentence and $A_i := i$ otherwise.

In our stage construction, Δ_a will stand for the set of formulas that we have put in our Henkin theory at stage a, $\Delta^{\perp}_a$ will stand for the set of formulas that we have decided at stage a to definitely keep out go our Henkin theory, ϕ will be a sequence of 0's, 1's and 2's that records the actions we take at each stage. We define:

stage 0 $\Delta_0 := \emptyset$, $\Delta^{\perp}_0 := \emptyset$.
stage $a + 1$ If A_a is a Γ^{h}-sentence:

- If $\Delta_a \cup \{A_a\} \nvdash_{\alpha,\Gamma^{\mathsf{h}}} \Delta^{\perp}_a$, we take $\phi(a) := 1$ and $\Delta_{a+1} := \Delta_a \cup \Phi(A_a)$ and $\Delta^{\perp}_{a+1} := \Delta^{\perp}_a$.
- If $\Delta_a \cup \{A_a\} \vdash_{\alpha,\Gamma^{\mathsf{h}}} \Delta^{\perp}_a$, we take $\phi(a) := 0$ and $\Delta_{a+1} := \Delta_a$ and $\Delta^{\perp}_{a+1} := \Delta^{\perp}_a \cup \Phi^{\perp}(A_a)$.

If A_a is not a Γ^{h}-sentence, we take $\phi(a) := 2$ and $\Delta_{a+1} := \Delta_a$ and $\Delta^{\perp}_{a+1} := \Delta^{\perp}_a$.

We note that our enumeration is chosen in such a way that $c[A_a]$ will be fresh w.r.t. Δ_a. Clearly, all information about the construction is contained in the function ϕ. The Henkin Theory $\Delta^{\star}$ will be the union of the Δ_a, where we let the variable a range over a suitable definable cut.

We address the details of the above construction. In S^1_2 we have codes for sequences and finite sets.

Consider a string σ over the alphabet $\{0, 1, 2\}$. Let us say that σ is *acceptable* when, for any $i < \mathsf{length}(\sigma)$, we have: if $\sigma_i \neq 2$, then i is a $\mathsf{form}^{\mathsf{h}}_{\Sigma}$-sentence. We note that initial substrings of acceptable strings are acceptable. We define, for acceptable σ:

- $\Theta_\sigma := \bigcup\{\Phi(A_a) \mid a < \mathsf{length}(\sigma) \text{ and } \sigma_a = 1\}$.
- $\Theta_\sigma^\perp := \bigcup\{\Phi^\perp(A_a) \mid a < \mathsf{length}(\sigma) \text{ and } \sigma_a = 0\}$.

Let $\ell := \mathsf{length}(\sigma)$. We note that each member of Θ_σ is estimated by $\omega_1(\ell)$. Moreover, the number of members of Θ_σ is at most 2ℓ. So Θ_σ itself is estimated by $(\omega_1(\ell))^{2\ell} = 2^{2|\ell|^2\ell} \leq 2^{2\ell^2}$ (for $\ell \geq 4$). Since ℓ is estimated by $|\sigma|$, it follows that Θ_σ is estimated by $\omega_1(\sigma)^2$. Thus, Θ_σ does indeed exist and, similarly, for $\Theta_\sigma^\perp$.

We note that our formulations are somewhat sloppy: there are different numerical codes for Θ_σ. A unique one can be singled out e.g. by choosing a fixed order of enumerating the elements of Θ_σ.

A 0,1,2-string σ is *adequate* iff σ is acceptable and, for all $i < \mathsf{length}(\sigma)$, we have:

- $\sigma_i = 1$ iff A_i is a Γ^{h}-sentence and $\Theta_{\sigma\restriction i} \cup \{A_i\} \nvdash_{\alpha,\Gamma^{\mathsf{h}}} \Theta^\perp_{\sigma\restriction i}$,
- $\sigma_i = 0$ iff A_i is a Γ^{h}-sentence and $\Theta_{\sigma\restriction i} \cup \{A_i\} \vdash_{\alpha,\Gamma^{\mathsf{h}}} \Theta^\perp_{\sigma\restriction i}$,
- $\sigma_i = 2$ iff A_i is not a Γ^{h}-sentence.

We use the length-first coding for strings so two 0,1,2-strings are equal iff they have the same numbers at the same places.

We claim that, for any a, there is at most one adequate string of length a. Consider two adequate strings σ and τ of length a. We prove by induction, that for all $i \leq a$, we have $\sigma \restriction i = \tau \restriction i$. The result is immediate noting that we have induction in S^1_2 for the property $\sigma \restriction i = \tau \restriction i$ seeing that a is a logarithmic in σ.

Consider the class I of the a such that, there is an adequate string σ of length a with $\Theta_\sigma \nvdash_{\alpha,\Gamma^{\mathsf{h}}} \Theta^\perp_\sigma$. We claim that I is closed under 0 and successor. We have $0 \in I$, since $\emptyset \nvdash_{\alpha,\Gamma^{\mathsf{h}}} \emptyset$. This is precisely our assumption $\Diamond_{\alpha,\Gamma}\top$, noting that we may replace the Henkin constants in any proof witnessing $\emptyset \vdash_{\alpha,\Gamma^{\mathsf{h}}} \emptyset$ by variables. A small argument is needed to guarantee that we can choose fresh variables in such a way that the result of substitution exists. Let π be our proof. We note that the number of fresh variables we need is bounded by $|\pi|$, the length of π. Moreover, to guarantee freshness it is sufficient to choose indices with length $|\pi|$. So our new proof will have length estimated by $|\pi|^2$, and, hence, the new proof itself is estimated by $\omega_1(\pi)$.

Suppose a is in I. Let σ be an adequate sequence of length a with $\Theta_\sigma \nvdash_{\alpha,\Gamma^{\mathsf{h}}} \Theta^\perp_\sigma$. Let $A := A_a$.

In case A is a Γ^{h}-sentence and $\Theta_\sigma, A \nvdash_{\alpha,\Gamma^{\mathsf{h}}} \Theta^\perp_\sigma$, the string $\tau := \sigma 1$ is adequate and $\Theta_\tau = \Theta_\sigma \cup \Phi(A)$ and $\Theta^\perp_\tau = \Theta^\perp_\sigma$. In case A is not of the form $\exists x\, Bx$ we are done. Suppose A is of the form $\exists x\, Bx$. Assume, to get a contradiction that $\Delta_\sigma, A, B(c[A]) \vdash \Delta^\perp_\sigma$. By construction $c[A]$ does not occur in $\Delta_\sigma, A, \Delta^\perp_\sigma$. So, by $\mathsf{L}\exists$, we find $\Delta_\sigma, A \vdash \Delta^\perp_\sigma$. Quod non.

In case A is a Γ^{h}-sentence and (a) $\Theta_\sigma, A \vdash_{\alpha,\Gamma^{\mathsf{h}}} \Theta^\perp_\sigma$, the string $\tau = \sigma 0$ is adequate and $\Theta_\tau = \Theta_\sigma$ and $\Theta^\perp_\tau = \Theta^\perp_\sigma \cup \Phi^\perp(A)$. Suppose A is not of the form $\forall x\, Bx$. To obtain a contradiction, suppose we have (b) $\Theta_\sigma \vdash_{\alpha,\Gamma} \Theta^\perp_\sigma, A$. Applying the cut rule to (a) and (b), we obtain: $\Theta_\sigma \vdash_{\alpha,\Gamma^{\mathsf{h}}} \Theta^\perp_\sigma$. Quod non. Now suppose A is of the form $\forall x\, Bx$. To obtain a contradiction, we assume $\Theta_\sigma \vdash_{\alpha,\Gamma^{\mathsf{h}}} \Theta^\perp_\sigma, A, B(c[A])$. We note that $c[A]$ does not occur in $\Theta_\sigma, A, \Theta^\perp_\sigma$. By $\mathsf{R}\forall$, we get $\Theta_\sigma \vdash_{\alpha,\Gamma^{\mathsf{h}}} \Theta^\perp_\sigma, A$. We again use the cut-rule to obtain a contradiction.

In case A is not a Γ^{h}-sentence, the string $\tau = \sigma 2$ is adequate. We have $\Theta_\tau = \Theta_\sigma$ and $\Theta_\tau^\perp = \Theta_\sigma^\perp$. So, we are immediately done.

Since I is inductive, we can shorten it to a ω_1-cut J. On this cut we have S^1_2. Note that J is closed under the syntactic operations including substitution. For $i \in J$, we define Δ_i as Θ_σ for the unique adequate σ of length i and, similarly, for $\Delta_i^\perp$. We define $\Delta^\star$ as the virtual class that is the union of the Δ_i such that $i \in J$. Similarly, $\Delta^{\perp\star}$ is the union of the $\Delta_i^\perp$ where i in J.

We first show that each Γ^{h}-sentence A in J is in $\Delta^\star \cup \Delta^{\perp\star}$. Suppose A is a Γ^{h}-sentence in J. Say $A = A_i$, where $i \in J$. Consider the unique adequate sequence σ of length $i+1$. In case $\sigma_i = 1$, we have $A \in \Theta_\sigma = \Delta_{i+1}$, in case $\sigma_i = 0$, we have $A \in \Theta_\sigma^\perp = \Delta_{i+1}^\perp$. Since we assumed A to be in Γ^{h}, by the definition of adequate sequence, we cannot have $\sigma_i = 2$. So either $A \in \Delta_{i+1} \subseteq \Delta^\star$ or $A \in \Delta_{i+1}^\perp \subseteq \Delta^{\perp\star}$.

The Δ_i will be monotonically increasing. Consider stages $i < j \in J$. Let σ be an adequate string of length i and let τ be an adequate string of length j. From the definition of adequacy, it is immediate that $\tau \restriction i$ is an adequate string of length i and, hence, we find $\sigma = \tau \restriction i$. Trivially, $\Theta_{\tau\restriction i} \subseteq \Theta_\tau$. By a similar argument the $\Delta_i^\perp$ are monotonically increasing.

We show that $\Delta^\star \cap \Delta^{\perp\star} = \emptyset$. Suppose $A \in \Delta^\star \cap \Delta^{\perp\star}$. Then, for some stages i and j, we will have $A \in \Delta_i$ and $A \in \Delta_j^\perp$. Let a be the maximum of i and j. It follows, by monotonicity, that $A \in \Delta_a \cap \Delta_a^\perp$. But this contradicts the fact that A is in J and hence in I, in combination with the deduction rules Share and W.

Consider a sentence A in $\Gamma^{\mathsf{h}} \cap J$ and any $i \in J$. We show that (†) if $\Delta_i \vdash_{\alpha,\Gamma^{\mathsf{h}}} \Delta_i^\perp, A$, then $A \in \Delta^\star$ and (‡) if $A, \Delta_i \vdash_{\alpha,\Gamma^{\mathsf{h}}} \Delta_i^\perp$, then $A \in \Delta^{\perp\star}$.

Suppose $\Delta_i \vdash_{\alpha,\Gamma^{\mathsf{h}}} \Delta_i^\perp, A$. Let $A := A_j$ and let $a := \mathsf{max}(i, j+1)$. In case $A_j \in \Delta_{j+1}^\perp$, then $A_j \in \Delta_a^\perp$. Hence, by Share and W, we find $\Delta_a \vdash_{\alpha,\Gamma^{\mathsf{h}}} \Delta_a^\perp$. But this contradicts the fact that $a \in J \subseteq I$. So $A \in \Delta_{j+1} \subseteq \Delta^\star$. Similarly, for the other case.

We note that it follows that $\forall A \in (\mathsf{sent}_\Sigma \cap \Gamma \cap J)\,(\Box_{\alpha,\Gamma} A \to A \in \Delta^\star)$.

Our next order of business is to verify the commutation conditions for Δ^*. Let $\mathcal{G}$ be the class of sentences of $\Gamma^{\mathsf{h}} \cap J$.

Consider $(B \vee C) \in \mathcal{G}$. Suppose $B = A_j$ and $C = A_k$ and $(B \vee C) = A_i$. Here $j < i$ and $k < i$.

Suppose $B \vee C$ is in $\Delta^\star$. It follows that $B \vee C$ is in Δ_{i+1}. In case neither B nor C is in $\Delta^\star$, it follows that both B and C are in $\Delta_{i+1}^\perp$. Hence $B, \Delta_{i+1} \vdash_{\alpha,\Gamma^{\mathsf{h}}} \Delta_{i+1}^\perp$ and $C, \Delta_{i+1} \vdash_{\alpha,\Gamma^{\mathsf{h}}} \Delta_{i+1}^\perp$. Ergo, by $\mathsf{L}\vee$, we have $B \vee C, \Delta_{i+1} \vdash_{\alpha,\Gamma^{\mathsf{h}}} \Delta_{i+1}^\perp$ and, hence, $\Delta_{i+1} \vdash_{\alpha,\Gamma^{\mathsf{h}}} \Delta_{i+1}^\perp$. Quod non. So, B is in $\Delta^\star$ or C is in $\Delta^\star$.

Conversely, if B is in $\Delta^\star$, we have $B \in \Delta_{j+1}$. Hence, by Share and W, we find $\Delta_{j+1} \vdash_{\alpha,\Gamma^{\mathsf{h}}} \Delta_{j+1}^\perp, B$. Ergo, by $\mathsf{R}\vee$, we obtain $\Delta_{j+1} \vdash_{\alpha,\Gamma^{\mathsf{h}}} \Delta_{j+1}^\perp, B \vee C$. So, $B \vee C$ is in $\Delta^\star$. Similarly, in case C is in $\Delta^\star$.

The case of conjunction is similar.

We treat the case of implication. Consider $(B \to C) \in \mathcal{G}$. Suppose $B = A_j$ and $C = A_k$ and $(B \to C) = A_i$. Here $j < i$ and $k < i$.

Suppose $B \to C$ is in $\Delta^\star$ and B is in $\Delta^\star$. It follows that $B \to C$ and B are in Δ_{i+1}. In case C is not in $\Delta^\star$, it follows that C is in $\Delta^{\perp\star}$ and, thus in $\Delta_{i+1}^\perp$. Ergo, $\Delta_{i+1} \vdash_{\alpha,\Gamma^{\mathsf{h}}}$

$\Delta^{\perp}_{i+1}$, B and C, $\Delta_{i+1} \vdash_{\alpha,\Gamma^{\mathsf{h}}} \Delta^{\perp}_{i+1}$. Thus, by $\mathsf{L}\!\rightarrow$, we have $B \rightarrow C, \Delta_{i+1} \vdash_{\alpha,\Gamma^{\mathsf{h}}} \Delta^{\perp}_{i+1}$ and, so, $\Delta_{i+1} \vdash_{\alpha,\Gamma^{\mathsf{h}}} \Delta^{\perp}_{i+1}$. Quod non. Hence, C is in $\Delta^{\star}$.

Next suppose that if B is in $\Delta^{\star}$, then C is in $\Delta^{\star}$. This means that either B is in $\Delta^{\perp\star}$ or C is in $\Delta^{\star}$. Hence, B is in $\Delta^{\perp}_{i+1}$ or C is in Δ_{i+1}. In both cases, we have, by Share and W, that $B, \Delta_{i+1} \vdash_{\alpha,\Gamma^{\mathsf{h}}} \Delta^{\perp}_{i+1}, C$. Hence, by $\mathsf{R}\!\rightarrow$, we find $\Delta_{i+1} \vdash_{\alpha,\Gamma^{\mathsf{h}}} \Delta^{\perp}_{i+1}, B \rightarrow C$. It follows that $(B \rightarrow C)$ is in $\Delta^{\star}$.

We treat the case of the existential quantifier. Suppose $\exists x\, Bx$ is in $\mathcal{G}$. Suppose $\exists x\, Bx$ is in $\Delta^{\star}$. Suppose $\exists x\, Bx$ is A_i. Since it is impossible that $\exists x\, Bx$ is in $\Delta^{\perp}_{i+1}$, we will have $\sigma_i = 1$, for an adequate σ of length $i + 1$. Hence, $B(c[\exists x\, B])$ is in Θ_{σ}, i.e. in Δ_{i+1}, and, hence in $\Delta^{\star}$.

Conversely, suppose that $B(c)$ is in Δ^{*}, for some Henkin constant c in J. Say, $B(c) = A_j$. Then, $B(c)$ is in Δ_{j+1}. It follows by Share and W that $\Delta_{j+1} \vdash_{\alpha,\Gamma^{\mathsf{h}}} \Delta^{\perp}_{j+1}, Bc$, and hence, by $\mathsf{R}\exists$ that $\Delta_{j+1} \vdash_{\alpha,\Gamma^{\mathsf{h}}} \Delta^{\perp}_{j+1}, \exists x\, Bx$. Ergo, by (†), $\exists x\, Bx$ is in $\Delta^{\star}$.

The proof for universal quantification is similar.

Now we are ready and set to define the Henkin-translation η. We take as our domain δ_{η} the Henkin constants in J. Note that these Henkin constants are not restricted to $c[A]$ where A is in Γ^{h}. We define:

- $P_{\eta}(x_0, \cdots, x_{n-1}) := \underline{\ulcorner P \urcorner} * \underline{\ulcorner (\urcorner} * x_0 * \underline{\ulcorner , \urcorner} * \cdots * \underline{\ulcorner , \urcorner} * x_{n-1} * \underline{\ulcorner) \urcorner} \in \Delta^{\star}$.[11]

Here $*$ stands for the arithmetization of concatenation. We include the relation of identity in the above definition.

We define $\mathsf{sat}_{\eta}(\sigma, A)$, where σ is an assignment from the free variables in A to elements of δ_{η} by: the result of substituting $\sigma(v)$ for v in A, for each free variable v of A, is in $\Delta^{\star}$.[12] It follows that we have the usual commutation conditions for the connectives for sat_{η} for formulas in $\Gamma \cap J$. E.g., for $\forall w\, B$ in $\Gamma \cap J$, we have: $\mathsf{sat}_{\eta}(\sigma, \forall w\, B)$ iff $\forall x \in \delta_{\eta}\ \mathsf{sat}_{\eta}(\sigma[w : x], B)$.

Using sat_{η}, we can reformulate the atomic definition of P_{η} as follows:

$$P_{\eta}(x_0, \cdots, x_{n-1}) \text{ is } \mathsf{sat}_{\eta}(\{\langle v_0, x_0\rangle, \cdots, \langle v_{n-1}, x_{n-1}\rangle\}, P(v_0, \cdots, v_{n-1})).$$

Finally consider a standard Σ-formula A in Γ. Since A is standard. A will also be in J. It follows from the commutation conditions that

$$\forall \vec{x} \in \delta_{\eta}\, (A^{\eta}(\vec{x}) \leftrightarrow \mathsf{sat}_{\eta}(\sigma_{\vec{x}}, \underline{\ulcorner A \urcorner}))$$

Here $\sigma_{\vec{x}}$ is a code of a function that assigns x_i to $\underline{\ulcorner x_i \urcorner}$.

Suppose $\underline{\ulcorner A \urcorner} \in \alpha$. It follows that $\Box_{\alpha,\Gamma} A$, and, hence, $\underline{\ulcorner A \urcorner} \in \Delta^{\star}$ or, equivalently, $\mathsf{sat}_{\eta}(\emptyset, \underline{\ulcorner A \urcorner})$. By the commutation conditions, we find: A^{η}.

---

[11] I choose for this representation of P_{η} since it contains no ambiguities. One would like to say that $P_{\tau}(x_0, \cdots, x_{n-1})$ is $P(c_0, \cdots, c_{n-1}) \in \Delta^{\star}$, were the c_i *are* the x_i.

[12] We are a bit sloppy, since the finite set need not have a unique code. The insight works for every code of the displayed set.

At this point we leave SF_Σ. We have shown $\mathsf{SF}_\Sigma \vdash \underline{\ulcorner A \urcorner} \in \alpha \rightarrow A^\eta$. Hence:

$$\begin{aligned} A \text{ is an axiom of } V &\Rightarrow \mathsf{sf}(V) \vdash \underline{\ulcorner A \urcorner} \in \alpha \\ &\Rightarrow \mathsf{sf}(V) \vdash A^\eta \end{aligned}$$

We may conclude that $H : \mathsf{sf}(V) \rhd V$, where H is the interpretation that is based on η. □

We note that η is not dependent of V. It is only in transmuting η into an interpretation of V rather than of predicate logic for Σ that the full theory $\mathsf{sf}(V)$ comes into play.

Remark 3.2 Where can the proof of Theorem 3.1 can be verified? We note that the reasoning up to and including the verification of the commutation conditions for sat_η just is one specific proof in SF_Σ. Thus, this part of the reasoning can be trivially executed in S^1_2 as metatheory.

The verification of the Tarski Biconditionals for sat_η involves a SF_Σ-external induction on syntax that is present in S^1_2 as metatheory.

The last step concerning $\mathsf{sf}(V)$ involves an arbitrary theory V, where the axioms of V may be any set X of Σ-sentences. Here the formulation of the result transcends the resources of S^1_2. Now let us assume that the axiom-set of V is given as a Δ^{b}_1-formula. Clearly, under this assumption, the verification that H is an interpretation just collects the previous resuls, and our whole verification is present in S^1_2.

What we have verified, under the extra assumption, is *axioms-interpretability*: for every axiom A of V we can prove A^η in $\mathsf{sf}(V)$. Can we also verify *theorems-interpretability*: for every theorem B of V we can prove B^η in sf(V)? Consider a V-proof π of B. What will the $\mathsf{sf}(V)$-verification, say $\pi^\star$, of B^η look like? Well, first we transform π into the translated proof π^η, where we η-translate every formula in π and add a bit of extra stuff to handle the domain relativization. This transformation is polynomial. The resulting proof is a $\mathsf{sf}(V)$-proof from 'axioms' A^η, where A is a V-axiom. The next step is to plug in proofs of A^η from $A \in \alpha \cap J$.

We remind the reader that this proof runs as follows: we start with the assumption $A \in \alpha \cap J$. From this we move to $\Box_{\alpha,\Gamma} A$ and from there to $\mathsf{sat}_\eta(\emptyset, A)$. Now we employ the commutation conditions to obtain A^η. The proof involving commutation is $\mathsf{sf}(V)$-external. Inspection shows that it is p-time in A.

The proofs that $A \in J$ are also p-time since the A are relatively standard. Note that, for each axiom A used in π. we have $|A| < |\pi|$. So, the various bits we added are each bounded by $|\pi|^n + k$, for standard n and k. The number of extra bits we added is bounded by $2|\pi|$. So the new proof will be bounded by $|\pi^{n'} + k'|$, for standard n' and k'. In other words the transformed proof is p-time in π. Thus, we have theorems interpretability.

Suppose we want to conclude $U \rhd V$ from $U \rhd \mathsf{sf}(V)$ inside S^1_2. In this case we have to be careful: even if $U \rhd V$ is given for theorems-interpretability, transitivity for theorems-interpretability may fail in some models of S^1_2. What we need for the transitivity of theorems-interpretability is Σ_1-collection. So, the most natural meta-theory for the formalization of Theorem 3.1 and its desired consequences, on the assumption of Δ_1^{b}-axiom sets, is S_2^1 plus Σ_1-collection. See [37], for further discussion of these subtle points.

Remark 3.3 We note that $\rho(\eta)$ is a fixed number, say $\mathfrak{h}$. Suppose $K : U \rhd \mathsf{sf}(V)$. Then $M := (H_V \circ K) : U \rhd V$, where $\rho(M) = \rho(K) + \mathfrak{h}$.

In our proof, we showed more than contained in the statement of Theorem 3.1. The next somewhat meandering theorem articulates some of that extra information. Let $\mathsf{SF}_{\Sigma,\mathsf{cut}}$ be the following theory. We expand the signature of SF_Σ by a new predicate J and add axioms to the effect that J is an ω_1-cut. The theory $\mathsf{sf}_{\mathsf{cut}}(V)$ is defined as $\mathsf{SF}_{\Sigma_V,\mathsf{cut}} + \mathsf{Sf}_{\mathsf{ax}_V}$.

Theorem 3.4 *Consider a signature Σ. There is a one-dimensional interpretation $H_\Sigma : \mathsf{SF}_\Sigma \rhd \mathsf{Pred}_\Sigma$ with underlying translation η_Σ. The domain of H_Σ is contained in a definable SF_Σ-cut J_Σ.*

From this point on we suppress the subscript Σ in the statement of the theorem.

In SF, we have a satisfaction predicate sat_η for H for formulas A in $\Gamma \cap J$. This predicate satisfies the commutation conditions for all formulas in its intended domain. The commutation conditions for the quantifiers involve relativization to δ_η and the commutation condition of an atomic predicate P is:

$$\mathsf{SF} \vdash \forall \vec{x} \in \delta_H\, (\mathsf{sat}_\eta(\vec{x}, \underline{\ulcorner P\vec{v}\urcorner}) \leftrightarrow P_\eta(\vec{x})).$$

We have:

$$\mathsf{SF} \vdash \forall A \in (\mathsf{sent}_\Sigma \cap \Gamma \cap J)\, (\Box_{\alpha,\Gamma} A \to \mathsf{sat}_\eta(\emptyset, A)).$$

Hence, a fortiori,

$$\mathsf{SF} \vdash \forall A \in (\alpha \cap J)\ \mathsf{sat}_\eta(\emptyset, A).$$

It follows that, for any Σ-theory V, we have an interpretation $H_V : \mathsf{sf}(V) \rhd V$, where H is based on η.

Expanding SF to $\mathsf{SF}_{\mathsf{cut}}$, we have the following. The restriction of δ_H to the ω_1-cut $J^* := J \cap \mathsf{J}$ gives us a $(\Gamma \cap J^*)$-elementary sub-interpretation H^* of H. This means that the restriction of sat_η to $J^* \times J^*$ is a satisfaction predicate sat_{η^*} that satisfies the commutation conditions for relativization to δ_{η^*} and for formulas from $\Gamma \cap J^*$. Here the atomic commutation condition is the one for H^*. We have again:

$$\mathsf{SF}_{\mathsf{cut}} \vdash \forall A \in (\mathsf{sent}_\Sigma \cap \Gamma \cap J^*)\, (\Box_{\alpha,\Gamma} A \to \mathsf{sat}_{\eta^*}(\emptyset, A)).$$

Proof The last bit of the theorem about the sub-cut is by observing that in our construction of $\Delta^\star$ any cut J^* that is shorter than the given J would do as well. □

We formulate a model theoretic consequence of Theorem 3.4.

Corollary 3.5 *Consider a model $\mathcal{N}$ of* $\mathsf{sf}(V)$. *Then there is an internal V-model $\mathcal{H}$ of $\mathcal{N}$ with underlying translation η. The domain of $\mathcal{H}$ is contained in an $\mathcal{N}$-definable ω_1-cut $\mathcal{J}$ of $\mathcal{N}$. Moreover, we have a satisfaction predicate sat_η for $\mathcal{H}$ for formulas A in $\Gamma_{\mathcal{N}} \cap \mathcal{J}$. This predicate satisfies the commutation conditions for all formulas in its intended domain.*

Consider any sub-ω_1-cut $\mathcal{J}^*$ of $\mathcal{J}$. The cut $\mathcal{J}^*$ need not be definable in $\mathcal{N}$. The restriction of $\delta_{\mathcal{H}}$ to $\mathcal{J}^*$ gives us a $(\Gamma_{\mathcal{N}} \cap \mathcal{J}^*)$-elementary sub-model $\mathcal{H}^*$ of $\mathcal{H}$.

3.3 Extending the Target Theory to a Sequential Theory

Since $\mathsf{sf}(V)$ is a sequential theory, we can strengthen Theorem 3.1 a bit in another direction. We define $\mathsf{seq}(V)$ as follows. We expand the language of V with a unary predicate D and a binary predicate $\in$. The theory $\mathsf{seq}(V)$ is axiomatized by the axioms of V relativized to D plus AS for $\in$. The theory $\mathsf{seq}(V)$ is clearly sequential.

Theorem 3.6 *Suppose U is (poly)sequential and $U \rhd V$. Then $U \rhd \mathsf{seq}(V)$.*

Proof We give the proof for the poly-sequential case. So we suppose that the direct interpretation of AS in U is m-dimensional.

In order not to get totally confused between the syntactically realized external sequences that are connected with the dimension of interpretations and the internal sequences of the theory U, we call an external sequence $(x_0, \cdots, x_{m-1})$ *an m-sequence* and an internal sequence simply *a sequence*.

Suppose $K : U \rhd V$. Since we can code any sequence in U using an m-sequence, we may without loss of generality assume that K is m-dimensional. We define the m-dimensional interpretation $M := \mathsf{seq}(K) : U \rhd \mathsf{seq}(V)$ with underlying translation μ as follows. Let $\alpha, \beta, \gamma, \ldots$ range over m-sequences.

- The domain δ_μ consists of m-sequences of the form $\langle 0, \alpha\rangle$ for arbitrary α and $\langle 1, \alpha\rangle$ for $\alpha \in \delta_K$. (Remember that in our context there may be many m-sequences implementing 0, implementing $\langle 0, \alpha\rangle$, etcetera.)
- $\mathsf{D}_\mu(\alpha)$ iff α is of the form $\langle 1, \beta\rangle$ and $\beta \in \delta_K$.
- $\alpha =_\mu \alpha'$ iff (α and α' are of the form $\langle 0, \beta\rangle$ respectively $\langle 0, \beta'\rangle$ and $\beta = \beta'$) or (α and α' are of the form $\langle 1, \beta\rangle$ respectively $\langle 1, \beta'\rangle$ and $\beta =_K \beta'$).
- $\alpha \in_\mu \alpha'$ iff α' is of the form $\langle 0, \beta'\rangle$ and ((α is of the form $\langle 0, \beta\rangle$ and $\beta \in \beta'$) or (α is of the form $\langle 1, \beta\rangle$ and for some γ with $\gamma =_\mu \beta$ we have $\gamma \in \beta'$)).
- $P_\mu(\alpha_0, \cdots, \alpha_{n-1})$ iff each α_i is of the form $\langle 1, \beta_i\rangle$ and $P_K(\beta_1, \cdots, \beta_{n-1})$.

As is easily seen μ supports the promised interpretation M. □

We note the following corollary.

Corollary 3.7 *If U is polysequential, then $U \equiv \mathsf{seq}(U)$. So every m-sequential theory is mutually interpretable with a 1-sequential theory.*

From the previous theorem, we immediately have:

Corollary 3.8 $\mathsf{sf}(V) \rhd \mathsf{seq}(V)$.

We note that $\mathsf{sf}(V)$ does not necessarily provide a satisfaction predicate for some reasonable formula-class for $\mathsf{seq}(V)$.

3.4 An Equivalent of sf(V)

Consider any finite signature Σ. We define the theory Comm_Σ as follows. We extend the signature of S^1_2 with three new predicate symbols Γ, sat and D. Our theory is axiomatized by axioms Sf_Σ1–7 plus the non-atomic clauses of the commutation conditions of sat for Γ-formulas, where the quantifiers are relativized to D. We define:

$$\mathsf{comm}(V) := \mathsf{Comm}_{\Sigma_V} + \{\alpha(A) \mid A \in \mathsf{ax}_V\}.$$

Let us write $\rhd_{\mathsf{cut}}$ for cut interpretability. This means interpretability by a parameter-free interpretation K such that (i) the domain formula is a definable cut and (ii) identity and the predicates corresponding to the arithmetical operations are translated identically but for the fact that we constrain their variables to be in the cut. Using this terminology, Theorem 3.4 tells us that $\mathsf{SF}_\Sigma \rhd_{\mathsf{cut}} \mathsf{Comm}_\Sigma$ and $\mathsf{sf}(V) \rhd_{\mathsf{cut}} \mathsf{comm}(V)$. Conversely, using several shortenings, we can prove that

$$\mathsf{Comm}_\Sigma \rhd_{\mathsf{cut}} \mathsf{SF}_\Sigma \text{ and } \mathsf{comm}(V) \rhd_{\mathsf{cut}} \mathsf{sf}(V).$$

Hence, $\mathsf{Comm}_\Sigma \equiv_{\mathsf{cut}} \mathsf{SF}_\Sigma$ and $\mathsf{comm}(V) \equiv_{\mathsf{cut}} \mathsf{sf}(V)$.

Open Question 3.9 We note that both comm and sf involve coding. Can we make a coding-free equivalent? (in Sect. 4, we will see some examples of coding-free variants for special cases.)

3.5 Treatment of Numerals

Consider H_V and suppose that $N : \mathsf{S}^1_2 \lhd V$. We can name $H_V \circ N$-numbers both using Henkin constants and using numerals. In this subsection we study how these two representations are related.

Since we are talking about interpretations, we need the relational version of numerals. We work for the moment in the relational version of the language of arithmetic. We select fixed, disjoint standard variables x, y, u, v, w. We define a sequence of formulas that will represent *dyadic numerals*.

- $C_0 x := \mathsf{Z}x$,
- $C_{2n+1} y := \exists x\, \exists u\, (C_n x \wedge \mathsf{A}xxu \wedge \mathsf{S}uy)$, if n is even,
  $C_{2n+1} x := \exists y\, \exists u\, (C_n y \wedge \mathsf{A}yyu \wedge \mathsf{S}ux)$, if n is odd,
- $C_{2n+2} y := \exists x\, \exists u\, \exists v\, (C_n x \wedge \mathsf{A}xxu \wedge \mathsf{S}uv \wedge \mathsf{S}vy)$, if n is even,
  $C_{2n+2} x := \exists y\, \exists u\, \exists v\, (C_n y \wedge \mathsf{A}yyu \wedge \mathsf{S}uv \wedge \mathsf{S}vx)$, if n is odd.
- For w disjoint from x, y, u, v, we write $C_n(w)$ for $C_n[y : w]$, if n is 1 or 2 modulo 4, and for $C_n[x : w]$, if n is 3 or 0 modulo 4.
- $D_n := \exists w\, C_n(w)$.

We clearly can implement this definition in S^1_2. We define the numerical constants by:

- $c_n := c[D_n^N]$. We call the c_n: *h-numerals* (w.r.t. V and N).

Clearly, the code of c_n is polynomial in n. To obtain our first insight in the behavior of h-numerals, we need a lemma.

Lemma 3.10 *Suppose $N : \mathsf{S}^1_2 \lhd V$. We work with a finite axiomatization of S^1_2, so there is one witnessing proof. Let this witnessing proof be π. (We assume that π includes the verification of the axioms of identity.) Let $n := \rho(\pi)$. Let $W := \mathsf{sf}(V) + (\Sigma^\star_n \subseteq \Gamma)$.*

Then:

1. $W \vdash \forall x\, \Box_{\alpha,\Gamma} \exists w \in \delta_N\, C^N_x(w)$.
2. $W \vdash \forall x\, \Box_{\alpha,\Gamma} \forall w, w' \in \delta_N\, ((C^N_x(w) \wedge C^N_x(w')) \to w =_N w')$.
3. $W \vdash \forall x\, \Box_{\alpha,\Gamma} \forall w \in \delta_N\, (C^N_x(w) \to C^N_{2x}(2w))$.
4. $W \vdash \forall x\, \Box_{\alpha,\Gamma} \forall w \in \delta_N\, (C^N_x(w) \to C^N_{2x+1}(2w+1))$.

Proof We assume the conditions of the theorem to be fulfilled. We reason in W. Let k be the code of S^1_2 and let k' be $\rho(\mathsf{S}^1_2)$. For any x, we have:

1'. $\Box_{\mathsf{S}^1_2,k,k'} \exists w\, C_x(w)$.
2'. $\Box_{\mathsf{S}^1_2,k,k'} \forall w\, \forall w'\, ((C_x(w) \wedge C_x(w')) \to w = w')$.
3'. $\Box_{\mathsf{S}^1_2,k,k'} \forall w\, (C_x(w) \to C_{2x}(2w))$.
4'. $\Box_{\mathsf{S}^1_2,k,k'} \forall w\, (C_x(w) \to C_{2x+1}(2w+1))$.

Let ν_i be the witnessing proof for (i'). We transform ν_i to ν_i^N from axioms $(\mathsf{S}^1_2)^N$. We have $\rho(\nu_i^N) = k' + \rho(N)$. Next we add, above the uses of the $(\mathsf{S}^1_2)^N$-axiom in ν_i^N, the proof π that verifies $(\mathsf{S}^1_2)^N$. Let the resulting proof be ν_i^*. We note that the conclusion of π has complexity $k' + \rho(N)$, so $\rho(\nu_i^*) = \rho(\pi) = n$. Thus the formulas of this proof are all in Γ. It follows that the ν_i^* witness:

1. $\Box_{\alpha,\Gamma} \exists w \in \delta_N\, C^N_x(w)$.
2. $\Box_{\alpha,\Gamma} \forall w, w' \in \delta_N\, ((C^N_x(w) \wedge C^N_x(w')) \to w =_N w')$.
3. $\Box_{\alpha,\Gamma} \forall w \in \delta_N\, (C^N_x(w) \to C^N_{2x}(2w))$.
4. $\Box_{\alpha,\Gamma} \forall w \in \delta_N\, (C^N_x(w) \to C^N_{2x+1}(2w+1))$.

□

Theorem 3.11 *Suppose $N : \mathsf{S}^1_2 \lhd V$. Let the witnessing proof be π. (We assume that π includes the verification of the axioms of identity.) Let $n := \rho(\pi)$. Let $W := \mathsf{sf}(V) + (\Sigma^\star_n \subseteq \Gamma)$. Consider the interpretation $H : W \rhd V$ based on η. Let J be the cut on which H was constructed; let J_0 be the common cut of J, and $H \circ N$ and let F be the definable isomorphism between J_0 and a cut of $H \circ N$. (See Theorem 2.5.)*

In W, we can shorten J_0 to an ω_1-cut J_1 such that, for all x in J_1, $F(x) =_{H \circ N} c_x$.

Proof We work in W. Let J_0 and F be as described in the formulation of the theorem.

Clearly, $F0 =_{H \circ N} c_0$.

Suppose $x \in J_0$ and $Fx =_{H \circ N} c_x$. Let k be the Gödel number of the axiom of S^1_2 and let $k' := \rho(\mathsf{S}^1_2)$.

By (1) of Lemma 3.10, we have in W that $\Delta^\star \vdash_{\alpha, \Gamma^{\mathsf{h}}} C_x(c_x)$ and $\Delta^\star \vdash C_{2x}(c_{2x})$. By (3) of Lemma 3.10, we obtain $\Delta^\star \vdash C_{2x}(2c_x)$. Hence, by (2) of Lemma 3.10, we may conclude: $\Delta^\star \vdash c_{2x} =_N 2c_x$. Thus, we have $(c_{2x} = 2c_x)^{NH}$. Since F is an isomorphism, it follows that $F(2x) =_{H \circ N} c_{2x}$. Similarly for $F(2x+1)$. We may conclude that the class X of x such that $F(x) =_{H \circ N} c_x$ is closed under 0, and the functions $\lambda x \cdot 2x$ and $\lambda x \cdot 2x + 1$. Hence, we can shorten X to the desired ω_1-cut J_1. ☐

Next we turn to the usual dyadic numerals. What does $A(\underline{\vec{n}}\,)$ mean? We interpret this using small scope elimination of terms. The transformation of the formula commutes with all connectives, so we only have to consider its meaning for atoms. Let us consider the case $P(x, \underline{n}, \underline{k})$. We translate this to:

$$\exists w \in \delta_N \, \exists w' \in \delta_N \, (C^N_n(w) \wedge C^N_k(w) \wedge P(x, w, w')).$$

Here we choose the w, w' fresh according to some predetermined rule. We have the following.

Theorem 3.12 *Suppose $N : \mathsf{S}^1_2 \lhd V$. Let the witnessing proof be π. (We assume that π includes the verification of the axioms of identity.) Let $n := \rho(\pi)$. Let $W := \mathsf{sf}(V) + (\Sigma^\star_n \subseteq \Gamma) + \forall A \in \Gamma\, (\Sigma^*_{\rho(A)} \subseteq \Gamma)$. Consider the interpretation $H : W \rhd V$ based on η. Let J be the cut on which H was constructed.*

In the present theorem, we write $A(\vec{v}\,)$ for the formula A with a selection *of its free variables displayed. So the real object we are talking about is something like $\langle A, \vec{v}\,\rangle$, where the elements of $\vec{v}$ are among the free variables of A.*

In W, we can shorten J to an ω_1-cut J_0 such that, for all formulas $A(\vec{v}\,)$ in J_0 and for all $\vec{a}$ in J_0, if $A(\underline{\vec{a}}\,)$ is in $\Gamma \cap J_0$, then $\mathsf{sat}_\eta(\sigma, A(\underline{\vec{a}}\,))$ iff $\mathsf{sat}_\eta(\sigma', A(\vec{v}\,))$, where σ' is σ extended with the function that sends v_i to c_{a_i}.

Proof We assume the conditions of the theorem. Note that under our assumptions, if $A(\underline{\vec{a}}\,)$ is in $\Gamma \cap J_0$, then so is $A(\vec{v}\,)$. We employ as measure of complexity ν that computes depth of connectives. This means that $\nu(C)$ is 0 if C is an atom and $\nu(D \wedge E) = \mathsf{max}(\nu(D), \nu(E)) + 1$, etcetera.

We consider the class X of all x in J such that, for all formulas $B(\vec{w}\,)$ in J of ν-complexity $\leq x$, and, for all $B(\underline{\vec{b}}\,)$, where the b_i are in J, if $B(\underline{\vec{b}}\,) \in \Gamma$, then, for all σ, $\mathsf{sat}_\eta(\sigma, B(\vec{\underline{b}}))$ iff $\mathsf{sat}_\eta(\sigma', B)$, where σ' is σ extended with the function that sends v_i to c_{a_i}.

Using Lemma 3.10(1)(2), for the atomic case and the commutation conditions for sat for the connectives, we may now show that X is closed under 0 and successor. We shorten X to an ω_1-cut J_0.

We find that, for all formulas $B(\vec{w})$ in J of ν-complexity in J_0, and, for all $B(\vec{\underline{b}})$, where the b_i are in J, if $B(\vec{\underline{b}}) \in \Gamma$, then, for all σ, $\mathsf{sat}_\eta(\sigma, B(\vec{\underline{b}}))$ iff $\mathsf{sat}_\eta(\sigma', B)$, where σ' is σ extended with the function that sends v_i to c_{a_i}.

Note that if $B(\vec{w})$ is in J_0, then $\nu(B)$ is in J_0 and $B(\vec{\underline{b}})$ is in J_0. It follows that J_0 has the property promised in the theorem. □

3.6 The Collapse

We can sharpen Theorem 3.4 further by providing 'normal forms' for the domain of the interpretation. I add these sharpenings for completeness of the presentation. They will not be used further in the paper. In the next theorem, we show that we can modify interpretation H to make it identity preserving and such that its domain is an initial segment.

Let Eqr be the following theory. We expand the signature of S^1_2 by a unary predicate symbol D and a binary predicate symbol E. The axioms of our theory are S^1_2 plus axioms stating that that E is an equivalence relation on D.[13]

Lemma 3.13 *In* Eqr, *we have an* ω_1*-cut* I *and a collapsing function* coll *that is a bijection between* $(\mathsf{D} \cap I)/(\mathsf{E} \restriction (\mathsf{D} \cap I))$ *and an initial segment of* I *(with ordinary identity). The bijection has the further property that for* $x \in \mathsf{D} \cap I$, *we have* $\mathsf{coll}(x) \leq x$.

Proof We work in Eqr. As a first step, we consider the class I_0 of x such that for all $y \leq x$, if $y \in \mathsf{D}$, then there is a smallest $z \in \mathsf{D}$ such that $y\mathsf{E}z$. We show that I_0 is inductive. If $x = 0$ and $y \leq x$ and $y \in \mathsf{D}$, then certainly $z = 0$ is the smallest $z \in \mathsf{D}$ such that $y = 0\,\mathsf{E}\,0 = z$. Suppose we have $x \in I_0$ and $y \leq \mathsf{S}x$. Then, $y \leq x$ or $y = \mathsf{S}x$. In the first case, we can find the desired z by applying the fact that $x \in I_0$ to y'. In the second case, suppose $y \in \mathsf{D}$. If there is a $y' < y$ with $y\mathsf{E}y'$, we have $y' \leq x$ and we can find z applying $x \in I_0$. In case, for no $y' < y$, we have $y\mathsf{E}y'$, we take $z := y$.

We shorten I_0 to an ω_1-cut I_1. We find that each equivalence class of $\mathsf{E} \restriction (\mathsf{D} \cap I_1)$ has a smallest element.

We proceed to work in I_1. By the usual tricks, we can assume that we have an extra element $*$. A sequence σ in I_1 is a *good* sequence, if for all $i < \mathsf{length}(\sigma)$ we have one of the following:

[13] We allow the domain D to be empty. We consider the the empty relation to be an equivalence relation on the empty set.

- $i \notin \mathsf{D}$ and $\sigma_i = *$.
- $i \in \mathsf{D}$ and the smallest i_0 that is E-equivalent to i is i itself. In this case, σ_i is the supremum of the $\sigma_j + 1$, for $j < i$ and $\sigma_j \neq *$.[14]
- $i \in \mathsf{D}$ and the smallest i_0 that is E-equivalent to i is strictly below i. In this case, $\sigma_i = \sigma_{i_0}$.

Consider any two good sequences σ and τ of length a. We may prove by induction on $i \leq a$ that $\sigma \restriction i = \tau \restriction i$ noting that a is logarithmic in σ.

Consider a good sequence σ of length a in I_1. Suppose, for $i_0, i_1 < a$ with $i_0, i_1 \in \mathsf{D}$, we have $i_0 \mathsf{E} i_1$. Then, by the definition of *good* it is immediate that $\sigma_i = \sigma_j$. Suppose, for $i_0, i_1 < a$ with $i_0, i_1 \in \mathsf{D}$, we have not $i_0 \mathsf{E} i_1$. Let i_0^* be the smallest element in the equivalence class of i_0 and let i_1^* be the smallest element in the equivalence class of i_1. Suppose e.g. that $i_0^* < i_1^*$. By the definition of good sequence, we find that $\sigma_{i_0} = \sigma_{i_0^*} < \sigma_{i_1^*} = \sigma_{i_1}$. So the sequence σ codes a partial injection from $(\mathsf{D} \cap I_1)/(\mathsf{E} \restriction (\mathsf{D} \cap I_1))$ to I_1-numbers.

Consider a good sequence σ of length a in I_1. By induction we find that, for all $i < a$ such that $\sigma_i \neq *$, we have $\sigma_i \leq i$, noting that a is logarithmic in σ.

Consider a good sequence σ of length a in I_1. By induction we find that, for all $i < a$ if $\sigma_i = j$ and $k < j$, then there is a $m < i$ such that $\sigma_m = k$, noting that all quantifiers are sharply bounded, since a is logarithmic in σ.

Consider the class I_2 of all a in I_1 such that a is the length of a good sequence. Clearly, I_2 is progressive. Let I be an ω_1-cut that is a shortening of I_2. We define, for x in $\mathsf{D} \cap \mathsf{I}$: $\mathsf{coll}(x) = y$ iff there is a good sequence σ of length $x + 1$ such that $\sigma_x = y$. It is easy to see that coll is a bijection between $(\mathsf{D} \cap I)/(\mathsf{E} \restriction (\mathsf{D} \cap I))$ and an initial segment of I. Moreover, $\mathsf{coll}(x) \leq x$, whenever $x \in \mathsf{D} \cap I$. □

Let EQ be the theory of identity.

Lemma 3.14 *Suppose* $K : V \lhd U$ *and* $M_0 : \mathsf{EQ} \lhd U$. *Let* $K_0 : \mathsf{EQ} \lhd U$ *be the reduct of* K *to the language of pure identity. This means that* K_0 *is the unique direct interpretation of* EQ *in* U. *Suppose there is a* U*-definable isomorphism* F *between* K_0 *and* M_0. *Then we can find an interpretation* $M : V \lhd U$, *such that* M_0 *is the reduct of* M *to the language of identity and* F *extends to an isomorphism between* K *and* M.

The simple proof is left to the reader.

Theorem 3.15 *There is a one-dimensional identity-preserving interpretation* H°_Σ : $\mathsf{SF}_{\mathsf{cut}} \rhd \mathsf{Pred}_\Sigma$ *with underlying translation* η°_Σ. *The domain of* H° *is an initial segment of a definable* $\mathsf{SF}_{\Sigma,\mathsf{cut}}$*-cut* J° *with* $J^\circ \subseteq \mathsf{J}$. *Moreover, we have a satisfaction predicate* $\mathsf{sat}_{\eta^\circ_\Sigma}$ *for* H°_Σ *for formulas* A *in* $\Gamma \cap J^\circ$. *This predicate satisfies the commutation conditions for all formulas in its intended domain with the usual understanding that commutation for the quantifiers is relativized to* $\delta_{\eta^\circ_\Sigma}$ *and that atoms are sent to their* η°_Σ*-translations. Our satisfaction predicate satisfies:*

[14] In S^1_2, the supremum of the elements of a sequence always exists. However, we do not need that insight for the present definition to be meaningful.

$$\mathsf{SF}_{\Sigma,\mathsf{cut}} \vdash \forall A \in (\mathsf{sent}_\Sigma \cap \Gamma \cap J^\circ)\,(\Box_{\alpha,\Gamma} A \to \mathsf{sat}_{\eta^\circ_\Sigma}(\emptyset, A)).$$

It follows that we have an interpretation $H^\circ_V : \mathsf{sf}_{\mathsf{cut}}(V) \rhd V$ based on $\eta^\circ_{\Sigma_V}$. Clearly, H°_V inherits the various properties of H°_Σ.

Proof We work in $\mathsf{sf}_{\mathsf{cut}}(V)$. We consider the interpretation $H' := H \restriction (J \cap \mathsf{J})$. If we interpret S^1_2 on $J \cap \mathsf{J}$ and D as $\delta_{H'}$ and E as $=_{H'}$, then we obtain an interpretation of Eqr. We take $J^\circ := I$, where I is as promised in Lemma 3.13. We use the function coll to construct the interpretation H° and the predicate $\mathsf{sat}_{\eta^\circ}$ using Lemma 3.14. □

3.7 The Second Incompleteness Theorem

This subsection is mainly intended to bring a question on the board. One of Feferman's discoveries is that a theory U can have a consistency statement for U that can be proved in U. This is reflected in our framework. In Sect. 4, we will see that restricted theories U interpret $\mathsf{sf}(U)$. The same holds for reflexive theories.

It would be interesting to see whether we can use the framework of this paper to get a more precise understanding of under what conditions a consistency statement can be proven.

We will briefly illustrate that at least not all theories U interpret $\mathsf{sf}(U)$. Let us say that a theory U is *introspective* iff $U \rhd \mathsf{sf}(U)$.

Theorem 3.16 *Suppose e.g. $\mathfrak{T}$ is a complete and consistent extension of* PA *in the language of* PA*. Then, $\mathfrak{T}$ is not introspective for any axiomatization.*[15]

Proof Let X be an axiomatization of $\mathfrak{T}$ and let T be the theory axiomatized by X. Suppose $K : T \rhd \mathsf{sf}(T)$. Consider any model $\mathcal{M}$ of T. Since T satisfies full induction $\mathcal{N} := \widetilde{K}(\mathcal{M})$ is an internal model with a definable initial embedding from $\mathcal{M}$ in $\mathcal{N}$. It is easy to see that as a consequence we will have $\mathcal{M}$-definable predicates α^* and Γ^* such that $\mathcal{M}(\alpha^*, \Gamma^*)$ satisfies $\mathsf{sf}(T)$.

Reason in $\mathcal{M}$. Suppose $\Box_{\alpha^*}\bot$. By cut-elimination it follows that we can find a cutfree-proof (in the sense of cutfree for predicate logic) of $\bot$. The formulas used in this cutfree proof are subformulas of formulas in α^* and hence these formulas will be in Γ^*. It follows that $\Box_{\alpha^*,\Gamma^*}\bot$. Quod non. However, since $\alpha^\star$ provides axioms for the complete theory $\mathfrak{T}$, we may now conclude that $\Box_{\alpha^*}$ is a truth-predicate for $\mathcal{M}$. But this is impossible. □

Consider a consistent finitely axiomatized sequential theory A. It has been shown by Jan Krajíček in [20] that we can find a function sending N to n_N such that $\mathsf{kraj}(A) := A + \{\Box^N_{A,n_N,n_N}\bot \mid N : \mathsf{S}^1_2 \lhd A\}$ is consistent. See also [38, 39]. (In fact, we can arrange it so that $A \rhd_{\mathsf{loc}} \mathsf{kraj}(A)$.) We assume that $\mathsf{kraj}(A)$ be axiomatized

15 We remind the reader that we allow axiomatizations of any complexity.

in the obvious way. As a consequence every standard formula of the language of A will be in $\mathsf{sf}(\mathsf{kraj}(A))$-verifiably in Γ.

Theorem 3.17 *Suppose A is a consistent sequential theory. Then, $\mathsf{kraj}(A)$ is not introspective.*

Proof Let A be consistent and finitely axiomatized. *Suppose* $K : \mathsf{kraj}(A) \rhd \mathsf{sf}(\mathsf{kraj}(A))$. We have $\mathsf{sf}(\mathsf{kraj}(A)) \xrightarrow{K} \mathsf{kraj}(A) \xrightarrow{H} \mathsf{sf}(\mathsf{kraj}(A))$. We note that all standard cuts of K in $\mathsf{kraj}(A)$ will be in Γ. We consider $\mathcal{J}$ with:

$$x \in \mathcal{J} :\leftrightarrow x \in \delta_{KH} \wedge$$
$$\forall I(v)\ (\omega_1\text{-}\mathsf{cut}_{KH}(\{y \in \delta_{KH} \mid \mathsf{sat}_\eta(\{\langle v, y\rangle\}, I(v))\}) \rightarrow$$
$$\mathsf{sat}_\eta(\{\langle v, x\rangle\}, I(v))).$$

Clearly $\mathcal{J}$ is an ω_1-cut in $\mathsf{sf}(\mathsf{kraj}(A))$. Also $\mathcal{J}$ is below every standard K-internal KH-ω_1-cut. So, we have $\mho(A) := \mathsf{S}^1_2 + \{\Diamond_{A,n,n}\top \mid n \in \omega\}$ on $\mathcal{J}$. Now we consider

$$\mathsf{sf}(\mathsf{kraj}(A)) \xrightarrow{K} \mathsf{kraj}(A) \xrightarrow{H} \mathsf{sf}(\mathsf{kraj}(A)) \xrightarrow{K} \mathsf{kraj}(A).$$

In the outer $\mathsf{kraj}(A)$ we find that $\mathcal{J}K$ satisfies S^1_2. The verification of this asks only a finite number of $\mathsf{kraj}(A)$ axioms over A; let their conjunction be B. We consider the disjunctive interpretation $N^* := \mathcal{J}K\langle B\rangle N_0$, where $N_0 : \mathsf{S}^1_2 \lhd A$. Clearly, $N^* : \mathsf{S}^1_2 \lhd A$. It follows that $\mathsf{kraj}(A) \vdash \Box^{N^*}_{A,n_{N^*},n_{N^*}}\bot$. Since $\mathsf{kraj}(A) \vdash B$ it follows that $\mathsf{kraj}(A) \vdash \Box^{\mathcal{J}K}_{A,n_{N^*},n_{N^*}}\bot$. But this contradicts the fact that we have $\mho(A)$ on $\mathcal{J}K$. $\Box$

Open Question 3.18 Is there an axiomatization of $\overline{\mathsf{kraj}(A)}$, for A consistent and sequential, that is introspective? My guess would be: *no*.

4 Characterization Theorems and The World of $\mho$

This section is an introduction to the $\mho$-family. This is a group of transformations of theories into theories that is closely connected to the main object of our paper $\mathsf{sf}(\cdot)$. The Friedman Characterization and the Orey–Hájek Characterization emerge naturally from the results on $\mho$.

4.1 $\mho_{*,n}$

In this subsection, we study restricted sequential theories. A theory is *n-restricted* if its axiom set is contained in Σ^*_n. A theory is *restricted* if it is n-restricted for some n.

We define $\mho_{*,n}(V)$ as the theory in the language of arithmetic extended with a unary predicate α given by:

- $\mho_{*,n}(V) := \mathsf{S}^1_2 + \forall x \in \alpha\, \mathsf{sent}_{\Sigma_V}(x) + \{\alpha(\ulcorner A \urcorner) \mid A \in \mathsf{ax}_V\} + \{\Diamond_{\alpha,k,n}\top \mid k \in \omega\}$.

We have the following theorem.

Theorem 4.1 *Suppose V is n-restricted then $V \lhd \mathsf{sf}(V) \equiv \mho_{*,n}(V)$. If V is also sequential, we have $V \rhd \mathsf{sf}(V)$, and hence, $V \equiv \mathsf{sf}(V) \equiv \mho_{*,n}(V)$.*

Proof Suppose V is n-restricted.

$\mathsf{sf}(V) \rhd V$. This is Theorem 3.1.

$\mathsf{sf}(V) \rhd \mho_{*,n}(V)$. We work in $\mathsf{sf}(V)$. We have $\Diamond_{\alpha,\Gamma}\top$. It clearly follows that $\Diamond_{\alpha,\Gamma\cap\Sigma^*_n}\top$. By Buss' result, we have cut-elimination for n-proofs π such that a tower of exponents of height $n + c$ exists, where c is a fixed standard constant. So, we can find a definable ω_1-cut I such that $\Diamond^I_{\alpha,\Sigma^*_n}\top$. So, a fortiori, $\Diamond^I_{\alpha,k,n}\top$. We interpret $\mho_{*,n}(V)$ taking I as our domain of interpretation and $\alpha \cap I$ as the interpretation of α.

$\mho_{*,n}(V) \rhd \mathsf{sf}(V)$. We take the identical interpretation for the arithmetical domain and predicates. We interpret Γ as Σ^*_n. Finally, we interpret α by, say α^* with $\alpha^*(A) :\leftrightarrow \alpha(A) \wedge \Diamond_{\alpha,A,n}\top$.[16] Suppose $\Box_{\alpha^*,\Sigma^*_n}\bot$. Let B be the largest axiom in the proof. It follows that $\Box_{\alpha,B,n}\bot$. Quod non, since $B \in \alpha^*$.

$V \rhd \mathsf{sf}(V)$, *if V is sequential.* In V we have an interpretation N of S^1_2 and a satisfaction predicate sat_n such that V proves that all Σ^*_n-formulas in N satisfy the commutation conditions. We write $\mathsf{true}_n(B)$ for $\mathsf{sat}_n(\emptyset, B)$. We can find an ω_1-cut I such that:

$$(\dagger)\quad V \vdash \forall A \in (\Sigma^\star_n \cap I)\, (\Box^I_{\mathsf{true}_n,n} A \to \mathsf{true}_n(A)).$$

It is now easily seen that interpreting S^1_2 by relativization to I and α as $\mathsf{true}_n \cap I$ and Γ as $\Sigma^\star_n \cap I$, we have an interpretation of $\mathsf{sf}(V)$. □

We note that $\mho_{*,n}(V)$ is dependent on the chosen axioms for V. However, by the above theorem, modulo mutual interpretability, we can find an extensionally equal variant $\overline{V}_n$ of V that is uniquely fixed by the theorems of V, to wit the theory axiomatised by the Σ^*_n-theorems of V.

For example, the theory of pure identity EQ is finitely axiomatized, and a fortiori restricted, but since EQ is decidable, we have $\mathsf{EQ} \not\rhd \mathsf{sf}(\mathsf{EQ})$.

We note that, for any V, the theory $\mathsf{sf}(V)$ is restricted and sequential. Hence, we have:

Corollary 4.2 *Let V be any theory. Then,* $\mathsf{sf}(\mathsf{sf}(V)) \equiv \mathsf{sf}(V)$.

In case we consider a finitely axiomatized sequential theory A, we find that the theory $\mho_{*,\rho(A)}(A)$ is ipso facto finitely axiomatised. The theory $\mho_{*,\rho(A)}(A)$ is clearly mutually interpretable with $\mathsf{S}^1_2 + \Diamond_{A,A,\rho(A)}\top$ (where the first subscript A is elliptic for the axiom set $x = \ulcorner A \urcorner$). Thus, we have: $A \equiv (\mathsf{S}^1_2 + \Diamond_{A,A,\rho(A)}\top)$. This insight

[16] We treat the A in the subscript of $\Diamond_{\alpha,A,n}$ as the number bounding the axiom of the theory A, which is *par abus de langage*, again A.

immediately leads to the Friedman Characterization of interpretability among finitely axiomatized sequential theories.

The Friedman Characterization was first published in [33]. For a careful analysis of its formalization, see [36]. An exploration of its consequences can be found in [45].

Theorem 4.3 (This result is due to Harvey Friedman around 1973) *Let A and B be finitely axiomatized theories. Then:* $A \rhd B$ *iff* $\mathsf{EA} \vdash \Diamond_{A,A,\rho(A)}\top \to \Diamond_{B,B,\rho(B)}\top$.

Proof Let A and B be finitely axiomatized sequential theories. We find, by Theorem 4.1:

$$(\dagger)\quad A \rhd B \text{ iff } (\mathsf{S}^1_2 + \Diamond_{A,A,\rho(A)}\top) \rhd (\mathsf{S}^1_2 + \Diamond_{B,B,\rho(B)}\top).$$

By a meta-theorem of Wilkie and Paris (see: [53]), we have

$$(\ddagger)\quad (\mathsf{S}^1_2 + \Diamond_{A,A,\rho(A)}\top) \rhd (\mathsf{S}^1_2 + \Diamond_{B,B,\rho(B)}\top) \text{ iff } \mathsf{EA} \vdash \Diamond_{A,A,\rho(A)}\top \to \Diamond_{B,B,\rho(B)}\top.$$

Combining (†) and (‡), we immediately have the Friedman Characterization. □

Reflection on the proof shows that the demand that B be sequential is superfluous. The characterization works for finitely axiomatized A and B with A sequential.

Open Question 4.4 Can we have something as nice as the Friedman Characterization for the restricted case?

4.2 $\mho_{*,*}$

We define three operations on theories:

- $\mho_{*,*}(V) := \mathsf{S}^1_2 + \forall x \in \alpha\, \mathsf{sent}_{\Sigma_V}(x) + \{\alpha(\underline{\ulcorner A \urcorner}) \mid A \in \mathsf{ax}_V\} + \{\Diamond_{\alpha,k,k}\top \mid k \in \omega\}$.
- $\mathsf{FC}_n(V)$ is a theory in the signature of V extended by second-order variables for all arities m with $1 \leq m \leq n$. We add to the axioms of V the axioms of first-order comprehension for m-ary relations, with $1 \leq m \leq n$, where the defining formulas are constrained to the original V-language, i.e., they contain no second-order variables. We write FC for FC_1.[17]
- $\mathsf{sf}^+(V) := (\mathsf{sf}(V) + \{(\Sigma^*_k \subseteq \Gamma) \mid k \in \omega\})$.

In terms of $\mho_{*,*}$ we can define the central notion of *reflexivity*.

- A theory V is *reflexive* iff $V \rhd \mho_{*,*}(V)$.

We have the following theorem.

[17] It would be more natural to work with the theory FC_∞ with class-variables for any arity. However, this would not fit with our policy to only consider finite signatures.

Theorem 4.5 *Consider any theory V. We have:*

$$\mathsf{FC}_n(V) \lhd \mathsf{sf}^+(V) \equiv \mho_{*,*}(V).$$

In case V is n-sequential we also have $\mathsf{FC}_n(V) \rhd \mathsf{sf}^+(V)$, and thus:

$$\mathsf{FC}_n(V) \equiv \mathsf{sf}^+(V) \equiv \mho_{*,*}(V).$$

Proof $\mathsf{sf}^+(V) \rhd \mathsf{FC}_n(V)$. We consider the Henkin interpretation $H : \mathsf{sf}^+(V) \rhd V$. For this interpretation we have a satisfaction predicate sat on $\Gamma \cap J$ for a certain ω_1-cut J. We extend H by interpreting our m-ary classes, for $1 \leq m \leq n$, as formulas y in $\Gamma \cap J$ with a sequence of length m of designated variables $\vec{v}$ free. We translate $\vec{x} \in y$ as $\mathsf{sat}(\{\langle v_0, x_0\rangle, \cdots, \langle v_{m-1}, x_{m-1}\rangle\}, y)$. It is easily seen that this gives us the desired interpretation of the first-order comprehension axioms for all $1 \leq m \leq n$, since all standard formulas are in Γ.

$\mathsf{sf}^+(V) \rhd \mho_{*,*}(V)$. This is immediate.

$\mho_{*,*}(V) \rhd \mathsf{sf}^+(V)$. We interpret $\mathsf{sf}^+(V)$ in $\mho_{*,*}(V)$ via τ by defining:

- $\delta_\tau(x) :\leftrightarrow \top$.
- τ is the identical translation on the arithmetical vocabulary.
- $\Gamma_\tau(x) :\leftrightarrow \mathsf{form}_{\Sigma_V}(x) \wedge \Diamond_{\alpha,\rho(x),\rho(x)}\top$.
- $\alpha_\tau(x) :\leftrightarrow \alpha(x) \wedge \Diamond_{\alpha,x,x}\top$.

We work in $\mho_{*,*}(V)$. We verify $\Diamond_{\alpha_\tau,\Gamma_\tau}\top$. Suppose we have an α_τ, Γ_τ-proof π of $\bot$. Let x be the largest axiom used in π and let y be the maximal complexity of a formula occurring in the proof. Let z be the maximum of x and y. So we have $\Diamond_{\alpha,z,z}\top$. On the other hand, π would be be certainly an α, z, z-proof of $\bot$. Quod non. The other axioms of $\mathsf{sf}^+(V)$ are easily verified.

If V is n-sequential, then $\mathsf{FC}_n(V) \rhd \mathsf{sf}^+(V)$. We treat the 1-sequential case. The general case requires only minor changes. Let V be a sequential theory. We fix a one-dimensional $N : \mathsf{S}^1_2 \lhd V$.

We work in $\mathsf{FC}(V)$. We suppress the mention of N in the proof, wherever it is clear that things are supposed to be coded in N.

Our first order of business is to construct an appropriate satisfaction predicate. Regrettably, the construction is a substantial amount of work, so, here, we can only sketch it. We can write down a formula $\Phi(X, Y, W, Z)$ with the following properties. Suppose Y is an ω_1-cut and X is a satisfaction predicate for $\Sigma^*_x \cap Y$, in the sense that X satisfies the commutation conditions on $\Sigma^*_x \cap Y$.

- Suppose that $\Phi(X, Y, W, Z)$. Then, Z is a sub-ω_1-cut of Y and W is a satisfaction predicate for $\Sigma^*_{x+1} \cap Z$. Moreover, X and W coincide on $\Sigma^*_x \cap Z$-formulas.
- Suppose that $\Phi(X, Y, W, Z)$ and $\Phi(X, Y, W', Z')$. Then W is extensionally equal to W' and Z is extensionally equal to Z'.

Using the fact that V has a good theory of sequences, we can represent a sequence of classes as a class of sequences. There is a small technical detail since the naive construction does not work when the empty class is an element of our sequence, but a small trick will solve that problem. A sequence S of pairs of classes is *good* if its first component is $\langle \emptyset, \delta_N \rangle$ and whenever $x + 1 < \mathsf{length}(S)$ and S_x is $\langle X, Y \rangle$ and S_{x+1} is $\langle W, Z \rangle$, then $\Phi(X, Y, W, Z)$.

We consider the virtual class $\mathcal{X}$ of all x such that:

i. There exists a good sequence of length x.
ii. For all $u \leq x$, all good sequences of length u are (extensionally) equal.
iii. Let $u < x$ and let S be a good sequence of length x. Let $S_x = \langle X, Y \rangle$. Then, Y is an ω_1-cut and X is a satisfaction predicate for $\Sigma^*_x \cap Y$.
iv. Let $u < v < x$ and let S be a good sequence of length x. Suppose S_u is $\langle X, Y \rangle$ and S_v is $\langle W, Z \rangle$. Then Z is a sub-ω_1-cut of Y and X and W coincide on $\Sigma^*_u \cap Z$.
v. If S is a good sequence of length x and $u < x$, then the restriction $S \restriction u$ of S to u exists. (We have to add this clause because of the limited amount of comprehension available.)

Clearly, we can prove, for each *standard* n, that $n \in \mathcal{X}$. We define:

- X_x is the first component of S_x, where $x + 1$ is in $\mathcal{X}$ and S is a good sequence of length $x + 1$.
- Y_x is the second component of S_x, where $x + 1$ is in $\mathcal{X}$ and S is a good sequence of length $x + 1$.
- $\mathcal{I}$ is the intersection of all Y_x for $x + 1 \in \mathcal{X}$.
- $\Sigma^*_{\mathcal{X}} := \bigcup_{x \in \mathcal{X}} \Sigma^*_x$.
- $\mathsf{sat}^*(\alpha, A)$ iff $A \in \Sigma^*_{\mathcal{X}} \cap \mathcal{I}$ and $X_{\rho(A)}(\alpha, A)$.

By our construction $\mathsf{sat}^*(\alpha, A)$ satisfies the commutation conditions on $\Sigma^*_{\mathcal{X}} \cap \mathcal{I}$. We define $\mathsf{true}^*(A)$ as $\mathsf{sat}^*(\emptyset, A)$. We can now find an ω_1-cut $\mathcal{J}$ such that:

$$(\dagger)\ \ \forall A \in (\Sigma^*_{\mathcal{X}} \cap \mathcal{J})\ (\Box^{\mathcal{J}}_{\mathsf{true}^*, \mathcal{X}} A \to \mathsf{true}^*(A)).$$

We define our translation ν.

- $\delta_\nu := \mathcal{J}$.
- The translation of the arithmetical vocabulary is the restriction of the vocabulary of N to $\mathcal{J}$,
- $\Gamma_\nu := \Sigma^*_{\mathcal{X}} \cap \mathcal{J}$.
- $\alpha_\nu := \mathsf{true}^* \cap \mathcal{J}$.

We can now verify the axioms of $\mathsf{sf}^+(V)$ under the translation ν. □

It is easy to see that FC preserves extensional identity $=_{\mathsf{ext}}$ of theories. So FC can as well be seen as a functor on theories qua sets of theorems. Similarly, it is easy to see that, if $U \rhd V$ via a 1-dimensional interpretation, then $\mathsf{FC}(U) \rhd \mathsf{FC}(V)$. Since every interpretation in a sequential theory can be transformed into a 1-dimensional interpretation, it follows that FC can be viewed as a functor on the degrees of interpretability of sequential theories. By Theorem 4.5, it follows that FC, sf^+ and $\mho_{*,*}$ define the same functor on the degrees of sequential theories.

Theorem 4.6 *The operations* FC, sf^+ *and* $\mho_{*,*}$ *define a closure operation on the interpretability degrees of sequential theories.*

Proof We have already seen that FC yields a functor on the degrees of sequential theories. Moreover, $V \lhd \mathsf{FC}(V)$. Finally, we check that $\mho_{*,*}(V) \rhd \mho_{*,*}(\mathsf{FC}(V))$. We construct a translation τ. This translation will be the identity on the arithmetical vocabulary. We take $\alpha_\tau := \alpha \cup \mathsf{fc}$, where fc is a Δ_1^{b}-representation of the comprehension axioms. We work in $\mho_{*,*}(V)$. Suppose π witnesses $\Box_{\alpha_\tau,n,n}\bot$. Suppose the comprehension axioms $\leq n$ involve the comprehension formulas $A_0, \cdots, A_{k-1}$. We interpret $x \in Y$ by $(y = 0 \wedge A_0(x)) \vee \cdots \vee (y = k-1 \wedge A_{k-1}(x))$, where the variable Y is translated to y and where the domain of classes is all numbers. Using this translation, say ν, we can transform π into a witness π' of $\Box_{\alpha,n,2n}\bot$, and a fortiori, of $\Box_{\alpha,2n,2n}\bot$. A contradiction. Hence, $\Diamond_{\alpha_\tau,n,n}\top$. □

Theorem 4.7 *Suppose* V *is sequential. The theory* $\mathsf{FC}(V)$ *is not mutually interpretable with a finitely axiomatized theory* A.

Proof Suppose V is sequential and $\mathsf{FC}(V)$ is mutually interpretable with the finitely axiomatized theory A. It follows that $A \equiv \mathsf{FC}(V) \equiv \mathsf{FC}(\mathsf{FC}(V)) \equiv \mathsf{FC}(A) \equiv \mho_{*,*}(A)$. So, for some $N : \mathsf{S}^1_2 \lhd A$, for all n, we have $A \vdash \Diamond^N_{A,n,n}\top$. If we take, n sufficiently large, it follows, by cut-elimination, that $A \vdash_n \Diamond^N_{A,n,n}\top$, where $\vdash_n$ stands for provability only involving Σ^*_n-formulas. This contradicts a version of the Second Incompleteness Theorem due to Pudlák. See [28]. □

The functor FC plays a role in the comparison of local and global interpretability. The next theorem will state the connection. We will use three notions.

- $V \restriction n$ is the theory axiomatized by the V-axioms with code $\leq n$.
- Let X be a set of numbers. Suppose $N : \mathsf{S}^1_2 \lhd V$. The theory V *N-binumerates* X iff, there is a V-formula $B(v)$ such that, for all $n \in X$, $V \vdash B(\underline{n})$ and, for all $n \notin X$, $V \vdash \neg B(\underline{n})$. Here $\underline{n}$ stands for the N-numeral of n. In case, V is sequential, and $N' : \mathsf{S}^1_2 \lhd V$, by Pudlák's Theorem 2.5, we find that V N-binumerates X iff V N'-binumerates X. So, for sequential theories we simply speak about *binumeration*.
- $\mho_{*,*}(U, V)$ is the result of extending $\mho_{*,*}(U)$ with a new predicate β and axioms $\{\beta(B) \mid B \in \mathsf{ax}_V\}$ and $\{\neg\beta(B) \mid B \notin \mathsf{ax}_V\}$.

Theorem 4.8 *Suppose* U *is a sequential theory and* U *binumerates the axioms of* V. *Then, the following are equivalent:*
(i) $U \rhd_{\mathsf{loc}} V$, *(ii)* $\mho_{*,*}(U, V) \vdash \mho_{*,*}(V)[\alpha := \beta]$, *(iii)* $\mathsf{FC}(U) \rhd V$.

Proof (i) ⇒ (ii). Suppose $U \rhd_{\mathsf{loc}} V$. Consider any n. We have, for some m and some K, that $K : (U \restriction m) \rhd (V \restriction n)$. Let $U \restriction m := \{A_0, \cdots A_{k-1}\}$ and let $V \restriction n := \{B_0, \cdots, B_{\ell-1}\}$. So, $K : \{A_0, \cdots A_{k-1}\} \rhd \{B_0, \cdots, B_{\ell-1}\}$. Let the witnessing proofs of the B_i^K from $\{A_0, \cdots A_{k-1}\}$ be ν_i. By Σ_1-completeness, we have, for each $i < \ell$:

$$\mho_{*,*}(U, V) \vdash (\nu_i : \{A_0, \cdots A_{k-1}\} \vdash B_i^K)^N.$$

We reason in $\mho_{*,*}(U, V)$. Suppose π is an n-proof witnessing the inconsistency of $\Box_{\beta,n,n}\bot$. By bi-numerability, π witnesses the inconsistency of $\{B_0, \cdots, B_{\ell-1}\}$. We can transform π into π^K a $n + \rho(K)$-proof of $\bot$ from the B_i^K. Adding the ν_i above the B_i^K, we get a p-proof of $\bot$, from axioms $\{A_0, \cdots A_{k-1}\}$, where p is a sufficiently large standard number. We may assume that $p \geq m$. It follows that $\Box_{\alpha,p,p}\bot$. Quod non. Hence $\Diamond_{\beta,n,n}\top$.

(ii) $\Rightarrow$ (iii). By Theorem 4.5, $\mathsf{FC}(U)$ interprets $\mho_{*,*}(U)$ say via N. Since U binumerates the axioms of V, it binumerates them in N. Hence, $\mathsf{FC}(U)$ interprets $\mho_{*,*}(U, V)$. It follows by (ii), that $\mathsf{FC}(U)$ interprets $\mho_{*,*}(U, V)$. Hence, by Theorems 4.5 and 3.1, $\mathsf{FC}(U) \rhd V$.

(iii) $\Rightarrow$ (i). It is easily seen that $U \rhd_{\mathsf{loc}} \mathsf{FC}(U)$. Hence $U \rhd_{\mathsf{loc}} V$. $\Box$

We note that some clause like our demand that U binumerates the axioms of V is necessary, since $\mathsf{FC}(U)$ interprets at most countably many incomparable theories in the language of V. On the other hand, the theory U locally interprets $2^{\aleph_0}$ incomparable theories. If we restrict ourselves to the recursively enumerable theories we can get rid of the extra clause.

The insight that, for sequential U in which the axioms of V are binumerable, we have $U \rhd_{\mathsf{loc}} V$ iff $\mho_{*,*}(U, V) \vdash \mho_*(V)[\alpha := \beta]$ is our version of the famous Orey–Hájek characterization.

Consider a recursively enumerable theory as set of theorems $\mathfrak{U}$. By Craig's trick we know that $\mathfrak{U}$ can be Δ_1^{b}-axiomatized. Consider two such axiomatizations α_0 and α_1. Then, it is easy to see that $\mathsf{S}_2^1 + \{\Diamond_{\alpha_0,n,n}\top \mid n \in \omega\}$ and $\mathsf{S}_2^1 + \{\Diamond_{\alpha_1,n,n}\top \mid n \in \omega\}$ prove the same theorems. Thus, from the perspective of theories as sets of theorems we are justified in writing $\mho(\mathfrak{U})$ for $\overline{\mathsf{S}_2^1 + \{\Diamond_{\alpha^*,n,n}\top \mid n \in \omega\}}$, where α^* is some Δ_1^{b}-axiomatization of $\mathfrak{U}$. Using these ideas, we get a simpler form of the Orey–Hájek characterization for the case of recursively enumerable theories qua sets of theorems.

Corollary 4.9 *Suppose $\mathfrak{T}$ and $\mathfrak{U}$ are recursively enumerable theories qua sets of theorems and $\mathfrak{T}$ is sequential. Then, $\mathfrak{T} \rhd_{\mathsf{loc}} \mathfrak{U}$ iff $\mho(\mathfrak{T}) \vdash \mho(\mathfrak{U})$ iff $\mathsf{FC}(\mathfrak{T}) \rhd \mathfrak{U}$.*

Corollary 4.10 FC *is the right adjoint of the projection functor of the degrees of global interpretability of recursively enumerable sequential theories on the degrees of local interpretability of recursively enumerable sequential theories.*

Open Question 4.11 The relation $V \lhd \mathsf{FC}(U)$ on the sequential theories induces a Kleisli pre-order category on the sequential global degrees. We have seen that the sub-category of this Kleisli category obtained by restricting ourselves to degrees of recursively enumerable theories is isomorphic to the category of local degrees of recursively enumerable sequential theories.

Clearly $V \lhd \mathsf{FC}(U)$ implies $V \lhd_{\mathsf{loc}} U$, but not vice versa.

Is there a characterization of the relation $V \lhd \mathsf{FC}(U)$ on the sequential theories that makes clear that it an alternative notion of interpretability?

4.3 $\mho_{*,\infty}$

We define:

- $\mho_{*,\infty}(V) := \mathsf{S}^1_2 + \forall x \in \alpha\, \mathsf{sent}_{\Sigma_V}(x) + \{\alpha(\ulcorner A \urcorner) \mid A \in \mathsf{ax}_V\} + \{\Diamond_{\alpha,k,\infty}\top \mid k \in \omega\}$.
- $\mathsf{PC}_n(V)$ is a theory in the signature of V extended by second-order variables for arities m with $1 \leq m \leq n$. We add to the axioms of V the axiom of predicative comprehension for m-ary relations, with $1 \leq m \leq n$, where the defining formulas contain no bound second-order variables. We write PC for PC_1.
  As is well known, if V is sequential (or even a pair theory) then $\mathsf{PC}(V)$ is finitely axiomatizable over V.[18]
- $\mathsf{sf}^{++}(V) := (\mathsf{sf}(V) + (\Gamma = \mathsf{form}_{\Sigma_V}))$.

We can define the important notion of strong reflexivity in terms of $\mho_{*,\infty}$.

- A theory V is *strongly reflexive* iff $V \rhd \mho_{*,\infty}(V)$.

In much of the literature the notion we call *strong reflexivity* is named *reflexivity*. This is due to the fact that this literature works over the basic theory EA rather than S^1_2.

We have the following theorem.

Theorem 4.12 *Consider any theory V. We have:*

$$\mathsf{PC}_n(V) \lhd \mathsf{sf}^{++}(V) \equiv \mho_{*,\infty}(V) \equiv (\mho_{*,*}(V) + \mathsf{exp}).$$

In case V is n-sequential, we also have $\mathsf{PC}_n(V) \rhd \mathsf{sf}^{++}(V)$, and thus:

$$\mathsf{PC}_n(V) \equiv \mathsf{sf}^{++}(V) \equiv \mho_{*,\infty}(V).$$

Proof $\mathsf{sf}^{++}(V) \rhd \mathsf{PC}_n(V)$. We consider the Henkin interpretation $H : \mathsf{sf}^{+}(V) \rhd V$. For this interpretation we have a satisfaction predicate sat on $\mathsf{form}_{\Sigma_V} \cap J$ for a certain ω_1-cut J. We extend H by interpreting our m-ary as formulas y in J with a designated sequence of variables $\vec{v}$ free. We translate $\vec{x} \in y$ as $\mathsf{sat}(\{\langle v_0, x_0\rangle, \cdots, \}\langle v_{m-1}, x_{m-1}\rangle, y)$. It is easily seen that this gives us the desired interpretation of the predicative comprehension axiom because of the good closure properties of $\mathsf{form}_{\Sigma_V} \cap J$.

$\mathsf{sf}^{++}(V) \rhd \mho_{*,\infty}(V)$. This is immediate.

$\mho_{*,\infty}(V) \rhd \mathsf{sf}^{++}(V)$. This is analogous to the proof that $\mho_{*,*}(V) \rhd \mathsf{sf}^{+}(V)$.

If V is n-sequential, then $\mathsf{PC}_n(V) \rhd \mathsf{sf}^{++}(V)$. We restrict ourselves to the 1-sequential case. We fix a one-dimensional $N : \mathsf{S}^1_2 \lhd V$. We work in $\mathsf{PC}(V)$. We suppress the mention of N in the proof, wherever it is clear that things are supposed to be coded in N.

---

[18] Regrettably, I do not have a reference for the argument in full generality. It is proven for the case of ACA_0 in [32, pp. 311–312]. However, the argument there does not easily lift to the general case.

We develop a satisfaction predicate. This has the same definition as the satisfaction predicate constructed in the proof that $\mathsf{FC}(V) \rhd \mathsf{sf}^{+}(V)$ (Theorem 4.5). The first point of divergence is that the virtual class $\mathcal{X}$ will be closed under successor. So, we have a definable ω_1-cut $\mathcal{J} \subseteq \mathcal{X}$. Thus, $\mathsf{sat}^{*}(\alpha, A)$ will satisfy the commutation conditions for all formulas on a suitable ω_1-cut $\mathcal{I}$. The rest of the proof follows the lines of the earlier proof that $\mathsf{FC}(V) \rhd \mathsf{sf}^{+}(V)$. □

It is easy to see that PC preserves extensional identity of theories. So PC can as well be seen as a functor on theories qua sets of theorems. Similarly, it is easy to see that, if $U \rhd V$ via a 1-dimensional interpretation, then $\mathsf{PC}(U) \rhd \mathsf{PC}(V)$. It follows that PC can be viewed as a functor on the degrees of interpretability of sequential theories. By Theorem 4.12, it follows that PC, sf^{++} and $\mho_{*,\infty}$ define the same functor on the degrees of sequential theories.

Theorem 4.13 *The operations* PC, sf^{++} *and* $\mho_{*,\mathrm{inf}}$ *define a closure operation on the interpretability degrees of sequential theories.*

The proof is analogous to the proof for the case of FC, sf^{+} and $\mho_{*,*}$

In contrast to FC, the functor PC does preserve finite axiomatizability for sequential theories. We do have:

Theorem 4.14 *Suppose $U \rhd \mho_{*,\infty}(U)$. Then, U is not mutually locally interpretable with a finitely axiomatized theory.*

Proof Suppose $U \rhd \mho_{*,\infty}(U)$ and $U \equiv_{\mathsf{loc}} A$, where A is finitely axiomatized. Let B be a finitely axiomatized subtheory of U such that $B \rhd A$. We have $U \rhd (\mathsf{S}^1_2 + \Diamond_B \top)$. So, for some finitely axiomatized subtheory C of U, we have $C \rhd (\mathsf{S}^1_2 + \Diamond_B \top)$. Hence:

$$A \rhd C \rhd (\mathsf{S}^1_2 + \Diamond_B \top) \vdash (\mathsf{S}^1_2 + \Diamond_A \top).$$

But this contradicts the Second Incompleteness Theorem. □

We end this subsection with a somewhat surprising theorem. Let exp be the axiom that expresses the totality of exponentiation.

Theorem 4.15 *Let U be sequential. Then $\mho_{*,\infty}(U) \equiv (\mho_{*,*}(U) + \mathsf{exp})$.*

Proof The theory $(\mho_{*,*}(U) + \mathsf{exp})$ interprets $\mho_{*,\infty}(U)$ on a superlogarithmic cut $\mathcal{J}$. The main ingredient is that $\mathsf{EA} = \mathrm{I}\Delta_0 + \mathsf{exp} \vdash \Diamond_{A,A,n} \top \to \Diamond_{A,A,\infty} \top$ for any (standardly) finitely axiomatized A by cut-elimination for predicate logic.

In the other direction, we note that $\mho_{*,\infty}(U)$ interprets U via, say H^{*}, with a satisfaction predicate sat^{*} that works for all formulas on some cut J^{*}. We fix some interpretation $N : \mathsf{S}^1_2 \lhd U$. We consider, in $\mho_{*,\infty}(U)$, the intersection $\mathcal{J}$ of all H^{*}-ω_1-cuts that are definable by a formula in J^{*} (using sat^{*}). Since, for every standard n, there is an N-ω_1-cut I, such that $U \vdash \Diamond^{I}_{U,n,n} \top$, we find that $\mho_{*,\infty}(U) \vdash \Diamond^{\mathcal{J}}_{U,n,n} \top$.

We show that $\mathcal{J}$ is closed under exponentiation. Reason in $\mho_{*,\infty}(U)$. Consider any x in $\mathcal{J}$. We show that 2^x is in every $\mathcal{H}$-ω_1-cut I that is definable by a formula in J^*. Consider such a cut I. We consider a logarithmic sub-ω_1-cut I' of I. The construction of such a subcut can be executed in J^*. By definition x is in I', and, hence, 2^x exists in I.

We now can use $\mathcal{J}$ to interpret $(\mho_{*,*}(U) + \mathsf{exp})$. □

Open Question 4.16 It is a bit strange that we needed sequentiality in the proof of Theorem 4.15. What happens when we drop the demand that U be sequential?

Remark 4.17 As we will see, when reflecting on the example of Sect. 5.1, one could imagine a somewhat more general kind of $\mho$. In our present treatment, we have one α and our consistency statements are concerned with the first n-axioms in α, say $\alpha \restriction n$, for increasing n. A more general idea would be to take α a two-place predicate $\alpha(x, y)$ or $\alpha_x(y)$, where the α_x are approximations to the final axiom set. In a nutshell, we replace $\alpha \restriction x$ by the more abstract α_x. As a result, we could also treat oracle provability as in Sect. 5.1 using a $\mho$-format.

5 Examples

In this section, we treat three examples of applications of the Interpretation Existence Lemma. These examples are more or less randomly chosen, but we hope they give a first impression of the wide range of applicability of the Lemma.

5.1 End-Extensions

Consider a model $\mathcal{M}$ of PA. Let α^* be the usual axiomatization of PA. We have $\mathcal{M} \models \mho_{*,\infty}(\mathsf{PA})[\alpha := \alpha^*]$. Hence, $\mathcal{M}$ has an internal model $\mathcal{N}$ that satisfies (the standard axioms of) PA. This model has a satisfaction predicate that works for all $\mathcal{M}$-formulas. Say the corresponding truth-predicate is true. If we consider the sentence L with $\mathcal{M} \models L \leftrightarrow \neg\,\mathsf{true}(L)$, we see that $\mathcal{M}$ and $\mathcal{N}$ are not elementary equivalent. By Pudlák's Lemma, $\mathcal{M}$ and $\mathcal{N}$ have a common $\mathcal{M}$-definable cut. Since $\mathcal{M}$ satisfies full induction, this cut must be $\mathcal{M}$ itself. We may conclude that $\mathcal{N}$ is a strict end-extension of $\mathcal{M}$.

Remark 5.1 What about *elementary* end-extensions? The McDowell-Specker Theorem tells us that every model of PA has an elementary strict end-extension.

Feferman, in [5], announces the following result: there is no arithmetically definable non-standard model of the theory of the standard model of arithmetic. In other words, no model can be an internal model of the standard model, non-standard, and elementarily equivalent to the standard model. A proof of this result was provided by

Scott in [31]. See also [1, Theorems 25.4a, c]. Inspecting the proof of the Feferman-Scott result, one can easily adapt it to show that no model $\mathcal{M}$ of PA has an internal model (possibly with parameters) $\mathcal{N}$ that is an elementary strict end-extension w.r.t. the unique definable initial embedding of $\mathcal{M}$ into $\mathcal{N}$. Thus, we cannot use the Interpretation Existence Lemma directly to create an elementary strict end-extension (w.r.t. the unique definable embedding).[19]

However, we can still use the Interpretation Existence Lemma as a proof ingredient to show the McDowell-Specker Theorem as was observed by Matt Kaufmann in a talk "Model theory for arithmetic and for set theory: a brief comparative survey" presented at the meeting of the Association for Symbolic Logic, University of Notre Dame, on April 1984. We refer the reader to Schmerl's paper [30] for a sketch of the proof.

As our example we reproduce a proof of a theorem due to Wilkie. See [30, 52]. This is Theorem 2.4.2 in Kossak and Schmerl's textbook [21]. We reproduce treatment in [21] for the case where we restrict ourselves to the language of PA. We follow the Kossak and Schmerl's presentation of the proof almost step by step except that we do not use their Lemma 2.4.1.

For a complete theory $\mathfrak{T}$, we define $\mathsf{Rep}(\mathfrak{T})$ as the set of representable sets of natural numbers of $\mathfrak{T}$. We write $\mathsf{SSy}(\mathcal{M})$ for the standard system of a model $\mathcal{M}$ of PA. The standard system consists of the intersections of the parametrically definable sets of the model with the standard numbers. See [18, 21] for information about *representable sets* and *standard system.*

Theorem 5.2 *Suppose that* $\mathcal{M}$ *is a model of* PA *and that* $\mathfrak{T}$ *is a complete theory as set of theorems that extends* $\overline{\mathsf{PA}}$, *the set of theorems of* PA. *Then* $\mathcal{M}$ *has an end-extension* $\mathcal{N} \models \mathfrak{T}$ *iff the following conditions hold:*

i. $\mathsf{Rep}(\mathfrak{T}) \subseteq \mathsf{SSy}(\mathcal{M})$.
ii. $\mathcal{M} \models \mathfrak{T} \cap \Pi_1$.

Proof Suppose $\mathcal{M}$ has an end-extension $\mathcal{N} \models \mathfrak{T}$. Then, $\mathsf{Rep}(\mathfrak{T}) \subseteq \mathsf{SSy}(\mathcal{N}) = \mathsf{SSy}(\mathcal{M})$. So, we have (i). Condition (ii) follows from the upwards preservation of Σ_1-sentences from models of PA to extensions and end-extensions again satisfying PA.

We assume $\mathcal{M}$ and $\mathfrak{T}$ satisfy Condition (i) and (ii) of the theorem.

We write $\mathcal{P} \sqsubseteq_i \mathcal{Q}$ for: $\mathcal{Q}$ is an end-extension of $\mathcal{P}$ and for every $A(\vec{x}\,)$ in Σ_i, and every $\vec{p}$ in $\mathcal{P}$, if $\mathcal{P} \models A(\vec{p}\,)$, then $\mathcal{Q} \models A(\vec{p}\,)$.

---

[19] We *do* have, by a direct application of the Interpretation Existence Lemma, that there is a large class of models $\mathcal{M}$ that have a parametric internal strict end-extension $\mathcal{N}$ that is elementarily equivalent to $\mathcal{M}$. In fact, the recursively saturated models are in this class. So, the difference between *elementary extension* and *extension that is elementarily equivalent* is essential in the extended Feferman-Scott result. An open end in our story is that it is unclear whether there is a model $\mathcal{M}$ with a *parameter-free* internal model $\mathcal{N}$ that is a strict end-extension and elementarily equivalent to $\mathcal{M}$.

We construct a sequence of models $\mathcal{N}_i$ such that: $\mathcal{N}_0 = \mathcal{M}$ and

$(\dagger)_i$ $\mathcal{N}_i \models \mathfrak{T} \cap \Pi_{i+1}$,
$(\ddagger)_i$ $\mathcal{N}_i \sqsubseteq_{i+1} N_{i+1}$.

The limit of the $\mathcal{N}_i$ will be the desired model $\mathcal{N}$. It is easy to see that $\mathcal{N}$ has the desired properties.

Clearly, $\mathcal{N}_0 := \mathcal{M}$ satisfies condition $(\dagger)_0$. Suppose we have constructed an end-extension $\mathcal{N}_i$ of $\mathcal{M}$ satisfying $(\dagger)_i$. We show how to find an $\mathcal{N}_{i+1} \sqsupseteq_i \mathcal{N}_i$ that satisfies $(\dagger)_{i+1}$.

Let C is any theorem of $\mathfrak{T}$. Since $\mathfrak{T}$ contains PA, we have, in $\mathfrak{T}$, that $\Diamond^{\Sigma_{i+1}} C$. Here $\Diamond^{\Sigma_{i+1}} C$ means that $C \cup \{D \mid \mathsf{true}_{\Sigma_{i+1}}(D)\}$ is consistent, where $\mathsf{true}_{\Sigma_{i+1}}$ is the usual Σ_{i+1}-truth-predicate. The set $\{D \mid \mathsf{true}_{\Sigma_{i+1}}(D)\}$ is to be understood $\mathfrak{T}$-internally. The statement $\Diamond^{\Sigma_{i+1}} C$ is Π_{i+1} (in the context of $\mathfrak{T}$), so, by $(\dagger)_i$, we have $\mathcal{N}_i \models \Diamond^{\Sigma_{i+1}} C$.

The set $\beta := \mathfrak{T} \cap \Pi_{i+2}$ is in $\mathsf{Rep}(\mathfrak{T})$ by the presence of a Π_{i+2}-truth-predicate. It follows that β is in $\mathsf{SSy}(\mathcal{M})$. Our model $\mathcal{N}_i$ is an end-extension of $\mathcal{M}$, so $\mathsf{SSy}(\mathcal{N}_i) = \mathsf{SSy}(\mathcal{M})$. So, β is coded in $\mathcal{N}_i$. Let α_0 be the set of axioms of PA. Clearly, $\gamma := \beta \cup \alpha_0$ is coded in $\mathcal{N}_i$, say by c. Let's write X_x for the set of 'elements' of c that are less than or equal to x.

We define $\alpha^*(x) := \mathsf{true}_{\Sigma_{i+1}}(x) \vee (x \in X_x \wedge \Diamond^{\Sigma_{i+1}}_{X_x} \top)$. It follows, by our above observation, that, for each C in γ, we have $\mathcal{N}_i \models \alpha^*(\underline{\ulcorner C \urcorner})$. Thus, we have an interpretation of $\mathsf{sf}(U)$, where U is the theory axiomatized by the axioms of PA plus $\mathfrak{T} \cap \Pi_2$, in $\mathcal{N}_i$, by setting $\alpha := \alpha^*$ and $\Gamma := \mathsf{form}_{\mathsf{PA}}$. We interpret the arithmetical part identically.

Since we are in an environment with full induction, all cuts involved in the development of Theorems 3.1, 3.11 and 3.12 are the identical cut. It follows that η^*, the interpretation of η, gives us an end-extension $\mathcal{N}_{i+1}$ of $\mathcal{N}_i$ such that the standard embedding of $\mathcal{N}_i$ in $\mathcal{N}_{i+1}$ sends a number to the value of its numeral and such that the satisfaction predicate behaves in a good way with respect to numerals.

By construction $\mathcal{N}_{i+1}$ is a model of PA that satisfies $\mathfrak{T} \cap \Pi_{i+2}$. We show that $\mathcal{N}_i \sqsubseteq_{i+1} \mathcal{N}_{i+1}$. Suppose $B(\vec{x})$ is in Σ_{i+1} and $\mathcal{N}_i \models B(\vec{b})$. We find that $\mathcal{N}_i \models \mathsf{true}_{\Sigma_{i+1}}(B(\underline{\vec{b}}))$. It follows that $\mathcal{N}_i \models (B(\underline{\vec{b}})) \in \alpha^*$, and hence $\mathcal{N}_i \models \mathsf{sat}_{\eta^*}(\emptyset, B(\underline{\vec{b}}))$. By, Theorem 3.12, we find:

$$\mathcal{N}_i \models \mathsf{sat}_{\eta^*}(\{\langle v_0, c_{b_0}\rangle, \cdots, \langle v_{n-1}, c_{b_{n-1}}\rangle\}, B(\vec{v})).$$

By Theorem 3.11, this means that $\mathcal{N}_{i+1} \models B(\vec{b})$, via the identification of elements implemented by the standard embedding of $\mathcal{N}_i$ in $\mathcal{N}_{i+1}$. □

5.2 *Properties of Degree Structures*

We can use the Interpretation Existence Lemma to prove all kinds of properties of degrees structures of interpretability. I choose one specific, more or less random, example that I have not seen before. Recently, Ali Enayat asked me the following question. If A is a finitely axiomatizable sequential theory that interprets PA, does A interpret ACA_0? The negative answer follows directly from the following theorem.

Theorem 5.3 *Let A and U be theories, where A is finitely axiomatized and U is recursively enumerable and sequential. Suppose $A \ntrianglerighteq U$. Then, there is a finitely axiomatized theory B such that $A \ntrianglerighteq B \ntrianglerighteq U$. Moreover, if A is sequential, B is sequential too.*

Proof We assume the conditions of the theorem.

We remind the reader of witness comparison notation. Let A be $\exists x\, A_0(x)$ and let B be of the form $\exists x\, B_0(x)$.

- $A \leq B :\leftrightarrow \exists x\, (A_0(x) \wedge \forall y < x \neg B_0(y))$. ($A$ is witnessed before or at the same time as B.)
- $A < B :\leftrightarrow \exists x\, (A_0(x) \wedge \forall y \leq x \neg B_0(y))$. ($A$ is witnessed strictly before B.)
- $(A \leq B)^\perp :\leftrightarrow B < A$, $(A < B)^\perp :\leftrightarrow B \leq A$. ($C^\perp$ is the opposite of C.)

We write $\Box^\star_U C$ for $\exists x\, \Box_{U,x,x} C$.

By the Gödel Fixed Point Lemma, we find a Σ_1-sentence R such that:

- $\mathsf{S}^1_2 \vdash R \leftrightarrow (B \rhd A) \leq (U \rhd B)$,
- $B = (A \boxplus (\mathsf{S}^1_2 + (R < \Box^\star_U \bot)))$.

We note that, in the definition of R, we have on the right-hand-side of the two occurrences $\rhd$ in each case a finitely axiomatized theory. We assume that such interpretability statements are written in the form $\exists x\, S_0(x)$, where S_0 is, say, $\Delta_0(\omega_1)$. Thus, R can indeed be taken to be a Σ_1-sentence.

Suppose R. In that case we have $B \rhd A$. Let r be a witness of R. Since U is sequential, we have $U \rhd (\mathsf{S}^1_2 + \Diamond_{U,\underline{r},\underline{r}} \top)$ and, hence, by Σ_1-completeness,

$$U \;\rhd\; (\mathsf{S}^1_2 + \underline{r} : R + \Diamond_{U,\underline{r},\underline{r}} \top) \;\vdash\; (\mathsf{S}^1_2 + (R < \Box^\star_U \bot)) \;\rhd\; B \;\rhd\; A.$$

So, $U \rhd A$. Quod non.

Suppose $R^\perp$. In that case we have $U \rhd B$. By Σ_1-completeness, it follows that:

$$U \;\rhd\; B \;\vdash\; (A \boxplus (\mathsf{S}^1_2 + R^\perp + (R < \Box^\star_U \bot))) \;\vdash\; (A \boxplus (\mathsf{S}^1_2 + R^\perp + R)) \;\rhd\; A.$$

Hence, $U \rhd A$. Quod non iterum.

Since R is false, it follows that $\mathsf{S}^1_2 + (R < \Box^\star_U \bot) \vdash \mho_{*,*}(U)$, since, internally, R cannot have a standard witness. Hence: (†) $(\mathsf{S}^1_2 + (R < \Box^\star_U \bot)) \rhd U$. We also have (‡) $A \rhd U$. So, by (†) and (‡), we find that:

$$B \;=\; (A \boxplus (\mathsf{S}^1_2 + (R < \Box^\star_U \bot))) \;\rhd\; U.$$

Thus, $A \rhd B \rhd U$. We show that none of the two steps can be reversed. Suppose $B \rhd A$. It then follows that R or $R^\perp$. Quod non. Suppose $U \rhd B$. In that case it again follows that R or $R^\perp$. Quod non. We may conclude that $A \ntrianglerighteq B \ntrianglerighteq U$, as promised.

Finally, note that B, as constructed, is the infimum of A and a sequential theory using the disjunction implementation of the infimum. So, if A is sequential, then B is also sequential. $\square$

Open Question 5.4 What happens when we drop the demand that U is sequential?

5.3 The Interpretability of Inconsistency

In my two papers [46, 49], I study Feferman's Theorem on the interpretability of inconsistency. Here I just present the basic result.

Theorem 5.5 *Suppose U is a recursively enumerable theory and suppose $N : \mathsf{S}^1_2 \lhd U$. Then $U \rhd (U + \Box^N_U \bot)$.*

Proof Clearly, $(U + \Box^N_U \bot) \rhd (U + \Box^N_U \bot)$. Moreover, using the formalized version of Gödel's Second Incompleteness Theorem, we have:

$$(U + \Diamond^N_U \top) \vdash (U + \Diamond^N_U \Box^N_U \bot) \rhd (U + \Box^N_U \bot).$$

So, using a disjunctive interpretation, we find $U \rhd (U + \Box^N_U \bot)$. $\square$

References

1. Boolos, G.S., Burgess, J.P., Jeffrey, R.C.: Computability and Logic, 4th edn. Cambridge University Press, Cambridge (2002)
2. Buss, S.R.: Bounded Arithmetic. Bibliopolis, Napoli (1986)
3. Buss, S.: Cut elimination in situ. In: Kahle, R., Rathjen, M. (eds.) Gentzen's Centenary, pp. 245–277. Springer International Publishing, Berlin (2015)
4. Ebbs, G.: Satisfying predicates: Kleene's proof of the Hilbert–Bernays theorem. Hist. Philos. Log. **36**(4), 1–21 (2014)
5. Feferman, S.: Arithmetically definable models of formalized arithmetic. Notices of the American Mathematical Society **5**, 679–680 (1958)
6. Feferman, S.: Arithmetization of metamathematics in a general setting. Fund. Math. **49**, 35–92 (1960)
7. Feferman, S.: My route to arithmetization. Theoria **63**(3), 168–181 (1997)
8. Gerhardy, P.: Refined complexity analysis of cut elimination. In: Baaz, M., Makowsky, J. (eds.) Proceedings of the 17th International Workshop CSL 2003. LNCS, vol. 2803, pp. 212–225. Springer, Berlin (2003)
9. Gerhardy, P.: The role of quantifier alternations in cut elimination. Notre Dame J. Form. Log. **46**(2), 165–171 (2005)
10. Gödel, K.: Die Vollständigkeit der Axiome des logischen Funktionenkalküls. Monatshefte für Mathematik **37**(1), 349–360 (1930)
11. Hájek, P.: On interpretability in set theories I. Comm. Math. Univ. Carolinae **12**, 73–79 (1971)
12. Hájek, P.: On interpretability in set theories II. Comm. Math. Univ. Carolinae **13**, 445–455 (1972)

13. Hilbert, D., Bernays, P.: Grundlagen der Mathematik II. Springer, Berlin (1939). Second edition: 1970
14. Henkin, L.: The completeness of the first-order functional calculus. J. Symb. Log. **14**(03), 159–166 (1949)
15. Hájková, M., Hájek, P.: On interpretability in theories containing arithmetic. Fund. Math. **76**(2), 131–137 (1972)
16. Hájek, P., Pudlák, P.: Metamathematics of First-Order Arithmetic. Perspectives in Mathematical Logic. Springer, Berlin (1993)
17. Jeřábek, E.: Sequence encoding without induction. Math. Log. Q. **58**(3), 244–248 (2012)
18. Kaye, R.: Models of Peano Arithmetic. Oxford Logic Guides. Oxford University Press, Oxford (1991)
19. Kleene, S.C.: Introduction to Metamathematics. North-Holland, Amsterdam (1952)
20. Krajíček, J.: A note on proofs of falsehood. Archiv für Mathematische Logik und Grundlagenforschung **26**(1), 169–176 (1987)
21. Kossak, R., Schmerl, J.: The Structure of Models of Peano Arithmetic. Clarendon Press, Oxford (2006)
22. Lindström, P.: Some results on interpretability. In: Jensen, F.V., Mayoh, B.H., Møller, K.K. (eds.) Proceedings of the 5th Scandinavian Logic Symposium 1979, pp. 329–361. Aalborg University Press, Aalborg (1979)
23. Löwenheim, L.: Über möglichkeiten im Relativkalkül. Math. Ann. **76**(4), 447–470 (1915)
24. Mycielski, J., Pudlák, P., Stern, A.S.: A lattice of chapters of mathematics (interpretations between theorems). Memoirs of the American Mathematical Society, vol. 84. AMS, Providence (1990)
25. Nelson, E.: Predicative Arithmetic. Princeton University Press, Princeton (1986)
26. Orey, S.: Relative interpretations. Zeitschrift für Mathematische Logik und Grundlagen der Mathematik **7**, 146–153 (1961)
27. Pudlák, P.: Some prime elements in the lattice of interpretability types. Trans. Am. Math. Soc. **280**, 255–275 (1983)
28. Pudlák, P.: Cuts, consistency statements and interpretations. J. Symb. Log. **50**(2), 423–441 (1985)
29. Putnam, H.: Trial and error predicates and a solution to a problem of Mostowski. J. Symb. Log. **30**(1), 146–153 (1965)
30. Schmerl, J.H.: End extensions of models of arithmetic. Notre Dame J. Form. Log. **33**(2), 216–219 (1992)
31. Scott, D.: On a theorem of Rabin. Indagationes Mathematicae (Proceedings) **63**, 481–484 (1960)
32. Simpson, S.G.: Subsystems of Second Order Arithmetic. Perspectives in Logic: Association for Symbolic Logic, 2nd edn. Cambridge University Press, Cambridge (2009)
33. Smoryński, C.: Nonstandard models and related developments. In: Harrington, L.A., Morley, M.D., Scedrov, A., Simpson, S.G. (eds.) Harvey Friedman's Research on the Foundations of Mathematics, pp. 179–229. North Holland, Amsterdam (1985)
34. Solovay, R.M.: Interpretability in set theories. Unpublished letter to P. Hájek. http://www.cs.cas.cz/hajek/RSolovayZFGB.pdf (1976)
35. Troelstra, A.S., Schwichtenberg, H.: Basic Proof Theory. Cambridge Tracts in Theoretical Computer Science, vol. 43. Cambridge University Press, Cambridge (2000)
36. Visser, A.: Interpretability logic. In: Petkov, P.P. (ed.) Mathematical logic. Proceedings of the Heyting 1988 Summer School in Varna, Bulgaria, pp. 175–209. Plenum Press, Boston (1990)
37. Visser, A.: The formalization of interpretability. Stud. Log. **51**(1), 81–105 (1991)
38. Visser, A.: The unprovability of small inconsistency. Arch. Math. Log. **32**(4), 275–298 (1993)
39. Visser, A.: Faith & Falsity: a study of faithful interpretations and false Σ_1^0-sentences. Ann. Pure Appl. Log. **131**(1–3), 103–131 (2005)
40. Visser, A.: Categories of theories and interpretations. In: Enayat, A., Kalantari, I., Moniri, M. (eds.) Logic in Tehran. Proceedings of the Workshop and Conference on Logic, Algebra and Arithmetic, held 18–22 October 2003. Lecture Notes in Logic, vol. 26, pp. 284–341. ASL, A.K. Peters Ltd., Wellesley, Mass (2006)

41. Visser, A.: Pairs, sets and sequences in first order theories. Arch. Math. Log. **47**(4), 299–326 (2008)
42. Visser, A.: Cardinal arithmetic in the style of Baron von Münchhausen. Review of Symbolic Logic **2**(3), 570–589 (2009). https://doi.org/10.1017/S1755020309090261
43. Visser, A.: Hume's principle, beginnings. Rev. Symb. Log. **4**(1), 114–129 (2011)
44. Visser, A.: What is sequentiality? In: Cégielski, P., Cornaros, Ch., Dimitracopoulos, C. (eds.) New Studies in Weak Arithmetics. CSLI Lecture Notes, vol. 211, pp. 229–269. CSLI Publications and Presses Universitaires du Pôle de Recherche et d'Enseingement Supérieur Paris-est, Stanford (2013)
45. Visser, A.: Interpretability degrees of finitely axiomatized sequential theories. Arch. Math. Log. **53**(1–2), 23–42 (2014)
46. Visser, A.: The interpretability of inconsistency, Feferman's theorem and related results. Logic Group Preprint Series 318, Faculty of Humanities, Philosophy, Utrecht University, Janskerkhof 13, 3512 BL Utrecht. http://www.phil.uu.nl/preprints/lgps/ (2014)
47. Visser, A.: Why the theory R is special. In: Tennant, N. (ed.) Foundational Adventures. Essays in honour of Harvey Friedman, pp. 7–23. College Publications, UK (2014). Originally published online by Templeton Press in 2012. http://foundationaladventures.com/
48. Visser, A.: On Q. Soft Comput. 1–18 (2016). Special issue in memoriam Franco Montagna (A. Di Nola, D. Mundici, C. Toffalori, A. Ursini, eds.)
49. Visser, A.: Transductions in arithmetic. Ann. Pure Appl. Log. **167**(3), 211–234 (2016)
50. Wang, H.: Arithmetic translations of axiom systems. Trans. Am. Math. Soc. **71**, 283–293 (1951)
51. Wang, H.: Arithmetical models for formal systems. Methodos **3**, 217–232 (1951)
52. Wilkie, A.J.: On the theories of end-extensions of models of arithmetic. Set Theory and Hierarchy Theory V, pp. 305–310. Springer, Berlin (1977)
53. Wilkie, A.J., Paris, J.B.: On the scheme of induction for bounded arithmetic formulas. Ann. Pure Appl. Log. **35**, 261–302 (1987)

Author Biography

Albert Visser was born on December 19, 1950 in Zwijndrecht. He studied Mathematics and Philosophy at the Utrecht University under Dirk van Dalen and Henk Barendregt. In 1981 he obtained his PhD with the thesis Aspects of Diagonalization & Provability. He continued to work at Utrecht University, where, in 1998, he became professor of logic, philosophy of mathematics and epistemology at the Department of Philosophy. In 2015, he became Faculty Professor of the Faculty of Humanities of Utrecht University. He retired in 2016. He is a member of the Royal Netherlands Academy of Arts and Sciences. His research is concerned with the metamathematics of interpretations, sequential theories, satisfaction predicates, schematic logics like provability and interpretability logic, constructive arithmetical theories and intuitionistic propositional logic. In addition, he works on a series of papers where he addresses various issues connected to philosophy and metaphysics. Finally, he has an enduring interest in the philosophy of language.

Tiered Arithmetics

Helmut Schwichtenberg and Stanley S. Wainer

To the memory of Sol Feferman, whose creative force at the centre of logic has been a constant inspiration

Abstract In his paper "Logics for Termination and Correctness of Functional Programs, II. Logics of Strength PRA" [5] Feferman was concerned with the problem of how to guarantee the feasibility (or at least the subrecursive complexity) of functions definable in certain logical systems. His ideas have influenced much subsequent work, for instance the final chapter of [13]. There, linear two-sorted systems LT(;) (a version of Gödel's T) and LA(;) (a corresponding arithmetical theory) of polynomial-time strength were introduced. Here we extend LT(;) and LA(;) in such a way that some forms of non-linearity are covered as well. This is important when one wants to deal on the proof level with particular algorithms, not only with the functions they compute. Examples are divide-and-conquer approaches as in treesort, and the first of two main sections here gives a detailed analysis of this. The second topic treated heads in a quite different direction, though again its roots lie in the final chapter of [13]. Instead of just two sorts we consider transfinite ramified sequences of them, or "tiers"; ordinally labelled copies of the natural numbers, respecting certain pointwise orderings. A hierarchy of infinitary arithmetical theories $\mathrm{EA}(I_\alpha)$ is devised, I_α designating the top tier. These are weak numerical analogues of the iterated inductive definitions underpinning much of Feferman's fundamental work over decades; see for example his survey [6] and the technical classic [3]. The computational strength of $\mathrm{EA}(I_\alpha)$ is summarized thus: it proves the totality of all functions elementary in the Fast-Growing F_α. A "pointwise" concept of transfinite induction

H. Schwichtenberg (✉)
Mathematisches Institut der Universität München, 80333 München, Germany
e-mail: schwicht@math.lmu.de

S.S. Wainer
School of Mathematics, University of Leeds, Leeds LS2 9JT, UK
e-mail: s.s.wainer@leeds.ac.uk

G. Jäger and W. Sieg (eds.), *Feferman on Foundations*, Outstanding Contributions to Logic 13, DOI 10.1007/978-3-319-63334-3_6

then provides an ordinal measure of strength, but this is a weak, finitistic analogue of the usual notions, related to the Slow-Growing hierarchy.

Keywords Polynomial time · Linear two-sorted arithmetic · Program extraction Tiered arithmetic · Fast and slow-growing hierarchies · Pointwise transfinite induction.

2010 Mathematics Subject Classification 03D20 · 03D15 · 03F05

1 Introduction

The principle of numerical induction:

$$A(0) \wedge \forall_a (A(a) \to A(a+1)) \to A(n)$$

may be viewed as being "impredicative", since establishing that n has property A might entail quantifying over all numbers, in particular n itself, even before it is completely understood (see Nelson [11]). "Tiering" is a way to unravel such impredicativities: one thinks of the input n as existing at a higher level (tier) than A's quantified variables – the output domain. Such tiered or predicative inductions severely restrict the computational strength of an arithmetical theory, and bring them more closely into the realms of "feasibility" (see Leivant [8, 9]).

For example, suppose we want to build a theory of elementary recursive strength. Then we will have the exponential function $E(n)$ representing 2^n:

$$E(0) := 1, \qquad E(S(n)) := D(E(n)).$$

However, its iteration

$$F(0) := 1, \qquad F(S(n)) := E(F(n))$$

should be avoided. Let us see how a proof of "computational strength" of F could arise. Let $F(n) \simeq a$ denote the graph of F, viewed as an inductively defined relation. We write $F(n){\downarrow}$ for $\exists_a (F(n) \simeq a)$. Then a proof of

$$\forall_n F(n){\downarrow}$$

should be disallowed. Such a proof would be by induction on n. In the step we would need to prove $\forall_n (F(n){\downarrow} \to F(S(n)){\downarrow})$. So assume n is given and we have an a such that $F(n) \simeq a$. We need to find b such that $E(F(n)) \simeq b$. Clearly $E(a)$ would be such a b, but here we have substituted the "output" variable a in the recursion (or "input") argument of E. We therefore distinguish input and output variables and argument positions in order to avoid superexponential strength. The underlying idea is that the inputs form a new "tier" lying over the domain of output values. While an input may be fed down to the output level and used as a bound on induction or recursion steps, this is a one-way process - for outputs may not be fed back as inputs.

If we want to even further restrict the computational strength to the subelementary level, we need a linearity restriction. Consider for example the function $B(a, n)$ representing $a + 2^n$:

$$B(a, 0) := S(a), \qquad B(a, S(n)) := B(B(a, n), n).$$

Let $B(a, n) \simeq b$ be the (inductively defined) graph of B, and consider the following proof of

$$\forall_{n,a} B(a, n)\downarrow,$$

by induction on n. In the step we have n and can assume $\forall_a B(a, n)\downarrow$; we need to show $\forall_a B(a, S(n))\downarrow$, i.e., $\exists_c (B(B(a, n), n) \simeq c)$. Given n and a, by induction hypothesis we have b such that $B(a, n) \simeq b$, and applying the induction hypothesis again, we have c such that $B(b, n) \simeq c$. This double use of the induction hypothesis is responsible for exponential growth, and hence we use a linearity restriction to stay within the subelementary realm.

2 Representing Algorithms in Linear Two-Sorted Arithmetic

In this section we define the constructive systems A(;) and LA(;), their intended use being to develop program specification proofs and then term extraction for practical algorithms. The computational strength of A(;) will be elementary recursive (going back to early developments of such theories by Leivant [9], based on the safe / normal discipline of Bellantoni and Cook [1] and earlier Simmons [14]). The subtheory LA(;) will be corresponding theory of polynomial strength, and therefore relevant for the development of feasible programs.

The main contents of this section will be a description of these theories and their basic properties, followed by examples illustrating their use. Our leading intuition is the Curry–Howard correspondence between terms in lambda-calculus (or more precisely in Gödel's T) and derivations in arithmetic. The restrictions needed to stay within elementary or polynomial strength will be incorporated in certain term systems T(;) and LT(;) corresponding to A(;) and LA(;). A two-sortedness restriction will allow to unfold the higher type recursion operator $\mathcal{R}$ in a controlled way to guarantee elementary complexity, and a further linearity restriction will ensure polynomial strength.

2.1 The Term Systems T(;) *and* LT(;)

We consider *types* built from base types ι by two forms $\rho \hookrightarrow \sigma$ and $\rho \to \sigma$ of arrow types, called input arrow and output arrow. A type is *safe* if it does not involve the input arrow $\hookrightarrow$.

As base types we have the type **B** of booleans tt, ff, the (unary) natural numbers **N** with constructors 0 and $S\colon \mathbf{N} \to \mathbf{N}$, products $\rho \times \sigma$ with constructor $\times^+\colon \rho \to \sigma \to \rho \times \sigma$ and lists $\mathbf{L}(\rho)$ with constructors [] and $::_\rho$ of type $\rho \to \mathbf{L}(\rho) \to \mathbf{L}(\rho)$. Note that all constructors have safe types.

Variables are typed, and come in two forms, input variables $\bar{x}^\rho$ and output variables x^ρ. Constants are (i) the constructors and (ii) the recursion operators for base types, for instance

$$\begin{aligned}\mathcal{R}^\tau_{\mathbf{N}}&\colon \mathbf{N} \hookrightarrow \tau \to (\mathbf{N} \hookrightarrow \tau \to \tau) \hookrightarrow \tau,\\ \mathcal{R}^\tau_{\mathbf{L}(\rho)}&\colon \mathbf{L}(\rho) \hookrightarrow \tau \to (\rho \hookrightarrow \mathbf{L}(\rho) \hookrightarrow \tau \to \tau) \hookrightarrow \tau,\end{aligned}$$

where the value type τ is required to be safe. This requirement is necessary because without it we could define the iterated exponential function F from the exponential function E via iteration with value type $\mathbf{N} \hookrightarrow \mathbf{N}$. We also have (iii) the cases operators for base types, for instance

$$\begin{aligned}\mathcal{C}^\tau_{\mathbf{N}}&\colon \mathbf{N} \to \tau \to (\mathbf{N} \hookrightarrow \tau) \to \tau,\\ \mathcal{C}^\tau_{\mathbf{L}(\rho)}&\colon \mathbf{L}(\rho) \to \tau \to (\rho \hookrightarrow \mathbf{L}(\rho) \hookrightarrow \tau) \to \tau,\\ \mathcal{C}^\tau_{\rho\times\sigma}&\colon \rho \times \sigma \to (\rho \hookrightarrow \sigma \hookrightarrow \tau) \to \tau,\end{aligned}$$

where again the value type τ is required to be safe.

Remark Recursion and cases operators are provided for all base types. However, with arbitrary base types, we may have more than one recursive call. If – as in Sect. 2.1 – we are to develop a theory based on linear ideas, we must disallow recursion operators with multiple recursive calls, since this would spoil the whole approach.

T(;)-*terms* are built from variables and the constants above by introduction and elimination rules for the two type forms $\rho \hookrightarrow \sigma$ and $\rho \to \sigma$:

$$\begin{aligned}&\bar{x}^\rho \mid x^\rho \mid C^\rho \text{ (constant)} \mid\\ &(\lambda_{\bar{x}^\rho} r^\sigma)^{\rho\hookrightarrow\sigma} \mid (r^{\rho\hookrightarrow\sigma} s^\rho)^\sigma \quad (s \text{ an input term}) \mid\\ &(\lambda_{x^\rho} r^\sigma)^{\rho\to\sigma} \mid (r^{\rho\to\sigma} s^\rho)^\sigma.\end{aligned}$$

A term s is an *input term* if all its free variables are input variables, or else s is of higher type and all its higher type free variables are input variables.

Remark The restriction for $(r^{\rho\hookrightarrow\sigma} s^\rho)^\sigma$ is more liberal here than in [13]: we now allow output variables of base type in case s is of higher type. This change does not affect the estimates ensuring elementary complexity. We also changed the type of step terms in recursion and cases operators, for instance $\rho \hookrightarrow \mathbf{L}(\rho) \hookrightarrow \tau \to \tau$ instead of $\rho \to \mathbf{L}(\rho) \to \tau \to \tau$ in $\mathcal{R}^{\tau}_{\mathbf{L}(\rho)}$. This makes is easier to construct step terms as lambda-abstractions, since now the abstracted variables corresponding to parts of the recursion argument are input variables and hence usable to build input terms. These changes do not affect the complexity estimates.

LT(;)-*terms* are built from variables and the constants above by introduction and elimination rules for the two type forms $\rho \hookrightarrow \sigma$ and $\rho \to \sigma$, but now with an additional linearity restriction:

$\bar{x}^{\rho} \mid x^{\rho} \mid C^{\rho}$ (constant) $\mid$	
$(\lambda_{\bar{x}^{\rho}} r^{\sigma})^{\rho \hookrightarrow \sigma} \mid (r^{\rho \hookrightarrow \sigma} s^{\rho})^{\sigma}$	(s an input term) $\mid$
$(\lambda_{x^{\rho}} r^{\sigma})^{\rho \to \sigma} \mid (r^{\rho \to \sigma} s^{\rho})^{\sigma}$	(higher type output variables in r, s distinct, r does not start with a cases operator $\mathcal{C}^{\tau}_{\iota}$) $\mid$
$\mathcal{C}^{\tau}_{\iota} t \vec{r}$	(higher type output variables in FV(t) not in $\vec{r}$)

with as many r_i as there are constructors of ι. The notion of an input term is the same as above. The restriction on output variables in the formation of $r^{\rho \to \sigma} s$ or $\mathcal{C}^{\tau}_{\iota} t \vec{r}$ ensures that every higher type output variable can occur at most once in a given LT(;)-term, except in the alternatives of a cases operator.

2.2 *The Theories* A(;) *and* LA(;)

We consider *formulas* built from atomic formulas by (i) the two forms $A \hookrightarrow B$ and $A \to B$ of implication, called input and output implication, and (ii) universal quantification either $\forall_{\bar{x}} A$ over an input variable $\bar{x}$ or $\forall_x A$ over an output variable x. *Atomic formulas* are either terms of the type **B** of booleans viewed as propositions or else inductively defined predicates – possibly with parameters – applied to argument terms. We view $\exists_{\bar{x}} A$ and $\exists_x A$ as atomic formulas, more precisely as (nullary) inductive predicates with predicates $\{ \bar{x} \mid A \}$ or $\{ x \mid A \}$ as parameter.

In this section proofs are in minimal logic, in natural deduction style. It is convenient to represent them as proof terms, as in Table 1. For quantification on input variables $\bar{x}$ we have similar rules, and also for the input implication $\hookrightarrow$. Assumption variables come in two forms, input ones $\bar{u}^A$ and output ones u^A. Axioms are the introduction and elimination axioms for $\exists$

$$\exists^{+}_{\{x \mid A\}} \colon \forall_x (A \to \exists_x A), \qquad \exists^{-}_{\{x \mid A(x)\}, P} \colon \exists_x A(x) \to \forall_{\bar{x}} (A(\bar{x}) \hookrightarrow P) \to P$$

and similarly for input variables $\bar{x}$. For every base type we have its induction and cases axioms, for instance

Table 1 Derivation terms for $\to$ and $\forall$

Derivation	Term
$u\colon A$	u^A
$\dfrac{\begin{array}{c}[u\colon A]\\ \mid M\\ B\end{array}}{A\to B}\ \to^+ u$	$(\lambda_{u^A} M^B)^{A\to B}$
$\dfrac{\begin{array}{c}\mid M\\ A\to B\end{array}\qquad \begin{array}{c}\mid N\\ A\end{array}}{B}\ \to^-$	$(M^{A\to B} N^A)^B$
$\dfrac{\begin{array}{c}\mid M\\ A\end{array}}{\forall_x A}\ \forall^+ x$ (with var.cond.)	$(\lambda_x M^A)^{\forall_x A}$ (with var.cond.)
$\dfrac{\begin{array}{c}\mid M\\ \forall_x A(x)\end{array}\qquad r}{A(r)}\ \forall^-$	$(M^{\forall_x A(x)} r)^{A(r)}$

$$\mathrm{Ind}_{\bar n,P}\colon \forall_{\bar n}(P(0)\to\forall_{\bar n}(P(\bar n)\to P(S(\bar n)))\hookrightarrow P(\bar n^{\mathbf{N}})),$$
$$\mathrm{Ind}_{\bar l,P}\colon \forall_{\bar l}(P([])\to\forall_{\bar x,\bar l}((P(\bar l)\to P(\bar x::\bar l))\hookrightarrow P(\bar l^{\mathbf{L}(\rho)}))$$

and

$$\mathrm{Cases}_{n,P}\colon \forall_n(P(0)\to\forall_{\bar n}P(S(\bar n))\to P(n^{\mathbf{N}})),$$
$$\mathrm{Cases}_{l,P}\colon \forall_l(P([])\to\forall_{\bar x,\bar l}P(\bar x::\bar l)\to P(l^{\mathbf{L}(\rho)})).$$

We call these *raw* proof terms. Note that when ignoring the annotations of implications and variables we obtain proofs terms in ordinary arithmetic. The raw proof terms need to be restricted to make up the theories A(;) and LA(;). To formulate these restrictions it is easiest to refer to the *extracted term* $\mathrm{et}(M)$ of a proof term M, which we introduce first. This requires some preparations.

Computational content in proofs arises from computationally relevant (c.r.) atomic formulas; in our setting the only ones are $\exists_{\bar{x}} A$ and $\exists_x A$. There are also non-computational (n.c.) atomic formulas, like equalities. Following Kolmogorov [7] we assign to every formula A an object $\tau(A)$, which is a type or the "nulltype" symbol $\circ$. The definition can be conveniently written if we extend the use of $\rho \hookrightarrow \sigma$, $\rho \to \sigma$ and $\rho \times \sigma$ to the nulltype symbol $\circ$:

$$(\rho \hookrightarrow \circ) := \circ, \quad (\circ \hookrightarrow \sigma) := \sigma, \quad (\circ \hookrightarrow \circ) := \circ \quad \text{and similarly for } \to,$$
$$(\rho \times \circ) := \rho, \quad (\circ \times \sigma) := \sigma, \quad (\circ \times \circ) := \circ.$$

With this understanding of $\rho \hookrightarrow \sigma$, $\rho \to \sigma$ and $\rho \times \sigma$ we can simply define

$$\begin{aligned}
\tau(A) &:= \circ \quad \text{if } A \text{ is an n.c. atomic formula,}\\
\tau(\exists_{\bar{x}^\rho} A) &:= \tau(\exists_{x^\rho} A) := \rho \times \tau(A),\\
\tau(A \hookrightarrow B) &:= (\tau(A) \hookrightarrow \tau(B)),\\
\tau(A \to B) &:= (\tau(A) \to \tau(B)),\\
\tau(\forall_{\bar{x}^\rho} A) &:= (\rho \hookrightarrow \tau(A)),\\
\tau(\forall_{x^\rho} A) &:= (\rho \to \tau(A)).
\end{aligned}$$

We introduce a special "nullterm" symbol ε to be used as a "realizer" for n.c. formulas, and extend term application to the nullterm symbol by

$$\varepsilon t := \varepsilon, \quad t\varepsilon := t, \quad \varepsilon\varepsilon := \varepsilon.$$

Now we can define the extracted term $\mathrm{et}(M)$ of a proof term M deriving A. It is relative to a fixed assignment of input variables $\bar{x}_{\bar{u}}$ of type $\tau(A)$ to input assumption variables $\bar{u}^A$, and similarly output variables x_u of type $\tau(A)$ to output assumption variables u^A. If A is n.c., then $\mathrm{et}(M) := \varepsilon$, else

$$\begin{aligned}
&\mathrm{et}(\bar{u}^A) && := \bar{x}_{\bar{u}}^{\tau(A)},\\
&\mathrm{et}(u^A) && := x_u^{\tau(A)},\\
&\mathrm{et}((\lambda_{\bar{u}^A} M)^{A \hookrightarrow B}) && := \lambda_{\bar{x}_{\bar{u}}^{\tau(A)}} \mathrm{et}(M),\\
&\mathrm{et}((\lambda_{u^A} M)^{A \to B}) && := \lambda_{x_u^{\tau(A)}} \mathrm{et}(M),\\
&\mathrm{et}(M^{A \hookrightarrow B} N) := \mathrm{et}(M^{A \to B} N) && := \mathrm{et}(M)\mathrm{et}(N),\\
&\mathrm{et}((\lambda_{\tilde{x}^\rho} M)^{\forall_{\tilde{x}} A}) && := \lambda_{\tilde{x}^\rho} \mathrm{et}(M),\\
&\mathrm{et}(M^{\forall_{\tilde{x}} A} r) && := \mathrm{et}(M) r,
\end{aligned}$$

with $\tilde{x}$ input or output variable. We also need to define extracted terms for the axioms, i.e., $\exists^+$, $\exists^-$ and for every base type its induction and cases axioms. The extracted terms are

$$\mathrm{et}(\exists^+_{\{x^\rho \mid A\}}) := \times^+ \qquad \text{of type } \rho \to \tau(A) \to \rho \times \tau(A)$$
$$\mathrm{et}(\exists^-_{\{x^\rho \mid A\},P}) := \mathcal{C}_{\rho\times\tau(A)} \text{ of type } \rho \times \tau(A) \to (\rho \hookrightarrow \tau(A) \hookrightarrow \tau(P)) \to \tau(P)$$

and for the induction and cases axioms the corresponding recursion and cases operators.

Now finally we are ready to define the theories A(;) and LA(;): a raw proof term M is in A(;) (or LA(;)) if $\mathrm{et}(M)$ is a term in T(;) (or LT(;)).

2.3 Treesort

In this section we extend LT(;) and LA(;) in such a way that some forms of non-linearity are covered as well. This is important when one wants to deal on the proof level with particular algorithms, not only with the functions they compute. Examples are divide-and-conquer approaches like in treesort. The method requires two recursive calls and hence is not covered by the linear setup in LT(;) and LA(;). However, one can show that the number of conversion steps in the parse-dag computation model still is a polynomial in the length of the list. Generally, one needs to extend LT(;) and LA(;) by constants defined by computation rules meeting certain criteria.

For the formulation of the treesort algorithm we use the base type $\mathbf{T}$ (branch labelled binary trees) with a nullary constructor $\diamond$ and a ternary constructor C of type $\mathbf{N} \to \mathbf{T} \to \mathbf{T} \to \mathbf{T}$. The treesort algorithm is given by the defined constants

$$\begin{aligned}
&\mathrm{TreeSort}(l) &&= \mathrm{Flatten}(\mathrm{MakeTree}(l)),\\
&\mathrm{MakeTree}([]) &&= \diamond,\\
&\mathrm{MakeTree}(a{::}l) &&= \mathrm{Insert}(a, \mathrm{MakeTree}(l)),\\
&\mathrm{Insert}(a, \diamond) &&= C_a(\diamond, \diamond),\\
&\mathrm{Insert}(a, C_b(u, v)) &&= \begin{cases} C_b(\mathrm{Insert}(a, u), v) & \text{if } a \le b\\ C_b(u, \mathrm{Insert}(a, v)) & \text{otherwise,}\end{cases}\\
&\mathrm{Flatten}(\diamond) &&= [],\\
&\mathrm{Flatten}(C_b(u, v)) &&= \mathrm{Flatten}(u) * (b{::}\mathrm{Flatten}(v))
\end{aligned}$$

where $*$ denotes the Append-function. Notice that the second defining equation of Flatten has two recursive calls. Therefore this "divide-and-conquer" algorithm is not covered by the treatment in [13]: the linearity restriction is violated. The point of the present section is to show how this problem can be overcome, by giving the non-linear Flatten-function a special treatment w.r.t. our parse dag computation model, which we describe next.

Let LT(;) + Flatten be the extension of LT(;) by the defined constant Flatten of type $\mathbf{T} \hookrightarrow \mathbf{L}(\mathbf{N})$. To obtain a polynomial upper bound on the time complexity of functions definable in LT(;) + Flatten, we need a careful analysis of the normalization process. Our time measurement is with respect to a computation model that fits well to the lambda-terms we have to work with, and is also close to actual computation.

We compute with terms represented as dags (directed acyclic graphs) where only nodes for terms of base type can have in-degree greater than one. Each graph is required to be connected and have a unique root (i.e., node with in-degree zero). Nodes can be (i) terminal nodes labelled by a variable or constant, (ii) abstraction nodes with one successor, labelled with a (typed input or output) variable and a pointer to the successor node, or (iii) application nodes with two successors, labelled with pointers to them. A *parse dag* is required to represent a parse tree for a term, i.e., the types must fit and all other conditions above on the formation of terms must be satisfied.

The *size* $\|d\|$ of a parse dag d is the number of nodes in it. A parse dag is *conformal* if (i) every node with in-degree greater than 1 is of base type, and (ii) every maximal path to a bound variable x passes through the same binding λ_x-node. A parse dag is *h-affine* if every higher type variable occurs at most once in the dag, except in the alternatives of a cases operator. We identify a parse dag with the term it represents.

In our computation model the following steps require one time unit.

(a) Creation of a node given its label and pointers to its successor nodes.
(b) Deletion of a node.
(c) Given a pointer to an interior node, to obtain a pointer to one of its successor nodes.
(d) Test on the type and the label of a node, and on the variable or constant in case the node is terminal.

We will estimate the number of steps it takes to reduce a term t to its normal form $\mathrm{nf}(t)$. For simplicity we fix an order of reduction, by requiring that the leftmost innermost redex is converted first. Let $\#t$ denote the total number of such reduction steps.

Lemma 2.1 *Let l be a numeral of type* $\mathbf{L}(\mathbf{N})$. *Then*

$$\#(l * l') = O(|l|).$$

Proof One easily proves $\#(l * l') \leq N \cdot (|l| + 1)$ by induction on $|l|$, for an appropriate N. □

To estimate $\#\mathrm{Flatten}(u)$ we use a size function for numerals u of type $\mathbf{T}$:

$$\|\diamond\| := 0,$$
$$\|C_n(u, v)\| := 2\|u\| + \|v\| + 3.$$

Lemma 2.2 *Let u be a numeral of type* $\mathbf{T}$. *Then*

$$\#\mathrm{Flatten}(u) = O(\|u\|).$$

Proof We prove $\#\mathrm{Flatten}(u) \leq N(\|u\| + 1)$ by induction on $\|u\|$, for an appropriate N, and only deal with the second defining equation of Flatten, which involves two recursive calls. Consider the parse dag for $\text{Flatten}(C_n(u, v))$. We can assume that it takes $\leq N$ steps to transform it into a the parse dag for $\text{Flatten}(u) * (a{::}\text{Flatten}(v))$. Then

$$\#\text{Flatten}(C_n(u, v)) \leq N + \#\text{Flatten}(u) + \#\text{Flatten}(v) + N(\|u\| + 1)$$

using $\#(l * l') \leq N(|l| + 1)$ (by Lemma 2.1) and $|\text{Flatten}(u)| \leq \|u\|$. Hence by induction hypothesis

$$\begin{aligned}\#\text{Flatten}(C_a(u, v)) &\leq N + N(\|u\| + 1) + N(\|v\| + 1) + N(\|u\| + 1)\\ &= N(2\|u\| + \|v\| + 4)\\ &= N(\|C_a(u, v)\| + 1).\end{aligned}$$

□

We show that all functions definable in LT(;) + Flatten are polynomial-time computable, by adapting the argument in [13] to the presence of defined constants, here the constant Flatten. Call a term $\mathcal{RD}$-free if it contains neither recursion constants $\mathcal{R}$ nor Flatten. A term is called *simple* if it contains no higher type input variables. Obviously simple terms are closed under reductions, taking of subterms, and applications. Every simple term is h-affine, due to the (almost) linearity of higher type output variables.

As in [13, pp. 416–418] we have

Lemma 2.3 (Simplicity) *Let t be a base type term whose free variables are of base type. Then* $\text{nf}(t)$ *contains no higher type input variables.*

Lemma 2.4 (Sharing normalization) *Let t be an $\mathcal{RD}$-free simple term. Then a parse dag for* $\text{nf}(t)$*, of size at most $\|t\|$, can be computed from t in time $O(\|t\|^2)$.*

Corollary 2.5 (Base normalization) *Let t be a closed $\mathcal{RD}$-free simple term of type* **N** *or* **L(N)**. *Then* $\text{nf}(t)$ *can be computed from t in time $O(\|t\|^2)$, and $\|\text{nf}(t)\| \leq \|t\|$.*

Lemma 2.6 ($\mathcal{RD}$-elimination) *Let $t(\vec{x})$ be a simple term of safe type. There is a polynomial P_t such that: if $\vec{r}$ are safe type $\mathcal{RD}$-free closed simple terms and the free variables of $t(\vec{r})$ are output variables, then in time $P_t(\|\vec{r}\|)$ one can compute an $\mathcal{RD}$-free simple term* $\text{rdf}(t; \vec{x}; \vec{r})$ *such that* $t(\vec{r}) \to^* \text{rdf}(t; \vec{x}; \vec{r})$.

Proof By induction on $\|t\|$, as in [13, pp. 418–20]. We only have to add a case for the defined constant Flatten.

Case Flatten. Then t is of the form $\text{Flatten}(l)$, because the term has safe type. Since l is an input term, all free variables of l are input variables – they must be in $\vec{x}$ since free variables of $t(\vec{r})$ are output variables. Therefore $l(\vec{r})$ is closed, implying $\text{nf}(l(\vec{r}))$ is a list. One obtains $\text{rdf}(l; \vec{x}; \vec{r})$ in time $P_l(\|\vec{r}\|)$ by the induction hypothesis. Then by base normalization one obtains the lists $\hat{l} := \text{nf}(\text{rdf}(l; \vec{x}; \vec{r}))$ in a further polynomial time. Now Lemma 2.2 implies the claim. □

Theorem 2.7 (Normalization) *Let t be a closed term in* LT(;) + Flatten *of type* $\mathbf{N} \twoheadrightarrow \dots \mathbf{N} \twoheadrightarrow \mathbf{N}$ *($\twoheadrightarrow \in \{\hookrightarrow, \rightarrow\}$). Then t denotes a poly-time function.*

Proof One must find a polynomial Q_t such that for all $\mathcal{RD}$-free simple closed terms $\vec{n}$ of type $\mathbf{N}$ one can compute $\mathrm{nf}(t\vec{n})$ in time $Q_t(\|\vec{n}\|)$. Let $\vec{x}$ be new variables of type $\mathbf{N}$. The normal form of $t\vec{x}$ is computed in an amount of time that may be large, but it is still only a constant with respect to $\vec{n}$. By the simplicity lemma $\mathrm{nf}(t\vec{x})$ is simple. By $\mathcal{RD}$-elimination one reduces to an $\mathcal{RD}$-free simple term $\mathrm{rdf}(\mathrm{nf}(t\vec{x}); \vec{x}\,; \vec{n})$ in time $P_t(\|\vec{n}\|)$. Since the running time bounds the size of the produced term, $\|\mathrm{rdf}(\mathrm{nf}(t\vec{x}); \vec{x}\,; \vec{n})\| \leq P_t(\|\vec{n}\|)$. By sharing normalization one can compute

$$\mathrm{nf}(t\vec{n}) = \mathrm{nf}(\mathrm{rdf}(\mathrm{nf}(t\vec{x}); \vec{x}\,; \vec{n}))$$

in time $O(P_t(\|\vec{n}\|)^2)$, so for Q_t one can take some constant multiple of $P_t(\|\vec{n}\|)^2$. □

2.4 *Treesort in* LA(;) + Flatten

We have seen that polynomial-time algorithms can be implemented as extracted terms of appropriate proofs in LA(;). This is developed in detail in [13, Sect. 8.4]. Here we describe that and how the same mechanism works when both LT(;) and LA(;) contain constants (like Flatten) defined by equations involving multiple recursive calls. As an example we treat the treesort algorithm in LA(;).

A tree u is called *sorted* if the list Flatten(u) is sorted. We recursively define a function I inserting an element a into a tree u in such a way that, if u is sorted, then so is the result of the insertion:

$$I(a, \diamond) := C_a(\diamond, \diamond),$$

$$I(a, C_b(u, v)) := \begin{cases} C_b(I(a, u), v) & \text{if } a \leq b, \\ C_b(u, I(a, v)) & \text{if } b < a \end{cases}$$

and, using I, a function S sorting a list l into a tree:

$$S([]) := \diamond, \qquad S(a::l) := I(a, S(l)).$$

We represent these functions by inductive definitions of their graphs. Thus, writing $I(a, u, u')$ to denote $I(a, u) = u'$ and similarly, $S(l, u)$ for $S(l) = u$, we have the following clauses:

$$I(a, \diamond, C_a(\diamond, \diamond)),$$
$$a \leq b \rightarrow I(a, u, u') \rightarrow I(a, C_b(u, v), C_b(u', v)),$$
$$b < a \rightarrow I(a, v, v') \rightarrow I(a, C_b(u, v), C_b(u, v')),$$

and

$$S([], \diamond),$$
$$S(l, u) \to I(a, u, u') \to S(a::l, u').$$

As an auxiliary function we use $\mathrm{tl}_i(l)$, which is the tail of the list l of length i, if $i < |l|$, and l otherwise. Its recursion equations are

$$\mathrm{tl}_i([]) := [], \quad \mathrm{tl}_i(a::l) := \begin{cases} \mathrm{tl}_i(l) & \text{if } i \le |l| \\ a::l & \text{else.} \end{cases}$$

We will need some easy properties of S and tl:

$$\begin{aligned}
&S([a], C_a(\diamond, \diamond)),\\
&S(l, C_b(u, v)) \to a \le b \to I(a, u, u') \to S(a::l, C_b(u', v)),\\
&S(l, C_b(u, v)) \to b < a \to I(a, v, v') \to S(a::l, C_b(u, v')),\\
&i \le |l| \to \mathrm{tl}_i(a::l) = \mathrm{tl}_i(l),\\
&\mathrm{tl}_{|l|}(l) = l, \qquad \mathrm{tl}_0(l) = [].
\end{aligned}$$

We would like to derive $\exists_u S(l, u)$ in LA(;). However, we shall not be able to do this. All we can achieve is $|l| \le n \to \exists_u S(l, u)$, with n an input parameter.

Lemma 2.8 (Tree insertion) LA(;) *proves* $\forall_{a,n,u}(|u| \le n \to \exists_{u'} I(a, u, u'))$.

Proof We fix a and use induction on n. In the base case we can take $u' := C_a(\diamond, \diamond)$. In the step assume the induction hypothesis and $|w| \le n+1$. If $|w| \le n$ use the induction hypothesis. Now assume $n < |w|$. Then w is of the form $C_b(u, v)$. If $a \le b$ pick u' by induction hypothesis and take $C_b(u', v)$. If $b < a$ pick v' by induction hypothesis and take $C_b(u, v')$. □

Lemma 2.9 LA(;) *proves* $\forall_{l,n,m}(m \le n \to \exists_u S(\mathrm{tl}_{\min(m,|l|)}(l), u))$.

Proof We fix l, n and use induction on m. In the base case we can take $u := \diamond$, using $\mathrm{tl}_0(l) = []$. In the step case assume the induction hypothesis and $m + 1 \le n$. If $|l| \le m$ we are done by the induction hypothesis. If $m < |l|$ we must show $\exists_u S(\mathrm{tl}_{m+1}(l), u)$. Now $\mathrm{tl}_{m+1}(l) = a::\mathrm{tl}_m(l)$ with $a := \mathrm{hd}(\mathrm{tl}_{m+1}(l))$, since $m < |l|$. If $m = 0$ take $u := C_a(\diamond, \diamond)$. If $0 < m$, by induction hypothesis we have w with $S(\mathrm{tl}_m(l), w)$, and w is of the form $C_b(u, v)$. If $a \le b$, pick u' by Lemma 2.8 such that $I(a, u, u')$. Take $S(a::\mathrm{tl}_m(l), C_b(u', v))$. If $b < a$, pick v' by Lemma 2.8 such that $I(a, v, v')$. Take $S(a::\mathrm{tl}_m(l), C_b(u, v'))$. □

Theorem 2.10 (Treesort) LA(;) *proves* $\forall_{l,n}(|l| \le n \to \exists_u S(l, u))$.

Proof Specializing Lemma 2.9 to l, n, n gives

$$\forall_{l,n} \exists_u S(\mathrm{tl}_{\min(n,|l|)}(l), u))$$

and hence also

$$\forall_{l,n}(|l| \leq n \to \exists_u S(l,u))$$

since $\mathrm{tl}_{|l|}(l) = l$. □

We have formalized these proofs in Minlog[1] and extracted their computational content, i.e., LT(;)-terms. For Lemma 2.8 we obtain a term involving the recursion operator $\mathcal{R}_{\mathbf{N}}^{\tau}\colon \mathbf{N} \hookrightarrow \tau \to (\mathbf{N} \hookrightarrow \tau \to \tau) \to \tau$ with $\tau := \mathbf{T} \to \mathbf{T}$. This term represents the function f of type $\mathbf{N} \to \mathbf{N} \hookrightarrow \mathbf{T} \to \mathbf{T}$ defined by

$$f(a,0,u) := C_a(\diamond,\diamond),$$

$$f(a,n+1,u) := \begin{cases} f(a,n,u) & \text{if } |u| \leq n, \\ C_{\mathrm{Lb}(u)}(f(a,n,L(u)), R(u)) & \text{if } n < |u| \text{ and } a \leq \mathrm{Lb}(u), \\ C_{\mathrm{Lb}(u)}(L(u), f(a,n,R(u))) & \text{if } n < |u| \text{ and } \mathrm{Lb}(u) < a \end{cases}$$

with $\mathrm{Lb}(u)$, $L(u)$, $R(u)$ label and left and right subtree of $u \neq \diamond$. For Lemma 2.9 we obtain the function g of type $\mathbf{L}(\mathbf{N}) \to \mathbf{N} \hookrightarrow \mathbf{N} \hookrightarrow \mathbf{T}$ with

$$g(l,n,0) := \diamond, \qquad g(l,n,m+1) :=$$

$$\begin{cases} u & \text{if } |l| \leq m, \\ C_{\mathrm{hd}(\mathrm{tl}_1(l))}(\diamond,\diamond), & \text{if } 0 = m < |l|, \\ C_{\mathrm{Lb}(u)}(f(a,m,L(u)), R(u)) & \text{if } 0 < m < |l| \text{ and } a \leq \mathrm{Lb}(u) \\ C_{\mathrm{Lb}(u)}(L(u), f(a,m,R(u))) & \text{if } 0 < m < |l| \text{ and } \mathrm{Lb}(u) < a \end{cases}$$

where $u := g(l,n,m)$ and $a := \mathrm{hd}(\mathrm{tl}_{m+1}(l))$.

Let $\bar{S}(l,l')$ express that l' is multiset-equal to l and sorted. One easily proves $S(l,u) \to \bar{S}(l, \mathrm{Flatten}(u))$ and thus gets an LA(;) + Flatten-derivation of

$$|l| \leq n \to \exists_{l'} \bar{S}(l,l'). \tag{1}$$

From Theorem 2.10 we get the function h of type $\mathbf{L}(\mathbf{N}) \to \mathbf{N} \hookrightarrow \mathbf{T}$ with $h(l,n) := g(l,n,n)$. For (1) we finally obtain an LT(;) + Flatten-term representing the function $\bar{h}$ of type $\mathbf{L}(\mathbf{N}) \to \mathbf{N} \hookrightarrow \mathbf{L}(\mathbf{N})$ with $\bar{h}(l,n) := \mathrm{Flatten}(h(l,n))$.

3 Transfinitely Iterated Tiering

This section investigates the effect of adding to an arithmetical base theory a transfinite hierarchy of number-theoretic tiers $\{I_\alpha\}$, indexed by countable "tree ordinals" α (that is, ordinals with fixed, chosen fundamental sequences $\{\lambda_i\}_{i \in N}$ assigned at

[1] www.minlog-system.de.

limits λ). I_0 will be the "output" domain, and unrestricted quantifiers are to be read as ranging over I_0. The first level of "inputs", I_1, controls the lengths of inductions on formulas of level/tier 0. Thus from the previous section and Chap. 8 of [13] one notes already that a (non-linear) tiered arithmetical theory incorporating just I_1 will already have elementary recursive computational strength. This section investigates what is gained by adding successively higher levels of input, and what conditions should be placed on those levels. The result is easily stated: *Each new tier gives access to the next level of the fast-growing, transfinitely extended Grzegorczyk hierarchy*. An informal explanation is also easy –

A version of the fast-growing hierarchy is:

$$F_0(n) = n + 1 \; ; \quad F_{\beta+1}(n) = F_\beta^{2^n}(n) \; ; \quad F_\lambda(n) = F_{\lambda_n}(n) \; .$$

The main principle here is that of *tiered induction*:

$$A(0) \wedge \forall_z (A(z) \to A(z+1)) \to \forall_z (I_\alpha(z) \to A(z))$$

where the "level" of A is $\prec \alpha$, i.e. every I_β occurring in A has $\beta \prec \alpha$.

Now let us assume inductively that we can prove $F_\beta \colon I_\beta \to I_\beta$, that is

$$\forall_x (I_\beta(x) \to \exists_y (I_\beta(y) \wedge F_\beta(x) \simeq y)) \; .$$

Then the formula $A(z) \equiv F_\beta^{2^z} {:} I_\beta \to I_\beta$ is inductive, and so by tiered induction: $\forall_z (I_{\beta+1}(z) \to A(z))$. The tiering also entails $I_{\beta+1}(z) \to I_\beta(z)$ and therefore by definition of $F_{\beta+1}(z)$,

$$\forall_z (I_{\beta+1}(z) \to \exists_y (I_\beta(y) \wedge F_{\beta+1}(z) \simeq y)) \; .$$

Now a further principle of tiering is that the level of a "computed" variable y, depending only on higher level inputs, may be lifted to the lowest level of non-zero input on which it depends. This rule was used in Cantini [4] and is a kind of Σ_1^0-reflection: in particular

$$\frac{I_{\beta+1}(z) \to \exists_y (I_\beta(y) \wedge F_{\beta+1}(z) \simeq y)}{I_{\beta+1}(z) \to \exists_y (I_{\beta+1}(y) \wedge F_{\beta+1}(z) \simeq y)}$$

and therefore $F_{\beta+1} \colon I_{\beta+1} \to I_{\beta+1}$.

At limits λ we would like to show $F_\lambda \colon I_\lambda \to I_\lambda$, but unfortunately the tiering principles mean that the best we can do is $F_\lambda \colon I_{\lambda+1} \to I_\lambda$. Assume inductively that this holds already at each stage λ_i of the fundamental sequence to λ, thus

$$\forall_z (I_{\lambda_i+1}(z) \to \exists_y (I_{\lambda_i}(y) \wedge F_{\lambda_i}(z) \simeq y)) \; .$$

The conditions placed on limit tiers must then entail, for each i, $I_\lambda(i) \to I_{\lambda_i+1}(i)$ for then, by the definition of $F_\lambda(i)$:

$$I_\lambda(i) \to \exists_y(I_{\lambda_i}(y) \wedge F_\lambda(i) \simeq y) \ .$$

Again, since the $\exists_y$ only depends upon i at level λ, the $I_{\lambda_i}(y)$ may be lifted to $I_\lambda(y)$. However, for technical reasons to do with the lifting rule below, the premise $I_\lambda(i)$ must then also be lifted up one level, to $I_{\lambda+1}(i)$. The final step $F_\lambda\colon I_{\lambda+1} \to I_\lambda$ then requires an ω-rule to universally quantify over i.

To summarize, the underlying mechanisms are (i) tiered induction, (ii) lifting or Σ^0_1 reflection, and (iii) diagonal: $I_\lambda(i) \to I_{\lambda_i}(i)$. These are formalized below, in a hierarchy of infinitary arithmetics $\mathrm{EA}(I_\alpha)$ whose computational strengths correspond exactly to the levels $\mathcal{E}^\alpha$ of the fast-growing, extended Grzegorczyk hierarchy.

3.1 The Infinitary Systems $\mathbf{EA}(I_\alpha)$

Given an ordinal α, the infinitary system $\mathrm{EA}(I_\alpha)$ derives Tait-style sequents with numerical input declarations:

$$n_1{:}I_{\beta_1}, \ldots n_j{:}I_{\beta_j} \vdash^\gamma \Gamma \quad \text{abbreviated } \vec{n}{:}\vec{I} \vdash^\gamma \Gamma$$

where $\alpha \succeq \beta_1 \succ \ldots \succ \beta_j$. Γ will be a finite set of closed formulas in the language of arithmetic augmented by elementary term constructors for coding sequences and suitable ordinal notations, and a new binary "input predicate" $I_b(m)$ representing $m{:}I_\beta$ when β is the tree ordinal denoted by the notation b. We will generally abuse notation and write $I_\beta(m)$ for $I_b(m)$, but no confusion should arise. The set Γ is said to be of "level" less than ξ (written $\mathrm{lev}(\Gamma) \prec \xi$) if every I_β occurring has $\beta \prec \xi$, An important convention will be that a declaration $n_j{:}I_{\beta_j}$ where $n_j = 0$ will be suppressed (i.e. assumed and not explicitly stated). Of the declared inputs, only finitely-many will be non-zero. An obvious principle is that $n{:}I_\beta \vdash^\gamma A$ means $\vdash^\gamma I_\beta(n) \to A$, and this is achieved by writing $\vdash^\gamma \neg I_\beta(n_1), \neg I_\beta(n_2), \ldots, \neg I_\beta(n_r), \Gamma$ instead as $n{:}I_\beta \vdash^\gamma \Gamma$ where $n = \max(n_1, n_2, \ldots, n_r)$. Thus declarations $n_j{:}I_{\beta_j}$ to the left of $\vdash^\gamma$ are kept distinct from the formulas $I_\beta(m)$ in Γ.

Both the ordinal bounds γ on the heights of derivations, and also the tier levels α, will come from a specified initial segment of tree-ordinals (e.g. ε_0 or Γ_0 etc.). There will be an associated elementary recursive notation-system allowing computation of e.g. successors, predecessors, and limit-projection $(\lambda, n) \mapsto \lambda_n$. It will be assumed that the ordering $\prec$ on tree ordinals will be the union of "pointwise" orderings $\prec_n$ where, for each n, $\prec_n$ is the transitive closure of $\delta \prec_n \delta + 1$ and $\lambda_n \prec_n \lambda$ for limits λ. Most "standard" notation systems satisfy $\delta \prec_n \gamma \Rightarrow \delta + 1 \preceq_{n+1} \gamma$ and this too will be a standing assumption here. The "n-descending chain" from γ is

$$0 \prec 1 \prec \ldots \prec \delta \prec \delta + 1 \prec \ldots \prec \lambda_n \prec \lambda \prec \ldots \prec \gamma$$

and this will therefore be contained in the $n+1$-descending chain. Alternative notation for $\gamma' + 1 \preceq_n \gamma$ is $\gamma' \in \gamma[n]$ – see part 2 of [13].

Logic Rules

To ensure appropriate levels of stratification, the ordinals γ' bounding the premises of all rules below, bear the following relationship to the ordinal bounds γ assigned to their conclusions: $\gamma' + 1 \preceq_n \gamma$ where *$n{:}I_\beta$ is a declared input at a level higher than the levels of all formulas in the premises or conclusion*. Thus if there is no explicit declaration, meaning just 0 is declared, or if there is no higher level, that is I_α already appears in Γ, then rules may still be applied, but only under the constraint $\gamma' + 1 \preceq_0 \gamma$.

The Axioms are $\vec{n}{:}\vec{I} \vdash^\gamma \Gamma$ where the set Γ contains a true atom (e.g. an equation or inequation between closed terms, or $s \neq s', t \neq t', \bar{I}_s(t), I_{s'}(t')$).

The Cut rule, with cut formula C, is

$$\frac{\vec{n}{:}\vec{I} \vdash^{\gamma'} \Gamma, \neg C \qquad \vec{n}{:}\vec{I} \vdash^{\gamma''} \Gamma, C}{\vec{n}{:}\vec{I} \vdash^{\gamma} \Gamma}.$$

The $\exists$-rules are:

$$\frac{\vec{n}{:}\vec{I} \vdash_c^{\gamma'} I_\beta(m) \qquad \vec{n}{:}\vec{I} \vdash^{\gamma''} A(m), \Gamma}{\vec{n}{:}\vec{I} \vdash^{\gamma} \exists x(I_\beta(x) \wedge A(x)), \Gamma}.$$

Here the left-hand premise "computes" witness m according to the computation rules given below.

The $\forall$-rules are versions of the ω-rule:

$$\frac{\ldots \max(n, m){:}I_\beta, \ldots \vdash^{\gamma'} A(m), \Gamma \ \text{ for every m in N}}{\ldots n{:}I_\beta, \ldots \vdash^{\gamma} \forall x(I_\beta(x) \to A(x)), \Gamma}$$

but note the fixed ordinal bound γ' on the premises, which does not vary with m. This helps to keep the theory "weak".

The $\vee, \wedge$ rules are unsurprising and we don't list them.

The final logic rule allows interaction with computation in the form:

$$\frac{\ldots n{:}I_\beta \vdash_c^{\gamma'} I_\beta(m) \qquad \ldots m{:}I_\beta, \ldots \vdash^{\gamma''} \Gamma}{\ldots n{:}I_\beta, \ldots \vdash^{\gamma} \Gamma}.$$

Computation Rules

The same conditions on tree-ordinal bounds γ apply.

The Computational Axioms are $\vec{n}{:}\vec{I}, n{:}I_\beta \vdash_c^\gamma I_{\beta'}(\ell), \Gamma$ provided $\ell \leq n + 1$ and there is a k declared in $\vec{n}{:}\vec{I}$ such that $\beta' \preceq_k \beta$. Thus by the ($\forall$)-rule, with any γ, $k{:}I_{\beta+1} \vdash^{\gamma+1} \forall_x(I_\beta(x) \to I_{\beta'}(x+1))$ provided $\beta' \preceq_k \beta$. Hence I_β is inductive and is contained in $I_{\beta'}$ whenever $\beta' \prec_0 \beta$.

The Lifting Rule, from $I_{\beta'}$ to I_β when $\beta' \prec_k \beta$, is:

$$\frac{\vec{n}{:}\vec{I},\, n{:}I_\beta \vdash_c^\gamma I_{\beta'}(m)}{\vec{n}{:}\vec{I},\, n{:}I_\beta \vdash_c^\gamma I_\beta(m)}$$

provided k is declared in $\vec{n}{:}\vec{I}$ at a level higher than β. Recall that, in the declaration, the blank after $n{:}I_\beta$ means zeros.

The Computation Rules (call-by-value) are:

$$\frac{\ldots n{:}I_\beta \vdash_c^{\gamma'} I_\beta(m) \qquad \ldots m{:}I_\beta \vdash_c^{\gamma''} I_\beta(\ell)}{\ldots n{:}I_\beta \vdash_c^\gamma I_\beta(\ell)} .$$

Alternative Ordinal Assignment

Alternatively, in each rule above one could simply take $\gamma = \max(\gamma_0, \gamma_1) + 1$ or $= \gamma' + 1$. But then, in order to make use of the $\forall$-rule, which requires a fixed bound on all premises, one needs to add an Accumulation Rule, as in Buchholz [2]: from $\vec{n}{:}\vec{I} \vdash^{\gamma'} \Gamma$ derive $\vec{n}{:}\vec{I} \vdash^{\gamma} \Gamma$ provided $\gamma' + 1 \preceq_n \gamma$, where n is declared at a level greater than the level of Γ. (This is also a suitably modified version of Mints' Repetition Rule [10].)

Basic Lemmas

Lemma 3.1 (Tiered Induction) *If the level of A is $\prec \beta$ and $\beta' \prec \beta$ and a appropriately measures the "size" of the formula A, so that $\vdash^a A, \neg A$, then*

$$\vdash^{a+2\omega+1} A(0) \wedge \forall_z (I_{\beta'}(z) \to A(z) \to A(z+1)) \to \forall_x (I_\beta(x) \to A(x)) .$$

Proof By repeatedly applying the $(\wedge)$ and $(\exists)$ rules, and the computational axioms, using $\vdash^a A, \neg A$, one obtains for each m,

$$m{:}I_\beta \vdash^{a+2m+1} \neg(A(0) \wedge \forall_z (I_{\beta'}(z) \to A(z) \to A(z+1))), A(m) .$$

Then $m : I_\beta \vdash^{a+2\omega} \neg(A(0) \wedge \forall_z (I_{\beta'}(z) \to A(z) \to A(z+1))), A(m)$ because for the tree ordinal ω we take the successor function as its "standard" fundamental sequence and so $m+1 \prec_m \omega$ and hence $a + 2m + 2 \prec_m a + 2\omega$. This holds for every m, so the result follows by applying the $(\forall)$-rule. □

Lemma 3.2 (Bounding) *Let $\{f_\beta(g)\}$ be the following functional version of the fast-growing hierarchy:*

$$\begin{aligned} f_0(g)(n; m) &= m + 1 \\ f_{\beta+1}(g)(n; m) &= f_\beta(g)(\max(n, m); -)^{2^{g(\max(n,m))}}(m) \\ f_\lambda(g)(n; m) &= f_{\lambda_n}(g)(n; m) . \end{aligned}$$

Let $G_\gamma(n)$ be the slow-growing function, giving the size of $\{\gamma' : \gamma' + 1 \preceq_n \gamma\}$.

Then if $\vec{n}{:}\vec{I} \vdash_c^\gamma I_\beta(m)$ by the computation rules alone, we have:

$$m \leq f_\beta(g)(n; -)^{2^{g(n)}}(\bar{m})$$

where $g = G_\gamma$, $\bar{m} = 1 + \max \vec{n}$ *and* n *is* $1+$ *the maximum of all declared inputs at levels* $\succ \beta$.

Proof Proceed by induction on β with a nested induction on γ. Let $g' = G_{\gamma'}$ and $g = G_\gamma$ and note that if $\gamma' + 1 \preceq_n \gamma$ then $g'(n) < g(n)$.

If $\vec{n}{:}\vec{I} \vdash_c^\gamma I_\beta(m)$ comes about by a computational axiom then $m < \bar{m} + 1 = f_0(g)(n; \bar{m})$ and the result is immediate.

If it arises by Lifting from $\vec{n}{:}\vec{I} \vdash_c^\gamma I_{\beta'}(m)$ where $\beta' \prec_{n-1} \beta$, then inductively one may assume that

$$m \leq f_{\beta'}(g)(\bar{m}; -)^{2^{g(\bar{m})}}(\bar{m}) = f_{\beta'+1}(g)(n; \bar{m}) .$$

Now since $\beta' + 1 \preceq_n \beta$, it follows that

$$m \leq f_{\beta'+1}(g)(n; \bar{m}) \leq f_\beta(g)(n; \bar{m}) \leq f_\beta(g)(n; -)^{2^{g(n)}}(\bar{m}) .$$

Suppose the given derivation comes about by the Computation Rule from premises $\vec{n}{:}\vec{I} \vdash_c^{\gamma'} I_\beta(\ell)$ and $\vec{n}{:}\vec{I}, \ell{:}I_\beta \vdash_c^{\gamma''} I_\beta(m)$. Note that in this case both $\gamma' + 1$ and $\gamma'' + 1$ are $\preceq_n \gamma$ so $g'(n), g''(n) < g(n)$. Then the induction hypothesis gives $m \leq f_\beta(g'')(n; -)^{2^{g''(n)}}(\max(\bar{m}, \ell))$ and also $\ell \leq f_\beta(g')(n; -)^{2^{g'(n)}}(\bar{m})$. Composing, and at the same time increasing $f_\beta(g')$ and $f_\beta(g'')$ to $f_\beta(g)$,

$$m \leq f_\beta(g)(n; -)^{2^{g'(n)} + 2^{g''(n)}}(\bar{m}) \leq f_\beta(g)(n; -)^{2^{g(n)}}(\bar{m})$$

as required. □

Lemma 3.3 (Cut elimination) *(i) Suppose* $\vec{n}{:}\vec{I} \vdash^\gamma \Gamma, \neg C$ *and* $\vec{n}{:}\vec{I} \vdash^\delta \Gamma, C$, *both with cut-rank (maximum size of cut formulas)* $\leq r$. *Suppose also that* C *is either an atom, or a disjunction* $D_0 \vee D_1$ *or of existential form* $\exists x(I_j(x) \wedge D(x))$ *with* D *of size* r *(the "size" of input predicates is defined to be zero). Then* $\vec{n}{:}\vec{I} \vdash^{\gamma+\delta} \Gamma$ *again with cut-rank* r.
(ii) Hence if $\vec{n}{:}\vec{I} \vdash^\gamma \Gamma$ *with cut-rank* $r+1$ *then* $\vec{n}{:}\vec{I} \vdash^{\omega^\gamma} \Gamma$ *with cut-rank* $\leq r$ *and (repeating this)* $\vec{n}{:}\vec{I} \vdash^{\gamma^*} \Gamma$ *with cut-rank* 0, *where* $\gamma^* = \exp_\omega^{r+1}(\gamma)$.

Proof The proofs are fairly standard. □

Note on Σ_1^0 Reflection

The $(\exists)$ and "lifting" rules combine to derive the following version of Σ_1^0-reflection:

Suppose one has a cut-free derivation of $n{:}I_\beta \vdash^\gamma \Gamma$ *where* Γ *is a set of* Σ_1^0 *formulas of level* $\beta' \prec_n \beta$. *Then* $n{:}I_{\beta+1} \vdash^{2\cdot\gamma} \Gamma'$ *where* Γ' *results from* Γ *by lifting (some or all) existential quantifiers to level* β.

The proof is by induction on γ. Briefly, suppose the premises of the last $\exists$-rule are $n{:}I_\beta \vdash_c^{\gamma'} I_{\beta'}(m)$ and $n{:}I_\beta \vdash^{\gamma''} \Gamma, B(m)$ where Γ contains the formula $\exists_x(I_{\beta'}(x) \wedge B(x))$. Then by the induction hypothesis, $n{:}I_{\beta+1} \vdash^{2\cdot\gamma''} \Gamma', B'(m)$, and by lifting, since $n{:}I_{\beta+1} \vdash_c^0 I_\beta(n)$, we have also $n{:}I_{\beta+1} \vdash_c^{2\cdot\gamma'+1} I_\beta(m)$. Then by reapplying the $\exists$-rule, $n{:}I_{\beta+1} \vdash^{2\cdot\gamma} \Gamma', \exists_x(I_\beta(x) \wedge B'(x))$, that is $n{:}I_{\beta+1} \vdash^{2\cdot\gamma} \Gamma'$.

3.2 *The Computational Strength of* EA(I_α)

To illustrate, we now fix attention on the segment $\alpha \prec \varepsilon_0$ and choose the "standard" notation system for it, based, say, on Cantor normal forms with base ω. The EA(I_α)'s thus provide a "tiering" of Peano Arithmetic.

The heights γ of derivations allowed in EA(I_α) was previously left open, but now we need to be specific. Thus, henceforth, the heights γ of derivations in EA(I_α) will also be restricted to $\gamma \prec \varepsilon_0$, allowing EA($I_\alpha$)-derivations to be closed under cut elimination.

Theorem 3.4 *The provably recursive functions of* EA(I_α) *are exactly those functions elementary recursive in* F_α *where* F *is the version of the fast-growing hierarchy defined earlier:*

$$F_0(n) = n + 1 \; ; \quad F_{\delta+1}(n) = F_\delta^{2^n}(n) \; ; \quad F_\lambda(n) = F_{\lambda_n}(n) \; .$$

(But note that any other standard version would do since they are elementarily interreducible.)

Proof EA(I_α) was devised in the first place, precisely in order to allow derivation of $\forall_n(I_\beta(n) \to \exists_y(I_{\prec\beta}(n, y) \wedge F_\beta(n) \simeq y))$ for each $\beta \preceq \alpha$, where $I_{\prec\beta}(n, y)$ is either $I_{\beta-1}(y)$ or $I_{\beta_n}(y)$ according as β is a successor or a limit. This is what we take as our notion of "provable recursiveness". Furthermore, examination of that argument (in the introduction to this section) would show that the height of this derivation is (of the order) $\omega \cdot \beta + 2$. The Computation rule will then allow finite compositions of these functions to be formed and derived in EA(I_α). Thus if f is elementary in F_β (i.e. computable in time bounded by some finite iterate of F_β) then there is an elementary relation $R(n, m)$ such that $f(n)$ may be computed by finding the least m satisfying $R(n, m)$, and furthermore this m is $\leq F_\beta^k(n)$ for a fixed k. To show that f is provably recursive in EA(I_α) it is therefore only necessary to prove $n{:}I_\beta \vdash^\gamma \exists_y(I_{\prec\beta}(n, y) \wedge R(n, y))$ with γ independent of n. But because m is bounded by $F_\beta^k(n)$ and this provably exists in EA(I_α), we have $n{:}I_\beta \vdash_c I_{\prec\beta}(n, m)$ with height independent of n, and also $n{:}I_\beta \vdash R(n, m)$ since this entails just the checking of bounded quantifiers. Application of the ($\exists$)-rule then gives, for some fixed γ and every n, $n{:}I_\beta \vdash^\gamma \exists_y(I_{\prec\beta}(n, y) \wedge R(n, y))$.

Conversely, suppose f is provably recursive in EA(I_α). This means there is an inductively-given or elementary relation R such that $R(n, m) \Rightarrow f(n) \simeq (m)_0$ holds, and at some level $I_\beta, \vdash^\gamma \forall_x(I_\alpha(x) \to \exists_y(I_\beta(y) \wedge R(x, y)))$. We may assume this to be cut-free, and by inversion, $n{:}I_\alpha \vdash^\gamma \exists_y(I_\beta(y) \wedge R(n, y))$ for every n. Therefore (inverting the $\exists_y$ several times if necessary) for each n the correct value m satisfying $R(n, m)$ is such that $n{:}I_\alpha \vdash_c^\gamma I_\beta(m)$. There are now two cases: first suppose $\beta \prec \alpha$. By the Bounding Lemma, $m \le f_\beta(g)(n; -)^{2^{g(n)}}(n)$ where $g = G_\gamma$. This g is elementary, because $G_\gamma(n)$ has the effect of replacing each ω in the Cantor normal form of γ by n or $n+1$. Therefore $g(n)$ is bounded by a fixed finite iterate of $F_1(n) = n + 2^n$. It is not difficult to see, by induction on β, that $f_\beta(g)(n; -) \le F_\beta(g(\max n, -) + \ell)$, and hence $m \le F_{\beta+1}(g(n) + \ell)$ for some fixed ℓ. This bound is, as a function of n, elementary in F_α, and so the function f, being given by bounded search (as the least m less than the bound such that $R(n, m)$) is also elementary in F_α.

Finally, suppose $\beta = \alpha$, so $n{:}I_\alpha \vdash_c^\gamma I_\alpha(m)$. The Bounding Lemma now gives $m \le f_\alpha(g)(0; -)^{2^{g(0)}}(n)$ which is a fixed finite iterate of $f_\alpha(g)(0; -)$ on n. But, as noted above, $f_\alpha(g)(0; -) \le F_\alpha(g(-) + \ell)$. Therefore $f(n) = m$ is bounded by a fixed finite iterate of $F_\alpha \circ g$ and this holds for every n. So again, the given provably recursive function is elementary in F_α. □

3.3 *Weak, Pointwise Transfinite Induction*

A basic version of transfinite induction up to γ is

$$A(0) \wedge \forall_\delta(A(\delta) \to A(\delta + 1)) \wedge \forall_\lambda(\forall i\, A(\lambda_i) \to A(\lambda)) \to A(\gamma)\ .$$

Weak, pointwise-at-x transfinite induction up to γ is the following principle:

$$A(0) \wedge \forall_\delta(A(\delta) \to A(\delta + 1)) \wedge \forall_\lambda(A(\lambda_x) \to A(\lambda)) \to A(\gamma)$$

where x is a numerical input variable. We denote this principle $PTI(x, \gamma, A)$ and write $PTI(x, \gamma)$ for the schema.

Using this, we can immediately prove, with only a small amount of basic coding apparatus, that the x-descending chain from γ exists. That is

$$\exists_s D(s, x, \gamma)$$

where $D(s, x, \gamma)$ is the bounded formula saying that s is the sequence number of ordinal notations such that $(s)_0 = 0$ and $(s)_{lh(s)-1} = \gamma$ and for each $i < lh(s) - 1$ either $(s)_{i+1}$ is a limit λ, in which case $(s)_i = \lambda_x$, or $(s)_{i+1}$ is a successor $\delta + 1$, in which case $(s)_i = \delta$.

Thus $\exists_s D(s, x, \gamma)$ expresses the pointwise-at-x well-foundedness of γ, and we often abbreviate it as $PWF(x, \gamma)$. The contrast between this Σ_1^0 notion and full Π_1^1

well-foundedness is stark, but even here there are interesting analogies to be drawn. Whereas the natural subrecursive hierarchies of proof-theoretic bounding functions are "fast" growing in the classical case, they are "slow" growing in the pointwise case. For detailed comparisons between the two, see [13], and Weiermann [16]. Schmerl [12] was the first to formulate such weak, pointwise induction schemes in the context of Peano Arithmetic.

Definition 3.5 The functions L_x and G_x are defined as follows:
$L_x(\gamma) = a$ iff $\exists_s(D(s, x, \gamma) \wedge a = lh(s) - 1)$
$G_x(\gamma) = a$ iff $\exists_s(D(s, x, \gamma) \wedge a = \#(s))$
where $\#(s)$ is the number of successors in the descending sequence s.

Lemma 3.6 *L_x and G_x satisfy the following recursive definitions:*
$L_x(0) = 0, \quad L_x(\delta + 1) = L_x(\delta) + 1, \quad L_x(\lambda) = L_x(\lambda_x) + 1$.
$G_x(0) = 0, \quad G_x(\delta + 1) = G_x(\delta) + 1, \quad G_x(\lambda) = G_x(\lambda_x)$.

These functions, being given "pointwise-at-x", are alternative versions of the slow growing hierarchy, and they are both provably defined as immediate consequences of pointwise well-foundedness. They each have their uses, though we favour G_x since, for each x, it more readily collapses the arithmetic of tree ordinals down onto ordinary arithmetic. Thus writing x as the subscript and γ as the argument (instead of the other way around) is often a more appropriate notation. We use both, depending on context. Under the assumption $\delta \prec_n \gamma \Rightarrow \delta + 1 \preceq_{n+1} \gamma$ it immediately follows that

$$G_n(\gamma) \leq L_n(\gamma) \leq G_{n+1}(\gamma) .$$

Of course, even to call $PTI(x, \gamma)$ a *transfinite* induction principle requires a stretch of the imagination, because it is really just a collection of finitary inductions indexed by x and uniformized by γ. The following lemma brings this out more clearly. The levels at which inputs and quantifiers are declared will, for the time being, be suppressed.

Lemma 3.7 *In any arithmetical theory containing the basic coding apparatus, $PTI(x, \gamma)$ implies Numerical Induction up to $G_x(\gamma)$, and conversely, Numerical Induction up to $L_x(\gamma)$ implies $PTI(x, \gamma)$.*

More precisely, given any formula $F(a)$, let $A(\delta) \equiv \forall a \leq G_x(\delta).F(a)$ where $\forall a \leq G_x(\delta).F(a)$ stands for $\exists_b(G_x(\delta) = b \wedge \forall_{a \leq b} F(a))$. Then one may prove (with x, γ declared at a level higher than that of $F(a)$ and $A(\delta)$)

$$PTI(x, \gamma, A) \rightarrow (\ F(0) \wedge \forall_b(F(b) \rightarrow F(b + 1)) \rightarrow \forall a \leq G_x(\gamma).F(a)\ ) .$$

Conversely, given any formula $A(\delta)$ let $F(b) \equiv \forall \delta \preceq_x \gamma(L_x(\delta) = b \rightarrow A(\delta))$ where $\delta \preceq_x \gamma$ means $\exists_s(D(s, x, \gamma) \wedge \exists_{i<lh(s)}((s)_i = \delta))$. Then

$$(\ F(0) \wedge \forall_b(F(b) \rightarrow F(b + 1)) \rightarrow \forall a \leq L_x(\gamma).F(a)\ ) \rightarrow PTI(x, \gamma, A) .$$

Proof We argue informally. For the first part, it is only necessary to show that the progressiveness of F implies $A(0)$ and $\forall_\delta(A(\delta) \to A(\delta+1))$ and $A(\lambda_x) \to A(\lambda)$ for limits λ. But $F(0)$ immediately implies $A(0)$. If $\forall_b(F(b) \to F(b+1))$ then $\forall a \le G_x(\delta).F(a) \to \forall a \le G_x(\delta+1).F(a)$ which gives $\forall_\delta(A(\delta) \to A(\delta+1))$. The limit case $A(\lambda_x) \to A(\lambda)$ is immediate since $G_x(\lambda_x) = G_x(\lambda)$. Therefore $PTI(x, \gamma, A)$ gives $A(\gamma) \equiv \forall a \le G_x(\gamma).F(a)$.

For the converse, assume A is progressive, i.e. $A(0)$ and $\forall\delta(A(\delta) \to A(\delta+1))$ and $A(\lambda_x) \to A(\lambda)$ at limits λ. Then one easily proves $F(0)$ and for any b, $F(b) \to F(b+1)$. For assume $F(b)$. Then if $\delta \preceq_x \gamma$ and $L_x(\delta) = b+1$, δ is either a successor or a limit and its immediate predecessor in the $\preceq_x$-sequence, call it δ', satisfies $L_x(\delta') = b$. Therefore $A(\delta')$ holds and, by the progressiveness of A one immediately gets $A(\delta)$. Hence $F(b+1)$, and so by numerical induction up to $L_x(\gamma)$ we then have $F(L_x(\gamma))$ and hence $A(\gamma)$. This implies $PTI(x, \gamma, A)$. □

The motto is: "In a theory of predicative, or tiered, numerical induction, $G_\gamma \downarrow$ witnesses the provability of pointwise transfinite induction up to γ."

Definition 3.8 Extend G_x to the third number-class by taking large sups to small sups. Thus: $G_x(0) = 0$; $G_x(\delta+1) = G_x(\delta)+1$; $G_x(\lambda) = G_x(\lambda_x)$ at small limits λ, and at large limits, $G_x(SUP_\xi\, \lambda_\xi) = \sup_i G_x(\lambda_i)$.

Note in particular, $G_x(\Omega) = \omega$.

Definition 3.9 For each α in the third number-class, define the function φ_α from countable tree ordinals to countable tree ordinals:

$$\varphi_\alpha(\beta) = \begin{cases} \beta+1 & \text{if } \alpha = 0 \\ \varphi_{\alpha-1}^{2^\beta}(\beta) & \text{if } \alpha \text{ is a successor} \\ \sup_i \varphi_{\alpha_i}(\beta) & \text{if } \alpha \text{ is a small limit} \\ \varphi_{\alpha_\beta}(\beta) & \text{if } \alpha \text{ is a large limit.} \end{cases}$$

Lemma 3.10 (Collapsing) *Provided each large limit $\lambda \preceq \alpha$ satisfies the condition $G_x(\lambda_\xi) = G_x(\lambda)_{G_x(\xi)}$, we have:*

$$G_x(\varphi_\alpha(\beta)) = F_{G_x(\alpha)}(G_x(\beta)) .$$

Proof As in Chap. 5 of [13]. □

Theorem 3.11 *For each $\alpha \prec \varepsilon_0$, let $\bar{\alpha}$ be the ordinal in the third number-class obtained by replacing ω by Ω throughout its Cantor normal form. Then $\varphi_{\bar{\alpha}+1}(\omega)$ is the supremum of the ordinals γ for which* EA(I_α) *proves pointwise transfinite induction up to γ.*

Proof Pointwise transfinite induction up to $\gamma = \varphi_{\bar{\alpha}+1}(\omega)$ cannot be proven in EA(I_α) because, by Collapsing, $G_\gamma(n) = F_{G_n(\bar{\alpha}+1)}(n) = F_{\alpha+1}(n)$ and this is not elementary in F_α. (Note: the proviso in the Collapsing lemma is satisfied for Cantor normal forms.) Hence Numerical Induction up to G_γ cannot be proven in EA(I_α). On the other hand, $\gamma = \sup_i \gamma_i$ where every $\gamma_i = \varphi^{2^i}_{\bar{\alpha}}(\omega)$. But pointwise transfinite induction up to each γ_i is provable in $\mathrm{EA}(I_\alpha)$ because G_{γ_i} is a finite iterate of F_α, therefore elementary in F_α. □

Corollary 3.12 *(i) The supremum of the ordinals γ for which* $\mathrm{EA}(I_{\prec\omega})$ *proves pointwise transfinite induction up to γ is $\varphi_\omega(\omega)$, the first prim-closed ordinal. (ii) The supremum of the ordinals γ for which* $\mathrm{EA}(I_\omega)$ *proves pointwise transfinite induction up to γ is Γ_0. (Since $\varphi_{\Omega+1}(\omega) = \sup_i \varphi^{2^i}_{\Omega}(\omega)$ and $\varphi_\Omega(\beta) = \varphi_\beta(\beta)$.)*

The φ functions used here are not the Bachmann-Veblen functions ϕ, but are closely related. Thus $\bigcup_{\alpha\prec\varepsilon_0}\mathrm{EA}(I_\alpha)$ is a tiered version of PA^∞ and its provable pointwise transfinite inductions hold up to all ordinals below the Bachmann-Howard $\varphi_{\varepsilon_{\Omega+1}}(\omega)$. Similarly one may further extend the methods to larger "cut-closed" initial segments beyond ε_0. For more on $\mathrm{EA}(I_\omega)$ and its predicative analogies, see [15].

References

1. Bellantoni, S., Cook, S.: A new recursion-theoretic characterization of the polytime functions. Comput. Complex. **2**, 97–110 (1992)
2. Buchholz, W.: An independence result for Π^1_1-CA+BI. Annals of Pure and Applied Logic **33**(2), 131–155 (1987)
3. Buchholz, W., Feferman, S., Pohlers, W., Sieg, W.: Iterated Inductive Definitions and Subsystems of Analysis: Recent Proof-Theoretical Studies. Lecture Notes in Mathematics. Springer, Berlin (1981)
4. Cantini, A.: Polytime, combinatory logic and positive safe induction. Arch. Math. Log. **41**(2), 169–189 (2002)
5. Feferman, S.: Logics for termination and correctness of functional programs, II. logics of strength PRA. In Aczel, P., Simmons, H., Wainer, S. (eds.) Proof Theory. A selection of papers from the Leeds Proof Theory Programme 1990, pp. 195–225. Cambridge University Press, Cambridge (1992)
6. Feferman, S.: The proof theory of classical and constructive inductive definitions. a forty year saga 1968-2008. In Schindler, R. (ed.) Ways of Proof Theory: Festschrift for W. Pohlers, pp. 7–30. Ontos Verlag (2010)
7. Kolmogorov, A.N.: Zur Deutung der intuitionistischen Logik. Math. Zeitschr. **35**, 58–65 (1932)
8. Leivant, D.: Intrinsic theories and computational complexity. In: Leivant, D. (ed.) Logic and Computational Complexity. International Workshop LCC '94, Indianapolis, IN, USA, October 1994, LNCS, vol. 960, pp. 177–194. Springer Verlag, Berlin, Heidelberg, New York (1995)
9. Leivant, D.: Ramified recurrence and computational complexity I: word recurrence and polytime. In: Clote, P., Remmel, J. (eds.) Feasible Mathematics II, pp. 320–343. Birkhäuser, Boston (1995)
10. Mints, G.: Finite investigations of transfinite derivations. J. Sov. Math. **10**:548–596, 1978. Translated from: Zap. Nauchn. Semin. LOMI. 49 (1975)
11. Nelson, E.: Predicative Arithmetic. Princeton University Press, Princeton (1986)
12. Schmerl, U.: A proof theoretical fine structure in systems of ramified analysis. Archiv für Mathematische Logik und Grundlagenforschung **22**, 167–186 (1982)
13. Schwichtenberg, H., Wainer, S.S.: Proofs and Computations. Perspectives in Logic. Association for Symbolic Logic and Cambridge University Press (2012)
14. Simmons, H.: The realm of primitive recursion. Arch. Math. Log. **27**, 177–188 (1988)

15. Wainer, S.S.: Pointwise transfinite induction and a miniaturized predicativity. In: Kahle, R., Strahm, T., Studer, T. (eds.) Advances in Proof Theory, pp. 409–425. Birkhäuser, Basel (2016)
16. Weiermann, A.: What makes a (pointwise) subrecursive hierarchy slow growing? In: Cooper, S., Truss, J. (eds) Sets and Proofs: Logic Colloquium '97, London Mathematical Society Lecture Notes, vol. 258, pp. 403–423. Cambridge University Press, Cambridge (1999)

Author Biographies

Helmut Schwichtenberg is an Emeritus Professor of Mathematics at Ludwig-Maximilians-Universität München. He has recently developed the "proof-assistant" Minlog, a computer-implemented logic system for proof/program development and extraction of computational content.

Stanley S. Wainer is an Emeritus Professor of Mathematics at the University of Leeds and a past-President of the British Logic Colloquiumis an Emeritus Professor of Mathematics at the University of Leeds and a past-President of the British Logic Colloquium.

Part II
Conceptual Expansions

Predicativity and Regions-Based Continua

Geoffrey Hellman and Stewart Shapiro

Abstract After recapitulating in summary form our basic regions-based theory of the classical one-dimensional continuum (which we call a semi-Aristotelian theory), and after presenting relevant background on predicativity in foundations of mathematics, we consider what adjustments would be needed for a predicative version of our regions-based theory, and then we develop them. As we'll see, such a predicative version sits between our semi-Aristotelian system and an Aristotelian one, as well as falling generally between fully constructive and fully classical theories. Finally, we compare the resulting predicative theory and our original semi-Aristotelian one with respect to their power and unity.

Keywords Continuity · Predicativity · Point-free · Constructive · Infinity

2010 Mathematics Subject Classification 03A05 · 03B20

1 Introduction

In earlier work [10, 11], we develop a so-called "classical theory" of a regions-based one-dimensional continuum and establish that it is mathematically equivalent to a standard Dedekind–Cantor, point-based theory. We have also shown that similar methods and results obtain for higher (finite-) dimensional continua in the contexts of Euclidean [12] and non-Euclidean geometries. These theories recapitulate the Aristotelian conception of the continuous as non-punctiform, reflecting the intuitive (and long-lived) idea that a true continuum is not constituted by points.

At several places, however, we make essential use of impredicative constructions as well as non-constructive appeals to "actual" as opposed to "potential" infini-

G. Hellman (✉)
Department of Philosophy, University of Minnesota, Heller Hall 831, Minneapolis, MN 55455, USA
e-mail: hellm001@umn.edu

S. Shapiro
Department of Philosophy, The Ohio State University, Columbus, OH 43210, USA

G. Jäger and W. Sieg (eds.), *Feferman on Foundations*, Outstanding Contributions to Logic 13, DOI 10.1007/978-3-319-63334-3_7

ties. The "actually infinite", famously, was not part of the Aristotelian conception of mathematical or geometric spaces. For this reason, we call our theories "semi-Aristotelian". We have elsewhere shown how our one-dimensional theory can be modified to accord with the restriction to potential infinity [13]. We dub that theory "Aristotelian" or "more Aristotelian".

It is interesting to consider how our semi-Aristotelian theories can be modified to accord with contemporary predicative foundations, as developed especially by Solomon Feferman and his co-workers.[1] Here there is no objection to actual infinities so long as they are countable and predicatively justifiable; and, moreover, classical logic is fully applicable.[2] So the modifications appropriate to our more Aristotelian setting are too harsh for a predicativist treatment. Our purpose here is to provide suitable modifications for a predicative theory and to assess the result. Predicative theories are sometimes called "semi-constructive". So the goal here is a semi-constructive, semi-Aristotelian theory of the continuous.

We will begin by reviewing our original semi-Aristotelian theory of a one-dimensional (non-punctiform) continuum. See [10] or [11] for more details and proofs. To motivate the present project, and to make the article self-contained, we provide relevant background on predicativity in general and then with respect in particular to continua. Then we will turn to the main task of modifying our regions-based continua theories to accord with demands of predicativity.

2 Recapitulation of the "Semi-Aristotelian" Continuum (with Emphasis On "Semi")

The formal background is classical, first-order logic with identity supplemented with a standard axiom system for second-order logic (or logic of plural quantification, with an unrestricted comprehension axiom for plurals). The more Aristotelian system does not invoke these higher-order resources and the predicative system, developed here, restricts them.

1a. Axioms on $x \leq y$ ("x is part of y"): reflexive, anti-symmetric, transitive.

As is standard, we define a binary relation called " overlaps": $x \circ y \Leftrightarrow^{df} \exists z(z \leq x \ \& \ z \leq y)$. And we write $x|y$ for $\neg x \circ y$, pronounced "x is discrete from y".[3]

1b. Axiom on $\leq$ **and** $\circ$: $x \leq y \leftrightarrow \forall z[z \circ x \rightarrow z \circ y]$.

---

[1]For an excellent account of predicativity in historical perspective, from Russell's ramified type theory ((over-)reacting to the set-theoretic paradoxes), through semi-constructive ideas of Poincaré and Weyl, up to modern developments, including work of Feferman and Strahm on *unfolding*, see [3, 4].

[2]Our semi-Aristotelian and more Aristotelian theories also invoke classical logic. It might be noted that Aristotle himself accepted the law of excluded middle. The differences between our Aristotelian and semi-Aristotelian theories concern the treatment of infinity.

[3]Following Nelson Goodman's usage, [8] we prefer the term 'discrete' to 'disjoint' here in order to emphasize the fact that our regions are not sets of points (indeed aren't sets at all).

This implies an extensionality principle, yielding so-called classical mereology:

Theorem 1 $x = y \leftrightarrow \forall z[z \circ x \leftrightarrow z \circ y]$.

2. Axiom of Fusion or Whole Comprehension:

$$\exists u \Phi(u) \rightarrow [\exists x \forall y\{y \circ x \leftrightarrow \exists z(\Phi(z) \& z \circ y)\}],$$

where Φ is a predicate of the second-order language (or language of plurals) lacking free x.

This is the only second-order axiom. In the plural locution, it says that if we are given any regions—no matter how many—there is a single region that is a fusion of them. To repeat, in the semi-Aristotelian theory, there are *no* restrictions on the number of regions that can be "fused".

This is the axiom that allows the play with infinity. In some cases, we take fusions of regions without bothering to check how many regions are so fused. In other cases, we explicitly fuse infinitely many regions. We take it that the "infinities" invoked by this axiom are "actual", since we conclude that there is a *single* region that is the fusion of *all* of the given regions.

Some initial examples illustrate the power of this principle. If x and y are regions, we write $x + y$ for the mereological sum or fusion of x and y. So a region ovelaps $x + y$ if and only if either it overlaps x or it overlaps y. If x and y are separated by some space, then $x + y$ is a region with discontiguous parts. Here, of course, we only fuse two regions, and we take this to be unproblematic for the Aristotelian and the predicativist.

If $x \circ y$, then we write $x \wedge y$ for the *meet* of x and y. It is the fusion of *all* regions that are part of both x and y. So $\forall z[z \leq x \wedge y \leftrightarrow z \leq x \,\&\, z \leq y]$ (and if x and y have no common part, then $x \wedge y$ is undefined). Here, then, is the first place where we take the fusion of some regions without bothering to make sure that there are only finitely many such.

Similarly, if $\exists z(z \circ x \,\&\, \neg(z \circ y))$, then we let $x - y$ be the sum of all regions z that are part of x but discrete from y (and if there is no such z, then $x - y$ is undefined). So $\forall z(z \leq x - y \leftrightarrow (z \leq x \,\&\, \neg(z \circ y)))$. Again, we do not check to see how many regions are so-fused.

We let G be the fusion of *all* regions. It is the entire line, as a single region. Since, as it happens, there are infinitely many pairwise discrete regions, this, too is not acceptable to an Aristotelian.

We next introduce a geometric primitive, $L(x, y)$, to mean "x is (entirely) to the left of y". The axioms for L specify that it is irreflexive, asymmetric, and transitive. And we define '$R(x, y)$', "x is (entirely) to the right of y", as $L(y, x)$.

This allows us to define *betweenness*: $Betw(x, y, z)$ for "y is (entirely) between x and z":

$$Betw(x, y, z) \Leftrightarrow^{df} [L(x, y) \,\&\, R(z, y)] \vee [R(x, y) \,\&\, L(z, y)]$$

It follows that $Betw(x, y, z) \leftrightarrow Betw(z, y, x)$.

$L(x, y)$ obeys the following axioms:
3a. $L(x, y) \vee R(x, y) \rightarrow x|y$.
3b. $L(x, y) \leftrightarrow \forall z, u[z \leq x \,\&\, u \leq y \rightarrow L(z, u)]$.
The following can now be inferred:
$Betw(x, y, z) \rightarrow x|y \,\&\, y|z \,\&\, x|z$, and
$Betw(x, y, z) \,\&\, Betw(u, x, z) \rightarrow Betw(u, y, z)$.
We next define the notion of a *connected* region, one that has no gaps:

$$Conn(x) \Leftrightarrow^{df} \forall y, z, u[z, u \leq x \,\&\, Betw(z, y, u) \rightarrow y \leq x]. \qquad \text{(Df } Conn\text{)}$$

So the entire line G is connected.

We also define what it means for a region to be *bounded* (on both sides):

$$Bounded(p) \Leftrightarrow^{df} \exists x, y\, Betw(x, p, y). \qquad \text{(Df } Bounded\text{)}$$

We call bounded, connected regions "*intervals* " and write '$Int(j)$', etc., when needed. This is the direct analogue of an interval in a standard, Dedekind–Peano punctiform space. Since the present space lacks points, and thus endpoints, there is no sense in which intervals are open, closed, or half-open.

Using L, we can impose a condition of dichotomy for discrete intervals:

4. Dichotomy axiom: $\forall i, j[i, j$ are two discrete intervals $\rightarrow (L(i, j) \vee L(j, i))]$.

It is straightforward to prove a linearity condition among intervals:

Theorem 2 (Linearity)*: Let x, y, z be any three pairwise discrete intervals; then exactly one of x, y, z is between the other two.*

To guarantee that arbitrarily small intervals exist everywhere, we adopt the following "gunkyness" axiom:
5. $\forall x \exists j[Int(j) \,\&\, j < x]$.

An important relation of two intervals is " adjacency", which is defined as follows:

$$Adj(j, k) \Leftrightarrow^{df} j|k \,\&\, \nexists m[Betw(j, m, k)]. \qquad \text{(Df } Adjacent\text{)}$$

Intuitively, adjacent intervals touch, but do not overlap. This is Aristotle's definition of contiguity. Since we do not recognize points, at least not as parts of regions, adjacent intervals are also continuous with each other, in Aristotle's sense.

The following equivalence relation is useful: "intervals *j and k are left-end equivalent*" just in case

$$\exists p[p \leq j \,\&\, p \leq k \,\&\, \nexists q(\{q \leq j \vee q \leq k\} \,\&\, L(q, p)].$$

"*Right-end equivalent*" is defined analogously. Left- (Right-) end equivalence means, intuitively, that the intervals "share their left (right) ends, or end-points, in common". Since our system does not recognize "ends" or " end-points", we need a definition like the above. Roughly, two intervals are Left- (Right-) end equivalent just in case any region that is Left (Right) of one of them is Left (Right) of the other.

Our final geometric primitive is *congruence*, a binary relation among intervals. We adopt the usual first-order axioms specifying that *Cong* is an equivalence relation. We will sometimes write $Cong(i, j)$ as $|i| = |j|$. Similarly, for intervals i, j, we can define, contextually, $|i| < |j|$ as meaning: $\exists j'[j'$ an interval & $j' < j$ & $\text{Cong}(i, j')]$; and we may write $|i| > |j|$ as equivalent to $|j| < |i|$. Say that $|i| \leq |j|$ just in case either $|i| < |j|$ or $|i| = |j|$.

The next axiom is crucial to our characterization:

6. Translation axiom: Given any two intervals, i and j, each is congruent both to a unique left-end-equivalent and to a unique right-end-equivalent of the other.

Intuitively, this means that any interval can be "transposed" or, perhaps better, instantiated, anywhere in the space G.

Lemma 1 *Given any two intervals i and j such that* $\neg Cong(i, j)$*, either there exists an interval* $i' < j$ *with* $Cong(i, i')$*; or there exists* i' *with* $j < i'$ *with* $Cong(i, i')$.

Theorem 3 (Trichotomy) *For any two intervals,* i, j*, either* $|i| = |j|$ *or* $|i| < |j|$ *or* $|i| > |j|$.

Our final axiom is that congruence respects nominalistic summation of adjacent intervals:

7. Additivity: Given intervals i, j, i', j' such that $Adj(i, j)$, $Adj(i', j')$, $Cong(i, i')$, $Cong(j, j')$, then $Cong(k, k')$, where $k = i + j$ and $k' = i' + j'$.

We now turn to the matter of the bi-infinitude of G. In fact, our axioms already guarantee this, as we can prove.

Theorem 4 (Bi-Infinity of G) *Let any interval i be given; then there exist exactly two intervals,* j, k*, such that* $Cong(i, j)$ & $Cong(i, k)$ & $Adj(i, j)$ & $Adj(i, k)$& *one of* j, k *is left of i and the other is right of i.*

In words, this says that given any interval i, there is an interval congruent to i and adjacent to i on its left, and there is another interval congruent to i and adjacent to i on its right. This obviously iterates. The proof of Theorem 4 also involves taking the fusion of some intervals and, here too, there is no reason to think, in advance, that the number of intervals so fused is finite.

Call an interval l an (*immediate*) *bi-extension* of interval i—$BiExt(l, i)$, or $biext(i) = l$—just in case $l = j + i + k$, where j, i, k behave as in the Bi-Infinity theorem.

Lemma 2 *Let i and j be intervals such that* $i < j$*; then* $\neg Cong(i, j)$.

One central result is the Archimedean property that, in effect, our entire space G is *exhausted* by iterating the process of flanking a given interval by two congruent ones as in Bi-Infinity.

Let X be any class (or plurality) of intervals such that an arbitrary but fixed interval $i \leq G$ is one of the X and such that if $k = biext(j)$ for j any of the intervals of X, then k is also in X. Call such X a "closure of i under *biext*".

Lemma 3 *There is an individual which is the common part of the fusions of each class X which is a closure of i under biext, which we call their meet or the minimal closure i*of i under biext. (Since i is stipulated to belong to any such X, the meet is non-null, as required in mereology.)*

Here we apply Axiom 2, of unrestricted fusions, this time to an *explicitly defined* infinite set (or plurality) of intervals. Moreover, the definition of this "minimal closure" is impredicative.

This is sufficient to establish an Archimedean property:

Theorem 5 (Characterization of G)*: Let G be the fusion of the objects in the range of the quantifiers of our axioms; then* $G = i^*$*, the fusion of the minimal closure of i under biext.*

Our next theorem is that any interval has a unique bisection:

Theorem 6 (Existence and uniqueness of bisections): *Given any interval i, there exist intervals* j, k *such that* $j < i \;\&\; k < i \;\&\; j|k \;\&\; j + k = i \;\&\; Cong(j, k)$*; and* j, k *are unique with these properties.*

Here, again, we take the fusion of some intervals, without checking to see how many intervals are fused.

It is perhaps interesting that Axiom 2, of unrestricted fusions, yields the existence of meets, differences, biextensions, and bisections; and it plays a crucial role in establishing the Archimedean property. These consequences are themselves acceptable to an Aristotelian, even though the background axiom of fusions is not. So in the more Aristotelian theory [13], the principles in question have to be added in by hand, as new axioms. As we shall soon see, the same goes for the predicativist.

We next sketch a way to embed the real numbers into G, given an arbitrary interval i to serve as a unit. Define a sequence $< j_i >$ of intervals increasing to the right (or left, for negative reals) to be *Cauchy* just in case, for any interval, ε, there exists a natural number N such that for any natural numbers $m > k > N$, $j_m - j_k$ is an interval, $R(j_m - j_k, j_k)$, and $|j_m - j_k| < |\varepsilon|$. The fusion of the members of such a sequence gives us an interval that can play the role of a given real number.

Let us illustrate this with a particular real number, say π. The sequence $< j_i >$ is defined as follows. j_0 is i, our unit interval; j_1 is congruent with i and adjacent to it on the right, and j_2 is congruent with i and adjacent to j_1 on its right. So the fusion of j_0, j_1, and j_2 is the interval that corresponds to the number 3.

Next consider the fractional part of π as a bicimal expansion. It starts 0010010000 111111... (see http://www.befria.nu/elias/pi/binpi.html). Consider the following procedure for adding intervals to the sequence $< j_i >$. For each natural number $n > 2$,

> If the $n - 2$nd member of the bicimal expansion of π is 0, then do nothing; go on to the next number; but
>
> if the $n - 2$nd member of the bicimal expansion of π is 1, then add an interval congruent to an $n - 2$-fold bisection of i that is adjacent to, and on the right of, the rightmost member of $< j_i >$ defined so far.

So j_5 is congruent to an eighth of i (i.e., the result of bisecting i three times), j_6 is congruent to a 64th of i, etc. The real number π corresponds to the fusion of the members of the sequence $< j_i >$.

But, of course, these "Cauchy sequences" of intervals are themselves infinite and so here, too, we take the fusion of an infinite set of regions. Fusions like this are not acceptable to an Aristotelian. And only some such fusions—those with appropriate definitions—are acceptable to the predicativist.

3 Background on Predicativity

In the first instance, the terms 'predicative', 'impredicative', refer to methods of introducing, defining, or specifying collections or pluralities[4] of given objects. For example, given the natural numbers, any arithmetical formula with a free variable, x, that is a formula containing bound variables (governed say by quantifiers) over natural numbers but not over sets or pluralities of them, counts as predicative, e.g. "even number", "prime number", etc.[5] However, if we try to introduce the whole collection of natural numbers as the intersection of all collections containing 0 and closed under the successor operation, we have a quantifier over collections of numbers (and possibly other objects), one of which turns out to be the very collection—viz. of exactly all the naturals—that we're trying to introduce. Such a definition or specification is paradigmatically *impredicative* (relative to the natural numbers, however they may be construed).

Now from a constructivist point of view,[6] it is clear that such impredicative definitions are problematic, since we have already to presuppose the existence of the totality we're trying to "construct" as a value of the universally quantified variable in order for the definition to be comprehended. But even if one is not requiring that all mathematical entities be "constructed" or "constructible"—whatever that means, precisely—it seems that one still has to regard as understood the notion "arbitrary collection of natural numbers", and, as we know from Cantor, this transcends the countably infinite and raises the prospect of transfinite iteration of operations on collections, e.g. that of taking power sets, as we've just done.[7]

Let us grant, then that there is natural motivation for accepting the countable infinity of natural numbers, or the integers, or even the rationals as unproblematic

---

[4]We use 'pluralities' as shorthand indicating the use of plural quantification and plural variables to avoid reference to higher-type objects.

[5]Below, we will refine this to be a *relative* distinction, i.e. the definition of "prime number" is predicative relative to the natural numbers, taken as already introduced or given. But for now, let us suppress mention of this relativity.

[6]This refers to programs based on intuitionistic logic, including Bishop constructivism as well as intuitionism itself.

[7]No wonder, then, that some great mathematicians such as Poincaré might look askance at the above logicist definition of the natural numbers system, preferring simply to presuppose the numbers (individually and plurally) as a starting point for analysis and higher mathematics.

totalities from which to launch the study of real and complex analysis and other advanced subjects. Clearly, however, this motivation does not call into question the free use of classical logic, including the law of excluded middle, double-negation elimination, etc. For a familiar example, the least number principle—from existence of a number with (even undecideable) property Φ to infer the existence of a least such—is a classical, non-constructive but predicatively acceptable consequence of mathematical induction.

The next step is to accept subsets or subpluralities of those countable domains which are extensions of formulas of mathematical language whose bound variables range over just the first-order objects of those domains, i.e. the first-order definable subsets or pluralities. Call those the level-1 subtotalities (sets or pluralitities). But then one should allow further subsets as extensions of formulas with bound variables ranging over either the first-order objects or the already accepted level-1 subtotalities. And this process clearly can be iterated, certainly finitely often, and perhaps into the transfinite.[8]

Turning to the crucial case of analysis, suppose we seek to introduce real numbers as (equivalence classes of) Cauchy sequences of rationals. At the first stage, we only consider such rational sequences specified by formulas with bound variables restricted to the rational numbers. However, the reasoning of Cantor's diagonal argument for uncountability of the reals implies that there must be further stages introducing more reals.[9] Indeed, at the next stage, the predicativist considers formulas with bound variables ranging over real numbers defined at the first stage. And on it goes. Thus one may develop a hierarchy of richer and richer systems as in ramified type theory, allowing at any stage quantifiers over reals specified at earlier stages in the defining formulae. Crucially, however, one lacks the machinery to specify the union of *all* such stages, for over what totality of ordinals would "all such stages" range? Proof-theoretic analysis (independently by Feferman and by Schütte) has shown that a certain countable limit ordinal, known as Γ_0, of the Veblen hierarchy qualifies as a limit to the available stages, but even to recognize this limit, one steps outside the predicativist framework proper (see [1, 7, 9]).[10]

[8]Transfinite iteration, of course, would have to be formulated carefully, and would involve a background theory of infinite ordinals. That in itself appears problematic from a predicativist point of view, since the very notion of " well-ordering" is impredicative (building in the requirement that *every* subset of the field of the well-ordering relation have a least element).

[9]Note that the immediate negative step in the Cantor diagonal argument, viz. that the altered diagonal sequence defined is a Cauchy sequence not on the given countable list, which therefore cannot be of *all* the reals, is constructively and predicatively acceptable. Thus the predicativist can see, at any stage, that more reals can be introduced. In this sense, "all the reals" makes predicative sense only as a *potential*, open-ended infinity, not an *actual* one.

[10]It should be noted that proponents of predicative mathematics need not espouse "predicativism" as a philosophical position, i.e. views that take only predicatively justified mathematics to be intelligible or legitimate. In particular, Feferman forswears such an "ism", explicitly in [5] and in practice, as illustrated by his work on specifying the outer limits of predicativity. Moreover, he has emphasized that "a little bit [of higher-order logical strength] goes a long way" [2]. Indeed, the exploration of various systems of predicative mathematics is, in our view, best understood as an integral part of the epistemology of mathematics.

Thus we seem to be saddled with a ramified hierarchy of levels or orders of reals, with all its awkwardness and complexities. For example, we still confront the problem of impredicativity of the least upper bound (or greatest lower bound) of predicatively acceptable sets of reals, in that those bounds will be of a level beyond the reals of the given set or plurality. Thus the predicativist program confronts a trade-off between the greater epistemic security offered by restrictions to predicatively definable subtotalitites, on the one hand, and recovering essential properties, such as continuity itself, on the other.[11]

Fortunately, however, work of Feferman et al. has provided systems that avoid the complexities of ramified type theory while still qualifying as "predicatively reducible", that is systems, such as Feferman's W [6], a "theory of flexible finite types" demonstrably conservative over first-order Peano Arithmetic, in which analysis can be quite naturally developed.

To sum up, thus far: predicative mathematics sits in between thoroughgoing constructive mathematics and full classical analysis and set theory. Regarding continua, it also sits between the semi-Aristotelian perspective of the previous section and that of the "more Aristotelian" systems developed in [13]. Like Aristotelian mathematics, predicative mathematics balks at the unrestricted use of actual infinity, as in the semi-Aristotelian framework, but does not reject the actually infinite outright. As already made clear, there is no objection to actual infinity as such, provided that any actual infinity recognized is countable and specifiable in a predicatively acceptable way. Also, it should be noted, demands of predicativity are no bar to punctiform systems of geometry or topology either. Although positive commitments to uncountable totalities are avoided, the mathematics of such spaces can be pursued in an "open-ended" fashion, as articulated in the program of *unfolding* [4].

Finally, to come full-circle, as noted above, predicativity is best understood as a relative notion; and the most common relativity is to the natural numbers as a countably infinite, though constructively generated, structure. One who is familiar, however, with the development of classical number systems from a logicist basis, à la Frege or Dedekind or Russell, may wonder about the decision to start with *sui generis* axioms for the natural numbers: for on those accounts, the class of the naturals is explicitly *defined* as the minimal closure of (the singleton of) the initial number designated '0' or '1', under the (defined) operation of successor—and this is a paradigmatically impredicative construction. So isn't the predicativist program, as usually presented, on shaky ground from the start? In response, one may begin with an elementary theory of finite sets[12] over an axiomatically provided, minimally

[11] Similar complexities led Weyl, eventually, to turn away from Brouwer's intuitionist program. But here we are raising the prospect of a slippery slope that calls into question the viability of a semi-constructivist or "definitionist" program (allowing free use of classical logic) of the sort that Weyl then pursued.

[12] Although classical theories of finite sets and of natural numbers may be formally interderivable, there is a fundamental difference to be noted regarding the objects of such theories. Arguably, numbers are identified by their "positions" in a structure; other aspects of their identity (if recognized at all) are mathematically irrelevant. In contrast, finite sets are self-standing, identified by their members, independently of any structure of which they are a part or to which they belong. In this

structured, countable domain (e.g. a pairing function and an urelement under pairing), thereby recovering a natural-numbers system (with induction for classes specified predicatively relative to finite sets), along with proofs of categoricity of the axioms and of recursion theorems, and a proof that the system is conservative relative to first-order Peano Arithmetic (again, see [1, 7]). This will prove useful in what follows here.

The goal now is to present a predicative, regions-based continuum. The situation is complicated by the fact that a theory of a regions-based continuum is, from the outset, aiming to describe an uncountable structure, one that is not predicated on, or constructed from, a prior countable one (such as the rational numbers). The semi-Aristotelian development sketched above begins with first-order logic enriched with a logic of plurals (or something equivalent), along with axioms of classical mereology, including an unrestricted comprehension scheme for plurals (or equivalent), and an axiom asserting the existence of the mereological sums or "fusions" of any (all) such pluralities. Such axioms are naturally regarded as impredicative and so must be restricted somehow. Sticking again, with the one-dimensional case, the task at hand is to determine what sort of predicative theory can be developed on the more restrictive basis and how that compares with the original classical theory.

4 Predicative Adjustments

The places where impredicativity threatens the semi-Aristotelian development are essentially the same as those where the Aristotelian demurs. In that case, there are several places where we took the fusion of some regions, without bothering to check whether the regions, so fused, are finite in number. The predicativist also balks at those places, since we did not check to make sure that we have a predicative definition of the relevant set or plurality. In other places, we explicitly took the fusion of infinitely many regions. Typically, these are also impredicative, often defined as minimal closures. As it happens, the predicativist can recapture some of those results.

Thus, from the predicativist standpoint, the troubling results from the semi-Aristotelian development are these:

(1) specification of arbitrary meets and differences among regions (via Axiom 2);

(2) specification of the universal region, G as the (unrestricted) fusion of *all* regions;

(3) in Theorem 4, the derivation of the existence of bi-extensions of any given interval (implying the bi-infinity of the space);

(4) the construction of minimal closure of $\{i\}$ under iterated bi-extension (Lemma 3), a paradigmatically impredicative construction; this yielded

(5) the Archimedean property;

(Footnote 12 continued)
way, recognizing numbers already raises questions about infinite totalitites (potential or actual), whereas such questions can be postponed in theorizing about finite sets, as reflected in Feferman and Hellman's theory which they call " Elementary Theory of Finite Sets and Classes" [1].

(6) the proof of Theorem 6, existence of bisections of intervals;

(7) the proof of Dedekind completeness of the interval structure (not mentioned above).

The task here is to formulate a regions-based theory that addresses the relevant issues as successfully as possible. The matter is complicated by the fact that, classically speaking, the individuals are regions of a geometric space, and even well-behaved regions such as intervals are uncountably many and so cannot enter into a predicatively acceptable totality or operation such as fusion. Thus arises the problem over the universal region G (2), and the break-down of proofs of existence of bi-extensions and bisections (problems (3) and (5)).

Since we are starting without any arithmetic axioms at all, it is useful to try to adapt the machinery of [1, 7] in their "predicative foundations of arithmetic", where axioms on ordered pairing provided for a countable set of individuals. Then "finite set" of such individuals is taken as primitive, subject to uncontroversial further axioms, and supplemented with a comprehension principle for "classes" of individuals (in which only individual and finite-set bound variables occur in the defining formulae). Hence the designation 'EFSC' of [1, 7], for "elementary theory of finite sets and classes".

In the present, more geometric setting, we will combine these ideas with axioms of atomless mereology (with $\leq$, part of, as primitive) and further axioms governing the special regions called "intervals", which enable us to dispense with primitive pairing. Instead of finite sets and classes of individuals, we will use plural variables *ff*, *gg*, *hh* etc., for finitely many regions or intervals, and *xx*, *yy*, *zz* for general plural reference; and we write $x \prec ff$, $y \prec xx$, etc., to express "x is one of, or is among, the *ff*", "y is one of, or is among, the *xx*", etc.

Our axioms start off the same as for the semi-Aristotelian theory described above, but we replace the unrestricted fusions Axiom 2 initially to allow only fusions of finitely many regions—as in the Aristotelian development. Later we broaden this to include certain countable fusions—those with proper definitions.

The other axioms of the semi-Aristotelian theory, through 7, carry over intact. Recall Axiom 5, the one that renders the system "gunky":

$$\forall x \exists j [Int(j) \,\&\, j < x].$$

Since the background logic is unfree, we assume the existence of at least one region. Axiom 5 thus gives us at least one interval. We designate one such interval to serve as a unit, for the purpose of a representation of the integers and, eventually, the rationals and predicatively specifiable reals.[13]

Also, we adopt another Axiom 8, for differences:

$$\text{If } \exists z(z \circ x \,\&\, \neg(z \circ y)) \text{ then } \exists w \forall z(z \leq w \leftrightarrow (z \leq x \,\&\, \neg(x \circ y)).$$

Then we can establish the existence of meets and biextensions (see Theorem A1 of [13]).

[13] If one prefers a free logic, then we'd need an axiom asserting the existence of an interval (or, given Axiom 5, a region).

Iterated biextensions, left and right, starting with our arbitrarily assumed unit interval, generate a bi-infinite linear structure isomorphic to the integers. In what follows, we thus assume such a bi-infinite sequence of sequentially adjacent intervals, all congruent to our unit, and we will designate these *canonical intervals*.

As noted, we cannot derive the existence of bisections (Theorem 6), as that involves arbitrary fusions, so instead we postulate it:

Axiom 9 *Bisections:* $\forall j \exists! k, m[Cong(k, m) \,\&\, Adj(k, m) \,\&\, j = k + m \,\&\, L(k, m)]$,

i.e. j is the fusion of k and m.

Finally, we have axioms governing the plural variables, both finitary and general:

Axiom 10 *Singleton finite plurality:* $\exists f\!f \exists! j[j \prec f\!f]$.

Axiom 11 *Adjunction:* $\forall f\!f \forall a \exists gg \forall x[x \prec gg \leftrightarrow x \prec f\!f \vee x = a]$.

Axiom 12 *Finite Separation:* $\forall f\!f \exists gg \forall x[x \prec gg \leftrightarrow x \prec f\!f \wedge \Phi(x)]$,

where Φ lacks free 'gg' and lacks bound general plural variables. (Thus, the individuals among the (finitely many) $f\!f$ satisfying such Φ are finitely many.)

Note that we do not need an axiom of induction governing the finite plural variables (the plurals analogue of finite-set induction), because the machinery so far already suffices to define minimal closures of finitely many given objects under the operations we need, providing theorems of induction for the formulae we need (cf. [1, 7]).

What about a comprehension principle for general plurals? Here, as already noted, we have to be careful lest we commit the predicativist to uncountable totalities. Our strategy will be to introduce certain interval sequences, already furnished by the axioms presented thus far, and then to allow into our general plurals comprehension axioms formulas with bound variables ranging over these intervals. They come in two sorts, depending on the operations of bi-extension and bisection, respectively, leading to them. The first sort was already described following the above introduction of bi-extensions: we're given a starting unit interval, call it j_0; then repeated applications of bi-extension right and left yield a bi-infinite sequence of mutually congruent intervals of the form

$$\ldots j_{-3}, j_{-2}, j_{-1}, j_0, j_{+1}, j_{+2}, j_{+3}, \ldots$$

with each interval adjacent to the ones immediately flanking it. (We designate this doubly infinite sequence with the plural constant $jj = \langle j_i \rangle$.) More formally, our biextensions theorem furnishes us with two 1-1 functions biext_L and biext_R yielding for any argument j_i its unique left-biextension and its unique right biextension, respectively. Then we can define the members of the above displayed canonical interval sequence by requiring that any finite plurality $f\!f$ such that (1) $j_{+k} \prec f\!f$ and (2) if $j_i \prec f\!f$, then $\mathrm{biext}_R^{-1}(j_i) \prec f\!f$ also satisfies that $j_0 \prec f\!f$, and likewise for any of the j_{-k},*mutatis mutandis.* These are our canonical intervals.

Next we specify intervals that can be called *binary fractions* of canonical intervals. These result from iterated bisection of subintervals, left or right at each stage, beginning with any of the canonical intervals. As in the case of biextensions, Axiom 9 on bisections furnishes two 1-1 functions, bisec_L and bisec_R, yielding for each interval j, its unique left half, j_L^{-2}, and its unique right half, j_R^{-2}, respectively. Thus, any sequence $\sigma = \langle \sigma_i \rangle$ obtained by iterating these functions is determined by requiring that any finite plurality ff such that (1) $\sigma_i \prec ff$ and (2) if $k \prec ff$, then either $\text{bisec}_L^{-1}(k)$ or $\text{bisec}_R^{-1}(k) \prec ff$ also satisfies that one of the canonical intervals $jj \prec ff$.

We are now in a position to state our comprehension principle governing general plural variables and fusions pertaining to them:

Axiom 13 *Plurals Fusion Comp:* $\exists u\Phi(u) \rightarrow [\exists x \forall y\{y \circ x \leftrightarrow \exists z(\Phi(z) \;\&\; z \circ y)\}]$, *where* Φ *lacks free* x *and the bound individual or finite or general plurals variables of* Φ *are restricted to canonical intervals or binary fractions thereof.*

Applying this in the case that $\Phi(u)$ is "u is a canonical interval of the sequence jj (displayed above)", we have existence of the fusion of the jj. To recover the classical characterization of G, we need to show that this is the same as the minimal closure of $\{j_0\}$ under iterated bi-extension, despite the impredicativity of the classical definition of that minimal closure as the intersection of all closures.

Here is where the method of [1, 7] can be adapted so that we can bypass the classical definition. In its stead, we simply require that the intervals starting with j_0 up to any j_i—proceeding in the sense $+$ or $-$ of i—form a finite plurality. This suffices to establish the existence of an isomorphism between the sequence jj and the integers, which collectively satisfy the minimal closure condition. Thus, just as [1, 7] recover the existence and uniqueness (up to isomorphism) of the natural-numbers-structure from the requirement that all initial segments of their defined privileged sequence (obtained by iterating the pairing operation, beginning with the posited urelement) be finite, so that requirement applied in the positive and negative directions of j suffices for uniqueness up to isomorphism of our bi-infinite linear interval structure.

We are not quite done, however, for we have yet to derive the Archimedean property. The above shows that the fusion of the sequence jj is part of the fusion of all intervals that are either to the right or to the left of our initial unit interval, j_0, but we need that the latter fusion be *identical* to the fusion of the jj. But that follows by the same proof given in the classical case, applying the Translation axiom, as no impredicative definitions arise there. Thus we have the identity, G is the fusion of the canonical interval sequence jj.

Furthermore, we observe that, with the machinery of binary fractions of (lengths of) canonical intervals together with mereology and plurals, the notion of *Cauchy interval sequence* is predicatively acceptable, which opens the way to introducing—as superstructure over the interval structure of G—a hierarchy of ever richer real-numbers structures, as in standard predicative analysis.

What about the *completeness* of the predicative continua, either at the geometric level of intervals or at the superstructural level? This depends on the notion of "completeness" involved. As already noted, *Dedekind completeness*—the least upper bound principle applied to *arbitrary* bounded pluralities or sets of real-length intervals—is paradigmatically impredicative, just as it is in the case of numbers. Indeed, predicative systems of analysis lack the means even to *express* the classical lub principle, as they are designed to avoid commitment to a totality of all (bounded) sets or pluralities of reals or of real-length intervals over which the principle quantifies. (Also, they are designed to avoid commitment to uncountable sets of reals or real-length intervals in the first place.) It is true that at any stage of introducing countably many (bounded, countable) pluralities of real-length intervals, the predicativist can derive a lub, by considering the fusion of the given bounded sequence. But, applying the reasoning of the Cantor diagonal argument, the predicativist can also pass to a proper extension. Thus one gets a hierarchy of lub principles. What cannot be done predicatively is to prove a formal result that expresses existence of a least bound of "any *predicatively definable* bounded totalities or pluralities of real-length intervals", since expressing " predicatively definable" would require stepping outside the framework, much as does proving the existence of the limit ordinal Γ_0 as a bound on predicatively specifiable countable ordinals. Thus, classical theorems of analysis that use full Dedekind completeness, such as the intermediate-value or extreme-value theorems, must be derived by other means, or else the statements of those theorems, must be suitably altered.

The situation with *Cauchy completeness*, that any given Cauchy sequence of reals (or real-length intervals) converges to a real (real-length interval) of the space, is favorable both to the constructivist and the predicativist, the main difference residing in their different criteria for what counts as a "Cauchy sequence of reals (or real-length intervals)". Classically, one allows arbitrary Cauchy sequences, while the constructivist demands that a function from ε to $N(\varepsilon)$, the term of the sequence beyond which differences must be within ε, be a *constructive* function. Similarly, the predicativist requires that such a function be predicatively definable.[14] Apart from such differences in the meanings of their terms, however, the argument for Cauchy completeness goes through. Here is a simple argument applicable to any stage of a hierarchy of predicative reals (or real-length intervals, which should be understood whenever we refer simply to reals, in the following).

Theorem P1 (Cauchy Completeness)*: Any Cauchy sequence, $\rho = \langle r_i \rangle$, of (predicative) reals converges.*[15]

Proof Each real r_i is determined (up to equivalence based on co-convergence) by a Cauchy sequence, $q_i = \langle q_{ij} \rangle, j = 1, 2, 3, \ldots$ of rationals (rational-length intervals). Without loss, we consider the unit interval and assume that each r_i of the sequence ρ is presented as a row, $r_{ij}, j = 1, 2, 3, \ldots$, in decimal notation, all such rows forming a countably infinite matrix M with entries r_{ij}. It suffices to transform M into a corresponding matrix M^q with rows representing rationals, q_i, such that the sequence of those rows, $\langle q_i \rangle$ co-converges with $\langle r_i \rangle$. This is accomplished simply by setting the entries of each row equal to those of the rational sequence $\langle q_{ij} \rangle$ determining r_i, for those values of $j \leq i$, setting $q_{ij} = 0$ for all $j > i$. It is easily checked that this

[14]To be sure, the predicativst cannot *state* this requirement, since, as already observed, the notion of predicative definability is not itself predicatively definable. See [1, 7].

[15]Inside a system of predicative analysis or geometry, the term 'predicative' would not occur. With it, the statement is classical. In all other respects, the proof is predicatively acceptable.

transforms M to M^q identical to M on the half left of the diagonal, including the diagonal entries, with 0's everywhere right of the diagonal, such that the sequence of rows $q = \langle q_i \rangle$ is co-convergent with $\rho = \langle r_i \rangle$. But by construction, $\langle q_i \rangle$ defines a real, as limit of both the rational sequence q and the real one ρ from which it is derived.[16] ■

As it turned out, in the Aristotelian system, we had to add an axiom for n-sections of intervals (where n is a natural number greater than 1). Here, we just add an axiom for bisections. The existence of n-sections follows from Theorem P1, as it is straightforward to define, for any given interval i, a predicative "Cauchy-sequence" of intervals that "converges" to the left-most n-section of i.

5 Conclusion

The above predicative theory of a regions-based continuum goes quite far, as indicated, including the derivation of the Archimedean property from Translation, using techniques of [1, 7]. But, lacking unrestricted fusions, it has to postulate differences and bisections outright, rather than deriving them; and it fails to derive Dedekind completeness in full generality. As to the extent of the intervals recognized, the predicativist could go on to introduce a hierarchy of increasingly rich theories of countably many real-length intervals, mimicking what it already does to approximate the classical point-based continuum. But here the loss seems greater: in the point-based classical theory, full Dedekind completeness is usually an extra axiom, one that the predicativist must forgo; but the regions-based classical theory (our "semi-Aristotelian" one) *derives that principle as a theorem*, whereas it remains out of reach for the predicativist. In sum, unrestricted, impredicative fusions, like its close cousin, unrestricted second-order comprehension, has great unifying power that predicative mathematics cannot access. This is a price the predicativist must pay for any epistemic gains it is ultimately in a position to claim.

References

1. Feferman, S., Hellman, G.: Predicative foundations of arithmetic. J. Philos. Log. **24**, 1–17 (1995)
2. Feferman, S.: Why a little bit goes a long way: logical foundations of scientifically applicable mathematics. In the Light of Logic, pp. 284–298. Oxford University Press, Oxford (1998)
3. Feferman, S.: Weyl indicated: Das Kontinuum seventy years later. In the Light of Logic, pp. 249–283. Oxford University Press, Oxford (1998)

---

[16] As a corollary, the predicativist can prove a LUB principle from Cauchy completeness, applicable to predicatively specifiable bounded sets of reals. This doesn't affect any of the substantive points in the paper, but it does call attention to the fact that the predicativist can prove a significant LUB principle, rather than having to take it as an axiom.

4. Feferman, S., Hellman, G.: Challenges to predicative foundations of arithmetic. In: Sher, G., Tieszen, R. (eds.) Between logic and intuition: essays in honor of Charles Parsons, pp. 317–338. Cambridge University Press, Cambridge (2000)
5. Feferman, S., Strahm, T.: The unfolding of non-finitist arithmetic. Ann. Pure Appl. Log. **104**, 75–96 (2000)
6. Feferman, S.: Comments on 'Predicativism as a philosophical position' by G. Hellman. Revue Internationale de Philosophie **58**, 313–323 (2004)
7. Feferman, S.: Predicativity. In: Shapiro, S. (ed.) The Oxford Handbook of Philosophy of Mathematics and Logic, pp. 590–624. Oxford University Press, Oxford (2005)
8. Goodman, N.: The Structure of Appearance, 3rd edn, p. 34 ff. Reidel, Dordrecht (1977)
9. Hellman, G.: Predicativism as a philosophical position. Revue Internationale de Philosophie **58**(3), 295–312 (2004)
10. Hellman, G., Shapiro, S.: Towards a point-free account of the continuous. Iyyun Jerus. Philos. Q. **61**, 263–287 (2012)
11. Hellman, G., Shapiro, S.: The classical continuum without points. Rev. Symb. Log. **6**, 488–512 (2013)
12. Hellman, G., Stewart, S.: Regions-based two dimensional continua: the Euclidean case. Logic and Logical Philosophy **24**, 499–534 (2015)
13. Linnebo, Ø., Shapiro, S., Hellman, G.: Aristotelian continua. Philosophia Mathematica **24**, 214–246 (2016)

Author Biographies

Geoffrey Hellman received his PhD in Philosophy in 1973 from Harvard University. He is currently Professor of Philosophy and Scholar of the College at the University of Minnesota, and he is a Fellow of the American Academy of Arts and Sciences. He has specialized in philosophy of mathematics and logic and in philosophy of quantum mechanics. In addition to many papers in these areas, he has published *Mathematics without Numbers: Towards a Modal-Structural Interpretation*, with Oxford University Press (1989) and has a second monograph, forthcoming with Oxford UP, *Varieties of Continua: From Regions to Points and Back* (jointly with Stewart Shapiro). In addition to courses in the above areas, he has taught musical aesthetics in the Schools of Music at Indiana University and the University of Minnesota. In addition, he has performed classical solo and chamber music recitals as a pianist at venues in the US and abroad.

Stewart Shapiro received an MA in mathematics in 1975, and a PhD in philosophy in 1978, both from the State University of New York at Buffalo. He is currently the O'Donnell Professor of Philosophy at The Ohio State University, and Professorial Fellow at the University of Oslo. He has specialized in philosophy of mathematics, logic, philosophy of logic, and philosophy of language, with a recent interest in semantics. He has published four research monographs, all with Oxford University Press: *Foundations without foundationalism: a case for second-order logic,Philosophy of mathematics: structure and ontology, Vagueness in context, and Varieties of logic*. He also has a forthcoming monograph, *Varieties of Continua: From Regions to Points and Back* (jointly with Geoffrey Hellman). He has taught courses in logic, philosophy of mathematics, philosophy of science, philosophy of religion, Marxism, aesthetics, Jewish philosophy, and medical ethics. He and his wife, Beverly Roseman-Shapiro, currently live in Columbus, Ohio, where they are enjoying the start of their empty nest and occasional visits to their grandchildren.

Unfolding Schematic Systems

Thomas Strahm

To Sol, with gratitude for his intellectual inspiration and friendship

Abstract The notion of unfolding a schematic formal system was introduced by Feferman in 1996 in order to answer the following question: *Given a schematic system* S, *which operations and predicates, and which principles concerning them, ought to be accepted if one has accepted* S? After a short summary of precursors of the unfolding program, we survey the unfolding procedure and discuss the main results obtained for various schematic systems S, including non-finitist arithmetic, finitist arithmetic, feasible arithmetic, and theories of inductive definitions.

Keywords Schematic systems · Unfolding · Finitist arithmetic · Non-finitist arithmetic · Feasible arithmetic · Inductive definitions

2010 Mathematics Subject Classification 03F03 · 03F30 · 03D75

1 Introduction

The search for new axioms which are *exactly as evident and justified as those with which you have started* was already advocated by Gödel in his program for new axioms, see Gödel [24], p. 151:

> Let us consider, e.g., the concept of demonstrability. It is well known that, in whichever way you make it precise by means of a formalism, the contemplation of this very formalism gives rise to new axioms which are exactly as evident and justified as those with which you started, and this process of extension can be iterated into the transfinite. So there cannot exist any formalism which would embrace all these steps, but this does not exclude that all these steps

T. Strahm (✉)
Institut für Informatik, Universität Bern, Neubrückstrasse 10,
CH-3012 Bern, Switzerland
e-mail: strahm@inf.unibe.ch

G. Jäger and W. Sieg (eds.), *Feferman on Foundations*, Outstanding Contributions to Logic 13, DOI 10.1007/978-3-319-63334-3_8

> (or at least all of them which give something new for the domain of propositions in which you are interested) could be described and collected together in some non-constructive way.

The first very natural candidates for new axioms to be added to an arithmetical system S are proof-theoretic reflection principles, roughly stating that everything which is provable in S is correct. More precisely, the local reflection schema is the collection of sentences

$$(\mathsf{Rfn}_\mathsf{S}) \qquad \mathsf{Prov}_\mathsf{S}(\ulcorner A \urcorner) \to A$$

for A being a sentence in the language of S. The generalization of this schema to arbitrary formulas uniformly in their free parameters is called the uniform reflection schema of S, in symbols, RFN_S. As was shown by Turing [36], one may iterate the addition of reflection principles (and consistency statements) along Kleene's constructive ordinal notations $\mathcal{O}$ in order to define for each $a \in \mathcal{O}$ a formal system S_a by adding the reflection principle of the previous system at successor stages and taking the union of the previous systems at limit stages. Turing called the so-obtained progressions of a given system S *ordinal logics*. These were taken up in Feferman [10] and there renamed to *transfinite recursive progressions of axiomatic theories*. While Turing obtained completeness results for Π^0_1 sentences by iteration of the consistency or local reflection principle Rfn_S, Feferman showed that one gets completeness for all arithmetic statements by iteration of the uniform reflection principle RFN_S. Both completeness results were considered to be problematic because they depend on clever choices of ordinal notations which were not justified on previously accepted grounds. Indeed, ordinal logics are far from being invariant under the choice of ordinal notation: Feferman and Spector [19] have shown that there are paths through $\mathcal{O}$ whose progression is not even complete for Π^0_1 sentences. For more information on ordinal logics, see Feferman [17] and Franzen [23].

The crucial condition which was missing in the previous proposals is the one of *autonomy* which guarantees that one is only allowed to advance to a system S_a for an $a \in \mathcal{O}$ in case the wellfoundedness of a has been established in a system S_b with b smaller than a; see Kreisel [26] and Feferman [11]. Thus we are led to the study of *all principles of proof and ordinals which are implicit in given concepts*, see Kreisel [28]. The most influential series of results in the "autonomy program" concerns the study of the limits of predicative provability by Feferman [11] and Schütte [31] who independently determined the so-called Feferman–Schütte ordinal Γ_0 as the limiting number of predicativity. The first system proposed for an analysis of predicativity was autonomous ramified analysis. After its ordinal Γ_0 had been found, Feferman developed (autonomous) progressions of hyperarithmetical analysis based on the hyperarithmetic comprehension rule and the uniform reflection principle as well as the system IR for inductive-recursive analysis.

One objection with the above-described approaches in the implicitness program one may have is the inclusion of the notion of ordinal or wellordering, which is not prima-facie implicit in our conception of the natural numbers or arithmetic. In his search for "a more perspicuous system for predicativity", Feferman [13] came up with a natural system capturing ramified analysis in levels less than Γ_0 without presupposing any notion of ordinal at the outset. Crucial in this system is the fact that arithmetic is treated as a *schematic system* with *free* predicate P and induction in schematic form,

$$P(0) \wedge (\forall x)[P(x) \rightarrow P(x')] \rightarrow (\forall x)P(x),$$

together with a rule of predicate substitution (Subst),

$$\text{(Subst)} \qquad A[P] \Rightarrow A[B/P],$$

expressing that whenever we derive a statement $A[P]$ possibly containing the free predicate P, we also accept all its substitution instances by any formula B. We note that a crucial feature of schemata as understood here is their *openendedness*, i.e. "they are not conceived of as applying to a specific language whose stock of basic symbols is fixed in advance, but rather as applicable to any language which one comes to recognize as embodying meaningful basic notions" (Feferman [15], p. 9).

A general notion of reflective closure of an arbitrary schematic formal system S was proposed by Feferman in a lecture for a meeting in 1979 on the work of Kurt Gödel, which was only published in 1991, see Feferman [14]. The basic observation underlying the reflective closure procedure in [14] is that the informal reasoning about what is implicit in S makes use of a notion of truth for S, which then leads us to also reason about statements involving truth and so on. The technical apparatus of the reflective closure is governed by what has become to be known as the Kripke-Feferman axioms of partial truth, rooted in Kripke's semantic theory of truth, see Kripke [29] and the article of Cantini, Fujimoto, and Halbach in this volume. The main result obtained in Feferman [14] is that the full reflective closure of Peano arithmetic, $\mathsf{Ref}^*(\mathsf{PA})$, is proof-theoretically equivalent to predicative analysis $\mathsf{RA}_{<\Gamma_0}$, where $\mathsf{Ref}^*(\mathsf{PA})$ includes a suitable version of the substitution rule (Subst).

As Feferman writes in [18], the axiomatic theory of truth "as an engine for the explanation of reflective closure still has an air of artificiality": It is at least questionable whether the axioms of truth are exactly as evident as those of the given system S. Also, the fact that some amount of arithmetic is presupposed in order to describe the coding machinery is not very pleasing. Given that a schematic system is formulated using function and predicate symbols in a given logical language, it is more attractive to expand and study the *operations* on individuals and predicates, which are implicit in the acceptance of S. This led Feferman [15] to his last proposal in the implicitness program, namely the notion of unfolding of an open-ended schematic system whose main aim is to answer the following question: *Given a schematic system* S, *which operations and predicates – and which principles concerning them – ought to be accepted if one has accepted* S?. The notion of unfolding has been applied to non-finitist-arithmetic (Feferman and Strahm [20]), finitist arithmetic (Feferman and Strahm [21]), feasible arithmetic (Eberhard and Strahm [9]), and theories of inductive definitions (Buchholtz [2]). The aim of this paper is to describe the unfolding procedure in detail and discuss the main results obtained for various schematic formal systems S.

2 The Unfolding of Non-finitist Arithmetic

The aim of this section is to spell out the unfolding procedure in detail for the case of non-finitist arithmetic NFA, and state the main results obtained in Feferman and Strahm [20]. We follow the presentation in Feferman and Strahm [21].

The schematic system for classical non-finitist arithmetic, NFA, is defined as follows. Its basic operations on individuals with the constant 0 are successor, Sc, and predecessor, Pd; the basic logical operations are $\neg$, $\wedge$, and $\forall$. It is given by the following axioms, where we write as usual, x' for $\mathsf{Sc}(x)$:

(1) $x' \neq 0$
(2) $\mathsf{Pd}(x') = x$
(3) $P(0) \wedge (\forall x)[P(x) \rightarrow P(x')] \rightarrow (\forall x)P(x)$.

Here P is a free predicate variable, and the intention is to use the induction scheme (3) in a wider sense than is limited by the basic language of NFA or any language fixed in advance. Namely, one applies the general rule of substitution

(Subst) $$A[P] \;\Rightarrow\; A[B/P]$$

to any formulas A and B that arise in the process of unfolding NFA.

In a first step, we shall describe the unfolding of a schematic system S informally by stating some general methodological "pre-axioms". Then we shall spell out these axioms in all detail for S being the schematic system NFA.

Underlying the idea of unfolding for arbitrary S are general notions of (partial) *operation* and *predicate*, belonging to a universe V extending the universe of discourse of S. These have to be thought of as intensional entities, given by rules of computation and defining properties, respectively. Operations have to be considered as pre-mathematical in nature and not bound to any specific mathematical domain. They can apply to other operations as well as to predicates. Some operations are universal and are naturally self-applicable as a result, like the identity operation or the pairing operation, while some are partial and presented to us on specific mathematical domains only. Operations satisfy the laws of a *partial combinatory algebra* with pairing, projections, and definition by cases. Predicates are equipped with a membership relation $\in$ to express that given elements satisfy the predicate's defining property.

For the formulation of the full unfolding $\mathcal{U}(\mathsf{S})$ of any given schematic axiom system S, we have the following axioms.

1. The universe of discourse of S has associated with it an additional unary relation symbol, $\mathsf{U_S}$, and the axioms of S are relativized to $\mathsf{U_S}$.

2. Each n-ary function symbol f of S determines an element $f^\star$ of our partial combinatory algebra, with $f(x_1, \ldots, x_n) = f^\star x_1 \ldots x_n$ on $\mathsf{U}^n_{\mathsf{S}}$ (or the domain of f in case f itself is given as a partial function).
3. Each relation symbol R of S together with $\mathsf{U_S}$ determines a predicate $R^\star$ with $R(x_1, \ldots, x_n)$ if and only if $(x_1, \ldots, x_n) \in R^\star$.
4. Operations on predicates, such as e.g. conjunction, are just special kinds of operations. Each logical operation l of S determines a corresponding operation $l^\star$ on predicates.
5. Sequences of predicates given by an operation f form a new predicate $\mathit{Join}(f)$, the disjoint union of the predicates from f.

Moreover, the free predicate variables $P, Q, \ldots$ used in the schematic formulation of S give rise to the crucial *rule of substitution* (Subst), according to which we are allowed to substitute any formula B for P in a previously recognized (i.e. derived) statement $A[P]$ depending on P.

The restriction $\mathcal{U}_0(\mathsf{S})$ of $\mathcal{U}(\mathsf{S})$ is obtained by dropping the axioms concerning predicates; $\mathcal{U}_0(\mathsf{S})$ is called the operational unfolding of S. Moreover, there is a natural intermediate predicate unfolding system $\mathcal{U}_1(\mathsf{S})$, which is simply $\mathcal{U}(\mathsf{S})$ without the predicate forming operation of *Join*.

The following spells out in detail the three unfolding systems $\mathcal{U}_0(\mathsf{S})$, $\mathcal{U}_1(\mathsf{S})$, and $\mathcal{U}(\mathsf{S})$ for $\mathsf{S} = \mathsf{NFA}$, the schematic system of non-finitist arithmetic introduced above. We begin with the operational unfolding $\mathcal{U}_0(\mathsf{NFA})$. Its language is first order, using variables $a, b, c, f, g, h, u, v, w, x, y, z \ldots$ (possibly with subscripts). It includes (i) the constant 0 and the unary function symbols Sc and Pd of NFA, (ii) constants for operations as individuals, namely sc, pd (successor, predecessor), k, s (combinators), p, $\mathsf{p_0}$, $\mathsf{p_1}$ (pairing and unpairing), d, tt, ff (definition by cases, true, false), and e (equality), and (iii) a binary function symbol $\cdot$ for (partial) term application. Further, we have (iv) a unary relation symbol $\downarrow$ (defined) and a binary relation symbol $=$ (equality), as well as (v) a unary relation symbol N (natural numbers). In addition, we have a symbol $\bot$ for the false proposition. Finally, a stock of free predicate symbols $P, Q, R, \ldots$ of finite arities is assumed.[1]

The *terms* $(r, s, t, \ldots)$ of $\mathcal{U}_0(\mathsf{NFA})$ are inductively generated from the variables and constants by means of the function symbols Sc, Pd, as well as $\cdot$ for application. In the following we often abbreviate $(s \cdot t)$ simply as (st), st or sometimes also $s(t)$; the context will always ensure that no confusion arises. We further adopt the convention of association to the left so that $s_1 s_2 \ldots s_n$ stands for $(\ldots(s_1 s_2) \ldots s_n)$. Further, we put $t' := \mathsf{Sc}(t)$ and $1 := 0'$. We define general n-tupling by induction on $n \geq 2$ as follows:

$$(s_1, s_2) := \mathsf{p} s_1 s_2, \qquad (s_1, \ldots, s_{n+1}) := ((s_1, \ldots, s_n), s_{n+1}).$$

Moreover, we set $(s) := s$ and $() := 0$.

---

[1]The constants sc and pd as well as the relation symbol N are used instead of the symbols $\mathsf{Sc}^\star$, $\mathsf{Pd}^\star$, and $\mathsf{U_{NFA}}$ mentioned in the informal description above.

The *formulas* $(A, B, C, \dots)$ of $\mathcal{U}_0(\mathsf{NFA})$ are inductively generated from the atomic formulas $\bot$, $s\downarrow$, $(s = t)$, $\mathsf{N}(s)$, and $P(s_1, \dots, s_n)$ by means of negation $\neg$, conjunction $\wedge$, and universal quantification $\forall$. The remaining logical connectives and quantifiers are defined as usual by making use of classical logic.

The sequence notation $\bar{u}$ and $\bar{t}$ is used in order to denote finite sequences of variables and terms, respectively. Moreover, we write $t[\bar{u}]$ to indicate a sequence $\bar{u}$ of free variables possibly appearing in the term t; however, t may contain other variables than those shown by using this bracket notation. Further, $t[\bar{s}]$ is used to denote the result of simultaneous substitution of the terms $\bar{s}$ for the variables $\bar{u}$ in the term $t[\bar{u}]$. The meaning of $A[\bar{u}]$ and $A[\bar{s}]$ is understood accordingly. Finally, we shall also use the sequence notation $\bar{A}$ in order to denote a finite sequence $\bar{A} = A_1, \dots, A_n$ of formulas.

$\mathcal{U}_0(\mathsf{NFA})$ is based on *partial* term application. Hence, it is not guaranteed that terms have a value, and $t\downarrow$ is read as "t is defined" or "t has a value". Accordingly, the *partial equality relation* $\simeq$ is introduced by

$$s \simeq t \;:=\; (s\downarrow \vee\, t\downarrow) \to (s = t).$$

Further, we shall use the following abbreviations concerning the predicate N for the natural numbers ($\bar{s} = s_1, \dots, s_n$):

$$\begin{aligned} \bar{s} \in \mathsf{N} &:= \mathsf{N}(s_1) \wedge \cdots \wedge \mathsf{N}(s_n), \\ (\exists x \in \mathsf{N})A &:= (\exists x)(x \in \mathsf{N} \wedge A), \\ (\forall x \in \mathsf{N})A &:= (\forall x)(x \in \mathsf{N} \to A). \end{aligned}$$

The logic of $\mathcal{U}_0(\mathsf{NFA})$ is the *classical logic of partial terms* LPT of Beeson [1], cf. also Feferman [12]. We recall that LPT embodies strictness axioms saying that all subterms of a defined compound term are defined as well. Moreover, if $(s = t)$ holds then both s and t are defined, and s is defined provided $\mathsf{N}(s)$ holds, and similarly for $P(\bar{s})$.

The axioms of $\mathcal{U}_0(\mathsf{NFA})$ are divided into three groups as follows.

I. Embedding of NFA

(1) The relativization of the axioms of NFA to the predicate N,[2]
(2) $(\forall x \in \mathsf{N})[\mathsf{Sc}(x) = \mathsf{sc}(x) \wedge \mathsf{Pd}(x) = \mathsf{pd}(x)]$.

II. Partial combinatory algebra, pairing, definition by cases

(3) $\mathsf{k}ab = a$,
(4) $\mathsf{s}ab\downarrow \wedge\ \mathsf{s}abc \simeq ac(bc)$,
(5) $\mathsf{p}_0(a, b) = a \wedge \mathsf{p}_1(a, b) = b$,
(6) $\mathsf{d}ab\,\mathsf{tt} = a \wedge \mathsf{d}ab\,\mathsf{ff} = b$.

III. Equality on the natural numbers N

[2] Note that this relativization also includes axioms such as $0 \in \mathsf{N}$ and $(\forall x \in \mathsf{N})(x' \in \mathsf{N})$.

(7) $(\forall x, y \in \mathsf{N})[\mathsf{e}xy = \mathtt{tt} \vee \mathsf{e}xy = \mathtt{ff}]$,
(8) $(\forall x, y \in \mathsf{N})[\mathsf{e}xy = \mathtt{tt} \leftrightarrow x = y]$.

Finally, crucial for the formulation of $\mathcal{U}_0(\mathsf{S})$ is the predicate substitution rule:

(Subst) $$A[\bar{P}] \;\Rightarrow\; A[\bar{B}/\bar{P}].$$

Here $\bar{P} = P_1, \ldots, P_m$ is a sequence of free predicate symbols possibly appearing in the formula $A[\bar{P}]$ and $\bar{B} = B_1, \ldots, B_m$ is a sequence of formulas. In the conclusion of this rule of inference, $A[\bar{B}/\bar{P}]$ denotes the formula $A[\bar{P}]$ with each subformula $P_i(\bar{t})$ replaced by $(\exists \bar{x})(\bar{t} = \bar{x} \wedge B_i[\bar{x}])$, where the length of $\bar{x}$ equals the arity of P_i.

We now turn to the full predicate unfolding $\mathcal{U}(\mathsf{NFA})$ and its restriction $\mathcal{U}_1(\mathsf{NFA})$.

The language of $\mathcal{U}(\mathsf{NFA})$ extends the language of $\mathcal{U}_0(\mathsf{NFA})$ by additional constants nat (natural numbers), eq (equality), pr_P (free predicate P), inv (inverse image), neg (negation), conj (conjunction), un (universal quantification), and join (disjoint unions). In addition, we have a new unary relation symbol Π for (codes of) predicates and a binary relation symbol $\in$ for expressing elementhood between individuals and predicates, i.e. satisfaction of those predicates by the given individuals. The *terms* of $\mathcal{U}(\mathsf{NFA})$ are generated as before but now taking into account the new constants. The *formulas* of $\mathcal{U}(\mathsf{NFA})$ extend the formulas of $\mathcal{U}_0(\mathsf{NFA})$ by allowing new atomic formulas of the form $\Pi(t)$ and $s \in t$.

The axioms of $\mathcal{U}(\mathsf{NFA})$ extend those of $\mathcal{U}_0(\mathsf{NFA})$, as follows.

IV. Basic axioms about predicates

(9) $\Pi(\mathsf{nat}) \wedge (\forall x)(x \in \mathsf{nat} \leftrightarrow \mathsf{N}(x))$,[3]
(10) $\Pi(\mathsf{eq}) \wedge (\forall x)(x \in \mathsf{eq} \leftrightarrow (\exists y)(x = (y, y)))$,
(11) $\Pi(\mathsf{pr}_P) \wedge (\forall \bar{x})((\bar{x}) \in \mathsf{pr}_P \leftrightarrow P(\bar{x}))$,
(12) $\Pi(a) \rightarrow \Pi(\mathsf{inv}(a, f)) \wedge (\forall x)(x \in \mathsf{inv}(a, f) \leftrightarrow fx \in a)$,
(13) $\Pi(a) \rightarrow \Pi(\mathsf{neg}(a)) \wedge (\forall x)(x \in \mathsf{neg}(a) \leftrightarrow x \notin a)$,
(14) $\Pi(a) \wedge \Pi(b) \rightarrow \Pi(\mathsf{conj}(a, b)) \wedge (\forall x)(x \in \mathsf{conj}(a, b) \leftrightarrow x \in a \wedge x \in b)$,
(15) $\Pi(a) \rightarrow \Pi(\mathsf{un}(a)) \wedge (\forall x)(x \in \mathsf{un}(a) \leftrightarrow (\forall y \in \mathsf{N})((x, y) \in a))$.

V. Join axiom

(16) $(\forall x \in \mathsf{N})\Pi(fx) \rightarrow \Pi(\mathsf{join}(f)) \wedge (\forall x)(x \in \mathsf{join}(f) \leftrightarrow J[f, x])$,

where $J[f, u]$ expresses that u is an element of the disjoint union of f over N, i.e.

$$J[f, u] \;:=\; (\exists y \in \mathsf{N})(\exists z)(u = (y, z) \wedge z \in fy).$$

In addition, $\mathcal{U}(\mathsf{NFA})$ contains the substitution rule (Subst), i.e. the rule $A[\bar{P}] \Rightarrow A[\bar{B}/\bar{P}]$, where now $\bar{B}$ denote arbitrary formulas in the language of $\mathcal{U}(\mathsf{NFA})$, but $A[\bar{P}]$ is required to be a formula in the language of $\mathcal{U}_0(\mathsf{NFA})$. This last restriction is due to the fact that predicates in general depend on the predicate parameters $\bar{P}$.

[3]Observe that nat is alternatively definable from the remaining predicate axioms by $x \in \mathsf{nat} \leftrightarrow (\exists y \in \mathsf{N})(x = y)$.

Finally, we obtain an intermediate predicate unfolding system $\mathcal{U}_1(\mathsf{NFA})$ by omitting axiom (16), i.e., $\mathcal{U}_1(\mathsf{NFA})$ is just $\mathcal{U}(\mathsf{NFA})$ without the *Join* predicate.

To state the proof-theoretic strength of the three unfolding systems $\mathcal{U}_0(\mathsf{NFA})$, $\mathcal{U}_1(\mathsf{NFA})$, and $\mathcal{U}(\mathsf{NFA})$, as usual we let $\mathsf{RA}_{<\alpha}$ denote the system of ramified analysis in levels less than α. In addition, Γ_0 is the so-called Feferman–Schütte ordinal, which was identified in the early sixties as the limiting number of predicative provability. As in Feferman and Strahm [20] we obtain the following proof-theoretic equivalences. In particular, the full unfolding of non-finitist arithmetic is equivalent to predicative analysis.

Theorem 1 *We have the following proof-theoretic equivalences:*

1. $\mathcal{U}_0(\mathsf{NFA}) \equiv \mathsf{PA}$.
2. $\mathcal{U}_1(\mathsf{NFA}) \equiv \mathsf{RA}_{<\omega}$.
3. $\mathcal{U}(\mathsf{NFA}) \equiv \mathsf{RA}_{<\Gamma_0}$.

In each case we have conservation with respect to arithmetic statements of the system on the left over the system on the right.

Let us give a few indications with respect to the proofs of these equivalences. In order to show that $\mathcal{U}_0(\mathsf{NFA})$ contains PA, one first shows by using the canonical fixed point operator of the underlying partial combinatory algebra that each primitive recursive function F can be represented by a term t_F in the language of $\mathcal{U}_0(\mathsf{NFA})$. Then one needs to show that these terms are well-typed on the natural numbers N, namely that $t_F(\bar{x}) \in \mathsf{N}$ for each $(\bar{x}) \in \mathsf{N}$: here one uses induction which follows by one application of the substitution rule to axiom (3) of NFA. A further application of (Subst) thus shows that the usual formulation of PA is directly contained in the operational unfolding $\mathcal{U}_0(\mathsf{NFA})$ of NFA. Indeed, $\mathcal{U}_0(\mathsf{NFA})$ does not go beyond PA, as is seen be formalizing its standard recursion-theoretic model in PA, see e.g. [20].

The full unfolding of NFA, $\mathcal{U}(\mathsf{NFA})$, derives the schema of transfinite induction along each initial segment of the Feferman–Schuütte ordinal Γ_0: Whenever we can derive in $\mathcal{U}(\mathsf{NFA})$ transfinite induction $\mathsf{TI}(\prec, P)$ along a primitive recursive ordering $\prec$, then we may substitute for P any formula and thus derive the existence of the predicate corresponding to the hyperarithmetic hierarchy along $\prec$, relative to any initial predicate p. Thus, using standard arguments from predicative wellordering proofs, whenever $\mathcal{U}(\mathsf{NFA})$ derives transfinite induction up to α, it also does so up to $\varphi\alpha 0$, hence the lower bound Γ_0. This bound is sharp according to Feferman and Strahm [20], see also Strahm [33].

Recall that in the intermediate unfolding $\mathcal{U}_1(\mathsf{NFA})$, the join principle is not available. We can still justify finite levels of the ramified analytical hierarchy, corresponding to the proof-theoretic ordinal $\varphi 20$, which is also the ordinal of the subsystem of second order arithmetic based on arithmetic comprehension and the bar rule, see Rathjen [30]. Indeed, in $\mathcal{U}_1(\mathsf{NFA})$, each application of the substitution rule lets us step from α to ε_α.

Let us close this section by mentioning that the original formulation of unfolding in Feferman [15] made use of a background theory of typed operations with general Least Fixed Point operator. The present formulation is a simplification of this approach. The upper bound computation in Feferman and Strahm [20] was done for this original formulation; it is worth mentioning that the proof-theoretic analysis of the unfolding of NFA in the present formulation is somewhat simpler and more elegant, since leastness for the fixed point operator is not present. A further difference is that predicates in the original formulation of unfolding were modeled as propositional functions using a truth predicate.

3 The Unfolding of Finitist Arithmetic

In this section we describe the unfolding of two schematic systems for finitist arithmetic, namely FA and FA plus a form of the Bar rule BR. The main results are that all three unfolding systems for FA are equivalent to Primitive Recursive Arithmetic PRA, while the three unfoldings of FA + BR reach precisely the strength of Peano arithmetic PA. These two characterizations of finitism are in accord with two prominent views about the limits of finitist reasoning due to Tait [35] and Kreisel [27]. In the sequel we follow Feferman and Strahm [21].

The logical operations of the basic schematic system FA of finitist arithmetic are restricted to $\wedge$, $\vee$, and $\exists$. In order to reason from such statements to new such statements given the above restriction of the logical operations of FA, we make use of a sequent formulation of our calculus, i.e. the statements proved are sequents Σ of the form $\Gamma \rightarrow A$, where Γ is a finite sequence (possibly empty) of formulas, and A may also be the false proposition $\bot$. Moreover, induction must now be given as a rule of inference involving such sequents. Accordingly, the basic axioms and rules of FA are as follows:

(1) $x' = 0 \rightarrow \bot$

(2) $\mathsf{Pd}(x') = x$

(3) $$\frac{\Gamma \rightarrow P(0) \qquad \Gamma,\, P(x) \rightarrow P(x')}{\Gamma \rightarrow P(x)}.$$

The substitution rule (Subst) may now generalized to incorporate sequent inference rules; the corresponding (meta) rule is called (Subst′) and will be spelled out in detail below.

Let us begin by describing the *operational unfolding* $\mathcal{U}_0(\mathsf{FA})$ of finitist arithmetic FA. That system tells us which operations from and to natural numbers, and which principles concerning them, ought to be accepted if we have accepted FA. It is seen that Skolem's system PRA of primitive recursive arithmetic is contained in $\mathcal{U}_0(\mathsf{FA})$. Indeed, the operational and even the full unfolding of finitist arithmetic do not go beyond PRA.

Large parts of the unfolding systems for FA and NFA are identical. Therefore, we shall confine ourselves in the sequel to mentioning the main differences in the specification of the unfolding systems for FA, beginning with its operational unfolding.

The *terms* of $\mathcal{U}_0(\mathsf{FA})$ are the same as the terms of $\mathcal{U}_0(\mathsf{NFA})$. Recall that FA is based on the logical operations $\wedge$, $\vee$, and $\exists$. Accordingly, the *formulas* of $\mathcal{U}_0(\mathsf{FA})$

are generated from the atomic formulas $\bot$, $s\downarrow$, $(s = t)$, $\mathsf{N}(s)$, and $P(\bar{s})$ by means of $\wedge$, $\vee$, and $\exists$; here P denotes an arbitrary free predicate variable of appropriate arity.

The underlying calculus of $\mathcal{U}_0(\mathsf{FA})$ is a Gentzen-type sequent system based on sequents of the form $\Gamma \rightarrow A$ for Γ being a finite sequence of formulas in the language of $\mathcal{U}_0(\mathsf{FA})$. In case Γ is empty, we shall write A for $\rightarrow A$. The logical axioms and rules of inference are the standard ones: apart from identity axioms, rules for $\bot$, cut and structural rules, these include the usual Gentzen-type rules for $\wedge$ and $\vee$ as well as introduction of $\exists$ on the left and on the right in the form

$$\frac{\Gamma \rightarrow A[t] \wedge t\downarrow}{\Gamma \rightarrow (\exists x)A[x]}, \qquad \frac{\Gamma,\, A[u] \rightarrow B}{\Gamma,\, (\exists x)A[x] \rightarrow B} \quad (u \text{ fresh})$$

Note that quantifiers range over defined objects only. Moreover, defined terms can be substituted for free variables according to the following rule of inference; here $\Gamma[t]$ stands for the sequence $(B[t] : B[u] \in \Gamma)$.

$$\frac{\Gamma[u] \rightarrow A[u]}{\Gamma[t],\, t\downarrow \rightarrow A[t]}$$

Finally, the equality and strictness axioms of our underlying logic of partial terms are given a Gentzen-style formulation in the obvious way.

The non-logical axioms and rules of $\mathcal{U}_0(\mathsf{FA})$ include the relativization of the axioms and rules of FA to the predicate N in the expected manner, as well as suitable formulations of the axioms (2)–(8) of $\mathcal{U}_0(\mathsf{NFA})$. We shall not spell out these axioms again, but instead give an example how to reformulate axiom (4) about the s combinator in our new setting. This now breaks into the following two axioms,

$$\mathsf{s}ab\downarrow \quad \text{and} \quad \mathsf{s}abc\downarrow \vee\, ac(bc)\downarrow \;\rightarrow\; \mathsf{s}abc = ac(bc).$$

What is still missing in $\mathcal{U}_0(\mathsf{FA})$ is a suitable version of the *substitution rule* (Subst), which is central to all unfolding systems. In order to fit this into our Gentzen-style setting, (Subst) has to be formulated in a somewhat more general form. For that purpose, we let $\Sigma, \Sigma_1, \Sigma_2, \dots$ range over sequents in the language of $\mathcal{U}_0(\mathsf{FA})$. A *rule of inference* for such sequents has the general form

$$\frac{\Sigma_1, \Sigma_2, \dots, \Sigma_n}{\Sigma},$$

which we simply abbreviate by $\Sigma_1, \Sigma_2, \dots, \Sigma_n \Rightarrow \Sigma$ in the sequel; we also allow n to be 0, i.e. rules with an empty list of premises are possible. As usual we call a rule of inference $\Sigma_1, \Sigma_2, \dots, \Sigma_n \Rightarrow \Sigma$ *derivable* from a collection of axioms and rules $\mathcal{T}$ (all in Gentzen-style), if the sequent Σ is derivable from $\mathcal{T} \cup \{\Sigma_1, \Sigma_2, \dots, \Sigma_n\}$.

In the following $\bar{P} = P_1, \dots, P_m$ denotes a finite sequence of free predicate symbols of finite arity and $\bar{B} = B_1, \dots, B_m$ a corresponding sequence of formulas in the language of $\mathcal{U}_0(\mathsf{FA})$. If $\Sigma[\bar{P}]$ is a sequent possibly containing the free predicates $\bar{P}$, then as above $\Sigma[\bar{B}/\bar{P}]$ denotes the sequent $\Sigma[\bar{P}]$ with each subformula of the form $P_i(\bar{t})$ replaced by $(\exists\bar{x})(\bar{t} = \bar{x} \wedge B[\bar{x}])$, where the length of $\bar{x}$ is equal to the arity of P_i.

We are now ready to state our (meta) substitution rule (Subst′). Its meaning is as follows: whenever the axioms and rules of inference at hand allow us to show that the rule $\Sigma_1, \Sigma_2, \ldots, \Sigma_n \Rightarrow \Sigma$ is *derivable*, then we can adjoin each of its substitution instances $\Sigma_1[\bar{B}/\bar{P}], \Sigma_2[\bar{B}/\bar{P}], \ldots, \Sigma_n[\bar{B}/\bar{P}] \Rightarrow \Sigma[\bar{B}/\bar{P}]$ as a new rule of inference to $\mathcal{U}_0(\mathsf{FA})$, for $B_i[\bar{x}]$ being formulas in the language of $\mathcal{U}_0(\mathsf{FA})$.[4] Symbolically,

$$\text{(Subst}'\text{)} \qquad \frac{\Sigma_1, \Sigma_2, \ldots, \Sigma_n \Rightarrow \Sigma}{\Sigma_1[\bar{B}/\bar{P}],\ \Sigma_2[\bar{B}/\bar{P}], \ldots, \Sigma_n[\bar{B}/\bar{P}] \ \Rightarrow\ \Sigma[\bar{B}/\bar{P}]}$$

It is not difficult to see that the primitive recursive functions can be introduced and proved total in $\mathcal{U}_0(\mathsf{FA})$. Indeed, the argument described in the previous section is readily seen to be formalizable in $\mathcal{U}_0(\mathsf{FA})$, since induction on N is available for equations and formulas of the form $t(\bar{x}) \in \mathsf{N}$. Thus, Primitive Recursive Arithmetic is interpretable in $\mathcal{U}_0(\mathsf{FA})$, see [21] for details.

The *full unfolding* $\mathcal{U}(\mathsf{FA})$ of finitist arithmetic FA is an extension of the operational unfolding $\mathcal{U}_0(\mathsf{FA})$ and is used, in addition, to answer the question of which operations on and to predicates, and which principles concerning them, are to be accepted if one has accepted FA. We shall see that $\mathcal{U}(\mathsf{FA})$ does not go beyond primitive recursive arithmetic PRA in proof-theoretic strength.

The language of $\mathcal{U}(\mathsf{FA})$ is an extension of the language of $\mathcal{U}_0(\mathsf{FA})$. It includes, in addition, the constants nat (natural numbers), eq (equality), pr_P (free predicate P), inv (inverse image), conj (conjunction), disj (disjunction), ex (existential quantification), and join (disjoint unions). Moreover, as above, we have a new unary relation symbol Π for (codes of) predicates and a binary relation symbol $\in$ for the elementhood relation. The *terms* of $\mathcal{U}(\mathsf{FA})$ are built as before. The *formulas* of $\mathcal{U}(\mathsf{FA})$ extend the formulas of $\mathcal{U}_0(\mathsf{FA})$ by allowing the new atomic formulas $\Pi(t)$ and $s \in t$.

[4]Observe that derivability of rules is a dynamic process as we unfold FA. In particular, new rules of inference obtained by (Subst′) allow us to establish new derivable rules, to which in turn we can apply (Subst′). In particular, the usual rule of induction

$$\frac{\Gamma \rightarrow A[0] \quad \Gamma,\ u \in \mathsf{N},\ A[u] \rightarrow A[u']}{\Gamma,\ v \in \mathsf{N} \rightarrow A[v]}$$

is an immediate consequence of (Subst′) applied to rule (3) of FA. Moreover, the substitution rule in its usual form as stated in Sect. 2,

$$\text{(Subst)} \qquad \frac{\Sigma[\bar{P}]}{\Sigma[\bar{B}/\bar{P}]}$$

is readily seen to be an admissible rule of inference of $\mathcal{U}_0(\mathsf{FA})$.

The axioms of $\mathcal{U}(\mathsf{FA})$ extend those of $\mathcal{U}_0(\mathsf{FA})$. In addition, we have the obvious defining axioms for the basic predicates of $\mathcal{U}(\mathsf{FA})$. These include straightforward reformulations using sequents of the axioms (9)–(12) and (14) of $\mathcal{U}(\mathsf{NFA})$ as well as the expected axiom about existentially quantified predicates; see Feferman and Strahm [21]. Further, axiom (16) of $\mathcal{U}(\mathsf{NFA})$ concerning join is now stated in terms of suitable inference rules; this is due to the absence of universal quantification in the framework of finitist arithmetic, see [21] for details.

Finally, $\mathcal{U}(\mathsf{FA})$ of course also includes the substitution rule (Subst') which we have spelled out for $\mathcal{U}_0(\mathsf{FA})$. The formulas $\bar{B}$ to be substituted for $\bar{P}$ are now in the language of $\mathcal{U}(\mathsf{FA})$; the rule in the premise of (Subst'), however, is required to be in the language of $\mathcal{U}_0(\mathsf{FA})$. This last restriction is imposed as before since predicates may depend on the free relation symbols $\bar{P}$. The intermediate unfolding system $\mathcal{U}_1(\mathsf{FA})$ for FA is obtained by dropping the rules about join.

It is shown in Feferman and Strahm [21] that all three unfolding systems for FA do not go beyond PRA in strength. This is obtained via a suitable recursion-theoretic interpretation into the subsystem Σ_1-IA of PA with induction on the natural numbers restricted to Σ_1 formulas; the latter system is known to be a Π_2 conservative extension of PRA by the well-known Mints-Parsons-Takeuti theorem, see Sieg [32] for a simple proof. The embedding essentially models the applicative axioms by means of partial recursive function application and the predicates by Σ_1 definable properties, where some special attention is required in order to validate the generalized substitution rule (Subst'). Thus we can summarize:

Theorem 2 $\mathcal{U}_0(\mathsf{FA})$, $\mathcal{U}_1(\mathsf{FA})$ *and* $\mathcal{U}(\mathsf{FA})$ *are all proof-theoretically equivalent to primitive recursive arithmetic* PRA.

In the remainder of this section we shall discuss an extension of FA by a Bar Rule BR and, correspondingly, three unfolding systems of $\mathsf{FA} + \mathsf{BR}$, all of strength Peano arithmetic.

Informally speaking, the Bar Rule BR says that if $\prec$ is a partial ordering provably satisfying $\mathsf{NDS}(\prec)$ (no infinite descending sequence property for $\prec$) then the principle $\mathsf{TI}(\prec, P)$ of transfinite induction on $\prec$ holds for arbitrary predicates P. It is sufficient to restrict this to provably decidable linear orderings $\prec$ in the natural numbers, with 0 as least element. But further restrictions have to be made in order to fit a version of BR to the language of FA. First of all, the statement that a given function f on N is descending in the $\prec$ relation, as long as it is not 0, is universal, so cannot be expressed as a formula of our language. Instead, we add a new function constant symbol f interpreted as an arbitrary (or "anonymous") function, and require that we establish a rule, $\mathsf{NDS}(\prec, \mathsf{f})$, that allows us to infer from the hypotheses that $\mathsf{f} : \mathsf{N} \to \mathsf{N}$ and that $\mathsf{f}(u') \prec \mathsf{f}(u)$ as long as $\mathsf{f}(u) \neq 0$ ('u' a free variable) the conclusion $(\exists x \in \mathsf{N})(\mathsf{f}(x) = 0)$. In addition, we must modify $\mathsf{TI}(\prec, P)$, since its standard formulation for a unary predicate P is of the form:

$$(\forall x)[(\forall u \prec x)P(u) \to P(x)] \to (\forall x)P(x).$$

Again, the idea is to treat this as a rule of the form:

$$\text{from} \quad (\forall u)[u \prec x \to P(u)] \to P(x) \quad \text{infer} \quad P(x).$$

But we still need an additional step to reformulate the hypothesis of this rule in the language of FA. For atomic A, B write $A \supset B$ for $(\neg A \vee B)$. Then the hypothesis is implied by

$$[t_1 \prec x \supset P(t_1)] \wedge \cdots \wedge [t_m \prec x \supset P(t_m)] \to P(x),$$

where the t_i are terms that have been proved to be defined. Now it may be that we cannot prove that $t_i \downarrow$ until we know that certain of its subterms $s_1, \ldots, s_n$ are defined and satisfy

$$[s_1 \prec x \supset P(s_1)] \wedge \cdots \wedge [(s_n \prec x \supset P(s_n)],$$

and so on. Indeed, as we shall see, that is necessary to establish closure under nested recursion on the $\prec$ ordering. This leads to the precise statement of BR in the language of FA augmented by a new function symbol f as follows.[5]

The rule NDS(f, $\prec$) says that for each possibly infinite descending chain f w.r.t. $\prec$ there is an x such that $\mathsf{f}x = 0$. Formally, it is given as follows:

$$\frac{\begin{array}{c} u \in \mathsf{N} \to \mathsf{f}u \in \mathsf{N}, \\ u \in \mathsf{N},\ \mathsf{f}u \neq 0 \to \mathsf{f}(u') \prec \mathsf{f}u, \\ u \in \mathsf{N},\ \mathsf{f}u = 0 \to \mathsf{f}(u') = 0 \end{array}}{(\exists x \in \mathsf{N})(\mathsf{f}x = 0)}$$

Next, the bar rule BR is spelled out in detail for the case of nesting level two and a predicate with one parameter. The general case for nesting of arbitrary level and number of parameters is analogous.

Let $\bar{s^r} = s_1^r, \ldots, s_n^r$ and $\bar{s^p} = s_1^p, \ldots, s_n^p$ be sequences of terms of length n, and let $\bar{t^r} = t_1^r, \ldots, t_m^r$ and $\bar{t^p} = t_1^p, \ldots, t_m^p$ be sequences of terms of length m. The superscripts'r' and'p' stand for recursion and parameter, respectively.

The bar rule BR now reads as follows. Whenever we have derived the four premises

(1) NDS(f, $\prec$)

(2) $x, y \in \mathsf{N} \to \bar{s^r} \in \mathsf{N} \wedge \bar{s^p} \in \mathsf{N}$

[5] In the formulation of the rules below we use a binary relation $\prec$ whose characteristic function is given by a closed term $t_\prec$ for which $\mathcal{U}_0(\mathsf{FA})$ proves $t_\prec : \mathsf{N}^2 \to \{0, 1\}$. We write $x \prec y$ instead of $t_\prec xy = 0$ and further assume that $\prec$ is a linear ordering with least element 0, provably in $\mathcal{U}_0(\mathsf{FA})$.

(3) $x, y \in \mathsf{N},\ \bigwedge_i [s_i^r \prec x \supset P(s_i^r, s_i^p)] \ \rightarrow\ \bar{t^r} \in \mathsf{N} \wedge \bar{t^p} \in \mathsf{N}$

(4) $x, y \in \mathsf{N},\ \bigwedge_i [s_i^r \prec x \supset P(s_i^r, s_i^p)],\ \bigwedge_j [t_j^r \prec x \supset P(t_j^r, t_j^p)] \ \rightarrow\ P(x, y)$

we can infer $x, y \in \mathsf{N} \ \rightarrow\ P(x, y)$.[6]

The new unfolding system $\mathcal{U}_0(\mathsf{FA} + \mathsf{BR})$ is the extension of $\mathcal{U}_0(\mathsf{FA})$ by this rule.

One of the crucial observations is that whenever we have derived $\mathsf{NDS}(\mathsf{f}, \prec)$ in $\mathcal{U}_0(\mathsf{FA} + \mathsf{BR})$, for a specific ordering $\prec$, then we can use the bar rule BR in order to justify function definitions by *nested* recursion along $\prec$, see Feferman and Strahm [21] for details.

Theorem 3 *Assume that* $\mathsf{NDS}(\mathsf{f}, \prec)$ *is derivable in* $\mathcal{U}_0(\mathsf{FA} + \mathsf{BR})$*. Then* $\mathcal{U}_0(\mathsf{FA} + \mathsf{BR})$ *justifies nested recursion along* $\prec$*.*

In the following let us assume that for each ordinal $\alpha < \varepsilon_0$ we have a standard primitive recursive wellordering $\prec_\alpha$ of ordertype α. Further, let us write $\mathsf{NDS}(\mathsf{f}, \alpha)$ for $\mathsf{NDS}(\mathsf{f}, \prec_\alpha)$. The crucial ingredient of the argument to show that $\mathcal{U}_0(\mathsf{FA} + \mathsf{BR})$ derives $\mathsf{NDS}(\mathsf{f}, \alpha)$ for each $\alpha < \varepsilon_0$ is the famous result by Tait [34] that nested recursion on $\omega\alpha$ entails ordinary recursion on ω^α or, more useful in our setting, nested recursion on $\omega\alpha$ entails $\mathsf{NDS}(\mathsf{f}, \omega^\alpha)$.

Theorem 4 *Provably in* $\mathcal{U}_0(\mathsf{FA} + \mathsf{BR})$*, nested recursion along* $\omega\alpha$ *entails* $\mathsf{NDS}(\mathsf{f}, \omega^\alpha)$*.*

Clearly, $\mathcal{U}_0(\mathsf{FA} + \mathsf{BR})$ proves $\mathsf{NDS}(\mathsf{f}, \omega 2)$ and hence we have nested recursion along $\omega 2$, which in turn entails $\mathsf{NDS}(\mathsf{f}, \omega^2)$; further, nested recursion on ω^2 gives us $\mathsf{NDS}(\mathsf{f}, \omega^\omega)$ and thus nested recursion along $\omega^\omega = \omega(\omega^\omega)$. Then we can derive $\mathsf{NDS}(\mathsf{f}, \omega^{\omega^\omega})$ and so on.

The upshot is that $\mathcal{U}_0(\mathsf{FA} + \mathsf{BR})$ derives $\mathsf{NDS}(\mathsf{f}, \omega_n)$ for each natural number n, where as usual we set $\omega_0 = \omega$ and $\omega_{n+1} = \omega^{\omega_n}$.

Corollary 5 *We have for each* $\alpha < \varepsilon_0$ *that* $\mathcal{U}_0(\mathsf{FA} + \mathsf{BR})$ *derives* $\mathsf{NDS}(\mathsf{f}, \alpha)$*.*

It is not difficult to see that this lower bound is sharp, see Feferman and Strahm [21].

Corollary 6 $\mathcal{U}_0(\mathsf{FA} + \mathsf{BR})$ *is proof-theoretically equivalent to Peano arithmetic* PA*.*

Even the full unfolding system with bar rule, $\mathcal{U}(\mathsf{FA} + \mathsf{BR})$, does not go beyond Peano arithmetic in strength.

Theorem 7 $\mathcal{U}_0(\mathsf{FA} + \mathsf{BR})$*,* $\mathcal{U}_1(\mathsf{FA} + \mathsf{BR})$*, and* $\mathcal{U}(\mathsf{FA} + \mathsf{BR})$ *are all proof-theoretically equivalent to Peano arithmetic* PA*.*

---

[6]In the formulation of this rule, we have used the shorthand $r \prec x \supset A$ for the formula $t_\prec rx = 1 \vee A$.

4 The Unfolding of Feasible Arithmetic

The aim of this section is to discuss the concept of unfolding in the context of a natural schematic system FEA for *feasible arithmetic*. We shall sketch various unfoldings of FEA and indicate their relationship to weak systems of explicit mathematics and partial truth. We follow Eberhard and Strahm [9].

Let us first introduce the basic schematic system FEA of feasible arithmetic. Its intended universe of discourse is the set $\mathbb{W} = \{0, 1\}^*$ of finite binary words and its basic operations and relations include the binary successors S_0 and S_1, the predecessor Pd, the initial subword relation $\subseteq$, word concatenation $\circledast$ as well as word multiplication $\boxtimes$.[7] The logical operations of FEA are conjunction ($\wedge$), disjunction ($\vee$), and bounded existential quantification ($\exists^{\leq}$). As in the case of finitist arithmetic FA, the statements proved in FEA are sequents of formulas in the given language, i.e. implication is allowed at the outermost level.

The language of FEA contains a countably infinite supply $\alpha, \beta, \gamma, \dots$ of variables (possibly with subscripts). These variables are interpreted as ranging over the set of binary words $\mathbb{W}$. We have a constant ϵ for the empty word, three unary function symbols S_0, S_1, Pd and three binary function symbols $\circledast$, $\boxtimes$, $\subseteq$.[8] Terms are defined as usual and are denoted by $\sigma, \tau, \dots$. Further, there is the binary predicate symbol $=$ for equality, and an infinite supply $P, Q, \dots$ of free predicate letters.

The atomic formulas of FEA are of the form $(\sigma = \tau)$ and $P(\sigma_1, \dots, \sigma_n)$. The formulas are closed under $\wedge$ and $\vee$ as well as under bounded existential quantification. In particular, if A is formula, then $(\exists \alpha \leq \tau)A$ is formula as well, where τ is not allowed to contain α. Further, as usual for theories of words, we use $\sigma \leq \tau$ as an abbreviation for $1 \boxtimes \sigma \subseteq 1 \boxtimes \tau$, thus expressing that the length of σ is less than or equal to the length of τ. As before, we use $\bar{\alpha}$, $\bar{\sigma}$, and $\bar{A}$ to denote finite sequences of variables, terms, and formulas, respectively.

FEA is formulated as a system of sequents Σ of the form $\Gamma \rightarrow A$, where Γ is a finite sequence of formulas and A is a formula. Hence, we have the usual Gentzen-type logical axioms and rules of inference for our underlying restricted language, see Eberhard and Strahm [9]. The non-logical axioms of FEA state the usual defining equations for the function symbols of its language, see, e.g., Ferreria [22] for similar axioms. Finally, we have the schematic induction rule formulated for a free predicate P as follows:

$$\frac{\Gamma \rightarrow P(\epsilon) \qquad \Gamma, P(\alpha) \rightarrow P(\mathsf{S}_i(\alpha)) \quad (i = 0, 1)}{\Gamma \rightarrow P(\alpha)}$$

In the various unfolding systems of FEA introduced below, we shall be able to substitute an arbitrary formula for an arbitrary free predicate letter P. Let us now quickly

---

[7] Given two words w_1 and w_2, the word $w_1 \boxtimes w_2$ denotes the length of w_2 fold concatenation of w_1 with itself.

[8] We assume that $\subseteq$ defines the characteristic function of the initial subword relation. Further, we employ infix notation for these binary function symbols.

review the *operational unfolding* $\mathcal{U}_0(\mathsf{FEA})$ of FEA. It tells us which operations from and to individuals, and which principles concerning them, ought to be accepted if one has accepted FEA.

In the operational unfolding, we make these commitments explicit by extending FEA by a partial combinatory algebra. Since it represents any recursion principle and thus any recursive function by suitable terms, it is expressive enough to reflect any ontological commitment we want to reason about. Using the notion of *provable totality*, we single out those functions and recursion principles we are actually committed to by accepting FEA.

The language of $\mathcal{U}_0(\mathsf{FEA})$ is an expansion of the language of FEA including new constants k, s, p, p_0, p_1, d, tt, ff, e, ϵ, s_0, s_1, pd, $\mathsf{c}_\subseteq$, $*$, $\times$, and an additional countably infinite set of variables $x_0, x_1, \ldots$.[9] The new variables are supposed to range over the universe of operations and are usually denoted by $a, b, c, x, y, z, \ldots$. The terms $(r, s, t, \ldots)$ are inductively generated from variables and constants by means of the function symbols of FEA and the application operator $\cdot$. We use the usual abbreviations for applicative terms as before. We have $(s = t)$, $s\downarrow$ and $P(\vec{s})$ as atoms. The formulas $(A, B, C, \ldots)$ are built from the atoms as before using $\vee$, $\wedge$ and the bounded existential quantifier, where as above the bounding term is a term of FEA not containing the bound variable. Finally, we write $s \leq \tau$ for $(\exists \beta \leq \tau)(s = \beta)$.[10]

The axioms and rules of $\mathcal{U}_0(\mathsf{FEA})$ are now spelled out in the expected manner, see Eberhard and Strahm [9] for details. In particular, $\mathcal{U}_0(\mathsf{FEA})$ includes the (meta) substitution rule (Subst$'$). Next we want to show that the polynomial time computable functions can be proved to be total in $\mathcal{U}_0(\mathsf{FEA})$. We call a function $F : \mathbb{W}^n \to \mathbb{W}$ provably total in a given axiomatic system, if there exists a closed term t_F such that (i) t_F defines F pointwise, i.e. on each standard word, and, moreover, (ii) there is a FEA term $\tau[\alpha_1, \ldots, \alpha_n]$ such that the assertion

$$t_F(\alpha_1, \ldots, \alpha_n) \leq \tau[\alpha_1, \ldots, \alpha_n]$$

is provable in the underlying system. Thus, in a nutshell, F is provably total iff it is provably and uniformly bounded.

We use Cobham's characterization of the polynomial time computable functions (cf. [4, 5]): starting off from the initial functions of FEA and arbitrary projections, the polynomial time computable functions can be generated by closing under composition and bounded recursion. In order to show closure under bounded recursion, assume that F is defined by bounded recursion with initial function G and step function H_i $(i = 0, 1)$, where τ is the corresponding bounding polynomial.[11] By the

[9]These variables are syntactically different from the FEA variables $\alpha_0, \alpha_1, \ldots$.

[10]It is important to note that we do not have a predicate W for binary words in our language, since this would allow us to introduce (hidden) unbounded existential quantifiers via formulas of the form $\mathsf{W}(t)$. Thus it is necessary to have two separate sets of variables for words and operations, respectively.

[11]We can assume that only functions built from concatenation and multiplication are permissible bounds for the recursion.

induction hypothesis, G and H_i are provably total via suitable terms t_G and t_{H_i}. Using the recursion or fixed point theorem of the partial combinatory algebra, we find a term t_F which provably in $\mathcal{U}_0(\mathsf{FEA})$ satisfies the following recursion equations for $i = 0, 1$:

$$\begin{aligned} t_F(\bar{\alpha}, \epsilon) &\simeq t_G(\bar{\alpha}) \mid \tau[\bar{\alpha}, \epsilon], \\ t_F(\bar{\alpha}, \mathsf{s}_i(\beta)) &\simeq t_{H_i}(t_F(\bar{\alpha}, \beta), \bar{\alpha}, \beta) \mid \tau[\bar{\alpha}, \mathsf{s}_i(\beta)] \end{aligned}$$

Here $\mid$ is the usual truncation operation such that $\alpha|\beta$ is α if $\alpha \leq \beta$ and β otherwise. Now fix $\bar{\alpha}$ and let $A[\beta]$ be the formula $t_F(\bar{\alpha}, \beta) \leq \tau[\bar{\alpha}, \beta]$ [12] and simply show $A[\beta]$ by induction on β. Thus F is provably total in $\mathcal{U}_0(\mathsf{FEA})$.

Next we shall describe the full predicate unfolding $\mathcal{U}(\mathsf{FEA})$ of FEA. It tells us, in addition, which predicates and operations on predicates ought to be accepted if one has accepted FEA. By accepting FEA one implicitly accepts an equality predicate and operations on predicates corresponding to the logical operations of FEA. Finally, we may accept the principle of forming the predicate for the disjoint union of a (bounded) sequence of predicates given by an operation.

The language of $\mathcal{U}(\mathsf{FEA})$ is an extension of the language of $\mathcal{U}_0(\mathsf{FEA})$ by new individual constants id (identity), inv (inverse image), con (conjunction), dis (disjunction), leq (bounded existential quantifier), and j (bounded disjoint unions); further new constants are of the form pr_P as combinatorial representations of free predicates. Finally, we have a new unary relation symbol Π in order to single out the predicates we are committed to as well as a binary relation symbol $\in$ for elementhood of individuals in predicates. The terms are generated as before but now taking into account the new constants. The formulas of $\mathcal{U}(\mathsf{FEA})$ extend the formulas of $\mathcal{U}_0(\mathsf{FEA})$ by allowing new atomic formulas of the form $\Pi(t)$ and $s \in t$.

The axioms of $\mathcal{U}(\mathsf{FEA})$ extend those of $\mathcal{U}_0(\mathsf{FEA})$ by the expected axioms about predicates, see Eberhard and Strahm [9] for details. Further, the full unfolding $\mathcal{U}(\mathsf{FEA})$ includes axioms stating that a bounded sequence of predicates determines the predicate of the disjoint union of this sequence. We use a set of three inference rules to express the join principle, see [9] for details. The rules of inference of $\mathcal{U}_0(\mathsf{FEA})$ are also available in $\mathcal{U}(\mathsf{FEA})$. In particular, $\mathcal{U}(\mathsf{FEA})$ contains the generalized substitution rule (Subst'): the formulas $\bar{B}$ to be substituted for $\bar{P}$ are now in the language of $\mathcal{U}(\mathsf{FEA})$; as above the rule in the premise of (Subst'), however, is required to be in the language $\mathcal{U}_0(\mathsf{FEA})$. This concludes the description of the predicate unfolding $\mathcal{U}(\mathsf{FEA})$ of FEA. We shall turn to its proof-theoretic strength at the end of this section.

Let us next discuss an alternative way to define the full unfolding of FEA. The truth unfolding $\mathcal{U}_{\mathsf{T}}(\mathsf{FEA})$ of FEA makes use of a truth predicate T which reflects

---

[12] Recall that by expanding the definition of the $\leq$ relation, the formula $A[\beta]$ stands for the assertion $(\exists \gamma \leq \tau[\bar{\alpha}, \beta])(t_F(\bar{\alpha}, \beta) = \gamma)$.

the logical operations of FEA in a natural and direct way. We shall see that the full predicate unfolding $\mathcal{U}(\mathsf{FEA})$ is directly contained in $\mathcal{U}_{\mathsf{T}}(\mathsf{FEA})$.[13]

As before, we want to make the commitment to the logical operations of FEA explicit. This is done by introducing a truth predicate for which truth biconditionals defining the truth conditions of the logical operations hold. The axiomatization of the truth predicate relies on a coding mechanism for formulas. In the applicative framework, this is achieved in a very natural way by using new constants designating the logical operations of FEA. The language of $\mathcal{U}_{\mathsf{T}}(\mathsf{FEA})$ extends the one of $\mathcal{U}_0(\mathsf{FEA})$ by new individual constants $\dot{=}$, $\dot{\wedge}$, $\dot{\vee}$, $\dot{\exists}$, as well as constants of the form pr_P. In addition, It includes a new unary relation symbol T. The terms and formulas are defined in the expected manner. Moreover, we shall use infix notation for $\dot{=}$, $\dot{\wedge}$ and $\dot{\vee}$.

The axioms of $\mathcal{U}_{\mathsf{T}}(\mathsf{FEA})$ extend those of $\mathcal{U}_0(\mathsf{FEA})$ by the following axioms about the truth predicate T:

$$(\dot{=}) \qquad \mathsf{T}(x \dot{=} y) \leftrightarrow x = y$$

$$(\dot{\wedge}) \qquad \mathsf{T}(x \mathbin{\dot{\wedge}} y) \leftrightarrow \mathsf{T}(x) \wedge \mathsf{T}(y)$$

$$(\dot{\vee}) \qquad \mathsf{T}(x \mathbin{\dot{\vee}} y) \leftrightarrow \mathsf{T}(x) \vee \mathsf{T}(y)$$

$$(\dot{\exists}) \qquad \mathsf{T}(\dot{\exists}\alpha x) \leftrightarrow (\exists \beta \le \alpha)\mathsf{T}(x\beta)$$

$$(\mathsf{pr}_P) \qquad \mathsf{T}(\mathsf{pr}_P(\bar{x})) \leftrightarrow P(\bar{x})$$

The generalized substitution rule (Subst′) can be stated in a somewhat more general form for $\mathcal{U}_{\mathsf{T}}(\mathsf{FEA})$, see Eberhard and Strahm [9] for a detailed discussion. It is easy to see that the full predicate unfolding $\mathcal{U}(\mathsf{FEA})$ is contained in the truth unfolding $\mathcal{U}_{\mathsf{T}}(\mathsf{FEA})$. The argument proceeds along the same line as the embedding of weak explicit mathematics into theories of truth in Eberhard and Strahm [8].

In Eberhard and Strahm [9] it is shown how to determine a suitable upper bound for $\mathcal{U}(\mathsf{FEA})$ and $\mathcal{U}_{\mathsf{T}}(\mathsf{FEA})$ thus showing that their provably total functions are indeed computable in polynomial time. There one proceeds via the weak truth theory T_{PT} introduced in Eberhard and Strahm [8] and Eberhard [6, 7], whose detailed and very involved proof-theoretic analysis is carried out in [7]. Thus we have:

Theorem 8 *The provably total functions of* $\mathcal{U}_0(\mathsf{FEA})$, $\mathcal{U}(\mathsf{FEA})$, *and* $\mathcal{U}_{\mathsf{T}}(\mathsf{FEA})$ *are exactly the polynomial time computable functions.*

[13] Recall that in Feferman's original definition of unfolding in [15], a truth predicate is used in order to describe the full unfolding of a schematic system.

5 The Unfolding of One Inductive Definition

Our last example for illustrating the unfolding program stems from a natural schematic system for arithmetical inductive definitions. We shall see that its unfolding corresponds to a generalization $\Psi(\Gamma_{\Omega+1})$ of Γ_0, which we shall describe below. The main result of this section is due to Buchholtz [2].

Let ID_1 be the usual system for one inductive definitions. In order to formulate it in schematic form as an extension of NFA, we have a new predicate constant $P_{\mathcal{A}}$ for each arithmetical operator form $\mathcal{A}[P, x]$ in which P occurs only positively. Then we obtain a schematic version of ID_1 as follows, with P denoting a free predicate variable:

(1) $(\forall x)(\mathcal{A}[P_{\mathcal{A}}, x] \to P_{\mathcal{A}}(x))$
(2) $(\forall x)(\mathcal{A}[P, x] \to P(x)) \to (\forall x)(P_{\mathcal{A}}(x) \to P(x))$.

The full unfolding $\mathcal{U}(\mathsf{ID}_1)$ of ID_1 is now defined according to the procedure described in detail for the case of NFA, with the only exception that the join axiom is formulated in a somewhat more general form: the family of predicates to which the join operation is applied, is not restricted to be indexed by the natural numbers N but by an arbitrary predicate p.

In order to describe the result about the strength of $\mathcal{U}(\mathsf{ID}_1)$ obtained in Buchholtz's thesis [2], let us review some basic ordinal theory needed to calibrate the proof-theoretic ordinal of $\mathcal{U}(\mathsf{ID}_1)$. Let Ω stand for $\aleph_1$. Then the sets $B_\Omega(\alpha)$ and ordinals $\Psi_\Omega(\alpha)$ are defined recursively as follows: $B_\Omega(\alpha)$ is the closure of $\{0, \Omega\}$ under $+$, the binary Veblen function φ, and $(\xi \mapsto \Psi_\Omega(\xi))_{\xi<\alpha}$; moreover,

$$\Psi_\Omega(\alpha) \simeq \min\{\xi < \Omega : \xi \notin B(\alpha)\}.$$

It is seen that $\Psi_\Omega(\alpha)$ is always defined and, hence, denotes an ordinal smaller than Ω. Finally, let $\Gamma_{\Omega+1}$ be the least Γ number beyond Ω. In the following, we are interested in the ordinal $\Psi_\Omega(\Gamma_{\Omega+1})$, which we simply denote by $\Psi(\Gamma_{\Omega+1})$. The main result of Buchholtz' thesis is the following

Theorem 9 *The proof-theoretic ordinal of* $\mathcal{U}(\mathsf{ID}_1)$ *is* $\Psi(\Gamma_{\Omega+1})$.

More recently, a number of natural systems have been identified whose proof-theoretic ordinal is $\Psi(\Gamma_{\Omega+1})$, see Buchholtz, Jäger, and Strahm [3]. Basically, those systems arise from natural systems of second order arithmetic of strength Γ_0 by allowing one generalized inductive definition at the bottom level, resulting in analogues of Δ^1_1 comprehension, Σ^1_1 choice and dependent choice, always with substitution rule, as well as Friedman's arithmetical transfinite recursion.

6 Concluding Remarks

One of the motivations in Feferman [15] for studying the unfolding of a schematic system S was to explicate some ideas that were initiated by Gödel regarding axioms for hierarchies of inaccessible and Mahlo cardinals. Gödel [24], p. 182 writes that "these axioms show clearly, not only that the axiomatic system of set theory as known today is incomplete, but also that it can be supplemented without arbitrariness by new axioms which are only the natural continuation of those set up so far".

Feferman [15] proposes a number of schematic systems for impredicative and admissible set theory. He further introduces a schematic reflection principle, so-called *Downward Reflection*, expressing that *whatever holds in the universe of sets already holds in arbitrary large transitive sets*. This principle entails a form of Bernays' downward second order reflection principle, from which the existence of hierarchies of Mahlo cardinals follows.

The unfolding systems for set theory mentioned above may also be directly expressed in the language of Feferman's operational set theory OST (cf. [16, 25]). We refer the reader to Feferman [18] for a number of interesting conjectures regarding various unfoldings of OST.

References

1. Beeson, M.J.: Foundations of Constructive Mathematics: Metamathematical Studies. Springer, Berlin (1985)
2. Buchholtz, U.: Unfolding of systems of inductive definitions. Ph.D. thesis, Stanford University (2013)
3. Buchholtz, U., Jäger, G., Strahm, T.: Theories of proof-theoretic strength $\Psi(\Gamma_{\Omega+1})$. In: Probst, D., Schuster, P. (eds.) Concepts of Proof in Mathematics, Philosophy, and Computer Science of Ontos Mathematical Logic. vol. 6, pp. 115–140. De Gruyter (2016)
4. Clote, P.: Computation models and function algebras. In: Griffor, E. (ed.) Handbook of Computability Theory, pp. 589–681. Elsevier (1999)
5. Cobham, A.: The intrinsic computational difficulty of functions. In: Logic, Methodology and Philosophy of Science II, pp. 24–30. North Holland, Amsterdam (1965)
6. Eberhard, S.: Weak Applicative Theories, Truth, and Computational Complexity. Ph.D. thesis, Universität Bern (2013)
7. Eberhard, S.: A feasible theory of truth over combinatory algebra. Ann. Pure Appl. Log. **165**(5), 1009–1033 (2014)
8. Eberhard, S., Strahm, T.: Weak theories of truth and explicit mathematics. In: Berger, U., Diener, H., Schuster, P., Seisenberger, M. (eds.) Logic, Construction, Computation, pp. 157–184. Ontos Verlag (2012)
9. Eberhard, S., Strahm, T.: Unfolding feasible arithmetic and weak truth. In: Achourioti, D., Galinon, H., Fujimoto, K., Martinez, J. (eds.) Unifying the Philosophy of Truth, pp. 153–167. Springer (2015)
10. Feferman, S.: Transfinite recursive progressions of axiomatic theories. J. Symb. Log. **27**, 259–316 (1962)
11. Feferman, S.: Systems of predicative analysis. J. Symb. Log. **29**(1), 1–30 (1964)
12. Feferman, S.: A language and axioms for explicit mathematics. In: Crossley, J. (ed.) Algebra and Logic of Lecture Notes in Mathematics, vol. 450, pp. 87–139. Springer (1975)

13. Feferman, S.: A more perspicuous system for predicativity. In: Konstruktionen vs. Positionen I. de Gruyter, pp. 87–139. Berlin (1979)
14. Feferman, S.: Reflecting on incompleteness. J. Symb. Log. **56**(1), 1–49 (1991)
15. Feferman, S.: Gödel's program for new axioms: Why, where, how and what? In: Hájek, P. (ed.) Gödel 1996 of Lecture Notes in Logic, vol. 6, pp. 3–22. Springer, Berlin (1996)
16. Feferman, S.: Operational set theory and small large cardinals. Inf. Comput. **207**, 971–979 (2009)
17. Feferman, S.: Turing's Thesis: Ordinal logics and oracle computability. In: Cooper, S.B., Van Leeuwen, J. (eds.) Alan Turing: His Work and Impact, pp. 145–150. Elsevier (2013)
18. Feferman, S.: The operational perspective: three routes. In: Kahle, R., Strahm, T., Studer, T. (eds.) Advances in Proof Theory of Progress in Computer Science and Applied Logic, vol. 28, pp. 269–289. Birkhäuser (2016)
19. Feferman, S., Spector, C.: Incompleteness along paths in progressions of theories. J. Symb. Log. **27**, 383–390 (1962)
20. Feferman, S., Strahm, T.: The unfolding of non-finitist arithmetic. Ann. Pure Appl. Log. **104**(1–3), 75–96 (2000)
21. Feferman, S., Strahm, T.: Unfolding finitist arithmetic. Rev. Symb. Log. **3**(4), 665–689 (2010)
22. Ferreira, F.: Polynomial time computable arithmetic. In: Sieg, W. (ed.) Logic and Computation, Proceedings of a Workshop held at Carnegie Mellon University, 1987, Contemporary Mathematics, vol. 106, pp. 137–156. American Mathematical Society, Providence, Rhode Island (1990)
23. Franzén, T.: Transfinite progressions: a second look at completeness. Bull. Symb. Log. **10**(3), 367–389 (2004)
24. Gödel, K.: Collected Works. In: Feferman, S. et al., (eds.) Vol. II. Oxford University Press, New York (1990)
25. Jäger, G.: On Feferman's operational set theory OST. Ann. Pure Appl. Log. **150**(1–3), 19–39 (2007)
26. Kreisel, G.: Ordinal logics and the characterization of informal concepts of proof. In: Proceedings International Congress of Mathematicians, 14–21 August 1958, pp. 289–299. Cambridge University Press, Cambridge (1960)
27. Kreisel, G.: Mathematical logic. In: Saaty, T.L. (ed.) Lectures on Modern Mathematics, vol. 3, pp. 95–195. Wiley, New York (1965)
28. Kreisel, G.: Principles of proof and ordinals implicit in given concepts. In: Kino, A., Myhill, J., Vesley, R.E. (eds.) Intuitionism and Proof Theory, pp. 489–516. North Holland, Amsterdam (1970)
29. Kripke, S.: Outline of a theory of truth. J. Philos. **72**(19), 690–716 (1975)
30. Rathjen, M.: The role of parameters in bar rule and bar induction. J. Symb. Log. **56**(2), 715–730 (1991)
31. Schütte, K.: Eine Grenze für die Beweisbarkeit der transfiniten Induktion in der verzweigten Typenlogik. Archiv für Mathematische Logik und Grundlagen der Mathematik **7**, 45–60 (1964)
32. Sieg, W.: Herbrand analyses. Arch. Math. Log. **30**(5+6), 409–441 (1991)
33. Strahm, T.: The non-constructive μ operator, fixed point theories with ordinals, and the bar rule. Ann. Pure Appl. Log. **104**(1–3), 305–324 (2000)
34. Tait, W.: Nested recursion. Mathematische Annalen **143**, 236–250 (1961)
35. Tait, W.: Finitism. J. Philos. **78**, 524–546 (1981)
36. Turing, A.: Systems of logic based on ordinals. Proc. London Math. Soc. 2nd Ser. **45**(I), 161–228 (1939)

Author Biography

Thomas Strahm is a teacher and researcher at the Institute of Computer Science in the Faculty of Science of the University of Bern, Switzerland. He is a member of the Logic and Theory Group. His principal research interests include mathematical logic and theoretical computer science, more specifically proof theory, explicit mathematics, the unfolding program, predicativity and metapredicativity, higher type functionals, as well as applicative theories and computational complexity. His teaching activities range from subjects in mathematical logic and theoretical computer science to applied courses in computer science.

Iterated Inductive Definitions Revisited

Wolfram Pohlers

Abstract In this paper we revisit our contribution to Lecture Note 897 and present a computation of the prooftheoretic ordinals of formal theories for iterated inductive definitions together with a characterization of their provably recursive functions. The techniques used here differ essentially from the original ones. Sections 1–5 contain a general survey of the connections between recursion theoretic and proof theoretic ordinals and roughly recap some basic facts on iterated inductive definitions. Beginning with Sect. 6 the paper becomes more technical. After introducing the semi–formal system for iterated inductive definition and the necessary ordinals we compute the *ordinal spectrum* of the formal theory for iterated inductive definitions and characterize their provably total functions in terms of a subrecursive hierarchy. The last section briefly discusses the foundational aspects of the paper.

Keywords Inductive definitions · Cut–elimination · Ordinal analysis
Provably recursive functions.

2010 Mathematics Subjects Classification 03F03 · 03F05 · 03F15 · 03D20 · 03D60 · 03D70

1 Introduction

My first meeting with Solomon Feferman was at a workshop in Tübingen 1973. This workshop was part of a series of logic workshops sponsored by the Volkswagenstiftung to foster mathematical logic in Germany. Sol's lectures on proof theory at this workshop impressed me lastingly. I had just passed my doctoral examination with a thesis based on Takeuti's treatment of second order arithmetic with Π^1_1–comprehension and I was eager to study impredicative theories further. Of course I did by far not understand everything Sol was talking about but I took notes. These

W. Pohlers (✉)
Institute for Mathematical Logic and Foundational Research Universität Münster,
Einsteinstrasses 62, D-48149 Münster, Germany
e-mail: pohlers@uni-muenster.de

G. Jäger and W. Sieg (eds.), *Feferman on Foundations*, Outstanding
Contributions to Logic 13, DOI 10.1007/978-3-319-63334-3_9

notes became my guidelines for a long time. To my remembrance it was the first time I learned about the proof theoretical aspects of iterated inductive definitions. I soon realized that my knowledge of Takeuti's methods could help to determine their prooftheoretic ordinals which lead to an upper bound first for finite iterations [27] and later also for transfinite iterations [32]. Jointly with Wilfried Buchholz we could show that these bounds are the precise ones [9]. It was again Sol Feferman—I think at a conference in Oxford—who gave me the essential hint that these results should also incorporate a reduction of the theories for inductive definitions based on classical logic to iterated accessibility inductive definitions based on intuitionistic logic.[1] All these results were then collected in my Habilitationsschrift.

There was, however, an essential drawback. Takeuti's reduction procedure, on which these results rested, was extremely opaque. Due to Sol Feferman's permanent nagging we in Munich looked therefore for more perspicuous methods to obtain these results. These efforts culminated in Lecture Note 897 by the four authors Buchholz, Feferman, Pohlers and Sieg.

Since the appearance of Lecture Note 897 in 1981 the techniques for ordinal analyses have been essentially improved. When being invited to contribute to the present volume with a paper around inductive definitions I revisited the old volume. This brought back to me all the technical difficulties with which I had to struggle in my then contribution and I realized that the paper is actually outdated and because of all its technicalities difficult to read. Therefore I decided to contribute a modernized version of—at least parts of—the old results with emphasis on its recursion theoretic aspects.

Following the suggestion of the anonymous referee I split my contribution into two parts. First a brief overview of the connection between abstractly defined recursion theoretic ordinals (GRT–ordinals) for an abstract structure $\mathfrak{M}$ and their prooftheoretic counterparts for axiomatizations for $\mathfrak{M}$. This includes a brief sketch of the abstract theory of generalized inductive definitions. We hope that this part will also be accessible for readers with a general background in abstract recursion theory who are non–experts in proof theory.

In the second part—beginning with Sect. 6—I will become more technical and recompute the prooftheoretic ordinals of the theory $\mathbf{ID}_\nu$ of ν–fold iterated positive inductive definitions. The main tool there are operator controlled semi–formal derivations, an improvement[2] of the technique of local predicativity which was the basis of the contribution in LN 897. Completely new is the technique to obtain a characterization of the provably recursive functions of $\mathbf{ID}_\nu$ which in contrast to the contribution in LN 897 does not need a formalization of the ordinal analysis but is obtained by an elimination of *all* abstract rules in semi–formal derivations of Σ^0_1–sentences with parameters as discussed in [38]. This technique roots in papers by Weiermann and Blankertz (cf. [3, 4]).

---

[1]A problem that was and apparently still is in the focus of interest. Cf. e.g. [1, 6, 14].

[2]This technique has been introduced by Wilfried Buchholz in [8].

The main emphasis of this revision is on some recursion–theoretic aspects of the proof theory of iterated inductive definitions. Their foundational aspect is widely neglected. In a last section I will therefore comment on some of the foundational aspects of this paper.

2 GRT–Ordinals and Their Prooftheoretic Equivalents

Given an abstract structure $\mathfrak{M}$ there is a series of abstractly defined ordinals which are—in a sense to be discussed later—characteristic for the structure $\mathfrak{M}$. Best known are the ordinals $\sigma^i_n(\mathfrak{M})$, $\pi^i_n(\mathfrak{M})$ and $\delta^i_n(\mathfrak{M})$ which are the suprema of the ordertypes of wellorderings which are definable by a Σ^i_n, Π^i_n or Δ^i_n–formulas in the language for $\mathfrak{M}$. Moreover these ordinals also return in completely different contexts, e.g. as ordinals which characterize the abstract recursion theory of "universes" above $\mathfrak{M}$.

The most prominent example for such an ordinal is ω_1^{CK}, the first ordinal that cannot be represented by a recursive wellordering on the natural numbers. It is well–known that $\omega_1^{CK} = \delta^0_0(\mathbb{N}) = \delta^1_0(\mathbb{N}) = \sigma^1_1(\mathbb{N})$ and that ω_1^{CK} characterizes the recursion theory of hyperarithmetical functions.

The wellfoundedness of a binary relation $\prec$ on an abstract structure $\mathfrak{M}$ is expressed by the Π^1_1–sentence

$$(\forall X) Wf(X, \prec) \;:\Leftrightarrow\; (\forall X)\big[(\forall z)[(\forall y \prec z) y \in X \to z \in X] \to (\forall x)[x \in X]\big]$$

where for simplicity we assume that the field of $\prec$ is $\mathfrak{M}$. By the *elementary language* of $\mathfrak{M}$ we understand its first order language where we generally allow free second order variables. In abstraction of the paradigmatic example $\mathbb{N}$ we call an abstract structure *acceptable* if it contains an elementarily definable copy of the natural numbers and an elementarily definable coding machinery.[3] Because of the coding machinery of an acceptable structure we will not take care of the arity of second order variables and just talk about *set variables*. It is a standard result in generalized recursion theory that defining

$$\delta^{\mathfrak{M}} := \sup\{\mathrm{otyp}(\prec) \mid \prec \text{ is elementarily definable in } \mathfrak{M} \wedge \mathfrak{M} \models (\forall X) Wf(X, \prec)\}$$

we obtain $\delta^{\mathfrak{M}} = \delta^1_0(\mathfrak{M}) = \sigma^1_1(\mathfrak{M})$ for countable acceptable structures $\mathfrak{M}$. In case that $\mathfrak{M}$ allows for a primitive recursive coding machinery we may even replace "elementarily definable" by "primitive–recursively (i.e., by a Δ^0_0–formula) definable" in the definition of $\delta^{\mathfrak{M}}$ and still obtain $\delta^{\mathfrak{M}} = \delta^0_0(\mathfrak{M})$.

Given an axiomatization T for the structure $\mathfrak{M}$ the obvious proof theoretic equivalent of $\delta^{\mathfrak{M}}$ is the ordinal

$$\delta^{\mathfrak{M}}(\mathsf{T}) := \sup\{\mathrm{otyp}(\prec) \mid \prec \text{ is elementarily definable in } \mathfrak{M} \wedge \mathsf{T} \vdash Wf(X, \prec)\}. \quad (1)$$

[3] Cf. [25] Exercise 1.6. and Chap. 8.

To stay within the realm of first order logic we have to replace the Π^1_1–sentence expressing the wellfoundedness of $\prec$ by the by the *pseudo–Π^1_1–sentence $Wf(X, \prec)$*. Pseudo Π^1_1–sentences are formulas in the first order language of $\mathfrak{M}$ in which at most set variables are allowed to appear freely. Semantically we treat pseudo–Π^1_1–sentences as Π^1_1–sentences, i.e.,

$$\mathfrak{M} \models F(X) \;:\Leftrightarrow\; \mathfrak{M} \models F(X)[S] \text{ for all } S \subseteq \mathfrak{M}.$$

In the paradigmatic case of the axiom system PA of Peano arithmetic for the standard structure $\mathbb{N}$ of arithmetic—which possesses a primitive recursive coding machinery—the ordinal $\delta^{\mathbb{N}}(\mathsf{PA})$ defined in (1) is apparently the familiar prooftheoretic ordinal of PA commonly denoted by $|\mathsf{PA}|$.

There is another abstractly defined ordinal $\pi^{\mathfrak{M}}$ that can be regarded as characteristic for a countable structure $\mathfrak{M}$.[4] Its definition needs the notion of a semi–formal system for $\mathfrak{M}$–logic. To give a first rough impression of a semi–formal system for $\mathfrak{M}$–logic—the details are in Sect. 6—it is a derivation in a, say one–sided, sequent calculus for the first order logic of $\mathfrak{M}$ augmented by the

($\mathfrak{M}$–rule) $\quad \vdash^{\alpha_m}_{\rho} \Gamma, F(\underline{m})$ and $\alpha_m < \alpha$ for all $m \in \mathfrak{M}$ implies $\vdash^{\alpha}_{\rho} \Gamma, (\forall x)F(x)$.

To take also syntactically care of set variables in the semi–formal calculus we need the

(X–rule) If s and t are closed terms in the language of $\mathfrak{M}$ such that $s^{\mathfrak{M}} = t^{\mathfrak{M}}$ then $\vdash^{\alpha}_{\rho} \Delta, s \in X, t \notin X$ holds true for all ordinals α and ρ.

The semi–formal system for $\mathfrak{M}$–logic derives finite sets Δ of pseudo–Π^1_1–sentences which should be interpreted as finite disjunctions. Its derivations are represented by countably branching wellfounded trees. The notion $\vdash^{\alpha}_{\rho} \Delta$ denotes that there is a semi–formal derivation whose length is bounded by α and whose cuts are all of complexity strictly less than ρ.

The central theorem of $\mathfrak{M}$–logic is the $\mathfrak{M}$–completeness theorem.

Theorem 2.1 *Let $(\forall X)F(X)$ be a Π^1_1–sentence in the language of a countable structure $\mathfrak{M}$. Then $\mathfrak{M} \models (\forall X)F(X)$ iff there is an ordinal $\alpha < \omega_1$ such that $\vdash^{\alpha}_{0} F(X)$.*[5]

Using Theorem 2.1 we can define a truth complexity for the pseudo–Π^1_1–sentences in the language of $\mathfrak{M}$ by

$$\mathrm{tc}(F) := \begin{cases} \min\{\xi \mid \vdash^{\xi}_{0} F\} & \text{if this exists} \\ \omega_1 & \text{otherwise.} \end{cases}$$

Putting

$$\mathrm{tc}((\forall X)F(X)) := \mathrm{tc}(F(X))$$

[4] For more details cf. [31].

[5] For a proof cf. e.g. [31] Theorem 3.5 or [35] Theorem 5.4.9 which treats the case $\mathfrak{M} = \mathbb{N}$.

the above definition extends to a definition of the truth complexities for the Π^1_1–sentences in the language of $\mathfrak{M}$. Using truth complexities we define the Π^1_1–ordinal of the structure $\mathfrak{M}$ by

$$\pi^{\mathfrak{M}} := \sup\{\mathrm{tc}(F) \mid \mathfrak{M} \models F\}. \tag{2}$$

Given an axiomatization T for a countable acceptable structure $\mathfrak{M}$ there is an obvious proof theoretic equivalent

$$\pi^{\mathfrak{M}}(\mathsf{T}) := \sup\{\mathrm{tc}(F) \mid \mathsf{T} \vdash F\} \le \pi^{\mathfrak{M}} \tag{3}$$

for the ordinal $\pi^{\mathfrak{M}}$. The distance between $\pi^{\mathfrak{M}}(\mathsf{T})$ and $\pi^{\mathfrak{M}}$ then serves as a measure of the performance of the axiom system T with respect to its target structure $\mathfrak{M}$. We say that an axiom system T' has an improved performance if this distance becomes smaller, i.e., if $\pi^{\mathfrak{M}}(\mathsf{T}) < \pi^{\mathfrak{M}}(\mathsf{T}')$.

The connection between the ordinals $\delta^{\mathfrak{M}}$ and $\pi^{\mathfrak{M}}$ is given by another central property of $\mathfrak{M}$–logic, the Abstract Boundedness Theorem.

Theorem 2.2 (Abstract Boundedness Theorem) *Let $\mathfrak{M}$ be a countable structure and $\prec$ a transitive wellfounded binary relation on $\mathfrak{M}$. Then*

$$\mathrm{otyp}(\prec) \le \mathrm{tc}((\forall X)\mathit{Wf}(X, \prec)).$$

If $\mathrm{otyp}(\prec)$ *is a limit ordinal we have equality.*

As an immediate consequence of the Boundedness Theorem we get $\delta^{\mathfrak{M}} \le \pi^{\mathfrak{M}}$ as well as $\delta^{\mathfrak{M}}(\mathsf{T}) \le \pi^{\mathfrak{M}}(\mathsf{T})$ for axiomatizations T of countable structures $\mathfrak{M}$. If $\mathfrak{M}$ allows for a coding machinery we also get the converse inequalities. Call an axiomatization T for an acceptable structure $\mathfrak{M}$ acceptable if its proves all the properties of the coding machinery of $\mathfrak{M}$.

Lemma 2.3 *For an acceptable countable structure $\mathfrak{M}$ we have $\delta^{\mathfrak{M}} = \pi^{\mathfrak{M}}$. If T is an acceptable axiomatization of $\mathfrak{M}$ we also get $\delta^{\mathfrak{M}}(\mathsf{T}) = \pi^{\mathfrak{M}}(\mathsf{T})$.*

The Boundedness Theorem goes back to a theorem of Gentzen in [15]. Though we do not want to reprove it here,[6] we mention (a coarsened form of) the lemma which is essential for its proof.

Lemma 2.4 *If $\mathop{\vdash}\limits^{\alpha}_{0} \neg(\forall x)[(\forall y)[y \prec x \rightarrow y \in X] \rightarrow x \in X], \Delta(X)$ for a finite set of X–positive pseudo Π^1_1–sentences $\Delta(X)$ then $\mathfrak{M} \models \bigvee\Delta(X)[\prec\restriction \alpha]$ where $\prec\restriction\alpha = \{x \in \mathit{field}(\prec) \mid |x|_\prec < \alpha\}$.*

The Boundedness Theorem is clearly an immediate consequence of Lemma 2.4. It has, however, another important consequence.

Theorem 2.5 *Let T be an acceptable axiomatization for a countable acceptable structure $\mathfrak{M}$ and T' an extension of T by a finite set of sentences that are true in $\mathfrak{M}$. Then $\delta^{\mathfrak{M}}(\mathsf{T}) = \delta^{\mathfrak{M}}(\mathsf{T}')$.*

---

[6]Proofs can be found in [2, 35], Sect. 6.6 and [31] Sect. 5.2.

Proof Let $\mathsf{T}' = \mathsf{T} \cup \{A\}$ for a true sentence A and assume $\mathsf{T}' \vdash Wf(X, \prec)$. Then $\mathsf{T} \vdash \neg A \vee Wf(X, \prec)$. Unraveling this proof into a semi–formal derivation yields

$$\vdash_0^{\alpha} \neg(\forall x)[(\forall y)[y \prec x \to y \in X] \to x \in X], (\forall x \in field(\prec))[x \in X], \neg A$$

for some ordinal $\alpha < \pi^{\mathfrak{M}}(\mathsf{T}) = \delta^{\mathfrak{M}}(\mathsf{T})$. According to Lemma 2.4 this yields

$$\mathfrak{M} \models (\forall x \in field(\prec))[|x|_{\prec} < \alpha] \vee \neg A.$$

Since $\neg A$ is false we have $\mathrm{otyp}(\prec) \leq \alpha \leq \delta^{\mathfrak{M}}(\mathsf{T})$. Hence $\delta^{\mathfrak{M}}(\mathsf{T}') \leq \delta^{\mathfrak{M}}(\mathsf{T})$ and the converse inequality holds trivially. □

It follows from Theorem 2.5 that augmenting an axiom system by true elementary sentences (even true Σ_1^1–sentences) in the language of $\mathfrak{M}$ will no improve its performance. To improve the performance of an axiom system for $\mathfrak{M}$ we thus need axioms talking about a universe above $\mathfrak{M}$. The most natural way—as we see it today—is to axiomatize a set universe above $\mathfrak{M}$. However, probably for historical reasons—going back to Hilbert's second problem in his famous 1900 list of mathematical problems and the fact that real numbers can be viewed as subsets of the natural numbers—first attempts aimed at axiomatizations of $\mathrm{Pow}(\mathfrak{M})$. Since there is no completeness theorem for full second order logic there is no hope to obtain an axiomatization for the full powerset of $\mathfrak{M}$. But even the restriction to weak second order logic in which the second order quantifiers are supposed to range over all definable subsets of the powerset of $\mathfrak{M}$ is—at least for our present knowledge—much too ambitious.[7] A viable attack is to axiomatize smaller universes above $\mathfrak{M}$ with sufficiently good closure properties.

Examples for such universes are Spector classes above $\mathfrak{M}$.[8] A Spector class Γ above an acceptable structure $\mathfrak{M}$ is a class of relations on $\mathfrak{M}$ satisfying:

- Γ is closed under positive Boolean operations, first order quantifications over $\mathfrak{M}$ and trivial combinatorial substitutions,
- Γ is parameterized, i.e., it contains a universal relation,
- Γ is normed, i.e., for every relation $R \in \Gamma$ there is an ordinal λ and a mapping $\sigma_R \colon R \longrightarrow \lambda$ onto λ such that the relations

$$x \preceq^* y \;:\Leftrightarrow\; x \in R \wedge [y \notin R \vee \sigma_R(x) \leq \sigma_R(y)]$$
$$x \prec^* y \;:\Leftrightarrow\; x \in R \wedge [y \notin R \vee \sigma_R(x) < \sigma_R(y)]$$

  are in Γ. We call the relations $\preceq^*$ and $\prec^*$ the prewellorderings of Γ.
- By Δ we denote the relations in $\Gamma \cap \neg\Gamma$, i.e. $R \in \Delta$ iff $R \in \Gamma$ and there is a relation $\breve{R} \in \Gamma$ such that $x \notin R \;\Leftrightarrow\; x \in \breve{R}$.

[7]The axiom system for weak second order logic above the axiomatization PA for the structure $\mathbb{N}$ of arithmetic, i.e. PA plus the full comprehension scheme, is also known as classical analysis.
[8]Cf. [25] Chap. 9.

For a Spector class Γ let

$$o^{\Gamma} := \sup\{\sigma_R(x) + 1 \,|\, x \in R \in \Gamma\}. \tag{4}$$

Then o^{Γ} is the first ordinal which cannot be represented by a relation in Δ.

Spector classes are structures which allow for an abstract computation theory.[9] Let Γ be a Spector class above a countable acceptable structure $\mathfrak{M}$. A partial function $f : \mathfrak{M} \longrightarrow \mathfrak{M}$ is Γ–partial recursive if its graph belongs to Γ. Without going into details we mention that the Γ–partial recursive functions are strongly connected to o^{Γ}–partial recursive functions. We will therefore call them occasionally also o^{Γ}–partial recursive functions.

Given an axiomatization T for a Spector class Γ we obtain the prooftheoretic equivalent of o^{Γ} as the ordinal

$$o^{\Gamma}(\mathsf{T}) := \sup\{\sigma_R(x) + 1 \,|\, R \in \Gamma \wedge \mathsf{T} \vdash x \in R\}.$$

If T axiomatizes a hierarchy Θ of Spector classes above a countable acceptable structure $\mathfrak{M}$ we can extend this to a spectrum of ordinals

$$\mathit{Spec}^{\mathfrak{M}}(\mathsf{T}) := \{o^{\Gamma}(\mathsf{T}) \,|\, \Gamma \subseteq \Theta \text{ is a Spector class above } \mathfrak{M}\} \cup \{\delta^{\mathfrak{M}}(\mathsf{T})\}. \tag{5}$$

We will see that the points $o^{\Gamma}(\mathsf{T})$ in $\mathit{Spec}^{\mathfrak{M}}(\mathsf{T})$ involve closure conditions for the Γ–recursive functions whose totality is provable in T.

Remark 2.6 The free set variables in the definition of the ordinals $\pi^{\mathfrak{M}}$ and $\pi^{\mathfrak{M}}(\mathsf{T})$ are mandatory. Any elementary sentence in the language of $\mathfrak{M}$ possesses a finite truth complexity. We thus would always get $\pi^{\mathfrak{M}} = \omega$ if we ban free set variables.

Also in the definition of the ordinal $\delta^{\mathfrak{M}}$ and $\delta^{\mathfrak{M}}(\mathsf{T})$ free set variables are mandatory in principle. However, one sometimes sees the definition

$$|\mathsf{T}| := \sup\left\{\mathrm{otyp}(\prec) \;\middle|\; \begin{array}{l} \prec \text{ is elementarily definable and} \\ \mathsf{T} \vdash \mathit{Wf}(F, \prec) \text{ for all formulas } F \text{ in the language of } \mathfrak{M} \end{array}\right\}$$

of the prooftheoretic ordinal of a formal theory T in which the pseudo Π^1_1–sentence $\mathit{Wf}(X, \prec)$ is replaced by the scheme $\mathit{Wf}(F, \prec)$. From a strict standpoint this is problematic. The GRT equivalent of the above definition would be

$$\sup\left\{\mathrm{otyp}(\prec) \;\middle|\; \begin{array}{l} \prec \text{ is elementarily definable and} \\ \mathfrak{M} \models \mathit{Wf}(F, \prec) \text{ for all } F \text{ in the language of } \mathfrak{M} \end{array}\right\},$$

i.e.

$$\sup\{\mathrm{otyp}(\prec) \,|\, \prec \text{ is elementarily definable } \wedge (\mathfrak{M}, \mathfrak{E}) \models (\forall X)\mathit{Wf}(X, \prec)\}$$

[9] Cf. [24].

where $\mathfrak{E}$ is the collection of all subsets of $\mathfrak{M}$ which are elementarily definable. The structure $(\mathfrak{M}, \mathfrak{E})$, however, is in general not a β–model, i.e., $(\mathfrak{M}, \mathfrak{E}) \models (\forall X)Wf(X, \prec)$ will in general not guarantee the wellfoundedness of $\prec$. To make the above definition of $|\mathsf{T}|$ sensible we have to anticipate that $\prec$ is a wellordering—which does in general not follow from the scheme $\mathsf{T} \vdash Wf(F, \prec)$. With this anticipation we get $\delta^{\mathfrak{M}}(\mathsf{T}) \leq |\mathsf{T}|$, since $\mathsf{T}(X) \vdash Wf(X, \prec)$ entails $\mathsf{T} \vdash Wf(F, \prec)$ for all elementary formulas of $\mathfrak{M}$.[10] On the other hand there is an elementary formula G and and a wellordering $\prec$ of ordertype $\geq \pi^{\mathfrak{M}}(\mathsf{T}) = \delta^{\mathfrak{M}}(\mathsf{T})$ such that $\mathsf{T} + Wf(G, \prec)$ proves the consistency of T. By Gödel's second incompleteness theorem we thus have $\mathrm{otyp}(\prec) < \delta^{\mathfrak{M}}(\mathsf{T})$ whenever $\mathsf{T} \vdash Wf(F, \prec)$. Hence $|\mathsf{T}| \leq \delta^{\mathfrak{M}}(\mathsf{T})$ and both ordinals coincide.

Nevertheless, even if the language of a formal theory T is supposed not to comprise free set variables we shall always assume that $\delta^{\mathfrak{M}}(\mathsf{T})$ is defined in the conservative extension $\mathsf{T}(X)$ of T in which set variables are axiomatized by $(\forall x)(\forall y)[x = y \to x \in X \to y \in X]$.

Finally we want to remark that the definition of the ordinals o^{Γ}, $o^{\Gamma}(\mathsf{T})$ and $Spec^{\mathfrak{M}}(\mathsf{T})$ do not need second order variables.

3 A Brief Summary of Abstract Inductive Definitions

The notion of a Spector class is too general as to be directly accessible to an ordinal analysis. There are too many examples for abstract Spector classes above an acceptable structure $\mathfrak{M}$.[11] In order to obtain ordinal analyses for axiomatizations of Spector classes we have to specify them further.

The least Spector class above a countable acceptable structure $\mathfrak{M}$ is the structure $\mathbf{ID}(\mathfrak{M}) = (\mathfrak{M}, \mathbf{\Gamma}(\mathfrak{M}))$ where $\mathbf{\Gamma}(\mathfrak{M})$ is the class of all relations which are positive inductively definable above $\mathfrak{M}$.[12] To clarify notations we give a brief summary of some basic facts of abstract inductive definitions. For a full account see [25].

Let $\mathfrak{M}$ be an acceptable abstract structure.[13] By an abstract inductive definition on $\mathfrak{M}$ we understand a monotone operator $\Phi\colon \mathrm{Pow}(\mathfrak{M}) \longrightarrow \mathrm{Pow}(\mathfrak{M})$.[14]

Such operators possess a least fixed–point I_Φ which is the intersection of all subsets of the domain of $\mathfrak{M}$ that are closed under Φ, i.e.,

$$I_\Phi = \bigcap \{X \subseteq \mathfrak{M} \mid \Phi(X) \subseteq X\}. \tag{6}$$

[10] Observe that the opposite direction is false in general. It is possible that $\mathsf{T} \vdash F(G)$ for every formula G with a proof depending on G whilst $\mathsf{T}(X) \nvdash F(X)$ since this would require a uniform proof. An effect which reminds of (and is connected to) the ω–defect of the first order language of T. Whilst $\mathsf{T} \vdash F(t)$ for any closed first order term t we cannot get $\mathsf{T} \vdash F(x)$ for a free variable x.

[11] Cf. [23].

[12] Cf. [25] Corollary 9A.3.

[13] In slight abuse of notation we will denote by $\mathfrak{M}$ both: the structure and its domain.

[14] Due to the fact that we have an elementary pairing in $\mathfrak{M}$ we sloppily talk about sets whenever [25] talks about relations.

Any fixed–point can be resolved into stages I_Φ^α which are recursively defined by

$$I_\Phi^\alpha := \Phi(I_\Phi^{<\alpha})$$

where $I_\Phi^{<\alpha}$ denotes the union of all stages less than α. Let $\overline{\overline{\mathfrak{M}}}$ denote the cardinality of the domain of $\mathfrak{M}$. By a simple cardinality argument there is an ordinal $\sigma < \overline{\overline{\mathfrak{M}}}^+$ such that $I_\Phi^{<\sigma} = I_\Phi^\sigma$. The least such ordinal is the *norm* $||\Phi||$ of the operator Φ. It is easy to see that

$$I_\Phi = I_\Phi^{<||\Phi||} = I_\Phi^{||\Phi||}. \tag{7}$$

For an element $n \in \mathfrak{M}$ and an inductive definition Φ on $\mathfrak{M}$ we define its *inductive norm* by

$$|n|_\Phi := \begin{cases} \min\{\xi \mid n \in I_\Phi^\xi\} & \text{if } n \in I_\Phi \\ \overline{\overline{\mathfrak{M}}}^+ & \text{otherwise.} \end{cases}$$

We then obtain

$$||\Phi|| = \sup\{|n|_\Phi + 1 \mid n \in I_\Phi\}. \tag{8}$$

A formula $F(X, x)$ in the elementary language of $\mathfrak{M}$ defines an operator

$$\Phi_F(S) = \{n \in \mathfrak{M} \mid \mathfrak{M} \models F(S, n)\}.$$

The operator Φ_F is monotone if X occurs only positively in $F(X, x)$. We then say that Φ_F is *positively definable* over $\mathfrak{M}$. Let

$$\kappa^{\mathfrak{M}} := \sup\{||\Phi_F|| \mid F(X, x) \text{ is an } X\text{–positive elementary formula in } \mathcal{L}_{\mathfrak{M}}\} \tag{9}$$

be the *closure ordinal* of the structure $\mathfrak{M}$ (cf. [25]) .

A set $S \subseteq \mathfrak{M}$ is positive–inductively definable over $\mathfrak{M}$ if it is a slice of a fixed–point of a positively definable operator over $\mathfrak{M}$, i.e., if

$$S = \{n \in \mathfrak{M} \mid \langle s, n\rangle \in I_{\Phi_F}\}$$

for an X–positive formula $F(X, x)$ in $\mathcal{L}_{\mathfrak{M}}$ and some $s \in \mathfrak{M}$. To be short we will briefly talk about *inductive sets* when we mean positive–inductively definable sets. A set $S \subseteq \mathfrak{M}$ is *hyperelementary* over $\mathfrak{M}$ iff S and its complement are inductive over $\mathfrak{M}$.

The elements of an inductive set $S \subseteq \mathfrak{M}$ that is an s–slice of an inductive definition Φ_F are well ordered by the relation

$$n \prec_S m \text{ iff } |\langle s, n\rangle|_{\Phi_F} < |\langle s, m\rangle|_{\Phi_F}.$$

The norm $||S||$ of the inductive set is the ordertype of the well ordering $\prec_S$. The inductive norm $|z|_S$ of an element z in the field of an inductively defined relation S is its ordertype in the relation $\prec_S$.

The closure ordinal of an abstract structure $\mathfrak{M}$ can also be expressed in terms of the norms of the elements that belong to the field of an inductively definable relation on the domain of $\mathfrak{M}$. We have

$$\kappa^{\mathfrak{M}} := \sup\{|z|_S + 1 \,|\, S \text{ is inductive above } \mathfrak{M} \wedge \mathfrak{M} \models z \in \mathit{field}(S).\} \tag{10}$$

One of the central results of the abstract theory of inductive definitions is the prewellordering theorem stating that for positive–inductively definable sets S and T the relations

$$s \preceq^* t \;:\Leftrightarrow\; s \in S \wedge (t \notin T \vee |s|_S \leq |t|_T)$$
$$s \prec^* t \;:\Leftrightarrow\; s \in S \wedge (t \notin T \vee |s|_S < |t|_T)$$

are inductive. The prewellordering theorem entails:

An inductive set S is hyperelementary over $\mathfrak{M}$ iff $||S|| < \kappa^{\mathfrak{M}}$.

That inductive sets are closed under "inductive in" is guaranteed by the Transitivity Theorem (cf. Theorem 1C.3 in [25]).

If $F(X, Y_1, \ldots, Y_n, x)$ is a formula in the language $\mathcal{L}_{\mathfrak{M}}$ in which X and all the variables Y_i occur positively and $P_1, \ldots, P_n$ are inductive over $\mathfrak{M}$ then the fixed point of the operator Γ_F defined by $F(X, P_1, \ldots, P_n, x)$ is again inductive over $\mathfrak{M}$.

As a consequence of the Transitivity Theorem we obtain:

If $P_1, \ldots, P_n$ is a tuple of sets that are hyperelementary over a structure $\mathfrak{M}$ and $\mathfrak{N}$ is the structure $(\mathfrak{M}, P_1, \ldots, P_n)$ then $\kappa^{\mathfrak{N}} = \kappa^{\mathfrak{M}}$.

However, if some of the sets $P_1, \ldots, P_n$ are inductive though not hyperelementary over $\mathfrak{M}$ we in general have $\kappa^{\mathfrak{M}} < \kappa^{\mathfrak{N}}$. Iteration of inductive definitions is among the main concerns of this paper.

Note 3.1 Without proof we mention that the class $\mathbf{\Gamma}(\mathfrak{M}) = \{R \,|\, R$ is inductive over $\mathfrak{M}\}$ is a Spector class above $\mathfrak{M}$. It is easy to check that $\mathbf{\Gamma}(\mathfrak{M})$ satisfies the closure conditions. A bit more difficult to see is that $\mathbf{\Gamma}(\mathfrak{M})$ is normed. This needs the prewellordering theorem proved in [25] Theorem 3A.3. It shows that the inductive norms $|z|_S$ induce the prewellorderings of $\mathbf{\Gamma}(\mathfrak{M})$. To prove that $\mathbf{\Gamma}(\mathfrak{M})$ is parameterized we have to use that $\mathfrak{M}$ is acceptable (cf. [25] Theorem 5D.1).

For an acceptable structure $\mathfrak{M}$ let $\mathbf{ID}(\mathfrak{M}) := (\mathfrak{M}, \mathbf{\Gamma}(\mathfrak{M}))$ denote the least Spector structure above $\mathfrak{M}$. Clearly $\mathbf{ID}(\mathfrak{M})$ is countable if $\mathfrak{M}$ is countable.

In view of Note 3.1 we have $\kappa^{\mathfrak{M}} = o^{\mathbf{ID}(\mathfrak{M})}$. To establish also the connection between $\kappa^{\mathfrak{M}}$ and $\pi^{\mathfrak{M}}$ we refer to the Π^1_1 characterization of fixed–points. For an X–positive formula $F(X, x)$ in the language of $\mathfrak{M}$ we have

$$m \in I_F \;\Leftrightarrow\; (\forall X)[(\forall x)[F(X,x) \to x \in X] \to m \in X].$$

Therefore $m \in I_F$ possesses a non–trivial truth complexity for which we obtain[15]

$$|m|_F \leq 2^{\mathrm{tc}(m \in I_F)}. \tag{11}$$

Given an elementarily definable ordering $\prec$ we obtain its *accessible part* as the fixed–point of the formula $A(X,x) \;:\Leftrightarrow\; (\forall y)[y \prec x \to y \in X]$. If $\prec$ is a wellordering we obtain

$$|m|_\prec \leq |m|_A \leq 2^{\mathrm{tc}(m \in I_A)} \text{ hence } \delta^{\mathfrak{M}} \leq \kappa^{\mathfrak{M}} \leq 2^{\pi^{\mathfrak{M}}}.$$

For acceptable countable structures $\mathfrak{M}$ the ordinal $\pi^{\mathfrak{M}}$ is an epsilon number, i.e., closed under exponentiations, and we obtain together with Lemma 2.3 the sharper result

$$\delta^{\mathfrak{M}} \leq \kappa^{\mathfrak{M}} \leq 2^{\pi^{\mathfrak{M}}} = \pi^{\mathfrak{M}} = \delta^{\mathfrak{M}}. \tag{12}$$

Given an axiomatization T of $\mathbf{ID}(\mathfrak{M})$ we obtain a prooftheoretic equivalent of $\kappa^{\mathfrak{M}}$

$$\kappa^{\mathfrak{M}}(\mathsf{T}) := \sup\{|z|_F + 1 \,|\, F(X,x) \text{ is an } X\text{–positive formula in } \mathfrak{M} \text{ and } \mathsf{T} \vdash z \in I_F\}$$

for which we have $\kappa^{\mathfrak{M}}(\mathsf{T}) = o^{\mathbf{ID}(\mathfrak{M})}(\mathsf{T})$.

Remark 3.2 According to Remark 2.6 we understand the ordinals $\delta^{\mathfrak{M}}(\mathsf{T})$ and $\pi^{\mathfrak{M}}(\mathsf{T})$ as defined for the conservative extension $\mathsf{T}(X)$ of the axiom system T by free set variables. If T contains the scheme of Mathematical Induction $\delta^{\mathfrak{M}}(\mathsf{T})$ is an epsilon number and Eq. (12) relativizes to

$$\delta^{\mathfrak{M}}(\mathsf{T}(X)) \leq \kappa^{\mathfrak{M}}(\mathsf{T}(X)) \leq 2^{\pi^{\mathfrak{M}}(\mathsf{T}(X))} = 2^{\delta^{\mathfrak{M}}(\mathsf{T}(X))} = \delta^{\mathfrak{M}}(\mathsf{T}(X)). \tag{13}$$

We will, however, give axiomatizations for inductive definitions which introduce constants for fixed–points. In these axiomatizations we can define $\kappa^{\mathfrak{M}}(\mathsf{T})$ without use of set variables. To save Eq. (13) also for such axiomatizations we have to check

$$\mathsf{T} \vdash z \in I_F \;\Leftrightarrow\; \mathsf{T}(X) \vdash (\forall x)[F(X,x) \to x \in X] \to z \in X. \tag{14}$$

Under this proviso we get

$$\delta^{\mathfrak{M}}(\mathsf{T}) = \pi^{\mathfrak{M}}(\mathsf{T}) = \kappa^{\mathfrak{M}}(\mathsf{T}) = o^{\mathfrak{M}_{\mu+1}}(\mathbf{ID}_\mu(T)). \tag{15}$$

[15]The proof is similar to that of the Boundedness Theorem. Cf. [31] Theorem 5.27. The 2–power is in general unavoidable.

4 Iterating Inductive Definitions

For a countable acceptable structure $\mathfrak{M}$ the Spector structure $\mathbf{ID}(\mathfrak{M})$ is again a countable acceptable structure. By countable iterations of the operation $\mathfrak{M} \mapsto \mathbf{ID}(\mathfrak{M})$ we thus obtain larger Spector classes above $\mathfrak{M}$. We define

$$\mathfrak{M}_0 := \mathfrak{M} =: \mathbf{ID}_0(\mathfrak{M})$$
$$\mathfrak{M}_{\nu+1} := \mathbf{ID}(\mathfrak{M}_\nu) =: (\mathfrak{M}, \boldsymbol{\Gamma}_{\nu+1}(\mathfrak{M})) =: \mathbf{ID}_{\nu+1}(\mathfrak{M}) \quad \text{and}$$
$$\mathfrak{M}_\lambda = \bigcup_{\xi<\lambda} \mathfrak{M}_\xi =: \bigcup_{\xi<\lambda} \mathbf{ID}_\xi(\mathfrak{M}) \text{ for limit ordinals } \lambda.$$

Definition 4.1 Let

$$\kappa_0^{\mathfrak{M}} = 0,$$
$$\kappa_{\nu+1}^{\mathfrak{M}} := \kappa^{\mathfrak{M}_\nu} = o^{\mathfrak{M}_{\nu+1}} \text{ and}$$
$$\kappa_\lambda^{\mathfrak{M}} := \sup\{\kappa_\xi^{\mathfrak{M}} \mid \xi < \lambda\} =: o^{\mathfrak{M}_\lambda} \text{ for limit ordinals } \lambda.$$

Observe that in the paradigmatic case $\mathfrak{M} = \mathbb{N}$ the ordinals $\kappa_\mu^{\mathbb{N}}$ are the initial ordinals of the constructive number classes, i.e. $\kappa_1^{\mathbb{N}} = \omega_1^{CK}$, $\kappa_2^{\mathbb{N}} = \omega_2^{CK}$, ...

Note 4.2 For reasons, which would be too far reaching to explain at that point, we restrict the iterations to countable ordinals ν below the first recursively inaccessible ordinal (Cf. Footnote 39 at the end of the paper.)

5 Axiomatization of Iterated Inductive Definitions

Note 5.1 In the axiomatizations of iterated inductive definitions we will here restrict ourselves to the case that the number of iteration steps is "given from outside". Though this is a restriction it suffices to demonstrate the main principles. Stronger versions as, e.g., autonomous iterations of inductive definitions are farer reaching. Their ordinal analyses, however, follow the same pattern.

To obtain an axiomatization of the structure $\mathbf{ID}_\nu(\mathfrak{M})$ for iterated positive inductive definitions we build on an axiomatization T of the initial structure $\mathfrak{M}$ which we assume to be Peano–like, i.e., T is an acceptable axiomatization such that all the axioms in T are sentences in the elementary language of $\mathfrak{M}$ and it includes the scheme of Mathematical Induction. In view of Remark 3.2 we aim at an axiomatization which does not use set variables. However, to develop the axiomatization for $\mathbf{ID}_\nu(\mathfrak{M})$ we have to work in the elementary language of $\mathfrak{M}$ extended by free set variables. Instead of $P(x)$ for P a predicate constant or set–variable we commonly write $x \in P$. Since there is an elementary pairing function we may be sloppy in distinguishing unary and n–ary predicates and set–variables.

For the rest of the paper let $\prec_0$ be an elementarily definable wellordering of ordertype ν. By lower case Greek letters $\rho, \mu, \nu_i, \mu_j, \dots$ we denote the members of the field of $\prec_0$ and write more intuitively $\rho < \mu$ instead of $\rho \prec_0 \mu$. We also mostly write $\mu < \nu$ instead of $\mu \in \mathit{field}(\prec_0)$.

For P a predicate or set–variable we abbreviate by $x \in P_y$ the formula $\langle x, y\rangle \in P$ where $\langle x, y\rangle$ is the elementary pairing function of $\mathfrak{M}$.

We use the "abbreviations" $x \in \{y \,|\, G(y)\}$ for $G(x)$ and $F(G)$ for the formula obtained from $F(X)$ by replacing all occurrences of $s \in X$ in $F(X)$ by $G(s)$.

For μ in the field of $\prec_0$ we denote by $P_{<\mu}$ the set $\{x \,|\, (\exists \rho)[\rho < \mu \wedge x \in P_\rho]\}$.

Definition 5.2 For every elementary X–positive formula $F(X, Y, x, y)$ in the language of $\mathfrak{M}$ which contains no further free variables we introduce a binary relation constant I_F.

The defining axioms for the constant I_F are the closure axiom

$$(\mathbf{ID}^1_\nu) \quad (\forall \mu)(\forall x)[F(I_{F_\mu}, I_{F_{<\mu}}, x, \mu) \to x \in I_{F_\mu}]$$

and the generalized induction scheme

$$(\mathbf{ID}^2_\nu) \quad (\forall \mu)\big[(\forall x)[F(G, I_{F_{<\mu}}, x, \mu) \to G(x)] \;\to\; (\forall x)[x \in I_{F_\mu} \to G(x)]\big]$$

where G is a any formula not containing free set variables.

Due to our understanding that the ordinal ν is given from outside we add the scheme of transfinite induction along $\prec_0$ to the axioms of $\mathbf{ID}_\nu$. i.e.,

$$(\mathbf{TI}_\nu) \quad (\forall \mu)\big[[(\forall \rho < \mu)G(\rho) \to G(\mu)] \;\to\; (\forall \mu)G(\mu)\big].$$

The axioms of $\mathbf{ID}_\nu(\mathsf{T})$ are the axioms in T augmented by the defining axioms for I_F with the scheme of Mathematical Induction extended to all formulas without set variables.

By $\mathbf{ID}_{<\nu}(\mathsf{T})$ we denote the union of the theories $\mathbf{ID}_\mu(\mathsf{T})$ for $\mu < \nu$.

Since we started with an axiom system T for $\mathfrak{M}$ the axiom system $\mathbf{ID}_\nu(\mathsf{T})$ is again an axiom system for $\mathfrak{M}$. According to its construction it does not use free set variables. Our aim is to compute its prooftheoretic ordinal $|\mathbf{ID}_\nu(\mathsf{T})| = \delta^{\mathfrak{M}}(\mathbf{ID}_\nu(\mathsf{T}))$. The axioms in $\mathbf{ID}_\nu(\mathsf{T})$, however, also axiomatize all the structures $\mathfrak{M}_\mu$ with $\mu < \nu$. Therefore it makes sense to ask also for the ordinals $\delta^{\mathfrak{M}_\mu}(\mathbf{ID}_\nu(\mathsf{T}))$ for $\mu < \nu$.

To compute these ordinals we want to draw on Eq. (15) and thus have to check the conditions in Eq. (14). By axiom $(\mathbf{ID}^2_\nu)$ we have

$$\mathbf{ID}_\nu(\mathsf{T})(X)) \vdash (\forall x)[F(X, I_{F_{<\mu}}, x, \mu) \to x \in X] \to z \in I_{F_\mu} \to z \in X.$$

So $\mathbf{ID}_\nu(\mathsf{T}) \vdash z \in I_{F_\mu}$ implies $\mathbf{ID}_\nu(\mathsf{T})(X)) \vdash (\forall x)[F(X, I_{F_{<\mu}}, x, \mu) \to x \in X] \to z \in X$ and the converse direction follows directly since $\mathbf{ID}_\nu(\mathsf{T}) \vdash (\forall x)[F(I_{F_\mu}, I_{F_{<\mu}}, x, \mu) \to x \in I_{F_\mu}]$ by axiom $(\mathbf{ID}^1_\nu)$.

According to Eq. (15) we thus have

$$|\mathbf{ID}_\nu(\mathsf{T})| := \delta^{\mathfrak{M}}(\mathbf{ID}_\nu(\mathsf{T})) = \kappa^{\mathfrak{M}}(\mathbf{ID}_\nu(\mathsf{T})) \tag{16}$$

and more general

$$\delta^{\mathfrak{M}_\mu}(\mathbf{ID}_\nu(\mathsf{T})) = \kappa^{\mathfrak{M}_\mu}(\mathbf{ID}_\nu(\mathsf{T})) = \kappa^{\mathfrak{M}}_{\mu+1}(\mathbf{ID}_\nu(\mathsf{T})) = o^{\mathfrak{M}_{\mu+1}}(\mathbf{ID}_\nu(\mathsf{T})), \tag{17}$$

hence

$$Spec^{\mathfrak{M}}(\mathbf{ID}_\nu(\mathsf{T})) = \{o^{\mathfrak{M}_{\mu+1}}(\mathsf{T}) \,|\, \mu < \nu\}.$$

To compute $Spec^{\mathfrak{M}}(\mathbf{ID}_\nu \mathsf{T})$ it therefore suffices to compute the inductive norms $|z|_{F_\mu}$ for $\mu < \nu$ for those z for which we have $\mathbf{ID}_\nu(\mathsf{T}) \vdash z \in I_{F_\mu}$.

6 Semi–formal Systems for Iterated Inductive Definitions

A meanwhile well established way to compute prooftheoretic ordinals of axiom systems is the use of semi–formal systems. By a semi–formal system we understand a deduction calculus which includes inferences with infinitely many premises.[16] As described in [31] semi–formal systems for axiom systems T are built around a *verification calculus* for their intended standard models.

To repeat briefly the notion of a verification calculus let $\mathfrak{S}$ be a countable structure. We divide the sentences of the elementary language of $\mathfrak{S}$—which is assumed to include names for all elements of $\mathfrak{S}$—into $\bigwedge$ –type and $\bigvee$ –type and decorate every sentence F with a *characteristic sequence* $\mathrm{CS}(F)$ of sentences such that

$$\begin{array}{l} \mathfrak{S} \models F \text{ if } \mathfrak{S} \models G \text{ for all } G \in \mathrm{CS}(F) \text{ for } F \in \bigwedge\text{–type}, \\ \mathfrak{S} \models F \text{ if } \mathfrak{S} \models G \text{ for some } G \in \mathrm{CS}(F) \text{ for } F \in \bigvee\text{–type}. \end{array} \tag{18}$$

The verification calculus $\models^{\alpha} \Delta$ for a finite set Δ of $\mathcal{L}_{\mathfrak{S}}$–sentences—which is to be read as a finite disjunction—is obtained by two rules

($\bigwedge$) If $F \in \Delta \cap \bigwedge$ –type and $\models^{\alpha_G} \Delta, G$ plus $\alpha_G < \alpha$ hold true for all $G \in \mathrm{CS}(F)$ then infer $\models^{\alpha} \Delta$

[16] Originally–and closer to formal systems–only infinite inferences whose premises could be enumerated primitive recursively were allowed in semi–formal systems. However, for the purpose of ordinal analysis the more liberalized version suffices.

and

($\bigvee$) If $F \in \Delta \cap \bigvee$ –type and $\models^{\alpha_0} \Delta, G$ holds true for some $G \in \mathrm{CS}(F)$ then infer $\models^{\alpha} \Delta$ for all $\alpha > \alpha_0$.

The verification calculus talks only about sentences. The language $\mathcal{L}_{\mathfrak{S}}$ is therefore supposed not to contain free variables.

To obtain a semi–formal systems which serves in the computation of $Spec^{\mathfrak{M}}(\mathbf{ID}_\nu(\mathsf{T}))$ we first have to fix the intended standard model for the axiom systems $\mathbf{ID}_\nu(\mathsf{T})$.

By recursion on $\mu < \nu$ we obtain the standard interpretation I_{F_μ} of I_{F_μ} as the least fixed–point of the operator Γ_{F_μ} defined by

$$\Gamma_{F\mu}(S) := \{x \in \mathfrak{M} | \mathfrak{M}_\nu \models F(X, Y, x, \mu)[S, I_{F_{<\mu}}]\}.$$

Since we aim at the computation of the stages of elements in I_{F_μ} we need a finer graining of the fixed–points I_{F_μ}. The stages of a μ–fold iterated inductive definition reach up to the closure ordinal $\kappa^{\mathfrak{M}}_{\mu+1}$. We thus extend the language of $\mathfrak{M}_\nu$ by constants for all the stages $I^{<\xi}_{F_\mu}$ for $\xi \leq \kappa^{\mathfrak{M}}_{\mu+1}$ of the fixed points I_{F_μ} for $\mu < \nu$ and call the extended model $\mathfrak{M}^*_\nu$.

The ordinals $\kappa^{\mathfrak{M}}_{\mu+1}$ are abstractly defined by their closure properties. Technically it is difficult to work with these closure properties. We will therefore introduce symbols Ω_μ for $0 \leq \mu \leq \nu$ whose intended interpretation are the ordinals $\kappa^{\mathfrak{M}}_\mu$ but allow also other interpretations and try to axiomatize them later by closure conditions which are easier to handle.[17]

More formally we obtain the language $\mathcal{L}^*_{\mathbf{ID}_\nu}$ by the following constants:

- for every ordinal $0 \leq \mu \leq \nu$ a constant Ω_μ,
- for every μ in the field of $\prec_0$, every X–positive elementary formula $F(X, Y, x, \mu)$ with not further free variables and every $\xi \leq \Omega_{\mu+1}$ a set–constant $I^{<\xi}_{F_\mu}$.[18]

We will use the language in Tait style. i.e. there is no negation symbol among our logical symbols but negation is defined via deMorgan's laws. This requires that for every set–constant there is also a constant for its complement. However, according to our convention to write $n \in P$ instead of $P(n)$ we can express by $n \notin P$ that n belongs to the complement. This makes extra symbols for the complements unnecessary.

The next step is to separate the $\mathcal{L}^*_{\mathbf{ID}}$–sentences into $\bigwedge$ –type and $\bigvee$ –type.

---

[17]The abstractly defined ordinals $\kappa^{\mathfrak{M}}_\mu$—symbolized by the constants" Ω_μ—are ideal objects as introduced in [38]. Their elimination in course of the ordinal analysis corresponds to Hilbert's "elimination of ideal objects" as discussed there.

[18]The notion $\xi \leq \Omega_{\mu+1}$ is to be read relative to an interpretation of the constants $\Omega_{\mu+1}$ which is not yet fixed. So the language $\mathcal{L}^*_{\mathbf{ID}_\nu}$ varies with different interpretations of $\Omega_{\mu+1}$.

The $\bigwedge$ –type comprises:	The $\bigvee$ –type comprises:
• All true atomic $\mathcal{L}(\mathfrak{M})$–sentences.	• All false atomic $\mathcal{L}(\mathfrak{M})$–sentences.
• All sentences of the form $(A \wedge B)$.	• All sentences of the form $(A \vee B)$.
• All sentences of the form $(\forall x)F(x)$.	• All sentences of the form $(\exists x)F(x)$.
• All sentences of the form $n \notin I_{F_\mu}^{<\xi}$.	• All sentences of the form $n \in I_{F_\mu}^{<\xi}$.
• All sentences of the form $n \notin I_{F_{<\mu}}$.	• All sentences of the form $n \in I_{F_{<\mu}}$.

To obtain the decoration for the language $\mathcal{L}^*_{\mathbf{ID}_\nu}$ we define:

- $\mathrm{CS}(A) = \emptyset$ if A is an atomic $\mathcal{L}(\mathfrak{M})$–sentence.
- $\mathrm{CS}(A \circ B) := \langle A, B\rangle$ for $\circ \in \{\wedge, \vee\}$.
- $\mathrm{CS}((\mathsf{Q}x)F(x)) := \big\langle F(\underline{n}) \big| n \in \mathfrak{M}\big\rangle$ for $\mathsf{Q} \in \{\forall, \exists\}$.
- $\mathrm{CS}((\mathsf{Q}\mu)F(\mu)) := \big\langle F(\rho) \big| \rho < \nu\big\rangle$ for $\mathsf{Q} \in \{\forall, \exists\}$.[19]
- $\mathrm{CS}(n \in I_{F_\mu}^{<\xi}) = \big\langle n \in I_{F_\mu}^{\eta} \big| \eta < \xi\big\rangle$, $\mathrm{CS}(n \notin I_{F_\mu}^{<\xi}) = \big\langle n \notin I_{F_\mu}^{\eta} \big| \eta < \xi\big\rangle$ where $n \in I_{F_\mu}^{\eta}$ is a shorthand for $F(I_{F_\mu}^{<\eta}, I_{F_{<\mu}}, n, \mu)$, $n \notin I_{F_\mu}^{\eta}$ a shorthand for $\neg F(I_{F_\mu}^{<\eta}, I_{F_{<\mu}}, n, \mu)$ and $I_{F_{<\rho}}$ a shorthand for $\{x \,|\, (\exists \mu)[\mu < \rho \wedge x \in I_{F_\mu}^{<\Omega_{\mu+1}}]\}$.

For every $\mathcal{L}^*_{\mathbf{ID}_\nu}$–sentence F we define its complexity by

$$\mathrm{rnk}(F) := \sup\{\mathrm{rnk}(G) + 1 | G \in \mathrm{CS}(F)\}.$$

Having a decoration of the language $\mathcal{L}^*_{\mathbf{ID}_\nu}$ we obtain a verification calculus $\models^{\alpha} \Delta$ for finite sets of $\mathcal{L}^*_{\mathbf{ID}_\nu}$–sentences.

Interpreting the constants Ω_μ as $\kappa_\mu^{\mathfrak{M}}$ and $I_{F_\mu}^{<\xi}$ standardly as the union of the stages $I_{F_\mu}^{\eta}$ below ξ shows that $\mathfrak{M}^*_\nu$ satisfies (18) for the given decoration. Therefore the verification calculus is sound for the model $\mathfrak{M}^*_\nu$. It is even complete for a fragment of $\mathfrak{M}^*_\nu$ to be defined below.

Definition 6.1 An $\mathcal{L}^*_{\mathbf{ID}_\nu}$–sentence G belongs to $\bigvee_1^{\Omega_{\mu+1}}$ –type if for any constant $I_{F_\rho}^{<\xi}$ occurring in G we have $\rho \le \mu$, $\xi \le \Omega_{\rho+1}$ and there are only positive occurrences of $I_{F_\mu}^{<\Omega_{\mu+1}}$. The class $\bigwedge_1^{\Omega_{\mu+1}}$ –type is the dual class, i.e., it contains at most negative occurrences of $I_{F_\mu}^{<\Omega_{\mu+1}}$.

Let $\bigvee_0^{\Omega_{\mu+1}}$ –type $= \bigwedge_0^{\Omega_{\mu+1}}$ –type $:= \bigvee_1^{\Omega_{\mu+1}}$ –type $\cap \bigwedge_1^{\Omega_{\mu+1}}$ –type.

Theorem 6.2 *The verification calculus is sound and complete for the sentences in* $\bigwedge_1^{\Omega_1}$ *–type.*

Proof Only completeness is to be checked. For F in $\bigwedge_1^{\Omega_1}$ –type such that $\mathfrak{M}^*_\nu \models F$ we get $\models^{\mathrm{rnk}(F)} F$ by an easy induction on $\mathrm{rnk}(F)$. Just to indicate three cases: If F is a true atomic formula then $F \in \bigwedge$ –type and $\mathrm{CS}(F) = \emptyset$ and we obtain $\models^{0} F$

[19] Observe that there are two types of ordinals. First ordinals $\le \nu$ which are represented by the elements in the field of $\prec_0$ and thus are just numbers, hence "quantifiable", and secondly the ordinal constants ξ occurring in the form of $I_{F_\mu}^{<\xi}$.

by a rule $(\bigwedge)$. If F is a sentence $n \in I_{F_0}^{<\xi}$ then $\xi < \Omega_1$ and there is an $\eta < \xi$ such that $n \in I_{F_0}^{\eta}$ is true. Hence $\models^{\alpha} n \in I_{F_0}^{\eta}$ by induction hypothesis for $\alpha = \mathrm{rnk}(n \in I_{F_0}^{\eta})$ which entails $\models^{\mathrm{rnk}(n \in I_{F_0}^{<\xi})} n \in I_{F_0}^{<\xi}$ by an inference $(\bigvee)$. If F is a sentence $n \notin I_{F_0}^{<\xi}$ then we have $\mathfrak{M}_{\nu}^{*} \models n \notin I_{F_0}^{\zeta}$ for all $\zeta < \xi$, hence $\models^{\alpha_\zeta} n \notin I_{F_0}^{\zeta}$ by induction hypothesis and the claim follows by an inference $(\bigwedge)$. □

We have now to check in how far the verification calculus is already a model for $\mathbf{ID}_{\nu}(\mathsf{T})$. It follows from Theorem 6.2 that all true elementary $\mathcal{L}(\mathfrak{M})$–sentences are verifiable. So it remains to check the axiom schemes of Mathematical Induction, $\mathbf{ID}_{\nu}^{1}$, $\mathbf{ID}_{\nu}^{2}$ and $\mathbf{TI}_{\nu}$. We postpone $\mathbf{ID}_{\nu}^{1}$ and try first to verify $\mathbf{ID}_{\nu}^{2}$, $\mathbf{TI}_{\nu}$ and Mathematical Induction. The sentence $Cl_{F_{\mu}}(G) :\Leftrightarrow (\forall x)[\neg F(G, I_{F_{<\mu}}, x, \mu) \vee G(x)]$ expresses that the set $\{x \,|\, G(x)\}$ is closed under the operator generated by $F_{\mu}(X) :\Leftrightarrow F(X, I_{F_{<\mu}}, x, \mu)$. Then

$$\models^{\alpha} \neg Cl_{F_{\mu}}(G), n \notin I_{F_{\mu}}^{<\xi}, G(n) \tag{19}$$

holds true for all n and all set terms $\{x \,|\, G(x)\}$ with $\alpha = 2 \cdot \mathrm{rnk}(G) + \Omega_{\mu} \cdot \xi + 1$. This is the standard proof by induction on ξ exploiting the monotonicity of the operator generated by F_{μ}.[20]

From (19) we easily obtain an $\alpha \in (2 \cdot \mathrm{rnk}(G) + \Omega_{\mu+1}, 2 \cdot \mathrm{rnk}(G) + \Omega_{\mu+1} + \omega)$ such that

$$\models^{\alpha} \neg Cl_{F_{\mu}}(G) \vee (\forall x)[x \notin I_{F_{\mu}}^{<\Omega_{\mu+1}} \vee G(x)].$$

Since this is true for all $\mu \in \mathit{field}(\prec_0)$ we finally obtain

$$\models^{\alpha} (\forall \mu)[\neg Cl_{F_{\mu}}(G) \vee (\forall x)[x \notin I_{F_{\mu}}^{<\Omega_{\mu+1}} \vee G(x)]] \tag{20}$$

for some $\alpha \in (\Omega_{\mu}, \Omega_{\nu} \cdot \omega)$ which is a translation of $\mathbf{ID}_{\nu}^{2}$ if we interpret $I_{F_{\mu}}$ by $I_{F_{\mu}}^{<\Omega_{\mu+1}}$.

Similar to (19) we obtain

$$\models^{\alpha} \neg(\forall \sigma)[\neg(\forall \rho < \sigma)G(\rho) \vee G(\sigma)], G(\mu) \tag{21}$$

for all $\mu < \nu$ for some ordinal $\alpha < 2 \cdot \mathrm{rnk}(G) + \omega \cdot \mu + 1$. This shows that the translation of $\mathbf{TI}_{\nu}$ is verifiable with an ordinal less than $\Omega_{\nu} \cdot \omega$. Mathematical Induction follows in the same vein.

There is, however, no chance to obtain a verification of the translations of the axiom $\mathbf{ID}_{\nu}^{1}$. There is no information about the meaning of the constants $\Omega_{\mu+1}$ in the verification calculus. We need axioms or defining rules—for which we are going to opt here—that fix the meaning of the constants $\Omega_{\mu+1}$. Therefore we extend the verification calculus $\models^{\alpha} \Delta$ to a semi–formal system $\vdash_{\rho}^{\alpha} \Delta$ by adding defining rules for the constants $\Omega_{\mu+1}$. However, adding new rules may spoil the admissibility of the cut rules why we also add a cut rule.

[20]The proof is similar to that of [35] Lemma 9.5.3.

Definition 6.3 We define the proof relation $\models^{\alpha}_{\rho} \Delta$ for a finite set Δ of $\mathcal{L}^*_{\mathbf{ID}_\nu}$–sentences by the rules:

($\bigwedge$) If $F \in \Delta \cap \bigwedge$–type and $\models^{\alpha_G}_{\rho} \Delta, G$ plus $\alpha_G < \alpha$ hold true for all $G \in \mathrm{CS}(F)$ then infer $\models^{\alpha}_{\rho} \Delta$.

($\bigvee$) If $F \in \Delta \cap \bigvee$–type and $\models^{\alpha_0}_{\rho} \Delta, G$ holds true for some $G \in \mathrm{CS}(F)$ then infer $\models^{\alpha}_{\rho} \Delta$ for all $\alpha > \alpha_0$.

(Cut) From $\models^{\alpha}_{\rho} \Delta, G$, $\models^{\alpha}_{\rho} \Delta, \neg G$ and $\mathrm{rnk}(G) < \rho$ infer $\models^{\beta}_{\rho} \Delta$ for any $\beta > \alpha$.

($\Omega_{\mu+1}$) If $(n \in I^{<\Omega_{\mu+1}}_{F_\mu}) \in \Delta$ for $\mu < \nu$ and $\models^{\alpha}_{\rho} \Delta, F(I^{<\Omega_{\mu+1}}_{F_\mu}, I_{F_{<\mu}}, n, \mu)$ then infer $\models^{\beta}_{\rho} \Delta$ for all $\beta > \alpha$.

We call the sentences F and $n \in I^{<\Omega_{\mu+1}}_{F_\mu}$, respectively, the *principal sentences* of the corresponding inference. A cut possesses no principal sentence but a *cut sentence*.

We clearly have $\models^{\alpha} \Delta \Rightarrow \models^{\alpha}_{0} \Delta$. However, $\models^{\alpha}_{0} \Delta \Rightarrow \models^{\alpha} \Delta$ holds only true if Δ is a subset of the class $\bigwedge^{\Omega_1}_1$–type.

The $\Omega_{\mu+1}$–rules enable us to obtain a semi–formal derivation also of the translation of $\mathbf{ID}^1_\nu$. To see that first observe that we have

$$\models^{2\mathrm{rnk}(H)}_{0} \Delta, \neg H, H$$

for any finite set Δ and any sentence H, hence especially

$$\models^{\alpha}_{0} \neg F(I^{<\Omega_{\mu+1}}_{F_\mu}, I_{F_{<\mu}}, n, \mu), F(I^{<\Omega_{\mu+1}}_{F_\mu}, I_{F_{<\mu}}, n, \mu)$$

for some ordinal less than $\Omega_{\mu+1} \cdot \omega$. Using an $\Omega_{\mu+1}$–inference this entails

$$\models^{\alpha'}_{0} \neg F(I^{<\Omega_{\mu+1}}_{F_\mu}, I_{F_{<\mu}}, n, \mu), n \in I^{<\Omega_{\mu+1}}_{F_\mu}$$

for α' still less than $\Omega_{\mu+1} \cdot \omega$. Since this holds true for all $\mu < \nu$ a ($\bigvee$)–inference followed by a ($\bigwedge$)–inference yield

$$\models^{\alpha}_{0} (\mathbf{ID}^1_\nu)^* \tag{22}$$

with an ordinal α less than $\Omega_\nu + \omega$.

Since all the translations of axioms of $\mathbf{ID}_\nu(\mathsf{T})$ are semi–formally provable with ordinals less than $\Omega_\nu \cdot \omega$ we finally obtain by an induction on the lengths of $\mathbf{ID}_\nu(\mathsf{T})$–derivations a translation theorem.

Theorem 6.4 *If* $\mathbf{ID}_\nu(\mathsf{T}) \vdash G(x_1, \ldots, x_n)$ *then there is an* $\alpha < \Omega_\nu \cdot \omega$ *and a finite ordinal* k *such that* $\vdash^{\alpha}_{\Omega_\nu + k} G(z_1, \ldots, z_n)^*$ *holds true for any choice of elements* $z_1, \ldots, z_n$ *in* $\mathfrak{M}$. *Here* G^* *is the* $\mathfrak{M}^*_\nu$*–sentence obtained from* G *replacing all occurrences of* I_{F_μ} *by* $I^{<\Omega_{\mu+1}}_{F_\mu}$.

Observe that Theorems 6.2 and 6.4 hold true independently of the interpretation of the constants Ω_μ. The $\Omega_{\mu+1}$–rules mimic the closure properties of the ordinals $\kappa^{\mathfrak{M}}_{\mu+1}$. If we interpret all Ω_μ standardly by $\kappa^{\mathfrak{M}}_\mu$ the $\Omega_{\mu+1}$–rules are apparently sound in the structure $\mathfrak{M}^*_\nu$. We thus have the following theorem.

Theorem 6.5 *When interpreting all* Ω_μ *standardly by* $\kappa^{\mathfrak{M}}_\mu$ *the proof–relation* $\vdash^{\alpha}_{\rho} F$ *is sound for the structure* $\mathfrak{M}^*_\nu$.

Corollary 6.6 *If* Δ *is finite set of sentences in* $\bigvee^{\Omega_{\mu+1}}_0$ *–type such that* $\vdash^{\alpha}_{\Omega_{\mu+1}} \Delta$ *and all constants* Ω_ρ *for* $\rho \le \mu$ *are interpreted standardly by* $\kappa^{\mathfrak{M}}_\rho$ *then* $\mathfrak{M}^*_\nu \models \bigvee \Delta$.

Proof Since $\Delta \subseteq \bigvee^{\Omega_{\mu+1}}_0$ –type the derivation $\vdash^{\alpha}_{\Omega_{\mu+1}} \Delta$ does not contain constants Ω_σ for $\sigma > \mu$. Therefore we have $\rho < \mu$ for all $(\Omega_{\rho+1})$–rules occurring in $\vdash^{\alpha}_{\Omega_{\mu+1}} \Delta$ and these rules are sound in $\mathfrak{M}^*_\nu$. □

Since the semi–formal system for $\mathfrak{M}^*_\nu$ derives only finite sets of *sentences* cut–elimination for this system comes nearly for free.

Theorem 6.7 (Semantical Cut Elimination)

$$\vdash^{\alpha}_{\rho} \Delta\,,\ \textit{implies}\ \vdash^{\alpha}_{0} \Delta\,.$$

Proof A simple induction on α shows that for a finite set Γ of sentences that are—standardly interpreted—all false in $\mathfrak{M}^*_\nu$ we get

$$\vdash^{\alpha}_{\rho} \Delta\,, \Gamma\ \textit{implies}\ \vdash^{\alpha}_{0} \Delta\,.$$

To emphasize that this proof is based on the fact that the semi–formal system only derives sentences let us look at the case that the last inference is a cut

$$\vdash^{\alpha_0}_{\rho} \Delta, \Gamma, G\,,\ \vdash^{\alpha_0}_{\rho} \Delta, \Gamma, \neg G\ \Rightarrow\ \vdash^{\alpha}_{\rho} \Delta, \Gamma\,. \tag{i}$$

Then either G or $\neg G$ is false in the standard interpretation. So we can pick the premise in which G or $\neg G$ is false and apply the induction hypothesis. All other cases follow from the induction hypothesis and the fact that all rules are sound in the standard interpretation. The claim follows from (i) for $\Gamma = \emptyset$. □

Remark 6.8 Since $\mathfrak{M} \models \neg F(X) \;\Leftrightarrow\; \mathfrak{M} \not\models F(X)$ is false for pseudo Π^1_1–sentences it is obvious that the above proof cannot work for semi–formal systems that derive finite sets of pseudo Π^1_1–sentences.

The theorem that enables us to use the semi–formal system for $\mathfrak{M}_\nu^*$ in the computation of $Spec^{\mathfrak{M}}(\mathbf{ID}_\nu(\mathsf{T}))$ is the following *Boundedness Theorem.*

Theorem 6.9 (Boundedness) *Let* $0 \le \mu < \nu$ *and assume* $\frac{\alpha}{\rho} \Delta(I_{F_\mu}^{<\xi})$ *for* $\alpha \le \Omega_{\mu+1}$. *Then we obtain* $\frac{\alpha}{\rho} \Delta(I_{F_\mu}^{<\zeta})$ *for any* $\zeta \in [\alpha, \xi]$ *and any positive occurrence of* $I_{F_\mu}^{<\xi}$ *in one of the sentences in* Δ.

Proof The proof is straightforward by induction on α. Nevertheless we give a rough sketch since it incorporates one of the crucial features of impredicative proof theory, *the elimination of ideal methods.*[21]

If the last inference is not an inference with principal sentence $n \in I_{F_\mu}^{<\xi}$ the claim follows immediately by the same inference from the induction hypothesis. So assume that the principal sentence of the last inference is $n \in I_{F_\mu}^{<\Omega_{\mu+1}}$. Then we are either in the case of an inference according to $(\bigvee)$ or an $(\Omega_{\mu+1})$–inference. In the prior case we have the premise $\frac{\alpha_0}{\rho} \Delta_0, n \in I_{F_\mu}^{<\xi}, n \in I_{F_\mu}^{\zeta_0}$ for some $\zeta_0 < \xi$. By induction hypothesis we obtain $\frac{\alpha_0}{\rho} \Delta_0, n \in I_{F_\mu}^{<\zeta}, n \in I_{F_\mu}^{\alpha_0}$ which entails $\frac{\alpha}{\rho} \Delta_0, n \in I_{F_\mu}^{<\zeta}$ by a $(\bigvee)$–inference since $\alpha_0 < \alpha \le \zeta$. In the latter case we have the premise

$$\frac{\alpha_0}{\rho} \Delta_0, F(I_{F_\mu}^{<\Omega_{\mu+1}}, I_{F_{<\mu}}, n, \mu), n \in I_{F_\mu}^{<\Omega_{\mu+1}}.$$

Since $I_{F_\mu}^{<\Omega_{\mu+1}}$ occurs positively we may apply the induction hypothesis to obtain $\frac{\alpha_0}{\rho} \Delta_0, n \in I_{F_\mu}^{\alpha_0}, n \in I_{F_\mu}^{<\zeta}$ and then an inference $(\bigvee)$ to get the claim. □

Observe that the additional "ideal" $(\Omega_{\mu+1})$–rule—which is not part of the verification calculus—is thus replaced by the $(\bigvee)$–rule of the verification calculus.

The Boundedness Theorem opens a strategy to compute the points $o^{\mathfrak{M}_{\mu+1}}(\mathbf{ID}_\nu(\mathsf{T}))$ in $Spec^{\mathfrak{M}}(\mathbf{ID}_\nu(\mathsf{T}))$. Given a formal proof $\mathbf{ID}_\nu(\mathsf{T}) \vdash n \in I_{F_\mu}$ for $\mu < \nu$ we unravel it into a semi–formal proof $\frac{\alpha}{\rho} n \in I_{F_\mu}^{<\Omega_{\mu+1}}$. If $\alpha, \rho < \Omega_{\mu+1}$ we interpret all Ω_ρ for $\rho \le \mu$ standardly and obtain by boundedness and Corollary 6.6 $\mathfrak{M}_\mu^* \models n \in I_{F_\mu}^{<\alpha}$, hence $|n|_{F_\mu} + 1 \le \alpha$. Finding a uniform bound below $\Omega_{\mu+1}$ for the lengths of the unravelled derivations thus gives also an upper bound for $o^{\mathfrak{M}_{\mu+1}}(\mathbf{ID}_\nu(\mathsf{T}))$.

For the first point in the spectrum we have $o^{\mathfrak{M}_1}(\mathbf{ID}_\nu(\mathsf{T})) = \kappa^{\mathfrak{M}_0}(\mathbf{ID}_\nu(\mathsf{T})) = \delta^{\mathfrak{M}}(\mathbf{ID}_\nu(\mathsf{T})) = |\mathbf{ID}_\nu(\mathsf{T})|$. It thus coincides with the familiar proof theoretic ordinal of the axiom system $\mathbf{ID}_\nu(\mathsf{T})$. In a derivation $\frac{\alpha}{\Omega_1} n \in I_{F_1}^{<\xi}$ with $\alpha < \Omega_{0+1}$ there are no occurrences of constants Ω_μ. We thus obtain the prooftheoretic ordinal of $\mathbf{ID}_\nu(\mathsf{T})$ by an elimination of all additional "ideal" constants Ω_μ. It is therefore independent of the interpretation of the constants Ω_μ.

The stumbling block in this strategy is that the ordinals α and ρ which arise in the unravelling procedure are in general too big. By (20) we know that ordinals $\alpha > \Omega_\nu$ are unavoidably needed for the embedding of the $\mathbf{ID}_\nu^2$ schemes; ordinals which are far too big as to yield reasonable upper bounds. On the other hand we know from (the proof of) Theorem 6.2 that for an element $z \in I_{F_0}$ with $|z|_{F_0} = \alpha$,

[21] Cf. the discussion in [38].

there is a verification, hence also a semi–formal proof $\vdash_0^{\xi} n \in I_{F_0}^{<\Omega_1}$ with an ordinal $\alpha \leq \xi < \kappa^{\mathfrak{M}}$. This indicates that there could exist a collapsing procedure for the semi–formal system. We are thus confronted with the problem:

- "Is there a procedure that collapses a derivation $\vdash_{\rho}^{\alpha} \Delta$ for $\Delta \subseteq \bigvee_{\Omega_1}^{1}$–type into a derivation of a length less than $\kappa^{\mathfrak{M}}$?"[22]

Note 6.10 It is the generalized induction axiom $\mathbf{ID}_{\nu}^{2}$ that forces us to use ordinals above $\kappa^{\mathfrak{M}}$. Therefore one could ask what happens if this axiom is lacking and $\mathbf{ID}_{\nu}^{1}$ is replaced by a fixed–point axiom. Such and similar theories in the language of set theory have been studied by Gerhard Jäger (c.f. [19]) and his school (see e. g. [20]). They turn out to be either predicative with prooftheoretic ordinals below Γ_0 or metapredicative with prooftheoretic ordinals between Γ_0 and the prooftheoretic ordinal of $\mathbf{ID}_1$. Since Eq. (11) is not applicable in this situation—it hinges on the fact that the fixed point is expressible by a Π_1^1–sentence—one cannot draw on (16) to obtain upper bounds for their prooftheoretic ordinals. It still needs pseudo Π_1^1–sentences in their computation and thus an (X)–rule for free set–variables which spoils semantical cut–elimination (cf. [38]). The prooftheoretic analysis of such systems therefore requires sophistications of methods of predicative proof theory, e.g., asymmetric interpretations.

7 Iterated Skolem Hulls

Collapsing derivations and keeping control over the lengths of the collapsed derivations needs also a collapsing function on the ordinals. Since ordinals as transitive sets are not collapsible this means that we must only use ordinals from a subclass of the ordinals which contains gaps that are large enough to allow for the necessary collapsing. A proven technique to generate such sets are iterations of Skolem hull operators.[23]

Let $\mathcal{F}$ be a set of functions on the ordinals of arities $n \geq 0$. The Skolem hull operator $\mathcal{H}^{\mathcal{F}}$ generated by $\mathcal{F}$ assigns to a set X of ordinals the least set $\mathcal{H}^{\mathcal{F}}(X)$ which contains X and is closed under all the functions in $\mathcal{F}$. Skolem hull operators are obviously inflationary, monotone and idempotent and thus not iterable. To obtain iterations of Skolem hull operators we have therefore to augment the generating functions in every iteration step.

Call an ordinal ξ inaccessible by $\mathcal{H}$ or a critical point of $\mathcal{H}$ if $\xi \notin \mathcal{H}(\xi)$ where ξ is viewed as the set of its predecessors. The idea is now to define by recursion on α the iterations $\mathcal{H}^{\alpha}$ together with a function ψ defined by $\psi^{\alpha} := \min\{\xi \,|\, \xi \notin \mathcal{H}^{\alpha}(\xi)\}$ and to augment the generators of $\mathcal{H}^{\alpha}$ by the function $\psi \restriction \alpha$. Doing so, it is not too hard

[22]Already Feferman in [12] Sect. 4.1 conjectured "*some collapsing argument my perhaps be possible*" in finding the "*provably recursive ordinals*" of $\mathbf{ID}_{\nu}$.

[23]A technique that goes back to work of many people among them Sol Feferman, Peter Aczel, Jane Bridge (Kister), Wilfried Buchholz and others.

to see that starting with $\mathcal{F} = \{0, +\}$ the least ordinal for which we have $\psi^\sigma = \sigma$ is ε_0, the prooftheoretic ordinal of Peano arithmetic.

Moreover we observe that for $\alpha \notin \mathcal{H}^\alpha(\xi)$ we get $\mathcal{H}^\alpha(\xi) = \mathcal{H}^{\alpha+1}(\xi)$ and thus $\psi^\alpha = \psi^{\alpha+1}$. To obtain 1–1 functions it is therefore reasonable to require the normal form condition $\alpha \in \mathcal{H}^\alpha(\xi)$ for critical points. In Sect. 8 we will recollect that predicative proof theory is governed by the Veblen functions φ_ξ. To refresh the definition of the Veblen functions φ_ξ recall that φ_0 enumerates the additively indecomposable ordinals, i.e., $\varphi_0(\eta) := \omega^\eta$, and for $\xi > 0$ the function φ_ξ enumerates the common fixed–points of the functions φ_ζ for $\zeta < \xi$. Though not absolutely necessary—$\lambda\xi.\,\omega^\xi$ would do—it is convenient to include the Veblen function in the generators of Skolem hull operators. We thus fix $\mathcal{F}_0 = \{0, +, \ \lambda\xi\eta.\ \varphi_\xi(\eta)\}$.

The operator generated by $\mathcal{F}_0$ and its iterations are transitive in the sense that they return a transitive set of ordinals when applied to a transitive set of ordinals. However, aiming at iterated inductive definition we have to incorporate the closure ordinals $\kappa_\mu^{\mathfrak{M}}$ as "initial points".

In Definition 4.1 the ordinals $\kappa_{\mu+1}^{\mathfrak{M}}$ are abstractly defined by their closure properties. These closure properties are technically difficult to handle. Instead of using these closure properties and to refer to the ordinals $\kappa_\mu^{\mathfrak{M}}$ directly we opt for an axiomatic approach and replace the ordinals $\kappa_\mu^{\mathfrak{M}}$ by ordinal constants Ω_μ together with defining axioms. We will have to check later in how far this axiomatization captures the intended interpretation.

We therefore extend the generator $\mathcal{F}_0$ to $\mathcal{F} := \mathcal{F}_0 \cup \{\Omega_\mu \,|\, 1 \le \mu \le \nu\}$ and denote the resulting Skolem hull operator by $\mathcal{H}$. We define

$$\begin{aligned} \mathcal{H}^0 &:= \mathcal{H}, \\ \Psi^\alpha_{\Omega_{\mu+1}} &:= \min\{\xi \,|\, \Omega_\mu \le \xi \notin \mathcal{H}^\alpha(\xi) \wedge \alpha \in \mathcal{H}^\alpha(\xi)\} \end{aligned}$$

and augment for $\alpha > 0$ the generators of $\mathcal{H}^\alpha$ by all the functions $\Psi_{\Omega_{\mu+1}} \restriction \alpha$ for $\mu < \nu$. Note that the functions $\Psi_{\Omega_{\mu+1}}$ are partial and we have $\alpha \in dom(\Psi_{\Omega_{\mu+1}})$ iff $\alpha \in \mathcal{H}^\alpha(\Psi^\alpha_{\Omega_{\mu+1}})$. When writing $\Psi^\alpha_{\Omega_{\mu+1}}$ we tacitly assume that $\alpha \in dom(\Psi_{\Omega_{\mu+1}})$.

Let for convenience $\Omega_0 := 0$. For $\mu > 0$ we axiomatize the ordinal constants Ω_μ by

$$(\mathbf{Ax}_\Omega) \qquad \left\{ \begin{array}{l} \Omega_\mu \text{ is strongly critical, i.e., closed under } + \text{ and } \varphi, \\ \rho < \mu \text{ implies } \Omega_\rho < \Omega_\mu, \\ \text{for } \alpha \in dom(\Psi_{\Omega_{\mu+1}}) \text{ it is } \Psi^\alpha_{\Omega_{\mu+1}} < \Omega_{\mu+1}, \\ \Omega_\lambda = \sup\{\Omega_\xi \,|\, \xi < \lambda\} \text{ for limit ordinals } \lambda. \end{array} \right.$$

Remark 7.1 The axiomatization of the ordinals Ω_μ is consistent. Interpreting the function $\lambda\xi.\,\Omega_\xi$ as the enumerating function of (an initial segment of) the class of uncountable cardinals augmented by 0, a simple cardinality argument shows that $(\mathbf{Ax}_\Omega)$ is satisfied.

According to this interpretation as the initial ordinals of uncountable number classes and the intended interpretation as the initial ordinals of the (relativized) constructible number classes we refer to the constants Ω_μ as *initial ordinals.*

Key properties of the iterated Skolem hull operators are

$$\begin{aligned}&\mathcal{H}^\alpha(\Psi^\alpha_{\Omega_{\mu+1}}) = \mathcal{H}^\alpha(\Omega_\mu),\\ &\Psi^\alpha_{\Omega_{\mu+1}} = \mathcal{H}^\alpha(\Omega_\mu) \cap \Omega_{\mu+1} = \min\{\xi \,|\, \alpha \in \mathcal{H}^\alpha(\Omega_\mu) \wedge \mathcal{H}^\alpha(\xi) \cap \Omega_{\mu+1} = \xi\}.\end{aligned} \tag{23}$$

The main point in showing the second claim in (23) is to prove the transitivity of the set $\mathcal{H}^\alpha(\Omega_\mu) \cap \Omega_{\mu+1}$ by induction on α.

According to Eq. (23) the normal–form condition $\alpha \in \mathcal{H}^\alpha(\Psi^\alpha_{\Omega_{\mu+1}})$ is satisfied iff $\alpha \in \mathcal{H}^\alpha(\Omega_\mu)$.

From Eq. (23) we get moreover

$$\begin{aligned}\Psi^\alpha_{\Omega_{\mu+1}} < \Psi^\beta_{\Omega_{\rho+1}} \;&\Leftrightarrow\; \Psi^\alpha_{\Omega_{\mu+1}} \in \mathcal{H}^\beta(\Psi^\beta_{\Omega_{\rho+1}}) \cap \Omega_{\rho+1} = \mathcal{H}^\beta(\Omega_\rho) \cap \Omega_{\rho+1}\\ &\Leftrightarrow\; \mu < \rho \vee\\ &\qquad (\mu = \rho \wedge \alpha \in \mathcal{H}^\beta(\Omega_\rho) \cap \beta)\end{aligned} \tag{24}$$

which in turn implies

$$\Psi^\alpha_{\Omega_{\mu+1}} = \Psi^\beta_{\Omega_{\rho+1}} \;\Leftrightarrow\; \mu = \rho \wedge \alpha = \beta. \tag{25}$$

The function $\Psi_{\Omega_{\mu+1}}$ is—by requirement—collapsing below $\Omega_{\mu+1}$. We are going to indicate that—independent of the interpretation of the constants Ω_μ—Ψ_{Ω_1} is even collapsing below $\kappa^{\mathfrak{M}}$.

The largest transitive segment of the ordinals we can reach by the iteration procedure is apparently $\mathcal{H}^{\Gamma_{\Omega_\nu+1}}(0) \cap \Omega_1 = \Psi^{\Gamma_{\Omega_\nu+1}}_{\Omega_1}$ where $\Gamma_{\Omega_\nu+1}$ denotes the first strongly critical ordinal above Ω_ν, i.e., the first ordinal above Ω_ν that is closed under the Veblen function viewed as a binary function.

We are going to indicate that the set $\mathcal{H}^{\Gamma_{\Omega_\nu+1}}(0)$ and also the less than relation restricted to it are elementarily definable. We define a set $\mathcal{T}$ of *ordinal terms* inductively by the following clauses.

($\mathcal{T}_0$) For $0 \le \mu \le \nu$ it is $\Omega_\mu \in \mathcal{T}$.
($\mathcal{T}_1$) If $\alpha =_{NF} \alpha_1 + \cdots + \alpha_n$ and $\{\alpha_1, \ldots, \alpha_n\} \subseteq \mathcal{T}$ then $\alpha \in \mathcal{T}$.[24]
($\mathcal{T}_1$) If $\alpha =_{NF} \varphi_{\alpha_1}(\alpha_2)$ and $\{\alpha_1, \alpha_2\} \subseteq \mathcal{T}$ then $\alpha \in \mathcal{T}$.[25]
($\mathcal{T}_3$) If $\alpha \in \mathcal{H}^\alpha(\Omega_\mu)$ and $\alpha \in \mathcal{T}$ then $\Psi^\alpha_{\Omega_{\mu+1}} \in \mathcal{T}$ for all $\mu < \nu$.

By induction on the definition of $\alpha \in \mathcal{T}$ we get $\alpha \in \mathcal{H}^{\Gamma_{\Omega_\nu+1}}(0)$ and conversely $\alpha \in \mathcal{H}^{\Gamma_{\Omega_\nu+1}}(0) \Rightarrow \alpha \in \mathcal{T}$ by induction on the definition of $\alpha \in \mathcal{H}^{\Gamma_{\Omega_\nu+1}}(0)$. Hence

---

[24]By $\alpha =_{NF} \alpha_1 + \cdots + \alpha_n$ we denote the Cantor normal form for α.

[25]$\alpha =_{NF} \varphi_{\alpha_1}(\alpha_2)$ means $\alpha = \varphi_{\alpha_1}(\alpha_2)$ and $\alpha_i < \alpha$ for $i = 1, 2$.

$$\mathcal{T} = \mathcal{H}^{\Gamma_{\Omega_\nu+1}}(0). \tag{26}$$

The only "non–syntactical" feature in the definition of $\mathcal{T}$ is the normal–form condition $\alpha \in \mathcal{H}^\beta(\Omega_\mu)$. To "syntactize" also this condition we define the sets

$$K_\mu\alpha := \begin{cases} \emptyset & \text{if } \alpha < \Omega_\mu \text{ or } \alpha = \Omega_\rho \text{ for some } \rho \leq \nu, \\ \bigcup_{i=1}^n K_\mu\alpha_i & \text{if } \alpha =_{NF} \alpha_1 + \cdots + \alpha_n, \\ K_\mu\alpha_1 \cup K_\mu\alpha_2 & \text{if } \alpha =_{NF} \varphi_{\alpha_1}(\alpha_2), \\ \{\beta\} & \text{if } \Omega_\mu < \alpha = \Psi^\beta_{\Omega_{\rho+1}}. \end{cases}$$

We then have

$$\alpha \in \mathcal{H}^\beta(\Omega_\mu) \;\Leftrightarrow\; K_\mu\alpha < \beta. \tag{27}$$

Now we can see by $(\mathbf{Ax}_\Omega)$, (24), (25), the definitions of $\mathcal{T}$ and $K_\mu\alpha$ and (27) that the set $\mathcal{T}$ and $<\restriction\mathcal{T}$ can be defined simultaneously by course of values recursion.[26] Together with the assumption that $\prec_0$ is elementarily decidable we thus obtain that $\mathcal{H}^\alpha(\emptyset) \cap \Omega_1$ has an ordertype less than $\delta^{\mathfrak{M}} = \kappa_1^{\mathfrak{M}}$. Hence $\Psi^\alpha_{\Omega_1} = \mathcal{H}^\alpha(0) \cap \Omega_1 < \kappa_1^{\mathfrak{M}}$ for any $\alpha \in \mathcal{T}$ which shows that Ψ_{Ω_1} collapses below $\kappa_1^{\mathfrak{M}}$.

This argument can be relativized by replacing $\mathfrak{M}$ by $\mathfrak{M}_\mu$. Then the set $\mathcal{H}^{\Gamma_{\Omega_\nu+1}}(\kappa_\mu^{\mathfrak{M}})$ and the less than relation restricted to it are elementarily definable in $\mathfrak{M}_\mu$. This implies that $\mathcal{H}^\alpha(\kappa_\mu^{\mathfrak{M}}) \cap \Omega_{\mu+1} < \delta^{\mathfrak{M}_\mu} = \kappa^{\mathfrak{M}_\mu} = \kappa_{\mu+1}^{\mathfrak{M}}$. Summing up we have the following theorem.

Theorem 7.2 *It is $\mathcal{H}^\alpha(\kappa_\mu^{\mathfrak{M}}) \cap \Omega_{\mu+1} < \kappa_{\mu+1}^{\mathfrak{M}}$ for any consistent interpretation of the constants Ω_σ for $\sigma > \mu$. Let $\Psi^\alpha_{\kappa_{\mu+1}^{\mathfrak{M}}} := \mathcal{H}^\alpha(\kappa_\mu^{\mathfrak{M}}) \cap \kappa_{\mu+1}^{\mathfrak{M}}$.*

Note 7.3 For a fixed $\mu < \nu$ a possible consistent interpretation of the constants $\Omega_{\mu+\sigma}$ for $\sigma > 0$ are the ordinals ω_σ where $\lambda\xi.\,\omega_\xi$ enumerates the infinite cardinals. The set $\mathcal{H}^\alpha(\kappa_\mu^{\mathfrak{M}})$ contains all the initial ordinals $\kappa_\rho^{\mathfrak{M}}$ for $\rho < \mu$. Aiming at $\Psi^\alpha_{\kappa_{\mu+1}^{\mathfrak{M}}}$ as a bound for a semi–formal derivation we can therefore assume that in all $\Omega_{\rho+1}$–rules with $\rho + 1 < \mu$ the constants $\Omega_{\rho+1}$ are interpreted standardly

8 Collapsing and Elimination of the Constants Ω_μ

For a set X of ordinals the set $\mathcal{H}^{\Gamma_{\Omega_\nu+1}}(X)$ is a set with gaps. To utilize these gaps in the collapsing procedures we introduce the notion of an *operator controlled* semi–formal calculus. A technique that has been developed by Wilfried Buchholz in [8] as a simplification of the technique of *local predicativity* which was the basis in

[26] To avoid a lengthy passage which is inessential for our result we suppressed to indicate that the less than relation between terms that are not both strongly critical, i.e., at least one of them is different from Ω_μ or $\Psi^\alpha_{\Omega_{\mu+1}}$, can also be obtained primitive recursively, i.e., elementarily in our setting. This has been widely handled in the literature. E.g [39] Sect. 14, [35] Definition 9.6.7.

[30]. The key idea is to allow in $\vdash^{\alpha}_{\rho} \Delta$ only ordinals $\alpha \in \mathcal{H}(\mathrm{par}(\Delta))$ as measures of derivation lengths where $\mathcal{H}$ is a Skolem hull operator.

To make this precise we first have to define what we understand by a *parameter* in an $\mathcal{L}^*_{\mathbf{ID}_\nu}$–sentence G. These are all the ordinals ξ that occur in the form $I^{<\xi}_{F_\mu}$ in the sentence G. Observe that the "ordinals" $\mu \in \mathit{field}(\prec_0)$ are not counted as parameters; these are only natural numbers in the field of $\prec_0$. For a finite set Δ of sentences $\mathrm{par}(\Delta)$ are all the parameters that occur in sentences in Δ.

Definition 8.1 A semi–formal derivation $\vdash^{\alpha}_{\rho} \Delta$ is controlled by an operator $\mathcal{H}$ (denoted by $\mathcal{H} \vdash^{\alpha}_{\rho} \Delta$) if $\alpha \in \mathcal{H}(\mathrm{par}(\Delta))$ and for all inferences

$$\mathcal{H} \vdash^{\alpha_\iota}_{\rho} \Delta_\iota \text{ for } \iota \in I \;\Rightarrow\; \mathcal{H} \vdash^{\alpha}_{\rho} \Delta$$

which are not according to $(\bigwedge)$ we have $\mathrm{par}(\Delta_\iota) \subseteq \mathcal{H}(\mathrm{par}(\Delta))$.

It is easy to see that operator controlled semi–formal derivations possess the following *weakening property*

(WP) $\alpha_0 \le \alpha_1,\ \rho_0 \le \rho_1,\ \Delta_0 \subseteq \Delta_1,\ \mathcal{H}_0(\mathrm{par}(\Delta_0)) \subseteq \mathcal{H}_1(\mathrm{par}(\Delta_1))$, $\alpha_1 \in \mathcal{H}_1(\mathrm{par}(\Delta_1))$ *and* $\mathcal{H}_0 \vdash^{\alpha_0}_{\rho_0} \Delta_0$ *imply* $\mathcal{H}_1 \vdash^{\alpha_1}_{\rho_1} \Delta_1$.

Remark 8.2 Operator controlled semi–formal derivations are "more formal" than the more liberal notion of semi–formal derivations, let alone that of the verification calculus. E.g. the proof of the completeness direction in Theorem 6.2 does not work for operator controlled derivations. If we, for instance, have $\mathfrak{M}^*_\nu \models z \in I^{<\xi}_{F_0}$ then there is an $\eta < \xi$ such that $\mathfrak{M}^*_\nu \models z \in I^{\eta}_{F_0}$. Even if we have $\mathcal{H} \vdash^{\alpha_\eta}_{\rho} z \in I^{\eta}_{F_0}$ by induction hypothesis we cannot apply an inference $(\bigvee)$ since in general we cannot secure that $\eta \in \mathcal{H}(\mathrm{par}(z \in I^{<\xi}_{F_0}))$.

Yet what we can do is to transfer formal derivations into operator controlled semi–formal derivations. The key observation is that for any Skolem hull operator we obtain

$$\mathcal{H} \vdash^{2\cdot\mathrm{rnk}(F)}_{0} \Delta, F, \neg F \tag{28}$$

by induction on $\mathrm{rnk}(F)$. The induction hypothesis gives

$$\mathcal{H} \vdash^{2\cdot\mathrm{rnk}(G)}_{0} \Delta, G, \neg G$$

for all $G \in \mathrm{CS}(F)$. Assuming that F is in $\bigvee$ –type we obtain $\mathcal{H} \vdash^{2\cdot\mathrm{rnk}(G)+1}_{0} \Delta, F, \neg G$ by an inference $(\bigvee)$, since here we have $\mathrm{par}(G) \subseteq \mathcal{H}(\mathrm{par}(\Delta, F, \neg G))$, and finally $\mathcal{H} \vdash^{2\cdot\mathrm{rnk}(F)}_{0} \Delta, F, \neg F$ by a clause $(\bigwedge)$, since there is no parameter condition for $(\bigwedge)$–clauses. □

So we can collect the following facts:

- Clearly all true atomic sentences are operator controlled derivable.
- Building on (28) we obtain that (20) and (21) are both operator controlled derivable provided that the operator is closed under ordinal addition, i.e., that $+$ is among its generators.
- Using the $(\Omega_{\mu+1})$–rule we obtain from (28) that also (22) is operator controlled provable for any Skolem hull operator $\mathcal{H}$. Here we need that for any generated Skolem hull operator we always have

$$Y \subseteq \mathcal{H}(X) \;\Rightarrow\; \mathcal{H}(Y) \subseteq \mathcal{H}(X) \tag{29}$$

  by monotonicity and idempotency.
- Putting all together we obtain by induction on the length m of a formal $\mathbf{ID}_\nu(\mathsf{T})$ derivation $\models^{m} G(x_1, \ldots, x_n)$ that $\mathcal{H} \,\big|\!\frac{\Omega_\nu \cdot \omega + m}{\Omega_\nu + k}\, G(z_1, \ldots, z_n)^*$ holds true for some $k < \omega$, any tuple $z_1, \ldots, z_n$ of elements in $\mathfrak{M}$ and any Skolem hull operator that is closed under ordinal addition.

Thus we obtain an operator controlled version of Theorem 6.4.

Theorem 8.3 *If* $\mathbf{ID}_\nu(\mathsf{T}) \vdash G(x_1, \ldots, x_n)$ *then there is an* $\alpha < \Omega_\nu \cdot \omega + \omega$ *and a finite ordinal k such that* $\mathcal{H} \,\big|\!\frac{\alpha}{\Omega_\nu + k}\, G(z_1, \ldots, z_n)^*$ *holds true for any choice of elements* $z_1, \ldots, z_n \in \mathfrak{M}$ *and any Skolem hull operator* $\mathcal{H}$ *closed under* $+$.

The central tool in predicative proof theory is cut elimination and the crucial theorem states

$$\big|\!\frac{\alpha}{\beta + \omega^\rho}\, \Delta \text{ implies } \big|\!\frac{\varphi_\rho(\alpha)}{\beta}\, \Delta\,.$$

In this sense predicative proof theory is ruled by the Veblen function. This theorem survives controlling operators provided that there are no initial ordinals in the interval $[\beta, \beta + \omega^\rho)$. Setting $\mathbb{K} = \{\Omega_\mu \,|\, \mu \leq \nu\}$ we thus have

$$\mathcal{H} \,\big|\!\frac{\alpha}{\beta + \omega^\rho}\, \Delta,\ \rho \in \mathcal{H}(\mathrm{par}(\Delta)) \textit{ and } [\beta, \beta + \omega^\rho) \cap \mathbb{K} = \emptyset \textit{ implies } \mathcal{H} \,\big|\!\frac{\varphi_\rho(\alpha)}{\beta}\, \Delta. \tag{30}$$

Since impredicative proof theory bases on predicative proof theory—in spite of semantical cut–elimination—these preliminary remarks show that we need operators that are closed under addition, the Veblen functions and allow for gaps. Though not absolutely necessary it is convenient for continuity reasons to have the Veblen function among the generation functions of the operator. That is why we have opted in Sect. 7 to generate $\mathcal{H}$ by $\{+,\ \varphi\} \cup \{\Omega_\mu \,|\, 0 < \mu \leq \nu\}$. From now on we fix this operator.

We next observe that also the Boundedness Property survives operator controlling with a slight modification the formulation of which needs the definition

$$\mathcal{H}[\mathcal{Y}](\mathcal{X}) := \mathcal{H}(\mathcal{Y} \cup \mathcal{X}). \tag{31}$$

With the hypotheses of the Boundedness Theorem (Theorem 6.9) the claim modifies to

$$\mathcal{H} \overset{\alpha}{\underset{\rho}{\vdash}} \Delta(I_{F_\mu}^{<\xi}) \;\Rightarrow\; \mathcal{H}[\{\xi\}] \overset{\alpha}{\underset{\rho}{\vdash}} \Delta(I_{F_\mu}^{<\zeta}). \tag{32}$$

The ordinal analysis of the theories $\mathbf{ID}_\nu(\mathsf{T})$ rests on the elimination of the initial ordinals which is obtained by a collapsing procedure on operator controlled derivations. To formulate the main theorem let

$$\Omega'_\mu = \begin{cases} \Omega_\mu & \text{if} \, 0 < \mu \in Lim \\ \Omega_\mu + 1 & \text{if} \, \mu \, \text{is a successor.} \end{cases}$$

Recall moreover the natural sum which for $\alpha =_{NF} \alpha_1 + \cdots + \alpha_n$ and $\beta =_{NF} \alpha_{n+1} + \cdots + \alpha_{n+k}$ returns $\alpha \# \beta = \alpha_{\pi(1)} + \cdots + \alpha_{\pi(n+k)}$ where π is a permutation of the numbers $1, \ldots, n+k$ such that $\alpha_{\pi(i)} \geq \alpha_{\pi(i+1)}$.

Theorem 8.4 *Assume* $\mathcal{H}^{\gamma+1} \overset{\alpha}{\underset{\Omega'_\rho}{\vdash}} \Delta$ *for a finite set* Δ *of* $\mathcal{L}^*_{\mathbf{ID}_\nu}$*–sentences in* $\bigvee_1^{\Omega_{\mu+1}}$*–type and* $\gamma \in \mathcal{H}^\gamma(par(\Delta))$ *such that* $par(\Delta) \subseteq \mathcal{H}^{\gamma+1}(\Omega_\mu)$*. Then we obtain*

$$\mathcal{H}^{\gamma \# \omega^{\alpha \# \Omega_\rho}+1} \overset{\Psi_{\Omega_{\mu+1}}^{\gamma \# \omega^{\alpha \# \Omega_\rho}}}{\underset{\Psi_{\Omega_{\mu+1}}^{\gamma \# \omega^{\alpha \# \Omega_\rho}}}{\vdash}} \Delta.$$

Proof The proof is by main–induction on ρ and side–induction on α. Let

(I) $\mathcal{H}^{\gamma+1} \overset{\alpha_\iota}{\underset{\Omega'_\rho}{\vdash}} \Delta_\iota$ for $\iota \in I \Rightarrow \mathcal{H}^{\gamma+1} \overset{\alpha}{\underset{\Omega'_\rho}{\vdash}} \Delta$

be the last inference. First we show

$$\mathrm{par}(\Delta_\iota) \subseteq \mathcal{H}^{\gamma+1}(\Omega_\mu). \tag{i}$$

If (I) is different from a $(\bigwedge)$–rule (i) is obvious from $\mathrm{par}(\Delta_\iota) \subseteq \mathcal{H}^{\gamma+1}(\mathrm{par}(\Delta)) \subseteq \mathcal{H}^{\gamma+1}(\Omega_\mu)$ and (29). In case of an $(\bigwedge)$–rule we either have $\mathrm{par}(\Delta_\iota) = \mathrm{par}(\Delta)$ or the principal sentence in Δ is $(n \notin I_{F_\sigma}^{<\xi})$ and $\Delta_\iota = \Delta_\eta := \Delta, (n \notin I_{F_\sigma}^{\eta})$ for $\eta < \xi$. Since $(n \notin I_{F_\sigma}^{<\xi})$ is in $\bigvee_1^{\Omega_{\mu+1}}$–type we have $\xi < \Omega_{\mu+1}$. Hence $\xi \in \mathcal{H}^{\gamma+1}(\Omega_\mu) \cap \Omega_{\mu+1} = \Psi_{\Omega_{\mu+1}}^{\gamma+1}$ which implies $\eta \in \mathcal{H}^{\gamma+1}(\Omega_\mu)$ and thus $\mathrm{par}(\Delta_\eta) = \mathrm{par}(\Delta) \cup \{\eta\} \subseteq \mathcal{H}^{\gamma+1}(\Omega_\mu)$.

If (I) is not a cut of rank $\geq \Omega_{\mu+1}$ we again have $\Delta_\iota \subseteq \bigvee_1^{\Omega_{\mu+1}}$–type and can apply the induction hypothesis to the premises of (I). Putting $\widehat{\delta} := \gamma \# \omega^{\delta \# \Omega_\rho}$ we thus have

$$\mathcal{H}^{\widehat{\alpha_\iota}+1} \overset{\Psi_{\Omega_{\mu+1}}^{\widehat{\alpha_\iota}}}{\underset{\Psi_{\Omega_{\mu+1}}^{\widehat{\alpha_\iota}}}{\vdash}} \Delta_\iota \text{ for all } \iota \in I. \tag{ii}$$

From $\alpha_\iota, \gamma \in \mathcal{H}^{\gamma+1}(\mathrm{par}(\Delta_\iota)) \subseteq \mathcal{H}^{\widehat{\alpha}}(\Omega_\mu)$ we get $\widehat{\alpha_\iota} \in \mathcal{H}^{\widehat{\alpha}}(\Omega_\mu) \cap (\widehat{\alpha})$, hence $\Psi_{\Omega_{\mu+1}}^{\widehat{\alpha_\iota}} < \Psi_{\Omega_{\mu+1}}^{\widehat{\alpha}}$ by (24). From $\alpha, \gamma \in \mathcal{H}^{\gamma+1}(\mathrm{par}(\Delta)) \subseteq \mathcal{H}^{\widehat{\alpha}+1}(\mathrm{par}(\Delta))$ we get

$\widehat{\alpha} \in \mathcal{H}^{\widehat{\alpha}+1}(\mathrm{par}(\Delta))$ and thus $\Psi^{\widehat{\alpha}}_{\Omega_{\mu+1}} \in \mathcal{H}^{\widehat{\alpha}+1}(\mathrm{par}(\Delta))$. So if (I) was not a cut of rank bigger than Ω_μ we obtain the claim by the same inference taking (ii) as premises.

So suppose that (I) is a cut of rank $\geq \Omega_\mu$. Then we have the premises

$$\mathcal{H}^{\gamma+1} \left|\frac{\alpha_0}{\Omega'_\rho}\right. \Delta, G \text{ and } \mathcal{H}^{\gamma+1} \left|\frac{\alpha_0}{\Omega'_\rho}\right. \Delta, \neg G \tag{iii}$$

with $\Omega_\mu \leq \mathrm{rnk}(G) < \Omega'_\rho$. If there is a τ such that $\Omega_\mu \leq \mathrm{rnk}(G) < \Omega_{\tau+1} < \Omega'_\rho$ then $\Delta, G, \neg G$ are in $\bigvee_1^{\Omega_{\tau+1}}$–type and $\mathrm{par}(\Delta, G, \neg G) \subseteq \mathcal{H}^{\gamma+1}(\Omega_{\tau+1})$. Applying the side induction hypothesis to (iii) thus yields

$$\mathcal{H}^{\widehat{\alpha_0}+1} \left|\frac{\Psi^{\widehat{\alpha_0}}_{\Omega_{\tau+1}}}{\Psi^{\widehat{\alpha_0}}_{\Omega_{\tau+1}}}\right. \Delta, G \text{ and } \mathcal{H}^{\widehat{\alpha_0}+1} \left|\frac{\Psi^{\widehat{\alpha_0}}_{\Omega_{\tau+1}}}{\Psi^{\widehat{\alpha_0}}_{\Omega_{\tau+1}}}\right. \Delta, \neg G. \tag{iv}$$

By cut we obtain from (iv)

$$\mathcal{H}^{\widehat{\alpha_0}+1} \left|\frac{\Psi^{\widehat{\alpha_0}}_{\Omega_{\tau+1}} \cdot 2}{\Omega'_\tau}\right. \Delta.$$

The main induction hypothesis then yields

$$\mathcal{H}^{\beta+1} \left|\frac{\Psi^{\beta}_{\Omega_{\mu+1}}}{\Psi^{\beta}_{\Omega_{\mu+1}}}\right. \Delta$$

for $\beta := \widehat{\alpha_0} \# \omega^{\Psi^{\widehat{\alpha_0}}_{\Omega_{\tau+1}} \cdot 2 \# \Omega_\tau} < \gamma \# \omega^{\alpha \# \Omega_\rho} = \widehat{\alpha}$. Hence $\beta \in \mathcal{H}^{\widehat{\alpha}}(\Omega_\mu) \cap \widehat{\alpha}$ which entails $\Psi^{\beta}_{\Omega_{\mu+1}} < \Psi^{\widehat{\alpha}}_{\Omega_{\mu+1}}$ and thus the claim by weakening.

It remains the case $\mathrm{rnk}(G) = \Omega_{\mu+1} = \Omega_\rho$. Then G has the form $n \in I_{F_\mu}^{<\Omega_{\mu+1}}$ and (iii) becomes

$$\mathcal{H}^{\gamma+1} \left|\frac{\alpha_0}{\Omega_{\mu+1}+1}\right. \Delta, n \in I_{F_\mu}^{<\Omega_{\mu+1}} \text{ and } \mathcal{H}^{\gamma+1} \left|\frac{\alpha_0}{\Omega_{\mu+1}+1}\right. \Delta, n \notin I_{F_\mu}^{<\Omega_{\mu+1}}. \tag{v}$$

Here we cannot apply the induction hypothesis to the right hand premise but with $\mathcal{H}^{\gamma+1} \left|\frac{\alpha_0}{\Omega_{\mu+1}+1}\right. \Delta, n \notin I_{F_\mu}^{<\Omega_{\mu+1}}$ we clearly also have

$$\mathcal{H}^{\gamma+1} \left|\frac{\alpha_0}{\Omega_{\mu+1}+1}\right. \Delta, n \notin I_{F_\mu}^{<\eta} \tag{vi}$$

for all $\eta < \Omega_{\mu+1}$ and $n \notin I_{F_\mu}^{<\eta}$ belongs to $\bigvee_1^{\Omega_{\mu+1}}$–type. The side induction hypothesis applied to the left hand premise and (vi) thus gives

$$\mathcal{H}^{\widehat{\alpha_0}+1} \left|\frac{\Psi^{\widehat{\alpha_0}}_{\Omega_{\mu+1}}}{\Psi^{\widehat{\alpha_0}}_{\Omega_{\mu+1}}}\right. \Delta, n \in I_{F_\mu}^{<\Omega_{\mu+1}} \text{ and } \mathcal{H}^{\widehat{\alpha_0}+1} \left|\frac{\Psi^{\widehat{\alpha_0}}_{\Omega_{\mu+1}}}{\Psi^{\widehat{\alpha_0}}_{\Omega_{\mu+1}}}\right. \Delta, n \notin I_{F_\mu}^{<\eta} \tag{vii}$$

By the Boundedness Theorem (as modified in (32)) we can lower $\Omega_{\mu+1}$ to $\Psi^{\widehat{\alpha_0}}_{\Omega_{\mu+1}}$ in the left derivation and pick $\eta = \Psi^{\widehat{\alpha_0}}_{\Omega_{\mu+1}} < \Omega_{\mu+1}$ in the right derivation. Cutting these derivations yields

$$\mathcal{H}^{\widehat{\alpha_0}+1} \Big|\frac{\Psi^{\widehat{\alpha_0}}_{\Omega_{\mu+1}}}{\Psi^{\widehat{\alpha_0}}_{\Omega_{\mu+1}}+1} \Delta. \tag{viii}$$

By predicative cut–elimination (30) we thus obtain

$$\mathcal{H}^{\widehat{\alpha_0}+1} \Big|\frac{\varphi_\beta(\beta)}{\Omega'_\mu} \Delta$$

with $\beta := \Psi^{\widehat{\alpha_0}}_{\Omega_{\mu+1}}$. The main–induction hypothesis finally leads to

$$\mathcal{H}^{\delta+1} \Big|\frac{\Psi^{\delta}_{\Omega_{\mu+1}}}{\Psi^{\delta}_{\Omega_{\mu+1}}} \Delta$$

for $\delta = \widehat{\alpha_0} \# \omega^{\varphi_\beta(\beta)\#\Omega_\mu} \in \mathcal{H}^{\delta+1}(\Omega_\mu) \cap \widehat{\alpha}$. Hence $\Psi^{\delta}_{\Omega_{\mu+1}} < \Psi^{\widehat{\alpha}}_{\Omega_{\mu+1}}$ which together with $\Psi^{\widehat{\alpha}}_{\Omega_{\mu+1}} \in \mathcal{H}^{\widehat{\alpha}+1}(\mathrm{par}(\Delta))$ entails the claim by weakening. □

Corollary 8.5 *Let* T *be a Peano–like axiomatization for an acceptable countable structure* $\mathfrak{M}$. *Then* $\kappa^{\mathfrak{M}}_{\mu+1}(\mathbf{ID}_\nu(\mathsf{T})) \leq \Psi^{\varepsilon_{\Omega_\nu+1}}_{\kappa^{\mathfrak{M}}_{\mu+1}} < \kappa^{\mathfrak{M}}_{\mu+1}$ *for* $\mu < \nu$.

Proof From $\mathbf{ID}_\nu(\mathsf{T}) \vdash n \in I_{F_\mu}$ we obtain by Theorem 8.3 $\mathcal{H} \Big|\frac{\alpha_0}{\Omega_\nu+k} n \in I^{<\Omega_{\mu+1}}_{F_\mu}$ and thence by predicative cut–elimination $\mathcal{H} \Big|\frac{\alpha_1}{\Omega'_\nu} n \in I^{<\Omega_{\mu+1}}_{F_\mu}$ for $\alpha_0, \alpha_1 \in \mathcal{H}(\emptyset) \cap \varepsilon_{\Omega_\nu+1}$. By Theorem 8.4 it follows $\mathcal{H}^{\alpha_2+1} \Big|\frac{\Psi^{\alpha_2}_{\Omega_{\mu+1}}}{\Psi^{\alpha_2}_{\Omega_{\mu+1}}} n \in I^{<\Omega_{\mu+1}}_{F_\mu}$ By Boundedness this implies $\mathcal{H}^{\alpha_2+1} \Big|\frac{\Psi^{\alpha_2}_{\Omega_{\mu+1}}}{\Psi^{\alpha_2}_{\Omega_{\mu+1}}} n \in I^{<\Psi^{\alpha_2}_{\Omega_{\mu+1}}}_{F_\mu}$. Since $n \in I^{<\Psi^{\alpha_2}_{\Omega_{\mu+1}}}_{F_\mu}$ belongs to $\bigvee^{\Omega_{\mu+1}}_0$–type we obtain $\mathfrak{M}^*_\nu \models n \in I^{<\Psi^{\alpha_2}_{\Omega_{\mu+1}}}_{F_\mu}$ by Corollary 6.6 if all Ω_ρ for $\rho \leq \mu$ are interpreted standardly by $\kappa^{\mathfrak{M}}_\rho$. Hence $|n|_{F_\mu} \leq \Psi^{\alpha_2}_{\Omega_{\mu+1}} < \Psi^{\varepsilon_{\Omega_\nu+1}}_{\Omega_{\mu+1}} = \mathcal{H}^{\varepsilon_{\Omega_\nu+1}}(\Omega_\mu) \cap \Omega_{\mu+1} = \mathcal{H}^{\varepsilon_{\Omega_\nu+1}}(\kappa^{\mathfrak{M}}_\mu) \cap \Omega_{\mu+1}$ $= \Psi^{\varepsilon_{\Omega_\nu+1}}_{\kappa^{\mathfrak{M}}_{\mu+1}} < \kappa^{\mathfrak{M}}_{\mu+1}$ by Theorem 7.2. □

Setting $\mu = 0$ in the above corollary we reobtain the upper bound for the prooftheoretic ordinal of $\mathbf{ID}_\nu(\mathsf{T})$.

Corollary 8.6 *Let* T *be a Peano–like axiomatization for an acceptable countable structure* $\mathfrak{M}$. *Then* $|\mathbf{ID}_\nu(\mathsf{T})| = \delta^{\mathfrak{M}}(\mathbf{ID}_\nu(\mathsf{T})) \leq \Psi^{\varepsilon_{\Omega_\nu+1}}_{\Omega_0+1} < \kappa^{\mathfrak{M}}$.

As first shown in [9] and later reproved in [5] this ordinal bound is the exact one. To make the paper not even longer we will not reprove this result. The main idea is to use the fact that in every $\mathfrak{M}_\mu$ for $\mu \leq \nu$ there is an elementarily definable copy of $\mathcal{H}^{\Gamma_{\Omega_\nu+1}}(\kappa^{\mathfrak{M}}_\mu)$. We thus obtain a hierarchy $<_\mu$ of orderings that are elementary in $\mathfrak{M}_\mu$. For every $\mu < \nu$ we then have the condensing property

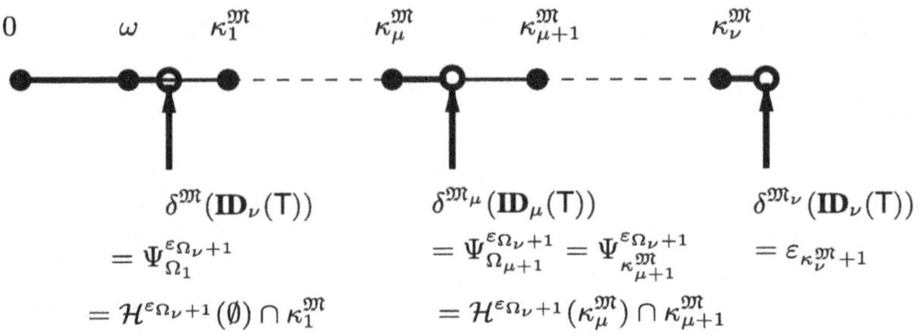

Fig. 1 *Spec*$^{\mathfrak{M}}(\mathbf{ID}_\nu(\mathsf{T}))$

$$\mathbf{ID}_\nu(\mathsf{T}) \vdash TI(<_\nu, \alpha) \;\Rightarrow\; \mathbf{ID}_\nu(\mathsf{T}) \vdash \Psi^\alpha_{\Omega_{\mu+1}} \in \mathrm{Acc}_{<_\mu}$$

where $TI(<_\nu, \alpha)$ expresses the wellfoundedness of $<_\nu\restriction\alpha$ and $\mathrm{Acc}_{<_\mu}$ denotes the accessible part of $<_\mu$, i.e., the fixed–point of the formula $(\forall y)[y <_\mu x \to y \in X]$.

Since $TI(<_\nu, \Omega_\nu)$ holds trivially and we have the full scheme of Mathematical Induction in $\mathbf{ID}_\nu(\mathsf{T})$ we can adapt the familiar Gentzen proof to obtain $TI(<_\nu, \alpha)$ for all $\alpha < \varepsilon_{\Omega_\nu+1}$. Putting $\Psi^\alpha_{\kappa_\nu^{\mathfrak{M}}} := \alpha$ and $\kappa_0^{\mathfrak{M}} = \omega$ we, together with the condensing property and Theorem 7.2, finally get our main theorem.

Theorem 8.7 *Let* T *be a Peano–like axiomatization for a countable acceptable structure* $\mathfrak{M}$. *Then*

$$Spec^{\mathfrak{M}}(\mathbf{ID}_\nu(\mathsf{T})) = \{\Psi^{\varepsilon_{\Omega_\nu+1}}_{\kappa^{\mathfrak{M}}_{\mu+1}} \,|\, \mu < \nu\} = \{\mathcal{H}^{\varepsilon_{\Omega_\nu+1}}(\kappa_\mu^{\mathfrak{M}}) \cap \kappa^{\mathfrak{M}}_{\mu+1} \,|\, \mu < \nu\}.$$

(Cf. Fig. 1).

Remark 8.8 In [30] Section δ.3 we already tried to give a definition of the *spectrum* of a formal theory. As mentioned there in Remark 3.7 a still tentative definition which was not completely felicitous. Instead of "the least set of ordinals which are needed to carry through the ordinal analysis" it should rather have been the "least operator" which is needed for the ordinal analysis. Then, however, the technique of operator controlled semi–formal derivations was not yet developed.

Indeed, it follows from Theorem 8.7 that *Spec*$^{\mathfrak{M}}(\mathbf{ID}_\nu(\mathsf{T}))$, as defined in Eq. (5), is generated by the operator $\mathcal{H}^{\varepsilon_{\Omega_\nu+1}}$ and this seems to be the "least" operator which is needed for the ordinal analysis of $\mathbf{ID}_\nu(\mathsf{T})$.

Actually, knowing $\mathcal{H}^{\varepsilon_{\Omega_\nu+1}}$ is more important and—as we will see in the next section—also more informative than just knowing the points in *Spec*$^{\mathfrak{M}}(\mathbf{ID}_\nu(\mathsf{T}))$.

Spector classes have been invented to generalize computation theory to abstract structures.[27] The Spector structures introduced in our context are the structures $\mathfrak{M}_{\mu+1} = (\mathfrak{M}, \mathbf{\Gamma}_{\mu+1}(\mathfrak{M}))$ with $o^{\mathfrak{M}_{\mu+1}} = \kappa^{\mathfrak{M}}_{\mu+1}$. We say that a partial function $f : \mathfrak{M} \longrightarrow \mathfrak{M}$

[27] Cf. [24].

is $\kappa^{\mathfrak{M}}_{\mu+1}$–recursive iff its graph G_f belongs to $\mathbf{\Gamma}_{\mu+1}(\mathfrak{M})$. The next theorem shows that the points in the spectrum of $\mathbf{ID}_\nu(\mathsf{T})$ characterize the $\kappa^{\mathfrak{M}}_{\mu+1}$–recursive functions whose totality is provable in $\mathbf{ID}_\nu(\mathsf{T})$. It is an immediate consequence of Theorem 8.4.

Theorem 8.9 *Let* T *be a Peano–like axiomatization of a countable acceptable structure* $\mathfrak{M}$ *and* $f\colon \mathfrak{M} \longrightarrow \mathfrak{M}$ *a* $\kappa^{\mathfrak{M}}_{\mu+1}$*–recursive function whose totality is provable in* $\mathbf{ID}_\nu(\mathsf{T})$*. Then* f *is bounded by* $\Psi^{\varepsilon_{\Omega_\nu+1}}_{\Omega_{\mu+1}}$*. i.e., for all* $\langle x, f(x)\rangle$ *we have* $|\langle x, f(x)\rangle|_{G_f} < \Psi^{\varepsilon_{\Omega_\nu+1}}_{\Omega_{\mu+1}}$.

Note 8.10 In this paper we have restricted ourselves to the simplest case of iterated inductive definitions, namely iteration along a given primitive–recursive wellordering. We did this by purpose in order not to obscure the—in principle— simple ideas behind the ordinal analysis. Generalizations to more complex theories, e.g., $\mathbf{ID}_{\prec^*}$ which allows iterations along the accessible part $\prec^*$ of a primitive–recursive relation $\prec$ or even AUT($\mathbf{ID}$), which allows for autonomous iterations of inductive definitions, are not too difficult to obtain. The additional work to handle these theories rests on a strengthening of the needed Skolem hull operator, e.g., augmenting the generating functions by a function $\xi \mapsto \Omega_\xi$. Once the iterations of this Skolem hull operator are studied the collapsing procedure for the semi–formal derivations follows the same pattern as described above.

9 Provably Recursive Functions

Note 9.1 In this section we will restrict ourselves to the case that the basis structure is the standard structure $\mathbb{N}$ of arithmetic which is axiomatized by (an extension of) the Peano axioms PA which includes the defining axioms for all primitive recursive functions and –relations. We put $\mathfrak{N}_\nu := \mathbf{ID}_\nu(\mathbb{N})$ and $\mathbf{ID}_\nu := \mathbf{ID}_\nu(\mathsf{PA})$.

The paradigmatic example of a Spector class is the class $\mathbf{\Sigma}^0_1$ of arithmetical relations that are Σ^0_1–definable.[28] We obtain $o^{\mathbf{\Sigma}^0_1} = \omega$ and the $\mathbf{\Sigma}^0_1$–partial recursive functions are the familiar partial recursive functions. Looking at Fig. 1 we notice that there is no arrow pointing below ω for the Spector class $\mathbf{\Sigma}^0_1$. So apparently a point in $Spec^{\mathbb{N}}(\mathbf{ID}_\nu)$ is lacking.

Clearly there cannot be a finite ordinal that satisfies an analogue of Theorem 8.9. Nevertheless we are going to show that the recursive functions the totality of which is provable in $\mathbf{ID}_\nu$ can be characterized by a subrecursive hierarchy which is canonically generated by the operator $\mathcal{H}^{\varepsilon_{\Omega_\nu+1}}$ and thus can be characterized by the ordinal $\mathcal{H}^{\varepsilon_{\Omega_\nu+1}}(0) \cap \Omega_1 = \Psi^{\varepsilon_{\Omega_\nu+1}}_{\Omega_1} = \delta^{\mathbb{N}}(\mathbf{ID}_\nu)$.

To become more specific let $\Phi\colon \mathbb{N} \longrightarrow \mathbb{N}$ be a strictly increasing primitive–recursive function and $\mathcal{T}$ a primitive–recursively definable set of ordinal–terms such

[28] "Closure under first order quantification" has to be replaced by "closure under bounded quantification" in $\mathbf{\Sigma}^0_1$.

that every $\alpha \in \mathcal{T}$ is equipped with a norm $N(\alpha) \in \mathbb{N}$. We assume that $N(\alpha)$ is primitive–recursively computable from α. Following an idea of Weiermann [10] we obtain a subrecursive hierarchy by defining a function $\phi\colon \mathcal{T} \longrightarrow \mathbb{N}$

$$\phi(\alpha) := \sup\{\phi(\beta) + 1 \,|\, \beta < \alpha \wedge N(\beta) \leq \Phi(N(\beta))\} \cup \{0\} \tag{33}$$

and then a family of functions $\Phi_\alpha\colon \mathbb{N} \longrightarrow \mathbb{N}$

$$\Phi_\alpha(n) := \phi(\omega \cdot \alpha + n). \tag{34}$$

If we denote by Φ^k the familiar iterated application of Φ this yields

$$\Phi_\alpha(n) \approx \Phi^\alpha(n)$$

for $\alpha < \omega$ and may thus be regarded as a generalization of the finite iterations of the function Φ. The subrecursive hierarchy $\{\Phi_\alpha \,|\, \alpha \leq \varepsilon_0\}$ is closely connected to the familiar fast growing hierarchy and matches the Hardy hierarchy (at least for an initial segment of the ordinals).

In Sect. 7 we have seen that the ordinals in $\mathcal{H}^{\Gamma_{\Omega_\nu+1}}(0)$ can be represented by the ordinal terms in $\mathcal{T}$. This set is inductively defined and we define the norm $N(\alpha)$ of an ordinal(term) α as the stage at which α appears in $\mathcal{T}$. This yields, e.g., $N(n) = n$ for $n < \omega$. In the sequel we tacitly assume $n < \omega$ whenever we write n.

Fixing a sufficiently increasing primitive–recursive start function Φ—requiring $\Phi(x) + 3n + 9 \leq \Phi(x + n)$ would do—and given a Skolem hull operator $\mathcal{H}$ we generalize (33) to

$$\phi^{\mathcal{H}(\mathcal{X})}(\alpha) := \sup\{\phi^{\mathcal{H}(\mathcal{X})}(\beta) + 1 \,|\, \beta \in \mathcal{H}(\mathcal{X}) \cap \alpha \wedge N(\beta) \leq \Phi(N(\alpha))\} \cup \{0\} \tag{35}$$

The crucial feature of the function $\phi^{\mathcal{H}}$—which is needed to prove nearly all its further properties—is

$$\phi^{\mathcal{H}(\mathcal{X})}(\alpha + \phi^{\mathcal{H}(\mathcal{X})}(\beta) + n) \leq \phi^{\mathcal{H}(\mathcal{X})}(\alpha \# \beta + n) \tag{36}$$

which follows by induction on β.

For a finite set $\mathcal{X}$ of ordinals let $|\mathcal{X}| := \max\{N(\alpha) \,|\, \alpha \in \mathcal{X}\} + 1$. Similarly we define for a finite set Δ of $\mathcal{L}^*_{\mathbf{ID}_\nu}$–sentences $|\Delta| = \max\{N(\alpha) \,|\, \alpha \in \mathrm{par}^*(\Delta)\} + 1$ where $\mathrm{par}^*(\Delta)$ are the parameters in the sentences of Δ where we also count constants for natural numbers as finite ordinals among the parameters.

For a finite set $\mathcal{X}$ of ordinals we generalize (34) to

$$\Phi^\alpha_{\mu+1}(\mathcal{X}) := \phi^{\mathcal{H}(\mathcal{X})}(\Psi^\alpha_{\Omega_{\mu+1}} + |\mathcal{X}|) \text{ and put } \Phi_\alpha(n) := \phi^{\mathcal{H}(\emptyset)}(\omega \cdot \alpha + n). \tag{37}$$

Observe that this yields $\Phi^\alpha_1(n) = \phi^{\mathcal{H}(n)}(\Psi^\alpha_{\Omega_1} + n) = \Phi_{\Psi^\alpha_{\Omega_1}}(n)$.

With these notions we define the *fragmentation* of an iterated Skolem hull operator.[29]

Definition 9.2 Let $f\mathcal{H}^{\alpha}_{\mu+1}(\mathcal{X}) := \{\xi \in \mathcal{H}^{\alpha}(\mathcal{X}) \,|\, N(\xi) \leq \Phi^{\alpha}_{\mu+1}(\mathcal{X})\}$.
We call $f\mathcal{H}^{\alpha}_{\mu+1}$ the $\mu+1$–fragmentation of the iterated operator $\mathcal{H}^{\alpha}$.

A derivation $\overset{\alpha}{\underset{\rho}{\vdash}} \Delta$ is *fragmented controlled* by $f\mathcal{H}^{\gamma}_{\mu+1}$—written as $f\mathcal{H}^{\gamma}_{\mu+1} \overset{\alpha}{\underset{\rho}{\vdash}} \Delta$—if $\{\gamma, \alpha\} \subseteq f\mathcal{H}^{\gamma}_{\mu+1}(\text{par}^*(\Delta))$ and $\text{par}^*(\Delta_\iota) \subseteq f\mathcal{H}^{\delta}_{\mu+1}(\text{par}^*(\Delta))$ holds true for the premises Δ_ι of all inferences with conclusion Δ different from inferences according to $(\bigwedge)$.

As a word of caution we want to mention that $\Psi^{\alpha}_{\Omega_{\mu+1}} + |\mathcal{X}| < \Psi^{\beta}_{\Omega_{\rho+1}} + |\mathcal{Y}|$ is not sufficient to obtain $\Phi^{\alpha}_{\mu+1}(\mathcal{X}) \leq \Phi^{\beta}_{\rho+1}(\mathcal{Y})$. What we need besides $\Psi^{\alpha}_{\Omega_{\mu+1}} < \Psi^{\beta}_{\Omega_{\rho+1}}$ is $N(\alpha) + |\mathcal{X}| \leq \Phi(N(\beta) + |\mathcal{Y}|)$.

The fragmented Skolem hull operators loose of course their idempotency, especially the property $\mathcal{Y} \subseteq \mathcal{H}(\mathcal{X}) \Rightarrow \mathcal{H}(\mathcal{Y}) \subseteq \mathcal{H}(\mathcal{X})$. A replacement is

$$\mathcal{Y} \subseteq f\mathcal{H}^{\gamma}_{\mu+1}(\mathcal{X}) \textit{ and } \beta \in f\mathcal{H}^{\gamma}_{\mu+1}(\mathcal{X} \cup \mathcal{Y}) \textit{ imply } \beta \in f\mathcal{H}^{\gamma+1}_{\mu+1}(\mathcal{X}) \tag{38}$$

which is easily checked using (36). The main theorem for fragmented controlled derivations is the following Witnessing Theorem.

Theorem 9.3 *Assume* $f\mathcal{H}^{\gamma}_{\mu+1} \overset{\alpha}{\underset{0}{\vdash}} (\exists x)F(x)$ *for a* Σ^0_1*–sentence with parameters. Then there is a natural number* $m \leq \Phi^{\gamma \# \omega^{\alpha+1}}_{\mu+1}$ *(par*((∃x)F(x))) such that* $\mathbb{N} \models F(m)$.

Proof The proof is by induction on α. The only possible premise is

$$f\mathcal{H}^{\gamma}_{\mu+1} \overset{\alpha_0}{\underset{0}{\vdash}} (\exists x)F(x), F(k) \tag{i}$$

for some natural number k. Let $\mathcal{X} := \text{par}^*((\exists x)F(x))$. We have $k \in f\mathcal{H}^{\gamma}_{\mu+1}(\mathcal{X})$, hence $k = N(k) \leq \Phi^{\gamma}_{\mu+1}(\mathcal{X}) \leq \Phi^{\gamma \# \omega^{\alpha+1}}_{\mu+1}(\mathcal{X})$. If $\mathbb{N} \models F(k)$ we are thus done. Otherwise a nested induction on α_0 proves

$$f\mathcal{H}^{\gamma}_{\mu+1}[\{k\}] \overset{\alpha_0}{\underset{0}{\vdash}} (\exists x)F(x)$$

for $f\mathcal{H}^{\gamma}_{\mu+1}[\mathcal{Y}](\mathcal{X}) := f\mathcal{H}^{\gamma}_{\mu+1}(\mathcal{X} \cup \mathcal{Y})$, hence

$$f\mathcal{H}^{\gamma+1}_{\mu+1} \overset{\alpha_0}{\underset{0}{\vdash}} (\exists x)F(x) \tag{ii}$$

by (38). By induction hypothesis we thus obtain an m such that

$$\mathbb{N} \models F(m) \text{ and } m \leq \Phi^{\gamma+1 \# \omega^{\alpha_0+1}}_{\mu+1}(\mathcal{X}). \tag{iii}$$

---

[29]This is based on an idea due to Jan Carl Stegert in [42].

Now we draw on $\alpha_0 \in f\mathcal{H}^{\gamma}_{\mu+1}(\mathcal{X} \cup \{k\})$, hence $\alpha_0 \in f\mathcal{H}^{\gamma+1}_{\mu+1}(\mathcal{X})$, and compute

$$\begin{aligned}\Phi^{\gamma+1\#\omega^{\alpha_0+1}}_{\mu+1}(\mathcal{X}) = \phi^{\mathcal{H}^{\gamma+1\#\omega^{\alpha_0+1}}(\mathcal{X})}(\Psi^{\gamma+1\#\omega^{\alpha_0+1}}_{\Omega_{\mu+1}} + |\mathcal{X}|) \leq \\ \phi^{\mathcal{H}^{\gamma+1\#\omega^{\alpha}}(\mathcal{X})}(\Psi^{\gamma+1\#\omega^{\alpha}}_{\Omega_{\mu+1}} + N(\alpha_0) + |\mathcal{X}|) \leq \\ \phi^{\mathcal{H}^{\gamma+1\#\omega^{\alpha}}(\mathcal{X})}(\Psi^{\gamma+1\#\omega^{\alpha}}_{\Omega_{\mu+1}} + \phi^{\mathcal{H}^{\gamma+1}(\mathcal{X})}(\Psi^{\gamma+1}_{\Omega_{\mu+1}} + |\mathcal{X}|) + |\mathcal{X}|) \leq \\ \phi^{\mathcal{H}^{\gamma+1\#\omega^{\alpha}}(\mathcal{X})}(\Psi^{\gamma+1\#\omega^{\alpha}}_{\Omega_{\mu+1}} \# \Psi^{\gamma+1}_{\Omega_{\mu+1}} + 2|\mathcal{X}|) \leq \Phi^{\gamma\#\omega^{\alpha+1}}_{\mu+1}(\mathcal{X}). \quad \square\end{aligned}$$

Once we succeed in finding a $f\mathcal{H}^{\gamma}_{1}$–controlled cut–free semi–formal derivation for the Σ^0_1–sentences that are provable in $\mathbf{ID}_\nu$ the Witnessing Theorem together with

$$\Phi^{\alpha}_{1}\{n_1, \ldots, n_k\} = \Phi_{\Psi^{\alpha}_{\Omega_1}}(\max\{n_1, \ldots, n_k\} + 1)$$

provides a characterization of the provable recursive functions in terms of a subrecursive hierarchy indexed by recursively countable ordinals.

A first step is the observation that the translations of the axioms $\mathbf{ID}^1_\nu$, $\mathbf{ID}^2_\nu$, $\mathbf{TI}_\nu$ and Mathematical Induction are $f\mathcal{H}^0_1$–controlled derivable. Based on this observation we can prove that for a formal $\mathbf{ID}_\nu$ proof $\frac{|m}{0}\, G\,(x_1, \ldots, x_n)$ of length m we find finite ordinals k and l such that

$$f\mathcal{H}^0_1[\{k\}] \,\Big|\frac{\Omega_\nu\cdot\omega+m}{\Omega_\nu+l}\; G(z_1, \ldots, z_n).$$

Since $f\mathcal{H}^0_1[\{k\}](\emptyset) \subseteq f\mathcal{H}^k_1(\emptyset)$ we may subsume this into

Theorem 9.4 *If* $\mathbf{ID}_\nu \vdash G(x_1, \ldots, x_n)$ *then there is a finite ordinal* m *such that* $f\mathcal{H}^m_1 \Big|\frac{\Omega_\nu\cdot\omega+m}{\Omega_\nu+m}\; G(z_1, \ldots, z_n)$ *holds true for all tuples* $z_1, \ldots, z_n$ *of numerals.*

The main lemma for predicative cut–elimination is the Reduction Lemma which because of (38) has to be slightly modified for fragmented controlled derivations.

Lemma 9.5 (Reduction Lemma) *Assume* $f\mathcal{H}^{\gamma}_{\mu+1} \Big|\frac{\alpha}{\rho}\; \Delta, F$ *and* $f\mathcal{H}^{\gamma}_{\mu+1} \Big|\frac{\beta}{\rho}\; \Gamma, \neg F$ *such that* $rnk(F) = \rho \notin \mathbb{K}$, $F \in \bigwedge$–type *and* $par^*(F) \subseteq f\mathcal{H}^{\gamma}_{\mu+1}(par^*(\Delta))$. *Then we obtain* $f\mathcal{H}^{\gamma+1}_{\mu+1} \Big|\frac{\alpha+\beta}{\rho}\; \Delta, \Gamma$.

This alteration affects the Predicative Elimination Theorem which becomes

Theorem 9.6 *If* $\rho \in f\mathcal{H}^{\gamma}_{\mu+1}(par(\Delta))$, $f\mathcal{H}^{\gamma}_{\mu+1} \Big|\frac{\alpha}{\beta+\omega^\rho}\; \Delta$ *and* $[\beta, \beta+\omega^\beta) \cap \mathbb{K} = \emptyset$ *then we get* $f\mathcal{H}^{\gamma\#\varphi_\rho(\alpha)+1}_{\mu+1} \Big|\frac{\varphi_\rho(\alpha)}{\beta}\; \Delta$.

Also the main theorem of impredicative proof theory, the Collapsing Theorem (Theorem 8.4), transfers to fragmented controlled derivations.

Theorem 9.7 *Assume* $f\mathcal{H}_{\mu+1}^{\gamma+1} \overset{\alpha}{\underset{\Omega'_\rho}{\vdash}} \Delta$ *for a finite set* Δ *of* $\mathcal{L}^*_{\mathbf{ID}_\nu}$*–sentences in* $\bigvee_1^{\Omega_{\mu+1}}$*–type such that* $par^*(\Delta) \subseteq \mathcal{H}^{\gamma+1}(\Omega_\mu)$. *Then we obtain* $f\mathcal{H}_{\mu+1}^{\gamma\#\omega^{\alpha+1\#\Omega_\rho}+1} \overset{\Psi_{\Omega_{\mu+1}}^{\gamma\#\omega^{\alpha+1\#\Omega_\rho}}}{\underset{\Psi_{\Omega_{\mu+1}}^{\gamma\#\omega^{\alpha+1\#\Omega_\rho}}}{\vdash}} \Delta.$

Proof The proof is that of Theorem 8.4 with some extra care on the finite parameters. We have all the hypotheses oft Theorem 8.4 and may thus draw the same conclusions. Let Δ_ι be the premises of the last inference. Then we again have $\mathrm{par}^*(\Delta_\iota) \subseteq \mathcal{H}^{\gamma+1}(\Omega_\mu)$ and in case that the last inference is not a cut of rank $\geq \Omega_{\mu+1}$, obtain by the induction hypothesis

$$f\mathcal{H}_{\mu+1}^{\gamma\#\omega^{\alpha_\iota+1\#\Omega_\rho}+1} \overset{\gamma\#\omega^{\alpha_\iota+1\#\Omega_\rho}}{\underset{\gamma\#\omega^{\alpha_\iota+1\#\Omega_\rho}}{\vdash}} \Delta_\iota.$$

In order to argue as in the proof of Theorem 8.4 we need

$$f\mathcal{H}_{\mu+1}^{\gamma\#\omega^{\alpha_\iota+1\#\Omega_\rho}+1}(\Delta_\iota) \subseteq f\mathcal{H}_{\mu+1}^{\gamma\#\omega^{\alpha+1\#\Omega_\rho}+1}(\Delta_\iota). \tag{i}$$

To secure (i) we draw on $\alpha_\iota \in f\mathcal{H}_{\mu+1}^{\gamma+1}(\mathrm{par}^*(\Delta_\iota))$, hence $N(\alpha_\iota) \leq \Phi_{\mu+1}^{\gamma+1}(\mathrm{par}^*(\Delta_\iota))$. Putting $\mathcal{X}_\iota := \mathrm{par}^*(\Delta_\iota)$ we compute—similarly as in the proof of Theorem 9.3—

$$\begin{aligned}
&\Phi_{\mu+1}^{\gamma\#\omega^{\alpha_\iota+1\#\Omega_\rho}+1}(\mathcal{X}_\iota) = \phi^{\mathcal{H}^{\gamma\#\omega^{\alpha_\iota+1\#\Omega_\rho}+1}(\mathcal{X}_\iota)}(\Psi_{\Omega_{\mu+1}}^{\gamma\#\omega^{\alpha_\iota+1\#\Omega_\rho}+1} + |\mathcal{X}_\iota|) \leq \\
&\phi^{\mathcal{H}^{\gamma\#\omega^{\alpha\#\Omega_\rho}+1}(\mathcal{X}_\iota)}(\Psi_{\Omega_{\mu+1}}^{\gamma\#\omega^{\alpha\#\Omega_\rho}+1} + N(\alpha_\iota) + |\mathcal{X}_\iota|) \leq \\
&\phi^{\mathcal{H}^{\gamma\#\omega^{\alpha\#\Omega_\rho}+1}(\mathcal{X}_\iota)}(\Psi_{\Omega_{\mu+1}}^{\gamma\#\omega^{\alpha\#\Omega_\rho}+1} + \phi^{\mathcal{H}^{\gamma+1}(\mathcal{X})}(\Psi_{\Omega_{\mu+1}}^{\gamma+1} + |\mathcal{X}_\iota|) + |\mathcal{X}_\iota|) \leq \\
&\phi^{\mathcal{H}^{\gamma\#\omega^{\alpha\#\Omega_\rho}+1}(|\mathcal{X}_\iota|)}(\Psi_{\Omega_{\mu+1}}^{\gamma\#\omega^{\alpha\#\Omega_\rho}+1} \# \Psi_{\Omega_{\mu+1}}^{\gamma+1} + 2|\mathcal{X}_\iota|) \leq \Phi_{\mu+1}^{\gamma\#\omega^{\alpha+1\#\Omega_\rho}+1}(|\mathcal{X}_\iota|).
\end{aligned}$$

Similar computations are needed in the cases of cuts with ranks $\geq \Omega_\mu$. Especially in the case of a cut formula G of rank $\Omega_{\mu+1}$ we have to apply the Predicative Elimination Theorem. In addition to the considerations in the proof of Theorem 8.4 we have to ensure that

$$f\mathcal{H}_{\mu+1}^{\delta+1}(\mathrm{par}^*(\Delta)) \subseteq f\mathcal{H}_{\mu+1}^{\tilde{\alpha}+1}(\mathrm{par}^*(\Delta)) \tag{ii}$$

for $\tilde{\alpha} := \gamma \# \omega^{\alpha+1\#\Omega_{\mu+1}}$, $\widetilde{\alpha_0} := \gamma \# \omega^{\alpha_0+1\#\Omega_{\mu+1}}$, $\beta := \Psi_{\Omega_{\mu+1}}^{\widetilde{\alpha_0}}$ and

$$\delta := \widetilde{\alpha_0} \# \omega^{\varphi_\beta(\beta)+1\#\Omega_\mu}.$$

Clearly $\delta < \tilde{\alpha}$. Let $\mathcal{X} := \mathrm{par}^*(\Delta)$. Drawing on $\gamma, \alpha_0 \in f\mathcal{H}_{\mu+1}^{\gamma+1}(\mathcal{X} \cup \mathcal{Y})$ for $\mathcal{Y} := \mathrm{par}^*(G) \subseteq f\mathcal{H}_{\mu+1}^{\gamma+1}(\mathcal{X})$ we compute

$$\Phi_{\mu+1}^{\delta+1}(\mathcal{X}) \le \phi^{\mathcal{H}^{\tilde\alpha}(\mathcal{X})}(\Psi_{\Omega_{\mu+1}}^{\gamma\#\omega^{\alpha\#\Omega_{\mu+1}}} + N(\delta) + 1 + |\mathcal{X}|) \le$$
$$\phi^{\mathcal{H}^{\tilde\alpha}(\mathcal{X})}(\Psi_{\Omega_{\mu+1}}^{\gamma\#\omega^{\alpha\#\Omega_{\mu+1}}} + N(\gamma) + N(\alpha_0) + 9 + |\mathcal{X}|) \le$$
$$\phi^{\mathcal{H}^{\tilde\alpha}(\mathcal{X})}(\Psi_{\Omega_{\mu+1}}^{\gamma\#\omega^{\alpha\#\Omega_{\mu+1}}} + \phi^{\mathcal{H}^{\tilde\alpha}(\mathcal{X})}(\Psi_{\Omega_{\mu+1}}^{\gamma+2} + |\mathcal{X}|) \cdot 2 + 9 + |\mathcal{X}|) \le$$
$$\phi^{\mathcal{H}^{\tilde\alpha}(\mathcal{X})}(\Psi_{\Omega_{\mu+1}}^{\gamma\#\omega^{\alpha\#\Omega_{\mu+1}}} \# (\Psi_{\Omega_{\mu+1}}^{\gamma+2} + |\mathcal{X}|) \cdot 2 + 9 + |\mathcal{X}|) \le$$
$$\phi^{\mathcal{H}^{\tilde\alpha}(\mathcal{X})}(\Psi_{\Omega_{\mu+1}}^{\gamma\#\omega^{\alpha+1\#\Omega_{\mu+1}}} + |\mathcal{X}| \cdot 3 + 9) \le \Phi_{\mu+1}^{\tilde\alpha}(\mathcal{X}).$$

Then (iii) together with $\delta < \tilde\alpha$ yield (ii). □

Definition 9.8 A computable function $\theta\colon \mathbb{N} \longrightarrow \mathbb{N}$ is a *dominant* for an arithmetical Π^0_2–sentence $(\forall x)(\exists y)F(x, y, \mathbf{k})$ if for any n there is an $m \le \theta(n)$ such that $\mathbb{N} \models F(n, m, \mathbf{k})$.[30] We say that a function $\Theta\colon \mathbb{N} \longrightarrow \mathbb{N}$ *dominates* the provable Π^0_2–sentences of an axiom system T if for every Π^0_2–sentence that is provable in T there is a dominant which is eventually majorized by Θ, i.e., if for every provable Π^0_2–sentence G there is an l such that $\lambda x\,.\,\Theta(x + l)$ is a dominant for G.

By Theorems 9.4, 9.6, 9.7 and again Theorem 9.6 we obtain for a Π^0_2–sentence $(\forall x)(\exists y)F(x, y, \mathbf{k})$ that is provable in $\mathbf{ID}_\nu$ ordinals $\alpha_0 \in \mathcal{H}^{\varepsilon_{\Omega_\nu+1}}(0) \cap \varepsilon_{\Omega_\nu+1}$ and $\alpha_1 < \Psi_{\Omega_1}^{\varepsilon_{\Omega_\nu+1}}$ such that $f\mathcal{H}_1^{\alpha_0} \overset{|\alpha_1}{\underset{0}{}} (\exists y)F(n, y, \vec{k})$ for all numerals n. By the Witnessing Theorem (Theorem 9.3) this implies that there is an $m \le \Phi_1^{\alpha_0}\{n, \vec{k}\} = \Phi_{\Psi_{\Omega_1}^{\alpha_0}}(\max\{n, \vec{k}\} + 1)$ such that $\mathbb{N} \models F(n, m, \vec{k})$. The function $n \mapsto \Phi_1^{\alpha_0}\{n, \vec{k}\}$ is thus a dominant for $(\forall x)(\exists y)F(x, y, \vec{k})$. Since $\Phi_{\Psi_1^{\alpha_0}}$ is eventually majorized by $\Phi_{\Psi_{\Omega_1}^{\varepsilon_{\Omega_\nu+1}}}$ we have the following result.

Theorem 9.9 *The function* $\Phi_{\Psi_1^{\varepsilon_{\Omega_\nu+1}}}$ *dominates the provable* Π^0_2*–sentences of* $\mathbf{ID}_\nu$.

Remark 9.10 From Theorem 9.9 we obtain for every Π^0_2–sentence G provable in $\mathbf{ID}_\nu$ a natural number l such that $\lambda x\,.\,\Phi_{\Psi_1^{\varepsilon_{\Omega_\nu+1}}}(x + l)$ is a dominant for G. As a word of warning we want to emphasize that the number l not only depends on the parameters of G but also on the norm of $\alpha \in \mathcal{H}^{\varepsilon_{\Omega_\nu+1}}(0) \cap \varepsilon_{\Omega_\nu+1}$ in Φ_1^α which in turn—as a consequence of Theorem 9.4—depends upon the $\mathbf{ID}_\nu$–proof of G.

Corollary 9.11 *The provably recursive functions of* $\mathbf{ID}_\nu$ *are eventually majorized by the function* $\Phi_{\Psi_{\Omega_1}^{\varepsilon_{\Omega_\nu+1}}}$.

Proof If $\mathbf{ID}_\nu \vdash (\forall x)(\exists y)T(e, x, y)$, where is T is the Kleene predicate, then $\{e\}(n) \le \Phi_{\Psi_{\Omega_1}^\alpha}(\max\{e, n\} + 1)$ and the latter is eventually majorized by $\Phi_{\Psi_{\Omega_1}^{\varepsilon_{\Omega_\nu+1}}}$. □

Corollary 9.12 *The provably recursive function of the theory* $\mathbf{ID}_\nu$ *are exactly the functions that are primitive–recursive in some* Φ_α *for* $\alpha < \Psi_{\Omega_1}^{\varepsilon_{\Omega_\nu+1}}$.

---

30 In [38] we call a dominant for $(\forall x)(\exists y)F(x, y, \mathbf{k})$ a testfunction since it gives an upper bound for testing instances for y in $(\exists y)F(n, y, \mathbf{k})$. More details on testfunctions and their impact on Hilbert's programme of elimination of ideal elements are discussed in [38].

Proof According to [9] the theory $\mathbf{ID}_\nu$ proves transfinite induction for every $\alpha < \Psi_{\Omega_1}^{\varepsilon_{\Omega_\nu+1}}$. Therefore we obtain $\mathbf{ID}_\nu \vdash (\forall x)(\exists y)[\Phi_\alpha(x) = y]$ for every $\alpha < \Psi_{\Omega_1}^{\varepsilon_{\Omega_\nu+1}}$. Since $\{e\}(m) = n \Leftrightarrow (\exists y \leq \Phi_\alpha(\max\{e, n\} + 1))[T(e, m, y) \wedge U(y) = n]$ the claim follows. □

Remark 9.13 1. Occasionally the Π_2^0–ordinal of an axiom system T is defined as the least stage in a subrecursive hierarchy at which the provably recursive function of T are eventually majorized. Then, according to Corollary 9.11, the Π_2^0–ordinal of $\mathbf{ID}_\nu$ coincides with its prooftheoretic ordinal $|\mathbf{ID}_\nu| = \delta^{\mathbb{N}}(\mathbf{ID}_\nu)$.

This definition of the Π_2^0–ordinal is, however, not as stable as the definition of the prooftheoretic ordinal. It depends on the definition of the subrecursive hierarchy. The hierarchy, as defined in this paper, depends on the starting function Φ and the norm of the ordinal(term)s. The dependency on Φ is harmless since for any different primitive–recursive starting function—which is sufficiently strong increasing—the hierarchies catch up very quickly. Therefore we may neglect this dependency.

More serious is the dependency on the definition of the norm. However, according to Sect. 7 the stages of the inductive definition of $\alpha \in \mathcal{T}$ define a canonical primitive–recursive norm on the ordinals in $\mathcal{H}^{\varepsilon_{\Omega_\nu+1}}(0)$. Given the operator $\mathcal{H}^{\varepsilon_{\Omega_\nu+1}}$ the construction in Sect. 9 of the subrecursive hierarchy $\{\Phi_\alpha \mid \alpha \in \mathcal{H}^{\varepsilon_{\Omega_\nu+1}}(\emptyset)\}$ is thus canonical.

If we—as discussed in Remark 8.8—consider the spectrum of $\mathbf{ID}_\nu$ as generated by the operator $\mathcal{H}^{\varepsilon_{\Omega_\nu+1}}$ it therefore also includes a canonical characterization of the provable recursive functions of $\mathbf{ID}_\nu$, i.e., it includes also the Spector class $\mathbf{\Sigma}_1^0$ above $\mathbb{N}$.

The arrows for the structures $\mathfrak{M}_{\mu+1}$ in Fig. 1 represent static bounds for the $\mathfrak{M}_{\mu+1}$–recursive functions whose totality is provable in $\mathbf{ID}_\nu(\mathsf{T})$. Such a static bound cannot exist for $\mathbf{\Sigma}_1^0$–recursive function. Instead of representing $\mathbf{\Sigma}_1^0$ by a fixed arrow in Fig. 1, pointing at some ordinal below ω, we need a "floating arrow". Given a provably recursive function f there is a number l such that for every mark $n < \omega$ bounding the inputs x for f the arrow has to flow to the point $\Phi_{\Psi_{\Omega_1}^{\varepsilon_{\Omega_\nu+1}}}(n + l) < \omega$ which marks a bound for $f(x)$.

2. In Theorem 9.4 we tacitly use the fact that all primitive recursive functions are eventually dominated by Φ_{ω^ω}. If we have an axiomatization T of $\mathbb{N}$ which includes more strongly increasing functions, e.g. functions more strongly increasing than $\Phi_{\Psi_{\Omega_0+1}^{\varepsilon_{\Omega_\nu+1}}}$ we have to alter the bounds in Theorem 9.4 and thus obtain a different Π_2^0–ordinal. Therefore Theorem 2.5 does not hold for the Π_2^0–ordinal of an axiom system. In contrast to the Π_1^1–ordinal $\pi^{\mathbb{N}}(\mathsf{T})$ the Π_2^0–ordinal of an axiom system T is sensitive to elementary sentences in T. Thus there are axiomatizations for which the Π_2^0–ordinal and the Π_1^1–ordinal differ.

10 Concluding Remarks

In revisiting the proof theory of iterated inductive definitions—viewed as special Spector classes—I have here put the emphasis on the recursion theoretic aspects. My belief is that such results enable us to study the independence of certain abstractions of combinatorial principles (e.g. generalizations of the hydra game, Kruskal sentences etc.) of the axiom systems $\mathbf{ID}_\nu$.

The original motivation to study generalized inductive definition were, however, problems in the foundations of mathematics. The interest in generalized inductive definitions from a proof theoretical point of view was launched by Kreisel in [21]. To cite Sol Feferman from [14] *"Kreisel's aim there was to assess the constructivity of Spector's consistency proof of full second order analysis* [41], *by means of a functional interpretation in the class of so–called bar recursive functionals.* [...]. *So Kreisel asked whether the intuitionistic theory of inductive definitions given by monotonic arithmetical closure conditions* [...] *serves to generate the class of (indices of) representing functions of the bar recursive functionals.* Although it rapidly turned out that this could not be fully obtained the interest in the prooftheoretic analysis of generalized inductive definitions and their iterations was raised. This is not the place to narrate the development of this interest. The story is much better told in [13] and its continuation [14] by Sol Feferman. I will therefore restrict myself to a few personal remarks.

My personal interest in the proof theory of iterated inductive definitions was raised by the conjectured prooftheoretic ordinals for systems of iterated generalized inductive definitions by Martin Löf in [22]. As already mentioned in the introduction this problem could be solved by "brut force" using Takeuti's reduction procedure for Π^1_1–comprehension.

. Since *"[a]ccessibility inductive definitions enjoy a privileged position [...]"* because *"[w]e have a direct picture of how the inductively defined sets are generated [...]"*[31] theories for accessibility inductive definition based on intuitionistic logic became the most favored candidates for constructivity. The reduction of the theories $\mathbf{ID}_\nu$ based on classical logic to theories $\mathbf{ID}^i_\nu(\mathrm{Acc})$ which are based on intuitionistic logic and only talk about the accessible part of elementarily definable orderings thus became a prominent problem. By work of Howard, Kreisel and Troelstra it was known that $\mathbf{ID}_1$ is reducible to $\mathbf{ID}^i_1(\mathcal{O})$ the intuitionistic theory of constructive ordinals which is an accessibility inductive definition. According to [11] this was achieved *"by a roundabout argument through a formal theory of choice sequences"* (which I admittedly never studied). Howard in [17][32] computed the prooftheoretic ordinal of $\mathbf{ID}^i_1(\mathcal{O})$ which via the just mentioned reduction also gave the prooftheoretic ordinal of $\mathbf{ID}_1$.

It was, however, not clear how to obtain similar reductions also for the iterated case. Zucker in [43] pointed out that for $\nu > 1$ there are definite obstacles for a

---

[31] Cited from [11].

[32] This paper is also in another context of seminal importance. It led Wilfried Buchholz to develop his Ω_μ–rules also presented in [5]. This is another story which I briefly touched in [29].

straightforward reduction of $\mathbf{ID}_\nu$ to $\mathbf{ID}^i_\nu(\mathrm{Acc})$ for positively defined accessibility inductive definitions.

It was Sol Feferman who called my attention to the fact that an ordinal analysis of $\mathbf{ID}_\nu$ should also entail a solution to the reduction problem. This led to a reduction of $\mathbf{ID}_\nu$ to $\mathbf{ID}^i_\nu$ for an accessibility inductive definition. The reduction was obtained using a formalization of the ordinal analysis by which it could be shown that $\mathbf{ID}_\nu$ is conservative for stable arithmetical sentences above Heyting arithmetic augmented by transfinite induction along all initial segments of the prooftheoretic ordinal of $\mathbf{ID}_\nu$. Since the wellfoundedness of all these initial segments is provable in accessibility $\mathbf{ID}^i_\nu$ this yields a reduction. In my Habiltationsschrift [28] this was obtained by embedding $\mathbf{ID}_\nu$ into iterated Π^1_1–comprehensions and then using Takeuti's reduction procedure.[33] The more perspicuous method by local predicativity was later presented in [5]. This volume contains also Wilfried Sieg's reduction of $\mathbf{ID}_{<\nu}$ to $\mathbf{ID}^i_{<\nu}(\mathcal{O})$ for limit ordinals ν and Wilfried Buchholz's (very elegant) reduction via his $\Omega_{\mu+1}$–rules.[34]

The remarkable fact is that Buchholz' and my reduction are based on ordinal analyses and that this is in an abstract sense also true for Sieg's reduction.[35] His aim was to give a reduction of classical $\mathbf{ID}_\nu$ to the mathematically meaningful intuitionistic theory $\mathbf{ID}^i_\nu(\mathcal{O})$ of constructive number classes—an especially distinguished accessibility theory. The reduction does not talk explicitly about prooftheoretic ordinals; rather, the reduction is obtained via a direct proof theoretic investigation of infinitary calculi PL_μ into which $\mathbf{ID}_\mu$ can be embedded. The obvious meta–theory is $\mathbf{ID}^i_{\mu+1}(\mathcal{O})$ as the infinitary deductions in PL_μ can be naturally viewed as elements of $\mathcal{O}_{\mu+1}$, the $(\mu+1)$–st constructive number class. This leads to a reduction in the limit case. Only recently Avigad and Townsner[36] presented a reduction of $\mathbf{ID}_1$ to $\mathbf{ID}^i_1(\mathcal{O})$ by a combination of functional interpretation with Sieg's result yielding a reduction of $\mathbf{ID}^i_1$ to $\mathbf{ID}^i_1(\mathcal{O})$ which seems to be generalizable to $\nu > 1$. This would improve Sieg's result in so far that a reduction of $\mathbf{ID}_\nu$ to $\mathbf{ID}^i_\nu(\mathcal{O})$ also holds true for successor ordinals.

Nonetheless let me mention that—though not explicitly stated—the present paper also contributes to the foundational problem. First we still obtain the reduction of $\mathbf{ID}_\nu$ to $\mathbf{ID}^i_\nu$. The wellfoundedness of all initial segments of $\mathcal{H}^{\varepsilon_{\Omega_\nu+1}}(\emptyset) \cap \Omega_{0+1}$ is provable within a system $\mathbf{ID}^i_\nu$ only talking about accessible parts of the relation $<\restriction\mathcal{T}$ as defined in Sect. 7. This guarantees that $\mathbf{ID}_\nu$ and $\mathbf{ID}^i_\nu$ have the same provably recursive functions and thus the same consistency strength. Although this gives no direct embedding of $\mathbf{ID}_\nu$ into $\mathbf{ID}^i_\nu$ it shows that the consistency problem of both theories is of the same complexity.

---

[33] This approach has the advantage that only the monotonicity of the inductive definition is used. All the results obtained there are (as explicitly stated there) valid for monotone inductive definitions. This is in contrast to the more perspicuous approach in [5] where the positivity of the inductive definition is needed.

[34] Cf. [7].

[35] Cf. [40].

[36] Cf. [1].

Moreover it should be observed that the collapsing theorems (Theorems 8.4 and 9.7) contribute to the aspect of "elimination of ideal objects" in Hilbert's programme as discussed in [38]. For sentences in $\bigvee_1^{\Omega_0+1}$–type all the "ideal" rules $(\Omega_{\mu+1})$ are eliminated in Theorem 8.4. As pointed out in [38] even sentences above a complexity Π^0_2 can be regarded as "ideal" in the sense of [16]. Theorem 9.7 shows that in a semi-formal proof of Π^0_2–sentences all ideal sentences are eliminable.

The surprising and striking point is that the solution of the elimination problem, which is so simple to phrase, needs infinitary methods manifested by the ordinal $\Psi_{\Omega_1}^{\varepsilon_{\Omega_\nu+1}}$. Since this bound is sharp no simpler methods are likely to work. It would be interesting to know if this is also true for the reduction of the axiom systems $\mathbf{ID}_\nu$ based on classical logic to its intuitionistic version $\mathbf{ID}^i_\nu$. Even if there is a form of reduction by functional interpretation I conjecture that the elimination of all "ideal methods" in the proof of the computability of the interpreting functionals will need infinitary methods leading to the known prooftheoretic ordinals. Such a result could indicate that, in spite of the equality of their prooftheoretic ordinals, there is in fact a profound difference in the constructive meaning of the classical and the intuitionistic version of iterated inductive definitions.

Finally a short comment on the aim to obtain axiom systems for second order arithmetic with increasing performance. Though second order arithmetic is the natural system to formulate classical Analysis—which here stands for the theory of real numbers and –functions—it is difficult to handle proof theoretically. Because of its apparent impredicativity the scheme of full comprehension seems to be untreatable. Also restrictions in the complexity of the comprehension scheme reaching essentially beyond Π^1_1 turned out to be untreatable directly. Approximations by increasingly growing Spector classes are likewise difficult to handle. Because of the lack of ordinals in the powerset of $\mathbb{N}$ their axiomatizations are complicated. Axiomatizing set universes above $\mathbb{N}$ which also contain ordinals have turned out to be more promising.[37] Here the roles of Spector classes above $\mathbb{N}$ are played by admissible universes above $\mathbb{N}$. The first point where this approach exceeds the approximation of Spector classes by iterated inductive definitions is a structure $\mathbb{A}$ which is an admissible limit of admissible universes.[38] I mention this structure since it is the first point at which the approach of Sect. 7 has to be altered. Since there we have $\Omega_I = I$ for $I = o^{\mathbb{A}}$ and we need a function Ψ_I^α which collapses below I the axiomatization of the ordinals Ω_μ becomes more complicated.[39]

Real gain in performance is obtained by axiomatizing reflection principles in set structures above $\mathbb{N}$. The strongest system for which there is a complete analysis including the characterization of the provable recursive functions is the theory of full reflection and a weak form of stability by Stegert (cf. [26, 42],). Ordinal analyses for a stronger form of stability which is equivalent to parameter free Π^1_2–comprehension have been developed by Rathjen in a series of papers (cf. [33, 36, 37].)

[37] Cf. [19].

[38] Cf. [18].

[39] This is the reason why we required in Note 4.2 to restrict the iterations to ordinals less than the first recursively inaccessible ordinal.

References

1. Avigad, J., Towsner, H.: Functional interpretation and inductive definitions. J. Symb. Logic **74**(4), 1100–1120 (2009)
2. Beckmann, A., Pohlers, W.: Application of cut-free infinitary derivations to generalized recursion theory. Ann. Pure Appl. Logic **94**, 1–19 (1998)
3. Blankertz, B., Weiermann, A.: How to characterize provably total functions by the Buchholz operator method. Lecture Notes in Logic, vol. 6. Springer, Heidelberg (1996)
4. Blankertz, B.: Beweistheoretischen Techniken zur Bestimmung von Π^0_2–Skolem Funktionen, Dissertation, Westfälische Wilhelms-Universität, Münster (1997)
5. Buchholz, W., Feferman, S., Pohlers, W., Sieg, W. (eds.): Iterated Inductive Definitions and Subsystems of Analysis: Recent Proof-theoretical Studies, Lecture Notes in Mathematics, vol. 897. Springer, Heidelberg (1981)
6. Buchholz, W.: Another reduction of classical ID_ν to constructive ID_ν. In: Schindler, R. (ed.) Ways of Proof Theory, Ontos Mathematical Logic, vol. 2, Ontos Verlag, Frankfurt, Paris, Lancaster, New Brunswick, pp. 183–197 (2010)
7. Buchholz, W.: The $\Omega_{\mu+1}$-rule, iterated inductive definitions and subsystems of analysis: recent proof-theoretical studies. In: Buchholz, W. et al. (eds.), Lecture Notes in Mathematics, vol. 897, pp. 188–233. Springer, Heidelberg (1981)
8. Buchholz, W.: A simplified version of local predicativity. In: Aczel, P. (ed.) Proof Theory, pp. 115–147. Cambridge University Press, Cambridge (1992)
9. Buchholz, W., Pohlers, W.: Provable well orderings of formal theories for transfinitely iterated inductive definitions. J. Symb. Logic **43**, 118–125 (1978)
10. Buchholz, W., Cichon, E.A., Weiermann, A.: A uniform approach to fundamental sequences and hierarchies. Math. Logic Quart. **40**, 273–286 (1994)
11. Feferman, S., Sieg, W.: Inductive definitions and subsystems of analysis. In: Buchholz, W. et al., (ed.), Iterated Inductive Definitions and Subsystems of Analysis: Recent Proof-theoretical Studies, Lecture Notes in Mathematics, vol. 897, pp. 16–77. Springer, Heidelberg (1981)
12. Feferman, S.: Formal theories for transfinite iteration of generalized inductive definitions and some subsystems of analysis. In: Kino, A. et al., (ed.) Intuitionism and Proof Theory, Studies in Logic and the Foundations of Mathematics, North-Holland Publishing Company, Amsterdam, pp. 303–326 (1970)
13. Feferman, S.: Preface: how we got from there to here. In: Buchholz W. et al., (eds.) Iterated Inductive Definitions and Subsystems of Analysis: Recent Proof-theoretical Studies, Lecture Notes in Mathematics, vol. 897, pp. 1–15. Springer, Heidelberg (1981)
14. Feferman, S.: The proof theory of classical and constructive inductive definitions. A forty year saga, 1968–2008. In: Schindler, R. (ed.) Ways of Proof Theory, Ontos Mathematical Logic, vol. 2, Ontos Verlag, Frankfurt, Paris, Lancaster, New Brunswick, pp. 79–95 (2010)
15. Gentzen, G.: Beweisbarkeit und Unbeweisbarkeit von Anfangsfällen der transfiniten Induktion in der reinen Zahlentheorie. Math. Ann. **119**, 140–161 (1943)
16. Hilbert, D., Die Grundlagen der Mathematik. Vortrag gehalten auf Einladung des Mathematischen Seminars im Juli 1927 in Hamburg. In: Hamburger Mathematische Einzelschriften, vol. 5, pp. 1–21. Heft (1928)
17. Howard, W.A.: A system of abstract constructive ordinals. J. Symb. Logic **37**, 355–374 (1972)
18. Jäger, G., Pohlers, W.: Eine beweistheoretische Untersuchung von (Δ^1_2 -CA) + (BI) und verwandter Systeme. Bayerische Akademie der Wissenschaften, Sitzungsberichte 1982, pp. 1–28 (1983)
19. Jäger, G.: Theories for Admissible Sets. A Unifying Approach to Proof Theory, Studies in Proof Theory, Lecture Notes, vol. 2. Bibliopolis, Naples (1986)
20. Jäger, G., Strahm, T.: Upper bounds for metapredicative Mahlo in explicit mathematics and admissible set theory. J. Symb. Logic **66**(2), 935–958 (2001)
21. Kreisel, G.: Generalized inductive definitions, Reports of seminars on the foundation of Analysis. (Stanford Report), mimeographed, vol. section III (1963)

22. Martin-Löf, P.: Hauptsatz for the intuitionistic theory of iterated inductive definitions. In: Fenstad, J.E. (ed.) Proceedings of the 2nd Scandinavian Logic Symposium, Studies in Logic and the Foundations of Mathematics, pp. 179–216, vol. 63. North-Holland Publishing Company, Amsterdam (1971)
23. Möllerfeld, M.: Systems of inductive definitions, Ph.D. thesis, Münster (2003)
24. Moschovakis, Y.N.: Axioms for computation theories – first draft. In: Gandy, R.O., Yates, C.M.E. (eds.), Logic colloquium '69 Studies in Logic and the Foundations of Mathematics, vol. 61, pp. 199–255. North-Holland Publishing Company, Amsterdam (1971)
25. Moschovakis, Y.N.: Elementary Induction on Abstract Structures, Studies in Logic and the Foundations of Mathematics, vol. 77. North-Holland Publishing Company, Amsterdam (1974)
26. Pohlers, W., Stegert, J.C.: Provably recursive functions of reflections. In: Berger, U. et al. (eds.), Logic, Construction, Computation, Ontos Mathematical Logic, vol. 3, pp. 381–474. Ontos Verlag, Frankfurt (2012)
27. Pohlers, W.: An upper bound for the provability of transfinite induction, $\models$. In: Diller, J., Müller, G.H. (eds.) ISILC Proof Theory Symposium, Lecture Notes in Mathematics, vol. 500, pp. 271–289. Springer, Heidelberg (1975)
28. Pohlers, W.: Beweistheorie der iterierten induktiven Definitionen. Ludwig Maximilians-Universität, München, München, Habilschrift (1977)
29. Pohlers, W.: From subsystems of analysis to subsystems of set theory. In: Kahle, R. et al., (eds.), Advances in Proof Theory, Progress in Computer Science and Applied Logic 28, pp. 319–338. Birkhäuser Verlag (2016)
30. Pohlers, W.: Proof-theoretical analysis of ID_ν by the method of local predicativity. In: Buchholz, W. et al., (eds.) Iterated Inductive Definitions and Subsystems of Analysis: Recent Proof-theoretical Studies, Lecture Notes in Mathematics, vol. 897, pp. 261–357. Springer, Heidelberg (1981)
31. Pohlers, W.: Semi–formal calculi and their applications. In: Kahle, R., Rathjen, M. (eds.), Gentzen's Centenary: The Quest for Consistency, pp. 195–232. Springer, Berlin (2015)
32. Pohlers, W.: Ordinals connected with formal theories for transfinitely iterated inductive definitions. J. Symb. Logic **43**, 161–182 (1978)
33. Pohlers, W.: Proof Theory, The First Step into Impredicativity. Universitext, Springer, Berlin (2009)
34. Pohlers, W.: Hilbert's programme and ordinal analysis. In Probst, D., Schuster, P. (eds.), Concepts of Proof in Mathematics, Philosophy and Computer Science, Ontos Mathematical Logic, DeGruyter, pp. 291–322 (2016)
35. Rathjen, M.: Proof theory of reflection. Ann. Pure Appl. Logic **68**, 181–224 (1994)
36. Rathjen, M.: An ordinal analyis of parameter free Π^1_2-comprehension. Arch. Math. Logic **48**(3), 263–362 (2005)
37. Rathjen, M.: An ordinal analyis of stability. Arch. Math. Logic **48**(2), 1–62 (2005)
38. Schindler, R. (ed.): Ways of Proof Theory, Ontos Mathematical Logic, vol. 2. Ontos Verlag, Frankfurt (2010)
39. Schütte, K.: Proof theory, Grundlehren der mathematischen Wissenschaften, vol. 225. Springer, Heidelberg (1977)
40. Sieg, W.: Inductive definitions, constructive ordinals and normal derivations. In: Buchholz, W. et al. (eds.), Iterated inductive Definitions and Subsystems of Analysis: Recent Proof-theoretical Studies, Lecture Notes in Mathematics, vol. 897, pp. 143–187. Springer, Heidelberg (1981)
41. Spector, C.: Provably recursive functionals of analysis: a consistency proof of analysis by an extension of principles formulated in current intuitionistic mathematics. In: Dekker, J.C.E. (ed.), Recursive Function Theory, Proceedings of Symposia in Pure Mathematics, vol. 5, pp. 1–27. American Mathematical Society, Providence (1962)
42. Stegert, J.-C.: Ordinal proof theory of Kripke–Platek set theory augmented by strong reflection principles, Ph.D. thesis, Westfälische Wilhelms-Universität, Münster (2011)
43. Zucker, J.I.: Iterated inductive definitions, trees, and ordinals. In: Troelstra, A.S. (ed.), Metamathematical Investigation of Intuitionistic Arithmetic and Analysis, Lecture Notes in Mathematics, vol. 344, pp. 392–453, Springer, Heidelberg (1973)

Author Biography

Wolfram Pohlers is a retired full professor of mathematics at the Westfälische Wilhelms University in Münster where he served as director of the Institute for Mathematical Logic and Foundational Research from 1985 until 2008. He received his scientific training at the Ludwigs-Maximilian University in München (LMU) under the supervision of Kurt Schütte. In 1980 he was appointed Associate Professor at the LMU. From 1989 to 1990 he was a visiting scholar at the MSRI in Berkeley and in the fall of 2005 he taught at the Ohio State University in Columbus.

The Operational Penumbra: Some Ontological Aspects

Gerhard Jäger

Abstract Feferman's explicit mathematics and operational set theory are two important examples of families of theories providing an operational approach to mathematics. My aim here is to survey some central developments in these two fields, to sketch some of Fefeman's main achievements, and to relate them to the work of others. The focus of my approach is on ontological questions.

Keywords Explicit mathematics · Operational set theory
Operational approach · Proof theory

2010 MSC MSC 03B30 · 03B40 · 03E20 · 03F03

1 Introduction

I first met Solomon Feferman at the 1978 Logic Colloquium meeting in Mons, Belgium. He gave a survey talk about various approaches to constructive mathematics and presented his own constructive theory of functions and classes. The written form [12] of his talk is published in the proceedings volume of that conference and is one of the three landmark papers about explicit mathematics. This was the time when I was working for my dissertation and, of course, Sol was already known to me very well by his many influential papers on proof theory and the foundations of mathematics. After that I had the privilege to learn from Sol in direct personal contact when we both spent the academic year 1979–1980 at the University of Oxford. We have been in close scientific and personal contact since then, including my visit as an assistant professor at Stanford University in the academic year 1982–1983.

Sol's influence on my scientific development has been manifold. One very important aspect is that he widened the range of my proof-theoretic interests and led me to work on new topics dealing with foundational questions, different from those I had previously studied. Maybe the most typical example along these lines is explicit

G. Jäger (✉)
Institut für Informatik, Universität Bern, Neubrückstrasse 10, 3012 Bern, Switzerland
e-mail: jaeger@inf.unibe.ch

G. Jäger and W. Sieg (eds.), *Feferman on Foundations*, Outstanding Contributions to Logic 13, DOI 10.1007/978-3-319-63334-3_10

mathematics, a subject that has never left me since then. It was characteristic of Sol that he was always asking for conceptual clarity and insisting on a clear methodological approach, not obscured by a self-satisfying technical machinery.

A general operational approach has been extremely successful in connection with the λ-calculus and combinatory logic leading, for example, to a variety of functional programming languages. The formal operational approach in mathematics, on the other hand, has not been so popular although many working mathematicians freely make use of operations and the operational machinery whenever convenient, but typically informally and without caring for its foundations. Church's approach to base the foundations of mathematics entirely on operations turned out to be inconsistent.

Feferman's explicit mathematics changed the picture. Motivated by the desire to set up a proper formal framework for Bishop's *Foundations of Constructive Analysis* [2] he proposed a new kind of formalism, baptized *explicit mathematics*. Bishop's book had enormous influence on the discussion of the foundations of mathematics. Bishop showed in his book, putting aside all ideological considerations, that most of the important theorems in real analysis can be established up to equivalence by constructive methods. The success of this book stimulated many logicians to develop formal frameworks for Bishop's approach, and Feferman's system (or family of systems) turned out to pave a very influential way. At the same time, his framework provided a way to account for predicative mathematics and descriptive set theory as well, which could not be done in the other approaches.

Soon after the first presentation of explicit mathematics in Feferman [9], its relevance for other parts of proof theory became evident. For example, systems of explicit mathematics – based on classical or intuitionistic logic – have their natural place in reductive proof theory and constitute a natural setting for studying various forms of abstract computability and recursion in higher type functionals.

In this article I will try to sketch some of the main lines in the research about explicit mathematics. A textbook by Solomon Feferman, Gerhard Jäger, and Thomas Strahm on the foundations of explicit mathematics is in preparation, aiming at providing a systematic approach to the topics mentioned above. In addition to that, Ulrik Buchholtz has set up an online bibliography of explicit mathematics and related topics at http://home.inf.unibe.ch/~ltg/em_bibliography.

The second main topic of this article is *operational set theory*, a further central stream in Feferman's operational approach. It goes back to Feferman [18] and is further elaborated in Feferman [19], where also much about the original ideology of operational set theory is explained. Further advances and technical results will be presented in Sect. 5 of this article.

Feferman's *unfolding program* is a third field under the operational perspective. However, we will not treat it in this article since Strahm's contribution for this volume is dedicated to unfolding. In addition, the reader will find a useful introduction to all three fields in Feferman [22].

My aim here is to survey some developments in explicit mathematics and operational set theory from a common operational perspective, to sketch some of Feferman's main achievements in these to fields, and to relate them to the work of others. The focus of my approach is on ontological questions, a point of view that has been neglected so far. But I am convinced that such ontological questions will play a crucial role in the further development of a general operational penumbra.

2 The General Operational Framework

Before turning to systems of explicit mathematics and operational set theory we set up the general operational framework. However, in contrast to Church (cf. [7, 8]), who wanted to base the foundations of mathematics solely on operations and whose approach turned out to be inconsistent, we confine ourselves to a consistent and relatively weak core operational theory. The basic idea is simple: The universe of discourse is a partial combinatory algebra; its elements are operations and share the following properties:

- Operations may be partial, they may freely be applied to each other, and self-application of operations is permitted.
- As a consequence, the general theory of operations is type-free. If needed sets or classes of operations can be added with the purpose to partly structure the universe.
- Operations are intensional objects; extensionality of operations is only assumed or claimed axiomatically in very special situations.

Since we will be dealing with possibly undefined objects, it is convenient to work with Beeson's logic of partial terms, see Beeson [1], rather than ordinary classical or intuitionistic logic. Terms are formed in this logic from the variables and constants of the language by simple term application, and we have atomic formulas of the form $(t\downarrow)$ to express that the term t has a value or is defined.

In his first articles [9, 11, 12] about explicit mathematics, Feferman did not make use of the logic of partial terms but worked with a three place relation $App[x, y, z]$ to express that operation x applied to y has value z.

Scott's [58] presents one of several alternative possibilities of dealing with existence and partiality in a logical context. In this E-logic we have a specific relation symbol E, where $E(t)$ has the intuitive interpretation "t exists". In the Beeson/Feferman approach all constants have a value and the free variables range over existing objects, in contrast to Scott's approach where they can also stand for possibly non-existing objects. In both approaches, quantifiers are supposed to range over existing objects only. In spite of this different philosophically attitude, both approaches are technically more or less equivalent; see Troelstra and van Dalen [61].

Any operational language L comprises the following primitive first order symbols:

(PS.1) Countably many individual variables $a, b, c, f, g, h, u, v, w, x, y, z$ (possibly with subscripts) and countably many individual constants, including k, s (combinators), p, p_0, p_1 (pairing and unpairing).

(PS.2) The binary function symbol $\circ$ for (partial) term application.

(PS.3) For every natural number n a countable (possibly empty) set of relation symbols, including the unary relation symbol $\downarrow$ for definedness and the binary relation symbol $=$ for equality.
(PS.4) The logical symbols $\neg$ (negation), $\vee$ (disjunction), and $\exists$ (existential quantification).

The *individual terms* $(r, s, t, r_0, s_0, t_0, \ldots)$ of an operational language L are inductively generated as follows:

(T.1) The individual variables and individual constants of L are individual terms of L.
(T.2) If r and s are individual terms of L, then so also is $\circ(r, s)$.

In the following $\circ(r, s)$ is usually written as $(r \circ s)$, as (rs), or – if no confusion arises – simply as rs. The convention of association to the left is also adopted so that $r_1 r_2 \ldots r_n$ stands for $(\ldots(r_1 r_2)\ldots)$, and we often also write $s(r_1, \ldots, r_n)$ for $sr_1 \ldots r_n$. General n-tupling is defined by induction on $n \geq 1$ as follows:

$$<r_1> := r_1 \quad \text{and} \quad <r_1, \ldots, r_{n+1}> := \mathsf{p}(<r_1, \ldots, r_n>, r_{n+1}).$$

Finally, the *formulas* $(A, B, C, A_0, B_0, C_0, \ldots)$ of L are inductively generated by the following three clauses:

(F.1) All expressions $(r\downarrow)$, $(rs\downarrow)$, and $(r = s)$ are (atomic) formulas of L.
(F.2) If L contains additional n-ary relation symbols R, then all expressions of the form $R(r_1, \ldots, r_n)$ are further (atomic) formulas of L.
(F.3) If A and B are formulas of L, then so also are $(\neg A)$, $(A \vee B)$, and $\exists x A$.

In this article we confine ourselves to classical logic. Hence the remaining logical connectives and the universal quantifier can be defined as usual. Also, $(r \neq s)$ is short for $\neg(r = s)$.

We will often omit parentheses and brackets whenever there is no danger of confusion. Moreover, we frequently make use of the vector notation $\vec{\mathcal{E}}$ as shorthand for a finite string $\mathcal{E}_1, \ldots, \mathcal{E}_n$ of expressions whose length is not important or is evident from the context. The set of free variables of a formula A is defined in the standard way. An L formula without free variables is called a closed L formula; the closed L terms are those without variables.

Suppose now that $\vec{a} = a_1, \ldots, a_n$ and $\vec{r} = r_1, \ldots, r_n$, where $a_1, \ldots, a_n$ are pairwise (syntactically) different variables. Then $A[\vec{r}/\vec{a}\,]$ is the L formula that is obtained from the L formula A by simultaneously replacing all free occurrences of the variables $\vec{a}$ by the L terms $\vec{r}$; in order to avoid collision of variables, a renaming of bound variables may be necessary. In case the L formula A is written as $A[\vec{a}\,]$, we often simply write $A[\vec{r}\,]$ instead of $A[\vec{r}/\vec{a}\,]$. Further variants of this notation will be obvious. The substitution of L terms for variables in L terms is treated accordingly.

As deduction system for the logic of partial terms we make use of a so-called *Hilbert calculus*, consisting of the following axioms and rules of inference.

Propositional Axioms and Propositional Rules. These comprise the usual axioms and rules of inference of some sound and complete Hilbert calculus for classical propositional logic

Quantifier Axioms and Quantifier Rules. The axioms for the existential quantifier consist of all L formulas

$$A[r] \wedge r\downarrow \;\rightarrow\; \exists x A[x],$$

where r may be an arbitrary L term. The rules of inference for the existential quantifier, on the other hand, are all configurations

$$\frac{A \rightarrow B}{\exists x A \rightarrow B}$$

for which the variable x does not occur free in B. Because of axiom (DE.1) below it is not necessary to claim in the premise that a is defined.

Definedness and Equality Axioms. For all constants r, all L terms s, all variables a, b, and all atomic formulas $A[u]$ of L:

(DE.1) $r\downarrow \wedge a\downarrow$.
(DE.2) $A[s] \rightarrow s\downarrow$.
(DE.3) $(a = a)$.
(DE.4) $(a = b) \wedge A[a] \rightarrow A[b]$.

The axioms (DE.2) are often referred to as *strictness axioms*. Important special cases are, for example, all assertions

$$(s = t) \rightarrow s\downarrow \wedge t\downarrow \quad \text{and} \quad (st)\downarrow \rightarrow s\downarrow \wedge t\downarrow,$$

stating that two terms can be equal only in case both have a value and that a compound term has a value only in case all its subterms have values as well. Thus the determination of the value of a term follows a *call-by-value* strategy. Observe that the axioms (DE.3) and (DE.4) are formulated for variables only. We must not claim $(r = r)$ in general since r may not have a value. However, we can introduce the notion of *partial equality* $\simeq$ à la Kleene,

$$(r \simeq s) \;:=\; (r\downarrow \vee s\downarrow) \rightarrow (r = s),$$

and obtain for all formulas $A[u]$ and terms r, s of L that $(r \simeq s)$ and $A[r]$ imply $A[s]$.

As mentioned above, it is an important aspect of the logic of partial terms that constants are defined and variables only range over defined objects. To point this out explicitly, we include axiom (DE.1). But observe that assertion $a\downarrow$ follows from (DE.2) and (DE.3).

The semantics of the logic of partial terms is based on *partial structures* consisting of a non-empty universe, interpretations of all constants within this universe, interpretations of all n-ary relation symbols as n-ary relations over this universe, and a partial binary function on this universe to take care of application. It is not difficult to shown that the above Hilbert system is sound and complete with respect to this semantics.

The basic theory BO(L) of operations for the language L comprises these axioms and rules of the logic of partial terms and axioms formalizing that the universe is a partial combinatory algebra and that pairing and projections are as expected.

Combinatory Axioms, Pairing and Projections

(Co.1) $\mathsf{k} \neq \mathsf{s}$.
(Co.2) $\mathsf{k}ab = a$.
(Co.3) $\mathsf{s}ab\downarrow \ \wedge \ \mathsf{s}abc \simeq (ac)(bc)$.
(Co.4) $\mathsf{p}_0{<}a, b{>} = a \ \wedge \ \mathsf{p}_1{<}a, b{>} = b$.

In general, totality of application is not assumed; but if it is required in a special situation we add the statement

$$\text{(Tot)} \qquad \forall x \forall y (xy\downarrow).$$

Two fundamental principles are an immediate consequence of the combinatory axioms: λ-abstraction and the fixed point theorem. In more detail: With any L term r, we associate an L term $(\lambda x.r)$ whose variables are those of r excluding x, such that BO(L) proves

$$(\lambda x.r)\downarrow \ \wedge \ (\lambda x.r)x \simeq r \ \wedge \ (s\downarrow \ \rightarrow \ (\lambda x.r)s \simeq r[s/x]).$$

As usual we can generalize λ-abstraction to several arguments by simply iterating abstraction for one argument. In addition, we have the following fixed point theorems for the partial and the total case.

Theorem 1 (Fixed points) *There exist closed* L *terms* fix *and* fix_t *such that* BO(L) *proves for any* f *and* a*:*

1. $\mathsf{fix}\ f\downarrow \ \wedge \ \mathsf{fix}(f, a) \simeq f(\mathsf{fix}\ f, a)$.
2. $\text{(Tot)} \ \rightarrow \ \mathsf{fix}_t\ f = f(\mathsf{fix}_t\ f)$.

This basic operational framework is an adaptation of λ-calculus and combinatory algebra to the partial case. According to my knowledge it has been set up in this form and in all details for the first time in Feferman [9].

We end this section with mentioning an interesting ontological relationship between full definition by cases and operational extensionality. For this purpose we assume that L contains an individual constant d and consider the additional axiom

$$\text{(d)} \qquad (u = v \ \rightarrow \ \mathsf{d}(a, b, u, v) = a) \ \wedge \ (u \neq v \ \rightarrow \ \mathsf{d}(a, b, u, v) = b).$$

This is "full definition by cases" since it tests for arbitrary elements of the universe whether they are equal. Later we will also introduce restricted versions of (d). Clearly, BO(L) + (d) is consistent.

Operational extensionality is the principle that claims that two operations are identical in case they have the same "input-output" behavior,

$$(\mathsf{Op\text{-}Ext}) \qquad \forall f \forall g(\forall x(fx = gx) \;\rightarrow\; f = g).$$

Also $\mathsf{BO}(\mathrm{L}) + (\mathsf{Op\text{-}Ext})$ is consistent. However, (d) and $(\mathsf{Op\text{-}Ext})$ as well as (d) and (Tot) are incompatible with each other.

Theorem 2 *Let* L *be an operational language with the constant* d. *Then* $\mathsf{BO}(\mathrm{L}) + (\mathsf{d}) + (\mathsf{Op\text{-}Ext})$ *and* $\mathsf{BO}(\mathrm{L}) + (\mathsf{d}) + (\mathsf{Tot})$ *are inconsistent.*

Proof To show the first inconsistency, set $r := \mathsf{fix}(\lambda yx.\mathsf{d}(\mathsf{k}, \mathsf{s}, y, \lambda z.\mathsf{s}))$. In view of the fixed point theorem we then have $r\downarrow$ and

$$(*) \qquad \forall x(rx \;\simeq\; \mathsf{d}(\mathsf{k}, \mathsf{s}, r, \lambda z.\mathsf{s})).$$

From $r = \lambda z.\mathsf{s}$ we would be able to deduce by (d) and $(*)$ that $\forall x(rx = \mathsf{k})$, in contradiction to the assumption $r = \lambda z.\mathsf{s}$. Hence $r \neq \lambda z.\mathsf{s}$. Since r and $\lambda z.\mathsf{s}$ are defined, (d) and $(*)$ now give us $\forall x(rx = \mathsf{s})$. But then operational extensionality $(\mathsf{Op\text{-}Ext})$ yields $r = \lambda z.\mathsf{s}$; again a contradiction.

To establish the second inconsistency, we work with the term fix_t and let r be the term $\mathsf{fix}_t(\lambda x.\mathsf{d}(\mathsf{k}, \mathsf{s}, x, \mathsf{s}))$. Now a simple calculation shows that $r = \mathsf{s}$ implies $r = \mathsf{k}$, and $r \neq \mathsf{s}$ yields $r = \mathsf{s}$; again a contradiction. □

3 Applicative Theories

Now I do not follow the historic timeline. Originally, Feferman's interest in the operational approach was triggered by his work on explicit mathematics to which we will turn in the following section. Most approaches to explicit mathematics choose a sort of second order operational approach that permits the formation of classes of operations and includes class formation principles of various strengths.

In this section we set a slower pace, stay first order and carefully extend the basic theory $\mathsf{BO}(\mathrm{L})$ by some elementary axioms for the natural numbers. Then we consider various forms of induction on the natural numbers, and later the numerical choice operator μ and the Suslin operator E_1. These theories constitute the first order part of explicit mathematics, and we call them *applicative theories*.

Let L_1 be an operational language that in addition to the primitive first order symbols mentioned in the previous section comprises constants 0 (zero), $\mathsf{s_N}$ (numerical successor), $\mathsf{p_N}$ (numerical predecessor), $\mathsf{d_N}$ (definition by numerical cases), $\mathsf{r_N}$ (primitive recursion), μ (unbounded search), E_1 (Suslin operator), and the unary relation symbol N for the collection of all natural numbers. Then we often use $(r \in \mathsf{N})$ interchangeably with $\mathsf{N}(r)$ and set

$$(r : \mathsf{N}^k \rightarrow \mathsf{N}) \;:=\; (\forall x_1, \ldots, x_k \in \mathsf{N})(r(x_1, \ldots, x_k) \in \mathsf{N}),$$

where k is supposed to be a positive natural number. Furthermore, in the following we generally write $(r : \mathsf{N} \to \mathsf{N})$ for $(r : \mathsf{N}^1 \to \mathsf{N})$ and r' for $\mathsf{s_N} r$.

The *basic theory of operations and numbers* BON is the extension of BO(L_1) by the following groups of axioms, dealing with the natural numbers.

Natural Numbers.

(Nat.1) $0 \in \mathsf{N} \wedge (a \in \mathsf{N} \to a' \in \mathsf{N})$.
(Nat.2) $a \in \mathsf{N} \to (a' \neq 0 \wedge \mathsf{p_N}(a') = a)$.
(Nat.3) $(a \in \mathsf{N} \wedge a \neq 0) \to (\mathsf{p_N} a \in \mathsf{N} \wedge (\mathsf{p_N} a)' = a)$.

Definition by Numerical Cases.

(Nat.4) $(a, b \in \mathsf{N} \wedge a = b) \to \mathsf{d_N}(u, v, a, b) = u$.
(Nat.5) $(a, b \in \mathsf{N} \wedge a \neq b) \to \mathsf{d_N}(u, v, a, b) = v$.

Primitive Recursion.

(Nat.6) $(a \in \mathsf{N} \wedge f : \mathsf{N}^2 \to \mathsf{N}) \to \mathsf{r_N}(a, f) : \mathsf{N} \to \mathsf{N}$.
(Nat.7) $(a, b \in \mathsf{N} \wedge f : \mathsf{N}^2 \to \mathsf{N} \wedge g = \mathsf{r_N}(a, f)) \to$
$(g0 = a \wedge g(b') = f(b, gb))$.

Axioms for the the constants μ and E_1 follow later. Thus far no induction principles are available, and this is the reason that the axioms (Nat.6) and (Nat.7) are needed for representing all primitive recursive functions within BON. But with these axioms at hand, it is straightforward to prove the following.

Theorem 3 (Primitive recursive functions) *For every k-ary primitive recursive function $\mathcal{F}$ there exists a closed term $\mathsf{prf}_{\mathcal{F}}$ of L_1 such that* BON *proves $\mathsf{prf}_{\mathcal{F}} : \mathsf{N}^k \to \mathsf{N}$ as well as the (canonical translations of the) defining equations of $\mathcal{F}$.*

Several forms of induction have been considered over BON. The weakest form, called *basic induction*, applies induction only to operations that are known to be total from N to N.

Basic Induction on N (B-$\mathsf{I_N}$).

$$(f : N \to N \wedge f0 = 0 \wedge (\forall x \in \mathsf{N})(fx = 0 \to f(x') = 0)) \to (\forall x \in \mathsf{N})(fx = 0).$$

The assumption $f : \mathsf{N} \to \mathsf{N}$ is central in this formulation and responsible for its relative weakness (see below): Basic induction allows us to prove properties of total operations from N to N; however, in general it cannot be employed to show that certain operations are total from N to N. Basic induction is, of course, a special case of the schema of induction on the natural numbers for arbitrary L_1 formulas.

L_1 **induction on** N (L_1-$\mathsf{I_N}$). For all L_1 formulas $A[u]$,

$$A[0] \wedge (\forall x \in \mathsf{N})(A[x] \to A[x']) \to (\forall x \in \mathsf{N})A[x].$$

The canonical model of $\mathsf{BON} + (\mathrm{L}_1\text{-}\mathsf{I_N})$ has the natural numbers $\mathbb{N}$ as universe and interprets application $\circ$ as the partial function $\circ_{\mathbb{N}}$ from $\mathbb{N} \times \mathbb{N}$ to $\mathbb{N}$ such that, for all $e, n \in \mathbb{N}$,

$$(e \circ_{\mathbb{N}} n) \simeq \{e\}(n),$$

where $\{e\}$ for $e = 0, 1, \ldots$ is the usual indexing of the partial recursive functions on $\mathbb{N}$. There are also partial and total term models of $\mathsf{BON} + (\mathrm{L}_1\text{-}\mathsf{I_N})$, see, e.g., Beeson [1], Feferman [12], and Troelstra and van Dalen [62]. Probably the simplest way to set up a model satisfying operational extensionality is to start off from a term model of the $\lambda\eta$-calculus (extended by reduction rules for the additional constants of L_1) and to use the standard translation of combinatory logic into the λ-calculus.

In the following theorem we summarize several consistency and inconsistency results concerning BON. The two stated inconsistencies are direct consequences of Theorem 2 since in the presence of $\forall x(x \in \mathsf{N})$ definition by numerical cases is full definition by cases.

Theorem 4 *We have the following consistency and inconsistency results:*

1. $\mathsf{BON} + (\mathsf{Tot}) + (\mathsf{Op\text{-}Ext}) + (\mathrm{L}_1\text{-}\mathsf{I_N})$ *is consistent.*
2. $\mathsf{BON} + \forall x(x \in \mathsf{N}) + (\mathsf{d}) + (\mathrm{L}_1\text{-}\mathsf{I_N})$ *is consistent.*
3. $\mathsf{BON} + \forall x(x \in \mathsf{N}) + (\mathsf{Op\text{-}Ext})$ *is inconsistent.*
4. $\mathsf{BON} + \forall x(x \in \mathsf{N}) + (\mathsf{Tot})$ *is inconsistent.*

The axioms of BON take care of the induction-free part of primitive recursive arithmetic PRA, equipped with the combinatorial machinery of $\mathsf{BO}(\mathrm{L}_1)$, which does not contribute to proof-theoretic strength. Depending on what form of induction we add to BON we thus obtain systems equivalent to primitive recursive arithmetic PRA or Peano arithmetic PA; there are also intermediate forms of induction that we omit.

Theorem 5 $\mathsf{BON} + (\mathsf{B\text{-}I_N})$ *is proof-theoretically equivalent to* PRA *and* $\mathsf{BON} + (\mathrm{L}_1\text{-}\mathsf{I_N})$ *to* PA.

Adding, for example, the assertion $\forall x(x \in \mathsf{N})$ would not spoil these two equivalences. However, the situation becomes much more interesting as soon as further axioms for the type-2 functionals μ and E_1 are taken into consideration. The numerical choice operator μ is characterized by the following two axioms.

Axioms for μ.

$(\mu.1)$ $f : \mathsf{N} \to \mathsf{N} \;\leftrightarrow\; \mu f \in \mathsf{N}$
$(\mu.2)$ $(f : \mathsf{N} \to \mathsf{N} \wedge (\exists x \in \mathsf{N})(fx = 0)) \;\to\; f(\mu f) = 0.$

μ is a non-constructive but predicatively acceptable operator, closely related to the well-known operator E_0 for quantification over the natural numbers. The relationship between μ and E_0 on the basis of BON has been studied in Kahle [47] in full detail: E_0 can be defined within BON from μ; for deriving μ from E_0 Kahle extends BON by specific (proof-theoretically irrelevant) strictness assertions.

The non-constructive operator μ and the functional E_0 are a well-studied objects in higher recursion theory, cf., for example, Feferman [10] and Hinman [27] for a comprehensive survey. It is known that the 1-sections of μ and E_0 are identical; they coincide with the set of natural numbers in the constructible hierarchy up to the first non-recursive ordinal ω_1^{ck}, and, hence, with the collection of hyperarithmetic sets of natural numbers. Consequently, the structure $(\mathbb{N}, 1\text{-}sec(\mathsf{E}_1), \ldots)$ is the minimal standard model of Δ^1_1 comprehension.

If one wants to speak about well-foundedness in this context, the natural step is to add the Suslin operator E_1 that tests for well-foundedness of binary relations on the natural numbers. For the formulation of the axioms of E_1 it is convenient to introduce the descending chain condition $DCC[f]$,

$$DCC[f] := (\exists g \in (\mathsf{N} \to \mathsf{N}))(\forall x \in \mathsf{N})(f(g(x'), gx) = 0),$$

stating that there exists a total operation g from N to N describing a descending chain $g0, g1, \ldots$ with respect to the binary relation coded by f.

Axioms for E_1.

(E_1.1) $f : \mathsf{N}^2 \to \mathsf{N} \;\leftrightarrow\; \mathsf{E}_1 f \in \mathsf{N}$.
(E_1.2) $f : \mathsf{N}^2 \to \mathsf{N} \;\to\; (DCC[f] \;\leftrightarrow\; \mathsf{E}_1 f = 0)$.

The recursion theory of E_1 is well established; see, for example, Hinman [27]. An important result states that the 1-section of E_1 coincides with the set of natural numbers in the constructible hierarchy up to the first recursively inaccessible ordinal ι_0. This ordinal is also the least ordinal not recursive in E_1. Also, Gandy showed that the 1-section of E_1 builds the least standard model of Δ^1_2 comprehension.

From the ontological point of view, the operators μ and E_1 behave as expected. μ takes care of quantification over the natural numbers. Therefore, if μ and the axioms for μ are available, every arithmetically definable set of natural numbers can be represented by a total operation from N to N. If, in addition, E_1 and the axioms for E_1 are at our disposal, we can operationally check for well-foundedness and thus the Π^1_1 normal form theorem allows us to represent all Π^1_1 sets of natural numbers as total operations from N to N.

In the following we write $\mathsf{BON}(\mu)$ for the extension of BON by the axioms (μ.1) and (μ.2). In spite of its "recursion-theoretic strength", $\mathsf{BON}(\mu) + (\mathsf{B}\text{-}\mathsf{I_N})$ is fairly weak proof-theoretically as shown in Feferman and Jäger [23].

Theorem 6 *We have the following proof-theoretic equivalences:*

1. $\mathsf{BON}(\mu) + (\mathsf{B}\text{-}\mathsf{I_N}) \;\equiv\; \mathsf{PA} \;\equiv\; \mathsf{ACA}_0 \;\equiv\; \Delta^1_1\text{-}\mathsf{CA}_0$.
2. $\mathsf{BON}(\mu) + (\mathsf{L}_1\text{-}\mathsf{I_N}) \;\equiv\; \Pi^0_1\text{-}\mathsf{CA}_{<\varepsilon_0} \;\equiv\; \Delta^1_1\text{-}\mathsf{CA}$.

In this theorem and whenever we mention subsystems of second order arithmetic or set theory later we follow the standard nomenclature and refrain from further explanations; see, for example, Buchholz, Feferman, Pohlers, and Sieg [3] or Simpson [59].

The theory $\mathsf{BON}(\mu) + (\mathsf{B}\text{-}\mathsf{I}_\mathsf{N})$ is particularly interesting in connection with Feferman's philosophical analysis of Weyl's *Das Kontinuum* and his reconstruction of the axiom system of *Das Kontinuum* in modern terms; see Feferman [16, 17]. One of his key results is that Weyl's approach can be developed within a conservative extension of PA. This system W can be easily reduced to $\mathsf{BON}(\mu) + (\mathsf{B}\text{-}\mathsf{I}_\mathsf{N})$.

Now we briefly turn to the proof theory of E_1. The applicative theory for E_1 is called SUS and consists of $\mathsf{BON}(\mu)$ and the additional axioms $(\mathsf{E}_1.1)$ and $(\mathsf{E}_1.2)$. The proof-theoretic analysis of SUS plus various forms of induction is carried through in detail in Jäger and Strahm [42] and Jäger and Probst [40].

Theorem 7 *We have the following proof-theoretic equivalences:*

1. $\mathsf{SUS} + (\mathsf{B}\text{-}\mathsf{I}_\mathsf{N}) \;\equiv\; \Pi^1_1\text{-}\mathsf{CA}_0 \;\equiv\; \Delta^1_2\text{-}\mathsf{CA}_0.$
2. $\mathsf{SUS} + (\mathsf{L}_1\text{-}\mathsf{I}_\mathsf{N}) \;\equiv\; \Pi^1_1\text{-}\mathsf{CA}_{<\varepsilon_0} \;\equiv\; \Delta^1_2\text{-}\mathsf{CA}.$

For the lower bounds of SUS plus various forms of induction we exploit the fact that the Suslin operator has the power to deal with Π^1_1 comprehension, provably in $\mathsf{SUS} + (\mathsf{B}\text{-}\mathsf{I}_\mathsf{N})$. Upper bounds are established in Jäger and Strahm [42] by making use of a very specific positive Δ^1_2 inductive definition in the framework of theories of admissible sets and by interpreting the application operation by a Σ definable fixed point of this inductive definition. A more direct approach to the computation of the upper bounds in question is provided in Jäger and Probst [40]; several theories featuring the Suslin operator are embedded into ordinal theories tailored to dealing with non-monotone inductive definitions that enable a smooth definition of the application relation.

4 Explicit Mathematics

As already mentioned above, Feferman [9] is the starting point of explicit mathematics. The two other "big elephants" are Feferman [11], in which explicit mathematics is discussed in the context of recursion theory, and Feferman [12], which discusses the relationship between explicit mathematics and several alternative approaches to constructive mathematics.

Originally, explicit mathematics was formulated in a single sorted first order language with a unary relation symbol Cl and a binary relation symbol η, where $Cl(u)$ expressed that u is a class and $(v\ \eta\ u)$ that v has the property described by u in case $Cl(u)$ holds. Later it turned out to be more convenient to formulate explicit mathematics in an extension of the logic of partial terms with class variables; see Jäger [29].

The underlying ontological idea is that we have two sorts of objects: the individuals as in the case of applicative theories and collections of such objects, called *classes*. The individuals form a partial combinatory algebra and are conceived as being given intensionally and explicitly as before, whereas the classes are subsets of the applicative universe and may even be considered to exist in a Platonic sense;

this is purposely left open. Membership of individuals in classes is as usual, and we have extensionality on the level of classes.

But also the classes can be addressed explicitly, though in an indirect way: We add a new binary relation $\mathfrak{R}$ to express that the individual x *represents* or *names* class X, written $\mathfrak{R}(x, X)$. Classes are explicitly generated with reference to their names in an operational way, and this process is made uniform in the parameters. For example, we will have a constant nat that names the class of natural numbers and a constant un such that $\mathsf{un}(u, v)$ is the name of the union of the classes U and V provided that u is the name of U and v the name of V.

A suitable language for our purpose is the extension L_2 of L_1 by class variables $U, V, W, X, Y, Z, \dots$ (possibly with subscripts), two new binary relation symbols $\in$ (membership) and $\mathfrak{R}$ (naming, representation) and the new individual constants nat (natural numbers), id (identity), co (complement), un (union), dom (domain), inv (inverse image), j (join), and i (inductive generation). The atomic formulas of L_2 are all expressions $r\downarrow$, $(r = s)$, $\mathsf{N}(r)$, $(r \in U)$, $(U = V)$, and $\mathfrak{R}(r, U)$, where r and s are individual terms of L_2.

The *formulas* $(A, B, C, A_0, B_0, C_0, \dots)$ of L_2 are generated from the atomic L_2 formulas by closing under the propositional connectives and quantification in both sorts. An L_2 formula is called *elementary* if it contains neither the relation symbol $\mathfrak{R}$ nor bound class variables. The *stratified* formulas are those L_2 formulas that do not contain the relation symbol $\mathfrak{R}$.

Some individual terms represent (or name) classes, and we write $(r \in \mathfrak{R})$ to express that r is a name,

$$(r \in \mathfrak{R}) \;:=\; \exists X \mathfrak{R}(r, X).$$

If r names class X, then X can be regarded as the *extension* of r, and in this sense we can transfer an element relation and extensional equality to the level of individuals:

$$(r \mathrel{\dot{\in}} s) \;:=\; \exists X (\mathfrak{R}(s, X) \wedge r \in X),$$
$$(r \mathrel{\dot{=}} s) \;:=\; \exists X (\mathfrak{R}(r, X) \wedge \mathfrak{R}(s, X)).$$

Clearly, $(r \notin U)$ and $(r \mathrel{\dot{\notin}} s)$ are short for $\neg(r \in U)$ and $\neg(r \mathrel{\dot{\in}} s)$, respectively. Since we have extensionality on the level of classes, the subclass relation on classes is as usual with the corresponding notion on the level of individual terms

$$(U \subseteq V) \;:=\; \forall x (x \in U \rightarrow x \in V),$$
$$(r \mathrel{\dot{\subseteq}} s) \;:=\; \exists X \exists Y (\mathfrak{R}(r, X) \wedge \mathfrak{R}(s, Y) \wedge X \subseteq Y).$$

Finally, if $\vec{r}$ is the string $r_1, \dots, r_n$ of individual terms and $\vec{U}$ the string $U_1, \dots, U_n$ of class variables of the same length, we set

$$\mathfrak{R}(\vec{r}, \vec{U}) \;:=\; \bigwedge_{i=1}^{n} \mathfrak{R}(r_i, U_i) \quad \text{and} \quad (\vec{r} \in \mathfrak{R}) \;:=\; \bigwedge_{i=1}^{n} (r_i \in \mathfrak{R}).$$

Observe that all formulas $\Re(\vec{r}, \vec{U})$, $(\vec{r} \in \Re)$, $(r \dot{\in} s)$, $(r \doteq s)$, and $(r \dot{\subseteq} s)$ are not stratified.

4.1 Elementary Explicit Comprehension

The logic of the first order part of our systems of explicit mathematics is still Beeson's logic of partial terms as in the previous sections, of course formulated now for the language L_2. In particular, the definedness axioms extend to atomic L_2 formulas and, therefore, $(r \in U)$ and $\Re(r, U)$ imply that the term r has a value. The logic for the second order part of the systems of explicit mathematics is classical predicate logic with equality.

The non-logical axioms of the *elementary theory* EC *of classes and names* comprises the non-logical axioms of BON plus the following two groups of class axioms for classes.

Explicit Representation and Extensionality.

(Cl.1) $\exists x \Re(x, U)$.
(Cl.2) $\Re(r, U) \wedge \Re(r, V) \;\rightarrow\; U = V$.
(Cl.3) $\forall x(x \in U \;\leftrightarrow\; x \in V) \;\rightarrow\; U = V$.

These axioms state that each class has a name, that there are no homonyms and that equality of classes is extensional. The second group of axioms for classes ensures the build-up of some basic classes, in parallel with a uniform naming process.

Basic Class Existence Axioms.

(Cl.4) $\mathsf{nat} \in \Re \;\wedge\; \forall x(x \dot{\in} \mathsf{nat} \;\leftrightarrow\; \mathsf{N}(x))$.
(Cl.5) $\mathsf{id} \in \Re \;\wedge\; \forall x(x \dot{\in} \mathsf{id} \;\leftrightarrow\; \exists y(x = \langle y, y \rangle))$.
(Cl.6) $r \in \Re \;\rightarrow\; (\mathsf{co}(r) \in \Re \;\wedge\; \forall x(x \dot{\in} \mathsf{co}(r) \;\leftrightarrow\; x \dot{\notin} r))$.
(Cl.7) $r, s \in \Re \;\rightarrow\; (\mathsf{un}(r, s) \in \Re \;\wedge\; \forall x(x \dot{\in} \mathsf{un}(r, s) \;\leftrightarrow\; (x \dot{\in} r \;\vee\; x \dot{\in} s)))$.
(Cl.8) $r \in \Re \;\rightarrow\; (\mathsf{dom}(r) \in \Re \;\wedge\; \forall x(x \dot{\in} \mathsf{dom}(r) \;\leftrightarrow\; \exists y(\langle x, y \rangle \dot{\in} r)))$.
(Cl.9) $r \in \Re \;\rightarrow\; (\mathsf{inv}(r, f) \in \Re \;\wedge\; \forall x(x \dot{\in} \mathsf{inv}(r, f) \;\leftrightarrow\; fx \dot{\in} r))$.

These axioms formalize that the natural numbers form a class and that there is the identity class; furthermore, classes are closed under complements, unions, domains and inverse images. It is important that the axioms (C.4)–(C.9) provide a finite axiomatization of uniform elementary comprehension.

Theorem 8 (Elementary comprehension) *For every elementary formula* $A[u, \vec{v}, \vec{W}]$ *with at most the indicated free variables there exists a closed term* t_A *such that* EC *proves:*

1. $\vec{z} \in \Re \rightarrow t_A(\vec{y}, \vec{z}) \in \Re$,
2. $\Re(\vec{z}, \vec{Z}) \rightarrow \forall x(x \dot{\in} t_A(\vec{y}, \vec{z}) \;\leftrightarrow\; A[x, \vec{y}, \vec{Z}])$.

Immediate obvious consequences of this theorem are, for example, the existence of the empty class $\emptyset$ and the universal class $\mathbf{V}$, the closure of the collection of all classes under complements, finite unions, finite intersections, finite Cartesian products, and the finitely iterated formation of function spaces.

By a model construction following Feferman [9, 12] it can be shown that EC is consistent with stratified comprehension, whereas a simple Russell-style argument shows that it is inconsistent with comprehension for arbitrary L_2 formulas.

Interesting induction principles in the context of EC are (B-I_N) and (L_1-I_N) as before plus two new forms of induction: class induction and the schema of induction for all L_2 formulas.

Class Induction on N (C-I_N).

$$\forall X(0 \in X \wedge (\forall x \in \mathsf{N})(x \in X \rightarrow x' \in X) \rightarrow (\forall x \in \mathsf{N})(x \in X)).$$

L_2 **induction on** N (L_2-I_N). For all L_2 formulas $A[u]$,

$$A[0] \wedge (\forall x \in \mathsf{N})(A[x] \rightarrow A[x']) \rightarrow (\forall x \in \mathsf{N})A[x].$$

All combinations of EC and its extension $\mathsf{EC}(\mu)$ by the type-2 functional μ with these forms of first and second order induction have been analyzed proof-theoretically; a detailed presentation will be given in Feferman, Jäger, and Strahm [24]. As illustration we mention three results.

Theorem 9 *We have the following proof-theoretic equivalences:*

1. $\mathsf{EC} + (\mathsf{C}\text{-}\mathsf{I}_\mathsf{N}) \equiv \mathsf{BON} + (L_1\text{-}\mathsf{I}_\mathsf{N}) \equiv \mathsf{ACA}_0 \equiv \mathsf{PA}$.
2. $\mathsf{EC} + (L_2\text{-}\mathsf{I}_\mathsf{N}) \equiv \mathsf{ACA}$.
3. $\mathsf{EC}(\mu) + (\mathsf{B}\text{-}\mathsf{I}_\mathsf{N}) \equiv \mathsf{BON}(\mu) + (\mathsf{B}\text{-}\mathsf{I}_\mathsf{N}) \equiv \mathsf{PA}$,

In Feferman [10, 12, 16] it is convincingly argued that EC-like systems provide a natural framework for dealing with large parts of predicative mathematics. In particular, the theory $\mathsf{EC}(\mu) + (\mathsf{B}\text{-}\mathsf{I}_\mathsf{N})$ is a natural extension of $\mathsf{BON}(\mu) + (\mathsf{B}\text{-}\mathsf{I}_\mathsf{N})$ and as such perfectly suited for developing Weyl's approach to the continuum. It is also shown in Feferman [10] that the intensional and extensional variants of finite type theories find their natural place within EC.

4.2 *Join and Inductive Generation*

Of course, the theorem about elementary comprehension tells us that in EC the classes are closed under the formation of finite unions and intersections. But in order to form the unions, intersections, and Cartesian products of general possibly infinite families of classes, Feferman introduced a further axiom, and here the constant j comes into play.

Join Axiom.

(J) $(a \in \Re \wedge (\forall x \mathrel{\dot{\in}} a)(f(x) \in \Re)) \rightarrow (\mathsf{j}(a, f) \in \Re \wedge DU[a, f, \mathsf{j}(a, f)])$,

where the formula $DU[a, f, b]$ is short for

$$\forall x(x \mathrel{\dot{\in}} b \leftrightarrow x = <(x)_0, (x)_1> \wedge (x)_0 \mathrel{\dot{\in}} a \wedge (x)_1 \mathrel{\dot{\in}} f((x)_0)).$$

This axiom states that given a class named by a and an operation f from this class to names, $\mathsf{j}(a, f)$ is the name of the disjoint union of the classes named by these $f(x)$ with $x \mathrel{\dot{\in}} a$. Clearly, the generation of $\mathsf{j}(a, f)$ is uniform in a and f.

Finally, let us turn to inductive generation and introduce an auxiliary abbreviation. Given an L_2 formula $A[u]$ we write $Prog[a, b, A]$ for

$$(\forall x \mathrel{\dot{\in}} a)(\forall y(<y, x> \mathrel{\dot{\in}} b \rightarrow A[y]) \rightarrow A[x]).$$

Moreover, $Prog[a, b, c]$ stands for $Prog[a, b, C]$ with $C[u]$ being $(u \mathrel{\dot{\in}} c)$. If we think of b coding a binary relation on the class named a, then $Prog[a, b, A]$ states that formula $A[u]$ is progressive on a with respect to b. Feferman's axioms about inductive generation guarantee the existence of accessible parts of classes with respect to binary relations.

Axioms for Inductive Generation.

(IG.1) $a, b \in \Re \rightarrow (\mathsf{i}(a, b) \in \Re \wedge Prog[a, b, \mathsf{i}(a, b)])$.
(IG.2) $(a, b \in \Re \wedge Prog[a, b, A]) \rightarrow (\forall x \mathrel{\dot{\in}} \mathsf{i}(a, b))A[x]$

for all L_2 formulas $A[u]$. Let a and b be names. According to (IG.1) then $\mathsf{i}(a, b)$ names a class and is progressive on a with respect to b. (IG.2) is an induction principle and states that the class named $\mathsf{i}(a, b)$ is minimal with respect to this property.

The most famous theory of explicit mathematics is called T_0 and extends EC by join, inductive generation and full induction on the natural numbers for arbitrary L_2 formulas,

$$\mathsf{T}_0 := \mathsf{EC} + (\mathsf{J}) + (\mathsf{IG}.1) + (\mathsf{IG}.2) + (\mathsf{L}_2\text{-}\mathsf{I}_\mathsf{N}).$$

Many subsystems of T_0 - obtained, for example, by restricting the induction principles or omitting inductive generation - have been introduced and studied in Chapter II (written by Feferman and Sieg), of Buchholz, Feferman, Pohlers, and Sieg [3].

As far as T_0 itself is concerned, [3] also provides an argument that it can be embedded into the system Δ^1_2-CA + (BI) of second order arithmetic. Then Jäger and Pohlers [39] determined the upper bound of the proof-theoretic strength the latter system via the theory KPi of iterated admissible sets, and Jäger [28] showed by a well-ordering proof within (even the intuitionistic version of) T_0 that this bound is sharp.

Theorem 10 $\mathsf{T}_0 \equiv \Delta^1_2\text{-}\mathsf{CA} + (\mathsf{BI}) \equiv \mathsf{KPi}$.

Recently Sato presented an interesting reduction of $\Delta^1_2\text{-}\mathsf{CA} + (\mathsf{BI})$ to T_0 without employing a well-ordering proof; see [56].

Feferman [9] also introduces the extension of T_0 by the non-constructive μ and baptizes it T_1. He makes a point that T_0 provides an elegant framework for Borelian and hyperarithmetic mathematics. In particular, he advocates studying a generalization of higher arithmetic model theory by means of formalization in T_0. Feferman [13] contains further conceptual work and technical results along similar lines.

Glaß and Strahm [25] mentiones that T_0 and T_1 are equiconsistent and determines the proof-theoretic strengths of many subsystems of T_1. Finally, ongoing work of Probst is about extensions of T_1 by the Suslin operator E_1. One of his observations is that the second order framework provides several ways of formulating E_1-like operators that may turn out not to be equivalent.

4.3 Monotone Inductive Definitions

A further interesting principle is introduced in Feferman [15]. It expresses that every monotone operation from classes to classes has a least fixed point. Define

$$\begin{aligned} Mon[f] &:= (\forall x, y \in \Re)(x \mathrel{\dot\subseteq} y \to fx \mathrel{\dot\subseteq} fy), \\ Lfp[f,a] &:= fa \doteq a \wedge (\forall x \in \Re)(fx \mathrel{\dot\subseteq} x \to a \mathrel{\dot\subseteq} x). \end{aligned}$$

In view of our definition of $(u \mathrel{\dot\subseteq} v)$, $Mon[f]$ implies that f maps names to names; similarly, $Lfp[f,a]$ implies that a is a name. Then (MID) is the axiom stating that every monotone operation has a least fixed point,

$$(\mathsf{MID}) \qquad\qquad \forall f(Mon[f] \to \exists a Lfp[f,a]).$$

The analysis of (MID) turned out to be very interesting. Adding (MID) to T_0 or a (natural) subsystem of T_0 leads to an enormous increase of its proof-theoretic strength. A first result in Takahashi [60] says that $\mathsf{T}_0 + (\mathsf{MID})$ is interpretable in $\Pi^1_2\text{-}\mathsf{CA} + (\mathsf{BI})$. Later Rathjen in a series of articles [50–53] and Glass, Rathjen Schlüter [26] managed to provide a complete proof-theoretic analysis of (MID) and the uniform version (UMID) of this principle over T_0 and some of its natural subsystems. They were able to determine the exact relationship between these systems of explicit mathematics and systems of second order arithmetic with Π^1_2 comprehension.

4.4 Universes

Universes have been introduced into explicit mathematics in Feferman [14], Marzetta [48], Jäger, Kahle, and Studer [46], and Jäger and Strahm [41] as a powerful method for increasing its expressive and proof-theoretic strength. Informally speaking, universes play a similar role in explicit mathematics as admissible sets in weak set theory and the sets V_κ (for regular cardinals κ) in full classical set theory; explicit universes are also closely related to universes in Martin-Löf type theory. More formally, universes in explicit mathematics are classes which consist of names only and reflect the theory $\mathsf{EC} + (\mathsf{J})$.

Let $\mathfrak{C}[U, a]$ be the closure condition that is formed by the disjunction of the following L_2 formulas:

(1) $a = \mathsf{nat} \vee a = \mathsf{id}$,
(2) $\exists x(a = \mathsf{co}(x) \wedge x \in U)$.
(3) $\exists x \exists y(a = \mathsf{un}(x, y) \wedge x \in U \wedge y \in U)$,
(4) $\exists x(a = \mathsf{dom}(x) \wedge x \in U)$.
(5) $\exists x \exists f(a = \mathsf{inv}(x, f) \wedge x \in U)$,
(6) $\exists x \exists f(a = \mathsf{j}(x, f) \wedge x \in U \wedge (\forall y \,\dot{\in}\, x)(fx \in U))$.

Thus the formula $\forall x(\mathfrak{C}[U, x] \rightarrow x \in U)$ states that U is a class that is closed under (the finite axiomatization of) elementary comprehension and join. If, in addition, all elements of U are names, we call U a *universe* and write $Univ[U]$ to express this fact,

$$Univ[U] \;:=\; \forall x(\mathfrak{C}[U, x] \rightarrow x \in U) \wedge (\forall x \in U)(x \in \mathfrak{R}).$$

Also, $\mathbb{U}[a]$ states that the individual a is a name of a universe,

$$\mathbb{U}[a] \;:=\; \exists X(\mathfrak{R}(a, X) \wedge Univ[X]).$$

It is an immediate consequence of the closure properties of universes that they satisfy elementary comprehension and join. The first important axiom in connection with universes is the *limit axiom*. We assume that L_2 contains a fresh individual constant ℓ and express this by

$$(\mathsf{Lim}) \qquad a \in \mathfrak{R} \;\rightarrow\; \mathbb{U}[\ell a] \wedge a \,\dot{\in}\, \ell a.$$

Hence this axiom states that the individual ℓ uniformly picks for each name x of a class the name ℓx of a universe containing x. Since universes are the explicit analogue of admissible sets, the axiom (Lim) is the explicit analog of the limit axiom in admissible set theory which enforces that any set is contained in an admissible set. The limit axiom (Lim) together with $\mathsf{EC} + (\mathsf{J})$ provides the explicit analogue of (recursive) inaccessibility.

There is a very natural way in explicit mathematics to go a step further and couch (recursive) Mahloness into this framework. To simplify the notation we set

$$(f \in \mathfrak{R} \rightarrow \mathfrak{R}) \;:=\; \forall x(x \in \mathfrak{R} \;\rightarrow\; fx \in \mathfrak{R}),$$
$$(f \,\dot{\in}\, a \rightarrow a) \;:=\; \forall x(x \,\dot{\in}\, a \;\rightarrow\; fx \,\dot{\in}\, a)$$

and let m be a further fresh individual constant of L_2. Then the *Mahlo axiom* is the assertion

$$\text{(Mahlo)}\quad \begin{array}{l} a \in \Re \wedge f \in (\Re \to \Re) \to \\ \qquad \mathbb{U}[\mathsf{m}(a, f)] \wedge a \,\dot{\in}\, \mathsf{m}(a, f) \wedge f \in (\mathsf{m}(a, f) \to \mathsf{m}(a, f)). \end{array}$$

This means that given a name a and an operation f from names to names the individual m uniformly picks a universe $\mathsf{m}(a, f)$ that contains a and is closed under f.

For the proof-theoretic analysis of (Lim) and (Mahlo) over the relevant metapredicative and impredicative systems of explicit mathematics we refer to Jäger, Kahle, and Studer [46] and Jäger and Strahm [24]. In all cases there is a direct correspondence to systems of iterated admissible sets, but space does not permit to go into details here. Jäger and Strahm [43] even explains how stronger reflection principles can be formulated within the explicit framework.

4.5 Names of Classes and Universes

One of the very central ontological observations is that the names of a class never form a class, no matter how simple this class may be. This theorem follows immediately from Jäger [31] and is proved in full detail in Jäger [30] and Feferman, Jäger, and Strahm [24].

Theorem 11 $\mathsf{EC} \vdash \forall X \neg \exists Y \forall z (z \in Y \leftrightarrow \Re(z, X))$.

In Sect. 2 we introduced the notion of operational extensionality. Clearly, there is also a corresponding notion of class extensionality:

$$\text{(Cl-Ext)} \qquad (\forall x, y \in \Re)(x \doteq y \to x = y),$$

claiming that two names are identical provided that they name the same class. Although at a first glance this principle may appear to be acceptable or even natural, we have to dismiss it since it is inconsistent with EC. The following theorem is a consequence of Theorem 11 above. An alternative proof, due to Gordeev, of a similar result is presented in Beeson [1].

Corollary 12 (Cl-Ext) *is inconsistent with* EC.

Proof Pick, for example, the class of natural numbers. From (Cl-Ext) we could conclude that all names of this class are identical to nat and thus form a class (in view of elementary comprehension), contradicting Theorem 11. □

Hence $\mathsf{T_0}$ + (Cl-Ext) is inconsistent as well, thus answering a question raised in Feferman [12]. Although the names of a class never form a class, it is consistent to claim that there exists the class of all names. This can be seen by extending the model construction for EC that is presented in detail in Feferman [12] and Feferman, Jäger, and Strahm [24].

Theorem 13 *The assertion* $\exists X \forall x(x \in X \leftrightarrow x \in \mathfrak{R})$ *is consistent with* EC, *but not provable in* EC.

With some additional effort even a strengthening of this result is possible: We can consistently assume in EC that all objects are names.

Let us now take a look at power classes. In principle, one could think of two forms of power classes. The *strong power class axiom* states that for every class X there exists a class Y such that Y contains exactly the names of all subclasses of X,

$$\text{(SP)} \qquad \forall X \exists Y \forall z(z \in Y \leftrightarrow \exists Z(\mathfrak{R}(z, Z) \wedge Z \subseteq X)).$$

On the other hand, the *weak power class axiom* asks for less. Then we only claim that for each class X there exists a class Y such that each element of Y names a subclass of X and for any subclass of X at least one of its names belongs to Y,

$$\text{(WP)} \quad \forall X \exists Y((\forall z \in Y)(\exists Z \subseteq X)(\mathfrak{R}(z, Z)) \wedge (\forall Z \subseteq X)(\exists z \in Y)\mathfrak{R}(z, Z)).$$

Clearly, each of these can be formulated uniformly by adjunction of suitable constants. Neither the strong nor the weak power class axiom is provable in EC. Much worse, by Theorem 11 we know that in EC the names of the empty class cannot form a class, and thus the strong power class of the empty class cannot exist.

Corollary 14 (SP) *is inconsistent with* EC.

As the following remark shows, the weak power class axiom is less problematic in this respect. Its consistency with EC is a consequence of Theorem 13.

Corollary 15 EC $+ \exists X \forall x(x \in X \leftrightarrow x \in \mathfrak{R})$ *proves*

$$\exists f(\forall a \in \mathfrak{R})((\forall b \mathrel{\dot\in} fa)(b \mathrel{\dot\subseteq} a) \wedge (\forall b \mathrel{\dot\subseteq} a)(\exists c \mathrel{\dot\in} fa)(b \doteq c).$$

Hence the (uniform version of the) weak power class axiom is provable in EC $+$ $\exists X \forall x(x \in X \leftrightarrow x \in \mathfrak{R})$ *and thus consistent with* EC.

Proof Let Z be the class of all names and let z be a name of Z. Also, let r be the closed term $\lambda xy.\mathsf{co}(\mathsf{un}(\mathsf{co}(x), \mathsf{co}(y)))$. This means that for all names a and b, $r(a, b)$ is a name of the intersection of the classes represented by a and b. Now we consider the elementary formula

$$A[u, v, W] := (\exists x \in W)(u = r(v, x))$$

and choose t_A according to Theorem 8. Then $t_A(v, w)$ is a name in case w is a name, and we have

$$\Re(w, W) \rightarrow \forall u(u \mathrel{\dot{\in}} t_A(v, w) \leftrightarrow (\exists x \in W)(u = r(v, x))).$$

Since Z is supposed to be the class of all names and z one of its names, this implies

$$\forall u(u \mathrel{\dot{\in}} t_A(v, z) \leftrightarrow (\exists x \in \Re)(u = r(v, x))).$$

Put $s := \lambda v.t_A(v, z)$. Clearly, $r(a, b) \mathrel{\dot{\subseteq}} a$ for all $a, b \in \Re$ and $s(a, b) \doteq b$ for any $b \mathrel{\dot{\subseteq}} a$. Hence s is a witness for the existential assertion we have to prove. □

However, we have to be careful. The join axiom (J) is incompatible with the weak power class axioms.

Theorem 16 EC + (J) *proves the negation of* (WP). *Also, in* EC + (J) *the names cannot form a class.*

Proof Working in EC + (J), we let a be a name of the universal class $\mathbf{V}$ and assume (WP). Then there exists an element $b \in \Re$ – namely a name of a weak power class of $\mathbf{V}$ – such that:

$$(\forall x \mathrel{\dot{\in}} b)(x \mathrel{\dot{\subseteq}} a), \tag{1}$$

$$(\forall x \in \Re)(\exists y \mathrel{\dot{\in}} b)(x \doteq y). \tag{2}$$

Assertion (1) implies that all elements of (the class represented by) b are names. Now we apply (J) to b and the operation $\lambda z.z$ and obtain that $\mathsf{j}(b, \lambda z.z) \in \Re$ and

$$\forall x(x \mathrel{\dot{\in}} \mathsf{j}(b, \lambda z.z) \leftrightarrow (\exists y_1 \mathrel{\dot{\in}} b)\exists y_2(x = <y_1, y_2> \wedge y_2 \mathrel{\dot{\in}} y_1).$$

By elementary comprehension we can thus form a class X satisfying

$$\forall x(x \in X \leftrightarrow <x, x> \mathrel{\dot{\notin}} \mathsf{j}(b, \lambda z.z)).$$

According to (2), X has a name $u \mathrel{\dot{\in}} b$. However, this implies

$$u \in X \leftrightarrow \neg(u \mathrel{\dot{\in}} b \wedge u \mathrel{\dot{\in}} u) \leftrightarrow u \mathrel{\dot{\notin}} u \leftrightarrow u \notin X.$$

This is a contradiction. Hence $\mathbf{V}$ cannot have a weak power class, and (WP) has been refuted. Therefore, it is also clear in view of the previous corollary that the names must not form a class. □

Now we turn to some remarkable ontological properties of universes. A first observation, proved in Marzetta [48], reveals that no universe may contain one of its names. We have mentioned already that the names of a class do not form a class. In connection with universes, a stronger result is possible: Each class has so many names that not all of them can be contained in a single universe; in other words, no universe is large enough to contain all names of a given type. For a proof of this result see Jäger, Kahle, and Studer [46] or Minari [49]. This result implies that in the presence of the limit axiom (Lim), a name a cannot have the same extensions as the universe represented by ℓa. Also, the operation ℓ does not preserve extensional equality; see [46] for details.

Theorem 17 *1.* $\mathsf{EC} \vdash \mathit{Univ}[U] \wedge \Re(a, U) \rightarrow a \notin U$.
2. $\mathsf{EC} + (\mathsf{J}) \vdash \mathit{Univ}[U] \rightarrow \exists x(\Re(x, V) \wedge x \notin U)$.
3. $\mathsf{EC} + (\mathsf{J}) + (\mathsf{Lim}) \vdash (\forall x \in \Re)(x \not\doteq \ell x) \wedge (\exists x, y \in \Re)(x \doteq y \wedge \ell x \not\doteq \ell x)$.

In this section several important ontological properties of explicit mathematics have been collected. For more along these lines consult Feferman [12], Jäger, Kahle, and Studer [46], and Jäger and Zumbrunnen [45].

5 Operational Set Theory

Feferman's original motivation for operational set theory was to provide a setting for the operational formulation of large cardinal statements directly over set theory in a way that seemed to him to be more natural mathematically than the metamathematical formulations using reflection and indescribability principles, etc. He saw operational set theory as a natural extension of the von Neumann approach to axiomatizing set theory. Another principal motivation was to relate formulations of classical large cardinal statements to their analogues in admissible set theory. However, in view of Jäger and Zumbrunnen [44] this aim of operational set theory has to be analyzed further; see below.

The central systems of present day operational set theory can be considered as an applicative (based) reformulation of systems of classical set theory ranging in strength from Kripke–Platek set theory to von Neumann–Bernays–Gödel set theory and a bit beyond.

The basic system OST has been introduced in Feferman [18] and further discussed in Feferman [19] and Jäger [32–35]. For a gentle introduction into operational set theory and some general motivation we refer to these articles, in particular to [19].

There is also an interesting relationship between some more constructive variants of operational set theory and constructive or semi-constructive set theory, but we will not discuss this line of research here. For a profound discussion of this topic and some interesting technical results see Cantini and Crosilla [5, 6], Cantini [4], and Feferman [20].

5.1 The Central Systems

Let $\mathcal{L}$ be a typical language of first order set theory with the binary symbols $\in$ and $=$ as its only relation symbols and countably many set variables $a, b, c, f, g, u, v, w, x, y, z, \ldots$ (possibly with subscripts). We further assume that $\mathcal{L}$ has a constant ω for the collection of all finite von Neumann ordinals. The formulas of $\mathcal{L}$ are defined as usual.

The language $\mathcal{L}^\circ$ of operational set theory extends $\mathcal{L}$ by the binary function symbol $\circ$ for partial term application, the unary relation symbol $\downarrow$ for definedness and a series of constants: (i) the combinators k and s, (ii) $\top$, $\bot$, **el**, **non**, **dis**, **e**, and **E** for logical operations, (iii) $\mathbb{D}$, $\mathbb{U}$, $\mathbb{S}$, $\mathbb{R}$, $\mathbb{C}$, and $\mathbb{P}$ for set-theoretic operations. The meaning of these constants will be specified by the axioms below.

$\mathcal{L}^\circ$ is an operational language in the sense of Sect. 2, and we define the terms and formulas of $\mathcal{L}^\circ$ exactly as there. To increase readability, we freely use standard set-theoretic terminology. For example, if $A[x]$ is an $\mathcal{L}^\circ$ formula, then $\{x : A[x]\}$ denotes the collection of all sets satisfying A; it may be (extensionally equal to) a set, but this is not necessarily the case. Special instances are

$$\mathbf{V} := \{x : x\downarrow\}, \quad \emptyset := \{x : x \neq x\}, \quad \text{and} \quad \mathbf{B} := \{x : x = \top \vee x = \bot\}$$

so that $\mathbf{V}$ denotes the collection of all sets (it is not a set itself), $\emptyset$ stands for the empty collection, and $\mathbf{B}$ for the unordered pair consisting of the truth values $\top$ and $\bot$ (it will turn out that $\emptyset$ and $\mathbf{B}$ are sets in OST). The following shorthand notation, for n an arbitrary natural number greater than 0,

$$(f : a^n \to b) := (\forall x_1, \dots, x_n \in a)(f(x_1, \dots, x_n) \in b)$$

expresses that f, in the operational sense, is an n-ary mapping from a to b. It does not say, however, that f is an n-ary function in the set-theoretic sense. In this definition the set variables a and b may be replaced by $\mathbf{V}$ and $\mathbf{B}$. So, for example, $(f : a \to \mathbf{V})$ means that f is total on a, and $(f : \mathbf{V} \to b)$ means that f maps all sets into b.

As in the case of explicit mathematics, all systems of operational set theory start off from the basic theory $\mathsf{BO}(\mathcal{L}^\circ)$. The additional non-logical axioms of OST comprise some basic set-theoretic axioms, the representation of elementary logical connectives as operations, and operational set existence axioms.

Basic Set-theoretic Axioms. They comprise: (i) the usual extensionality axiom; (ii) assertions that give the appropriate meaning to the constant ω; (iii) $\in$-induction for arbitrary formulas $A[u]$ of $\mathcal{L}^\circ$,

$$\forall x((\forall y \in x)A[y] \to A[x]) \to \forall x A[x].$$

Logical Operations Axioms.

(L.1) $\top \neq \bot$,
(L.2) $(\mathbf{el} : \mathbf{V}^2 \to \mathbf{B}) \wedge \forall x \forall y(\mathbf{el}(x, y) = \top \leftrightarrow x \in y)$,
(L.3) $(\mathbf{non} : \mathbf{B} \to \mathbf{B}) \wedge (\forall x \in \mathbf{B})(\mathbf{non}(x) = \top \leftrightarrow x = \bot)$,
(L.4) $(\mathbf{dis} : \mathbf{B}^2 \to \mathbf{B}) \wedge (\forall x, y \in \mathbf{B})(\mathbf{dis}(x, y) = \top \leftrightarrow (x = \top \vee y = \top))$,
(L.5) $(f : a \to \mathbf{B}) \to (\mathbf{e}(f, a) \in \mathbf{B} \wedge (\mathbf{e}(f, a) = \top \leftrightarrow (\exists x \in a)(fx = \top)))$.

Set-theoretic Operations Axioms.

(S.1) Unordered pair: $\mathbb{D}(a, b)\downarrow \wedge \forall x(x \in \mathbb{D}(a, b) \leftrightarrow x = a \vee x = b)$.
(S.2) Union: $\mathbb{U}(a)\downarrow \wedge \forall x(x \in \mathbb{U}(a) \leftrightarrow (\exists y \in a)(x \in y))$.

(S.3) Separation for definite operations:

$$(f : a \to \mathbf{B}) \to (\mathbb{S}(f, a)\downarrow \wedge \forall x(x \in \mathbb{S}(f, a) \leftrightarrow (x \in a \wedge fx = \top))).$$

(S.4) Replacement:

$$(f : a \to \mathbf{V}) \to (\mathbb{R}(f, a)\downarrow \wedge \forall x(x \in \mathbb{R}(f, a) \leftrightarrow (\exists y \in a)(x = fy))).$$

(S.5) Choice: $\exists x(fx = \top) \to (\mathbb{C}f\downarrow \wedge f(\mathbb{C}f) = \top)$.

This finishes our description of the system OST. OST($\mathbb{P}$) is OST + ($\mathbb{P}$) and OST($\mathbf{E}, \mathbb{P}$) is OST + ($\mathbb{P}$) + ($\mathbf{E}$), where ($\mathbb{P}$) and ($\mathbf{E}$) are axioms providing for the operational form of power set and unbounded existential quantification, respectively:

$$(\mathbb{P}) \qquad (\mathbb{P} : \mathbf{V} \to \mathbf{V}) \wedge \forall x \forall y(x \in \mathbb{P}y \leftrightarrow x \subset y),$$

$$(\mathbf{E}) \qquad (f : \mathbf{V} \to \mathbf{B}) \to (\mathbf{E}(f) \in \mathbf{B} \wedge (\mathbf{E}(f) = \top \leftrightarrow \exists x(fx = \top))).$$

Finally, OST^r($\mathbf{E}, \mathbb{P}$) is obtained from OST($\mathbf{E}, \mathbb{P}$) by restricting the schema of $\in$-induction for arbitrary $\mathcal{L}^\circ$ formulas to $\in$-induction for sets.

Theorem 18 *We have the following proof-theoretic equivalences:*

1. OST $\equiv$ KP.
2. OST($\mathbb{P}$) $\equiv$ KP(P).
3. OST^r($\mathbf{E}, \mathbb{P}$) $\equiv$ ZFC.
4. OST($\mathbf{E}, \mathbb{P}$) $\equiv$ NBG$^+$.

KP is Kripke–Platek set theory with infinity, KP(P) is Power Kripke–Platek set theory as in Rathjen [55], ZFC is Zermelo-Fraenkel set theory with the axiom of choice, and NBG$^+$ is von Neumann–Bernays–Gödel theory NBG for sets and classes extended by a suitable form of (Σ^1_1-AC) for classes and $\in$-induction for all formulas.

For proofs of the first equivalence see Feferman [18, 19] and Jäger [32], the second equivalence is due to Rathjen (see his [54, 55] and private communication); it should also be provable via an adaptation of the method in Sato and Zumbrunnen [57]. The third equivalence is proved in Jäeger [32], and the fourth follows from Jäger [33] together with Jäger and Krähenbühl [38].

5.2 *Operational Closure*

With respect to ontological properties, it is a natural question to ask what it means for a set to be operationally closed. As it turns out, this has a very direct relationship to the concept of stability. More precisely, operationally closed sets behave like Σ_1 substructures of the universe. The detailed proof-theoretic analysis of the concept of operational closure is carried through in Jäger [35].

Definition 19 1. A set d is called *operationally closed*, in symbols $Opc[d]$, iff d is transitive, contains the constants of $\mathcal{L}^\circ$ as elements, and satisfies

$$(\forall x, f \in d)(fx\downarrow \rightarrow fx \in d).$$

2. The *operational limit axiom* states that every set is an element of an operational closed set,

$$\text{(OLim)} \qquad \forall x \exists y(x \in y \wedge Opc[y]).$$

An immediate consequence of this definition is that all closed terms of $\mathcal{L}^\circ$ that have a value are contained in every operationally closed set. Also, if we have a λ-term of $\mathcal{L}^\circ$ whose free variables belong to an operationally closed d, then this term belongs to d as well. The strength of the concept of operational closure and its connection to Σ_1 substructures becomes evident by the following observation.

Theorem 20 *For any Δ_0 formula $A[\vec{u}, v]$ of the language $\mathcal{L}$ with at most the variables $\vec{u}, v$ free, the theory* OST *proves that*

$$Opc[d] \wedge \vec{a} \in d \wedge \exists x A[\vec{a}, x] \rightarrow (\exists x \in d)A[\vec{a}, x].$$

Recall that a transitive set d with $\omega \in d$ is called a Σ_1*-elementary substructure of the transitive class* $\mathbf{M}$ iff $d \in \mathbf{M}$ and for all Σ_1 formulas $A[\vec{u}]$ with parameters $\vec{u}$ and all $\vec{a} \in d$,

$$d \models A[\vec{a}] \iff \mathbf{M} \models A[\vec{a}].$$

Hence the preceding theorem says that any operationally closed set is an Σ_1-elementary substructure of the universe $\mathbf{V}$. Also it implies that all instances of

$$(\Sigma_1\text{-Sep}) \qquad \forall x \exists y \forall z(z \in y \rightarrow z \in x \wedge A[z]),$$

where $A[u]$ is a Σ_1 formula of $\mathcal{L}$, are provable in OST + (OLim).

Theorem 21 KP + (Σ_1-Sep) *is contained in* OST + (OLim).

On the other hand, an ordinal α is called *stable* (in symbols $Stab[\alpha]$) iff $\mathbf{L}_\alpha$ is a Σ_1-elementary substructure of the constructible universe $\mathbf{L}$. Then KP + ($\mathbf{V}$=$\mathbf{L}$) + (Σ_1-Sep) proves that every ordinal α is majorized by a stable ordinal,

$$\mathsf{KP} + (\mathbf{V}{=}\mathbf{L}) + (\Sigma_1\text{-Sep}) \vdash \forall \alpha \exists \beta(a < \beta \wedge Stab[\beta]).$$

Since by means of the inductive model construction presented in Jäger and Zumbrunnen [44] the theory OST + (OLim) can be reduced to KP + ($\mathbf{V}$=$\mathbf{L}$) + $\forall \alpha \exists \beta(a < \beta \wedge Stab[\beta])$, and adding ($\mathbf{V}$=$\mathbf{L}$) to KP + ($\Sigma_1$-Sep) does not increase its proof-theoretic strength, we obtain the following characterization.

Theorem 22 *We have the following proof-theoretic equivalences:*

$$\mathsf{OST} + (\mathsf{OLim}) \;\equiv\; \mathsf{KP} + (\mathbf{V}{=}\mathbf{L}) + \forall\alpha\exists\beta(\alpha < \beta \wedge \mathit{Stab}[\beta]) \;\equiv\; \mathsf{KP} + (\Sigma_1\text{-}\mathsf{Sep}).$$

In Jäger [35] it is also shown that $\mathsf{OST} + \exists x\,\mathit{Opc}[x]$ is equiconsistent to KP plus parameter-free Σ_1 separation on ω.

So we notice that the concept of operational closure is proof-theoretically very powerful, lifting OST to a new dimension. However, from a operational perspective, this notion is somewhat problematic: The uniform version of (OLim), with a new constant OC,

$$\forall x(x \in \mathsf{OC}(x) \wedge \mathit{Opc}[\mathsf{OC}(x)]),$$

is easily seen to lead to inconsistency.

5.3 Relativizing Operational Set Theory

A further motivation for operational set theory, formulated in Feferman [18, 19], was to use his general applicative framework for explaining the admissible analogues of various large cardinal notions. Everything works out fine as along as only one (classically or recursively) regular universe is concerned. However, in view of Jäger and Zumbrunnen [44] this aim of OST had to be analyzed further. It is shown in [44] that a direct relativization of operational reflection leads to theories that are significantly stronger than theories formalizing the admissible analogues of classical large cardinal axioms. This refutes the conjecture 14(1) on p. 977 of Feferman [19].

The main reason is that simply restricting quantifiers to specific sets and operations to operations from and to those sets does not affect the global application relation and thus substantial strength may be imported – so to say – through the back door. Hence relativizing operational set theory requires a more cautious approach.

In a nutshell: The applicative structure must also be relativized when explaining the notion of relativized regularity in the context of OST. In contrast to the usual way of relativizing formulas with respect to a given set d, we now relativize our formulas A with respect to a set d and a set $e \subseteq d^3$ to formulas $A^{(d,e)}$; then d is the new universe and e takes care of application in the sense described below. This way of relativizing operational set theory is worked out in all details in Jäger [37].

First we add to $\mathcal{L}^\circ$ a fresh binary relation symbol Reg to express relativized regularity and a fresh constant **reg** for the operational representation of Reg in the sense of the following axiom that has to be added to the logical operations axioms,

(L.6) $(\mathbf{reg} : \mathbf{V}^2 \to \mathbf{B}) \wedge \forall x\forall y((\mathbf{reg}(x, y) = \top \leftrightarrow \mathsf{Reg}(x, y)).$

Then we turn to relativizing application: For all $\mathcal{L}^\circ$ terms r and variables e we define the formula $(r \,\partial\, e)$ by induction on the complexity of r as follows:

1. If r is a variable or a constant of $\mathcal{L}^\circ$, then $(r \ \partial \ e)$ is the formula $(r = r)$.
2. If r is the $\mathcal{L}^\circ$ term $r_1 r_2$, then choose some variable x not appearing in r_1, r_2 and different from e and let $(r \ \partial \ e)$ be the formula

$$(r_1 \ \partial \ e) \ \wedge \ (r_2 \ \partial \ e) \ \wedge \ \exists x(\langle r_1, r_2, x\rangle \in e).$$

Think of e as a ternary relation; then $(r \ \partial \ e)$ formalizes that the term r is defined if application within r is treated according to e. For us only such relations are interesting that are compatible with the real term application. To single those out, we set

$$Comp[e] \ := \ \forall x \forall y \forall z(\langle x, y, z\rangle \in e \ \rightarrow \ xy = z).$$

Clearly, if $Comp[e]$ and $(r \ \partial \ e)$, then $r\downarrow$. However, observe that in general we may have $Comp[e]$ and $r\downarrow$, but not $(r \ \partial \ e)$; so it is possible that term r has a value without being defined in the sense of e.

In a next step this form of relativizing application via e is combined with restricting the universe of discourse to d. For all $\mathcal{L}^\circ$ formulas A we define the relativized formula $A^{(d,e)}$ by induction on the complexity of A as follows:

$$\begin{aligned}
(r = s)^{(d,e)} \ &:= \ (r \ \partial \ e) \ \wedge \ (s \ \partial \ e) \ \wedge \ r = s, \\
(r \in s)^{(d,e)} \ &:= \ (r \ \partial \ e) \ \wedge \ (s \ \partial \ e) \ \wedge \ r \in s, \\
(r\downarrow)^{(d,e)} \ &:= \ (r \ \partial \ e) \ \wedge \ r \in d, \\
\mathsf{Reg}(r, s)^{(d,e)} \ &:= \ (r \ \partial \ e) \ \wedge \ (s \ \partial \ e) \ \wedge \ \mathsf{Reg}(r, s), \\
(\neg A)^{(d,e)} \ &:= \ \neg A^{(d,e)}, \\
(A \vee B)^{(d,e)} \ &:= \ (A^{(d,e)} \vee C^{(d,e)}), \\
((\exists x \in r)A)^{(d,e)} \ &:= \ (r \ \partial \ e) \ \wedge \ (\exists x \in r)A^{(d,e)}, \\
(\exists x A)^{(d,e)} \ &:= \ (\exists x \in d)A^{(d,e)},
\end{aligned}$$

Now the relation Reg comes into play. $\mathsf{Reg}(d, e)$ is supposed to state that set d is regular with respect to e, and has the following intuitive interpretation: (i) d is a transitive set containing all constants of $\mathcal{L}^\circ$ as elements and e is a ternary relation on d compatible with the general application relation; (ii) if application is interpreted in the sense of e, then d satisfies the axioms of OST; (iii) we claim a linear ordering of those pairs $\langle d, e\rangle$ for which $\mathsf{Reg}(d, e)$ holds. To make this precise, we add to OST additional so-called Reg-axioms. Here $TranCon[d]$ is short for the $\mathcal{L}^\circ$ formula stating that d is transitive and contains all constants of $\mathcal{L}^\circ$.

Axioms for Reg.

(Reg.1) $\mathsf{Reg}(d, e) \ \rightarrow \ (TranCon[d] \ \wedge \ e \subseteq d^3 \ \wedge \ Comp[e])$.

(Reg.2) If A is an applicative axiom, logical operations axiom, or set-theoretic operations axiom with at most the variables $\vec{x}$ free such that neither the variables d, e do not appear in the list $\vec{x}$, then

$$\mathsf{Reg}(d,e) \;\rightarrow\; (\forall \vec{x} \in d)A^{(d,e)}.$$

(Reg.3) $\mathsf{Reg}(d_1,e_1) \wedge \mathsf{Reg}(d_2,e_2) \;\rightarrow\; d_1 \in d_2 \vee d_1 = d_2 \vee d_2 \in d_1.$
(Reg.4) $\mathsf{Reg}(d_1,e_1) \wedge \mathsf{Reg}(d_2,e_2) \wedge d_1 \in d_2 \;\rightarrow\; e_1 \in d_2 \wedge e_1 \subseteq e_2.$

In the following we write $\mathsf{OST(LR)}$ for the extension of OST by the axioms (Reg.1)-(Reg.4) and the limit axiom for (relativized) regular sets,

(Lim-Reg) $$\forall x \exists y \exists z(x \in y \wedge \mathsf{Reg}(y,z)).$$

One of the central results of Jäger [37] is that $\mathsf{OST(LR)}$ is proof-theoretically equivalent to the theory KPi of iterated admissible sets and thus describes an recursively inaccessible universe from an operational perspective.

Theorem 23 $\mathsf{OST(LR)} \equiv \mathsf{KPi}$.

As can be seen from the proof of this equivalence, our notion of relativized regularity is the operational analogue of admissibility and thus provides a first essential step in capturing recursive analogues of large cardinal assertions. There is no intrinsic reason to stop at inaccessibility, and it seems that we can deal with, for example, Mahloness in an analogous way. The hope is that also the recursive versions of very strong forms of reflection can be handled in this way.

6 Future Work

Explicit mathematics and operational set theory are couched in an operational framework and as such have a lot in common. However, there are also significant differences. The article Jäger and Zumbrunnen [45] tries to clarify this relationship more systematically, especially from an ontological perspective.

A basic and significant difference is that in explicit mathematics we deal with individuals and classes, whereas operational set theory is completely first order. Hence it is an interesting question whether there exist natural operational theories of sets and classes. Feferman's draft notes [21] present some first ideas and Jäger [36] discusses several technical and conceptual problems; it also presents a "technically working" system that, however, does not satisfy the criterion of naturalness.

In explicit mathematics we can take a given applicative structure and build the universe of classes above this structure without being forced to change the underlying applicative structure; no new individuals are created. In an operational theory of sets and classes the situation is different: Again we may start off from the applicative universe, which now models set-theoretic axioms. However, building classes above this universe may force us to generate new sets, in particular if we want the "Aussonderungsprinzip" to be satisfied: given a set x and a class Y, the intersection $x \cap Y$ is a set. Therefore, a sort of strong impredicativity makes the interplay between sets and classes very delicate.

In spite of such difficulties it is worthwhile to search for "good" operational theories of sets and classes, even if they can only cope with systems of very high consistency strength. If successful, this framework is likely to be very useful in studying strong reflection principles from an operational perspective.

The analysis of strong forms of reflection is also a topic in explicit mathematics. This together with the development of a convincing operational descriptive set theory are major tasks for the future.

References

1. Beeson, M.J.: Foundations of Constructive Mathematics: Metamathematical Studies. Ergebnisse der Mathematik und ihrer Grenzgebiete, vol. 3/6. Springer, Berlin (1985)
2. Bishop, E.: Foundations of Constructive Analysis. McGraw-Hill Series in Higher Mathematics. McGraw-Hill, New York (1967)
3. Buchholz, W., Feferman, S., Pohlers, W., Sieg, W.: Iterated Inductive Definitions and Subsystems of Analysis: Recent Proof-theoretical Studies. Lecture Notes in Mathematics, vol. 897. Springer, Berlin (1981)
4. Cantini, A.: Extending constructive operational set theory by impredicative principles. Mathematical Logic Quarterly **57**(3), 299–322 (2011)
5. Cantini, A., Crosilla, L.: Constructive set theory with operations. In: Andretta, A., Kearnes, K., Zambella, D. (eds.) Logic Colloquium 2004. Lecture Notes in Logic, vol. 29, pp. 47–83. Cambridge University Press, Cambridge (2007)
6. Cantini, A., Crosilla, L.: Elementary constructive operational set theory. In: Schindler, R. (ed.) Ways of Proof Theory. Ontos Mathematical Logic, vol. 2, pp. 199–240. De Gruyter, New York (2010)
7. Church, A.: A set of postulates for the foundation of logic. Ann. Math. **33**(2), 346–366 (1932)
8. Church, A.: A set of postulates for the foundation of logic (second paper). Ann. Math. **34**(4), 839–864 (1933)
9. Feferman, S.: A language and axioms for explicit mathematics. In: Crossley, J.N. (ed.) Algebra and Logic. Lecture Notes in Mathematics, vol. 450, pp. 87–139. Springer, Berlin (1975)
10. Feferman, S.: Theories of finite type related to mathematical practice. In: Barwise, K.J. (ed.) Handbook of Mathematical Logic. Studies in Logic and the Foundations of Mathematics, vol. 90, pp. 913–9711. Elsevier, Amsterdam (1977)
11. Feferman, S.: Recursion theory and set theory: a marriage of convenience. In: Fenstad, J.E., Gandy, R.O., Sacks, G.E. (eds.) Generalized Recursion Theory II, Oslo 1977. Studies in Logic and the Foundations of Mathematics, vol. 94, pp. 55–98. Elsevier, New York (1978)
12. Feferman, S.: Constructive theories of functions and classes. In: Boffa, M., van Dalen, D., McAloon, K. (eds.) Logic Colloquium '78. Studies in Logic and the Foundations of Mathematics, vol. 97, pp. 159–224. Elsevier, Amsterdam (1979)
13. Feferman, S.: Generalizing set-theoretical model theory and an analogue theory on admissible sets. In: Hintikka, J., Niiniluoto, I., Saarinen, E. (eds.) Essays on Mathematical and Philosophical Logic. Synthese Library, vol. 22, pp. 171–195. Reidel, Dordrecht (1979)
14. Feferman, S.: Iterated inductive fixed-point theories: application to Hancock's conjecture. In: Metakides, G. (ed.) Patras Logic Symposion. Studies in Logic and the Foundations of Mathematics, vol. 109, pp. 171–196. Elsevier, New York (1982)
15. Feferman, S.: Monotone inductive definitions. In: Troelstra, A.S., van Dalen, D. (eds.) The L.E.J. Brouwer Centenary Symposium. Studies in Logic and the Foundations of Mathematics, vol. 110, pp. 77–89. Elsevier, New York (1982)
16. Feferman, S.: In the light of logic; chapter 13: Weyl vindicated: Das Kontinuum seventy years later. Logic and Computation in Philosophy. Oxford University Press, Oxford (1998)

17. Feferman, S.: The significance of Hermann Weyl's Das Kontinuum. In: Hendricks, V.F., Pedersen, S.A., Jørgensen, K.F. (eds.) Proof Theory: History and Philosophical Significance. Synthese Library Series, pp. 179–194. Kluwer Academic Publishers, Dordrecht (2000)
18. Feferman, S.: Notes on operational set theory, I. generalization of "small" large cardinals in classical an admissible set theory, Technical notes, Stanford University (2001)
19. Feferman, S.: Operational set theory and small large cardinals. Inf. Comput. **207**, 971–979 (2009)
20. Feferman, S.: On the strength of some semi-constructive theories. In: Berger, U., Diener, H., Schuster, P., Seisenberger, M. (eds.) Logic, Construction, Computation. Ontos Mathematical Logic, vol. 3, pp. 201–225. De Gruyter, New York (2012)
21. Feferman, S.: An operational theory of sets and classes, Technical Notes, Stanford University (2013)
22. Feferman, S.: The operational perspective: three routes. In: Kahle, R., Strahm, T., Studer, T. (eds.), Advances in Proof Theory. Progress in Computer Science and Applied Logic. Birkhäuser, Basel (2016)
23. Feferman, S., Jäger, G.: Systems of explicit mathematics with non-constructive μ-operator. part I. Ann. Pure Appl. Logic **65**(13), 243–263 (1993)
24. Feferman, S., Jäger, G., Strahm, T.: Foundations of explicit mathematics, in preparation
25. Glaß, T., Strahm, T.: Systems of explicit mathematics with non-constructive μ-operator and join. Ann. Pure Appl. Logic **82**(2), 193–219 (1996)
26. Glaß, T., Rathjen, M., Schlüter, A.: On the proof-theoretic strength of monotone induction in explicit mathematics. Ann. Pure Appl. Logic **85**(1), 1–46 (1997)
27. Hinman, P.G.: Recursion-Theoretic Hierarchies. Perspectives in Mathematical Logic, vol. 9. Springer, Berlin (1978)
28. Jäger, G.: A well-ordering proof for Feferman's theory T_0. Archiv für mathematische Logik und Grundlagenforschung **23**(1), 65–77 (1983)
29. Jäger, G.: Induction in the elementary theory of types and names. In: Börger, E., Kleine Büning, H., Richter, M.M. (eds.) CSL '87. Lecture Notes in Computer Science, vol. 329, pp. 118–128. Springer, Berlin (1987)
30. Jäger, G.: Applikative Theorien und explizite Mathematik. Vorlesungsskript. Universität Bern, Bern (1996)
31. Jäger, G.: Power types in explicit mathematics?. J. Symb. Logic **62**(4), 1142–1146 (1997)
32. Jäger, G.: On Feferman's operational set theory OST. Ann. Pure Appl. Logic **150**(1–3), 19–39 (2007)
33. Jäger, G.: Full operational set theory with unbounded existential quantification and power set. Ann. Pure Appl. Logic **160**(1), 33–52 (2009)
34. Jäger, G.: Operations, sets and classes. In: Glymour, C., Wei, W., Westerståhl, D. (eds.), Logic, Methodology and Philosophy of Science - Proceedings of the Thirteenth International Congress, pp. 74–96. College Publications, Norcross (2009)
35. Jäger, G.: Operational closure and stability. Ann. Pure Appl. Logic **164**(7–8), 813–821 (2013)
36. Jäger, G.: Operational set theory with classes?. Scientific Talk (2013)
37. Jäger, G.: Relativizing operational set theory. Bull. Symb. Logic **22**, 332–352 (2016)
38. Jäger, G., Krähenbühl, J.: Σ^1_1 choice in a theory of sets and classes. In: Schindler, R. (ed.) Ways of Proof Theory. Ontos Mathematical Logic, vol. 2, pp. 199–240. De Gruyter, New York (2010)
39. Jäger, G., Pohlers, W.: Eine beweistheoretische Untersuchung von $(\Delta^1_2\text{-CA}) + (\text{BI})$ und verwandter Systeme. Sitzungsberichte der Bayerischen Akademie der Wissenschaften, Mathematisch-Naturwissenschaftliche Klasse **1**, 1–28 (1982)
40. Jäger, G., Probst, D.: The Suslin operator in applicative theories: Its proof-theoretic analysis via ordinal theories. Ann. Pure Appl. Logic **162**(8), 647–660 (2011)
41. Jäger, G., Strahm, T.: Upper bounds for metapredicative Mahlo in explicit mathematics and admissible set theory. J. Symb. Logic **66**(2), 935–958 (2001)

42. Jäger, G., Strahm, T.: The proof-theoretic strength of the Suslin operator in applicative theories. In: Sieg, W., Sommer, R., Talcott, C. (eds.), Reflections on the Foundations of Mathematics: Essays in Honor of olomon Feferman, Lecture Notes in Logic, Association for Symbolic Logic, vol. 15, pp. 270–292 (2002)
43. Jäger, G., Strahm, T.: Reflections on reflection in explicit mathematics. Ann. Pure Appl. Logic **136**(1–2), 116–133 (2005)
44. Jäger, G., Zumbrunnen, R.: About the strength of operational regularity. In: Berger, U., Diener, H., Schuster, P., Seisenberger, M. (eds.) Logic, Construction, Computation. Ontos Mathematical Logic, vol. 3, pp. 305–324. De Gruyter, New York (2012)
45. Jäger, G., Zumbrunnen, R.: Explicit mathematics and operational set theory: some ontological comparisons. Bull. Symb. Logic **20**(3), 275–292 (2014)
46. Jäger, G., Kahle, R., Studer, T.: Universes in explicit mathematics. Ann. Pure Appl. Logic **109**(3), 141–162 (2001)
47. Kahle, R.: Applikative Theorien und Frege-Strukturen, Ph.D. thesis, Institut für Informatik und angewandte Mathematik, Universität Bern (1997)
48. Marzetta, M.: Predicative theories of types and names, Ph.D. thesis, Institut für Informatik und angewandte Mathematik, Universität Bern (1994)
49. Minari, P.: Axioms for universes, Handwritten Notes?
50. Rathjen, M.: Monotone inductive definitions in explicit mathematics. J. Symb. Logic **61**(1), 125–146 (1996)
51. Rathjen, M.: Explicit mathematics with the monotone fixed point principle. J. Symb. Logic **63**(2), 509–542 (1998)
52. Rathjen, M.: Explicit mathematics with the monotone fixed point principle. ii: models. J. Symb. Logic **64**(2), 517–550 (1999)
53. Rathjen, M.: Explicit mathematics with monotone inductive definitions: a survey. In: Sieg, W., Sommer, R., Talcott, C. (eds.) Reflections on the Foundations of Mathematics: Essays in Honor of Solomon Feferman, Lecture Notes in Logic, Association for Symbolic Logic, vol. 15, pp. 329–346 (2002)
54. Rathjen, M.: Relativized ordinal analysis: the case of power Kripke-Platek set theory. Ann. Pure Appl. Logic **165**(1), 316–339 (2014)
55. Rathjen, M.: Power Kripke-Platek set theory and the axiom of choice, Annals of Pure and Applied Logic (submitted)
56. Sato, K.: A new model construction by making a detour via intuitionistic theories II: interpretability lower bound of Feferman's explicit mathematics T_0. Ann. Pure Appl. Logic **166**(7–8), 800–835 (2015)
57. Sato, K., Zumbrunnen, R.: A new model construction by making a detour via intuitionistic theories I: operational set theory without choice is $\prod_1$-equivalent to KP. Ann. Pure Appl. Logic **166**(2), 121–186 (2015)
58. Scott, D.S.: Identity and existence in formal logic. In: Fourman, M., Mulvey, C., Scott, D. (eds.), Applications of Sheaves. Lecture Notes in Mathematics, vol. 753, pp. 660–696. Springer, Berlin (1979)
59. Simpson, S.G.: Subsystems of Second Order Arithmetic. Perspectives in Logic, Association for Symbolic Logic, 2nd edn. Cambridge University Press, Cambridge (2009)
60. Takahashi, S.: Monotone inductive definitions in a constructive theory of functions and classes. Ann. Pure Appl. Logic **42**(3), 255–297 (1989)
61. Troelstra, A.S., van Dalen, D.: Constructivism in Mathematics, I. Studies in Logic and the Foundations of Mathematics, vol. 121. Elsevier, Amsterdam (1988)
62. Troelstra, A.S., van Dalen, D.: Constructivism in Mathematics, II. Studies in Logic and the Foundations of Mathematics, vol. 123. Elsevier, Amsterdam (1988)

Author Biography

Gerhard Jäger is Professor of Logic and Theoretical Computer Science at the University of Bern. He received his mathematical education at the Ludwig-Maximilians-Universität in Munich. His principal research interests include proof theory (in particular the proof-theoretic analysis of subsystems of set theory and second order arithmetic), explicit mathematics and operational set theory, non-classical logics, philosophical foundations of mathematics (more specifically the relationship between predicative, metapredicative, and impredicative mathematics), as well as applications of logical methods in computer science.

Part III
Axiomatic Foundations

Feferman and the Truth

Andrea Cantini, Kentaro Fujimoto and Volker Halbach

Abstract We outline some of Feferman's main contributions to the theory of truth and the motivations behind them. In particular, we sketch the role truth can play in the foundations of mathematics and in the formulation of reflection principles, systems of ramified truth, several variants of the Kripke–Feferman theory, a deflationist theory in an extension of classical logic, and the system for determinate truth.

Keywords Axiomatic theories of truth · Ramified truth · Kripke–Feferman theory · Reflective closure · Determinate truth

2010 Mathematics Subject Classification 03F03 · 03F25 · 03F35 · 3F40 · 03A05.

1 Truth and the Foundations of Mathematics

Theories of comprehension and satisfaction are closely intertwined. To say that a is an element of the class $\{x : \varphi(x)\}$ for a formula $\varphi(x)$ seems tantamount to saying that the formula $\varphi(x)$ is satisfied by a or that the formula $\varphi(x)$ is true of a. Using this observation, one can reduce class theories to theories of satisfaction or those

---

Dedicated to the memory of Solomon Feferman.

A. Cantini (✉)
Dipartimento di Lettere e Filosofia, Universitá di Firenze, via Bolognese 52, 50139 Firenze, Italy
e-mail: andrea.cantini@unifi.it

K. Fujimoto
Department of Philosophy and School of Mathematics, University of Bristol, Cotham House, Bristol BS6 6JL, England
e-mail: fujimotokentaro@gmail.com

V. Halbach
University of Oxford, New College, Oxford OX1 3BN, England
e-mail: volker.halbach@philosophy.ox.ac.uk

G. Jäger and W. Sieg (eds.), *Feferman on Foundations*, Outstanding Contributions to Logic 13, DOI 10.1007/978-3-319-63334-3_11

of truth: Quantification over classes is translated to quantification over formulae or propositional functions. The basic strategy can be traced back at least to Russell [51] (see Schindler [53]).

In the good old days of logical positivism, however, the concept truth was considered with suspicion. It had been stained by too many dodgy philosophical theories. So the reduction of mathematical theories to theories of satisfaction or truth didn't look too attractive. Why should one want to replace a respectable mathematical theory with a 'philosophical' theory of truth or satisfaction? As foundational concepts membership and classes seemed much more suitable than truth and satisfaction.

More desirable was a reduction in the other direction, that is, the reduction of truth and satisfaction to a theory of classes. Consequently many philosophers were much more interested in Tarski's [58] definition of truth in a type theory over a theory of syntax. This reduction vindicated truth as a respectable notion for philosophers like Popper (see Leitgeb [43]).[1] Tarski added further reasons not to treat truth or satisfaction as a primitive notions to the neopositivist qualms: He thought that either the theory of truth will lack 'deductive power', if truth is axiomatized by the typed T-sentences, or the axioms for truth will have an 'accidental character' (see Halbach [33, Sect. 7]). Hence mathematical theories with comprehension axioms retained their conceptual priority over axiomatic theories of truth in foundational discussion. Of course some work on truth was done, but truth wasn't much used in the foundations of mathematics.

It took a long time for truth to recover from the neopositivist qualms and Tarski's verdict that any axiomatization of truth is either weak or arbitrary. Even philosophers beyond any suspicion of neopositivist convictions have endorsed the conceptual priority of some class theory over a theory of truth or satisfaction. For instance, the type-free truth theories advanced by Kripke [42], Gupta [30] or Herzberger [35] follow the pattern of Tarski's theory in a crucial aspect. On their accounts set theory is employed as the basic framework; then a semantics for a language with the truth predicate is defined within set theory. The concept of truth these authors are interested in is more worrisome than Tarski's, because type-free truth is prone to paradox. Consequently it is not a surprise that these authors rely on set theory as ultimate framework and show that semantics for languages with a type-free truth predicate can be developed within set theory. Again a theory of classes serves as the bedrock foundational framework on which a theory of truth can rest.

Truth, however, conceived as a primitive, undefined notion, does have a potential for use in the foundations of mathematics and the formal sciences. One use where the need for a truth predicate is obvious are the proof-theoretic reflection principles. As Gödel [29] had shown, a system S cannot prove even very weak consequences of the soundness of the theory. In particular, S cannot prove the consistency of S, if the latter is expressed in a natural way. However, by endorsing or accepting a system S, one is also committed to the soundness of S and thus to all consequences of its soundness such as the consistency of S. One can try to add statements expressing the soundness

[1] For those who have doubts about the success of Tarski's reduction, we add that worries had been raised early on and more recently by Field [22] and Putnam [48].

of S to the system S in order to obtain a system that features commitments implicit in the acceptance of S as explicit theorems. As a very early reaction to Gödel's incompleteness theorems, Turing [59] had tried to add such principles to S and to iterate this procedure even along transfinite ordinals.

Kreisel and Lévy [41, p. 98] describe a reflection principle for a system S as 'the formal statement stating the soundness of S'. As a soundness principle the consistency statement is very weak. More powerful are the local reflection principle $\mathrm{Bew}_{\mathsf{S}}(\ulcorner\varphi\urcorner) \rightarrow \varphi$ and the uniform reflection principle $\forall x\, (\mathrm{Bew}_{\mathsf{S}}(\ulcorner\varphi(\dot{x})\urcorner) \rightarrow \varphi(x))$. But even these schemata fall short of expressing full soundness.

Full soundness is the statement that all (closed) theorems of S are true. Kreisel and Lévy [41, p. 98] call a version of this the *global reflection principle*. They write:

> Literally speaking, the intended reflection principle cannot be formulated in S itself by means of a single statement. This would require a *truth definition* T_{S} [...]

They go on to point out that such a truth definition doesn't exist because of Tarski's theorem on the undefinability of truth. Hence, under the usual assumptions, the 'intended reflection' cannot be expressed in the language of S. It is also hardly an option to pass from S to another stronger system containing S with the resources for defining a predicate T_{S}, because it is only soundness that is to be added to the system and not, for instance, more comprehension axioms or other axioms sufficient for defining T_{S}.

In order to avoid the addition of new mathematical resources for the definition of T_{S}, one can add a primitive symbol for truth in S. This approach goes directly against the neopositivist qualms against truth. Thus it doesn't come as a surprise that initially a primitive formal truth predicate wasn't used in the discussion about reflection principles, reflective closure, recursive progressions and related topics. Turing's early work was continued without the use of a truth predicate in the object language. Of course, Feferman is the single most influential author in the wake of Turing's approach. In particular, Feferman [9] contained amazing results on the iterated addition of proof-theoretical reflection principles in the sense of Kreisel and Lévy [41] to Peano arithmetic. These progressions of theories were all formulated in the language of arithmetic without any additions.

In the 1970s more logicians were less restrained by worries about truth and started working on formal truth theories with truth as a predicate in the object language. They also became bolder and turned to type-free notions of truth. They explored other ways of solving or blocking the paradoxes than Tarski's restrictive method involving the distinction between an object and a metalanguage. Especially after the publication of Kripke [42], more philosophers and logicians turned their attention to truth. Feferman recognized their potential for adding strong soundness claims and making explicit assumptions implicit in the acceptance of a theory.

2 Ramified Truth and Reflection

In 1979 Feferman gave a talk entitled *Gödel's incompleteness theorems and the reflective closure of theories* at the meeting of the Association of Symbolic Logic in San Diego in 1979 ([45] and [17, p. 3]). This paper, which was to become the paper [17], had been circulated for quite some time and other logicians including Reinhardt [50] and Cantini [5] had published on Feferman's ideas. Feferman's talk and the subsequent papers by him and others mark the return of truth as an undefined, primitive notion of foundational significance. Let $\mathcal{L}_{\mathsf{PA}}$ be the language of first-order arithmetic.

The starting point is the 'compositional' axioms for truth. Given a *system* satisfying certain natural conditions, axioms for a truth predicate T can be added that correspond to the clauses in Tarski's definition of satisfaction. When one works in a theory that can encode every object as a closed term within it one can dispense with satisfaction and use a unary truth predicate – at least if the theory satisfies certain further assumptions. Feferman [17] gave fairly general account that applies to a variety of theories such as set theory for instance (see Fujimoto [28]). This generality is significant and not just a trivial extension of the special case of Peano arithmetic as starting theory.

However, here we use Peano arithmetic as our base theory. This will allow us to keep the presentation sufficiently simple. Let $\mathcal{L}_T$ be $\mathcal{L}_{\mathsf{PA}} \cup T$. For PA the compositional axioms can be chosen as follows:

T1 $\forall s\, \forall t\, \big(T(s \mathbin{\underset{\cdot}{=}} t) \leftrightarrow s^\circ = t^\circ)\big)$ and similarly for other predicates other than $=$, except for the special predicate T

T2 $\forall x\, \big(\mathrm{Sent}_{\mathsf{PA}}(x) \rightarrow (T(\underset{\cdot}{\neg} x) \leftrightarrow \neg T(x))\big)$

T3 $\forall x\, \forall y\, \big(\mathrm{Sent}_{\mathsf{PA}}(x \mathbin{\underset{\cdot}{\wedge}} y) \rightarrow (T(x \mathbin{\underset{\cdot}{\wedge}} y) \leftrightarrow T(x) \wedge T(y))\big)$

T4 $\forall v\, \forall x\, \big(\mathrm{Sent}_{\mathsf{PA}}(\underset{\cdot}{\forall}\, vx) \rightarrow (T(\underset{\cdot}{\forall}\, vx) \leftrightarrow \forall t\, T(x[t/v]))\big)$

Here and in what follows we use quantifiers $\forall s$ and $\forall t$ to range over the codes (or Gödel numbers) of closed $\mathcal{L}_{\mathsf{PA}}$-terms; namely, the expression $\forall t$ is short for $\forall x\, (\mathrm{ClTerm}(x) \rightarrow \ldots)$, where $\mathrm{ClTerm}(x)$ represents the set of the codes of closed $\mathcal{L}_{\mathsf{PA}}$-terms. The symbol $^\circ$ is a representation in PA of a recursive function that takes a code of a closed $\mathcal{L}_{\mathsf{PA}}$-term and returns its value in the standard model; e.g., $\ulcorner 0 + 0 \urcorner^\circ = 0$. Hence the axiom T1 expresses that a closed equation $s = t$ is true iff the values of the closed terms s and t coincide. There won't be a function symbol $^\circ$ in the language, but the function can be expressed using suitable formulae. The formula $\mathrm{Sent}_{\mathsf{PA}}(x)$ represents that x is a code of a sentence of the language of PA. The symbol $\underset{\cdot}{\neg}$ is a representation in PA of a recursive function that takes a code of a sentence and returns the code of its negation; e.g., $\underset{\cdot}{\neg} \ulcorner 0 = 0 \urcorner = \ulcorner 0 \neq 0 \urcorner$. Hence the axiom T2 expresses that the negation of a sentence of PA is true iff the sentence is not true. The symbol $\underset{\cdot}{\wedge}$ is a representation in PA of a recursive function that takes two codes of sentences and returns the code of their conjunction; e.g., $\ulcorner 0 = 0 \urcorner \mathbin{\underset{\cdot}{\wedge}} \ulcorner 0 \neq 0 \urcorner = \ulcorner 0 = 0 \wedge 0 \neq 0 \urcorner$. Hence the axiom T3 expresses that the conjunction of two sentences of PA is true iff both of the conjuncts are true. The symbol $\underset{\cdot}{\forall}$ is a representation in PA of a recursive function that takes a code of a variable and a code of a formula and returns the universal quantification of the formula with respect to the variable; e.g., $\underset{\cdot}{\forall}(\ulcorner v \urcorner, \ulcorner \varphi \urcorner) = \ulcorner \forall v \varphi \urcorner$ for a variable v and a formula φ. The operation $x[y/z]$ yields the code of the result of substituting a term encoded by y for a variable encoded by z in a formula encoded by x; e.g., $\ulcorner \varphi(v) \urcorner[\ulcorner t \urcorner / \ulcorner v \urcorner] = \ulcorner \varphi(t) \urcorner$ for a formula φ, a term t, and a variable v. Hence axiom T4 expresses that a universally quantified sentence of PA is true iff all its substitution instances (with closed terms) are true. A more detailed explanation of the notation can be found in Halbach [33]. But we hope that our notation should be largely self-explanatory. There are also some minor deviations from Feferman's definition of this theory in [17, p. 14].

The axioms T1–T4 are adjoined to those of Peano arithmetic. Crucially, the induction schema is expanded to the new language with the truth predicate. The resulting theory T(PA) or also called CT (for 'compositional truth') proves the soundness of Peano arithmetic, that is, it proves the global reflection principle

$$\forall x \left(\mathrm{Sent}_{\mathsf{PA}}(x) \wedge \mathrm{Bew}_{\mathsf{PA}}(x) \rightarrow T(x) \right) \tag{2.1}$$

Thus the expressive resources needed for stating the soundness of PA already imply the soundness of PA.[2] The global reflection principle doesn't have to be added as an additional axiom. As we have discussed, the uniform reflection principle is derived from (2.1) with the help of the axioms T1–T4; furthermore, CT is strong enough to derive the iterated reflection principle along any constructive ordinal α provably well-founded in CT in the sense that all the $\mathcal{L}_T$-instances of transfinite induction along α are derivable in CT.

If the commitment to the soundness of PA is implicit in the acceptance of PA, then also the resources needed for expressing the soundness claim are implicit in the acceptance of PA. Thus the acceptance of PA commits one to CT. Thus CT makes explicit some implicit commitments of the acceptance of PA. However, the implicit commitment in the acceptance of PA is not exhausted by CT and (2.1), and the iteration procedure further continues. For, once CT is explicitly accepted, one is then committed to the soundness of CT and thus to a truth predicate for CT, which is needed to express the soundness of CT. To this end one can add a further truth predicate T_1 that applies to all sentences formulated in the language of arithmetic expanded by T. The new predicate T_1 is then axiomatized in the same way as T except that T is treated as one of the non-special predicate symbols in T1. Moreover in T2–T4 quantification over sentences of the arithmetical language is replaced with quantification over sentences of the arithmetical language with T. This procedure can be iterated and an axiomatization of Tarski's hierarchy of languages is obtained. The exact specification of the procedure requires some more detail; but a general

[2] CT contains the expanded induction schema, and this expansion is indeed crucial in deriving (2.1), since CT without the expanded induction schema is conservative over PA and thus does not yield (2.1). The question whether the expanded induction schema is an essential part of 'the expressive resources needed for stating the soundness of PA' is a subtle issue and gave rise to lively debates in the context of deflationism; see a debate between Shapiro [56] and Field [23] for instance.

recipe for the transition from one level to the next can be specified. The procedure fits Feferman's [9, p. 274] description of a reflection principle:

> By a *reflection principle* we understand a description of a procedure for adding to any set of axioms A certain new axioms whose validity follow from the validity of the axioms A and which formally express, within the language of A, evident consequences of the assumption that all the theorems of A are valid.

This characterization of reflection principles precedes Feferman's work on truth by more than one and a half decades.

The addition of truth predicates and accompanying axioms can be iterated into the transfinite. A truth predicate T_λ at a limit level λ is axiomatized in the same way as the truth predicates at successor level with the exception that an axiom is added that says that a sentence is true iff it is true at one of the previous levels. For details see Halbach [33, Sect. 9.1] and Fujimoto [28].

The need for an ordinal notation system is obvious: The truth predicates need to be indexed and all levels of the hierarchy need to be axiomatized. Technically it is no problem to define theories up to any recursive ordinal. To this end one can pick a path through Kleene's $\mathcal{O}$. However, this would betray the original motivation for considering these iterated theories of truth: their purpose is to make explicit assumptions implicit in the acceptance of PA. Using our general mathematical machinery we can prove that there is a well-founded ordering of natural numbers whose order-type is any ordinal below the first non recursive ordinal ω_1^{CK}. But this theorem is not implicit in the acceptance of PA in any way. In particular, PA doesn't prove that these orderings are well-founded. Kreisel [40] suggested to employ autonomous iterations in situations of this kind. That is, one iterates a procedure of this kind in unfolding the implicit commitments in accepting a theory only if the theory in question can prove transfinite induction up to that ordinal level.

Since PA proves transfinite induction for any ordinal up to ε_0, the truth theories are iterated up to that point. This new theory, however, proves transfinite induction for longer wellorderings. Hence the truth theories are iterated even further until a point is reached where the hierarchy of truth theories is iterated to a point Γ_0 that proves transfinite induction for all ordinals smaller than Γ_0. This ordinal Γ_0 is the so-called Feferman–Schütte ordinal that had figured prominently in Feferman's earlier work [10] on predicativity.

The iterated truth theories very much resemble the systems of predicative analysis, which had been studied thoroughly by Feferman [10] and Schütte [54] in the 1960s. Thus, in a sense, the results on iterated truth theories are formally not extremely exciting. Perhaps this is the reason why they figure less prominently in the published paper [17] than in the draft version [16].

From a foundational point of view, however, we think that the iterated truth theories are significant. They are a very convincing way of carrying out the programme of determining the reflective closure of PA, that is, of characterizing the theory that makes explicit what is implicit in the acceptance of PA.

The formulation of the systems of iterated truth is technically awkward. The specification of the language already requires an ordinal notation system. Then the motivation of the terminal ordinal ε_0 or Γ_0 relies on some deeper results. Moreover, it is highly specific to PA.

Feferman has made various attempts at characterizing the reflective closure of theories in a more elegant way. The reasons for seeking a more succinct characterization are not only of an aesthetic nature. A method of defining the reflective closure of a theory that is less reliant on ordinal notation systems and an explicit appeal to proof-theoretic techniques and notions, should also be more generally applicable; moreover, it would also be philosophically less prone to the objection that it depends on arbitrary stipulation; a more elegant system would depend on a 'natural' ordinal notation system and arithmetization.

Feferman has tried various methods for characterizing the reflective closure of a theory. The various approaches should not be seen so much as competing but rather as different characterization of the same concept. The situation is similar to that in recursion theory: The different characterizations of computability do not exclude each another, rather their equivalence assures us that we have found a stable concept.

3 Kripke–Feferman

The first method of characterizing the reflective closure of PA is fairly close to the iterated truth theories. But it shuns already the need for an ordinal notation system. Somewhat metaphorically speaking, the ordinals emerge from the theory itself and are not imposed on it from the outside.

The theory has been dubbed the *Kripke–Feferman theory* or KF for short. Presumably Reinhardt was the first to publish on the theory in [49, 50] and by the time Feferman published his paper [17], the label KF had already been established. Unfortunately, since no authoritative version had been published by Feferman, different authors formulated KF in slightly different ways. Here we try to stick to the formulation chosen by Feferman; but we use a slightly different notation.

Feferman [17] formulated KF in the language of arithmetic augmented with two new unary predicates T and F for truth and falsity. Let $\mathcal{L}_{\mathsf{KF}} := \mathcal{L}_{\mathsf{PA}} \cup \{T, F\}$ denotes the language of KF. As the axiom K4 below indicates, the falsity predicate F is actually not necessary and could be understood as defined notion, because the falsity of a sentence coincides with the truth of its negation. Hence for the sake of simplicity, we will identify $\mathcal{L}_{\mathsf{KF}}$ with $\mathcal{L}_T$. The axioms of KF comprise those of PA and the following truth-theoretic axioms:

K1 $\forall s \forall t \left((T(s \mathbin{\dot{=}} t) \leftrightarrow s^\circ = t^\circ)) \wedge (F(s \mathbin{\dot{=}} t) \leftrightarrow s^\circ \neq t^\circ)\right)$, and similarly for other predicates other than $=$, except for the special predicate T;

K2 $\forall s \left((T(\dot{T}\, s) \leftrightarrow T(s^\circ)) \wedge (F(\dot{T}\, s) \leftrightarrow F(s^\circ))\right)$;

K3 $\forall s \left((T(\dot{F}\, s) \leftrightarrow F(s^\circ)) \wedge (F(\dot{F}\, s) \leftrightarrow T(s^\circ))\right)$;

K4 $\forall x \left(\mathrm{Sent}_{\mathsf{KF}}(x) \rightarrow (T(\dot{\neg} x) \leftrightarrow F(x)) \wedge (F(\dot{\neg} x) \leftrightarrow T(x))\right)$;

K5 $\forall x \forall y \left(\mathrm{Sent}_{\mathsf{KF}}(x \mathbin{\dot{\wedge}} y) \rightarrow (T(x \mathbin{\dot{\wedge}} y) \leftrightarrow T(x) \wedge T(y)) \wedge (F(x \mathbin{\dot{\wedge}} y) \leftrightarrow F(x) \vee F(y))\right)$;

K6 $\forall v \forall x \left(\mathrm{Sent}_{\mathsf{KF}}(\underset{\cdot}{\forall} vx) \rightarrow (T(\underset{\cdot}{\forall} vx) \leftrightarrow \forall t\, T(x[t/v])) \wedge (F(\underset{\cdot}{\forall} vx) \leftrightarrow \exists t\, F(x [t/v]))\right)$.

The axiom K1, exactly like T1, says that a closed equation $s = t$ is true (false) iff the value of the terms s and t agree (disagree, resp.). The formula $\mathrm{Sent}_{\mathsf{KF}}(x)$ expresses that x is a sentence of the language of (i.e. $\mathcal{L}_T$); hence the axiom K5 and K6 express essentially the same compositional axioms as T3 and T4 but extended to sentences of the larger language $\mathcal{L}_T$. The axioms K2 and K3 describes the iterative self-applicative characteristic of the truth and falsity predicates in KF; K2 says that it is true (false) that a sentence is true iff the sentence is true (false, resp.) and K3 says its dual. Finally, K4 defines the falsity of a sentence to be the truth of its negation. There are different ways to motivate the axioms of KF and Halbach [33] develops a fuller picture. KF can be seen as a generalization of CT, which is a subtheory of KF, or, more naturally, as a generalization of a theory PT of a positive inductive definition of truth and falsity [33, Sect. 8.7].

Although KF derives transfinitely iterated uniform reflection principles for its base theory (i.e., PA in the current setting), KF cannot derive its own soundness due to Gödel's incompleteness theorem: namely, there is an $\mathcal{L}_T$-sentence φ (e.g., $0 = 1$) such that

$$\mathsf{KF} \nvdash \left(\mathrm{Bew}_{\mathsf{KF}}(\ulcorner\varphi\urcorner) \rightarrow \varphi\right).$$

Hence KF doesn't derive the global reflection principle for itself:

$$\mathsf{KF} \nvdash \forall x \left(\mathrm{Sent}_{\mathsf{KF}}(x) \wedge \mathrm{Bew}_{\mathsf{KF}}(x) \rightarrow T(x)\right).$$

This fact may suggest one to iterate KF-truth as in the case of RT_α; c.f., [27]. However, Feferman [17] gave an argument against such iteration, and thereby explain why KF (or $Ref^*(\mathsf{PA}(P))$ defined below in Sect. 3.2.3) is to be called reflective *closure*, that is, why KF is to be seen as exhausting 'what notions and principles one ought to accept if one accepts the basic notions and principles of the theory' [18, p.205]. Iterating KF would mean to adopt principles that go beyond what is implicit in the acceptance of the base theory, that is, PA in the case considered here.

We sketch Feferman's argument against using iterations of KF in order to define the reflective closure of PA. If one were to add axioms for a further truth predicate T', that is, a truth predicate for the language of KF including the truth predicate T, one would specify axioms for T' analogous to those for T. To this end, one would now quantify over sentences in the full language with T' in axioms K4–K6; moreover, one would treat T as just like another predicate of the base language and therefore add the following axiom in analogy to K1 for the predicate T [17, p.40]:

$$\forall s \left((T'(\underset{\cdot}{T}\, s) \leftrightarrow T(s^\circ)) \wedge (F'(\underset{\cdot}{T}\, s) \leftrightarrow \neg T(s^\circ))\right) \tag{3.1}$$

From the logical truth $\forall s\ (T(s^\circ) \vee \neg T(s^\circ)$ and this axiom, we can derive the following 'totality' claim:

$$\forall s\,\big(T'(\underset{\cdot}{T}\, s) \vee F'(\underset{\cdot}{T}\, s)\big) \tag{3.2}$$

Therefore, by endorsing the new axiom (3.1), we would treat T as a 'total' or 'determined' predicate, just like the predicates of the base language. Feferman [17, p.40] thereby concludes that 'when iterating reflective closure we thus vitiate the informal idea behind the use of partial predicates of truth and falsity.' Hence, for Feferman, KF is closed under the process of making explicit what are implicitly accepted in accepting the basic notions and principles.

KF is a quite rich theory. Indeed, it is intimately related to an approach to predicativity, to the theme of unfolding (see Feferman and Strahm [20], Feferman and Strahm [21]), and it can be regarded as characteristic of a fruitful interaction between the logical *development of non-extensional concepts* (classification, operation) and *semantical investigations*. Further, it has a sort of mathematical appeal: it hinges upon well-known lattice theoretical facts (Knaster–Tarski theorem), and it naturally implies the existence of fixed points of arithmetically definable monotone operators. One can show that KF and the standard fixed point theory $\widehat{\mathsf{ID}}_1$ (see Feferman [14]) are mutually interpretable, thus reducing the classification of proof theoretic strength of KF to that of $\widehat{\mathsf{ID}}_1$. $\widehat{\mathsf{ID}}_1$ leads towards the foundations of intuitionistic type theory and predicatively reducible subsystems of analysis. Below we summarize a few important facts and results on KF.

3.1 Inner Logic and Outer Logic

The theory KF is sometimes criticized for yielding a discrepancy of its outer and inner logic. Let us call $\{\varphi \mid \varphi$ is an $\mathcal{L}_T$–sentence and $\mathsf{KF} \vdash \varphi\}$ the *outer theory* of KF and $\{\varphi \mid \varphi$ is an $\mathcal{L}_T$ -sentence and $\mathsf{KF} \vdash T(\ulcorner \varphi \urcorner)\}$ the *inner theory* of KF, and also call the logic governing the former the outer logic of KF and the logic governing the latter the inner logic of KF. [3] The inner logic of KF is strong Kleene logic whereas its outer logic is classical logic; also the inner theory of KF fails to coincide with the outer theory of KF. Hence, KF does not meet one of the desiderata that Leitgeb [44] suggests for a theory of truth.

In order to develop a theory that is an axiomatization of Kripke's theory with strong Kleene logic that avoids this discrepancy, Halbach and Horsten [34] presented a system based on strong Kleene logic. The resulting theory PKF has exactly the same inner theories and outer theories. Halbach and Horsten [34] showed that PKF proves significantly less arithmetical statements than the classical system KF. They suggested that thus KF should not be seen as a formal device for generating theorems in strong Kleene logic: the use of the classical outer logic is indispensable for proving

[3] The notion of inner logic thus defined is ambiguous, because it is not clear enough how to extract logic from a given set of sentences, and one sometimes simply identify outer/inner theories and outer/inner logics. At any rate, the intended inner logic of KF is strong Kleene logic, and Halbach and Horsten's [34] result can be construed to have 'shown' that the inner logic of KF is indeed strong Kleene logic.

certain arithmetical theorems. Hence it is unlikely that a purely 'instrumentalist' interpretation of KF in the spirit of Reinhardt [50] is a viable option.

The difference between inner and outer logic is adumbrated in the Sect. 6.1.2 of Feferman [17, p.40] on the informal interpretation of partial-self-applicable truth predicate. There, considering the question as to whether Kripke's [42] construction corresponds to a (more or less) clear informal notion, Feferman finds it 'reasonably convincing' that in the formulation of KF, $T(\ulcorner\varphi\urcorner)$ expresses that φ is a *grounded truth*, that is, that the denotation of T is given by Kripke's least fixed-point construction, understood in its full generality.[4] As a consequence, the analogy with partial computable predicates becomes helpful: Truth should be relative to *given rules of computation* or *inductive rules*, and ought to be distinguished from our everyday informal notion of truth with which *classical logic* is justified.

On the formal side, this leads to define for an $\mathcal{L}_T$-sentence φ to have a *determined* truth-value, when $\mathsf{KF} \vdash T(\ulcorner\varphi\urcorner) \leftrightarrow \neg F(\ulcorner\varphi\urcorner)$ holds, and to choose D to be the set of all sentences with a determined truth-value, that is the set of all sentences that are true or false, but not both (see the next subsection for a formal elaboration). This set D is interpreted to apply to those sentences whose truth-value is determinable by the rules of *internal* truth and falsity, while the complement of D is interpreted to apply to those whose truth-value is *not determinable* by those rules, but still *definite* in the sense that classical logic is justified with it. Hence, for instance, the statement $\lambda \vee \neg\lambda$ is not an internal truth but true in our informal and, say, naïve platonistic sense.

Those sentences with determined truth-value in this sense enjoy some nice properties. First, the laws of classical logic are provably true for all determined $\mathcal{L}_T$-sentences[5]: namely, if $\varphi_1, \ldots, \varphi_k$ have determined truth-value, then each logical axiom of classical logic involving these sentences is provably true in KF. Second, more importantly, the T-schema holds when restricted to those sentences even with the untyped truth predicate of KF: namely, it holds that, for all $\mathcal{L}_T$-sentence φ, $\mathsf{KF} \vdash D(\ulcorner\varphi\urcorner) \to \big(T(\ulcorner\varphi\urcorner) \leftrightarrow \varphi\big)$.

3.2 *Kripke–Feferman: The Proof Theoretic Side*

The main result which is proven in Feferman's paper on reflective closure is the following:

Theorem 1

$$\mathsf{KF} \equiv (\Pi^0_1\text{-}CA)_{<\varepsilon_0} \tag{3.3}$$

[4]He adds that the 'facts', on which T and F are grounded, may be representable as true (false) sentences of *any system* (arithmetic, set theory, etc.) we come to accept as basic.

[5]As well as for total predicates, see Lemma 1.

In general $(\Pi^0_1\text{-CA})_{<\alpha}$ is the theory of iterated jump up to any $\beta < \alpha$.[6]
Let us outline the proof of the theorem, possibly following different routes.

3.2.1 Lower Bound

We exploit the foundational side of truth, as a way of interpreting non-extensional predicate application, in order to show the main result of Feferman [17].

In order to enhance readability, we adopt the following shortenings:

(i) $a(x)$ stands for the the term representing the operation of substituting the first free variable of the expression encoded by a with the xth-numeral.
(ii) $D(a) := \mathsf{Sent_{KF}}(a) \wedge (T(a) \vee T(\dot\neg a)) \wedge \neg(T(a) \wedge T(\dot\neg a))$; $D(a)$ means 'a is determined';
(iii) non-extensional membership: $x \in a := T(a(x)) \wedge \neg T(\dot\neg a(x)))$; non-extensional co-membership: $x \bar\in a := T(\dot\neg a(x)) \wedge \neg T(a(x))$
(iv) the class(ification) predicate $Cl(a) := \forall x D(a(x))$.

Then it is easily seen that the language $\mathcal{L}_2$ of second order arithmetic can be regarded as a sublanguage of $\mathcal{L}_T$.

In order to relate KF with standard subsystems of second order arithmetic, we need a few additional definitions:

(i) A formula $\varphi(x, \vec{z})$ of $\mathcal{L}_T$ is *elementary in* $\vec{z} := z_1, \ldots z_n$ iff it is inductively generated from atoms of the form $t = s$, $t \in z_i$, $(1 \le i \le n)$ by means of $\wedge$, $\neg$, $\forall u$, with $u \notin \{z_1, \ldots z_n\}$.
(ii) We say that Cl is closed under *elementary comprehension* iff for every $\varphi(x, \vec{z})$ elementary in $\vec{z}$, then there exists a primitive recursive term $s_\varphi(\vec{z})$ such that, provably in KF,

$$Cl(\vec{z}) \to Cl(s_\varphi(\vec{z})) \wedge \forall x(\varphi(x, \vec{z}) \leftrightarrow x \in s_\varphi(\vec{z})). \tag{3.4}$$

(iii) *A family* g *of classes indexed by a class* a is simply an index g of a partial recursive function $\lambda n.\{g\}(n)$,[7] such that $\forall y(y \in a \to Cl(\{g\}(y))$.
(iv) Finally, we say that Cl is *closed under join* iff, whenever f is a family of classes indexed by a class a, there exists a term $j(a, f)$ whose elements are exactly those ordered pairs (u, v) such that that $u \in a$ and $v \in \{f\}(u)$.

Lemma 1 *The collection of classes is closed – provably in* KF- *under elementary comprehension and join.*[8]

---

[6]For a precise definition, see Feferman [17]. One can replace $(\Pi^0_1\text{-CA})_{<\alpha}$ in the statement of the theorem and below with ramified analysis up to any level $< \alpha$ in a fixed formalization, provided α has the form ω^β, $\beta \ge \omega$.

[7]We assume a standard formalization of standard recursion theory via the Kleene bracket relation.

[8]The statement can be used to interpret into KF a basic system of Feferman's Explicit Mathematics, see Feferman [13].

Proposition 1

$$(\Pi^0_1\text{-}CA)_{<\varepsilon_0} \leq \mathsf{KF} \tag{3.5}$$

Proof First of all, KF proves, for each $\alpha < \varepsilon_0$, the transfinite induction scheme $TI(\varphi, < \alpha)$, for arbitrary formulas φ in the full language. This is well-known (Gentzen–Schütte–Feferman) and it follows insofar as KF has the full number theoretic induction schema (see Lemma 4.3.2 in Feferman [17]). Then apply $TI(\varphi, < \alpha)$ in conjunction with the previous lemma. □

The fact that the jump hierarchy is available in KF up to any $\alpha < \varepsilon_0$ is crucial in order to carry out a wellordering proof for each initial segment of the standard wellordering of $\phi\varepsilon_0 0$ (see Feferman [17], appendix and Feferman [11]).

Lastly, it is not difficult to prove that KF can directly interpret the fixed point theory $\widehat{\mathsf{ID}}_1$ (see Cantini [5, Sect. 3.11] and Halbach [33, Sect. 19.5]).

3.2.2 Upper Bound

We sketch two strategies for proving the upper bound direction of the theorem.

The first route: Kripke–Feferman as a fixed point theory. This route is essentially Feferman's route in Feferman [17] and consists of a reduction of KF to classical subsystems of known strength.

The theory $\widehat{\mathsf{ID}}_1$ contains– besides the usual axioms for Peano arithmetic and induction schema for the whole language– fixed point axioms FP asserting the existence of fixed points I_φ for arbitrary elementary positive operators $\varphi(x, P)$[9]

$$\forall x(\varphi(x, I_\varphi) \leftrightarrow I_\varphi(x)) \tag{3.6}$$

but $\widehat{\mathsf{ID}}_1$ has no minimality schema.

By Aczel [2] it is known that $\widehat{\mathsf{ID}}_1$ has the same arithmetical theorems as PA with the schema of transfinite induction for each initial segment of the canonical primitive recursive wellordering of type $\phi\varepsilon_0 0$. The idea is to show that there are (not necessarily minimal) Σ^1_1-solutions to the fixed point Eq. (3.6) by standard diagonalization, provably in the subsystem Σ^1_1-AC_0.[10] The argument works in second order arithmetic with arithmetical comprehension except that Σ^1_1-AC_0 is also required in order to show that the result of replacing the parameter P by means of a Σ^1_1-predicate in a positive elementary operator can be still made equivalent to a Σ^1_1-formula (see Feferman [17], 4.2). The argument also holds when induction is restricted to arith-

[9] So $\varphi(x, P)$ is a formula in the language of PA expanded with the new predicate symbol P and positive in P.

[10] Concerning this subsystem and the corresponding one with full induction, see Simpson [57]. There are in the literature several strategies for classifying its proof-theoretic strength, which apply either non-standard models (as in H. Friedman's original proof) or some kind of proof theoretic machinery (in papers by several authors, among them Feferman himself).

metical formulae. For any system S, let $\mathsf{S}\restriction$ be the system *with the induction axioms restricted to the language of arithmetic*.

Theorem 2

(i) $\mathsf{KF} \leq \Sigma^1_1\text{-}\mathsf{AC}_0$
(ii) $\mathsf{KF}\restriction \leq \Sigma^1_1\text{-}\mathsf{AC}_0\restriction$

This is sufficient for calibrating KF and its subsystems.

The second route. The previous proof is simple but it has a disadvantage: it cannot be adapted to deal with *truth consistency* (or minimality), i.e. if we want to consider KF plus CONS:

$$\forall x(\mathrm{Sent}_{\mathsf{KF}}(x) \to \neg T(x) \vee \neg F(x)) \tag{CONS}$$

Of course, Axiom CONS rules out truth value gluts. That is, the axiom excludes models where the extension and the antiextension of the truth predicate overlap and sentences are simultaneously true and false. This restriction to consistent fixed points is in line with Kripke's [42] original account.

In order to analyze KF plus CONS, we can easily apply methods from predicative proof theory. Thus we devise a suitable sequent calculus version of KF, $\mathsf{KF}\restriction$, ..., choosing to rephrase $\mathsf{KF}\restriction$ as *a fixed point theory with consistency*. This means that we can easily find a formula $\mathcal{T}(x, P)$, formalizing the closure properties of the truth predicate T (see Halbach [33], p. 281), i.e. such that $\forall x(\mathcal{T}(x, T) \leftrightarrow T(x))$.

Since we like to formalize KF and $\mathsf{KF}\restriction$ as Tait calculi, the basic *positive atoms* have the form: $t = s$, $T(t)$, and the *negative atoms* are obtained by negating the positive ones. An *atom* is simply a positive or a negative atom and we stipulate that $\neg\neg\varphi := \varphi$ (φ atom). Formulas are inductively generated from atoms by closing under disjunction, conjunction, unbounded quantification. If φ is an arbitrary formula, $\neg\varphi$ is the formula which results from the negation normal form of $\neg\varphi$ by erasing each even sequence of occurrences of negation in front of atoms.

Let us expand $\mathcal{L}_T$ with a new predicate symbol P. If $Q := P, T$, a formula φ of $\mathcal{L}_T$ is *Q-positive* (*Q-negative*) if every occurrence of Q in φ occurs within positive (negative) atoms of the form $Q(t)$ ($\neg Q(t)$). A formula φ is *Q-separated* if φ is Q-positive or Q-negative. A formula φ is *Q-free* if Q does not occur in φ. A Q-free formula can be regarded as both Q-positive and Q-negative.

Definition 1 The system $\mathsf{TKF}^{\mathsf{c}}$ consists of:

- logical axioms of the form

$$\Gamma, \neg\varphi, \varphi$$
$$\Gamma, \neg t = s, \varphi[x := t], \varphi[x := s]$$

  where φ is an atom (according to the previous definitions);
- axioms of the form Γ, Δ where Δ is an e-atom or a finite set of e-atoms; Δ formalizes the standard axioms for zero, successor, or the defining equations for the function symbols of $\mathcal{L}_T$;

- standard logical rules (see Schwichtenberg and Wainer [55]) for introducing $\wedge$, $\vee$, $\forall$, $\exists$ and the cut rule:

$$\frac{\Gamma, \varphi \quad \Gamma, \neg\varphi}{\Gamma}$$

- T-consistency:

$$\Gamma, \neg\,\mathsf{Sent}_{\mathsf{KF}}(t), \neg T(t), \neg T(\dot{\neg} t)$$

- induction rule for *classifications*: if t is an arbitrary term,

$$\frac{\Gamma, Cl(a) \quad \Gamma, 0 \in a \quad \Gamma, \forall x(x \in a \to (x+1) \in a)}{\Gamma, t \in a}$$

- T-closure:

$$\frac{\Gamma, \mathcal{T}(t, T)}{\Gamma, T(t)}$$

- T-soundness:

$$\frac{\Gamma, \neg\mathcal{T}(t, T)}{\Gamma, \neg T(t)}$$

TKF is the calculus which is obtained from $\mathsf{TKF}^{\mathsf{c}}$ by replacing the induction rule for *classes* by full number-theoretic induction rule, that is, if $\varphi(x)$ is an arbitrary formula of $\mathcal{L}_T$ an arbitrary term,

$$\frac{\Gamma, \varphi(0) \quad \Gamma, \forall x(\varphi(x) \to \varphi(x+1))}{\Gamma, \varphi(t)}$$

If $S := \mathsf{TKF}^{\mathsf{c}}, \mathsf{TKF}$, we can inductively define a derivability relation with explicit tree height and suitable cut complexity, so that the following holds:

Lemma 2 (Partial Cut-elimination) *Every* $\mathsf{TKF}^{\mathsf{c}}$*-derivation* $\mathcal{D}$ *can be effectively transformed into a* $\mathsf{TKF}^{\mathsf{c}}$*-derivation of the same end-sequent, where cut–formulas are* T*-separated.*

The argument is standard; it essentially depends on the fact that the active formulas in the axioms and in the conclusions of the mathematical inferences (in particular, number theoretic induction) are T-separated. It is also obvious that the system $\mathsf{TKF}^{\mathsf{c}}$ (TKF) proves the sequents corresponding[11] to the theorems of $\mathsf{KF}\!\restriction$($\mathsf{KF}$) plus CONS.

Approximating truth by its finite levels. By Lemma 2 it is then possible to approximate truth by its finite levels and hence to eliminate truth from $\mathcal{L}_T$.

[11] Under the obvious translation of the language of KF, $\mathsf{KF}\!\restriction$ into the Tait framework.

Definition 2 Let $\bot = 0 = 1$; then

$$T^0(t) = 0 = 1 \tag{3.7}$$
$$T^{m+1}(t) = \mathcal{T}(t, T^m)) \tag{3.8}$$

Clearly *each formula in the sequence belongs to the language* $\mathcal{L}_{\mathsf{PA}}$.

If φ is any formula in negation normal form, let $\varphi[m, n]$ be obtained from φ by replacing each atom of the form $T(t)$ $(\neg T(t))$ by $T^n(t)$ $(\neg T^m(t))$.

Finally, a derivation $\mathcal{D}$ of $\mathsf{TKF^c}$ is *quasi-normal* provided all cut–formulas in $\mathcal{D}$ are T-separated.

Theorem 3 *Let $\mathcal{D}$ be a quasi-normal* $\mathsf{TKF^c}$*-derivation of Γ with height k. Then, provably in* PA, *for every $m > 0$,*[12] *if $H(m) := m + 2^k$:*

$$\Gamma[m, H(m)] \tag{3.9}$$

The proof is by induction on the height k of the given derivation. The soundness of the asymmetric interpretation – with respect to PA-provability, relies on the fact that cut formulas are always T-separated and on standard persistence properties.[13] On the other hand, restriction to classes has the effect that number theoretic induction for $\mathcal{L}_{\mathsf{PA}}$-formulas is enough, as a consequence of the fact that for each given m, $T^m(t)$ is an arithmetical formula[14] As to the verification of consistency, it reduces to verify by outer induction on m

$$\forall x(\mathrm{Sent}_{\mathsf{KF}}(x) \to \neg T^m(x) \vee \neg T^m(\dot{\neg} x)) \tag{3.10}$$

Since the transform $\varphi \mapsto \varphi[m, n]$ is the identity function on formulas of $\mathcal{L}_{\mathsf{PA}}$, we eventually conclude that

Corollary 1 *If* $\mathsf{TKF^c} \vdash \varphi$ *and* $\varphi \in \mathcal{L}_{\mathsf{PA}}$, *then* $\mathsf{PA} \vdash \varphi$.

Theorem 4

(i) $\mathsf{PA} \equiv \mathsf{KF}\restriction + \mathsf{CONS}$
(ii) $(\Pi^0_1\text{-}CA)_{<\varepsilon_0} \equiv \mathsf{KF} + \mathsf{CONS}$

As to $\mathsf{KF} + \mathsf{CONS} \leq (\Pi^0_1\text{-CA})_{<\varepsilon_0}$: by lifting the previous method to the case of systems with full number theoretic induction. This can be carried out by standard embedding into systems with ω-rule (see Schwichtenberg and Wainer [55]).

---

[12] In general, if $\Gamma := \{\varphi_1, \ldots, \varphi_q\}$, $\Gamma[m, n] := \{\varphi_1[m, n], \ldots, \varphi_q[m, n]\}$.

[13] This means that, if $0 < m_2 \leq m_1 \leq k_1 \leq k_2$, Γ is a set of formulas such that $\Gamma[m_1, k_1], \Delta$ is derivable in PA, then $\Gamma[m_2, k_2], \Delta$ is also PA-derivable, leaving height and cut complexity unchanged.

[14] Note, however, that its logical complexity increases with m.

3.2.3 From Ordinary Reflective Closure to Schematic Reflective Closure

Equation 3.4 means that the ordinary reflective closure of arithmetic is in fact equivalent to predicative analysis of any level up to the first ϵ-number ε_0 i.e. roughly to the fragment of second order arithmetic, which is based upon transfinite iteration up to ε_0 of the standard arithmetical comprehension

$$(\exists X)(\forall u)(u \in X \leftrightarrow \varphi(u)),$$

φ being a formula containing only number theoretic quantifiers and possibly free second order variables.

The notion of ordinary reflective closure can then be extended *by a suitable substitution rule* in order to answer the problem of which *schemata* can be regarded as implicit in accepting a given list of *schematic axioms and rules*. The substitution rule has the form

$$\frac{\varphi(P)}{\varphi(\hat{x}\psi(x))}$$

where φ is a formula of $\mathcal{L}_{\mathsf{PA}}$ with an additional predicate symbol P, ψ is a formula of the language $\mathcal{L}_T$ expanded with P, and $\varphi(\hat{x}\psi(x))$ is the formula obtained by replacing the atoms of the form $P(t)$ in $\varphi(P)$ by $\psi(t)$.[15] Informally, the rule allows us to make inferences from schemata accepted in the original arithmetical language to schemata of the language with reflective means (i.e. self-referential truth). It turns out that the resulting notion of *schematic reflective closure* $Ref^*(\mathsf{PA}(P))$ yields *an alternative unramified characterization of predicative analysis* (see Feferman [17]):

$$Ref^*(\mathsf{PA}(P)) \equiv (\Pi^0_1\text{-CA})_{<\Gamma_0};$$

this proof-theoretic[16] equivalence still holds even when we add consistency (CONS) to $Ref^*(\mathsf{PA}(P))$. The theorem witnesses that investigations deriving from formal semantics and paradoxes have reached a high level of integration with different areas of foundational investigations; for instance, while the lower bound result is actually a refinement of typical predicative well-ordering techniques, the upper bound theorem can be achieved by techniques and results from reductive proof theory (see Feferman [12]).

Remarks

(i) The main result about reflective closure (either ordinary or schematic) still holds once consistency (CONS) is replaced by the *completeness* axiom, i.e.

---

[15] With the obvious proviso ensuring that no clash of variables occurs.

[16] Γ_0 is the first strongly critical ordinal, which is known to be the limit of predicative provability in the sense of Feferman and Schütte.

$$(\forall x)(\mathrm{Sent}_{\mathsf{KF}}(x) \rightarrow (T(x) \vee F(x)).$$

This axiom, which is true in the greatest fixed point of the Kripkean operator for self-referential truth, rules out truth value gaps in the same way as the consistency axiom (CONS) rules out truth value gluts. Kripke [42] considered only consistent fixed point (that is, fixed points without overlapping extension and antiextension of truth). But later Visser [60] and others generalized the approach to fixed points where extension and antiextension are allowed to overlap and thus a sentence can be simultaneously true and false.

(ii) It turns out that *suitable variants of 'reflective closure' can be profitably applied as tools for proof-theoretic investigations*: for instance, as an intermediate step for computing the proof theoretic strength of transfinitely iterated fixed point theories (see Jäger et al. [38]).

3.2.4 Digression: KF with Minimality

Many philosophers think that the minimal fixed point model of Kripke's [42] theory is the most natural. It gives a picture of grounded truth: The truth and falsity of any sentence ultimately depends on the truth and falsity of non-semantic sentences, that is, sentences that do not contain the truth or falsity predicates. In other fixed points that are not minimal ungrounded sentences such as a truth teller sentence can be true and false as well.

KF is the theory of *all fixed point models* of the monotone operator specifying the clauses for reflective truth. Consequently, KF doesn't decide the truth teller sentence. If one aims at a theory of the minimal fixed point and thus of grounded truth, one can try to add axioms that exclude fixed point models that are not minimal. Of course, this is conceptually relevant for the whole enterprise of truth: think of the important distinctions that arise from considering the plurality of truth predicates (grounded sentence, paradoxical sentence, intrinsic, etc. see Kripke [42]).

Definition 3

(i) KF + GID: see Cantini [5]. GID is the schema

$$\forall x(\varphi(x, \hat{x}\psi(x)) \rightarrow \psi(x)) \rightarrow \forall x(x \in I_\varphi \rightarrow \psi(x)) \tag{3.11}$$

where ψ is an arbitrary formula, I_φ is obtained by diagonalization (Lemma 3.9, [5]), $\varphi(x, P)$ is an elementary positive (in P) operator in the language of $\mathcal{L}_T$ with an additional predicate variable, P positive.[17]

(ii) KF_μ (Truth with minimality, see Burgess [4]): it is the fragment of $\mathsf{KF}\restriction$ with

 (i) only the *composition principles*, e.g. $\forall x(\mathcal{T}(x, T) \rightarrow T(x))$;

[17] Of course, represented in $\mathcal{L}_T$, so that $P(t)$ is translated into $t \in p$, p fresh variable; $\varphi(x, \psi)$ is obtained by the substitution $t \in y \mapsto \psi(t)$. The schema (3.11) claims that I_φ represents the least fixed point of the monotone operator defined by $\varphi(x, P)$ in a given arithmetical model.

(ii) the schema: if ψ is an arbitrary formula,

$$\forall x(T(x, \hat{x}\psi(x)) \rightarrow \psi(x)) \rightarrow \forall x(T(x) \rightarrow \psi(x)) \tag{3.12}$$

Then KF_μ proves the decomposition axioms and the consistency axiom. Also, it explicitly refutes statements that fail in the least fixed point model. Clearly $\mathsf{KF}_\mu\upharpoonright$ has an inner model in $\mathsf{KF}\upharpoonright + GID$ (see proposition 3.12, Cantini [5]).

As to the upper bound, we can apply the proof-theoretic methods of Cantini [7] to $\mathsf{KF}_\mu\upharpoonright$.

As to the lower bound, let $\mathsf{ID}_1^{\mathsf{acc}}$ be the theory of accessibility inductive definitions over PA, which is known to be proof-theoretically equivalent to the theory of elementary inductive definitions. Then we can lift to the present context a suitable version of the so-called *bar-induction* schema, which goes back to Kreisel:

Theorem 5 *If* $\prec$ *is a binary relation which is* determined,[18] *then the schema of transfinite induction on the largest well-founded part* $W(\prec)$ *of* $\prec$ *holds for arbitrary formulas* φ*, provably in* KF_μ.

The proof takes inspiration from an analogous result of Friedman and Sheard [25]; it is also exploited by Burgess [4]. Hence we have by the previous theorem:

Corollary 2 $\mathsf{ID}_1^{\mathsf{acc}}$ *is interpretable in* KF_μ.

3.3 *Related Works Inspired by* KF

KF is nowadays considered to have brought the birth to the subject called axiomatic theories of truth as an area of logical research in its own right. A variety of formal theories of truth not in a semantic form but in an axiomatic form have been presented since KF. Some try to axiomatize known semantic theory or construction of truth as KF does for Kripke's theory of truth; others take more purely axiomatic approach by considering what combinations of truth-theoretic principles are possible (consistent) and plausible. Friedman and Sheard [25] listed nine natural and naïvely correct postulates of truth, which are inconsistent altogether, and determined the maximal consistent combinations of them. The other two theories of Feferman's own presented below would be counted in the latter 'purely axiomatic' type of axiomatic theories of truth as well. Cantini [6] presented a system that can be regarded as an axiomatization of Kripke's least fixed-point truth with supervaluation schema. Horsten et al. [36] made an attempt to axiomatize Gupta–Belnap–Herzberger revision theoretic truth. There are far more examples that we could list here, and the subject is still lively developing.

[18] I.e. total in the sense of T, F, see Sect. 3.2.1.

4 Type-Free T-Schema with Reinterpreted Biconditional

The second theory of truth by Feferman [15] was originally intended to provide a uniform type-free treatment for set-comprehension and truth-predication, both of which notoriously yield a contradiction with their naïve formulations. Feferman [18] described the purpose of the theory as 'pragmatic'; particularly, on the side of set-comprehension, the theory was presented for the purpose of suitably and consistently dealing with natural type-free statements that mathematicians would like to make, for instance, in category theory. However, he later gave it a reformation into a theory of deflationist truth in Feferman [19].

The core idea behind the theory is to formally interpret the informal biconditional 'if and only if' in the Comprehension Axiom and T-schema, not as the material biconditional $\leftrightarrow$, but as another new primitive logical connective $\equiv$ so that $\equiv$ preserves certain essential connotations or characteristics of the informal biconditional but still avoids a contradiction. Let us restrict ourselves to his theory of truth. The T-schema is expressed in English as

$$\ulcorner\varphi\urcorner \text{ is true if and only if } \varphi \tag{4.1}$$

for arbitrary $\varphi \in \mathcal{L}_T$. If we interpret 'if and only if' by the standard material biconditional $\leftrightarrow$, then the resulting translation of (4.1) is inconsistent. Hence, Feferman introduces a new binary connective $\equiv$ into the syntax of a theory, and add the following as a new axiom schema that meant to interpret (4.1):

$$T(\ulcorner\varphi\urcorner) \equiv \varphi,$$

for arbitrary $\varphi \in \mathcal{L}_T$.

Since $\equiv$ is a newly introduced connective separate and independent of any other standard connectives such as $\rightarrow$ and $\neg$, a suitable axiomatic characterization should be provided for it, and also such a characterization should be so made that $\equiv$ is seen as representing the informal biconditional 'if and only if' at least in the context of our discourse involving truth.

Feferman [15] initially gave a semantic characterization of $\equiv$, where he describes how to define the extension of a truth predicate and the semantic evaluation rule for the new connective $\equiv$ over a given model of a base theory. The construction goes as follows. We start with an arbitrary model $\mathfrak{M}$ with a domain M of a base theory B over a language $\mathcal{L}$. For the simplicity of argument, we take $\mathcal{L}_{\mathsf{PA}}$ and PA to be $\mathcal{L}$ and B respectively in what follows. Recall that we defined $\mathcal{L}_T = \mathcal{L}_{\mathsf{PA}} \cup \{T\}$. Let $\mathcal{L}_T(\equiv)$ be a language obtained by augmenting $\mathcal{L}_T$ with the connective $\equiv$. We first give the Kripkean least fixed-point construction of truth over $\mathfrak{M}$ with a suitable partial logic such as strong Kleene or 3-valued Łukasiewicz logic, and let $X \subset M$ be the extension of the predicate T in thus constructed fixed-point semantics over $\mathcal{L}_T$ with the partial logic we have chosen. Now, for each $\mathcal{L}_T(\equiv)$-formula φ, let us denote by φ^* the $\mathcal{L}_T$-formula obtained by replacing all the occurrences of $\equiv$ in φ with the material biconditional $\leftrightarrow$. Then, we expand the base model $\mathfrak{M}$ to an $\mathcal{L}_T(\equiv)$-structure $\mathfrak{M}(\equiv)$ by putting

- $\mathfrak{M}(\equiv) \models T(a)$, iff $a = \ulcorner\sigma\urcorner$ for some $\mathcal{L}_T(\equiv)$-sentence σ and σ^* is contained in thus constructed Kripkean truth, and
- $\mathfrak{M}(\equiv) \models \varphi \equiv \psi$, iff φ^* and ψ^* have the same semantic value in the fixed-point semantics;

note that $\equiv$ is now changed to $\leftrightarrow$ in φ^* and ψ^* and it is evaluated according to the partial logic we chose in the fixed-point semantics; the evaluation of $\mathcal{L}_{\mathsf{PA}}$-atomics in $\mathfrak{M}(\equiv)$ are the same as $\mathfrak{M}$, and the other logical connectives and quantifiers are classically evaluated in $\mathfrak{M}(\equiv)$. Hence, an informal reading of $\varphi \equiv \psi$ is 'the semantic values of φ and ψ coincides in the Kripke's least fixed-point semantics', and thus since the Kripkean construction gives the same semantic value to $T(\ulcorner\varphi\urcorner)$ and φ for all $\mathcal{L}_T$-sentences φ it holds that

$$\mathfrak{M}(\equiv) \models T(\ulcorner\varphi\urcorner) \equiv \varphi, \text{ for all } \mathcal{L}_T(\equiv)\text{–sentences } \varphi. \tag{4.2}$$

Under this interpretation $\equiv$ enjoys several desired properties. Let $\mathbf{t} := 0 = 0$ and $\mathbf{f} := 0 \neq 0$: namely, $\mathbf{t}$ and $\mathbf{f}$ respectively stand for definitely true and false sentences. Then we define $\mathcal{D}(\varphi)$ to abbreviate $(\varphi \equiv \mathbf{t}) \vee (\varphi \equiv \mathbf{f})$, which informally expresses 'φ is determinate.' Then, we can show that $\mathfrak{M}(\equiv)$ is a model of the following.

S1 $T\ulcorner\varphi\urcorner \equiv \varphi$ for all $\mathcal{L}_T(\equiv)$-sentences φ
S2 $\equiv$ is an equivalence relation
S3 $\mathbf{t} \not\equiv \mathbf{f}$
S4 $\equiv$ preserves all the connectives and quantifiers including $\equiv$ itself; e.g., $((\varphi \equiv \psi) \wedge (\sigma \equiv \chi)) \rightarrow ((\varphi \equiv \sigma) \equiv (\psi \equiv \chi)) \wedge ((\varphi \vee \sigma) \equiv (\psi \vee \chi))$.
S5 $\mathcal{D}(\varphi)$ for all $\mathcal{L}_{\mathsf{PA}}$-sentences.
S6 $\mathcal{D}$ is closed under $\neg$, $\vee$, and $\forall$.
S7 $(\varphi \equiv \mathbf{t} \rightarrow \varphi) \wedge (\varphi \equiv \mathbf{f} \rightarrow \neg\varphi)$.

Proof S1 immediately follows from (4.2). S2 and S4 are obvious from the evaluation for $\equiv$. S3 and S5 hold since the fixed-point semantics does not change the evaluations of $\mathcal{L}_{\mathsf{PA}}$-sentences. S6 is due to the evaluation schema of the logic we chose. Finally we can show S7 by induction on φ. □

Later Feferman [19] adopted these seven properties of $\equiv$ as the axioms of a deflationist theory of truth which he calls S. There he interprets deflationism in the form that the proposition $T\ulcorner\varphi\urcorner$ is evaluated in the same way as φ and they are thereby equated even if they do lack a determined truth value[19]; hence the aforementioned informal reading of $\equiv$ well fits in this interpretation. Furthermore, since any model of PA can be expanded to a model of S, S is conservative over B; it is even 'semantically

[19]This intuition yields a model in a suitable infinitary combinatory logic, as detailed in Aczel and Feferman [3].

conservative' over B in the sense of Craig and Vaught [8]. Thus it satisfies the 'conservativeness requirement' for deflationist theories of truth.[20]

Remark 1 Several extra axioms can be still conservatively added to S. For instance, we may consider the following further axioms for S.

S8 $(\varphi \vee \psi) \equiv \mathbf{t} \leftrightarrow \varphi \equiv \mathbf{t} \vee \psi \equiv \mathbf{t}$.
S9 $\exists x \varphi(x) \equiv \mathbf{t} \leftrightarrow \exists x (\varphi(x) \equiv \mathbf{t})$.
S10 $(\varphi \equiv \psi) \equiv \mathbf{t} \leftrightarrow \varphi \equiv \psi$.
S11 $(\varphi \equiv \psi) \equiv \mathbf{t} \leftrightarrow \mathcal{D}(\varphi) \wedge \mathcal{D}(\psi) \wedge (\varphi \equiv \psi)$.
S12 $\mathcal{D}$ is strongly compositional with respect to the connectives and quantifiers except $\equiv$; e.g., $\mathcal{D}(\varphi \vee \psi) \leftrightarrow \mathcal{D}(\varphi) \wedge \mathcal{D}(\psi)$.

Which of them can be conservatively added to S depends on which partial logic we choose in the construction of the Kripkean fixed-point semantics over $\mathfrak{M}$. For instance, if we choose 3-valued Łukasiewicz logic, then we can show that the resulting semantics $\mathfrak{M}(\equiv)$ is a model of S8–S10 and thus they can be conservatively added to S; if we choose strong Kleene logic instead, S8–S9 and S11 are satisfied in $\mathfrak{M}(\equiv)$ and thus can be conservatively added to S; by choosing weak Kleene logic, we can also see that S11–S12 can be conservatively added to S.

5 Determinate Truth

In the system S of the previous section one has the unrestricted T-schema. However, the biconditional in the T-schema is no longer the classical biconditional $\leftrightarrow$ but rather the nonclassical $\equiv$. This is the price for the unrestricted T-schema. Instead of having the unrestricted T-schema and a nonclassical biconditional, one can restrict the T-schema to well-behaved sentences and then have the full classical T-schema and all other truth-theoretic axioms one might want to have. Feferman [18] pursued this strategy with his system DT.

Of course the restriction needs to be well motivated. According to Russell [52], every predicate P has a *domain of significance* D and it makes sense to apply P only to objects in D and the principles which are supposed to characterize (or axiomatize) the concept expressed by P are meant to be only about the objects in D. Hence, in the case of truth, if such a domain is appropriately given and if no contradictory sentences, such as the liar sentence, are contained in D, the naïve truth-theoretic principles, such as Tarski's T-schema, can be safely postulated. For Feferman such

---

[20]Deflationism is a claim that truth is a 'metaphysically thin and insubstantial' notion and a merely logico-linguistic device for generalization and implicit endorsement. It is often argued that deflationist theory of truth should be conservative over a base theory, since otherwise the addition of a truth predicate would yield something that was not obtained without the help of it in the base theory. See also footnote 2 above. Shapiro [56] and McGee [47] discuss model-theoretic or semantic conservativeness.

a domain D for the case of truth consists of the sentences that are *meaningful* and *determinate*,[21] that is, have a definite truth value, true of false.

Let us assume that our base theory is PA as before. A theory of truth based on the above view have two predicate T and D representing truth and its domain of significance; namely, the language $\mathcal{L}_{\mathsf{DT}}$ of DT is $\mathcal{L}_{\mathsf{PA}} \cup \{T, D\}$. It is plausible for him to require that the condition on the domain D of significance for any predicate P should be prior to the conditions on P; in the current context, an axiomatic characterization of D should be given without appealing to T. Feferman takes *strong compositionality* as the axiomatic characterization of the domain D of significance for T. Namely, D is closed under the propositional operations and quantifiers, and, conversely, a sentence is meaningful only if all of (the substitution instances of) its parts are meaningful. Thus conceived *strong compositionality* can be axiomatized as follows:

D1 $\forall x\,\big(\mathrm{AtSent}_{\mathsf{PA}}(x) \to D(x)\big)$
D2 $\forall s\,\big(D(\underset{.}{T}\, s) \leftrightarrow D(s^{\circ})\big)$
D3 $\forall x\,\big(\mathrm{Sent}_{\mathsf{DT}}(x) \to \big(D(\underset{.}{\neg}x) \leftrightarrow D(x)\big)$
D4 $\forall x \forall y\,\big(\mathrm{Sent}_{\mathsf{DT}}(x \underset{.}{\vee} y) \to \big(D(x \underset{.}{\vee} y) \leftrightarrow (Dx \wedge Dy)\big)\big)$
D5 $\forall v \forall x\,\big(\mathrm{Sent}_{\mathsf{DT}}(\underset{.}{\forall} v.x) \to \big(D(\underset{.}{\forall} v.x) \leftrightarrow \forall t\, D(x[t/v])\big)\big)$

The formula $\mathrm{AtSent}_{\mathsf{PA}}$ stands for the set of all atomic $\mathcal{L}_{\mathsf{PA}}$-sentences, and $\mathrm{Sent}_{\mathsf{DT}}$ for the set of sentences of $\mathcal{L}_{\mathsf{DT}}$. The axiom D1 does not come from the strong compositionality requirement in question but from the common and plausible view that the sentences of the base language, which contains no semantic predicates, should be non-paradoxical, determinate, and counted in the domain of the significance of T.

The T-schema and other truth-theoretic principles are accordingly restricted to such sentences. For instance, the T-schema is formulated as follows:

$$D(\ulcorner\varphi\urcorner) \to (T(\ulcorner\varphi\urcorner) \leftrightarrow \varphi), \quad \text{for all } \mathcal{L}_{\mathsf{DT}}\text{-sentences } \varphi. \tag{5.1}$$

Feferman points out that thus formulated T-schema yields the following intimate connection of D and T as a special phenomenon particular to the case of truth:

$$D(\ulcorner\varphi\urcorner) \leftrightarrow \big(T(\ulcorner\varphi\urcorner) \vee T(\ulcorner\neg\varphi\urcorner)\big) \tag{5.2}$$

Its proof goes as follows. Suppose $D(\ulcorner\varphi\urcorner)$ for one direction. By D3 we have $D(\ulcorner\neg\varphi\urcorner)$. Hence, (5.1) yields $T(\ulcorner\varphi\urcorner) \leftrightarrow \varphi$ and $T(\ulcorner\neg\varphi\urcorner) \leftrightarrow \neg\varphi$. Finally, the law of excluded middle $\varphi \vee \neg\varphi$ entails $T(\ulcorner\varphi\urcorner) \vee T(\ulcorner\neg\varphi\urcorner)$. Suppose $(T(\ulcorner\varphi\urcorner) \vee T(\ulcorner\neg\varphi\urcorner))$ for the converse. Since T only applies to its domain of significance D, we have $D(\ulcorner\varphi\urcorner) \vee D(\ulcorner\neg\varphi\urcorner)$. Hence $D(\ulcorner\varphi\urcorner)$ follows from D3. Feferman then generalizes (5.2) to a universal statement and postulates

D0 $\forall x\big(D(x) \leftrightarrow (T(x) \vee T(\underset{.}{\neg} x))\big)$

[21] Feferman [17] adopts the term 'determined' for a similar notion; see p. 20.

Thereby, specially in the case of truth, the domain predicate D becomes definable in terms of T and redundant. Indeed, Feferman [18] does not take a predicate D as a primitive predicate for this reason. Instead he formulates DT over $\mathcal{L}_T$ and introduces Dx as a mere abbreviation of $T(x) \vee T(\dot{\neg} x)$. Therefore let us assume $\mathcal{L}_{\mathsf{DT}} = \mathcal{L}_T$ in what follows.

Given these characterizations of D, Feferman formulates the axioms for T as the the compositional Tarskian clauses restricted to D:

D6 $\forall s\, \forall t\, \big(T(s \mathbin{\dot{=}} t) \leftrightarrow s^\circ = t^\circ)\big)$, and similarly for other predicates of $\mathcal{L}_{\mathsf{PA}}$
D7 $\forall s\, \big(D(s^\circ) \rightarrow \big(T(\dot{T}\, s) \leftrightarrow T(s^\circ)\big)\big)$
D8 $\forall x\, \big(\mathrm{Sent}_{\mathsf{DT}}(x) \wedge D(x) \rightarrow \big(T(\dot{\neg} x) \leftrightarrow \neg T(x)\big)$
D9 $\forall x \forall y\, \big(\mathrm{Sent}_{\mathsf{DT}}(x \mathbin{\dot{\vee}} y) \wedge D(x \mathbin{\dot{\vee}} y) \rightarrow \big(T(x \mathbin{\dot{\vee}} y) \leftrightarrow (T(x) \vee T(y))\big)\big)$
D10 $\forall v \forall x\, \big(\mathrm{Sent}_{\mathsf{DT}}(\dot{\forall} v.x) \wedge D(\dot{\forall} v.x) \rightarrow \big(T(\dot{\forall} v.x) \leftrightarrow \forall t T(x[t/v])\big)\big)$

Let us denote by DT^{w} the axiomatic system comprising PA plus D0–D10 augmented by the expanded induction for $\mathrm{Sent}_{\mathsf{DT}}$. We can easily show by induction on formulae that (5.1) is a theorem of DT. This DT^{w} is a straightforward axiomatization of Feferman's view described so far and is already proof-theoretically pretty strong; DT^{w} is proof-theoretically equivalent to KF and indeed identical as theories with $\mathsf{WKF} + \mathsf{CONS}$, the Kripke–Feferman system for weak Kleene logic plus the axiom of consistency; see [26] for more details. However, he sees some defects in DT^{w} as a theory of truth and slightly amends it.

We can standardly show that DT^{w} proves the global reflection principle as KF does: i.e.,

$$\mathsf{DT}^{\mathsf{w}} \vdash \forall x \big(\mathrm{Bew}_{\mathsf{PA}}(x) \rightarrow T(x)\big).$$

However, in contrast to KF, DT^{w} does not prove *the truth of the global reflection principle*. Feferman requires for an adequate theory of truth to derive it and thereby suggests to amend DT^{w}.

Let us first see how its proof goes in KF. In addition to the global reflection principle, both KF and DT^{w} derive $\forall s \big(T(\dot{T}\, s) \leftrightarrow T(s^\circ)\big)$ as well as

$$\forall x\, \Big(\big(T(\ulcorner \mathrm{Bew}_{\mathsf{PA}}(\dot{x})\urcorner) \leftrightarrow \mathrm{Bew}_{\mathsf{PA}}(x)\big) \wedge \big(T(\ulcorner \neg \mathrm{Bew}_{\mathsf{PA}}(\dot{x})\urcorner) \leftrightarrow \neg \mathrm{Bew}_{\mathsf{PA}}(x)\big)\Big).$$

Hence, it follows from these that both KF and DT^{w} derive

$$\forall x\, \big(T(\ulcorner \mathrm{Bew}_{\mathsf{PA}}(\dot{x})\urcorner) \rightarrow T(\dot{T}\, \dot{x})\big).$$

By using the equivalence of $\varphi \rightarrow \psi$ and $\neg\varphi \vee \psi$, we obtain

$$\forall x\, \big(T(\ulcorner \neg \mathrm{Bew}_{\mathsf{PA}}(\dot{x})\urcorner) \vee T(\dot{T}\, \dot{x})\big). \tag{5.3}$$

However, one has to verify in $\mathsf{DT^w}$ that $\forall x D(Tx)$, in order to proceed from (5.3) to $\forall x(T\ulcorner\neg\mathrm{Bew}_{\mathsf{PA}}(\dot{x}) \vee Tx\urcorner)$, but $\forall x D(Tx)$ is incompatible with the view behind $\mathsf{DT^w}$ and indeed underivable in $\mathsf{DT^w}$; in contrast, KF can immediately derive it via the axioms K4 and K5 and thereby the truth of the global reflection principle. Given this observation, Feferman proposes to treat $\rightarrow$ as separate basic propositional connectives from $\vee$ and $\neg$, not identifying $\varphi \rightarrow \psi$ with $\neg\varphi \vee \psi$, and then postulates an independent axiom for the meaningfulness and determinateness of the sentences of the form $\varphi \rightarrow \psi$. Namely, he postulate the following two axioms for the conditional:

$$\text{D11}\quad \forall x \forall y \Big(\mathrm{Sent}_{\mathsf{DT}}(x \dot{\rightarrow} y) \rightarrow \big(D(x \dot{\rightarrow} y) \leftrightarrow \big(D(x) \wedge (T(x) \rightarrow D(y))\big)\big)\Big)$$

$$\text{D12}\quad \forall x \forall y \Big(\mathrm{Sent}_{\mathsf{DT}}(x \dot{\rightarrow} y) \wedge D(x \dot{\rightarrow} y) \rightarrow \big(T(x \dot{\rightarrow} y) \leftrightarrow (T(x) \rightarrow T(y))\big)\Big)$$

It is true, as Feferman himself points out, strong compositionality in the aforementioned sense is not met with respect to the conditional '$\rightarrow$', since D11 indicates that if φ is false then $D(\ulcorner\varphi \rightarrow \psi\urcorner)$ may hold even when $D(\ulcorner\psi\urcorner)$ fails, but he argues that it is rather natural since we do not care whether $D(\ulcorner\varphi \rightarrow \psi\urcorner)$ holds when φ is definitely false. Thereby the system DT of determinate truth is defined to be PA plus D0–D12 augmented by the expanded induction schema for $\mathcal{L}_{\mathsf{DT}}$.

A closer inspection reveals that DT is very intimately related to KF. KF is an "axiomatization" of Kripke's fixed-point semantics with strong Kleene logic. Naturally we may consider a similar "axiomatization" of the Kripkean fixed-point semantics with other logics. We have already mentioned such a theory for weak Kleene logic, which is related to $\mathsf{DT^w}$. As a matter of fact, DT is identical with such a KF-style axiomatization of the Kripkean fixed-point semantics with what we call Aczel–Feferman logic, plus the axiom of consistency CONS; namely, the inner logic of the KF-style system is Aczel–Feferman logic; see Fujimoto [27] for details.[22] Here we mean by Aczel–Feferman logic a variant of weak Kleene logic with an independent evaluation for the conditional '$\rightarrow$' than that for $\vee$ and $\neg$; in Aczel–Feferman logic, '$\rightarrow$' is not defined away in terms of '$\neg$' and '$\vee$' and the evaluation of '$\rightarrow$' is determined by the following separate truth table:

$\rightarrow$	T	U	F
T	T	U	F
U	U	U	U
F	T	T	T

Hence, from a purely technical point of view, Feferman's theory of determinate truth essentially boils down to his own axiomatization of Kripkean fixed-point semantics, and therefore DT is proof-theoretically equivalent to KF. Also, if we extend DT in the same way that we extend the reflective closure of PA, i.e., KF, to its *schematic reflective closure*, the resulting theory has the strength of the predicative limit Γ_0.

[22]What Aczel [1] does in his construction of Frege structure essentially amounts to the construction of Kripkean fixed-point semantics with Aczel–Feferman logic.

Remark 2 Feferman [18] raised two conjectures about the strength of the theories of determinate truth over PA, and the conjectures have been first settled by constructing a direct interpretation of them in KF and $Ref^*(\mathsf{PA}(P))$ in Fujimoto [26], but indeed the construction of Sect. 3.2.2 can also be adapted to the system DT (over PA), in order to produce an alternative proof of the first conjecture.

6 The Impact of Feferman's Work on Truth

We have mentioned already some further research inspired by Feferman's work. However, we don't aim at a survey of the impact of his work, because the impact is far too wide-ranging to be summarized in a few pages. Moreover, an assessment of the reception of Feferman's work is made harder by the fact that his work has been studied by different communities: In mathematical logic proof theorists have worked most extensively on Feferman's system and their variants; but set theorists have also availed themselves to Feferman's ideas (see Koellner [39]).

Also in philosophy Feferman's work has been fruitful not only for one topic. For once, Feferman advanced the use of truth as a primitive concept in the foundations of mathematics. Of course, this was his main motivation in the seminal *Reflecting on Incompleteness*. Generally Feferman didn't put so much emphasis on his theories as attempts to resolve the liar paradox; rather he was after theories of truth (and comprehension) that prove useful for foundational purposes. Of course, the paradoxes have to be addressed in some way, but the analysis of the paradoxes themselves doesn't figure very prominently in Feferman's work. Philosophers – among them Reinhardt [50], McGee [46] and Field [24] – have discussed Feferman's systems from the perspective of the theory of paradoxes. In this respect they also proved very fruitful.

One aspect of Feferman's work that hasn't been fully exploited is its relevance for the discussion about deflationism. Of course in some way, deflationists have employed ideas from Feferman's papers. In particular, Field's [24] recent research programme is very much motivated by deflationist or disquotationalist considerations and Field acknowledges Feferman's influence.

In particular, much of the work of Field and the work inspired by it can be seen as a continuation of Feferman's [15]: The challenge is to obtain the full T-schema in an extension of classical logic with a reasonably behaved conditional (or biconditional).

Deflationists have often seen truth as a device for expressing generalizations and infinite conjunctions (see Horwich [37] and Halbach [31]). Feferman's proof-theoretic analysis – more specifically the proof of the lower bound in Feferman [17, Sects. 4.3 and 5.3] – proceeds in terms of infinite conjunctions. While the discussion about the purpose of truth as a device of expressing infinite conjunctions usually remains somewhat metaphorical in the literature on deflationist, Feferman's analysis offers a precise conceptual analysis on the usability of truth for expressing infinite conjunctions. This is just one aspect of Feferman's work that offers scope for future philosophical research.

References

1. Aczel, P.: Frege structures and the notions of proposition, truth and set. In: Barwise, J., Keisler, H., Kunen, K. (eds.) The Kleene Symposium, pp. 31–59. North Holland, Amsterdam (1980)
2. Aczel, P.: The strength of Martin-Löf's intuitionistic type theory with one universe. In: Mietissen, S., Väänänen, J. (eds.) Proceedings of the Symposiums in Mathematical Logic in Oulu 1974 and Helsinki 1975, pp. 1–32. Department of Philosophy, University of Helsinki, Report no. 2 (1977)
3. Aczel, P., Feferman, S.: Consistency of the unrestricted abstraction principle using an intensional equivalence operator. In: Seldin, J.P., Hindley, J.R. (eds.) To H.B. Curry: Essays on Combinatory Logic, Lambda Calculus and Formalism, pp. 67–98. Academics Press, New York (1980)
4. Burgess, J.P.: Friedman and the axiomatization of Kripke's theory of truth. In: Tennant, N. (ed.) Foundational Adventures: Essays in Honor of Harvey M. Friedman, pp. 125–148. College Publications, London (2014)
5. Cantini, A.: Notes on formal theories of truth. Zeitschrift für mathematische Logik und Grundlagen der Mathematik **35**, 97–130 (1989)
6. Cantini, A.: A theory of formal truth arithmetically equivalent to ID_1. J. Symb.Logic **55**, 244–259 (1990)
7. Cantini, A.: Logical Frameworks for Truth and Abstraction: An Axiomatic Study. Number 135 in Studies in Logic and the Foundations of Mathematics. Elsevier, Amsterdam, Lausanne, New York, Oxford, Shannon and Tokyo (1996)
8. Craig, W., Vaught, R.: Finite axiomatizability using additional predicates. J. Symb. Logic **23**, 289–308 (1958)
9. Feferman, S.: Transfinite recursive progressions of axiomatic theories. J.Symb. Logic **27**, 259–316 (1962)
10. Feferman, S.: Systems of predicative analysis I. J. Symb.Logic **29**, 1–30 (1964)
11. Feferman, S.: Systems of predicative analysis II. J. Symb. Logic **33**(2), 193–220 (1968)
12. Feferman, S.: A more perspicuous formal system for predicativity. In: Lorenz, K. (ed.) Konstruktionen versus Positionen, vol. I, pp. 68–93. de Gruyter, Berlin (1978)
13. Feferman, S.: Constructive theories of functions and classes. In: Boffa, M., van Dalen, D., McAloon, K. (eds.) Logic Colloquium '78: Proceedings of the Colloquium held in Mons, August 1978, pp. 159–224. North-Holland, Amsterdam, New York and Oxford (1979)
14. Feferman, S.: Iterated inductive fixed-point theories: application to Hancock's conjecture. In: Metakides, G. (ed.) Patras Logic Symposion, pp. 171–195, Amsterdam, North Holland (1982)
15. Feferman, S.: Towards useful type-free theories I. J. Symb. Logic **49**, 75–111 (1984)
16. Feferman, S.: Reflecting on incompleteness (outline draft). handwritten notes (1987)
17. Feferman, S.: Reflecting on incompleteness. J. Symb. Logic **56**, 1–49 (1991)
18. Feferman, S.: Axioms for determinateness and truth. Rev. Symb. Logic **1**, 204–217 (2008)
19. Feferman, S.: Axiomatizing truth: Why and how. In: Berger, U., Diener, H., Schuster, P. (eds.) Logic, Construction, Computation, 185–200. Ontos, Frankfurt (2012)
20. Feferman, S., Strahm, T.: The unfolding of non-finitist arithmetic. Rev. Symb. Logic **3**, 665–689 (2000)
21. Feferman, S., Strahm, T.: Unfolding finitist arithmetic. Ann. Pure Appl. Logic **104**, 75–96 (2010)
22. Field, H.: Tarski's theory of truth. J. Philos. **69**, 347–375 (1972)
23. Field, H.: Deflating the conservativeness argument. J. Philos. **96**(10), 533–40 (1999)
24. Field, H.: Saving Truth From Paradox. Oxford University Press, Oxford (2008)
25. Friedman, H., Sheard, M.: An axiomatic approach to self-referential truth. Ann. Pure Appl. Logic **33**, 1–21 (1987)
26. Fujimoto, K.: Relative truth definability of axiomatic truth theories. Bull. Symb. Logic **16**, 305–344 (2010)
27. Fujimoto, K.: Autonomous progression and transfinite iteration of self-applicable truth. J. Symb. Logic **76**, 914–945 (2011)

28. Fujimoto, K.: Classes and truths in set theory. Ann. Pure Appl. Logic **163**, 1484–1523 (2012)
29. Gödel, K.: Über formal unentscheidbare Sätze der Principia Mathematica und verwandter Systeme I. Monatshefte für Mathematik **38**, 173–198 (1931)
30. Gupta, A.: Truth and paradox. J. Philos. Logic **11**, 1–60 (1982)
31. Halbach, V.: Disquotationalism and infinite conjunctions. Mind **108**, 1–22 (1999)
32. Halbach, V.: How not to state the T-sentences. Analysis **66**, 276–280 (2006). Correction of printing error in 67, 268
33. Halbach, V.: Axiomatic Theories of Truth, revised edition. Cambridge University Press, Cambridge (2014). first edition 2011
34. Halbach, V., Horsten, L.: Axiomatizing Kripke's theory of truth. J. Symb. Logic **71**, 677–712 (2006)
35. Herzberger, H.G.: Notes on naive semantics. J. Philos. Logic **11**, 61–102 (1982)
36. Horsten, L., Leigh, G., Leitgeb, H., Welch, P.: Revision revisited. Rev. Symb. Logic **5**, 642–664 (2012)
37. Horwich, P.: Truth, 2nd edn. Oxford University Press, Oxford (1998). first edition 1990
38. Jäger, G., Kahle, R., Setzer, A., Strahm, T.: The proof-theoretic analysis of transfinitely iterated fixed point theories. J. Symb. Logic **64**(1), 53–67 (1999)
39. Koellner, P.: On reflection principles. Ann. Pure Appl. Logic **157**, 206–219 (2009)
40. Kreisel, G.: Ordinal logics and the characterization of informal notions of proof. In: Todd, J.A. (ed.) Proceedings of the International Congress of Mathematicians, Edinburgh 1958, pp. 289–299, Cambridge University Press, Cambridge (1960)
41. Kreisel, G., Lévy, A.: Reflection principles and their use for establishing the complexity of axiomatic systems. Zeitschrift für mathematische Logik und Grundlagen der Mathematik **14**, 97–142 (1968)
42. Kripke, S.: Outline of a theory of truth. J. Philos. **72**, 690–712 (1975). reprinted in Martin (1984)
43. Leitgeb, H.: Poppers Wahrheitstheorie(n). In: Morscher, E (ed.) Was wir Karl R. Popper und seiner Philosophie verdanken: Zu seinem 100. Geburtstag, ProPhil · Projekte zur Philosophie, pp. 185–217. Academia Verlag, Sankt Augustin (2002)
44. Leitgeb, H.: What theories of truth should be like (but cannot be). In: Blackwell Philosophy Compass 2/2, pp. 276–290. Blackwell (2007)
45. Manaster, A.B., Payne, T.H., Harrah, D.: Meeting of the Association of Symbolic Logic, San Diego 1979. J. Symb. Logic **46**, 199 (1981)
46. McGee, V.: Truth, Vagueness, and Paradox: An Essay on the Logic of Truth. Hackett Publishing, Indianapolis and Cambridge (1991)
47. McGee, V.: In praise of the free lunch: Why disquotationalists should embrace compositional semantics. In: Hendricks, V.F., Pedersen, S.A. (eds.) Self-Reference, vol. 178, pp. 95–120. CSLI Publications, Stanford (2006)
48. Putnam, H.: A comparison of something with something else. New Lit. Hist. **17**(1), 61–79 (1985). Reprinted in (Putnam 1994, 330-50)
49. Reinhardt, W.: Remarks on significance and meaningful applicability. In: de Alcantara, L.P. (ed.) Mathematical Logic and Formal Systems: A Collection of Papers in Honor of Professor Newton C.A. Da Costa. Lecture Notes in Pure and Applied Mathematics, vol. 94, pp. 227–242. Marcel Dekker inc. (1985)
50. Reinhardt, W.: Some remarks on extending and interpreting theories with a partial predicate for truth. J. Philos. Logic **15**, 219–251 (1986)
51. Russell, B.: On some difficulties in the theory of transfinite numbers and order types. Proc. Lond. Math. Soc. **4**, 29–53 (1906)
52. Russell, B.: Mathematical logic as based on the theory of types. Am. J. Math. **30**, 222–262 (1908)
53. Schindler, T.: Type-free truth. Ph.D. Thesis, Ludwig-Maximilians-Universität München (2014)
54. Schütte, K.: Proof Theory. Springer, Berlin (1977)
55. Schwichtenberg, H., Wainer, S.: Proofs and Computations. Perspectives in Logic. Association of Symbolic Logic and Cambridge University Press, Cambridge (2012)

56. Shapiro, S.: Proof and truth: through thick and thin. J. Philos. **95**, 493–521 (1998)
57. Simpson, S.: Subsystems of Second Order Arithmetic. Springer, Berlin (1998)
58. Tarski, A.: Der Wahrheitsbegriff in den formalisierten Sprachen. Studia Philosophica Commentarii Societatis Philosophicae Polonorum **1**, 261–405 (1935). translation of Tarski (1936) by L. Blaustein, translated as 'The Concept of Truth in Formalized Languages' in (Tarski 1956, 152–278)
59. Turing, A.: Systems of logic based on ordinals. Proc. Lond. Math. Soc. **45**, 161–228 (1939)
60. Visser, A.: Four-valued semantics and the liar. J. Philos. Logic **13**, 181–212 (1984)

Author Biographies

Andrea Cantini is Professor of Logic at the University of Florence. His research interests include proof theory, axiomatic theories of truth, theories of operations and classes, history and philosophy of contemporary logic. Among his publications, the monograph *Logical Frameworks for Abstraction and Truth* (Elsevier/North–Holland 1996).

Kentaro Fujimoto is Lecturer in Mathematical and Philosophical Logic at the University of Bristol. His research interests include proof theory, axiomatic theories of truth, philosophy of logic and mathematics. He earned his master's degree from Stanford University under the supervision of Professor Solomon Feferman.

Volker Halbach is Professor of Philosophy at the University of Oxford and Fellow of New College. His fields of research are logic, philosophy of mathematics, philosophy of language, and epistemology. Among his publications is the monograph *Axiomatic Theories of Truth* (Cambridge University Press, 2011).

Feferman's Forays into the Foundations of Category Theory

Ali Enayat, Paul Gorbow and Zachiri McKenzie

Abstract This paper is primarily concerned with assessing a set-theoretical system, S^*, for the foundations of category theory suggested by Solomon Feferman. S^* is an extension of NFU, and may be seen as an attempt to accommodate unrestricted categories such as the category of all groups (without any small/large restrictions), while still obtaining the benefits of ZFC on part of the domain. A substantial part of the paper is devoted to establishing an improved upper bound on the consistency strength of S^*. The assessment of S^* as a foundation of category theory is framed by the following general desiderata (R) and (S). (R) asks for the unrestricted existence of the category of all groups, the category of all categories, the category of all functors between two categories, etc., along with natural implementability of ordinary mathematics and category theory. (S) asks for a certain relative distinction between large and small sets, and the requirement that they both enjoy the full benefits of the ZFC axioms. S^* satisfies (R) simply because it is an extension of NFU. By means of a recursive construction utilizing the notion of strongly cantorian sets, we argue that it also satisfies (S). Moreover, this construction yields a lower bound on the consistency strength of S^*. We also exhibit a basic positive result for category theory internal to NFU that provides motivation for studying NFU-based foundations of category theory.

Keywords Category theory · Stratified set theory · Zermelo-Fraenkel set theory · Small/large distinction

2010 Mathematics Subject Classification 03E70 · 18A15

A. Enayat (✉) · P. Gorbow · Z. McKenzie
Department of Philosophy, Linguistics, and Theory of Science,
University of Gothenburg, Box 200, 405 30 Göteborg, Sweden
e-mail: ali.enayat@gu.se

P. Gorbow
e-mail: paul.gorbow@gu.se

Z. McKenzie
e-mail: zach.mckenzie@gmail.com

G. Jäger and W. Sieg (eds.), *Feferman on Foundations*, Outstanding Contributions to Logic 13, DOI 10.1007/978-3-319-63334-3_12

1 Introduction

The foundations of category theory has been a source of many perplexities ever since the groundbreaking 1945-introduction of the subject by Eilenberg and Mac Lane; e.g., how is one to avoid Russell-like paradoxes and yet have access to objects that motivate the study in the first place, such as the category of all groups, or the category of all topological spaces? Solomon Feferman has grappled with such perplexities for over 45 years, as witnessed by his six papers on the subject during the period 1969–2013 [6, 7, 9–11, 13]. Our focus in this paper is on two important, yet quite different set-theoretical systems proposed by Feferman for the implementation of category theory: the ZF-style system ZFC/S [6] and the NFU-style system S^* [7, 11]; where NFU is Jensen's urelemente-modification of Quine's New Foundations system NF of set theory.[1]

Our assessment of Feferman's systems ZFC/S and S^* will be framed by the following general desiderata (R) and (S) of the whole enterprise of building a set-theoretical foundation for category theory. (R) is derived from Feferman's work, especially [7, 11], while (S) is extracted from Shulman's excellent survey [25]. Both (R) and (S) will be elaborated in Sect. 2.1.

(R) asks for the *unrestricted existence* of the category of all groups, the category of all categories, the category of all functors between two categories, etc., along with *natural implementability of ordinary mathematics and category theory.*

(S) asks for a certain *relative distinction between large and small sets,* and the requirement that they *both enjoy the full benefits of the* ZFC *axioms.*

Feferman's choice for a system to meet the demands of (R) was motivated by the fact that in contrast with ZFC-style systems, NFU accommodates a universal set of all sets, as well as the category of all groups, the category of all categories, etc. However, in order to deal with the fact that NFU is not powerful enough to handle some parts of ordinary mathematics and category theory, Feferman proposed an extension of NFU, called S^*, which directly interprets ZFC by a constant, and he established the consistency of S^* assuming the consistency of ZFC + $\exists\kappa\exists\lambda$ "$\kappa < \lambda$ are inaccessible cardinals" [7]. In our expository account of Feferman's work on S^*, we refine Feferman's results in two ways: we show that S^* interprets a significant strengthening of the Kelley-Morse theory of classes; and we also demonstrate the consistency of a natural strengthening S^{**} of S^* within ZFC + $\exists\kappa$ "κ is an inaccessible cardinal".

A widely accepted partial (partial because it is absolute, not relative) solution to (S) was chosen by Saunders Mac Lane in his standard reference "Categories for the Working Mathematician" [20], namely the theory ZFC + $\exists\kappa$ "κ is an inaccessible cardinal". Here the sets of rank less than a fixed inaccessible cardinal are regarded as small. A similar (but full, i.e. relative) solution to (S) due to Alexander Grothendieck,

[1] Feferman [9, 13] has put forward poignant criticisms of the general case of using category theory as an autonomous foundation for mathematics. Moreover, he suggested that a theory of *operations and collections* should also be pursued as a viable alternative platform for category theory; e.g. systems of Explicit Mathematics [8] and Operational Set Theory [12].

requires the existence of arbitrarily large inaccessible cardinals. However, as demonstrated by Feferman [6], the needs of category theory in relation to (S) can already be met in a conservative extension ZFC/S of ZFC, thus fully satisfying (S), albeit with the consequence that, while working with categories and functors in this foundational system, one often needs to verify that they satisfy a property called small-definability.

Besides the aforementioned fine-tuning of Feferman's results on S^*, our paper also has two other innovative features that were inspired by Feferman's work. One has to do with the observation that the category **Rel** in NFU has products and coproducts indexed by $\{\{i\} \mid i \in V\}$ (the same existence result also holds for coproducts in **Set**). The importance lies in that it marks a divergence from the analogous category of small sets with relations (respectively functions) as morphisms in a ZFC setting; this situation in ZFC is clarified by a theorem of Peter Freyd, which we prove in detail (Freyd's theorem also serves to exemplify why the large/small distinction is important for category theory). We consider this positive result on the existence of some limits for **Rel** and **Set** in NFU, to strengthen the motivation for exploring NFU-based foundations of category theory. Thus, another innovative feature is our proposal to champion the system NFUA (a natural extension of NFU) in meeting the demands imposed by both (R) and (S). It is known that NFUA is equiconsistent with ZFC + the schema: "for each natural number n there is an n-Mahlo cardinal".[2] This puts NFUA in a remarkably close relationship to the system ZMC/S, which is a strengthening of ZFC/S suggested by Shulman [25] as a complete solution to (S).

2 Requirements on Foundations for Category Theory

2.1 *The Problem Formulations (R) and (S)*

We will now elaborate each of the desiderata (R) and (S) mentioned in the introduction by more specific demands. The following requirements have been obtained from Feferman [7, 11].

(R1) For any given kind of mathematical structure, such as that of groups or topological spaces, the associated category of all such structures exists.
(R2) For any given categories A and B, the category B^A of all functors from A to B with natural transformations as morphisms, exists.
(R3) Ordinary mathematics and category theory, along with its distinction between large and small, are naturally implementable.

From a purely formal point of view, a foundational system is only required to facilitate implementation of the mathematical theories it is founding. But in practice,

[2]This equiconsistency result is due to Robert Solovay, whose 1995-proof is unpublished. The first-named-author used a different proof in [3] to establish a refinement of Solovay's equiconsistency result.

it is also desirable for this implementation to be user friendly. This is what the informal notion "naturally implementable" above seeks to capture. For the purposes of this paper, it suffices to distinguish three levels of decreasing user friendliness within the admittedly vague class of natural implementations of category theory in set-theory: (1) The usual set-theoretical description of category theoretic notions can be accommodated directly with the membership relation symbol of the underlying language, and with no restriction. (2) As above, but restricted to a set or class. (3) The notions can be accommodated with another well motivated defined membership relation, and restricted to a set or class. For example, the system NFUA facilitates a level (3) implementation of category theory by means of the set of equivalence classes of pointed extensional well-founded structures. But we will see that an equiconsistent extension of NFUA facilitates a level (2) implementation.

Feferman actually denotes his requirements as "(R1)–(R4)". His (R1)–(R2) are as above. His (R4) asks for a consistency proof of the system relative to some standard system of set theory; which we take as an implicit requirement. His (R3) requires that the system allows "us to establish the existence of the usual basic mathematical structures and carry out the usual set-theoretical operations" [11]. Feferman proposed a system S^*, which will be looked at in detail in the present paper, as a partial solution to his problem (R). But at least on one natural reading of Feferman's (R), Michael Ernst has recently shown in [5] that Feferman's (R3) is inconsistent with (R1) and (R2). Ernst uses a theorem of William Lawvere, which formulates Cantor's diagonalization technique in a general category theoretic context [21]. Although Ernst argues generally for the inconsistency of Feferman's (R), it is worth noting that there is a more direct result that applies to NFU: Feferman's (R3) may be understood to require that the category **Set** of sets and functions is cartesian closed, but two different proofs of the failure of cartesian closedness for **Set** in NF (also applicable to NFU) are provided in [14, 23].

Our formulation of (R3) is weaker in the sense that we only require *natural implementability*. We will show that Feferman's system S^* (as well as NFUA), solve this version of (R).

Shulman [25] does not provide an explicit list of requirements, but we extract the following from his exposition.

(S1) Ordinary category theory, along with its distinction between large and small, is naturally implementable.
(S2) The large/small-distinction is relative, in the sense that for any x, there is a notion of smallness such that x is small.
(S3) ZFC is interpreted by $\in$, both when quantifiers are restricted to large sets, and when quantifiers are restricted to small sets.

In order to avoid confusion, let us just clarify that Shulman does not directly advocate that a foundation of category ought to satisfy all of these conditions. Rather, in his concluding section he leaves that "to the reader's aesthetic and mathematical judgement" [25]. We have just extracted these conditions from the motivations Shulman gives for adopting one or another of the many ZFC-based foundations for category theory he considers. On our reading, Shulman's story-line for ZFC-based approaches culminates with systems aimed at fully satisfying (S).

2.2 The Distinction Between Large and Small

In ZFC, the familiar categories **Set**, **Top**, **Grp** can only be treated as proper classes. Still, it is perfectly possible to state and prove some theorems about a particular category, e.g. "for any two groups, there is a group, which is their product". However, if we want to prove that something holds for every category that satisfies some condition, say "for every category with binary products, we have that ...", and if we want this result to be applicable to **Set**, **Top**, **Grp**, ..., then ZFC is insufficient.

Several extensions of ZFC have been proposed for dealing with this situation, that are concerned with a distinction between large and small sets. In the standard reference on category theory by Saunders Mac Lane, "Categories for the Working Mathematician" [20], ZFC $+ \exists\kappa$ "κ is an inaccessible cardinal", is chosen as the foundational system. The sets of rank less than κ are considered small, and all sets are considered large. The set of small sets, V_κ, is then a model of ZFC. It is also common to say that V_κ is a Grothendieck universe, but we will not use that terminology in the present paper. The point is that any mathematical theory that can be developed within ZFC, can be developed exclusively by means of small sets. In ZFC $+ \exists\kappa$ "κ is an inaccessible cardinal", we can easily construct **Set**, **Top**, **Grp**, ... as the categories of all small sets, small groups, small topological spaces, ...(these categories will not be small however). In fact, we can conveniently state and prove many things about categories, as witnessed by Mac Lane's comprehensive book.

One might suppose that the category theory unfolding from this approach would only utilize the notion of smallness to obtain a formal foundation for stating and proving results. But, as Shulman observes [25], many results of category theory are actually concerned with this notion in a mathematically interesting way: For a simple example, we will prove in detail that (in such a ZFC-based approach) any small (large) category which has small (large) limits, is just a preorder. As a consequence, it is more interesting to study the situation of a large category having small limits, a condition satisfied by **Set**, **Top**, **Grp**. Important well-known examples making essential use of smallness are the Yoneda lemma and the adjoint functor theorems.

3 A Quick Dip into Category Theory

3.1 Limits

We have definitions of groups and group homomorphisms, topological spaces and continuous functions, and so on for many types of objects and morphisms in many areas of mathematics. One of the ways in which category theory is helpful, is that it proceeds abstractly from these basic data of objects and morphisms, to define standard notions like subobjects, quotients and products, for all categories at once. But category theory does not only provide guidelines for defining familiar notions in new categories. It also guides us when introducing new notions (beyond the familiar product and so on) in familiar categories. As an example, we will consider the category theoretic notion of limit, of which the notion of product is just one of the simplest instances.

Throughout the paper, we will occasionally use "$X \in \mathbf{C}$", as shorthand for "X is an object of the category $\mathbf{C}$". Suppose that $\mathbf{F} : \mathbf{J} \to \mathbf{C}$ is a functor. A *cone* to $\mathbf{F}$ is an object N of $\mathbf{C}$ together with a set of morphisms $\{n_X : N \to \mathbf{F}(X) \mid X \in \mathbf{J}\}$, such that for any $f : X \to_{\mathbf{J}} Y$, the following diagram commutes.

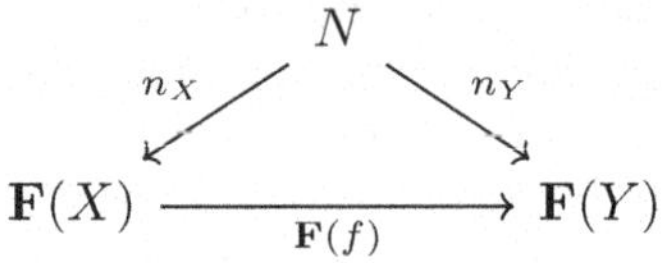

A *limit* to $\mathbf{F}$ is a universal cone, i.e. a cone $U, \{u_X\}$ to $\mathbf{F}$, such that for any cone $N, \{n_X\}$ to $\mathbf{F}$, there is a unique $v : N \to_{\mathbf{C}} U$, such that for any $f : X \to_{\mathbf{J}} Y$, the following diagram commutes.

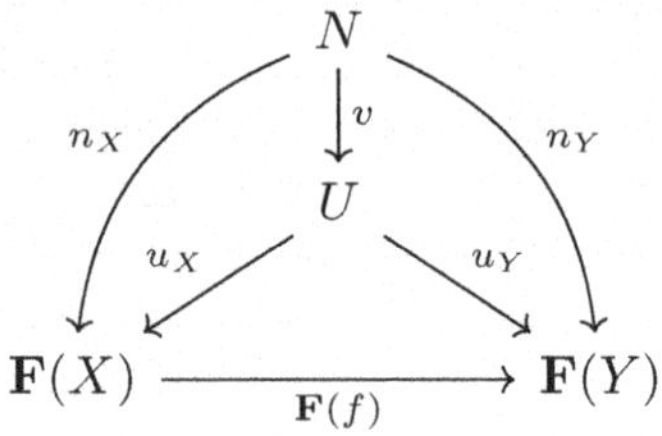

Since the morphism v (if it exists) is uniquely determined by the cone $N \ldots$ and the limit $U \ldots$, we call it the *universal morphism from the cone* $N \ldots$ *to the limit* $U \ldots$. Moreover, $\mathbf{J}$ is referred to as an *index* category, and $\mathbf{F} : \mathbf{J} \to \mathbf{C}$ is called a *diagram*. Of course, an index category can be any category, and a diagram can be any functor. We say that $\mathbf{C}$ has *limits indexed by* $\mathbf{J}$, if for each diagram $\mathbf{F} : \mathbf{J} \to \mathbf{C}$, it has a limit to $\mathbf{F}$.

Small index categories and diagrams are of particular importance. We say that $\mathbf{C}$ has *small limits*, if for any small index category $\mathbf{J}$ and any small diagram $\mathbf{F} : \mathbf{J} \to \mathbf{C}$, we have a limit in $\mathbf{C}$ to $\mathbf{F}$.

If the category $\mathbf{C}$ is a preorder, i.e. a category with at most one morphism between any two objects, and $a \geq b$ corresponding to an arrow $a \to b$, then a limit to $\mathbf{F}$ is the same as a least upper bound of the image of the object function of $\mathbf{F}$ (so the morphisms in $\mathbf{J}$ will be inconsequential). *Hence, the notion of the existence of limits for categories generalizes the notion of completeness for preorders.* We will see further below that with a ZFC-based foundation, every small category with small limits is a preorder.

However, many large non-preordered categories, such as **Set**, **Top**, **Grp**, have small limits.

We now show how a product is an example of a limit. Let **2** be the category consisting of only two objects, say 0 and 1, along with their identity morphisms. Suppose **C** is any category, and $A, B \in \mathbf{C}$. Then the map sending 0 to A and 1 to B determines a functor $\mathbf{F} : \mathbf{2} \to \mathbf{C}$. A *product* of A and B is (defined as) a limit to **F**. We may familiarly write this limit as $A \times B$, $\pi_0 : A \times B \to A$, $\pi_1 : A \times B \to B$, where π_0, π_1 are called the *projection* morphisms. The universality property now spells out, that for any object $N \in \mathbf{C}$ and any morphisms $f : N \to A$ and $g : N \to B$ (i.e. for any cone to **F**), there is a unique morphism $\langle f, g\rangle : N \to A \times B$, such that the following diagram commutes.

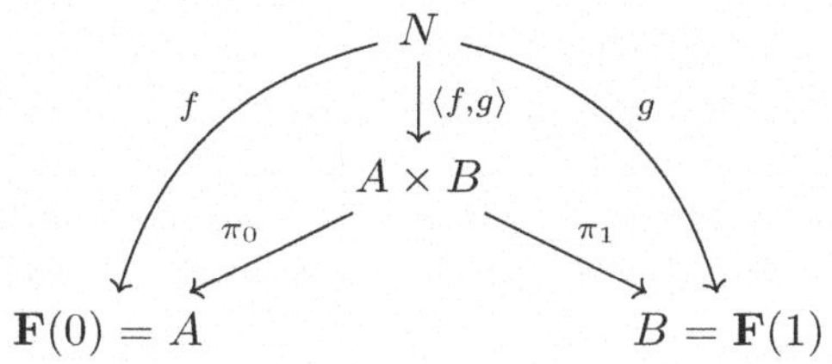

If $\mathbf{C} = \mathbf{Set}$, then the standard cartesian product of A, B together with the standard projection functions, is a limit to **F**. Moreover, we would have $\langle f, g\rangle(x) = \langle f(x), g(x)\rangle$, for each $x \in N$, explaining why the ordered pair notation $\langle f, g\rangle$ is commonly abused for the purpose of denoting that morphism.

Of course, such a limit may not exist. But if, for every functor $\mathbf{F} : \mathbf{2} \to \mathbf{C}$, a limit to **F** exists, then we say that **C** has *binary products*. In this sense, given the notion of limit, we may think of the index category **2** as providing all of the data for the notion of binary products. Similarly, a limit to a diagram from the index category with two objects and two morphisms $A \rightrightarrows B$ (in addition to the identity morphisms) is an *equalizer*; and a limit to a diagram from the index category with three objects and two morphisms $A \to C$, $B \to C$ (in addition to the identity morphisms) is a *pullback*. Thus, we may informally think of the notion of limit as defining a map from index categories to notions. The notions of product, equalizer and pullback would be three particular values of this map.

3.2 The Limits of a Proper Category (in ZFC)

Recall that the category **2** used to define binary products, only has identity morphisms. By considering index categories only having identity morphisms, more generally, we obtain stronger notions of product.

Theorem 1 *Assume that our set-theoretic foundation for category theory satisfies ZFC. Suppose that* **C** *is a category, and let* $\mathrm{Arr}(\mathbf{C})$ *be the category with the arrows* *of* **C** *as objects and only identity morphisms. If* **C** *has limits (i.e. products) indexed by* $\mathrm{Arr}(\mathbf{C})$*, then* **C** *is a preorder.*

Proof Assume towards a contradiction, that we have two different morphisms $f, g : A \to_{\mathbf{C}} B$. Let $P, \{p_X \mid X \in \mathrm{Arr}(\mathbf{C})\}$ be a limit (product) to the constant diagram mapping every object of $\mathrm{Arr}(\mathbf{C})$ to B. In suggestive notation, we may write $P = \Pi_{X \in \mathrm{Arr}(\mathbf{C})} B$. By the universality property of limits, we now obtain a function $\mathcal{U} : 2^{\mathrm{Arr}(\mathbf{C})} \to \mathbf{C}(A, P)$, defined by sending each function $S : \mathrm{Arr}(\mathbf{C}) \to 2$ to the universal morphism from the cone

$$A, \{a_X^S \mid X \in \mathrm{Arr}(\mathbf{C})\}, \text{ where } \forall X \in \mathrm{Arr}(\mathbf{C}).[(a_X^S = f \leftrightarrow S(X) = 0) \wedge (a_X^S = g \leftrightarrow S(X) = 1)],$$

to the limit $P, \{\pi_X \mid X \in \mathrm{Arr}(\mathbf{C})\}$.

Since $\mathbf{C}(A, P)$ is a subset of (the objects of) $\mathrm{Arr}(\mathbf{C})$, it suffices to prove that $\mathcal{U}$ is injective, for then Cantor's theorem yields a contradiction. So suppose that $S, S' : \mathrm{Arr}(\mathbf{C}) \to 2$ are different. Then there is $X \in \mathrm{Arr}(\mathbf{C})$, such that $S(X) \neq S'(X)$. Thus, we may assume that $a_X^S = f$ and $a_X^{S'} = g$. Since $P \ldots$ is a limit, the following diagram commutes.

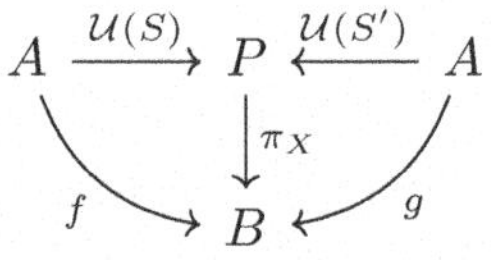

Finally, using the assumption $f \neq g$, we conclude that $\mathcal{U}(S) \neq \mathcal{U}(S')$. □

Corollary 2 *Assume that our set-theoretic foundation satisfies ZFC. Then every small category, which has small limits, is a preorder; and every large category, which has large limits, is a preorder.*

Proof A small category has a small set of morphisms, and a large category has a large set of morphisms. □

So in terms of large and small, the strongest notion of "having all limits", that may still be interesting for a category theory based on a foundation satisfying ZFC, is that of a large category having small limits. Therefore, some authors plainly call such a category *complete*. It turns out that many of the most familiar categories are complete, including **Set**, **Top**, **Grp**. The category of small metric spaces and the category of small fields are examples of large categories which are not complete.

4 Approaches to Satisfying (S)

This section is a brief summary of Sects. 8–11 in Shulman's exposition [25], where various ZFC-based approaches to the foundations of category theory are assessed and compared in detail.

4.1 Approaches Based on Inaccessible Cardinals

We have already mentioned the system ZFC $+\exists\kappa$ "κ is an inaccessible cardinal", which seems to work for most applications. We take Mac Lane's book [20] as sufficient evidence that it satisfies (S1). It fully satisfies (S3): Firstly, the set of small sets, V_κ, is a model of ZFC. Secondly, since every set is large, and ZFC is a subset of the theory, the large sets satisfy ZFC. This is an important difference from the system KMC (Kelley-Morse set theory with choice) – which has also been suggested as a foundation for category theory – where the classes (considered as large sets) do not satisfy ZFC.[3]

But ZFC $+\exists\kappa$ "κ is an inaccessible cardinal" imposes a rigid notion of smallness, fixed once and for all. Hence, it does not satisfy requirement (S2). The results of category theory involving the notion of smallness, do not depend on any fixed collection of objects. Rather, it is the relative smallness that is important. This is already evident in the proposal ZFC $+\exists\kappa$ "κ is an inaccessible cardinal": It is not important to choose, say the least inaccessible, as the demarcation of smallness, but only to pick some inaccessible.

Grothendieck proposed a stronger system, ZFC + "there are arbitrarily large inaccessible cardinals", to the following effect. For any set A, there is a notion of smallness such that A is small, and the set of small sets is a model of ZFC: Pick an inaccessible κ larger than the rank of A, and stipulate that a set is small if and only if it is an element of V_κ. Thus, this theory, which is known as Tarski-Grothendieck set theory (or just TG), satisfies (S2). It also satisfies (S1) and (S3) for the same reasons as given for ZFC $+\exists\kappa$ inaccessible above.

A worry with Mac Lane's approach, which is even more worrisome in Grothendiek's approach, is that we increase the consistency strength of the theory. The motivations for passing to these theories have had more to do with obtaining a useful notion of smallness, than with a pressing need from category theory for more consistency strength, so one might wonder if it is really necessary to introduce such large cardinal axioms. There are at least two concerns with increasing the consistency strength of the foundational system.

1. The risk that the theory is inconsistent increases.
2. The theory becomes inconsistent with other axioms that may turn out to be of interest to category theory.

As for the first concern, the large cardinal axioms we have looked at so far, are much weaker than other ones, which have also been quite thoroughly studied by set theorists. Hence, the risk that e.g. TG is inconsistent, is thought to be small.

---

[3] KM $+\ \Pi^1_\infty$-AC is bi-interpretable with $T :=$ ZFC $-$ Power Set $+\exists\kappa$ "κ is an inaccessible cardinal, and $\forall x\ |x| \leq \kappa$", where Π^1_∞-AC is the schema of Choice whose instances are of the form $\forall s\ \exists X\ \phi(s, X) \rightarrow \exists Y\ \forall s\ \phi(s, (Y)_s)$, where $\phi(s, X)$ is a formula of class theory with set variable s and class variable Y, and $(Y)_s$ is the "s-th slice of Y", i.e., $(Y)_s = \{x \mid \langle s, x\rangle \in Y\}$. This bi-interpretability was first noted by Mostowski; a modern account is given in a recent paper of Antos and Friedman [1, 2], where KM $+\ \Pi^1_\infty$-AC is referred to as MK*, and T is referred to as SetMK*.

As for the second concern, the authors are not aware of any principles that are both attractive to category theory and consistent with ZFC, but inconsistent with ZFC + a large cardinal axiom. However, we will of course consider the principle (R1) in this paper, which is attractive to category theory but inconsistent with ZFC.

4.2 Feferman's ZFC/S as a Solution to (S)

Now, even though we do not have strong reasons against including large cardinal axioms in a foundational system for category theory, a solution to (S) which is conservative over ZFC would yield insight on the large/small distinction, and is therefore of interest to the foundations of category theory. So how might we satisfy (S), while remaining at the consistency strength of ZFC? This is essentially the question which Feferman answers with his system ZFC/S, "ZFC with smallness". The idea of ZFC/S is to utilize the Reflection principle (provable in ZFC).

Theorem 3 (Reflection principle in ZFC) *If Φ is a finite set of formulas, and A is a set, then there is a $V_\alpha \supset A$, such that for any $\phi(x_0, \ldots, x_{n-1}) \in \Phi$, and any $a_0, \ldots, a_{n-1} \in V_\alpha$,*

$$\phi^{V_\alpha}(a_0, \ldots, a_{n-1}) \leftrightarrow \phi(a_0, \ldots, a_{n-1}).$$

V_α may be understood as the model obtained in ZFC by restricting $\in$ to the set V_α. If $\hat{\phi}$ is a name for ϕ in ZFC, then the formula $V_\alpha \models \hat{\phi}$, is defined recursively in the familiar Tarskian way of the semantics of first order logic. It naturally turns out that it is equivalent to the formula ϕ^{V_α}, which is obtained from ϕ by restricting each quantifier to V_α. We say that the model V_α *reflects* the formulas Φ. Hence, we implicitly assume that Φ consists of formulas in the language of set theory.

The system ZFC/S is defined as follows. Add an extra constant symbol $\mathbb{S}$ to the language of set theory. ZFC/S is the theory in this language consisting of the axioms of ZFC, the axiom saying that $\mathbb{S}$ is transitive and supertransitive, i.e. the axioms

$$x \in y \in \mathbb{S} \rightarrow x \in \mathbb{S}$$

$$x \subset y \in \mathbb{S} \rightarrow x \in \mathbb{S},$$

plus the axiom schema that, for each natural number n and each formula $\phi(x_0, \ldots, x_{n-1})$ in the language of set theory, adds the axiom

$$\forall x_0 \in \mathbb{S} \ldots \forall x_{n-1} \in \mathbb{S}.\big[\phi^{\mathbb{S}}(x_0, \ldots, x_{n-1}) \leftrightarrow \phi(x_0, \ldots, x_{n-1})\big].$$

Working in ZFC/S, we interpret "small" as "$\in \mathbb{S}$" (every set is considered large). By the Reflection schema, we may interpret ZFC by restricting quantifiers to $\mathbb{S}$, so (S3) is satisfied.

Theorem 4 *If* ZFC *is consistent, then* ZFC/S *is consistent.*

Proof Suppose that $\lightning_{\mathrm{ZFC/S}}$ is a proof of $\Psi \vdash \bot$, for some finite $\Psi \subset \mathrm{ZFC/S}$. Note that $\Psi = \Psi_{\mathrm{ZFC}} \cup \Psi_{\mathrm{Trans}} \cup \Psi_{\mathrm{Refl}}$, where $\Psi_{\mathrm{ZFC}} \subset \mathrm{ZFC}$, $\Psi_{\mathrm{Trans}} \subset$ {Transitivity, Supertransitivity} and $\Psi_{\mathrm{Refl}} \subset$ Reflection schema. Each $\psi \in \Psi_{\mathrm{refl}}$ is of the form $\forall x \in \mathbb{S}.\big(\phi^{\mathbb{S}}(\vec{x}) \leftrightarrow \phi(\vec{x})\big)$. By reflection,

$$\mathrm{ZFC} \vdash \exists V_\alpha. \bigwedge_{\psi \in \Psi_{\mathrm{Refl}}} \psi[\mathbb{S} := V_\alpha]. \tag{†}$$

Pick a witness V_α of (†). Then $\mathrm{ZFC} \vdash \Psi_{\mathrm{Refl}}[\mathbb{S} := V_\alpha]$ is just (†). $\mathrm{ZFC} \vdash \Psi_{\mathrm{ZFC}}[\mathbb{S} := V_\alpha]$ is trivial, and $\mathrm{ZFC} \vdash \Psi_{\mathrm{Trans}}[\mathbb{S} := V_\alpha]$ is basic. So there is a proof p of $\mathrm{ZFC} \vdash \Psi[\mathbb{S} := V_\alpha]$. Concatenating p with $\lightning_{\mathrm{ZFC/S}}[\mathbb{S} := V_\alpha]$ yields a proof $\lightning_{\mathrm{ZFC}}$ of $\mathrm{ZFC} \vdash \bot$. □

The technique used in the proof can be applied quite generally: Suppose we are building a proof from ZFC, in which we want to apply a finite set of theorems T of ZFC/S, and in which we would like to consider all the elements of some set A to be small. As in the proof of Con(ZFC) → Con(ZFC/S), the Reflection principle gives us

$$\mathrm{ZFC} \vdash \exists V_\alpha \supset A. \bigwedge_{\tau \in T} \tau[\mathbb{S} := V_\alpha]. \tag{‡}$$

Thus, given any finite set of theorems T of ZFC/S and any set A, all of whose elements we want to consider small, we can choose $V_\alpha \supset A$ and translate each theorem $\tau \in T$ of ZFC/S into the theorem $\tau[\mathbb{S} := V_\alpha]$ of ZFC. Then we can apply the translated versions in a proof from ZFC.

On the other hand, suppose T is an infinite theorem schema in ZFC/S. By a generalized version of the Reflection principle, the same technique works if T is Σ_n (in Levy's hierarchy of set-theoretic formulae) for some finite n, but it *does not work* for an arbitrary T.

We conclude that ZFC/S gives us a relative large/small-distinction, so that (S2) is satisfied, albeit with the limitation on T stated above.

What about (S1)? Can standard results of category theory be obtained in ZFC/S? The evidence 'on the ground' seems to indicate a positive answer; e.g., Andreas Blass has informed the authors that he used ZFC/S as a foundational platform in a course he taught on category theory at the University of Michigan, and that everything worked smoothly, except possibly for Kan extensions. The main issue with using ZFC/S in practice, seems to be that we often need to ensure that categories, functors, etc. are "small-definable", i.e. that they can be defined by a formula of ZFC/S all of whose parameters are elements of $\mathbb{S}$. For example, consider restricting a properly large locally small category **C** to a small subset A of the objects of **C**. If **C** is not small-definable, then the replacement axiom on $\mathbb{S}$ (obtained from the corresponding instance of the Reflection schema) cannot be applied. Therefore, we cannot conclude that the restriction of **C** to A is small. But categories, functors, etc. that are definable as classes over ZFC, obviously have direct counterparts that are small-definable in ZFC/S. Therefore, in applications this condition is almost always satisfied.

We have now justified the claim that ZFC/S is a conservative extension of ZFC satisfying (S), albeit with minor limitations.

5 Approaches to Satisfying (R)

5.1 A Very Short Introduction to Stratified Set Theory

(R1) asks for the category of all sets, the category of all groups, the category of all categories, etc. A natural place to look for a solution to (R), is therefore set theories proving the existence of a universal set. An approach to a set theory with this property is the simple theory of types, developed by Bertrand Russell and Alfred North Whitehead, and simplified by Leon Chwistek and Frank Ramsey, independently. In the usual form, called TST, the theory has types indexed by $\mathbb{N}$. There are countably infinitely many variables of each type, and an element- and equality-symbol for each type. A formula is well-formed only if each atomic subformula is of the form $x^i =^i y^i$ or $x^i \in^i y^{i+1}$, where $i \in \mathbb{N}$ denotes a type. For each type the theory has a corresponding axiom of extensionality and an axiom schema of comprehension. Russell's paradox is avoided because the formula '$x \in y$' becomes well-formed only if x is assigned one type lower than y, thus banishing any formula of the form '$x \in x$' from the language. But '$x^i =^i x^i$' is well-formed, so by comprehension there is a universal set of type $i + 1$ for each type $i \in \mathbb{N}$.

It is easily seen that $\text{TST} \vdash \phi \leftrightarrow \text{TST} \vdash \phi^+$, where ϕ^+ is obtained by raising all the type indices in ϕ by 1: A proof of ϕ can be turned into a proof of ϕ^+ simply by raising all the type indices in the proof by 1, and vice versa. From reflection on this fact, Quine was led to suggest the theory NF [24], which is obtained from TST simply by forgetting about the types. The result is an untyped first-order theory with the axiom of extensionality and the axiom schema of stratified comprehension. A formula is stratified if its variables can be assigned types so as to become a well-formed formula of TST. The stratified comprehension schema only justifies formation of sets which are the extensions of stratified formulas. But the well-formed formulas of NF are as usual in set theory. For example, in contrast to TST, the statement $\text{NF} \vdash \exists x . x \in x$ makes sense, and is in fact witnessed true by the universal set V obtained from the instance $\exists y . \forall x . (x \in y \leftrightarrow x = x)$ of the stratified comprehension axiom schema.

It has turned out to be very difficult to prove the consistency of NF relative to a set theory in the Zermelo-Fraenkel tradition.[4] However, if the extensionality axiom is weakened to allow for atoms, then the system NFU is obtained. This system was

[4]Two claimed proofs have fairly recently been announced: first by Randall Holmes [18], and then by Murdoch Gabbay [15].

proved consistent relative to Mac Lane set theory[5] in 1969 by Ronald Jensen [19]. Jensen's proof combines two techniques due to Ernst Specker and Frank Ramsey respectively. Specker had showed essentially that NF is equiconsistent with TST plus an automorphism between the types, and Jensen used Ramsey's theorem to obtain the automorphism at the cost of weakened extensionality. Jensen's proof also gives the consistency of NFU plus the axiom of choice and the axiom of infinity. This system is equiconsistent with NFU + the axiom of choice + the axiom of type-level pair. It is standard in the literature to refer to this latter system simply as NFU, and we do the same in this paper. A thorough introduction to NFU is found here [17].

The move to allow for atoms in NFU can be accomplished either by having extensionality only for non-empty sets, or by introducing a predicate of sethood and restricting extensionality to sets. We opt for the former alternative. It is then helpful to have a designated empty set; i.e. we pick an atom once and for all, and denote it by $\varnothing$. (The existence of atoms, now defined simply as sets without elements, is provided by applying stratified comprehension to the formula $x \neq x$.)

5.2 *Feferman's S* as a Solution to (R)*

Jensen's proof of the consistency of NFU appeared in 1969; it did not take long for Feferman to use NFU, in the early 1970's, as the source of a viable solution to (R). Indeed, it is easy to see that NFU satisfies (R1) and (R2). A detailed verification is found in [7, 3.4]. But since NFU is equiconsistent with Mac Lane set theory, which is proved consistent in ZFC, it follows from Gödel's second incompleteness theorem that NFU does not interpret ZFC. Let us now consider (R3). If ZFC is taken as the standard for "ordinary mathematics", then it is clear that NFU does not meet that standard. However, the equiconsistency of NFU with Mac Lane set theory, which arguably suffices for ordinary mathematics and category theory, suggests a path for advocating NFU as a foundation for category theory.

Feferman suggests the system S^*, which extends NFU with a constant symbol $\bar{V}$, and axioms ensuring that $\langle \bar{V}, \in \rangle$ is a model of ZFC (actually with stronger replacement and foundation schemata than required for that) [7, 11].

The language of S^*, denoted $\mathcal{L}^*$, is the two-sorted extension of the language of set theory with set variables $x, y, z, \ldots$, class variables $X, Y, Z, \ldots$, a constant symbol $\bar{V}$ and a binary function symbol P. As we alluded to above, the function symbol P, which will act as a type-level pairing function, is inherited from the language of NFU, and the constant symbol $\bar{V}$ will be used to obtain a notion of smallness. The terms of $\mathcal{L}^*$ are generated from $\bar{V}$, and both set and class variables using the function symbol P. The atomic formulae of $\mathcal{L}^*$ are all of the formulae in the form $s \in t$ and $s = t$ where s and t are $\mathcal{L}^*$-terms. In order to state the axioms of S^* we first need to extend the notion of stratification to $\mathcal{L}^*$-formulae.

[5]The axioms of Mac Lane set theory are specified in Sect. 6.

Definition 5 An $\mathcal{L}^*$-formula ϕ is said to be stratified if the following conditions can be satisfied:

1. Each term s in ϕ can be assigned a natural number that we will call the type assigned to s;
2. the type assigned to a term s is the same as the type assigned to every class variable occurring in s;
3. each class variable of ϕ has the same type assigned to all of its occurrences;
4. for each subformula of ϕ of the form $s = t$, the type assigned to the term s is the same as the type assigned to the term t;
5. for each subformula of ϕ of the form $s \in t$, if s is assigned the type n then t is assigned the type $n + 1$.

The theory S^* is the $\mathcal{L}^*$-theory that is axiomatised by the universal closures of the following:

1. (Stratified Comprehension) for all stratified $\mathcal{L}^*$-formulae $\phi(X, \vec{Z})$,
$$\exists Y \forall X (X \in Y \iff \phi(X, \vec{Z}))$$
2. (Weak Extensionality)
$$\mathcal{S}(X) \wedge \mathcal{S}(Y) \Rightarrow (X = Y \iff \forall Z (Z \in X \iff Z \in Y))$$
3. (Pairing) $P(X_1, X_2) = P(Y_1, Y_2) \Rightarrow X_1 = Y_1 \wedge X_2 = Y_2$
4. (Sets and Classes)
   (a) $\forall x \exists X (x = X)$
   (b) $X \in \bar{V} \iff \exists x (x = X)$
   (c) $X \in x \Rightarrow X \in \bar{V}$
5. (Empty Set) $\exists! z \forall y (y \notin z)$ — we use $\varnothing$ to denote the unique $z \in \bar{V}$ such that $\forall y (y \notin z)$
6. (Operations on Sets)
   (a) $\{x, y\} \in \bar{V}$
   (b) $\bigcup x \in \bar{V}$
   (c) $\mathcal{P}(x) \in \bar{V}$
   (d) $P(x, y) = \{\{x\}, \{x, y\}\}$
7. (Infinity)
$$\exists a (\exists z (z \in a \wedge \forall y (y \notin z)) \wedge \forall x (x \in a \Rightarrow x \cup \{x\} \in a))$$
8. (Replacement) for all $\mathcal{L}^*$-formulae $\phi(x, y, \vec{Z})$,
$$\forall x \forall y_1 \forall y_2 (\phi(x, y_1, \vec{Z}) \wedge \phi(x, y_2, \vec{Z}) \\ \Rightarrow y_1 = y_2) \Rightarrow \forall a \exists b \forall y (y \in b \iff \exists x (x \in a \wedge \phi(x, y, \vec{Z})))$$

9. (Foundation) for all $\mathcal{L}^*$-formulae $\phi(x, \vec{Z})$,

$$\exists x \phi(x, \vec{Z}) \Rightarrow \exists x(\phi(x, \vec{Z}) \wedge (\forall y \in x) \neg \phi(y, \vec{Z}))$$

10. (Universal Choice)

$$\exists C \begin{pmatrix} \forall X \forall Y_1 \forall Y_2 (P(X, Y_1) \in C \wedge P(X, Y_2) \in C \Rightarrow Y_1 = Y_2) \wedge \\ \forall X (\exists Y (Y \in X) \Rightarrow \exists Y (Y \in X \wedge P(X, \{Y\}) \in C)) \end{pmatrix}$$

Since S^* has a two-sorted language with set- and class-variables, we will call arbitrary objects *classes*, and we will call elements of $\bar{V}$ *sets*. This contrasts with the practice in ZFC, where one talks of proper classes, but cannot prove their existence as objects. In S^*, just as in NBG or KMC, classes exist as objects. It follows immediately from the definition of S^*, as an extension of NFU, that it satisfies (R1) and (R2). Feferman proves that S^* is consistent from ZFC $+ \exists\kappa\exists\lambda$ "$\kappa < \lambda$ are inaccessible cardinals" [7] using a technique from [19] based on the Erdős–Rado theorem. In this paper, the combinatorics of Feferman's proof is adjusted so as to prove the consistency of S^* already from ZFC $+ \exists\kappa$ "κ is an inaccessible cardinal".

S^* satisfies (R3) in a stronger sense than NFU does, and such considerations motivate its strong replacement and foundation schemata. Let us work in S^*, interpreting "small" as "$\in \bar{V}$" and (provisionally) "large" as "anything". Let $\mathbf{V}$ denote the category of all sets with functions as morphisms. In [7], Feferman proves a version of the Yoneda lemma from S^*:

Lemma 6 (Yoneda in S^*) *If all the objects and morphisms of* $\mathbf{C}$ *are small, and* $\mathbf{F}$ *is a functor from* $\mathbf{C}$ *to* $\mathbf{V}$*, then for each object* A *of* $\mathbf{C}$*, there is a bijection* $\mathrm{yon}_A : \mathbf{F}(A) \to \mathbf{Nat}(\mathbf{C}(A, -), \mathbf{F})$.

$\mathbf{C}(A, -)$ denotes the covariant hom functor from $\mathbf{C}$ to $\mathbf{V}$ given by A, and $\mathbf{Nat}(\mathbf{C}(A, -), \mathbf{F})$ denotes the set of natural transformations from $\mathbf{C}(A, -)$ to $\mathbf{F}$. This differs in two ways from the standard Yoneda lemma. Firstly, we only require that the objects and morphisms are small, not the stronger condition that the category is locally small. Secondly, the co-domain of $\mathbf{F}$ is the category $\mathbf{V}$, not the category of small sets. But of course, local smallness of $\mathbf{C}$ is exactly what is needed for the hom functors to be small-valued, so this readily specializes to the standard statement.

This is a first step towards satisfying the category theory part of (R3), but it is a daunting task to verify all the standard results of category theory involving the large/small-distinction. We will instead indicate why we feel confident that almost all such results would go through. In the family of systems including some form of the stratified comprehension schema, the following notions turn out to be quite important.

Definition 7 A is *cantorian* if there is a bijection from A to $\{\{a\} \mid a \in A\}$. A is *strongly cantorian* if there is a bijection from A to $\{\{a\} \mid a \in A\}$ that maps each $a \in A$ to $\{a\}$. We adopt $\mathrm{stcan}(A)$ as shorthand for "A is strongly cantorian".

The following basic results are easily verified. S^* proves that $\bar{V}$ is strongly cantorian. Moreover, NFU (and therefore S^*) proves that the powerset of a strongly cantorian set is strongly cantorian. NFU also proves that if B is a set of bijections, each witnessing strong cantorianicity of its domain, then $\cup B$ is a bijection witnessing strong cantorianicity of its domain.

Working with the von Neumann ordinals $\mathrm{Ord}^{\bar{V}}$ in $\bar{V}$, and relying on the strong foundation axiom schema, we now wish to recursively define $\bar{V}_0 = \bar{V}$, $\bar{V}_{\alpha+1} = \mathcal{P}(\bar{V}_\alpha)$ and (if alpha is a limit ordinal) $\bar{V}_\alpha = \cup\{\bar{V}_\beta \mid \beta < \alpha\}$. Our intention is also to show by induction that $\bar{V}_\alpha$ is strongly cantorian, for any $\alpha \in \mathrm{Ord}^{\bar{V}}$. As we shall see, that will lend support to the claim that S^* satisfies (R3) sufficiently. We will need the recursion theorem for NFU, which may be found here [17, 15.1]. The foundation axiom schema of S^* implies that $\mathrm{Ord}^{\bar{V}}$ forms a well-founded class. So we obtain the following special case of the recursion theorem for recursion over $\mathrm{Ord}^{\bar{V}}$.

Theorem 8 (Recursion in NFU and S^*) *Suppose that*

$$\mathcal{G} : \{H : \alpha \to V \mid \alpha \in \mathrm{Ord}^{\bar{V}}\} \to \{\{x\} \mid x \in V\}.$$

Then there is a unique function $F : \mathrm{Ord}^{\bar{V}} \to V$, *such that* $\{F(\alpha)\} = \mathcal{G}(F \restriction \alpha)$, *for each* $\alpha \in \mathrm{Ord}^{\bar{V}}$.

We now wish to apply this to obtain a hierarchy, recursively defined by $\bar{V}_0 = \bar{V}$, $\bar{V}_{\alpha+1} = \mathcal{P}(\bar{V})$, and $\bar{V}_\alpha = \cup\{\bar{V}_\beta \mid \beta < \alpha\}$ (for α a limit). Note that in NFU, we define $\mathcal{P}$ so that only the designated empty class, and no other atoms, are elements of a powerclass. The main obstacle to defining the required $\mathcal{G}$ of the recursion theorem, is that $\mathcal{P}$ is not a function in NFU.

Theorem 9 *In* S^*, *there are unique functions* $R, C : \mathrm{Ord}^{\bar{V}} \to V$, *such that for all* $\alpha \in \mathrm{Ord}^{\bar{V}}$

$$R(\alpha) = \begin{cases} \bar{V} & \text{if } \alpha = 0 \\ \mathcal{P}(R(\beta)) & \text{if } \alpha \text{ is a successor } \beta + 1 \\ \cup\{R(\beta) \mid \beta < \alpha\} & \text{if } \alpha \text{ is a limit} \end{cases}$$
$$C(\alpha) = \text{A bijection } R(\alpha) \to \{\{x\} \mid x \in R(\alpha)\} \text{ witnessing } \mathrm{stcan}(R(\alpha)). \tag{†}$$

Proof We need to restate (†) in a stratified form. First note that stratified comprehension allows us to form the function

$$\mathcal{P}' : \{X \mid \forall Y \in X. \exists Z. Y = \{Z\}\} \to V,$$

defined by $\mathcal{P}'(X) = \mathcal{P}(\cup X)$, which is stratified. Let $\bar{B}$ be the bijection from $\bar{V}$ to $\{\{X\} \mid X \in \bar{V}\}$, such that $\forall X \in \bar{V}. \bar{B}(X) = \{X\}$. We can now reformulate (†) as follows.

$$R(\alpha) = \begin{cases} \bar{V} & \text{if } \alpha = 0 \\ \mathcal{P}'\big(\{C(\beta)(X) \mid X \in R(\beta)\}\big) & \text{if } \alpha \text{ is a successor } \beta + 1 \\ \cup\{R(\beta) \mid \beta < \alpha\} & \text{if } \alpha \text{ is a limit} \end{cases}$$

$$C(\alpha) = \begin{cases} \bar{B} & \text{if } \alpha = 0 \\ \big\{\langle S, \{\cup T\}\rangle \mid S \subset R(\beta) \wedge T = \{C(\beta)(X) \mid X \in S\}\big\} & \text{if } \alpha \text{ is a successor } \beta + 1 \\ \cup\{C(\beta) \mid \beta < \alpha\} & \text{if } \alpha \text{ is a limit} \end{cases}$$

For applying the recursion theorem, we wish to show the existence of the function F defined on $\text{Ord}^{\bar{V}}$ by $\alpha \mapsto \langle R(\alpha), C(\alpha)\rangle$. The definition of the needed $\mathcal{G}$ is implicit from the expression above. $\mathcal{G}$ is obtained by comprehension if the expression is stratified, which is easily verified with this type-assignment:

$$\begin{array}{ll} R, C, \mathcal{P}' & \mapsto 2 \\ R(\alpha), R(\beta), C(\alpha), C(\beta), S, T, \bar{V} & \mapsto 1 \\ X, C(\beta)(X) & \mapsto 0 \end{array}$$

Therefore, unique functions R and C exist, satisfying the reformulated version of (†). But it remains to show that our reformulated version of (†) is equivalent to (†). I.e. it remains to clarify that $R(\alpha + 1) = \mathcal{P}(R(\alpha))$, that $C(\alpha)(X) = \{X\}$ and that $\text{dom}(C(\alpha)) = R(\alpha)$, for all $\alpha \in \text{Ord}^{\bar{V}}$ and all $X \in \text{dom}(C(\alpha))$. We do this through an induction argument, justified by the strong foundation axiom schema of S^*. So let $\alpha \in \text{Ord}^{\bar{V}}$ and let $X \in \text{dom}(C(\alpha))$.

Successor case, $\alpha = \beta + 1$: Assume that $C(\beta)(Y) = \{Y\}$, for all $Y \in \text{dom}(C(\beta))$, and assume that $\text{dom}(C(\beta)) = R(\beta)$. So by definition of C, we have that $X \subset R(\beta)$, and

$$C(\alpha)(X) = \big\{\cup\{C(\beta)(Y) \mid Y \in X\}\big\} = \big\{\cup\{\{Y\} \mid Y \in X\}\big\} = \{X\}.$$

Moreover, by definition of R,

$$R(\alpha) = \mathcal{P}'\big(\{C(\beta)(Y) \mid Y \in R(\beta)\}\big) = \mathcal{P}(\cup\{\{Y\} \mid Y \in R(\beta)\}) = \mathcal{P}(R(\beta)),$$

and it follows that $\text{dom}(C(\alpha)) = R(\alpha)$.

Limit case: Assume that $C(\beta)(Y) = \{Y\}$, and that $\text{dom}(C(\beta)) = R(\beta)$, for all $\beta < \alpha$ and $Y \in \text{dom}(C(\beta))$. By definition of C and R, we have

$$\text{dom}(C(\alpha)) = \cup\{\text{dom}(C(\beta)) \mid \beta < \alpha\} = \cup\{R(\beta) \mid \beta < \alpha\} = R(\alpha),$$

where the sets $\{\text{dom}(C(\beta)) \mid \beta < \alpha\}$ and $\{R(\beta) \mid \beta < \alpha\}$ are obtained from stratified comprehension and the existence of R and C. Therefore,

$$\langle X, Z\rangle \in C(\alpha) \Leftrightarrow \exists \beta < \alpha . C(\beta)(X) = Z \Leftrightarrow Z = \{X\}.$$

So C and R turn out to be the desired functions. □

Thanks to this result, we have the class $\bar{V}_\alpha := R(\alpha)$, for each $\alpha \in \mathrm{Ord}^{\bar{V}}$. By induction, these are all transitive (and supertransitive) classes, so it is easily seen that they are hereditarily strongly cantorian. Since we obtained a function $\alpha \mapsto \bar{V}_\alpha$ on $\mathrm{Ord}^{\bar{V}}$, we can take the union $\bar{\bar{V}} := \bigcup\{\bar{V}_\alpha \mid \alpha \in \mathrm{Ord}^{\bar{V}}\}$. We can also take the union $\bar{\bar{B}} := \bigcup\{C(\alpha) \mid \alpha \in \mathrm{Ord}^{\bar{V}}\}$, to obtain a bijection witnessing that $\bar{\bar{V}}$ is strongly cantorian.

Theorem 10 *In* S^*, $\bar{\bar{V}}$ *is a model of*

$$\begin{aligned}&\mathrm{ZC} + \text{“}\mathrm{Ord}^{\bar{V}} \textit{ is an inaccessible cardinal”}\\ &\quad + \text{“}\textit{for every ordinal } \alpha, V_\alpha \textit{ exists”}\\ &\quad + \bar{V} = V_{\mathrm{Ord}^{\bar{V}}}\\ &\quad + \text{“}\textit{replacement schema for domains of cardinality in } \mathrm{Ord}^{\bar{V}}\text{”}.\end{aligned}$$

Proof [6] $\mathrm{Ord}^{\bar{V}}$ is an inaccessible cardinal: It is clearly transitive and totally ordered by $\in$, and by foundation well-ordered by $\in$. Hence, it is an ordinal. Suppose towards a contradiction that it were not a cardinal. Then there would exist a bijection $f \in \bar{V}$ from $\kappa \in \mathrm{Ord}^{\bar{V}}$ to $\mathrm{Ord}^{\bar{V}}$. So by replacement, we would have $\mathrm{Ord}^{\bar{V}} \in \bar{V}$, which contradicts foundation. So $\mathrm{Ord}^{\bar{V}}$ is a cardinal. By the same argument (augmented with the powerset axiom), 2^κ does not stand in surjection to $\mathrm{Ord}^{\bar{V}}$, for any $\kappa \in \mathrm{Ord}^{\bar{V}}$. Lastly, suppose that there is $\kappa \in \mathrm{Ord}^{\bar{V}}$ and a function f from κ to $\mathrm{Ord}^{\bar{V}}$ that is unbounded. By replacement, its image I is then an element of $\bar{V}$, and $\bigcap I = \mathrm{Ord}^{\bar{V}} \in \bar{V}$, a contradiction. We conclude that $\mathrm{Ord}^{\bar{V}}$ is an inaccessible cardinal.

Existence of V_α, for ordinals α, and $\bar{V} = V_{\mathrm{Ord}^{\bar{V}}}$: Since S^* interprets ZFC by restricting quantifiers to $\bar{V}$, the analogous interpretation is satisfied in $\bar{\bar{V}}$. So in $\bar{\bar{V}}$, by transitivity of $\bar{V}$, V_α exists and is an element and a subset of $\bar{V}$, for each $\alpha < \mathrm{Ord}^{\bar{V}}$. Conversely, for each $x \in \bar{V}$, there is $\alpha < \mathrm{Ord}^{\bar{V}}$ such that $x \in V_\alpha$. Therefore, $V = \bigcup_{\alpha < \mathrm{Ord}^{\bar{V}}} V_\alpha = V_{\mathrm{Ord}^{\bar{V}}}$. Hence, recalling the construction of $\bar{\bar{V}}$, from S^*, $\bar{\bar{V}}$ is a union of V_α:s, for $\mathrm{Ord}_{\bar{V}} \leq \alpha < \mathrm{Ord}_{\bar{V}} + \mathrm{Ord}_{\bar{V}}$. It is easily seen that these V_α relativize appropriately to $\bar{\bar{V}}$, in the sense that $S^* \vdash$ “$\alpha \in \bar{\bar{V}}$ is an ordinal” $\leftrightarrow$ (“α is an ordinal”)$^{\bar{\bar{V}}}$, and $S^* \vdash$ (“$\alpha \in \bar{\bar{V}}$ is an ordinal and $X = V_\alpha$”) $\rightarrow (X = V_\alpha)^{\bar{\bar{V}}}$. So by Theorem 9, we have in $\bar{\bar{V}}$ that V_α exists for each ordinal.

[6]The proof given, for the separation and the weakened replacement schemata, quantifies over formulas in the meta-theory. So what is proved is actually a theorem schema about $\bar{\bar{V}}$, where $\bar{\bar{V}}$ is considered externally as a submodel of a model of S^*. However, Roland Hinnion's development of the semantics of first-order logic in [16] applies to NFU, and provides a means to internalize this proof to S^*.

Extensionality: None of the sets in $\bar{\bar{V}}$ contains any atom other than $\varnothing$.
Infinity: $\omega \in \bar{V}$.
Let $A, B \in \bar{\bar{V}}$. Then we may fix $\alpha + 2 \in \mathrm{Ord}^{\bar{V}}$, such that $A, B \in \bar{V}_{\alpha+2}$.
Union: Since $\bar{V}_{\alpha+2} = \mathcal{P}(\mathcal{P}(\bar{V}_{\alpha}))$, we have $\cup A \subset \bar{V}_{\alpha}$, so $\cup A \in \bar{V}_{\alpha+1}$.
Pair: $\{A, B\} \in \bar{V}_{\alpha+3}$.
Powerset: Since $A \subset \bar{V}_{\alpha+1}$, we have $\mathcal{P}(A) \subset \mathcal{P}(\bar{V}_{\alpha+1})$, so $\mathcal{P}(A) \in \bar{V}_{\alpha+3}$.
Choice: Follows from global choice.

Foundation: Let β be the least ordinal in $\mathrm{Ord}^{\bar{V}}$, such that there is $C \in \bar{V}_{\beta} \cap A$. For each $U \in C$, there is $\gamma < \beta$, such that $U \in \bar{V}_{\gamma}$. Hence, for each $U \in C$, we have that $U \notin A$.

Separation schema: Let $\phi(X, Y)$ be a formula. We need to show that $\{X \in A \mid \phi^{\bar{\bar{V}}}(X, B)\}$ exists in $\bar{\bar{V}}$. Since $\bar{\bar{V}}$ and B are strongly cantorian, the formula $X \in A \wedge \phi^{\bar{\bar{V}}}(X, B)$ is equivalent to a stratified formula. (This is done by using the bijections witnessing that the sets are strongly cantorian, to shift the types in the formula. For more details, see the Subversion theorem in [17, 17.5].) Hence, by stratified comprehension, we have that $\{X \in A \mid \phi^{\bar{\bar{V}}}(X, B)\}$ exists. Since it is a subset of A, it is a subset of $\bar{V}_{\alpha+1}$, and therefore an element of $\bar{V}_{\alpha+2}$.

Replacement schema for sets of cardinality in $\mathrm{Ord}^{\bar{V}}$: Suppose $|A| = \kappa \in \mathrm{Ord}^{\bar{V}}$. It suffices to show that if $\phi(\xi, Z, B)$ is a formula such that $[\forall \xi \in \kappa . \exists^! Z \phi(\xi, Z, B)]^{\bar{\bar{V}}}$, then there is $C \in \bar{\bar{V}}$ such that $[\forall Z(Z \in C \leftrightarrow \exists \xi \in \kappa . \phi(\xi, Z, B))]^{\bar{\bar{V}}}$. Consider the formula $\psi(\xi, \zeta, B)$ defined as "ζ is the least ordinal in $\mathrm{Ord}^{\bar{V}}$ such that there exists $Z \in \bar{V}_{\zeta}$ for which $\phi^{\bar{\bar{V}}}(\xi, Z, B)$". By the strong replacement axiom schema of S^*, the image of κ under ψ exists in $\bar{V}$. Let λ be the least ordinal in $\mathrm{Ord}^{\bar{V}}$, which is not in that image. Then we have $[\forall Z(Z \in \bar{V}_{\lambda} \leftarrow \exists \xi \in \kappa . \phi(\xi, Z, B))]^{\bar{\bar{V}}}$. Now, by separation in $\bar{\bar{V}}$, we obtain $C := \{Z \in \bar{V}_{\lambda} \mid [\exists \xi \in \kappa . \phi(\xi, Z, B)]^{\bar{\bar{V}}}\}$, as desired. □

We also extract the following general fact about ZC.

Proposition 11 *ZC $\vdash$ Replacement schema for functional formulas with bounded image.*

Proof Suppose $\alpha \in \mathrm{Ord}$ and $\phi(X, Z, B)$ is a formula such that $\forall X \in A . \exists^! Z \phi(X, Z, B)$ and $\forall Z(Z \in V_{\alpha} \leftarrow \exists X \in A . \phi(X, Z, B))$. Then, by separation, we obtain $C := \{Z \in V_{\alpha} \mid \exists X \in A . \phi(X, Z, B)\}$, as desired. □

We now suggest to interpret "small" as "$\in \bar{V}$" and to re-interpret "large" as "$\in \bar{\bar{V}}$". We then have quite a workable set theory for the large categories. The kind of operations on large categories, that would require more than what we have in $\bar{\bar{V}}$, are unusual in ordinary mathematics and category theory. For example, we cannot form the union $\cup\{\bar{V}_{\alpha} \mid \alpha \in \mathrm{Ord}^{\bar{V}}\}$ in the model $\bar{\bar{V}}$. On the other hand, if for some unusual application we would need to form that union, then we can make use of the fact that $\bar{\bar{V}}$ is strongly cantorian, and apply the iterated powerclass argument above to obtain

$\bar{\bar{\bar{V}}}_0, \bar{\bar{\bar{V}}}_1, \ldots, \bar{\bar{\bar{V}}}_\alpha, \ldots$, as well as their limit, the model $\bar{\bar{\bar{\bar{V}}}}$, in which $\cup\{\bar{V}_\alpha \mid \alpha \in \mathrm{Ord}^{\bar{V}}\}$ is obtained. Of course, we can also iterate that argument. We conclude that S^* is a level (2) solution to (R3), and thereby to all of (R).

We end this section with two remarks. Firstly, so far we only used the strong replacement axiom schema of S^* in the proof of Theorem 10, to show that $\bar{\bar{\bar{V}}}$ is a model of "the replacement schema for sets of cardinality in $\mathrm{Ord}^{\bar{V}}$". Another benefit of the strong replacement schema for the foundations of category theory, is that the small-definability concerns encountered in ZFC/S disappear: Working in S^*, strong replacement immediately yields the following: If f is a function and $A \in \bar{V}$ is a subset of $\mathrm{dom}(f)$, such that $f(A) \subset \bar{V}$, then $f(A) \in \bar{V}$. So for example, if $\mathbf{C}$ is a category, A is a small subset of its set of objects, and for all $X, Y \in A$, the set $\mathbf{C}(X, Y)$ of all $\mathbf{C}$-morphisms from X to Y is small, then the natural restriction of $\mathbf{C}$ to A is small. Without strong replacement, this argument requires the additional small-definability assumption that the natural restriction of $\mathbf{C}$ to $\bar{V}$ is definable over $\bar{V}$, which is not necessarily the case.

Secondly, on this interpretation of small and large, the categories enabling us to satisfy (R1) and (R2), are neither large nor small. The notions of small and large have come to be associated with various results of category theory, that do not necessarily hold for those categories. We will look at one example of this phenomenon shortly, but first let us consider the theory NFUA as a foundation of category theory.

5.3 NFUA as a Solution to both (R) and (S)

NFUA is defined as NFU + "every cantorian set is strongly cantorian". NFUA satisfies (R1) and (R2) simply because NFU does. In an unpublished proof, Robert Solovay showed in 1995 that NFUA is equiconsistent with ZFC + the schema "for each $n \in \mathbb{N}$, there is an n-Mahlo cardinal". A cardinal κ is (defined as) 0-Mahlo, if the set of inaccessible cardinals below κ is stationary below κ. And κ is $n+1$-Mahlo, if the set of n-Mahlo cardinals below κ is stationary below κ, where $n \in \mathbb{N}$. A set $S \subset \kappa$ is (defined as) stationary below κ, if S intersects every subset of κ that is unbounded and closed under suprema below κ. The proof of a refinement of Solovay's equiconsistency result is presented in [3].

For each $n \in \mathbb{N}$, the theory ZFC + "there is an n-Mahlo cardinal" is interpretable in NFUA by means of the set of equivalence classes of pointed extensional well-founded structures. This technique, invented by Roland Hinnion [16], is often used to obtain lower bounds on the consistency strength of variants of NFU and NF. We may of course solve (S) within this interpretation, for example through the Grothendieck approach of Sect. 4.1. In the terminology of Sect. 2.1, that approach yields a level (3) solution to (R3) and (S1). The proof of the converse direction of the equiconsistency result shows that we can add a relatively consistent axiom to NFUA that provides us with a level (2) implementation. We will briefly explain how this works for a certain refined solution to (S), that is similarly motivated as ZFC/S.

Just below ZFC + $\exists\kappa$ "κ is a Mahlo cardinal", in consistency strength, we have the theory ZMC defined roughly as ZFC + "$|V|$ is a Mahlo-cardinal". Of course, we cannot state the cardinality of V since it is not a set, but we can add the schema "for each formula $\phi(\xi)$, such that the class of ordinals satisfying ϕ is unbounded and closed under suprema, there is an inaccessible cardinal λ satisfying ϕ". This schema essentially says that $|V|$ is a Mahlo cardinal.

As explained by Shulman [25], we can form ZMC/S, in analogy with Feferman's ZFC/S. The same proof as with ZFC/S, gives us that ZMC/S is a conservative extension of ZMC. It turns out that ZMC/S is equivalent to ZFC/S + "$|\mathbb{S}|$ is inaccessible". This allows us to obtain a stronger replacement axiom schema in relation to $\mathbb{S}$: If $A \in \mathbb{S}$ and $\phi(x, y, u)$ is a formula with u an arbitrary parameter (i.e. not necessarily an element of $\mathbb{S}$), such that $\forall x \in A.\exists^! y \in \mathbb{S}.\phi(x, y, u)$, then there is $B \in \mathbb{S}$, such that $\forall y.(y \in B \leftrightarrow \exists x \in A.\phi(x, y, u))$. This means that we do not need to worry about small-definability. Hence, ZMC/S is quite a convenient solution to (S).

To interpret ZMC/S in NFUA, directly with the element relation of the language, restricted to a set, we may proceed as follows. In [3] a model of NFUA is constructed from a countable model of the theory T = ZFC + the schema "for each $n \in \mathbb{N}$, there is an n-Mahlo cardinal". Performing the "over S" transformation to T (just as from ZFC to ZFC/S in Sect. 4.2), gives us the theory T/S. As in the proof of Theorem 4, $\text{Con}(T) \Rightarrow \text{Con}(T/\text{S})$. So we obtain a countable model of T/S. Then, since $T \subset T/\text{S}$, the construction of the model $\mathcal{M}$ of NFUA goes through as before. In $\mathcal{M}$, ZMC/S is directly interpretable by restricting the element relation of the language to a set, yielding a level (2) solution to (R3) and (S1).

NFUA solves (R) and (S) somewhat separately: (S) may be solved within the interpretation of ZMC/S in NFUA (this also yields (R3)), while (R1) and (R2) are solved by means of categories (typically built from the universal set V) that do not fit within the interpretation of ZMC/S. For someone who is only interested in foundations solving (S), it would obviously be more economical to work in ZMC/S than in NFUA. So it is natural to wonder: Is there a benefit for category theory in enabling the existence of categories yielding (R1) and (R2), apart from the intuitive appeal of obtaining e.g. the category of all groups without any small/large restrictions? It seems that little research has been directly aimed at answering this question. In the next subsection we will just scratch the surface, by expressing and proving a basic fact of stratified set theory in category theoretic form. This result contrast with Theorem 1, showing that proper (R1)-categories can, in a sense, be more complete than what is possible in a purely ZFC-based setting.

5.4 *The Limits of a Proper Category (in NFU)*

We already saw that in ZFC-based foundations, a category **C** which is not a preorder, can only have products indexed by a set of lower cardinality than the cardinality of the set of morphisms of **C**. The proof in the ZFC setting uses Cantor's theorem, which holds for cantorian sets in NFU. But since $V \supset \mathcal{P}(V)$, the universe V is a counterexample to Cantor's theorem, and the categories yielding (R1) will typically have sets of objects and morphisms derived from V. Cantor's theorem has the following type-shifted version in NFU, essentially proved the same way.

Theorem 12 NFU $\vdash \forall x.\left|\left\{\{y\} \mid y \in x\right\}\right| < |\mathcal{P}(x)|$.

Again observing that $\mathcal{P}(V) \subset V$, we now see that $\left|\left\{\{y\} \mid y \in V\right\}\right| < |V|$. As an example of how categories in NFU satisfying (R1) can behave quite differently from their counterparts in ZFC, we will now show that the category **Rel**, with all sets as objects and binary relations as morphisms, has products and coproducts indexed by the set of all singletons. Similarly, the category **Set** has coproducts indexed by the set of all singletons. Compare this with the fact (a consequence of Theorem 1) that the corresponding locally small categories in a ZFC-based setting, do not have such limits indexed by the set/class of all small singletons. Note that in order to accommodate the weaker extensionality axiom of NFU, we stipulate that neither an object nor a morphism of **Rel** or **Set** may be an atom other than $\varnothing$.

Before getting into this proof, we would like to take the opportunity to point out that Thomas Forster, Adam Lewicki and Alice Vidrine have done unpublished work on the category theory of sets in stratifiable set theories at Cambridge University. This work establishes several category theoretic properties of NF: For example, the category **Set** in NF is exhibited as an "almost topos" and as a category of classes. Moreover, the status of the Yoneda lemma internal to NF is clarified. Being a category of classes entails, among other properties, having finite limitsco-limit and finite coproducts. In the present paper we are simply being explicit about how large the index category can be; there is no technical novelty in the proof. The point is to show possibilities in NFU-based category theory that are not available in purely ZFC-based approaches.

Proposition 13 NFU $\vdash$ **Rel** *has products indexed by* $\left\{\{i\} \mid i \in V\right\}$.

Proof Let **Sing** be the category with $\left\{\{i\} \mid i \in V\right\}$ as its set of objects and only identity morphisms, and let $\mathbf{F} : \mathbf{Sing} \to \mathbf{Rel}$ be a diagram. Set

$$P := \left\{\langle x, i\rangle \mid x \in \mathbf{F}(\{i\})\right\},$$

and define $\pi_{\{i\}} : P \to_{\mathbf{Rel}} \mathbf{F}(\{i\})$ by

$$\pi_{\{i\}} := \{\langle\langle x, i\rangle, x\rangle \mid x \in \mathbf{F}(\{i\})\},$$

for each $\{i\} \in \mathbf{Sing}$. Note that the definition of the map $\{i\} \mapsto \pi_{\{i\}}$ is stratified, and therefore realized by a function.

Consider an arbitrary cone to $\mathbf{F}$ in **Rel**, i.e. a set A and a relation $R_{\{i\}} : A \to_{\mathbf{Rel}} \mathbf{F}(\{i\})$, for each $\{i\} \in \mathbf{Sing}$. Again, the map-definition $\{i\} \mapsto R_{\{i\}}$ is stratified and therefore realized by a function. We may now define $u : A \to P$ by

$$u := \{\langle a, \langle x, i\rangle\rangle \mid a R_{\{i\}} x\}.$$

We just need to check that for an arbitrary $\{i\} \in \mathbf{Sing}$, we have $R_{\{i\}} = \pi_{\{i\}} \circ u$, i.e. that

$$\begin{array}{ccc} A & \xrightarrow{\ u\ } & P \\ & \searrow^{R_{\{i\}}} & \downarrow{\scriptstyle \pi_{\{i\}}} \\ & & \mathbf{F}(\{i\}) \end{array}$$

commutes. So suppose $a \in A$ and $x \in \mathbf{F}(\{i\})$. Then,

$$a(\pi_{\{i\}} \circ u)x \iff \exists p \in P.(aup \wedge p\pi_{\{i\}}x) \iff au\langle x, i\rangle \iff aR_{\{i\}}x.$$

Since a morphism of **Rel** cannot be any other atom than $\varnothing$, we have by extensionality that $R_{\{i\}} = \pi_{\{i\}} \circ u$. □

For any relation $R : A \to B$, we have a converse relation $R^\dagger : B \to A$ defined by $xRy \Leftrightarrow yR^\dagger x$. (**Rel** is a dagger symmetric monoidal category, hence the notation.) Clearly, $(Q \circ R)^\dagger = R^\dagger \circ Q^\dagger$, for any morphisms Q, R in **Rel**. Thus, by replacing u, the $\pi_{\{i\}}$ and the $R_{\{i\}}$ by their converses, in the proof above, we see:

Proposition 14 NFU $\vdash$ **Rel** *has coproducts indexed by* $\big\{\{i\} \mid i \in V\big\}$.

Note that the $\pi^\dagger_{\{i\}}$ are functions. Moreover, if we assume that the $R^\dagger_{\{i\}}$ are functions, then it follows that $u^\dagger$ is a function. Therefore:

Proposition 15 NFU $\vdash$ **Set** *has coproducts indexed by* $\big\{\{i\} \mid i \in V\big\}$.

Since we did not use choice, these results hold in NFU $\setminus$ {AC} and its extension NF (with full extensionality) as well.

As explained above, the proof given for Theorem 1 in ZFC, is blocked in NFU. But since $\big|\big\{\{y\} \mid y \in V\big\}\big| < |V|$, these propositions are not counterexamples to Theorem 1 in NFU. So the question whether NFU proves Theorem 1 or not, remains open, at least modulo the present paper. The point of Propositions 13–15 is that we have positive category theoretic results in NFU (assuming that limits are generally desirable), whose obvious translations into ZFC-based foundations are falsified by Theorem 1. We take this to be an indication that the (R1)-categories in NFU-based foundations may be of value to category theory, thus strengthening the motivation for exploring category theory internal to NF and NFU, as well as extensions like S^* and NFUA.

6 The Consistency of S^*

In this section we present Feferman's consistency proof of his system S^*. By carefully keeping track of the cardinals used to build a model of S^* we are able to improve on the result reported in [7, 11] by showing that ZFC + $\exists\kappa$"κ is an inaccessible cardinal" proves the consistency of S^*.

We will make reference to two subsystems of ZFC studied in [22]. Mac Lane Set Theory (Mac) is the subsystem of ZFC axiomatised by: extensionality, pair, emptyset, union, powerset, infinity, Δ_0-separation, transitive containment, regularity and the axiom of choice. The set theory Kripke-Platek with Ranks (KPR) is obtained from Mac by deleting the axiom of choice and adding Δ_0-collection, Π_1-foundation and an axiom asserting that for every ordinal α, the set V_α exists.[7] If $\mathcal{L}'$ is an extension of the language of set theory and $\mathcal{M}$ is an $\mathcal{L}'$-structure then for $a \in M$ we write a^* for the class $\{x \in M \mid \mathcal{M} \models (x \in a)\}$.

Throughout this section we work in the theory ZFC + $\exists\kappa$ "κ is an inaccessible cardinal". Let κ be an inaccessible cardinal. Let $A = V_\kappa \cup \{V_\kappa\}$. Jensen's consistency proof of NFU [19] reveals that a model of NFU can be built from model $\mathcal{M}$ of Mac that admits a non-trivial automorphism j and such that there exists a point $c \in M$ with

$$\mathcal{M} \models (c \cup \mathcal{P}(c) \subseteq j(c)). \tag{1}$$

In [7] Feferman obtains a model of S^* (in the theory ZFC + $\exists\kappa\exists\lambda$ "$\kappa < \lambda$ are inaccessible cardinals") by building a model $\mathcal{M}$ of ZFC that admits a non-trivial automorphism (the existence of the rank function $\alpha \mapsto V_\alpha$ in ZFC ensures that every non-trivial automorphism satisfies (1)) and such that $\mathcal{M}$ is an end-extension of A. This ensures that there is an isomorphic copy A in the well-founded part of the resulting model $\mathcal{N}$ of NFU, and moreover every point in this isomorphic copy of A is strongly Cantorian in $\mathcal{N}$. This allows Feferman to interpret the constant symbol $\bar{V}$ using the point corresponding to V_κ in $\mathcal{N}$. The fact that $\bar{V}$ is isomorphic to V_κ where κ is an inaccessible cardinal and $\mathcal{N}$ exists as a set ensures that $\bar{V}$ satisfies axioms (5–9) of S^*.

Feferman builds a model of ZFC that end-extends A and admits a non-trivial automorphism using tools from infinitary logic. The fine-tuned proof we present here uses the same techniques as [7], however the model $\mathcal{M}$ that we build, which is an end-extension of A and admits a non-trivial automorphism, will satisfy a fragment of ZFC that is sufficient to ensure that $\mathcal{M}$ equipped with its non-trivial automorphism still yields a model of NFU. Let $\mathcal{L}^A$ be the extension of the language of set theory obtained by adding new constant symbols $\hat{a}$ for every $a \in A$. We use $\mathcal{L}^A_{\infty\omega}$ to denote the infinitary language obtained from $\mathcal{L}^A$ that permits arbitrarily long conjunctions and disjunctions, but only finite blocks of quantifiers. Let T_A be the $\mathcal{L}^A_{\infty\omega}$-theory with axioms:

[7]Mathias omits powerset and transitive containment from his axiomatisation of KPR in [22], but both of these axioms follow from the existence of V_α for every ordinal α.

$$\forall x \left(x \in \hat{a} \iff \bigvee_{b \in a} (x = \hat{b}) \right) \text{ for each } a \in A.$$

Note that T_A asserts that an $\mathcal{L}^A$-structure is an end-extension of A. Let $\mathcal{L}_A$ be the least Skolem fragment of $\mathcal{L}^A_{\infty\omega}$ that contains T_A. So, in addition to the symbols of $\mathcal{L}^A$, $\mathcal{L}_A$ contains an n-ary function symbol $F_{\exists x\phi}$ for each n-ary $\mathcal{L}_A$-formula $\exists x \phi(x, y_1, \ldots, y_n)$. We write Sk_A for the $\mathcal{L}_A$-theory with axioms:

$$\forall \vec{y}(\exists x \phi(x, \vec{y}) \Rightarrow \phi(F_{\exists x \phi}(\vec{y}), \vec{y})) \text{ for each } \mathcal{L}_A\text{-formula } \phi(x, \vec{y}).$$

Let Fm_A be the set of $\mathcal{L}_A$-formulae. Note that $|\mathrm{Fm}_A| = \kappa$ and so there 2^κ many $\mathcal{L}_A$-theories.

We obtain a model that admits a non-trivial automorphism from a model equipped with an infinite class of order indiscernibles.

Definition 16 Let $\mathcal{M}$ be an $\mathcal{L}_A$-structure. We say that a linear order $\langle I, < \rangle$ with $I \subseteq M$ is a set of n-variable indiscernibles for $\mathcal{M}$ if for all $\mathcal{L}_A$-formulae $\phi(x_1, \ldots, x_n)$ and for all $a_1 < \cdots < a_n$ and $b_1 < \cdots < b_n$ in I,

$$\mathcal{M} \models \phi(a_1, \ldots, a_n) \text{ if and only if } \mathcal{M} \models \phi(b_1, \ldots, b_n).$$

If for all $n \in \omega$, $\langle I, < \rangle$ with $I \subseteq M$ is a set of n-variable indiscernibles for $\mathcal{M}$ then we say that $\langle I, < \rangle$ is a set of indiscernibles for $\mathcal{M}$.

The following result from [2] allows us to build $\mathcal{L}_A$-structures with indiscernibles:

Lemma 17 (Barwise–Kunen) *Suppose that for all $n \in \omega$, $\mathcal{M}_n$ is model of Sk_A and $\langle I_n, <_n \rangle$ with $I_n \subseteq M_n$ is a set of n-variable indiscernibles for $\mathcal{M}_n$. If for all $n \in \omega$, for all $a_1 <_n \cdots <_n a_n$ in I_n and for all $b_1 <_{n+1} \cdots <_{n+1} b_n$ in I_{n+1},*

$$\langle \mathcal{M}_n, a_1, \ldots, a_n \rangle \equiv_{\mathcal{L}_A} \langle \mathcal{M}_{n+1}, b_1, \ldots, b_n \rangle$$

then for any linear order $\langle I, < \rangle$ there is an $\mathcal{L}_A$-structure $\mathcal{M}$ with $I \subseteq M$ such that I is a set of indiscernibles for $\mathcal{M}$ and for all $n \in \omega$, for all $a_1 <_n \cdots <_n a_n$ in I_n and for all $b_1 < \cdots < b_n$ in I,

$$\langle \mathcal{M}_n, a_1, \ldots, a_n \rangle \equiv_{\mathcal{L}_A} \langle \mathcal{M}, b_1, \ldots, b_n \rangle.$$

We now turn to building a model of T_A with indiscernibles. The following definition appears in [19].

Definition 18 Let I be a set and let $\langle f_n \mid n \in \omega \rangle$ be a family of partitions such that for all $n \in \omega$, f_n has domain $[I]^n$. We say that $\langle c_n \mid n \in \omega \rangle$ is realizable for $\langle f_n \mid n \in \omega \rangle$ if for all $n \in \omega$ and for all $\beta < |I|$, there exists an $I_n \subseteq I$ such that $|I_n| \geq \beta$ and $f_k([I_n]^k) = \{c_k\}$ for all $k \leq n$.

The n-variable indiscernibles on the $\mathcal{L}_A$-structures required so that we can apply Lemma 17 are obtained by considering a sequence of partitions $\langle f_n \mid n \in \omega \rangle$ that admit a realizable sequence $\langle c_n \mid n \in \omega \rangle$. If λ is a cardinal then the generalized beth operation is defined by recursion: $\beth_0(\lambda) = \lambda$, $\beth_{\alpha+1}(\lambda) = 2^{\beth_\alpha(\lambda)}$ and (if α is a limit ordinal) $\beth_\alpha(\lambda) = \sup\{\beth_\beta(\lambda) \mid \beta < \alpha\}$. If λ is a cardinal then the function $\alpha \mapsto \beth_\alpha(\lambda)$ is inflationary and continuous. It follows from this observation that the function $\alpha \mapsto \beth_\alpha(\lambda)$ has arbitrarily large fixed points. We will use the Erdős–Rado Theorem [4] to produce realizable sequences.

Lemma 19 (Erdős–Rado) *Let λ be an infinite cardinal. If $f : [X]^{n+1} \longrightarrow \lambda$ is a partition with $|X| \geq \beth_n(\lambda)^+$ then there exists $H \subseteq X$ with $|H| \geq \lambda^+$ and $\gamma \in \lambda$ such that $f``[H]^{n+1} = \{\gamma\}$.*

We now turn to building a model of KPR that end-extends A and admits a non-trivial automorphism. Let η_0 be the least ordinal such that $\beth_{\eta_0}(2^\kappa) = \eta_0$. Now, recursively define

$$\eta_{\alpha+1} \text{ to be the least ordinal } > \eta_\alpha \text{ such that } \beth_{\eta_{\alpha+1}}(2^\kappa) = \eta_{\alpha+1},$$

$$\eta_\alpha = \sup_{\beta<\alpha} \eta_\beta \text{ for limit } \alpha.$$

Since the function $\alpha \mapsto \beth_\alpha(2^\kappa)$ is continuous it follows that for all ordinals α,

$$\eta_\alpha = \beth_{\eta_\alpha}(2^\kappa).$$

Therefore, for all ordinals α, $V_{\eta_\alpha} = H_{\eta_\alpha}$ and so $\langle V_{\eta_\alpha}, \in \rangle$ is a model of KPR. Let

$$I = \{V_{\eta_\alpha} \mid \alpha < \eta_{(2^\kappa)^+}\}.$$

The membership relation ($\in$) linearly orders I and we will often abuse notation and use $<$ to denote this linear order. Now, $|I| = \eta_{(2^\kappa)^+}$ and $\mathrm{cf}(\eta_{(2^\kappa)^+}) = (2^\kappa)^+ > 2^\kappa$. The argument used to prove the following Lemma can be found in [19].

Lemma 20 *If $\langle f_n \mid n \in \omega \rangle$ is a family of partitions such that for all $n \in \omega$,*

$$f_n : [I]^{n+1} \longrightarrow 2^\kappa$$

then there exists a sequence $\langle c_n \mid n \in \omega \rangle$ that is realizable for $\langle f_n \mid n \in \omega \rangle$.

Proof We inductively construct $\langle c_n \mid n \in \omega \rangle$. Suppose that we have $c_0, \ldots, c_{n-1} \in 2^\kappa$ such that for all $\beta < |I|$, there exists $D \subseteq I$ with $|D| \geq \beta$ and

$$f_k([D]^{k+1}) = \{c_k\} \text{ for all } k < n.$$

Let $\mathcal{U}_{n-1}$ be the set of all $D \subseteq I$ such that

$$f_k([D]^{k+1}) = \{c_k\} \text{ for all } k < n.$$

Note that our inductive hypothesis ensures that $\mathcal{U}_{n-1}$ contains arbitrarily large subsets of I. Now, let $\mathcal{U}'$ be the set of all $B \subseteq I$ such that there exists a $D \in \mathcal{U}_{n-1}$ and $c \in 2^\kappa$ with $B \subseteq D$ and

$$f_n([B]^{n+1}) = \{c\}.$$

Lemma 19 ensures that $\mathcal{U}'$ contains arbitrarily large subsets of I. Using the fact that $\mathrm{cf}(|I|) > 2^\kappa$ we can find $c_n \in 2^\kappa$ and $\mathcal{U}_n \subseteq \mathcal{U}'$ which contains arbitrarily large subsets of I such that for all $B \in \mathcal{U}_n$,

$$f_n([B]^{n+1}) = \{c_n\}.$$

Therefore, for all $\beta < |I|$, there exists $D \subseteq I$ with $|D| \geq \beta$ and

$$f_k([D]^{k+1}) = \{c_k\} \text{ for all } k \leq n.$$

Therefore we can build $\langle c_n \mid n \in \omega\rangle$ that is realizable for $\langle f_n \mid n \in \omega\rangle$ by induction.

□

Let $\lhd$ be a well-ordering of $V_{\eta_{\eta_{(2^\kappa)^+}}}$. Define an $\mathcal{L}_A$-structure

$$\mathcal{M} = \langle M, \in, (\hat{a}^{\mathcal{M}})_{a\in A}, (F^{\mathcal{M}}_{\exists x\phi})_{\exists x\phi \in \mathcal{L}_A}\rangle$$

such that:

- $M = V_{\eta_{\eta_{(2^\kappa)^+}}}$,
- for all $a \in A$, $\hat{a}^{\mathcal{M}} = a$,
- for all $\mathcal{L}_A$-formulae $\exists x\phi(x, \vec{y})$,

$$F^{\mathcal{M}}_{\exists x\phi}(\vec{y}) = \begin{cases} \varnothing & \text{if } \mathcal{M} \models \neg\exists x\phi(x, \vec{y}) \\ \lhd\text{-least } a \text{ s.t. } \mathcal{M} \models \phi(a, \vec{y}) & \text{otherwise} \end{cases}$$

Therefore, we have

$$\mathcal{M} \models (\mathrm{KPR} + \mathrm{Sk}_A + T_A) \text{ and } I \subseteq M.$$

Lemma 21 *There exists a family $\langle I_n \mid n \in \omega\rangle$ of infinite subsets of I such that for all $n \in \omega$, I_n is a set of n-variable indiscernibles for $\mathcal{M}$, and for all $a_0 < \cdots < a_n$ in I_n and for all $b_0 < \cdots < b_n$ in I_{n+1},*

$$\langle \mathcal{M}, a_0, \ldots, a_n\rangle \equiv_{\mathcal{L}_A} \langle \mathcal{M}, b_0, \ldots, b_n\rangle.$$

Proof For each $n \in \omega$, define $f_n : [I]^{n+1} \longrightarrow \mathcal{P}(\mathrm{Fm}_A)$ by

$$f_n(\{a_0 < \cdots < a_n\}) = \{\phi \mid \phi \text{ is an } n+1\text{-ary } \mathcal{L}_A\text{-formula and } \mathcal{M} \models \phi(a_0, \ldots, a_n)\}.$$

Using Lemma 20 we can find a sequence $\langle c_n \mid n \in \omega \rangle$ that is realizable for $\langle f_n \mid n \in \omega \rangle$. Therefore we can inductively build a sequence $\langle I_n \mid n \in \omega \rangle$ such that for all $n \in \omega$, $|I_n| \geq \omega$ and

$$f_k([I_n]^{k+1}) = \{c_k\} \text{ for all } k \leq n.$$

Let $n \in \omega$. It is clear that I_n is a set of n-variable indiscernibles for $\mathcal{M}$. Let $a_0 < \cdots < a_n$ be in I_n and let $b_0 < \cdots < b_n$ be in I_{n+1}. We have

$$f_n(\{a_0 < \cdots < a_n\}) = c_n = f_n(\{b_0 < \cdots < b_n\}).$$

Therefore, for all $\mathcal{L}_A$ formulae $\phi(x_0, \ldots, x_n)$,

$$\mathcal{M} \models \phi(a_0, \ldots, a_n) \text{ if and only if } \mathcal{M} \models \phi(b_0, \ldots, b_n).$$

Therefore

$$\langle \mathcal{M}, a_0, \ldots, a_n \rangle \equiv_{\mathcal{L}_A} \langle \mathcal{M}, b_0, \ldots, b_n \rangle.$$

□

Let $\langle I_n \mid n \in \omega \rangle$ be the sequence whose existence is guaranteed by Lemma 21. We can now apply Lemma 17 to obtain an $\mathcal{L}_A$-structure $\mathcal{M}'$ and a linear order $\langle Q, < \rangle$ with $Q \subseteq M'$ satisfying

(I) the order-type of $\langle Q, < \rangle$ is $\mathbb{Z}$— we write $Q = \{q_i \mid i \in \mathbb{Z}\}$,
(II) Q is a set of indiscernibles for $\mathcal{M}'$,
(III) $\mathcal{M}' \equiv_{\mathcal{L}_A} \mathcal{M}$,
(IV) for all $\mathcal{L}_A$-formulae $\phi(x_0, \ldots, x_n)$, if

$$\mathcal{M} \models \phi(a_0, \ldots, a_n) \text{ for all } a_0 < \cdots < a_n \text{ in } I \text{ then}$$

$$\mathcal{M}' \models \phi(q_{i_0}, \ldots, q_{i_n}) \text{ for all } i_0 < \cdots < i_n \text{ in } \mathbb{Z}.$$

Let $\mathcal{M}''$ be the $\mathcal{L}_A$-substructure of $\mathcal{M}'$ generated by Q, and the constants and functions of $\mathcal{L}_A$. Therefore $\mathcal{M}'' \prec_{\mathcal{L}_A} \mathcal{M}'$. Consider the order automorphism $j' : Q \longrightarrow Q$ defined by $j'(q_i) = q_{i+1}$ for all $i \in \mathbb{Z}$. Since every element of $\mathcal{M}''$ is the result of applying an $\mathcal{L}_A$-Skolem function to a finite tuple of elements of Q, the bijection j' can be raised to an automorphism $j : \mathcal{M}'' \longrightarrow \mathcal{M}''$. Note that for all $a < b$ in I,

$$\mathcal{M} \models (a \cup \mathcal{P}(a) \subseteq b).$$

Therefore, it follows from (IV) above that

$$\mathcal{M}'' \models (q_0 \cup \mathcal{P}(q_0) \subseteq j(q_0)). \tag{2}$$

Define the $\mathcal{L}^*$-structure $\mathcal{N} = \langle N, N_{\text{sets}}, \in^{\mathcal{N}}, P^{\mathcal{N}}, \bar{V}^{\mathcal{N}} \rangle$ by

- $N = q_0^*$,
- $N_{\text{sets}} = (\hat{V}_\kappa^{\mathcal{M}''})^*$,
- for all $x, y \in q_0^*$,

$$x \in^{\mathcal{N}} y \text{ if and only if } \mathcal{M}'' \models (j(y) \subseteq q_0 \wedge x \in j(y)),$$

$$P^{\mathcal{N}}(x, y) = \{\{x\}, \{x, y\}\} \in q_0^*,$$

- $\bar{V}^{\mathcal{N}} = \hat{V}_\kappa^{\mathcal{M}''}$

where the set variables $x, y, z, \ldots$ range over the domain N_{sets} and the class variables $X, Y, Z, \ldots$ range over the domain N.

Lemma 22 $\mathcal{N} \models S^*$.

Proof Note that since every element of N_{sets} interprets a constant symbol in $\mathcal{M}''$, the automorphism j fixes every element of N_{sets}. This means that the structure $\langle \bar{V}^*, \in^{\mathcal{N}} \rangle$ is isomorphic to $\langle V_\kappa, \in \rangle$. The arguments in [19] show that since q_0 satisfies (2) and j fixes every element of N_{sets}, $\mathcal{N}$ satisfies axiom 1 and axiom scheme 2 of S^*. The fact that Universal Choice is an $\mathcal{L}^*$-stratified sentence that holds in every element of I implies axiom 10 of S^* holds in $\mathcal{N}$. The fact that axiom 3 of S^* holds in $\mathcal{N}$ follows immediately from the definition of $P^{\mathcal{N}}$. The fact that axioms 4–7 and axiom scheme 9 of S^* hold in $\mathcal{N}$ follow immediately from the fact that $\langle \bar{V}^*, \in^{\mathcal{N}} \rangle$ is isomorphic to $\langle V_\kappa, \in \rangle$. Since the structure $\mathcal{N}$ is a set (in the metatheory) and $\langle \bar{V}^*, \in^{\mathcal{N}} \rangle$ is isomorphic to $\langle V_\kappa, \in \rangle$ it follows that axiom scheme 8 of S^* holds in $\mathcal{N}$. □

Since the structure $\mathcal{N}$ is a set in the theory ZFC $+ \exists\kappa$"κ is an inaccessible cardinal" we have shown:

Theorem 23 ZFC $+ \exists\kappa.$"*κ is an inaccessible cardinal*" $\vdash$ Con(S^*).

Conversely, recall that by Theorem 10:

Theorem 24 $S^* \vdash$ Con(ZC $+ \exists\kappa.$"*κ is an inaccessible cardinal*" $+ \forall\alpha \in$ Ord."*V_α exists*").

This strengthens Feferman's result that S^* proves the consistency of Morse-Kelley set theory. Within the above extension of ZC, if α is an ordinal, then $V_{\kappa+\alpha}$ yields a model of α:th order Morse-Kelley set theory. Having provided improved upper and lower bounds on the consistency strength of S^*, we ask:

Question 25 What is the exact consistency strength of S^* relative to an extension of ZFC?

Motivated by Question 25 Feferman and the first author of this paper proposed an extension of S^* which is called S^{**} in [11]. The theory S^{**} is obtained from S^* by adding the universal closure of the following axiom scheme:

11. for all $\mathcal{L}^*$-formulae $\phi(x, \vec{Z})$,

$$\exists X (\forall x \in \bar{V})(x \in X \iff \phi(x, \vec{Z})).$$

We will conclude this section by sketching how our proof that the theory ZFC $+ \exists\kappa$ "κ is an inaccessible cardinal" proves the consistency of S^* can be modified to show that the theory ZFC $+ \exists\kappa$ "κ is an inaccessible cardinal" also proves the consistency of S^{**}. Work in the theory ZFC $+ \exists\kappa$ "κ is an inaccessible cardinal" and let κ be and inaccessible cardinal. Let $A' = V_{\kappa+1}$. We can then define a fragment of infinitary logic $\mathcal{L}_{A'}$ that is equipped with Skolem functions, and capable of asserting that a structure is an end-extension of A', with the property that $|\mathrm{Fm}_{A'}| = 2^\kappa$. Therefore there are $2^{2^\kappa} = \beth_2(\kappa)$ many $\mathcal{L}_{A'}$-theories. We now define

$$I = \{V_{\eta_\alpha} \mid \alpha < \eta_{\beth_2(\kappa)^+}\}.$$

Therefore $|I| = \eta_{\beth_2(\kappa)^+}$ and $\mathrm{cf}(|I|) = \beth_2(\kappa)^+ > \beth_2(\kappa)$. We can then modify the definition of the structure $\mathcal{M}$ above by setting $M = V_{\eta_{\eta_{\beth_2(\kappa)^+}}}$ and expanding the interpretations to all symbols in $\mathcal{L}_{A'}$. The resulting $\mathcal{L}_{A'}$-structure is an end-extension of A', and satisfies KPR and the Skolem theory of $\mathcal{L}_{A'}$. Using the same argument that we used above we can use $\mathcal{M}$ to build an $\mathcal{L}_{A'}$-elementary equivalent structure $\mathcal{M}''$ that admits a non-trivial automorphism $j : \mathcal{M}'' \longrightarrow \mathcal{M}''$. We define an $\mathcal{L}^*$-structure $\mathcal{N}$ from $\mathcal{M}''$ and j in the same way as we did above. The structure $\mathcal{N}$ satisfies all of the axioms of S^*. Since $\mathcal{M}''$ is an end-extension of A', the structure $\langle \mathcal{P}(\bar{V})^*, \in^{\mathcal{N}} \rangle$ will be isomorphic to $\langle V_{\kappa+1}, \in \rangle$. This is enough to ensure that axiom scheme 11 holds in $\mathcal{N}$. Thus we have:

Theorem 26 ZFC $+ \exists\kappa$ "*κ is an inaccessible cardinal*" $\vdash \mathrm{Con}(S^{**})$.

Question 27 What is the exact consistency strength of S^{**} relative to an extension of ZFC?

References

1. Antos, C., Friedman, S.: Hyperclass forcing in Morse-Kelley class theory (2015). http://arxiv.org/abs/1510.04082
2. Barwise, J., Kunen, K.: Hanf numbers for frgments of $L_{\infty\omega}$. Isr. J. Math. **10**, 306–320 (1971)
3. Enayat, A.: Automorphisms, Mahlo cardinals, and NFU. In: Enayat, A., Kossak, R. (Eds.), Nonstandard Models of Arithmetic and Set Theory, Contemporary Mathematics, vol. 361, American Mathematical Society (Providence) (2004)
4. Erdős, P., Rado, R.: A partition calculus in set theory. Bull. Am. Mat. Soc. **62**, 427–488 (1956)
5. Ernst, M.: The prospects of unlimited category theory: doing what remains to be done. Rev. Symb. Log. **8**(2), 549–575 (2015)

6. Feferman, S.: Set-theoretical foundations for category theory (with an appendix by G. Kreisel). In: Barr, M. et al. (eds.) Reports of the Midwest Category Seminar III. Lecture Notes in Mathematics, vol. 106, pp. 201–247. Springer, Berlin (1969)
7. Feferman, S.: Some formal systems for the unlimited theory of structures and categories (1974). http://math.stanford.edu/~feferman/papers/Unlimited.pdf
8. Feferman, S.: A language and axioms for explicit mathematics. In: Algebra and Logic. Lecture Notes in Mathematics, vol. 450, pp. 87–139. Springer, Berlin (1975)
9. Feferman, S.: Categorical foundations and foundations of category theory. In: Butts, R.E., Hintikka, J. (eds.) Logic, Foundations of Mathematics, and Computability Theory, pp. 149–169. Reidel, Dordrecht (1977)
10. Feferman, S.: Typical ambiguity. Trying to have your cake and eat it too. In: Link, G. (Ed.) One Hundred Years of Russell's Paradox, pp. 135–151. Walter de Gruyter, Berlin (2004)
11. Feferman, S.: Enriched stratified systems for the foundations of category theory. In: Sica, G. (ed.) What is category theory? Polimetrica, Milan (2006). http://math.stanford.edu/~feferman/papers/ess.pdf
12. Feferman, S.: Operational set theory and small large cardinals. Inf. Comput. **207**, 971–979 (2009)
13. Feferman, S.: Foundations of unlimited category theory: what remains to be done. Rev. Symb. Logic **6**, 6–15 (2013)
14. Forster, T.E.: Why the Sets of NF do not form a Cartesian-closed Category (2007). https://www.dpmms.cam.ac.uk/~tf/cartesian-closed.pdf
15. Gabbay, M.J.: Consistency of Quine's new foundations using nominal techniques (2014). http://arxiv.org/abs/1406.4060
16. Hinnion, R.: Sur la théorie des ensembles de Quine. Ph.D. thesis, ULB Brussels. English translation by Thomas E. Forster (1975). http://www.logic-center.be/Publications/Bibliotheque/hinnionthesis.pdf
17. Holmes, M.R.: Elementary Set Theory with a Universal Set. Cahiers du Centre de logique (vol. 10). Louvain-la-Neuve, Academia (1998). http://math.boisestate.edu/~holmes/holmes/head.pdf
18. Holmes, M.R.: The consistency of NF (2015). http://arxiv.org/abs/1503.01406
19. Jensen, R.B.: On the consistency of a slight (?) modification of Quine's NF. Synthese **19**, 250–263 (1969)
20. Lane, S.M.: Categories for the Working Mathematician. Springer, Berlin (1998)
21. Lawvere, F.W.: Diagonal arguments and cartesian closed categories. Lecture Notes in Mathematics, vol. 92, pp. 134–145 (1969). Reprinted: Reprints in Theory and Applications of Categories, vol. 15, pp. 1–13 (2006)
22. Mathias, A.R.D.: The strength of Mac Lane set theory. Ann. Pure Appl. Logic **110**, 107–234 (2001)
23. McLarty, C.: Failure of cartesian closedness in NF. J. Symb. Logic **57**, 555–556 (1992)
24. Quine, W.V.: New foundations for mathematical logic. Am. Math. Monthly **44**, 111–115 (1937)
25. Shulman, M.A.: Set theory for category theory (2008). http://arxiv.org/abs/0810.1279

Author Biographies

Ali Enayat is currently Professor of Logic at the University of Gothenburg, Sweden. His research is focused on the model-theoretic investigation of various foundational axiomatic systems, which range in strength from weak arithmetical theories in which exponentiation of finite numbers need not be a total operation, all the way up to systems of set theory with large cardinals.

Paul Gorbow is currently a PhD student in Logic at the University of Gothenburg. He does research on stratified set theory (e.g. NF and NFU), its interplay with category theory and conventional set theory, as well as applications to philosophy.

Zachiri McKenzie held a postdoctoral research position in Logic at the University of Gothenburg in the period 2014–2016, during which this article was prepared. He has worked on the model theory of subsystems and extensions of ZFC, and on alternative set-theoretic foundational theories including the Simple Theory of Types, and variants of NF and NFU.

On Some Semi-constructive Theories Related to Kripke–Platek Set Theory

Fernando Ferreira

Abstract We consider some very robust semi-constructive theories related to Kripke–Platek set theory, with and without the powerset operation. These theories include the law of excluded middle for bounded formulas, a form of Markov's principle, the unrestricted collection scheme and, also, the classical contrapositive of the bounded collection scheme. We analyse these theories using forms of a functional interpretation which work in tandem with the constructible hierarchy (or the cumulative hierarchy, if the powerset operation is present). The main feature of these functional interpretations is to treat bounded quantifications as "computationally empty." Our analysis is extended to a second-order setting enjoying some forms of class comprehension, including strict-Π^1_1 reflection. The key idea of the extended analysis is to treat second-order (class) quantifiers as bounded quantifiers and strict-Π^1_1 reflection as a form of collection. We will be able to extract some effective bounds from proofs in these systems in terms of the constructive tree ordinals up to the Bachmann–Howard ordinal.

Keywords Intuitionistic Kripke–Platek set theory · Functional interpretations · Σ-ordinal · Strict-Π^1_1 reflection · Power Kripke–Platek set theory

MSC (2010) 03F10 · 03F35 · 03F65

F. Ferreira—I would like to thank the editors of this volume, Gerhard Jäger and Wilfried Sieg, for inviting me to contribute to this volume dedicated to Solomon Feferman and his work. It is an honor and a joy to participate. Over several decades, Sol was a role model for a certain approach to mathematical logic and the foundations of mathematics, combining high technical proficiency with sustained philosophical reflection. On a more personal level, he was always supportive of my work, ever since he told me more than a quarter of century ago that he had studied with interest my (then recent) doctoral dissertation. Thank you Sol! You'll be remembered. [The financial support of FCT, by way of grant PEst-OE/MAT/UI0209/2013 to the research center CMAFIO, is hereby duly acknowledged.].

F. Ferreira (✉)
Faculdade de Ciências, Departamento de Matemática, Universidade de Lisboa,
Lisboa, Portugal
e-mail: fjferreira@fc.ul.pt

G. Jäger and W. Sieg (eds.), *Feferman on Foundations*, Outstanding Contributions to Logic 13, DOI 10.1007/978-3-319-63334-3_13

1 Introduction

In recent writings, Solomon Feferman was urging the study of semi-construtive theories. His papers "On the strength of some semi-constructive theories" [8] and "Logic, mathematics, and conceptual structuralism" [9] (especially Sect. 6) are examples of these writings. These theories are a blend of intuitionistic and classical logic and their philosophical rationale can be described succinctly: according to some philosophical conceptions, there are good reasons to treat certain collections as constituting a definite totality (and, therefore, membership in them as abiding by the law of excluded middle) and others as open ended. For instance, one may want to see sets defined by bounded formulas as definite, and unbounded set-theoretic quantifications as open ended (and, as a consequence, treated intuitionistically). The consideration of intuitionistic subsystems of set theory with the law of excluded middle for bounded formulas was apparently first given by Lawrence Pozsgay in [16] (see [8] for more information on this regard). Even though the present paper studies semi-constructive theories of this kind, its rationale is mainly technical. It is an exploitation of a form of functional interpretation that was introduced in the classical setting in [10] and whose roots can be found in a seminal paper of Jeremy Avigad and Henry Towsner [2]. This form of functional interpretation works in tandem with Gödel's constructible hierarchy (or with the cumulative hierarchy, in case the powerset operation is present) and treats bounded and second-order (class) quantifications as "computationally empty." As it turns out, the theory of these functional interpretations is very natural and satisfying.

The layout of this paper is the following. In Sect. 2 we introduce and draw some simple but fundamental consequences of the basic semi-constructive theory that we will analyse. Section 3 studies with some detail the term calculus of the primitive recursive functionals, introduced by William Howard in [14]. The Ω-type tree terms q of this calculus give the means to refer to the various (countable) stages L_q of the constructible hierarchy (or of the cumulative hierarchy V_q). The most important section is the fifth, where the main functional interpretation is defined and where a pertinent soundness theorem is proved. Departing from tradition, the verifications of the functional interpretations of this paper do not take place within formal theories, but are rather seen to hold semantically. A previous Sect. 4 introduces the two basic semantical structures with which we will be working with. We will be able to extract constructive information from proofs of sentences of the form $\forall x \exists y\, \phi(x, y)$, where ϕ can take various forms. The constructive information is given by a closed term t of type $\Omega \to \Omega$ such that $\forall c^{\Omega} \forall x \in L_c \exists y \in L_{tc}\, \phi(x, y)$ holds. In particular, if the semi-constructive theory proves a Σ_1-sentence then it follows that this sentence already holds in L_α, for α an ordinal less than the Bachmann–Howard ordinal. Our methods are able to provide a Σ-ordinal analysis in this sense.

We extend the analysis to a second-order setting with a form of bounded comprehension and with strict-Π^1_1 reflection. In a first study, the extension keeps the original separation scheme. This is done in Sect. 6 and the conclusion is that an ordinary Σ-ordinal analysis still goes through. In the last section, we allow second-order parameters in the separation scheme. This simple modification entails a major change because, as Vincenzo Salipante has observed in [19], with this form of separation one can prove the totality of the powerset operation. Nevertheless, a functional interpretation of the second-order theory with the extended separation scheme can still be made. There is a crucial difference, though. We now need the (countable) stages of the *cumulative* hierarchy. We are also able to extract some constructive information and to perform a *relativized* Σ-ordinal analysis (in the sense of Michael Rathjen in [17]). As a last item, we recuperate Rathjen's relativized Σ-ordinal analysis of the classical theory dubbed *power Kripke–Platek set theory*.

2 Intuitionistic Kripke–Platek Set Theory

Kripke–Platek set theory with infinity, with acronym $\mathsf{KP}\omega$, is a well-known theory framed in the language of set theory. It is a theory of classical logic whose axioms are extensionality, (unordered) pair, union, infinity, and the schemata of Δ_0-separation, Δ_0-collection and (unrestricted) foundation. The reader can consult [5] for a precise formulation. Since we are interested in (semi) intuitionistic versions of $\mathsf{KP}\omega$, the primitive logical symbols are absurdity, conjunction, disjunction, implication, and the universal and existential quantifiers. It is also convenient in our setting (as it was in [10]) to include bounded quantifiers as a primitive syntactic device. Both $\forall x \in z\, \phi$ and $\exists x \in z\, \phi$ are part of the primitive syntactic apparatus and are not considered as abbreviations of $\forall x\, (x \in z \to \phi)$ and $\exists x\, (x \in x \wedge \phi)$, respectively. Instead, our axioms include the corresponding equivalences between these formulas. The class of bounded or Δ_0-formulas is the smallest class of formulas that includes the atomic formulas (including the absurdity) and which is closed under propositional connectives and bounded quantifiers. For the record (and because of its importance), we state the scheme Δ_0-Coll of bounded collection:

$$\forall y \in w\, \exists x\, \phi(x, y) \to \exists z\, \forall y \in w\, \exists x \in z\, \phi(x, y),$$

where $\phi(x, y)$ is a bounded formula, possibly with parameters. The scheme of foundation is formulated in its inductive form (which is the form appropriate for intuitionistic studies):

$$\forall x\, (\forall y \in x\, \phi(y) \to \phi(x)) \to \forall x\, \phi(x),$$

for *every* formula $\phi(x)$, possibly with parameters. Since the scheme is unrestricted, it is easy to see that the scheme is (intuitionistically) equivalent to the rule

$$\frac{\forall x\, (\forall y \in x\, \phi(y) \to \phi(x))}{\forall x\, \phi(x)}$$

where $\phi(x)$ is any formula (possibly with other free variables besides x). The proof of the soundness theorem (Theorem 2) simplifies if we consider the rule instead of the axiom scheme.

Following [8], let $\mathsf{IKP}\omega$ be the system $\mathsf{KP}\omega$ with the logic restricted to be intuitionistic. Let Δ_0-LEM be the scheme $\phi \vee \neg\phi$ of excluded middle for bounded formulas ϕ. Our basic intuitionistic theory is $\mathsf{IKP}\omega + \Delta_0$-$\mathsf{LEM}$. In the remaining of this section we present a series of four definitions that introduce principles that the functional interpretation of Sect. 5 is able to interpret.

Definition 1 Markov's principle MP is the scheme $\neg\forall x\, \phi(x) \to \exists x\, \neg\phi(x)$, for ϕ a bounded formula (possibly with parameters).

Proposition 1 $\mathsf{IKP}\omega + \Delta_0$-$\mathsf{LEM} + \mathsf{MP}$ *proves the following scheme for bounded formulas ϕ and ψ (parameters are allowed):* $(\forall x\, \phi(x) \to \psi) \to \exists x\, (\phi(x) \to \psi)$.

Proof Assume $\forall x\, \phi(x) \to \psi$. By Δ_0-LEM, there are two cases to consider. If ψ holds then any x will do. Otherwise, we have $\neg\forall x\, \phi(x)$. By MP take x_0 such that $\neg\phi(x_0)$. Of course, $\phi(x_0) \to \psi$. We are done. □

A Σ_1-formula ϕ is a formula of the form $\exists z\, \psi(z)$, where $\psi(z)$ is a bounded formula (possibly with parameters). Up to provability in $\mathsf{IKP}\omega$, this class of formulas is closed under conjunction, disjunction, bounded quantifications and existential quantifications. This is well-known in the classical setting, but the argument also goes through in $\mathsf{IKP}\omega$. We defer the discussion of the dual class Π_1 until the end of this section. A Π_2-formula is a formula of the form $\forall w\, \phi(w)$, where $\phi(w)$ is a Σ_1-formula. We have only defined these formulas with a single universal quantifier '$\forall w$', but it is clear (using the pair axiom and the closure properties of bounded formulas) that a tuple of universal quantifications yields a formula equivalent (in $\mathsf{IKP}\omega$) to a Π_2-formula. The following theorem shows that the theory $\mathsf{IKP}\omega + \Delta_0$-$\mathsf{LEM} + \mathsf{MP}$ has a certain robustness.

Theorem 1 *The theory* $\mathsf{KP}\omega$ *is* Π_2*-conservative over* $\mathsf{IKP}\omega + \Delta_0$-$\mathsf{LEM} + \mathsf{MP}$.

Proof This is an easy consequence of the (Gödel and Gentzen) negative translation. The translation is extended to the bounded quantifiers in the natural way: $(\forall x \in z\, \phi(x))^g$ is $\forall x \in z\, \phi^g(x)$, and $(\exists x \in z\, \phi(x))^g$ is $\neg\neg\exists x \in z\, \phi^g(x)$. Note that the translation of a bounded formula is still a bounded formula. Therefore, a bounded formula is equivalent to its negative translation in $\mathsf{IKP}\omega + \Delta_0$-$\mathsf{LEM}$. From this it is clear that the translations of the axioms of extensionality, pair, union, infinity and Δ_0-separation are theorems of $\mathsf{IKP}\omega + \Delta_0$-$\mathsf{LEM}$. The negative translation of an instance of the scheme of foundation is still an instance of foundation. In order to argue that the negative translation of $\mathsf{KP}\omega$ is contained in $\mathsf{IKP}\omega + \Delta_0$-$\mathsf{LEM} + \mathsf{MP}$, it remains to study the scheme of Δ_0-collection. Well, the negative translation of an instance of Δ_0-collection has the form $\forall x \in z \neg\neg\exists y\, \phi^g(x, y) \to \neg\neg\exists w \forall x \in z \neg\neg\exists y \in w\, \phi^g(x, y)$, where ϕ is a bounded formula. In the presence of MP, the antecedent of the above implication is equivalent to $\forall x \in z \exists y \phi^g(x, y)$. Now, by an application of bounded collection, we get something stronger than the consequent of the implication.

We are ready to prove the theorem. Suppose that $\mathsf{KP}\omega$ proves $\forall x \exists y\, \phi(x, y)$, with ϕ a bounded formula. By the properties of the negative translation, the theory $\mathsf{IKP}\omega + \Delta_0\text{-}\mathsf{LEM} + \mathsf{MP}$ proves $\forall x \neg\neg \exists y\, \phi(x, y)$. Using MP, we get the desired conclusion. □

Definition 2 The independence of premises principle bIP_{Π_1} is the scheme

$$(\forall x\, \phi(x) \to \exists y\, \psi(y)) \to \exists y\, (\forall x\, \phi(x) \to \exists z \in y\, \psi(z)),$$

where ϕ is a bounded formula and ψ is any formula (parameters are allowed).

This principle is reminiscent of the independence of premises principle of Gödel's *dialectica* interpretation (cf. [1]). The analogy is not total because of the intrusion of a bounded quantification in the consequent of the existential consequent above. This is a crucial feature and it is in line with the bounded functional interpretation [11]. However, when the formula ψ is bounded then the bounded quantifier is not needed.

Lemma 1 $\mathsf{IKP}\omega + \Delta_0\text{-}\mathsf{LEM} + \mathsf{MP} + \mathsf{bIP}_{\Pi_1}$ *proves the following scheme for bounded formulas* ϕ *and* ψ *(parameters are allowed):* $(\forall x\, \phi(x) \to \exists y\, \psi(y)) \to \exists y \exists x\, (\phi(x) \to \psi(y))$.

Observation 1 Note that $(\forall x\, \phi(x) \to \exists y\, \psi(y)) \to \exists y\, (\forall x\, \phi(x) \to \psi(y))$ follows intuitionistically.

Proof of Lemma 1 Assume $\forall x\, \phi(x) \to \exists y\, \psi(y)$. By bIP_{Π_1}, take y_0 so that $\forall x\, \phi(x) \to \exists z \in y_0\, \psi(z)$. Since the consequent of the latter formula is bounded, by Proposition 1 there is x_0 such that $\phi(x_0) \to \exists z \in y_0\, \psi(z)$. By Δ_0-LEM, there are two cases to consider. If $\phi(x_0)$ holds, let z_0 be an element of y_0 such that $\psi(z_0)$. Otherwise, let z_0 be $\emptyset$. Clearly, $\phi(x_0) \to \psi(z_0)$. □

Proposition 2 *The theory* $\mathsf{IKP}\omega + \Delta_0\text{-}\mathsf{LEM} + \mathsf{MP} + \mathsf{bIP}_{\Pi_1}$ *proves* Δ_1*-separation, i.e., it proves* $\forall x\, (\forall u\, \phi(u, x) \leftrightarrow \exists v\, \psi(v, x)) \to \forall z \exists y \forall x\, (x \in y \leftrightarrow (x \in z \wedge \exists v\, \psi(v, x)))$, *for bounded formulas* ϕ *and* ψ *(possibly with parameters).*

Proof Suppose that $\forall x\, (\forall u\, \phi(u, x) \leftrightarrow \exists v\, \psi(v, x))$ and fix z. From the left-to-right direction and the above lemma, $\forall x \in z \exists u, v\, (\phi(u, x) \to \psi(v, x))$. By bounded collection, there is w such that

$$\forall x \in z \exists u, v \in w\, (\phi(u, x) \to \psi(v, x)).$$

It is easy to see that we can take $y = \{x \in z : \exists v \in w\, \psi(v, x)\}$. □

Corollary 1 *The theory* $\mathsf{IKP}\omega + \Delta_0\text{-}\mathsf{LEM} + \mathsf{bIP}_{\Pi_1} + \mathsf{MP}$ *proves the* Δ_1 *law of excluded middle, i.e., it proves* $(\forall u\, \phi(u) \leftrightarrow \exists v\, \psi(v)) \to (\forall u\, \phi(u) \vee \neg\forall u\, \phi(u))$, *for bounded formulas* ϕ *and* ψ *(possibly with parameters).*

Proof Let $y_0 = \{x \in \{0, 1\} : (x = 0 \wedge \exists u \neg\phi(u)) \vee (x = 1 \wedge \exists v\, \psi(v))\}$. This set exists by the previous proposition. It is clear that $1 \in y_0 \leftrightarrow \forall u\, \phi(u)$. We are done. □

In a personal communication, Makoto Fujiwara observed that the Δ_1 law of excluded middle is a consequence of $\mathsf{IKP}\omega + \Delta_0\text{-}\mathsf{LEM} + \mathsf{bIP}_{\Pi_1}$ (MP is not needed). We will see this in the similar situation of Proposition 6.

One of the characteristic principles of the bounded functional interpretation [11] is a generalization of weak Kőnig's lemma. In the second-order setting of Sect. 6 ahead, the classical contrapositive of (a restriction of) strict-Π^1_1 reflection (see Chap. VIII of [5]) takes the place of weak Kőnig's lemma. However, at a more fundamental level, the above mentioned characteristic principle of the bounded functional interpretation is better seen as the classical contrapositive of a collection scheme (it was dubbed bounded *contra-collection* scheme in [11]). Functional interpretations which treat bounded quantifications as computationally empty, as it is the case of the bounded functional interpretation [11] and the functional interpretation of this paper (see Sect. 5), enjoy the novelty of interpretating a bounded contra-collection scheme. As a consequence, they are able to interpret a semi-intuitionism that is able to accomodate principles like the lesser limited principle of omniscience LLPO of Errett Bishop (see [6] and also [7]). It also gives us a good theory of Δ_1-predicates. Let us first look at these matters in our first-order Kripke–Platek framework.

Definition 3 The principle of bounded contra-collection $\Delta_0\text{-}\mathsf{CColl}$ is the scheme

$$\forall z \exists y \in w \forall x \in z\, \phi(x, y) \to \exists y \in w \forall x\, \phi(x, y),$$

where ϕ is a bounded formula (possibly with parameters).

Note that, classically, this is just the bounded collection scheme. Bounded contra-collection easily generalizes for a tuple of z's. The proof of the following lemma presents the argument for a pair of z's:

Lemma 2 *For each bounded formula ϕ, the theory* $\mathsf{IKP}\omega + \Delta_0\text{-}\mathsf{CColl}$ *proves*

$$\forall x, z \exists y \in w \forall u \in x \forall v \in z\, \phi(u, v, y) \to \exists y \in w\, \forall u, v\, \phi(u, v, y).$$

Proof Suppose that $\forall x, z \exists y \in w\, \forall u \in x \forall v \in z\, \phi(u, v, y)$. We claim that

$$\forall s \exists y \in w \forall r \in s \forall u \in r \forall v \in r\, \phi(u, v, y).$$

Let s be given. Using the assumption with x and z taking the common value $\bigcup s$, we get $\exists y \in w \forall u \in \bigcup s\, \forall v \in \bigcup s\, \phi(u, v, y)$ and, as a consequence, the claim. By $\Delta_0\text{-}\mathsf{CColl}$, $\exists y \in w \forall r \forall u \in r\, \forall v \in r\, \phi(u, v, y)$. The result follows using the pair axiom. □

Proposition 3 *For bounded formulas ϕ and ψ, the theory* $\mathsf{IKP}\omega + \Delta_0\text{-}\mathsf{CColl}$ *proves*

$$\forall x, z\, (\forall u \in x\, \phi(u) \vee \forall v \in z\, \psi(v)) \to \forall u\, \phi(u) \vee \forall v\, \psi(v).$$

Proof Suppose that $\forall x, z\, (\forall u \in x\, \phi(u) \vee \forall v \in z\, \psi(v))$. It clearly follows that

$$\forall x, z \exists y \in \{0, 1\} \forall u \in x\, \forall v \in z\, ((y = 0 \wedge \phi(u)) \vee (y = 1 \wedge \psi(v))).$$

The result follows from the previous lemma. □

We are now able to derive the analogue of the lesser limited principle of omniscience in our setting (parameters are allowed):

Corollary 2 *For bounded formulas* ϕ *and* ψ*, the theory* $\mathsf{IKP}\omega + \Delta_0\text{-}\mathsf{LEM} + \Delta_0\text{-}\mathsf{CColl}$ *proves*

$$\forall u, v\, (\phi(u) \vee \psi(v)) \to \forall u\, \phi(u) \vee \forall v\, \psi(v).$$

Proof In the presence of Δ_0-LEM, it can easily be argued that $\forall u, v\, (\phi(u) \vee \psi(v))$ entails $\forall x, z\, (\forall u \in x\, \phi(x) \vee \forall v \in z\, \psi(z))$. Now, apply the previous proposition. □

The theory $\mathsf{IKP}\omega + \Delta_0\text{-}\mathsf{LEM} + \mathsf{MP} + \mathsf{bIP}_{\Pi_1} + \Delta_0\text{-}\mathsf{CColl}$ is a rather robust theory. For instance, this theory is able to prove the results described by Jon Barwise between Sects. 3 and 6 of chapter I of [5]. These results include the existence of ordered pairs, cartesian products and transitive closures as well as various forms of reflection and replacement. More importantly, this theory is well behaved regarding the introduction of Δ_1-relation symbols and Σ_1-function symbols. The arguments of the referred sections of [5] rely crucially on the fact that Σ_1-formulas are closed under conjunctions, disjunctions, bounded quantifications and existential quantifications. As observed, this is also the case in our intuitionistic setting. It *also* relies crucially on corresponding dual properties of Π_1-formulas. This is immediate in the classical setting. However, in our semi-constructive setting, one must proceed with some care. A Π_1-formula is a formula of the form $\forall z\, \psi(z)$, where ψ is a bounded formula. Note that a negation of a Π_1-formula is (equivalent to) a Σ_1-formula, thanks to MP. We claim that Π_1-formulas are closed (up to equivalence in $\mathsf{IKP}\omega + \Delta_0\text{-}\mathsf{LEM} + \mathsf{MP} + \mathsf{bIP}_{\Pi_1} + \Delta_0\text{-}\mathsf{CColl}$) under conjunctions, disjunctions, bounded quantifications and universal quantifications. This is clear for conjunctions and universal quantifications (bounded and unbounded). Corollary 2 entails that Π_1-formulas are close under disjunctions. The closure under existential bounded quantifications follows from Δ_0-CColl. One last word, the theory $\mathsf{IKP}\omega + \Delta_0\text{-}\mathsf{LEM} + \mathsf{MP} + \mathsf{bIP}_{\Pi_1} + \Delta_0\text{-}\mathsf{CColl}$ also allows definitions by Σ-recursion.

The last principle of this section is the *unrestricted* collection scheme. The functional interpretation of Sect. 5 is able to interpret it. This is possible because we are in an intuitionistic setting (this is analogous to the functional interpretation of [8]). In the classical setting of [10], only the bounded collection scheme is interpretable.

Definition 4 The principle of (unrestricted) collection Coll is the scheme

$$\forall y \in w \exists x \, \phi(x, y) \to \exists z \forall y \in w \exists x \in z \, \phi(x, y),$$

where ϕ is *any* formula (possibly with parameters).

3 On the Term Calculus of the Primitive Recursive Tree Functionals

It was said in the introduction that the functional interpretations of this paper uses the combinatory term calculus $\mathcal{L}_\Omega$ of the primitive recursive tree functionals. This term calculus is due to Howard (cf. [14]), and in [10] we have presented a streamlined version of it. Let us briefly go through $\mathcal{L}_\Omega$. We expand Gödel's language of "primitive recursive functions of finite-type" (see [1]) with a new ground type Ω for the countable constructive tree ordinals. The ground type of the natural numbers is denoted by N. The complex types are obtained from the ground types by closing under arrow. We use the Greek letters $\rho, \tau, \sigma, \ldots$ to denote the types. The language has a denumerable stock of variables $a, b, c, \ldots$ for each type. We follow [1] for notations and conventions concerning omission of parentheses. The constants of $\mathcal{L}_\Omega$ are:

(a) *Logical constants* or *combinators*. For each pair of types ρ, τ there is a combinator of type $\rho \to \tau \to \rho$ denoted by $\Pi_{\rho,\tau}$. For each triple of types δ, ρ, τ there is a combinator of type

$$(\delta \to \rho \to \tau) \to (\delta \to \rho) \to (\delta \to \tau)$$

denoted by $\Sigma_{\delta,\rho,\tau}$.

(b) *Arithmetical constants*. The constant 0_N of type N. The *successor* constant S of type $N \to N$. For each type ρ, a *(number) recursor* constant of type

$$N \to \rho \to (N \to \rho \to \rho) \to \rho$$

denoted by R^N_ρ.

(c) *Tree constants*. The constant 0_Ω of type Ω. The *supremum* constant *Sup* of type $(N \to \Omega) \to \Omega$. For each type ρ, a *tree recursor* constant of type

$$\Omega \to \rho \to ((N \to \Omega) \to (N \to \rho) \to \rho) \to \rho$$

denoted by R^Ω_ρ.

The above treatment is not completely rigorous because the recursors must operate simultaneously on tuples of variables (simultaneous recursion), and not only on a single variable. A rigorous treatment for arithmetic with simultaneous recursions is given in [15]. Another option would be to permit product types $\rho \times \tau$. We will not worry in this paper about these fine details. In any circumstance, a combinatory calculus with a notion of weak equality $=_w$ can be associated with the above. The conversions for the tree recursors can be found in [10], but can also be read from their set-theoretical interpretation (definition) in the next section. In this paper we follow the treatment of [13], including the way of defining lambda terms (abstration). Let us introduce some important terms (see [10] for more information).

i. $q^{\Omega} + 1 :\equiv Sup(\lambda x^N.q)$.
ii. It is possible to define by number recursion a closed term $d^{\Omega\to\Omega\to N\to\Omega}$ such that $d(q, s, 0) =_w q$ and $d(q, s, Sn) =_w s$. Let $\max(a, b) + 1 :\equiv Sup(\lambda x^N.d(a, b, x))$ (the notation '$\max(a, b) + 1$' should be viewed syncategorematically).
iii. We can define by number recursion a closed term $q^{N\to\Omega}$ such that $q0 =_w 0_\Omega$ and $q(Sn) =_w S(qn)$. We write n_Ω instead of qn.
iv. $\omega_\Omega :\equiv Sup(\lambda x^N.x_\Omega)$.
v. It is possible to define by tree recursion a term Sup^{-1} of type $\Omega \to (N \to \Omega)$ such that $Sup^{-1}(Sup(t)) =_w t$ for each term t of type $N \to \Omega$. We abbreviate $Sup^{-1}(q)(n)$ by $q\langle n\rangle$, for terms q^Ω and n^N. Clearly, $Sup(t)\langle n\rangle =_w tn$. We have also $(q + 1)\langle n\rangle =_w q$, $(\max(q, s) + 1)\langle 0\rangle =_w q$ and $(\max(q, s) + 1)\langle Sn\rangle =_w s$.
vi. Fix p a pairing term of type $N \to (N \to N)$ with inverse functions l and r, both of type $N \to N$. Hence, $p(l(n), r(n)) =_w n$, $l(p(m, k)) =_w m$ and $r(p(m, k)) =_w k$, for terms n, m, k of type N. Given $t^{N\to\Omega}$, we define $\bigsqcup t :\equiv Sup\,(\lambda y^N.(t(ly))\langle ry\rangle)$. This is a term of type Ω. An important particular case of this "square union" is $q \sqcup s :\equiv \bigsqcup(\lambda x^N.d(q, s, x))$, where q and s are of type Ω and d is the term introduced in (ii) above.

The next definition is a refinement of a similar definition in [2]:

Definition 5 Let t, q be terms of type Ω and r a term of type $N \to N$. We say that $t \sqsubseteq_r q$ if $t\langle x\rangle =_w q\langle rx\rangle$, where x is a fresh variable of type N.

Sometimes we write only $t \sqsubseteq q$ when the witnessing term r is presupposed.

Lemma 3 *Let t be a term of type $N \to \Omega$ and k^N. Then, for each term n of type N, $(\bigsqcup t)\langle p(k, n)\rangle =_w (tk)\langle n\rangle$ and (therefore) $tk \sqsubseteq_{\lambda x^N.p(k,x)} \bigsqcup t$. (Here, p is the pairing function of* (vi) *above.) In particular, for q and s of type Ω, $q \sqsubseteq q \sqcup s$ and $s \sqsubseteq q \sqcup s$.*

Proof $(\bigsqcup t)\langle p(k, n)\rangle =_w (\lambda y^N.(t(ly))\langle ry\rangle)(p(k, n)) =_w (tk)\langle n\rangle$. □

In order to deal with the functional interpretation of Sect. 5, we need to lift some of the above notions from the type Ω to so-called *pure* Ω-types (i.e., types obtained only from the ground type Ω by means of the arrow). The lifting is done pointwise. As a consequence, we need to have a stronger notion of equality: one that incorporates some extensionality. We use *combinatory extensional equality*, as explained in [13]. This notion allows for the so-called ζ-rule: from $tx = qx$ infer $t = q$, where x is a fresh variable (of appropriate type). This rule is equivalent to the ξ-rule: from $t = q$ infer $\lambda x.t = \lambda x.q$. This is proved in [13]. Note that the ξ-rule is automatic in a direct treatment (i.e., not via combinators) of the lambda calculus. In a direct treatment of the lambda calculus, the extensionality needed is given by the η-axiom scheme: $\lambda x.qx = q$, where q is a term in which the variable x does not occur free. On the other hand, in combinatory logic, with a proper definition of abstraction (as in [13], and which we follow), the η-axiom scheme is automatic. These are somewhat technical issues relating the combinatory calculus with the lambda calculus. To cut through the fog, the proper way to state extensional equality in the combinatory calculus is the above ζ-rule. For more information, consult [13]. Risking confusion (but essentially following the notation of [13]), we use the notation $=_{\beta\eta}$ for combinatory extensional equality. Of course, if $t =_w q$ then $t =_{\beta\eta} q$.

vii. If q is a term of pure Ω-type τ of the form $\tau_1 \to \cdots \to \tau_k \to \Omega$, one defines $q + 1 :\equiv \lambda \underline{z}.((q\underline{z}) + 1)$, where $\underline{z}$ abbreviates a tuple of variables $z^{\tau_1}, \ldots, z^{\tau_k}$. If t is a term of type $N \to \tau$, we let $Sup(t) :\equiv \lambda \underline{z}.Sup(\lambda n.tn\underline{z})$.
viii. If q is as above and n is of type N, let $q\langle n\rangle :\equiv \lambda \underline{z}.((q\underline{z})\langle n\rangle)$. We claim that $(q+1)\langle n\rangle =_{\beta\eta} q$. Therefore, $(q+1)\langle n\rangle =_{\beta\eta} (\lambda \underline{z}.((q\underline{z})+1))\langle n\rangle =_{\beta\eta} \lambda \underline{z}.(((q\underline{z})+1)\langle n\rangle) =_{\beta\eta} \lambda \underline{z}.q\underline{z} =_{\beta\eta} q$. It is in the penultimate equality that extensionality is used (a ξ-rule application). If t is a term of type $N \to \tau$, where τ is the pure Ω-type above, we have $Sup(t)\langle n\rangle =_{\beta\eta} \lambda \underline{z}.tn\underline{z}$. To see this, notice that $Sup(t)\langle n\rangle =_{\beta\eta} \lambda \underline{z}.((Sup(\lambda n.tn\underline{z}))\langle n\rangle) =_{\beta\eta} \lambda \underline{z}.tn\underline{z}$. We are using (v) and the ξ-rule application in the last equality.
ix. Again, if t is a term of type $N \to \tau$, where τ is the pure Ω-type above, we define $\bigsqcup t :\equiv \lambda \underline{z}.(\bigsqcup \lambda x^N.(tx\underline{z}))$. If q and s are of type τ, we define $q \sqcup s$ pointwise in analogy to (vi) above: $q \sqcup s :\equiv \lambda \underline{z}.\bigsqcup \lambda x^N.d(q\underline{z}, s\underline{z}, x)$.

Definition 6 Let t, q be terms of pure Ω-type τ and let r a term of type $N \to N$. We say that $t \sqsubseteq_r q$ if $t\langle x\rangle =_{\beta\eta} q\langle rx\rangle$, where x is a fresh variable of type N.

One should see this definition as superseding Definition 5 and the next lemma as superseding Lemma 3:

Lemma 4 *Let t be a term of type $N \to \tau$, where τ is a pure Ω-type. Let k and n be terms of type N. Then $(\bigsqcup t)\langle p(k,n)\rangle =_{\beta\eta} (tk)\langle n\rangle$ and (therefore) $tk \sqsubseteq_{\lambda x^N.p(k,x)} \bigsqcup t$. In particular, for q and s of type τ, $q \sqsubseteq q \sqcup s$ and $s \sqsubseteq q \sqcup s$.*

Proof Suppose τ is $\tau_1 \to \cdots \to \tau_r \to \Omega$. Then, if $\underline{z}$ stands for a tuple of variables $z_1^{\tau_1}, \ldots, z_r^{\tau_r}$, we have

$$(\bigsqcup t)\langle p(k,n)\rangle =_{\beta\eta} (\lambda \underline{z}.\bigsqcup \lambda x^N.(tx\underline{z}))\langle p(k,n)\rangle =_{\beta\eta} \lambda \underline{z}.((\bigsqcup x^N.(tx\underline{z}))\langle p(k,n)\rangle) =_{\beta\eta}$$

$$\lambda \underline{z}.((tk\underline{z})\langle n\rangle) =_{\beta\eta} (tk)\langle n\rangle$$

Extensionality is used in the penultimate equality. □

We were somewhat careful (perhaps even pedantic) in the discussion of extensionality because there is a *faux pas* in Sect. 6 of [10]. In that paper, we discussed a so-called internalization of an interpretation introduced in a previous section (in the present paper we do not discuss internalizations). The referred internalization is given by the intensional model of Sect. 9.3 of [1] (a version of the hereditarily recursive operations for our setting). However, the conclusion of the soundness theorem is not verified in this structure because of a lack of extensionality: the above discussed $\beta\eta$ equalities do not hold in the intensional model. The problem is easily fixed, though. The internalization should have been done with the analogue of the hereditarily effective operations for our setting (also discussed in 9.3 of [1]).

4 Brief Semantical Considerations

In the next section we define a functional interpretation of the theory $\mathsf{IKP}\omega + \Delta_0\text{-}\mathsf{LEM} + \mathsf{MP} + \mathsf{bIP}_{\Pi_1} + \Delta_0\text{-}\mathsf{CColl} + \mathsf{Coll}$ and prove an appropriate soundness theorem. The conclusion of the soundness theorem is not verified in a formal theory (as it is the tradition in functional interpretations) but it is rather semantically verified, i.e., seen to be true in a certain structure. This is a simplificatory option of this paper. In principle, one can refine the soundness theorem in order to have a verification in a suitable formal theory, as we did in Sect. 6 of [10] with an "internalization" of the semantical interpretation presented there in Sects. 2 and 3. With verifications in formal theories, one can obtain stronger results (viz., conservation results) but, as noted at the end of the previous section, one has to proceed carefully and this is would make the present paper too long and perhaps even distracting from its main objective. So, in this paper, we will proceed semantically in the verification of the soundness theorems.

The interpretation of the term calculus of this paper is common to all the interpretations of this paper. It is the (full and extensional) set-theoretical structure $\langle S_\rho \rangle$ of Sect. 9.4 of [1] (see also [10]). The variables of each type ρ of $\mathcal{L}_\Omega$ range over S_ρ. These sets are defined thus:

1. $S_N = \omega$
2. S_Ω is the smallest set W which contains 0 and is such that, whenever f is a function that maps ω into W, then the ordered pair $(1, f)$ is in W.
3. $S_{\rho\to\tau} = \{f : f \text{ is a set-theoretic function that maps } S_\rho \text{ into } S_\tau\}$

It is clear how the terms of $\mathcal{L}_\Omega$ are interpreted in the set-theoretical model (see [10] for some details). We only note that the constant *Sup* is interpreted by the function which, on input $f \in S_{N\to\Omega}$, outputs the element $(1, f)$ of W. We also remark that, to each element c of W, we can associate a countable set-theoretical ordinal $|c|$ so that $|0| = 0$ and, for a function $f : \omega \to W$, $|Sup(f)| = \sup\{|f(n)| + 1 : n \in \omega\}$. Observe that $|f(n)| < |Sup(f)|$, for each natural number n. It is well-known that the first uncountable ordinal ω_1 is the supremum of all $|c|$, with $c \in W$. The previous discussion also permits to define, by classical ordinal recursion, the interpretations of the tree recursors R^{Ω}_{ρ} so that:

$$R^{\Omega}_{\rho}(0_{\Omega}, a, F) = a \text{ and } R^{\Omega}_{\rho}(Sup(f), a, F) = F(f, \lambda x^N.R^{\Omega}_{\rho}(f(x), a, F)),$$

for all $a \in S_\rho$ and F a function that maps $S_{N\to\rho}$ into $(S_\rho)^{S_{N\to\rho}}$.

Clearly, the equalities (both $=_w$ and $=_{\beta\eta}$) established in the previous section between terms of $\mathcal{L}_\Omega$ give rise to set-theoretical equalities in $\langle S_\rho \rangle$.

Lemma 5 *(i) If $c \in W$ and $c \neq 0$, then $|c| = Sup_{n\in\omega}(|c\langle n\rangle| + 1)$.*
(ii) If $c, d \in W$ and $c \sqsubseteq d$ then $|c| \leq |d|$.

Proof (ii) is an immediate consequence of (i). Given $c \neq 0$, let $c = \sup f$, for a certain $f : \omega \to W$. Then $|c| = \sup\{|f(n)| + 1 : n \in \omega\} = \sup\{|c\langle n\rangle| + 1 : n \in \omega\}$.□

Lemma 6 *Let $b, c : \rho \to \tau$ and $a : \rho$ be term variables of pure Ω-types and f be a term variable of type $N \to N$. Then the implication $b \sqsubseteq_f c \to ba \sqsubseteq_f ca$ is true in the set-theoretical structure $\langle S_\rho \rangle$ for any values of the variables.*

Proof Let τ be $\tau_1 \to \cdots \to \tau_k \to \Omega$. Take n a natural number. Since $b \sqsubseteq_f c$, then $b\langle n\rangle = c\langle fn\rangle$ holds set-theoretically. Note that $\rho \to \tau$ is $\rho \to \tau_1 \to \cdots \to \tau_k \to \Omega$. Hence, $\lambda w^\rho, \underline{z}.((bw\underline{z})\langle n\rangle) = \lambda w^\rho, \underline{z}.((cw\underline{z})\langle fn\rangle)$ holds set-theoretically (where $\underline{z}$ is a tuple of variables $z_1^{\tau_1}, \ldots, z_k^{\tau_k}$). Therefore, $(\lambda w^\rho, \underline{z}.((bw\underline{z})\langle n\rangle))a = (\lambda w^\rho, \underline{z}.((cw\underline{z})\langle fn\rangle))a$ holds set-theoretically and, hence, $\lambda\underline{z}.((ba\underline{z})\langle n\rangle) = \lambda\underline{z}.((ca\underline{z})\langle fn\rangle)$ holds set-theoretically. In sum, $ba \sqsubseteq_f ca$. □

Finally, before discussing our so-called mixed structures, observe that the interpretation of a *closed* term t of ground type Ω is an element of W and, therefore, has an associated set-theoretical ordinal which, with abuse of notation, we denote by $|t|$. The Bachmann–Howard ordinal is the supremum of all these ordinals.

The functional interpretation of the next section translates a formula ϕ of the language of set theory into formulas ϕ^{B} and ϕ_{B} of a mixed language $\mathcal{L}_\Omega^{\mathrm{mix}}$. A version of this language was introduced in [10]. Let us briefly describe it. The mixed language $\mathcal{L}_\Omega^{\mathrm{mix}}$ has three kinds of terms: the terms of $\mathcal{L}_\Omega$ (including, of course, the variables a, b, c, etc. of $\mathcal{L}_\Omega$), the set-theoretical variables x, y, z, etc. and (set) terms M_t, where t is a term of $\mathcal{L}_\Omega$. The *atomic formulas* of $\mathcal{L}_\Omega^{\mathrm{mix}}$ are the formulas of the form $x = y$, $x \in y$ or $x \in M_t$, for x and y set-theoretical variables and t a term of $\mathcal{L}_\Omega$ of type Ω. The *bounded mixed formulas* are generated from the atomic formulas by means of the propositional logical connectives $\neg$ and $\wedge$ and quantifications of the form $\forall x \in y$, $\forall x \in M_t$ and $\forall n^N$ (note that this last quantifier is classified as bounded). Since our semantics is classical, there is no need to introduce other connectives and quantifiers, as they can be defined. The *formulas* of $\mathcal{L}_\Omega^{\mathrm{mix}}$ are generated from the bounded mixed formulas by means of propositional connectives and quantifications of the form $\forall a^\rho$, where a is a term variable (of a certain type ρ) of the term language $\mathcal{L}_\Omega$. Observe that we do not need unbounded set-theoretic quantifications.

The two basic interpretations for $\mathcal{L}_\Omega^{\text{mix}}$ are the structures $L_{\omega_1}^{\text{mix}}$ and $V_{\omega_1}^{\text{mix}}$. In both of these structures, the terms of $\mathcal{L}_\Omega$ (and, therefore, the range of term variables) are interpreted set-theoretically in $\langle S_\rho \rangle$ (as described by points 1, 2 and 3 above). The set-theoretic variables range over L_{ω_1}, respectively V_{ω_1}. The terms M_t are interpreted as $L_{|t|}$ in $L_{\omega_1}^{\text{mix}}$ and as $V_{|t|}$ in $V_{\omega_1}^{\text{mix}}$. Abusing language, we often replace the term M_t by the notations L_t or V_t according to the interpretation that we have in mind.

5 The Main Functional Interpretation

We are going to associate to each formula $\phi(x_1, \ldots, x_n)$ of the language of set theory (free variables as shown) a bounded mixed formula ϕ_{B} of the form

$$\phi_{\text{B}}(a_1, \ldots, a_k, b_1, \ldots, b_m, x_1, \ldots, x_n),$$

with the free variables as shown (the a's and the b's are variables of $\mathcal{L}_\Omega$ of pure Ω-type) and also the following formula $\phi^{\text{B}}(x_1, \ldots, x_n)$ of the language $\mathcal{L}_\Omega^{\text{mix}}$:

$$\exists a_1 \cdots \exists a_k \forall b_1 \cdots \forall b_m \, \phi_{\text{B}}(a_1, \ldots, a_k, b_1, \ldots, b_m, x_1, \ldots, x_n).$$

Note that k or m (or both) can be zero. For notational simplicity, we simply write $\phi(x)$, $\phi^{\text{B}}(x)$ and $\phi_{\text{B}}(a, b, x)$, instead of carrying the tuple notation. Many times we also omit the parameters x.

Definition 7 To each formula ϕ of the language of set theory (possibly with parameters), we assign formulas ϕ^{B} and ϕ_{B} so that ϕ^{B} is of the form $\exists a \forall b \, \phi_{\text{B}}(a, b)$, with $\phi_{\text{B}}(a, b)$ a bounded mixed formula, according to the following clauses:

1. ϕ^{B} and ϕ_{B} are simply ϕ, for bounded formulas ϕ of the language of set theory.

For the *remaining* cases, if we have already interpretations for ϕ and ψ given by $\exists a \forall b \, \phi_{\text{B}}(a, b)$ and $\exists d \forall e \, \psi_{\text{B}}(d, e)$ (respectively), then we define:

2. $(\phi \wedge \psi)^{\text{B}}$ is $\exists a, d \forall b, e \, [\phi_{\text{B}}(a, b) \wedge \psi_{\text{B}}(d, e)]$,
3. $(\phi \vee \psi)^{\text{B}}$ is $\exists a, d \forall b, e \, [\forall n \, \phi_{\text{B}}(a, b\langle n \rangle) \vee \forall m \, \psi_{\text{B}}(d, e\langle m \rangle)]$,
4. $(\phi \to \psi)^{\text{B}}$ is $\exists B, D \forall a, e \, [\forall n \, \phi_{\text{B}}(a, (Bae)\langle n \rangle) \to \psi_{\text{B}}(Da, e)]$,
5. $(\forall x \in z \, \phi(x, z))^{\text{B}}$ is $\exists a \forall b \, [\forall x \in z \, \phi_{\text{B}}(a, b, x, z)]$,
6. $(\exists x \in z \, \phi(x, z))^{\text{B}}$ is $\exists a \forall b \, [\exists x \in z \forall n \, \phi_{\text{B}}(a, b\langle n \rangle, x, z)]$,
7. $(\forall x \phi(x))^{\text{B}}$ is $\exists A \forall c^{\Omega}, b \, [\forall x \in L_c \, \phi_{\text{B}}(Ac, b, x)]$,
8. $(\exists x \phi(x))^{\text{B}}$ is $\exists c^{\Omega}, a \forall b \, [\exists x \in L_c \forall n \, \phi_{\text{B}}(a, b\langle n \rangle, x)]$.

The lower B-translations are displayed inside the square parentheses. Note that they are bounded mixed formulas. The following lemma is an immediate consequence of the definitions.

Lemma 7 *Let ϕ be a Δ_0-formula. Then:*

(i) $(\exists x \phi(x))^B$ *is* $\exists c^{\Omega}[\exists x \in L_c\, \phi(x)]$.
(ii) $(\forall x \exists y \phi(x, y))^B$ *is* $\exists A^{\Omega \to \Omega} \forall c^{\Omega}[\forall x \in L_c \exists y \in L_{Ac}\, \phi(x, y)]$.

As it is usual with this kind of functional interpretations, we have the following crucial property:

Lemma 8 (Monotonicity property) *Let $\phi(x)$ be a formula of the language of set theory. In $L_{\omega_1}^{\text{mix}}$ one has the implication $a \sqsubseteq_f c \wedge \phi_B(a, b, x) \to \phi_B(c, b, x)$.*

Proof It is clear that the clauses 2, 3, 5 and 6 preserve this property. By Lemma 6, clause 7 preserves the property, and by Lemma 5 (ii), so does clause 8. Let us look now at clause 4. Suppose that $\forall n\, \phi_{\text{B}}(a, (Bae)\langle n \rangle) \to \psi_{\text{B}}(Da, e)$ holds in $L_{\omega_1}^{\text{mix}}$ and that $B \sqsubseteq_f B'$ and $D \sqsubseteq_g D'$. Assume that $\forall n\, \phi_{\text{B}}(a, (B'ae)\langle n \rangle)$. Let k be an arbitrary natural number. Since $B \sqsubseteq_f B'$, by two applications of Lemma 6, we get $Bae \sqsubseteq_f B'ae$. Hence, $(Bae)\langle k \rangle = (B'ae)\langle fk \rangle$. By the assumption, we have $\phi_{\text{B}}(a, (B'ae)\langle fk \rangle)$. Now, by the arbitrariness of k, we may conclude that $\forall k\, \phi_{\text{B}}(a, (Bae)\langle k \rangle)$. By hypothesis we may infer $\psi_{\text{B}}(Da, e)$ and, therefore by Lemma 6 and the induction hypothesis, $\psi_{\text{B}}(D'a, e)$. We are done. □

We are now ready to state and prove the soundness theorem of the functional interpretation.

Theorem 2 (Soundness Theorem) *Let ϕ be a sentence of the language of set theory. Suppose that* $\mathsf{IKP}\omega + \Delta_0\text{-}\mathsf{LEM} + \mathsf{MP} + \mathsf{bIP}_{\Pi_1} + \Delta_0\text{-}\mathsf{CColl} + \mathsf{Coll} \vdash \phi$. *Then there are closed terms t of $\mathcal{L}_{\Omega}$ such that, for appropriate types ρ,*

$$L_{\omega_1}^{\text{mix}} \models \forall b^{\rho}\, \phi_B(t, b).$$

Proof The proof is by induction on the length of the derivation. We show that if a formula $\phi(w)$ is provable in the theory of the theorem, then there are closed terms t of $\mathcal{L}_{\Omega}$ such that, for appropriate types ρ, we have

$$L_{\omega_1}^{\text{mix}} \models \forall c^{\Omega} \forall b^{\rho}\, [\forall w \in L_c\, \phi_{\text{B}}(tc, b, w)],$$

where $\phi(w)^{\text{B}}$ is $\exists a \forall b\, \phi_{\text{B}}(a, b, w)$.

For ease of reading, in the following we ignore parameters that do not play an important role in the proof of the theorem. For the logical part of the theory, we rely on the formalization of intuitionistic logic given in [1]. The verification of these axioms and rules has some rough similarities with the verifications of [11]. In the following, we take ϕ and ψ as in Definition 7, and γ with γ^{B} given by $\exists u \forall v\, \gamma_{\text{B}}(u, v)$. Let us now discuss each rule and axiom:

1. $\phi, \phi \to \psi \Rightarrow \psi$. By induction hypothesis, there are terms t, r and s such that $\forall b\, \phi_{\text{B}}(t, b)$ and $\forall a, e\, [\forall n\, \phi_{\text{B}}(a, (rae)\langle n \rangle) \to \psi_{\text{B}}(sa, e)]$ hold in $L_{\omega_1}^{\text{mix}}$. Let $q :\equiv st$. It is clear that we can conclude $\forall e\, \psi_{\text{B}}(q, e)$.

2. $\phi \to \psi, \psi \to \gamma \Rightarrow \phi \to \gamma$. By hypothesis we have terms t, s, r and q such that the following holds in $L^{\text{mix}}_{\omega_1}$: (i) $\forall a, e\, [\forall n\, \phi_{\text{B}}(a, (sae)\langle n\rangle) \to \psi_{\text{B}}(ta, e)]$ and (ii) $\forall d, v\, [\forall m\, \psi_{\text{B}}(d, (rdv)\langle m\rangle) \to \gamma_{\text{B}}(qd, v)]$. Take $l :\equiv \lambda a, v. \bigsqcup_m s(a, (r(ta, v))\langle m\rangle)$ and $o :\equiv \lambda a.q(ta)$. We show that $L^{\text{mix}}_{\omega_1} \models \forall a, v\, [\forall k\, \phi_{\text{B}}(a, (lav)\langle k\rangle) \to \gamma(oa, v)]$. Take a and v and suppose that $\forall k\, \phi_{\text{B}}(a, (lav)\langle k\rangle)$. Then, for every n, m, we have $\phi_{\text{B}}(a, (lav)\langle p(m, n)\rangle)$ (here p is the pairing term). By Lemma 4, $(lav)\langle p(m, n)\rangle = (s(a, (r(ta, v))\langle m\rangle))\langle n\rangle$. Hence, fixing m, we have $\forall n\, \phi_{\text{B}}(a, (s(a, (r(ta, v))\langle m\rangle))\langle n\rangle)$. Particularizing (i) with e as $(r(ta, v))\langle m\rangle$, we get $\psi_{\text{B}}(ta, (r(ta, v))\langle m\rangle)$. By the arbitrariness of m and (ii), we conclude $\gamma(q(ta), v)$. That is what we want.

3a. $\phi \vee \phi \to \phi$. A simple computation of $(\phi \vee \phi \to \phi)^{\text{B}}$ shows that we must find terms q_1, q_2 and t such that, for all a, a', b'' the following holds in $L^{\text{mix}}_{\omega_1}$:

$$\forall k, k' (\forall n\, \phi_{\text{B}}(a, (q_1 aa'b'')\langle k\rangle\langle n\rangle) \vee \forall m\, \phi_{\text{B}}(a', (q_2 aa'b''')\langle k'\rangle\langle m\rangle) \to \phi_{\text{B}}(taa', b'').$$

We claim that the above is true with $t :\equiv \lambda a, a'.(a \sqcup a')$, and with both q_1 and q_2 as $\lambda a, a', b''.(b'' + 2)$. In effect, the antecedent above entails $\phi_{\text{B}}(a, b'') \vee \phi_{\text{B}}(a', b'')$. By the monotonicity property, we get $\phi(a \sqcup a', b'')$, as wanted.

3b. $\phi \to \phi \wedge \phi$. We must find terms t_1, t_2 and q such that

$$\forall a, b', b''\, [\forall n\, \phi_{\text{B}}(a, (qab'b'')\langle n\rangle) \to \phi_{\text{B}}(t_1 a, b') \wedge \phi_{\text{B}}(t_2 a, b'')]$$

holds in $L^{\text{mix}}_{\omega_1}$. Let t_1 and t_2 be the term $\lambda a.a$ and $q :\equiv \lambda a, b', b''.((b' + 1) \sqcup (b'' + 1))$. If $\forall n\, \phi_{\text{B}}(a, ((b' + 1) \sqcup (b'' + 1))\langle n\rangle)$ we have, in particular $\phi_{\text{B}}(a, b')$ and $\phi_{\text{B}}(a, b'')$ because $((b' + 1) \sqcup (b'' + 1))\langle p(0, n)\rangle = b'$ and $((b' + 1) \sqcup (b'' + 1))\langle p(1, n)\rangle = b''$, where p is the pairing function.

4a. $\phi \to \phi \vee \psi$. We must find terms q, t and r such that

$$\forall a, b', e\, [\forall k\, \phi_{\text{B}}(a, (rab'e)\langle k\rangle) \to \forall n\, \phi_{\text{B}}(qa, b'\langle n\rangle) \vee \forall m\, \psi_{\text{B}}(ta, e\langle m\rangle)]$$

holds in $L^{\text{mix}}_{\omega_1}$. It is clear that $q :\equiv \lambda a.a$, $t :\equiv \lambda a.0$, and $r :\equiv \lambda a, b', e.b'$ works (here, 0 is the usual zero constant of appropriate pure Ω-type).

4b. $\phi \wedge \psi \to \phi$. We must find terms q, t and r such that

$$\forall a, d, b'\, [\forall n \forall m\, (\phi_{\text{B}}(a, (tadb')\langle n\rangle) \wedge \psi_{\text{B}}(d, (radb')\langle m\rangle)) \to \phi_{\text{B}}(qad, b')]$$

holds in $L^{\text{mix}}_{\omega_1}$. Clearly, $q :\equiv \lambda a, d.a$, $t :\equiv \lambda a, d, b'.(b' + 1)$ and $r :\equiv \lambda a, d, b'.0$ works.

5a. $\phi \vee \psi \to \psi \vee \phi$. We must find terms q, t, r and s such that

$$\forall a, b, b', e'\, [\forall k \forall k'\, (\forall n\, \phi_{\text{B}}(a, (qabb'e')\langle k\rangle\langle n\rangle) \vee \forall m\, \psi_{\text{B}}(d, (tabb'e')\langle k'\rangle\langle m\rangle)) \to$$

$$\forall n\, \phi_{\text{B}}(rad, b'\langle n\rangle) \vee \forall m\, \psi_{\text{B}}(sad, e'\langle m\rangle)]$$

holds in $L^{\text{mix}}_{\omega_1}$. It is clear that $r :\equiv \lambda a, d.a$, $s :\equiv \lambda a, d.d$, $q :\equiv \lambda a, b, b', e'.(b' + 1)$ and $t :\equiv \lambda a, b, b', e'.(e' + 1)$ work.

5b. $\phi \wedge \psi \rightarrow \psi \wedge \phi$. Clear.

6. $\phi \rightarrow \psi \Rightarrow (\gamma \vee \phi \rightarrow \gamma \vee \psi)$. By hypothesis we have terms t and q such that $\forall a, e\, [\forall k\, \phi_{\mathrm{B}}(a, (tae))\langle k\rangle \rightarrow \psi_{\mathrm{B}}(qa, e)]$. We must show that there are terms o, l, r and s such that

$$\forall u, a, v, e\, [\forall k \forall k' (\forall n\, \gamma_{\mathrm{B}}(u, (ruave)\langle k\rangle\langle n\rangle) \vee \forall m\, \phi_{\mathrm{B}}(a, (suave)\langle k'\rangle\langle m\rangle)) \rightarrow$$

$$\forall n\, \gamma(oua, v\langle n\rangle) \vee \forall m\, \psi(lua, e\langle m\rangle)]$$

holds in $L^{\mathrm{mix}}_{\omega_1}$. Let us define $o :\equiv \lambda u, a.u$, $l :\equiv \lambda u, a.qa$, $r :\equiv \lambda u, a, v, e.(v+1)$ and $s :\equiv \lambda u, a, v, e.((\bigsqcup_i t(a, e\langle i\rangle)) + 1)$. Suppose the antecedent. Particularizing for $k = k' = 0$, we get $\forall n\, \gamma_{\mathrm{B}}(u, v\langle n\rangle) \vee \forall m\, \phi_{\mathrm{B}}(a, (\bigsqcup_i t(a, e\langle i\rangle))\langle m\rangle)$. If we breakhave the first disjunct, we are done. Otherwise $\forall k, m\, \phi_{\mathrm{B}}(a, (\bigsqcup_i t(a, e\langle i\rangle))\langle p(m, k)\rangle)$, where p is the pairing term. Therefore, by Lemma 4, $\forall k, m\, \phi_{\mathrm{B}}(a, t(a, e\langle m\rangle)\langle k\rangle)$. The hypothesis now entails $\forall m\, \psi_{\mathrm{B}}(qa, e\langle m\rangle)$.

7a. $\phi \wedge \psi \rightarrow \gamma \Rightarrow \phi \rightarrow (\psi \rightarrow \gamma)$. By hypothesis, there are terms t, q and r such that $\forall a, d, v\, [\forall n \forall m\, (\phi_{\mathrm{B}}(a, (tadv)\langle n\rangle) \wedge \psi_{\mathrm{B}}(d, (qadv)\langle m\rangle)) \rightarrow \gamma_{\mathrm{B}}(rad, v)]$ holds in $L^{\mathrm{mix}}_{\omega_1}$, and we must obtain terms t', q' and r' such that the following also holds:

$$\forall a, d, v\, [\forall n\, \phi_{\mathrm{B}}(a, (t'adv)\langle n\rangle) \rightarrow (\forall m\, \psi_{\mathrm{B}}(d, (q'adv)\langle m\rangle) \rightarrow \gamma_{\mathrm{B}}(rad, v))].$$

Of course, $t' :\equiv t$, $q' :\equiv q$ and $r' :\equiv r$ work.

7b. $\phi \rightarrow (\psi \rightarrow \gamma) \Rightarrow \phi \wedge \psi \rightarrow \gamma$. Similar to 7a.

8. $\bot \rightarrow \phi$. Clear.

9. $\phi \rightarrow \psi(w) \Rightarrow \phi \rightarrow \forall w\, \psi(w)$, where w does not occur free in ϕ. By hypothesis, there are terms t and q such that

$$\forall c^{\Omega}, a, e\, [\forall w \in L_c\, (\forall n\, \phi_{\mathrm{B}}(a, (tcae)\langle n\rangle) \rightarrow \psi_{\mathrm{B}}(qca, e, w))]$$

holds in $L^{\mathrm{mix}}_{\omega_1}$. The interpretation asks for terms r and s such that

$$\forall c^{\Omega}, a, e\, [\forall n\, \phi_{\mathrm{B}}(a, (race)\langle n\rangle) \rightarrow \forall w \in L_c\, \psi_{\mathrm{B}}(sac, e, w)].$$

This is clear.

10. $\forall x\, \phi(x) \rightarrow \phi(w)$. A computation of the upper-B translation of this formula shows that we must find terms t, q and r such that for all $c \in W$ and A and b' of appropriate types in the set-theoretical structure $\langle S_\rho \rangle$,

$$\forall w \in L_c\, (\forall n, m \forall x \in L_{(qcAb')\langle n\rangle}\, \phi_{\mathrm{B}}(A((qcAb')\langle n\rangle), (rcAb')\langle m\rangle, x) \rightarrow \phi_{\mathrm{B}}(tcA, b', w))$$

holds in $L^{\mathrm{mix}}_{\omega_1}$. It is clear that the terms $t :\equiv \lambda c, A.Ac$, $q :\equiv \lambda c, A, b'.(c+1)$ and $r :\equiv \lambda c, A, b'.(b'+1)$ do the job.

11. $\phi(w) \to \exists x\, \phi(x)$. We must find terms t, q and r such that

$$\forall c^{\Omega}, a, b\, [\forall w \in L_c\, (\forall n\, \phi_{\mathrm{B}}(a, (qcab)\langle n\rangle, w) \to \exists x \in L_{tca} \forall n\, \phi_{\mathrm{B}}(rca, b\langle n\rangle, x))]$$

holds in $L^{\mathrm{mix}}_{\omega_1}$. It is clear that $t :\equiv \lambda c, a.c$, $q :\equiv \lambda c, a, b.b$ and $r :\equiv \lambda c, a.a$ do the job.

12. $\phi(w) \to \psi \Rightarrow \exists w\, \phi(w) \to \psi$, where w does not occur free in ψ. By hypothesis there are terms t and q such that

$$\forall c^{\Omega}, a, e\, [\forall w \in L_c\, (\forall n\, \phi_{\mathrm{B}}(a, (tcae)\langle n\rangle, w) \to \psi_{\mathrm{B}}(qca, e))]$$

holds in $L^{\mathrm{mix}}_{\omega_1}$. The interpretation asks for terms r and s such that

$$\forall c^{\Omega}, a, e\, [\forall m \exists w \in L_c\, \forall n\, \phi_{\mathrm{B}}(a, (rcae)\langle m\rangle\langle n\rangle, w) \to \psi_{\mathrm{B}}(sca, e)].$$

It is clear that $s :\equiv q$ and $r :\equiv \lambda c, a, e.((tcae) + 1)$ work.

If we do not count the axioms of equality of $\mathsf{IKP}\omega$, we are done with the logical part. The axioms of equality pose no problem because they can be taken as universal formulas and, hence, are interpreted by themselves. Before proceeding to the mathematical axioms of $\mathsf{IKP}\omega$, we must still pay attention to the four axioms that regulate the primitive bounded quantifiers $\forall x \in z\, (\ldots)$ and $\exists x \in z\, (...)$.

In order to study universal bounded quantification, we compute the upper-B translations of $\forall x \in z\, \phi(x, z)$ and $\forall x\, (x \in z \to \phi(x, z))$. They are $\exists a \forall b\, [\forall x \in z\, \phi_{\mathrm{B}}(a, b, x, z)]$ and $\exists A \forall c^{\Omega}, b\, [\forall x \in L_c\, (x \in z \to \phi_{\mathrm{B}}(Ac, b, x, z))]$, respectively.

<u>The axiom</u> $\forall x \in z\, \phi(x, z) \to \forall x\, (x \in z \to \phi(x, z))$. It is easy to see that we must obtain terms t and q such that for all $e, c \in W$ and a and b of appropriate types in the set-theoretical structure $\langle S_\rho \rangle$, and for all $z \in L_{|e|}$, the statement

$$\forall n \forall x \in z\, \phi_{\mathrm{B}}(a, (tecab)\langle n\rangle, x, z) \to \forall x \in L_c\, (x \in z \to \phi_{\mathrm{B}}(qeac, b, x, z))$$

holds in $L^{\mathrm{mix}}_{\omega_1}$. This is clearly the case with $t :\equiv \lambda e, c, a, b.(b + 1)$ and $q :\equiv \lambda e, a, c.a$.

<u>The axiom</u> $\forall x\, (x \in z \to \phi(x, z)) \to \forall x \in z\, \phi(x, z)$. We must obtain terms t, q and s such that for all for all $e \in W$ and A, b of appropriate types in the set-theoretical structure $\langle S_\rho \rangle$, and for all $z \in L_{|e|}$, the statement

$$\forall n, m \forall x \in L_{(teAb)\langle n\rangle} (x \in z \to \phi_{\mathrm{B}}(A((teAb)\langle n\rangle), (qeAb)\langle m\rangle, x, z)) \to \forall x \in z\, \phi_{\mathrm{B}}(seA, b, x, z)$$

holds in $L^{\mathrm{mix}}_{\omega_1}$. It is clear that the terms $s :\equiv \lambda e, A.Ae$, $t :\equiv \lambda e, A, b.(e + 1)$ and $q :\equiv \lambda e, A, b.(b + 1)$ work.

Now we discuss the existential bounded quantifier. The upper B-translations of $\exists x \in z\, \phi(x, z)$ and $\exists x\, (x \in z \wedge \phi(x, z))$ are, $\exists a \forall b\, [\exists x \in z \forall n\, \phi_{\mathrm{B}}(a, b <n>, x, z)]$ and $\exists c^{\Omega}, a \forall b\, [\exists x \in L_c \forall n\, (x \in z \wedge \phi_{\mathrm{B}}(a, b\langle n\rangle, x, z))]$ respectively.

The axiom $\exists x \in z\, \phi(x, z) \to \exists x (x \in z \wedge \phi(x, z))$. We must find terms t, q and s such that, for all $e \in W$ and a and b of appropriate types in the set-theoretical structure $\langle S_\rho \rangle$, and for all $z \in L_{|e|}$,

$$\forall m \exists x \in z \forall n\, \phi_B(a, (teab)\langle m \rangle\langle n \rangle, x, z) \to \exists x \in L_{qea} \forall n\, (x \in z \wedge \phi_B(sea, b\langle n \rangle, x, z))$$

holds in $L^{\text{mix}}_{\omega_1}$. Clearly, $t :\equiv \lambda e, a, b.(b+1)$, $q :\equiv \lambda e, a.e$ and $s :\equiv \lambda e, a.a$ work.

The axiom $\exists x\, (x \in z \wedge \phi(x, z)) \to \exists x \in z\, \phi(x, z)$. We must obtain terms t and q such that, for all $e, c \in W$ and for all a, b of appropriate types in the set-theoretical structure $\langle S_\rho \rangle$, and for all $z \in L_{|e|}$, the statement

$$\forall m \exists x \in L_c \forall n\, (x \in z \wedge \phi_B(a, (tecab)\langle m \rangle\langle n \rangle, x, z)) \to \exists x \in z \forall n\, \phi_B(qeca, b\langle n \rangle, x, z)$$

holds in $L^{\text{mix}}_{\omega_1}$. This is clearly the case with $t :\equiv \lambda e, a, c, b.(b+1)$ and $q :\equiv \lambda e, c, a.a$.

Let us now turn to the mathematical axioms of $\mathsf{IKP}\omega$. Extensionality poses no problem because it is the universal closure of a bounded formula. The verification of the pairing, union and infinity axioms is like the verification done in [10]. For completeness, for these three axioms we need (respectively) closed terms t, q and r such that $\forall c, e \forall x \in L_c \forall y \in L_e \exists z \in L_{tce}\, (x \in z \wedge y \in z)$, $\forall c \forall x \in L_c \exists z \in L_{qc}$ $\forall y \in x \forall w \in y\, (w \in z)$ and $\exists x \in L_r\, Lim(x)$, where $Lim(x)$ is a bounded formula which expresses that x is a limit ordinal. The terms $t :\equiv \lambda c, e.(\max(c, e) + 1)$, $q :\equiv \lambda c.c$ and $r :\equiv \omega_\Omega + 1$ do the job.

The separation scheme is $\forall w \forall y \exists z \forall x\, (x \in z \leftrightarrow x \in y \wedge \phi(x, w))$, where ϕ is a bounded formula in which the variable z does not occur. Note that the inner universal statement can be considered bounded. A straightforward computation of the upper B-translation of this formula shows that we need a closed term t of type $\Omega \to (\Omega \to \Omega)$ such that

$$\forall c^\Omega, e^\Omega\, [\forall y \in L_c \forall w \in L_e \exists z \in L_{tce} \forall x\, (x \in z \leftrightarrow x \in y \wedge \phi(x, w))]$$

holds in $L^{\text{mix}}_{\omega_1}$. It is clear that the term $t :\equiv \lambda c, e.(\max(c, e) + 1)$ does the job. To see this, just take z to be $\{x \in L_\alpha : x \in y \wedge \phi(x, w)\}$, where $\alpha = \max(|c|, |e|)$.

The bounded collection scheme is a sub-scheme of the scheme of unrestricted collection Coll that will be discussed later. Let us now study the foundation rule. A computation of the upper-B translation of the premise of the induction rule shows that, by induction hypothesis, there are terms t and q such that

$$(*) \quad \forall c^\Omega, a, b\, [\forall x \in L_c\, (\forall n \forall y \in x\, \phi_B(a, (tcab)\langle n \rangle, y) \to \phi_B(qca, b, x))]$$

holds in $L^{\text{mix}}_{\omega_1}$. We want to find a term r such that

$$(\star) \quad L^{\text{mix}}_{\omega_1} \models \forall c^\Omega, b\, [\forall x \in L_c\, \phi_B(rc, b, x)]$$

Define by tree recursion the term r as follows: $r0_\Omega = 0$ and

$$r(Sup(f)) = \bigsqcup_k q(fk, r(fk)).$$

We now check ($\star$) by transfinite induction on $|c|$. If $|c| = 0$ there is nothing to prove. Suppose that $c = Sup(f)$. Take b appropriate of type in the set-theoretical structure $\langle S_\rho \rangle$ and $x \in L_{|c|}$. Take $y \in x$. Since $L_{|c|} = \bigcup_k L_{|f(k)|+1}$, there is a natural number k_0 (which we identify with the corresponding type N term) such that $x \in L_{|f(k_0)|+1}$. By transfinite induction hypothesis,

$$L^{\text{mix}}_{\omega_1} \models \forall b \forall y \in L_{f(k_0)}\, \phi_{\text{B}}(r(f(k_0)), b, y)$$

because $|f(k_0)| < |c|$. In particular,

$$L^{\text{mix}}_{\omega_1} \models \forall b \forall y \in L_{f(k_0)} \forall n\, \phi_{\text{B}}(r(f(k_0)), (t(f(k_0), r(f(k_0)), b))\langle n \rangle, y).$$

Using the hypothesis ($*$), we may conclude that

$$L^{\text{mix}}_{\omega_1} \models \forall b\, \phi_{\text{B}}(q(f(k_0), r(f(k_0))), b, x).$$

But, by Lemma 4, $q(f(k_0), r(f(k_0))) \sqsubseteq \bigsqcup_k q(f(k), r(f(k)))$ holds in $L^{\text{mix}}_{\omega_1}$. By the definition of r and c, we have $L^{\text{mix}}_{\omega_1} \models q(f(k_0), r(f(k_0))) \sqsubseteq rc$. Now, ($\star$) follows using the monotonicity property of the existencial entry of ϕ_{B}.

It remains to check the principles Δ_0-LEM, MP, bIP$_{\Pi_1}$, Δ_0-CColl and Coll. Of course, Δ_0-LEM is trivially interpreted by itself. We now check these four principles:

MP. We must find a term t so that $\forall c^\Omega [\neg \forall n \forall x \in L_{c\langle n \rangle}\, \phi(x) \to \exists x \in L_{tc}\, \neg \phi(x)]$ holds in $L^{\text{mix}}_{\omega_1}$. The identity term in type Ω works.

bIP$_{\Pi_1}$. The upper B-translation of the antecedent of this principle is

$$\exists c^\Omega, d, B \forall e\, [\forall k \forall x \in L_{(Be)\langle k \rangle}\, \phi(x) \to \exists y \in L_c \forall n\, \psi_{\text{B}}(d, e\langle n \rangle, y)],$$

whereas the upper B-translation of the consequent is

$$\exists c^\Omega, d, B \forall e\, [\exists y \in L_c \forall m\, (\forall k \forall x \in L_{(B(e\langle m \rangle))\langle k \rangle}\, \phi(x) \to \exists z \in y \forall n\, \psi_{\text{B}}(d, e\langle m \rangle\langle n \rangle, z))].$$

Therefore, we must find terms t, q, r and s such that for all $c \in W$ and d, B and e of appropriate types in the set-theoretical structure $\langle S_\rho \rangle$,

$$\forall m\, (\forall k \forall x \in L_{(B((tcdBe)\langle m \rangle))\langle k \rangle}\, \phi(x) \to \exists z \in L_c \forall n\, \psi_{\text{B}}(d, (tcdBe)\langle m \rangle\langle n \rangle, z)) \to$$

$$\exists y \in L_{qcdB} \forall m\, (\forall k \forall x \in L_{((scdB)(e\langle m \rangle))\langle k \rangle}\, \phi(x) \to \exists z \in y \forall n\, \psi_{\text{B}}(rcdB, e\langle m \rangle\langle n \rangle, z))$$

holds in $L^{\text{mix}}_{\omega_1}$. It does hold with $t :\equiv \lambda c, d, B, e.e$, $s :\equiv \lambda c, d, B.B$, $q :\equiv \lambda c, d, B.(c + 1)$ and $r :\equiv \lambda c, d, B.d$. To see this we have to check that

$$\forall c^{\Omega}, d, B, e\, [\forall m\, (\forall k \forall x \in L_{(B(e\langle m\rangle))\langle k\rangle}\, \phi(x) \to \exists z \in L_c \forall n\, \psi_{\mathrm{B}}(d, e\langle m\rangle\langle n\rangle, z)) \to$$

$$\exists y \in L_{c+1} \forall m\, (\forall k \forall x \in L_{(B(e\langle m\rangle))\langle k\rangle}\, \phi(x) \to \exists z \in y \forall n\, \psi_{\mathrm{B}}(d, e\langle m\rangle\langle n\rangle, z))]$$

holds in $L^{\mathrm{mix}}_{\omega_1}$. In effect, if we assume the antecedent, then the consequent is seen to immediately hold with $y = L_c$.

Δ_0-CColl. According to the upper B-translation of this principle, we need a term t such that

$$\forall d^{\Omega}, c^{\Omega}\, [\forall w \in L_d\, (\forall n \forall z \in L_{(tdc)\langle n\rangle} \exists y \in w \forall x \in z\, \phi(x, y) \to \exists y \in w \forall n \forall x \in L_{c\langle n\rangle}\, \phi(x, y))]$$

holds in $L^{\mathrm{mix}}_{\omega_1}$. It is easy to argue that $t :\equiv \lambda d, c.(c + 2)$ works. One just has to instantiate the antecedent with $z = L_c$ (and $n = 0$, say) to see that the consequent holds.

Coll. The upper B-translation of the antecedent of this principle is

$$\exists c^{\Omega}, a \forall b\, [\forall y \in w \exists x \in L_c \forall n\, \phi_{\mathrm{B}}(a, b\langle n\rangle, x, y)],$$

and the upper B-translation of the consequent is

$$\exists d^{\Omega}, a \forall b\, [\exists z \in L_d \forall k \forall y \in w \exists x \in z \forall n\, \phi_{\mathrm{B}}(a, b\langle k\rangle\langle n\rangle, x, y)].$$

Therefore, we must obtain terms t, q and r such that

$$\forall e^{\Omega}, c^{\Omega}, a, b\, [\forall w \in L_e\, (\forall k \forall y \in w \exists x \in L_c \forall n\, \phi_{\mathrm{B}}(a, (tecab)\langle k\rangle\langle n\rangle, x, y) \to$$

$$\exists z \in L_{qeca} \forall k \forall y \in w \exists x \in z \forall n\, \phi_{\mathrm{B}}(reca, b\langle k\rangle\langle n\rangle, x, y))]$$

holds in $L^{\mathrm{mix}}_{\omega_1}$. Just take $t :\equiv \lambda e, c, a, b.b$, $q :\equiv \lambda e, c, a.(c + 1)$ and $r :\equiv \lambda e, c, a.a$. Obviously, the consequent becames true (given the antecedent) with $z = L_c$. □

The following proposition is an immediate consequence of the Soundness Theorem and of (ii) of Lemma 7:

Proposition 4 *If* $\mathsf{IKP}\omega + \Delta_0\text{-}\mathsf{LEM} + \mathsf{MP} + \mathsf{bIP}_{\Pi_1} + \Delta_0\text{-}\mathsf{CColl} + \mathsf{Coll} \vdash \forall x \exists y\, \phi(x, y)$*, where* ϕ *is a bounded formula (*x *and* y *are the only free variables), then there is a closed term* t *of type* $\Omega \to \Omega$ *such that*

$$L^{\mathrm{mix}}_{\omega_1} \models \forall c^{\Omega} \forall x \in L_c \exists y \in L_{tc}\, \phi(x, y).$$

Moreover, $L_{\mathrm{BH}} \models \forall x \exists y\, \phi(x, y)$*, where* BH *is the Bachmann–Howard ordinal.*

In the last conclusion, one uses the absoluteness of bounded formulas. The same for the following:

Corollary 3 *If* $\mathsf{IKP}\omega + \Delta_0\text{-}\mathsf{LEM} + \mathsf{MP} + \mathsf{bIP}_{\Pi_1} + \mathsf{Coll} + \Delta_0\text{-}\mathsf{CColl} \vdash \exists x\, \phi(x)$, *where* ϕ *is a bounded formula (x is the only free variable), then there is an ordinal* α *smaller than the Bachmann–Howard ordinal such that* $L_\alpha \models \exists x\, \phi(x)$.

A part of Proposition 4 can be much improved because the translations that define the functional interpretation are correct in $L_{\omega_1}^{\text{mix}}$ (however, the improvement does not seem susceptible to an internalization like the one in Sect. 6 of [10]). In order to discuss this improvement it is convenient to permit also unbounded set-theoretic quantifications in the mixed language. With these quantifications, we refer to the language as the *extended* mixed language.

Lemma 9 *For every formula* ϕ *of the language of set theory,* $L_{\omega_1}^{\text{mix}} \models \phi \leftrightarrow \phi^{\text{B}}$.

Proof The proof is by induction on the complexity of ϕ. The result is clear for ϕ bounded, for the conjunction and also for the disjunction. For the latter one, just use the fact that for every $b \in S_\rho$, the equality $(b+1)\langle n\rangle = b$ holds set-theoretically. The remaining cases follow from the following fact:
<u>Fact</u>. Let $z \in L_{\omega_1}$, ρ a pure Ω-type and ϕ a formula of the extended mixed language. Then

$$L_{\omega_1}^{\text{mix}} \models \forall x \in z\, \exists a^\rho\, \phi(a, x, z) \to \exists a\, \forall x \in z\, \exists n\, \phi(a\langle n\rangle, x, z).$$

The proof of the fact is easy. Let $z \in L_{\omega_1}$. Suppose $L_{\omega_1}^{\text{mix}} \models \forall x \in z\, \exists a^\rho\, \phi(a, x, z)$. Since z is countable, we can take an enumeration $(x_n)_{n\in\omega}$ of the elements of z. For each $n \in \omega$, choose $a_n \in S_\rho$ be such that $L_{\omega_1}^{\text{mix}} \models \phi(a_n, x_n, z)$. Call f this function $n \rightsquigarrow a_n$. By (viii) of Sect. 3, $L_{\omega_1}^{\text{mix}} \models \forall n^N((Sup\, f)\langle n\rangle = a_n)$. It is now clear that we can take for a the element $Sup\, f$. □(end of proof of fact)

Let us study the universal bounded quantifier. We need to show that

$$L_{\omega_1}^{\text{mix}} \models \forall x \in z\, \exists a \forall b\, \phi_{\text{B}}(a, b, x, z) \leftrightarrow \exists a \forall b \forall x \in z\, \phi_{\text{B}}(a, b, x, z).$$

Only the left-to-right direction needs to be argued. Assume the antecedent. By the Fact, $\exists a \forall x \in z\, \exists n \forall b\, \phi_{\text{B}}(a\langle n\rangle, b, x, z)$. By Lemma 4 and the monotonicity lemma, we easily get $\forall b \forall x \in z\, \phi_{\text{B}}(\tilde{a}, b, x, z)$, where $\tilde{a}$ is $\bigsqcup \lambda x^N.(a\langle x\rangle)$.

Regarding the existential bounded quantifier, we must show that

$$L_{\omega_1}^{\text{mix}} \models \exists x \in z \exists a \forall b\, \phi_{\text{B}}(a, b, x, z) \leftrightarrow \exists a \forall b\, \exists x \in z \forall n\, \phi_{\text{B}}(a, b\langle n\rangle, x, z).$$

This time, the left-to-right direction is trivial. So, assume the right-hand side and take an element a such that $\forall b\, \exists x \in z \forall n\, \phi_{\text{B}}(a, b\langle n\rangle, x, z)$. We use the contrapositive of the Fact in order to we get what we want.

Let us consider the universal quantifier. We must show that

$$L_{\omega_1}^{\text{mix}} \models \forall x \exists a^\rho \forall b\, \phi_{\text{B}}(a, b, x) \leftrightarrow \exists A \forall c^\Omega, b \forall x \in L_c\, \phi_{\text{B}}(Ac, b, x).$$

Suppose the left-hand side. Hence, for all $c \in W$, $\forall x \in L_c \exists a \forall b\, \phi_B(a, b, x)$. By the discussion of the universal bounded quantifier case, we get that for all $c \in W$ there is $a \in S_\rho$ such that $\forall x \in L_c \forall b\, \phi_B(a, b, x)$. Take a function $A : W \to S_\rho$ (that is, an element of $S_{\Omega \to \rho}$) such that, for all elements c in W, $\forall b \forall x \in L_c\, \phi_B(Ac, b, x)$. This A works. Now, assume the right-hand side. Let $A \in S_{\Omega \to \rho}$ such that, for all $c \in W$ one has $\forall b \forall x \in L_c\, \phi_B(A(c), b, x)$. Take x an arbitrary element of L_{ω_1}. Then there is $c \in W$ such that $x \in L_{|c|}$ (we are using the fact that $\omega_1 = \sup_{c \in W} |c|$). It is clear that if we take a to be $A(c)$, we get $\forall b\, \phi_B(a, b, x)$.

For the existential quantifier we must show that

$$L_{\omega_1}^{\text{mix}} \models \exists x \exists a^\rho \forall b\, \phi_B(a, b, x) \leftrightarrow \exists c^\Omega, a \forall b \exists x \in L_c \forall n\, \phi_B(a, b\langle n\rangle, x).$$

The left-to-right direction follows from the fact $L_{\omega_1} = \bigcup_{c \in W} L_{|c|}$. To see the other direction, take $c \in W$ and $a \in S_\rho$ such that $\forall b \exists x \in L_c \forall n\, \phi_B(a, b\langle n\rangle, x)$. By the contrapositive of the Fact, we obtain $\exists x \in L_c \forall b\, \phi_B(a, b, x)$ and, therefore, the left-hand side.

It remains to check the implication. We must argue that the following holds in $L_{\omega_1}^{\text{mix}}$:

$$(\exists a \forall b\, \phi_B(a, b) \to \exists d \forall e\, \psi_B(d, e)) \leftrightarrow \exists B, D \forall a, e\, (\forall n\, \phi_B(a, (Bae)\langle n\rangle) \to \psi_B(Da, e)).$$

Note that $\forall b\, \phi_B(a, b)$ is equivalent to $\forall b \forall n\, \phi_B(a, b\langle n\rangle)$. With this in mind, an appropriate (partial) prenexification of the left-hand side of the equivalence yields

$$\forall a \exists d \forall e \exists b\, (\forall n\, \phi_B(a, b\langle n\rangle) \to \psi_B(d, e)).$$

The equivalence follows using two applications of the axiom of choice in the structure (S_ρ). □

Proposition 5 *If* $\mathsf{IKP}\omega + \Delta_0\text{-}\mathsf{LEM} + \mathsf{MP} + \mathsf{bIP}_{\Pi_1} + \Delta_0\text{-}\mathsf{CColl} + \mathsf{Coll} \vdash \forall x \exists y\, \phi(x, y)$, *where* ϕ *is an arbitrary formula of the language of set theory (*x *and* y *are the only free variables), then there is a closed term* t *of type* $\Omega \to \Omega$ *such that*

$$\forall c \in W \forall x \in L_{|c|} \exists y \in L_{|t(c)|}\ L_{\omega_1} \models \phi(x, y).$$

Proof Let $\phi(x, y)^B$ be $\exists a^\tau \forall b^\rho \phi_B(a, b, x, y)$. It is easy to see that $(\forall x \exists y\, \phi(x, y))^B$ is

$$\exists A, D \forall c^\Omega, b \forall x \in L_c \exists y \in L_{Dc} \forall n\, \phi_B(Ac, b\langle n\rangle, x, y).$$

By the Soundness theorem, there are closed terms t and q such that

$$\forall c^\Omega \forall x \in L_c \forall b\, \exists y \in L_{tc} \forall n\, \phi_B(qc, b\langle n\rangle, x, y)$$

holds in $L^{\mathrm{mix}}_{\omega_1}$. By the contrapositive of the Fact of the previous lemma, we get:

$$\forall c^{\Omega} \forall x \in L_c \exists y \in L_{tc} \forall b\, \phi_{\mathrm{B}}(qc, b, x, y).$$

In particular, $\forall c \forall x \in L_c \exists y \in L_{tc} \exists a \forall b\, \phi_{\mathrm{B}}(a, b, x, y)$, i.e., $\forall c \forall x \in L_c \exists y \in L_{tc}$ $\phi^{\mathrm{B}}(x, y)$. By the previous lemma, we are done. □

6 Adding Strict-Π^1_1 Reflection

Admissibility is the playground where finiteness, recursive enumerability and other recursion-theoretic notions find a fertile ground for generalization (see, for instance, the preface of [18]). Weak Kőnig's lemma is an important (second-order) principle in recursion theory (viz., in relation with low degrees and compactness) as well as in subsystems of second-order arithmetic and reverse mathematics (cf. [20]). The principle of strict-Π^1_1 reflection, introduced by Barwise in [3, 4], is a natural generalization of weak Kőnig's lemma from the arithmetical setting to the admissible setting. Strict-Π^1_1 reflection can be stated as follows:

$$\forall X \exists x\, \phi(x, X) \rightarrow \exists z \forall X \exists x \in z\, \phi(x, X),$$

where ϕ is a bounded formula (for details see chapter VIII of [5]). In the second part of [10], we extended the Σ-ordinal analysis of $\mathsf{KP}\omega$ to a second-order theory with the principle of strict-Π^1_1 reflection. We proved the novel result that the Σ-ordinal of this second-order theory is still the Bachmann–Howard ordinal. The main idea for this analysis was an extension of the functional interpretation in which second-order quantifications are treated as bounded quantifications. It may sound surprising at first that this treatment works because the transformations of formulas underlying the functional intepretation of the second-order quantifiers are not truth-preserving in our semantics (more about this later). However, on second thought – for a person familiar with the bounded functional interpretation of [11] – the idea is compelling.

In this paper we are interested in analysing theories based on intuitionistic logic. Some phenomena absent in the classical setting emerge in the intuitionistic theories. As we saw, the bounded collection scheme of $\mathsf{KP}\omega$ is inflated into unrestricted collection Coll (while the non-intuitionistic classical contrapositive of bounded collection Δ_0-CColl is kept). A similar sort of thing happens with strict-Π^1_1 reflection (see Definitions 12 and 13). The inflation also extends to the notion of bounded formula, now that second-order quantifications are regarded as bounded. There is no apparent reason to stick any longer to the usual notion of bounded formula.

Let us set up the basic second-order theories $\mathsf{IKP}\omega_2{\restriction}$ and $\mathsf{IKP}\omega_2$ (the latter is discussed in the next section). The language of second-order set theory is the enlargement of the language of set theory (as described in Sect. 2) with monadic second-order quantification (we have both universal and existential quantifiers). We use capital letters X, Y, Z, ...for the monadic predicates and call them *classes*. As it is common usage, we write '$x \in X$' instead of the (syntactically correct) '$X(x)$'. This is an abuse of notation because the membership sign in the expression '$x \in X$' is not the membership sign of the language of (first-order) set theory. It is just a harmless and felicitous notational device, and we read '$x \in X$' as saying that x is a member of (the class) X.

Definition 8 The class of $\Delta_0^{\mathcal{C}}$*-formulas* of the language of second-order set theory is the smallest class of formulas that contains the atomic formulas $x \in y, x = y, x \in X$, the absurdity, and which is closed under propositional connectives, the bounded quantifiers $\forall x \in y$ and $\exists x \in y$, and the second-order quantifiers $\forall X$ and $\exists X$.

In [19], Salipante uses the notation Δ_0^{C}, with a roman letter 'C', for a more restricted version of bounded formula, namely for Δ_0-formulas in which second-order parameters are permitted (second-order quantifications are *not* allowed). In order to distinguish our notion from Salipante's, we use a caligraphic '$\mathcal{C}$' instead. This caligraphic notation has also the advantage of cohering with a notation of Rathjen in [17] that we will be needing in the next section.

We denote the law of excluded middle restricted to $\Delta_0^{\mathcal{C}}$-formulas by $\Delta_0^{\mathcal{C}}$-LEM.

Definition 9 The second-order theory $\mathsf{IKP}\omega_2\!\upharpoonright$ is the intuitionistic theory of the language of second-order set theory that contains $\mathsf{IKP}\omega$ and extends the scheme of foundation in order to permit *all* the formulas of the new language.

Some comments are in order. In the above, we did not change the original schemata of separation and bounded collection. They remain exactly as in $\mathsf{IKP}\omega$. We could have opted for allowing in the collection scheme the wider class of $\Delta_0^{\mathcal{C}}$-formulas and, also, second-order parameters. This would be more in line with the definition of $\mathsf{KP}\omega_2\!\upharpoonright$ in [10]. However, this enlarged collection scheme is a particular case of the unrestricted collection scheme $\mathsf{Coll}^{\mathcal{C}}$ given in (d) of Definition 11, and our functional interpretation is able to realize it. The *point of attention* is really the separation scheme of Definition 9. It is the *original* formulation of the separation scheme, without second-order parameters and with the original bounded (i.e., Δ_0) formulas. That explains the restriction sign in the acronym of the theory. As we will see, the Σ-ordinal of the restricted theory together with the $\Delta_0^{\mathcal{C}}$-LEM, some comprehension for second-order class formation and strict-Π_1^1 reflection is still the Bachmann–Howard ordinal. On the other hand, Salipante showed that if *second-order parameters* are allowed in the separation scheme (never mind allowing for $\Delta_0^{\mathcal{C}}$-matrices) then, in the presence of suitable comprehension and strict-Π_1^1 reflection, one is able to prove the powerset axiom. This will be discussed in the next section (see Theorem 4). Antecipating the results of that section, we may add that with the help of other interpretable principles Salipante's result gives rise to a very strong theory, namely to (an intuitionistic version of) the so-called power Kripke–Platek set theory $\mathsf{KP}\omega\,(\mathcal{P})$, as described in [17].

Next, we introduce the principles of class comprehension that will be of our interest (notice the analogy with similar principles in subsystems of second-order arithmetic). A $\Sigma_1^{\mathcal{C}}$-formula is a formula of the form $\exists x \varphi(x)$, where $\varphi(x)$ is a $\Delta_0^{\mathcal{C}}$-formula. The notion of $\Pi_1^{\mathcal{C}}$-formula is defined dually.

Definition 10 The following schemata are defined in the second-order language of set theory (first and second-order parameters are allowed):

I. The scheme $\Delta_0^{\mathcal{C}}$-CA is $\exists X \forall x\, (x \in X \leftrightarrow \phi(x))$, where $\phi(x)$ is a $\Delta_0^{\mathcal{C}}$-formula (X is a fresh variable).

II. The scheme $\Delta_1^{\mathcal{C}}$-CA is $\forall x\, (\phi(x) \leftrightarrow \psi(x)) \rightarrow \exists X \forall x\, (x \in X \leftrightarrow \phi(x))$, where $\phi(x)$ is a $\Sigma_1^{\mathcal{C}}$-formula and $\psi(x)$ is a $\Pi_1^{\mathcal{C}}$-formula (X is a fresh variable).

We now list some principles which our functional interpretation is able to realize:

Definition 11 The following schemata are defined in the second-order language of set theory (in all the schemata below, both first and second-order parameters are allowed):

(a) *Markov's principle* $\mathsf{MP}^{\mathcal{C}}$ is the scheme $\neg \forall x\, \phi(x) \rightarrow \exists x\, \neg \phi(x)$, for ϕ a $\Delta_0^{\mathcal{C}}$-formula.

(b) The *independence of premises principle* $\mathsf{bIP}_{\Pi_1^{\mathcal{C}}}$ is the scheme

$$(\forall x\, \phi(x) \rightarrow \exists y\, \psi(y)) \rightarrow \exists y\, (\forall x\, \phi(x) \rightarrow \exists z \in y\, \psi(z)),$$

where ϕ is a $\Delta_0^{\mathcal{C}}$-formula and ψ is *any* formula of the second-order language.

(c) The principle of *bounded contra-collection* $\Delta_0^{\mathcal{C}}$-CColl is the scheme

$$\forall z \exists y \in w \forall x \in z\, \phi(x, z) \rightarrow \exists y \in w \forall x \phi(x, y),$$

where ϕ is a $\Delta_0^{\mathcal{C}}$-formula.

(d) The principle of (unrestricted) *collection* $\mathsf{Coll}^{\mathcal{C}}$ is the scheme

$$\forall y \in w \exists x\, \phi(x, y) \rightarrow \exists z \forall y \in w \exists x \in z\, \phi(x, y),$$

where ϕ is *any* formula of the second-order language.

Some of the results of Sect. 2 adapt to the new setting. The following result is analogous to the law of excluded middle of Corollary 1. The proof of that corollary used a separation result that we do not have in our present setting, due to the restrictions of the separation scheme discussed above. However a direct proof is forthcoming (moreover, Markov's principle is not needed, as observed by Fujiwara):

Proposition 6 *The theory* $\mathsf{IKP}\omega_2^{\restriction} + \Delta_0^{\mathcal{C}}$-LEM $+\, \mathsf{bIP}_{\Pi_1^{\mathcal{C}}}$ *proves the* $\Delta_1^{\mathcal{C}}$ *law of excluded middle, i.e., it proves* $(\forall u\, \phi(u) \leftrightarrow \exists v\, \psi(v)) \rightarrow (\forall u\, \phi(u) \vee \neg \forall u\, \phi(u))$, *for* $\Delta_0^{\mathcal{C}}$-*formulas* ϕ *and* ψ *(possibly with first and second-order parameters).*

Proof Suppose that $\forall u\, \phi(u) \leftrightarrow \exists v\, \psi(v)$. Applying $\mathsf{bIP}_{\Pi_1^c}$ to the left-to-right direction of the equivalence, there is v_0 such that $\forall u\, \phi(u) \to \exists v \in v_0\, \psi(v)$. If $\exists v \in v_0\, \psi(v)$, our supposition entails $\forall u\, \phi(u)$. If $\neg\exists v \in v_0\, \psi(v)$, we directly conclude that $\neg\forall u\, \phi(u)$. □

A version of the lesser limited principle of omniscience also holds in the present setting (first and second-order parameters are allowed in the following, of course). The proof is analogous to the proof of Corollary 2.

Proposition 7 *If ϕ and ψ are Δ_0^C-formulas, then*

$$\mathsf{IKP}\omega_2\!\restriction + \Delta_0^C\text{-}\mathsf{LEM} + \Delta_0^C\text{-}\mathsf{CColl} \vdash \forall u, v\, (\phi(u) \vee \psi(v)) \to \forall u\, \phi(u) \vee \forall v\, \psi(v).$$

The appropriate version of the strict-Π_1^1 reflection scheme in our intuitionistic setting comes in two installments. The reader will notice that they are like collection schemes. The first one is a version of the contrapositive of strict-Π_1^1 reflection. The second one is a vast generalization, only possible because we are in an intuitionistic setting. Of course, each installment entails strict-Π_1^1 reflection (the first one classically, the second intuitionistically).

Definition 12 The principle of *bounded class contra-collection* Δ_0^C-CColl_2 is the scheme

$$\forall z \exists X \forall x \in z\, \phi(x, X) \to \exists X \forall x\, \phi(x, X),$$

where ϕ is a Δ_0^C-bounded formula (possibly with first and second-order parameters).

It was argued in [10] that with the aid of strict-Π_1^1 reflection, the bounded comprehension scheme upgrades to a Δ_1-comprehension scheme. A similar result holds in the present setting.

Proposition 8 $\mathsf{IKP}\omega_2\!\restriction + \Delta_0^C\text{-}\mathsf{LEM} + \Delta_0^C\text{-}\mathsf{CA} + \Delta_0^C\text{-}\mathsf{CColl}_2 \vdash \Delta_1^C\text{-}\mathsf{CA}$.

Proof The proof is *mutatis mutandis* the argument for lemma 5.3 of [10]. Suppose that $\forall u\, (\exists y\, \phi(u, y) \leftrightarrow \forall z\, \psi(u, z))$, where ϕ and ψ are Δ_0^C-formulas. Then,

$$\forall w \exists X \forall x \in w \forall u, y, z \in x\, ((\phi(u, y) \to u \in X) \wedge (u \in X \to \psi(u, v))).$$

It is easy to argue this. Given w, take $\tilde{w}$ its transitive closure. Clearly, we can take X to be $\{u : \exists y \in \tilde{w}\, \phi(u, y)\}$. Note that this class exists by Δ_0^C-CA.

By Δ_0^C-CColl, we get

$$\exists X \forall x \forall u, y, z \in x\, ((\phi(u, y) \to u \in X) \wedge (u \in X \to \psi(u, z))).$$

Clearly, this X is formed by the elements u that satisfy $\exists y\, \phi(u, y)$. □

We have said that Δ_0^C-CColl_2 is like a collection scheme. In fact, it generalizes Δ_0^C-CColl:

Proposition 9 $\mathsf{IKP}\omega 2\!\upharpoonright + \Delta_0^{\mathcal{C}}\text{-}\mathsf{LEM} + \Delta_0^{\mathcal{C}}\text{-}\mathsf{CA} + \Delta_0^{\mathcal{C}}\text{-}\mathsf{CColl}_2 \vdash \Delta_0^{\mathcal{C}}\text{-}\mathsf{CColl}$.

Proof Let ϕ be a $\Delta_0^{\mathcal{C}}$-formula and suppose that $\forall z \exists y \in w \forall x \in z\, \phi(x, y)$. We claim that $\forall z \exists X \forall x \in z\, [\exists y \in w\, (y \in X \wedge \forall u \in w\, (u \in X \to u = y) \wedge \phi(x, y))]$. Given z, by the supposition there is $y_0 \in w$ such that $\forall x \in z\, \phi(x, y_0)$. We just have to take X to be the singleton class formed by y_0 (it exists by $\Delta_0^{\mathcal{C}}$-CA). Since the formula between square parenthesis is a $\Delta_0^{\mathcal{C}}$-formula, we can apply Δ_0-CColl_2 in order to get

$$\exists X \forall x\, [\exists y \in w\, (y \in X \wedge \forall u \in w\, (u \in X \to u = y) \wedge \phi(x, y))].$$

Clearly, $X \cap w$ must be a singleton (i.e., this *class* has only one element). Let y_0 be the only element of this class. We get $y_0 \in w$ and $\forall x\, \phi(x, y_0)$. □

The second installment of strict-Π_1^1 reflection is the following:

Definition 13 The principle of (unrestricted) *class collection* $\mathsf{Coll}_2^{\mathcal{C}}$ is the scheme

$$\forall X \exists x\, \phi(x, X) \to \exists z \forall X \exists x \in z\, \phi(x, X),$$

where ϕ is *any* formula (possibly with first and second-order parameters).

At this point, the following proposition should not be surprising:

Proposition 10 $\mathsf{IKP}\omega 2\!\upharpoonright + \Delta_0^{\mathcal{C}}\text{-}\mathsf{LEM} + \Delta_0^{\mathcal{C}}\text{-}\mathsf{CA} + \mathsf{Coll}_2^{\mathcal{C}} \vdash \mathsf{Coll}^{\mathcal{C}}$.

Proof Let ϕ be a $\Delta_0^{\mathcal{C}}$-formula and suppose that we have $\forall y \in w \exists x\, \phi(x, y)$. We claim that $\forall X \exists x\, [\forall y \in w\, (y \in X \wedge \forall u \in w\, (u \in X \to u = y) \to \phi(x, y))]$. Given X, either there is $y \in w$ such that $y \in X \wedge \forall u \in w\, (u \in X \to u = y)$ or not (the previous formula is bounded and, hence, we can apply $\Delta_0^{\mathcal{C}}$-LEM). In the first case, by supposition, there is x such that $\phi(x, y)$ and we are done. If not, the assertion is trivially true.

By $\mathsf{Coll}_2^{\mathcal{C}}$, there is z_0 such that

$$\forall X \exists x \in z_0\, [\forall y \in w\, (y \in X \wedge \forall u \in w\, (u \in X \to u = y) \to \phi(x, y))].$$

Given $y \in w$, take X to be the singleton class formed by y. Clearly, $\exists x \in z_0\, \phi(x, y)$. □

The theory $\mathsf{IKP}\omega 2\!\upharpoonright + \Delta_0^{\mathcal{C}}\text{-}\mathsf{LEM} + \Delta_0^{\mathcal{C}}\text{-}\mathsf{CA} + \mathsf{MP}^{\mathcal{C}} + \mathsf{bIP}_{\Pi_1^{\mathcal{C}}} + \Delta_0^{\mathcal{C}}\text{-}\mathsf{CColl}_2 + \mathsf{Coll}_2^{\mathcal{C}}$ (which, as we saw, includes $\Delta_0^{\mathcal{C}}$-CColl and $\mathsf{Coll}^{\mathcal{C}}$) is very robust. For instance, modulo this theory, the $\Sigma_1^{\mathcal{C}}$-formulas and the $\Pi_1^{\mathcal{C}}$-formulas enjoy strong closure properties and this permits the smooth introduction of $\Delta_1^{\mathcal{C}}$-relation symbols and of $\Sigma_1^{\mathcal{C}}$-function symbols. In effect, modulo the above theory, the $\Sigma_1^{\mathcal{C}}$-formulas are closed under conjunctions, disjunctions, bounded quantifications, second-order quantifications and existential (first-order) quantifications. The closure under conjunctions, disjunctions and bounded, second-order and unbounded existential quantifications is clear. The closure under bounded and second-order universal quantifications follows from $\mathsf{Coll}^{\mathcal{C}}$ and $\mathsf{Coll}_2^{\mathcal{C}}$, respectively. Dually, the $\Pi_1^{\mathcal{C}}$-formulas are closed under conjunctions, disjunctions, bounded quantifications, second-order quantifications and universal (first-order) quantifications. The closure under conjunctions, bounded, second-order and unbounded universal quantifications is clear. The closure under disjunction is a consequence of Proposition 7. The closure under bounded and second-order existential quantifications follows from $\Delta_0^{\mathcal{C}}$-CColl and $\Delta_0^{\mathcal{C}}$-CColl_2, respectively. The introduction of the powerset operation in the next section uses these facts crucially. They are essential for the interpretation of the power Kripke–Platek set theory in a theory based on $\mathsf{IKP}\omega_2$.

The functional interpretation given in Definition 7 is extended to the second-order language by the following two clauses.

9. $(\forall X\, \phi(X))^{\mathrm{B}}$ is $\exists a \forall b\, [\forall X\, \phi_{\mathrm{B}}(a, b, X)]$,
10. $(\exists X\, \phi(X))^{\mathrm{B}}$ is $\exists a \forall b\, [\exists X \forall n\, \phi_{\mathrm{B}}(a, b\langle n\rangle, X)]$.

Notice that now the lower B-translations of formulas include second-order quantifications. The notion of bounded mixed formula of Sect. 4 has to be generalized to the notion of *second-order bounded mixed formula* in which closure under second-order quantifications is also allowed. The formulas of the second-order mixed language $\mathcal{L}_\Omega^{\mathrm{mix}^{\mathcal{C}}}$ are defined accordingly, as those that are generated from the second-order bounded formulas by means of propositional connectives and quantifications of the form $\forall a^\rho$, where a is a term variable (of a certain type ρ) of the term language $\mathcal{L}_\Omega$. Notice that, as before, we only need a set of classical connectives (our verifying semantics – described in the next paragraph – is classical) and that unbounded set-theoretic quantifiers are not present in the language $\mathcal{L}_\Omega^{\mathrm{mix}^{\mathcal{C}}}$.

As with the soundness theorem of Sect. 2, the soundness theorem of this section is verified semantically. There are several ways of extending the structures $L_{\omega_1}^{\mathrm{mix}}$ and $V_{\omega_1}^{\mathrm{mix}}$ to the second-order setting. The first semantics that we consider is $(L_{\omega_1}^{\mathrm{mix}^{\mathcal{C}}}, \mathcal{P}(L_{\omega_1})\!\upharpoonright)$. In this semantics, the second-order variables range over $\mathcal{P}(L_{\omega_1})\!\upharpoonright$, i.e., over the sets of the form $x \cap L_\alpha$, where $x \subseteq L_{\omega_1}$ and $\alpha < \omega_1$. There is a subtlety here. The intuitive second-order semantics is $\mathcal{P}(L_{\omega_1})$, not the subsets of L_{ω_1} truncated at a certain level α ($\alpha < \omega_1$) of the construtible hierarchy. However, the truncated semantics is enough. A second semantics that we will briefly consider is $(L_{\omega_1}^{\mathrm{mix}^{\mathcal{C}}}, L_{\omega_1})$. In this semantics the values of second-order variables range over *elements* of L_{ω_1}, i.e., over the sets in L_{ω_1}. This semantics is even subtler because both first-order set variables and second-order class variables range over the same domain, viz. L_{ω_1}. It was, in fact, the semantics used in [10]. In the next section we will consider the structure $(V_{\omega_1}^{\mathrm{mix}^{\mathcal{C}}}, V_{\omega_1})$. In this structure, the terms t of $\mathcal{L}_\Omega$ of type Ω index the (countable) stages V_t of the cumulative hierarchy. Of course, by this is meant that V_t is interpreted as $V_{|t|}$, as discussed at the end of Sect. 4. On the other hand, the second-order class variables range over elements of V_{ω_1} (the same range as the first-order set variables). Note that this range can also be described as being constituted by the sets of the form $x \cap V_\alpha$, where $x \subseteq V_{\omega_1}$ and $\alpha < \omega_1$.

Theorem 3 (Second-order soundness theorem I) *Let ϕ be a sentence of the language of second-order set theory. Suppose that*

$$\mathsf{IKP}\omega_2{\restriction} + \Delta_0^{\mathcal{C}}\text{-}\mathsf{LEM} + \Delta_0^{\mathcal{C}}\text{-}\mathsf{CA} + \mathsf{MP}^{\mathcal{C}} + \mathsf{bIP}_{\Pi_1^{\mathcal{C}}} + \Delta_0^{\mathcal{C}}\text{-}\mathsf{CColl}_2 + \mathsf{Coll}_2^{\mathcal{C}} \vdash \phi.$$

Then there are closed terms t of $\mathcal{L}_\Omega$ such that, for appropriate types ρ,

$$(L_{\omega_1}^{\mathrm{mix}^{\mathcal{C}}}, \mathcal{P}(L_{\omega_1}){\restriction}) \models \forall b^\rho\, \phi_B(t, b).$$

Proof The proof is by induction on the length of the derivation. We show that if a formula $\phi(w, W)$ is provable in the theory of the theorem, then there are closed terms t of $\mathcal{L}_\Omega$ such that, for appropiate types ρ, we have

$$L_{\omega_1}^{\mathrm{mix}} \models \forall c^\Omega \forall b^\rho\, [\forall W \forall w \in L_c\, \phi_{\mathrm{B}}(tc, b, w, W)],$$

where $\phi(w, W)^{\mathrm{B}}$ is $\exists a \forall b\, \phi_{\mathrm{B}}(a, b, w, W)$.

The various verifications are, *mutatis mutandis*, the ones given in the proof of Theorem 2. We only need to complement the verifications with the study of the logical rules for the second-order quantifiers and the principles $\Delta_0^{\mathcal{C}}$-CColl_2, $\mathsf{Coll}_2^{\mathcal{C}}$ and $\Delta_0^{\mathcal{C}}$-CA. We use the same layout as in the proof of Theorem 2. Let us start with the four new axioms and rules for second-order quantifiers.

13. $\phi \to \psi(W) \Rightarrow \phi \to \forall W\, \psi(W)$, where W is not free in ϕ. By induction hypothesis, there are terms q and r such that

$$\forall a, e\, [\forall W\, (\forall n\, \phi_{\mathrm{B}}(a, (qae)\langle n\rangle) \to \psi_{\mathrm{B}}(ra, e, W))]$$

holds in $(L_{\omega_1}^{\mathrm{mix}^{\mathcal{C}}}, \mathcal{P}(L_{\omega_1}){\restriction})$. We must obtain terms t and s such that

$$\forall a, e\, [\forall n\, \phi_{\mathrm{B}}(a, (sae)\langle n\rangle) \to \forall W\, \psi_{\mathrm{B}}(ta, e, W)]$$

holds in $(L_{\omega_1}^{\mathrm{mix}^{\mathcal{C}}}, \mathcal{P}(L_{\omega_1}){\restriction})$. Just take $t :\equiv r$ and $s :\equiv q$.

14. $\forall X\, \phi(X) \to \phi(W)$. A computation of the upper-B translation of this formula shows that we must find terms q and r such that

$$\forall a, b'\, [\forall W\, (\forall n \forall X\, \phi_{\mathrm{B}}(a, (qab')\langle n\rangle, X) \to \phi_{\mathrm{B}}(ra, b', W))]$$

holds in $(L_{\omega_1}^{\mathrm{mix}^{\mathcal{C}}}, \mathcal{P}(L_{\omega_1}){\restriction})$. Just put $q :\equiv \lambda a, b'.(b' + 1)$ and $r :\equiv \lambda a.a$.

15. $\phi(W) \to \exists X\, \phi(X)$. A computation of the upper-B translation of this formula shows that we must find terms q and r such that

$$\forall a, b'\, [\forall W\, (\forall n\, \phi_{\mathrm{B}}(a, (qab')\langle n\rangle, W) \to \exists X \forall n\, \phi_{\mathrm{B}}(ra, b'\langle n\rangle, X))]$$

holds in $(L_{\omega_1}^{\mathrm{mix}^{\mathcal{C}}}, \mathcal{P}(L_{\omega_1}){\restriction})$. Just put $q :\equiv \lambda a, b'.b'$ and $r :\equiv \lambda a.a$.

16. $\phi(W) \to \psi \Rightarrow \exists W\, \phi(W) \to \psi$, where W is not free in ψ. By induction hypothesis, there are terms q and r such that

$$\forall a, e[\forall W\, (\forall n\, \phi_{\mathrm{B}}(a, (qae)\langle n\rangle, W) \to \psi_{\mathrm{B}}(ra, e))]$$

holds in $(L_{\omega_1}^{\mathrm{mix}^C}, \mathcal{P}(L_{\omega_1})\lceil)$. We must obtain terms t and s such that

$$\forall a, e\, [\forall n \exists W \forall k\, \phi_{\mathrm{B}}(a, (sae)\langle n\rangle\langle k\rangle, W) \to \psi_{\mathrm{B}}(ta, e)]$$

holds in $(L_{\omega_1}^{\mathrm{mix}^C}, \mathcal{P}(L_{\omega_1})\lceil)$. Just take $t :\equiv r$ and $s :\equiv \lambda a, e.((qae) + 1)$.

Let us now discuss the principle $\Delta_0^{\mathcal{C}}$-CColl_2. According to its upper-B translation, we must find a term q such that

$$\forall c^{\Omega}\, [\forall n \forall z \in L_{(qc)\langle n\rangle} \exists X \forall x \in z\, \phi(x, X) \to \exists X \forall n \forall x \in L_{c\langle n\rangle}\, \phi(x, X)]$$

holds in $(L_{\omega_1}^{\mathrm{mix}^C}, \mathcal{P}(L_{\omega_1})\lceil)$. Well, it does hold with $q :\equiv \lambda c.(c + 2)$. To see this, note that the hypotehsis above entails $\forall z \in L_{c+1} \exists X \forall x \in z\, \phi_{\mathrm{B}}(x, X)$. In particular, this holds with z particularized as L_c, and we get what we want.

In order to discuss the principle $\mathsf{Coll}_2^{\mathcal{C}}$, we compute the upper-B translations of its antecedent and consequent. They are, $\exists c^{\Omega}, a \forall b\, [\forall X \exists x \in L_c \forall n\, \phi_{\mathrm{B}}(a, b\langle n\rangle, x, X)]$ and $\exists c^{\Omega}, a \forall b\, [\exists z \in L_c \forall k \forall X \exists x \in z \forall n\, \phi_{\mathrm{B}}(a, b\langle k\rangle\langle n\rangle, x, X)]$, respectively. Therefore, we need term q, r and t such that

$$\forall c^{\Omega}, a, b\, [\forall k \forall X \exists x \in L_c \forall n\, \phi_{\mathrm{B}}(a, (qcab)\langle k\rangle\langle n\rangle, x, X) \to$$

$$\exists z \in L_{rca} \forall k \forall X \exists x \in z \forall n\, \phi_{\mathrm{B}}(tca, b\langle k\rangle\langle n\rangle, x, X)]$$

holds in $(L_{\omega_1}^{\mathrm{mix}^C}, \mathcal{P}(L_{\omega_1})\lceil)$. Take $q :\equiv \lambda c, a, b.b$, $r :\equiv \lambda c, a.(c + 1)$ and $t :\equiv \lambda c, a.a$. With these data, the above holds (just let z be L_c).

Finally, we study the comprehension principle $\Delta_0^{\mathcal{C}}$-CA. An instance of this principle has the form $\forall W, w \exists X \forall x\, (x \in X \leftrightarrow \phi(x, w, W))$, where ϕ is a $\Delta_0^{\mathcal{C}}$-formula in which X does not occur. The upper B-translation of this instance is:

$$\forall d^{\Omega}, b^{\Omega}[\forall W \forall w \in L_d \exists X \forall n^N \forall x \in L_{b\langle n\rangle}(x \in X \leftrightarrow \phi(x, w, W))].$$

This statement holds with the set $X :\equiv \{x \in L_\alpha : \phi(x, w, W)\}$, where $\alpha = |b|$. $\square$

The proof of the previous theorem goes through in $(L_{\omega_1}^{\mathrm{mix}^C}, L_{\omega_1})$ except for one single step. It is in the verification of the scheme $\Delta_0^{\mathcal{C}}$-CA. If instead of $\Delta_0^{\mathcal{C}}$-CA one had the scheme $\exists X \forall x\, (x \in X \leftrightarrow \phi(x, w, W))$, where $\phi(x)$ is restricted to be a Δ_0-formula, then the set $\{x \in L_\alpha : \phi(x, w, W)\}$ is in L_{ω_1}. In effect, the parameter W (and, also, w) of the structure $(L_{\omega_1}^{\mathrm{mix}^C}, L_{\omega_1})$ takes a value in a certain L_β, for a certain $\beta < \omega_1$. Therefore, the previous set is an element of L_γ, for $\alpha, \beta < \gamma$. We think that this is worth remarking (specially because this was the strategy adopted in [10]).

Proposition 11 *If* $\mathsf{IKP}\omega_2\!\restriction + \Delta_0^C\text{-}\mathsf{LEM} + \Delta_0^C\text{-}\mathsf{CA} + \mathsf{MP}^C + \mathsf{bIP}_{\Pi_1^C} + \Delta_0^C\text{-}\mathsf{CColl}_2 + \mathsf{Coll}_2^C$ *proves* $\forall x \exists y\, \phi(x, y)$, *where* ϕ *is a* Δ_0*-formula (x and y are the only free variables), then there is a closed term t of type* $\Omega \to \Omega$ *such that*

$$L_{\omega_1}^{\mathrm{mix}} \models \forall c^{\Omega} \forall x \in L_c \exists y \in L_{tc}\, \phi(x, y).$$

Moreover, $L_{\mathrm{BH}} \models \forall x \exists y\, \phi(x, y)$, *where* BH *is the Bachmann–Howard ordinal.*

The above proposition is an immediate consequence of Theorem 3. Note that the formulas ϕ are restricted to Δ_0-formulas.

Corollary 4 *If* $\mathsf{IKP}\omega_2\!\restriction + \Delta_0^C\text{-}\mathsf{LEM} + \Delta_0^C\text{-}\mathsf{CA} + \mathsf{MP}^C + \mathsf{bIP}_{\Pi_1^C} + \Delta_0^C\text{-}\mathsf{CColl}_2 + \mathsf{Coll}_2^C$ *proves* $\exists x\, \phi(x)$, *where* ϕ *is a* Δ_0*-formula (x is the only free variable), then there is an ordinal* α *smaller than the Bachmann–Howard ordinal so that* $L_\alpha \models \exists x\, \phi(x)$.

Lemma 9 does not generalize to the functional interpretation extended to the second-order language, neither when the semantics is $(L_{\omega_1}^{\mathrm{mix}^C}, \mathcal{P}(L_{\omega_1})\!\restriction)$, nor when the semantics is $(L_{\omega_1}^{\mathrm{mix}^C}, L_{\omega_1})$. For instance, consider the following instance of Coll^C:

$$\forall X \exists x \forall z\, (X = \{z\} \to Ord(x) \wedge z \in L_x) \to \exists w \forall X \exists x \in w \forall z\, (X = \{z\} \to Ord(x) \wedge z \in L_x).$$

It is clear that this sentence is false in both structures above. The reason why the proof of Lemma 9 does not generalize is, of course, the fact that the transformations (9) and (10) of the definition of the functional interpretation are not truth preserving (in the said structures). However, as it was shown in Lemma 9, the remaining transformations are truth preserving. Hence, as long as we restrict ourselves to first-order formulas ϕ, we have the following:

Proposition 12 *If* $\mathsf{IKP}\omega_2\!\restriction + \Delta_0^C\text{-}\mathsf{LEM} + \Delta_0^C\text{-}\mathsf{CA} + \mathsf{MP}^C + \mathsf{bIP}_{\Pi_1^C} + \Delta_0^C\text{-}\mathsf{CColl}_2 + \mathsf{Coll}_2^C$ *proves* $\forall x \exists y\, \phi(x, y)$, *where* ϕ *is any formula (x and y are the only free variables)* without *second-order quantifications, then there is a closed term t of type* $\Omega \to \Omega$ *such that*

$$\forall c \in W \forall x \in L_{|c|} \exists y \in L_{|t(c)|}\ L_{\omega_1} \models \phi(x, y).$$

The proof is like the one of Proposition 5.

7 Salipante's Result and Power Kripke–Platek Set Theory

The theory $\mathsf{IKP}\omega_2$ is like the theory $\mathsf{IKP}\omega_2\!\restriction$ of Definition 9 except that we now permit Δ_0^C-formulas in the separation scheme *with (first and) second-order parameters*. In [19], Salipante observed the following (he worked in a classical theory, but the argument is the same):

Theorem 4 (Salipante) *The theory* $\mathsf{IKP}\omega_2 + \Delta_0\text{-}\mathsf{LEM} + \Delta_0\text{-}\mathsf{CA} + \mathsf{s}\Pi^1_1\text{-}\mathsf{ref}$ *proves the powerset axiom, i.e., it proves the sentence* $\forall y \exists z \forall x\, (x \in z \leftrightarrow x \subseteq y)$.

Proof Let y be given. The theory $\mathsf{IKP}\omega_2$ proves $\forall X \exists w\, (w = X \cap y)$, where $X \cap y$ abbreviates the *set* $\{u \in y : u \in X\}$. This set exists by separation (using the second-order parameter X). By $\mathsf{s}\Pi^1_1$-ref, $\exists z \forall X \exists w \in z\, (w = X \cap y)$. Let z_0 be such a set. We claim that $\forall x\, (x \subseteq y \to x \in z_0)$. To see this, take x a subset of y. By Δ_0-CA, let $X_0 = \{u : u \in x\}$. By the choice of z_0, there is $w \in z_0$ such that $w = X_0 \cap y$. Since $X_0 \cap y = x$, we conclude that $x \in z_0$, as wanted. The powerset of y can now be obtained from z_0 by ordinary separation. □

The above proof also holds if the separation scheme applies only to Δ_0-formulas (that is how Salipante stated his theorem). The *crucial thing* is to allow second-order parameters in the separation scheme.

In [17] Rathjen introduced the theory $\mathsf{KP}\omega\,(\mathcal{P})$ of power Kripke–Platek set theory. In order to formulate this theory we need the following definition:

Definition 14 The class of $\Delta^{\mathcal{P}}_0$-formulas is the smallest class of formulas of the language of set theory containing the atomic formulas (including $\bot$) and closed under $\wedge$, $\vee$, $\to$ and the quantifications

$$\forall x \in z, \ \exists x \in z, \ \forall x \subseteq z, \ \exists x \subseteq z,$$

where the last two quantifications abbreviate $\forall x\, (x \subseteq z \to \ldots)$ and $\exists x\, (x \subseteq z \wedge \ldots)$, respectively. $\Sigma^{\mathcal{P}}_1$-formulas are formulas of the form $\exists z\, \psi(z)$, where $\psi(z)$ is a $\Delta^{\mathcal{P}}_0$-formula (possibly with parameters). $\Pi^{\mathcal{P}}_1$-formulas are defined dually. A $\Pi^{\mathcal{P}}_2$-formula is a formula of the form $\forall w\, \phi(w)$, where $\phi(w)$ is a $\Sigma^{\mathcal{P}}_1$-formula.

The theory $\mathsf{KP}\omega\,(\mathcal{P})$ is a classical theory in the language of set theory with the following axioms: extensionality, pairing, union, infinity, powerset, $\Delta^{\mathcal{P}}_0$-separation, $\Delta^{\mathcal{P}}_0$-collection and unrestricted foundation. The transitive models of $\mathsf{KP}\omega\,(\mathcal{P})$ are the *power admissible* sets introduced by Harvey Friedman in [12]. As Rathjen observes, the theory $\mathsf{KP}\omega\,(\mathcal{P})$ can also be described as the theory $\mathsf{KP}\omega$ framed in the language of set theory extended with a new primitive unary function symbol $\mathcal{P}$ for the powerset operation, the axiom $\forall x\, (x \in \mathcal{P}(y) \leftrightarrow x \subseteq y)$, and the schemata of Δ_0-separation and Δ_0-collection extended to the Δ_0-formulas of this new language. It should be noticed, as Rathjen warns us, that the theory $\mathsf{KP}\omega\,(\mathcal{P})$ is *not* the same theory as $\mathsf{KP}\omega$ with the powerset axiom. This latter theory is much weaker than $\mathsf{KP}\omega\,(\mathcal{P})$, as Rathjen discusses in [17].

In this paper we are interested in semi-constructive theories. The natural theories to consider are the theories of Sect. 2 in which the Δ_0-formulas are replaced by the wider class of $\Delta^{\mathcal{P}}_0$-formulas. We are naturally led to consider the theory

$$\mathsf{IKP}\omega\,(\mathcal{P}) + \Delta^{\mathcal{P}}_0\text{-}\mathsf{LEM} + \mathsf{MP}^{\mathcal{P}} + \mathsf{bIP}_{\Pi^{\mathcal{P}}_1} + \Delta^{\mathcal{P}}_0\text{-}\mathsf{CColl} + \mathsf{Coll}^{\mathcal{P}},$$

where it should be clear what the acronyms above stand for. This is also a very robust theory, and it is clear that we have the analogue of Proposition 1:

Proposition 13 *The theory* $\mathsf{KP}\omega\,(\mathcal{P})$ *is* $\Pi_2^{\mathcal{P}}$*-conservative over* $\mathsf{IKP}\omega\,(\mathcal{P}) + \Delta_0^{\mathcal{P}}$*-*$\mathsf{LEM} + \mathsf{MP}^{\mathcal{P}}$.

We could adapt the analysis that we have made of the semi-constructive theory $\mathsf{IKP}\omega + \Delta_0\text{-}\mathsf{LEM} + \mathsf{MP} + \mathsf{bIP}_{\Pi_1} + \Delta_0\text{-}\mathsf{CColl} + \mathsf{Coll}$ to the powerset version mentioned above (this adaptation requires the cumulative hierarchy instead of the constructible hierarchy). The next theorem provides an illuminating alternative. We need a lemma first:

Lemma 10 *The theory* $\mathsf{IKP}\omega_2 + \Delta_0^{\mathcal{C}}\text{-}\mathsf{LEM} + \Delta_0^{\mathcal{C}}\text{-}\mathsf{CA} + \mathsf{MP}^{\mathcal{C}} + \mathsf{bIP}_{\Pi_1^{\mathcal{C}}} + \Delta_0^{\mathcal{C}}\text{-}\mathsf{CColl}_2 + \mathsf{Coll}_2^{\mathcal{C}}$ *proves* $\Delta_1^{\mathcal{C}}$*-separation, i.e., it proves*

$$\forall x\,(\forall u\,\phi(u,x) \leftrightarrow \exists v\,\psi(v,x)) \rightarrow \forall z\exists y\forall x\,(x \in y \leftrightarrow (x \in z \wedge \exists v\,\psi(v,x))),$$

for $\Delta_0^{\mathcal{C}}$*-formulas* ϕ *and* ψ *(possibly with first and second-order parameters).*

The above lemma is proven like Proposition 2. The proof is possible because we now permit $\Delta_0^{\mathcal{C}}$-formulas and parameters (first and second-order) in the separation scheme.

Theorem 5 *The theory* $\mathsf{IKP}\omega\,(\mathcal{P}) + \Delta_0^{\mathcal{P}}\text{-}\mathsf{LEM} + \mathsf{MP}^{\mathcal{P}} + \mathsf{bIP}_{\Pi_1^{\mathcal{P}}} + \Delta_0^{\mathcal{P}}\text{-}\mathsf{CColl} + \mathsf{Coll}^{\mathcal{P}}$ *is a subtheory of the second-order theory*

$$\mathsf{IKP}\omega_2 + \Delta_0^{\mathcal{C}}\text{-}\mathsf{LEM} + \Delta_0^{\mathcal{C}}\text{-}\mathsf{CA} + \mathsf{MP}^{\mathcal{C}} + \mathsf{bIP}_{\Pi_1^{\mathcal{C}}} + \Delta_0^{\mathcal{C}}\text{-}\mathsf{CColl}_2 + \mathsf{Coll}_2^{\mathcal{C}}.$$

Proof This result hinges on two facts. The first is that the relation $z = \mathcal{P}(y)$ is given by the $\Delta_0^{\mathcal{C}}$-formula $\forall x \in z\,(x \subseteq y) \wedge \forall X\,(X \cap y \in z)$. Without the abbreviation, it reads

$$\forall x \in z\,(x \subseteq y) \wedge \forall X \exists x \in z \forall w\,(w \in x \leftrightarrow w \in X \wedge w \in y).$$

Let us denote this $\Delta_0^{\mathcal{C}}$-formula by $P(y,z)$. By Theorem 4 (and its argument), the second-order theory of the theorem proves $\forall y, z\,(P(y,z) \leftrightarrow \forall x\,(x \in z \leftrightarrow x \subseteq y))$ and $\forall y \exists^1 z\, P(y,z)$. Therefore, it proves the powerset axiom. The second important fact is that the second-order theory of the theorem has a good theory for introducing $\Sigma_1^{\mathcal{C}}$-function symbols (see the comments after Definition 13). In particular, $\Delta_0^{\mathcal{C}}$-formulas in the new language with the extra function symbols translate into $\Delta_1^{\mathcal{C}}$-formulas of the original language (the equivalence between the corresponding pair of $\Sigma_1^{\mathcal{C}}$-formulas and $\Pi_1^{\mathcal{C}}$-formulas is proven in the second-order theory of the theorem, of course). Therefore, we can introduce a $\Sigma_1^{\mathcal{C}}$-function symbol that satisfies the defining axiom of the powerset operation and, as a consequence, $\Delta_0^{\mathcal{P}}$-formulas are rendered by $\Delta_1^{\mathcal{C}}$-formulas. The above lemma entails that $\Delta_0^{\mathcal{P}}$-separation is provable in the second-order theory. The theorem should now be clear. □

The structure $(V_{\omega_1}^{\mathrm{mix}^{\mathcal{C}}}, V_{\omega_1})$ for the second-order mixed language $\mathcal{L}_{\Omega}^{\mathrm{mix}^{\mathcal{C}}}$ was introduced just before Theorem 3. Remember that in this structure both the first-order set variables and second-order (class) variables of $\mathcal{L}_{\Omega}^{\mathrm{mix}^{\mathcal{C}}}$ take values in V_{ω_1}.

Theorem 6 (Second-order soundness theorem II) *Let ϕ be a sentence of the language of second-order set theory. Suppose that*

$$\mathsf{IKP}\omega_2 + \Delta_0^C\text{-}\mathsf{LEM} + \Delta_0^C\text{-}\mathsf{CA} + \mathsf{MP}^C + \mathsf{bIP}_{\Pi_1^C} + \Delta_0^C\text{-}\mathsf{CColl}_2 + \mathsf{Coll}_2^C \vdash \phi.$$

Then there are closed terms t of $\mathcal{L}_\Omega$ such that, for appropriate types ρ,

$$(V_{\omega_1}^{\mathrm{mix}^C}, V_{\omega_1}) \models \forall b^\rho\, \phi_B(t, b).$$

Proof The proof is *mutatis mutandis* the proof of Theorem 3 (see also the note after that proof), but one must replace systematically the construtible hierarchy by the cumulative hierarchy. We need the cumulative hierarchy because the separation axiom of $\mathsf{IKP}\omega_2$ has second-order parameters. Let us see in detail why this is so. The separation axiom with (first and) second-order parameters is

$$\forall w, W \forall y \exists z \forall x\, (x \in z \leftrightarrow x \in y \wedge \phi(x, w, W)),$$

where ϕ is a Δ_0^C-formula in which the variable z does not occur. As before, notice that the inner universal statement can be considered Δ_0^C. Hence, the upper B-translation of this formula shows that we need a closed term t of type $\Omega \to (\Omega \to \Omega)$ such that

$$\forall c^\Omega, e^\Omega\, [\forall y \in V_c \forall w \in V_e \forall W \exists z \in V_{tce} \forall x\, (x \in z \leftrightarrow x \in y \wedge \phi(x, w, W))]$$

holds in $(V_{\omega_1}^{\mathrm{mix}^C}, V_{\omega_1})$. It is clear that the term $t :\equiv \lambda c, e.(c + 1)$ does the job. To see this, just take z to be $\{x \in V_\alpha : x \in y \wedge \phi(x, w, W)\}$, where $\alpha = |c|$.
[Note that the proof does not go through in the constructible hierarchy because the term t is only allowed to depend on y and w, via c and e (respectively), but *not* on W.]
□

As usual, we can draw some consequences regarding Π_2^C and Σ_1^C consequences of the second-order theory of the theorem. We are going to present them in a particlular fashion, with an eye to their application to power Kripke–Platek set theories.

We have the following absoluteness property:

Lemma 11 *If α and β are ordinals and $\phi(x_1, \ldots, x_n)$ is a Δ_0^C-formula with its free variables as shown (they are all first-order), then*

$$x_1, \ldots, x_n \in V_\alpha \wedge \alpha < \beta \to [(V_\alpha, V_{\alpha+1}) \models \phi(x_1, \ldots, x_n) \leftrightarrow (V_\beta, V_\beta) \models \phi(x_1, \ldots, x_n)].$$

Observation 2 In the above, in a structure of the form (V, W), with $W \subseteq \mathcal{P}(V)$, the first-order variables take values in V and the second-order variables take values in W.

Proof We show a bit more in order to get an induction on Δ_0^C-formulas going. We prove by induction on Δ_0^C-formulas $\phi(x_1, \ldots, x_n, X_1, \ldots, X_k)$ that for all $x_1, \ldots, x_n \in V_\alpha$ and $X_1, \ldots, X_k \in V_\beta$ we have

$$(V_\alpha, V_{\alpha+1}) \models \phi(x_1, \dots, x_n, X_1 \cap V_\alpha, \dots, X_k \cap V_\alpha) \leftrightarrow (V_\beta, V_\beta) \models \phi(x_1, \dots, x_n, X_1, \dots, X_k)$$

Note that we are abusing notation by confusing variables with the sets that take their values. For ease of reading, we will also omit tuples. The proof by induction is straightforward except for the case of second-order quantifications. We study the universal second-order quantifier (the case of the existential second-order quantifier follows immediately because $\Delta_0^{\mathcal{C}}$-formulas are closed under negation).

Consider the formula $\forall W\, \phi(x, X, W)$, with ϕ a $\Delta_0^{\mathcal{C}}$-formula. Let $x \in V_\alpha$, $X \in V_\beta$ and assume that $(V_\alpha, V_{\alpha+1}) \models \forall W\, \phi(x, X \cap V_\alpha, W)$. Let Y be an arbitrary element of V_β. Since $Y \cap V_\alpha \in V_{\alpha+1}$, we have $(V_\alpha, V_{\alpha+1}) \models \phi(x, X \cap V_\alpha, Y \cap V_\alpha)$. By induction hypothesis, we get $(V_\beta, V_\beta) \models \phi(x, X, Y)$. By the arbitrariness of Y, we conclude $(V_\beta, V_\beta) \models \forall W\, \phi(x, X, W)$. To prove the converse, let $x \in V_\alpha$, $X \in V_\beta$ and assume that $(V_\beta, V_\beta) \models \forall W\, \phi(x, X, W)$. Let $Y \in V_{\alpha+1}$ be arbitrary. In particular, $Y \in V_\beta$. Hence, $(V_\beta, V_\beta) \models \phi(x, X, Y)$. By induction hypothesis, we get $(V_\alpha, V_{\alpha+1}) \models \phi(x, X \cap V_\alpha, Y \cap V_\alpha)$. Since $Y \cap V_\alpha = Y$, we have $(V_\alpha, V_{\alpha+1}) \models \phi(x, X \cap V_\alpha, Y)$. Therefore, by the arbitrariness of Y, we conclude that $(V_\alpha, V_{\alpha+1}) \models \forall W\, \phi(x, X \cap V_\alpha, W)$. □

Proposition 14 *If* $\mathsf{IKP}\omega_2 + \Delta_0^{\mathcal{C}}\text{-}\mathsf{LEM} + \Delta_0^{\mathcal{C}}\text{-}\mathsf{CA} + \mathsf{MP}^{\mathcal{C}} + \mathsf{bIP}_{\Pi_1^{\mathcal{C}}} + \Delta_0^{\mathcal{C}}\text{-}\mathsf{CColl}_2 + \mathsf{Coll}_2^{\mathcal{C}}$ *proves* $\forall x \exists y\, \phi(x, y)$, *where* ϕ *is a* $\Delta_0^{\mathcal{C}}$*-formula (*x *and* y *are the only free variables), then there is a closed term* t *of type* $\Omega \to \Omega$ *such that*

$$(V_{\omega_1}^{\mathrm{mix}^{\mathcal{C}}}, V_{\omega_1}) \models \forall c^{\Omega} \forall x \in V_c \exists y \in V_{tc}\, \phi(x, y).$$

Moreover, $(V_{\mathrm{BH}}, V_{\mathrm{BH}}) \models \forall x \exists y\, \phi(x, y)$, *where* BH *is the Bachmann–Howard ordinal.*

The above proposition is an immediate consequence of Theorem 6. The final conclusion follows from two applications of Lemma 11. With this lemma, we also get:

Corollary 5 *If* $\mathsf{IKP}\omega_2 + \Delta_0^{\mathcal{C}}\text{-}\mathsf{LEM} + \Delta_0^{\mathcal{C}}\text{-}\mathsf{CA} + \mathsf{MP}^{\mathcal{C}} + \mathsf{bIP}_{\Pi_1^{\mathcal{C}}} + \Delta_0^{\mathcal{C}}\text{-}\mathsf{CColl}_2 + \mathsf{Coll}_2^{\mathcal{C}} \vdash \exists x\, \phi(x)$, *where* $\phi(x)$ *is a* $\Delta_0^{\mathcal{C}}$*-formula (*x *is its only free variable), then there is an ordinal* α *smaller than the Bachmann–Howard ordinal such that* $(V_\alpha, V_{\alpha+1}) \models \exists x\, \phi(x)$.

An analysis of the theory $\mathsf{IKP}\omega\,(\mathcal{P}) + \Delta_0^{\mathcal{P}}\text{-}\mathsf{LEM} + \mathsf{MP}^{\mathcal{P}} + \mathsf{bIP}_{\Pi_1^{\mathcal{P}}} + \Delta_0^{\mathcal{P}}\text{-}\mathsf{CColl} + \mathsf{Coll}^{\mathcal{P}}$ is now forthcoming. The strategy is clear: Use Theorem 5 to reduce the analysis of this theory to the analysis of the second-order theory of that proposition. In the following lemma, $P(y, z)$ is the $\Delta_0^{\mathcal{C}}$-formula of the proof of Theorem 5:

Lemma 12 *Let* $\phi(x_1, \dots, x_n)$ *be a* $\Delta_0^{\mathcal{P}}$*-formula, with its free variables as shown. Then there is a* Δ_0*-formula* $\phi^*(x_1, \dots, x_n, z)$ *such that the second-order theory*

$$\mathsf{IKP}\omega_2 + \Delta_0^{\mathcal{C}}\text{-}\mathsf{LEM} + \Delta_0^{\mathcal{C}}\text{-}\mathsf{CA} + \mathsf{MP}^{\mathcal{C}} + \mathsf{bIP}_{\Pi_1^{\mathcal{C}}} + \Delta_0^{\mathcal{C}}\text{-}\mathsf{CColl}_2 + \mathsf{Coll}_2^{\mathcal{C}}$$

proves the equivalence

$$\phi(x_1, \ldots, x_n) \leftrightarrow \exists z\, (P(\mathrm{tc}(x_1 \cup \ldots \cup x_n), z) \wedge \phi^*(x_1, \ldots, x_n, z)).$$

Moreover, for all $x_1, \ldots, x_n \in V_{\omega_1}$,

$$V_{\omega_1} \models \phi(x_1, \ldots, x_n) \text{ if, and only if, } V_{\omega_1} \models \phi^*(x_1, \ldots, x_n, \mathcal{P}(\mathrm{tc}(x_1 \cup \ldots \cup x_n))).$$

Here, $\mathrm{tc}(w)$ *stands for the transitive closure of* w.

Proof The proof is by induction on the complexity of ϕ. We will only study negation and the universal quantifications $\forall w \subseteq x\, (\ldots)$ and $\forall w \in x\, (\ldots)$. Negation is clear because, using the induction hypothesis, the equivalence

$$\neg\phi(x_1, \ldots, x_n) \leftrightarrow \exists z\, (P(\mathrm{tc}(x_1 \cup \ldots \cup x_n), z) \wedge \neg\phi^*(x_1, \ldots, x_n, z))$$

is provable in the second-order theory of the lemma. Of course, $(\neg\phi)^*$ is defined as being $\neg(\phi^*)$. Let us now consider the formula $\forall w \subseteq x\, \phi(w, x, x_1, \ldots, x_n)$, with $\phi \in \Delta_0^{\mathcal{P}}$. By induction hypothesis, the second-order theory of the lemma proves the equivalence of the above formula with $\forall w \subseteq x \exists z\, (P(\mathrm{tc}(w \cup x \cup x_1 \cup \ldots \cup x_n), z) \wedge \phi^*(w, x, x_1, \ldots, x_n, z))$. This is equivalent to

$$\exists z\, (P(\mathrm{tc}(x \cup x_1 \cup \ldots \cup x_n), z) \wedge \forall w \in z\, (w \subseteq x \rightarrow \phi^*(w, x, x_1, \ldots, x_n, z))).$$

This is due to the fact that $w \cup x = x$ (and the uniqueness of the z). The argument for the second part of the lemma is similar. The situation is clear now.

The treatment of the usual bounded quantification $\forall w \in x\, (\ldots)$ is analogous. Here one takes notice that $\mathrm{tc}(w \cup x \cup x_1 \cup \ldots \cup x_n) = \mathrm{tc}(x \cup x_1 \cup \ldots \cup x_n)$ when $w \in x$. □

We are ready to prove the following proposition and corollary:

Proposition 15 *If* $\mathsf{IKP}\omega\,(\mathcal{P}) + \Delta_0^{\mathcal{P}}\text{-}\mathsf{LEM} + \mathsf{MP}^{\mathcal{P}} + \mathsf{bIP}_{\Pi_1^{\mathcal{P}}} + \Delta_0^{\mathcal{P}}\text{-}\mathsf{CColl} + \mathsf{Coll}^{\mathcal{P}} \vdash \forall x \exists y\, \phi(x, y)$, *where* ϕ *is a* $\Delta_0^{\mathcal{P}}$*-formula (x and y are the only free variables), then there is a closed term t of type* $\Omega \rightarrow \Omega$ *such that*

$$\forall c \in W \forall x \in V_{|c|} \exists y \in V_{|t(c)|}\; V_{\omega_1} \models \phi(x, y).$$

Moreover, $V_{\mathrm{BH}} \models \forall x \exists y\, \phi(x, y)$, *where* BH *is the Bachmann–Howard ordinal.*

Proof Suppose that $\mathsf{IKP}\omega\,(\mathcal{P}) + \Delta_0^{\mathcal{P}}\text{-}\mathsf{LEM} + \mathsf{MP}^{\mathcal{P}} + \mathsf{bIP}_{\Pi_1^{\mathcal{P}}} + \Delta_0^{\mathcal{P}}\text{-}\mathsf{CColl} + \mathsf{Coll}^{\mathcal{P}}$ proves $\forall x \exists y\, \phi(x, y)$. By Theorem 5, so does the theory of Proposition 14. By the previous lemma, this second-order theory proves $\forall x \exists y \exists z\, (P(\mathrm{tc}(x \cup y), z) \wedge \phi^*(x, y, z))$. Using Proposition 14, it can be argued that there is a closed term t such that

$$\forall c \in W \forall x \in V_{|c|} \exists y \in V_{|t(c)|} \; (V_{\omega_1}, V_{\omega_1}) \models \exists z \, (P(\mathrm{tc}(x \cup y, z)) \wedge \phi^*(x, y, z)).$$

Since $(V_{\omega_1}, V_{\omega_1}) \models P(\mathrm{tc}(x \cup y), z) \leftrightarrow z = \mathcal{P}(\mathrm{tc}(x \cup y))$, we obtain the desired conclusion using the second part of previous lemma.

It also follows that $V_{\mathrm{BH}} \models \forall x \exists y \, \phi(x, y)$ because $\Delta_0^{\mathcal{P}}$-formulas are absolute between the various levels of the cumulative hierarchy. □

Corollary 6 *If* $\mathsf{IKP}\omega\,(\mathcal{P}) + \Delta_0^{\mathcal{P}}\text{-}\mathsf{LEM} + \mathsf{MP}^{\mathcal{P}} + \mathsf{bIP}_{\Pi_1^{\mathcal{P}}} + \Delta_0^{\mathcal{P}}\text{-}\mathsf{CColl} + \mathsf{Coll}^{\mathcal{P}} \vdash \exists x \, \phi(x)$, *where* $\phi(x)$ *is a* $\Delta_0^{\mathcal{P}}$*-formula (x is its only free variable), then there is an ordinal* α *smaller than the Bachmann–Howard ordinal such that* $V_\alpha \models \exists x \, \phi(x)$.

Using Proposition 13, we can give a Σ-ordinal analysis (in the relativized sense of [17]) of the *classical* theory $\mathsf{KP}\omega\,(\mathcal{P})$:

Proposition 16 *If* $\mathsf{KP}\omega\,(\mathcal{P}) \vdash \forall x \exists y \, \phi(x, y)$, *where* ϕ *is a* $\Delta_0^{\mathcal{P}}$*-formula (x and y are the only free variables), then there is a closed term t of type* $\Omega \to \Omega$ *such that*

$$\forall c \in W \forall x \in V_{|c|} \exists y \in V_{|t(c)|} \; V_{\omega_1} \models \phi(x, y).$$

Moreover, $V_{\mathrm{BH}} \models \forall x \exists y \, \phi(x, y)$, *where* BH *is the Bachmann–Howard ordinal.*

Corollary 7 *If* $\mathsf{KP}\omega\,(\mathcal{P}) \vdash \exists x \, \phi(x)$, *where* $\phi(x)$ *is a* $\Delta_0^{\mathcal{P}}$*-formula (x is its only free variable), then there is an ordinal* α *smaller than the Bachmann–Howard ordinal such that* $V_\alpha \models \exists x \, \phi(x)$.

These two results are due to Rathjen in [17]. We obtained them in a very roundabout way, via second-order semi-constructive theories. A direct way, using our kind of functional interpretations, would be just to adapt – replacing in a straightforward manner the constructible hierarchy by the cumulative hierarchy – the analysis of $\mathsf{KP}\omega$ provided in [10].

References

1. Avigad, J., Feferman, S.: Gödel's functional ("Dialectica") interpretation. In: Buss, S.R. (ed.) Handbook of Proof Theory. Studies in Logic and the Foundations of Mathematics, vol. 137, pp. 337–405. North Holland, Amsterdam (1998)
2. Avigad, J., Towsner, H.: Functional interpretation and inductive definitions. J. Symb. Log. **74**(4), 1100–1120 (2009)
3. Barwise, J.: Implicit definability and compactness in infinitary languages. In: Barwise, J. (ed.) The Syntax and Semantics of Infinitary Logic. Lecture Notes in Mathematics, vol. 72, pp. 1–35. Springer, Berlin (1968)
4. Barwise, J.: Application of strict Π_1^1 predicates to infinitary logic. J. Symb. Log. **34**, 409–423 (1969)
5. Barwise, J.: Admissible Set Theory and Structures: An Approach to Definability Theory. Perspectives in Mathematical Logic. Springer, Berlin (1975)

6. Bishop, E.: Schizophrenia in contemporary mathematics. In: Bishop, E. (ed.) Reflections on Him and His Research. Contemporary Mathematics, pp. 1–32. American Mathematical Society (1985). First published in 1973
7. Bridges, D.S., Richman, F.: Varieties of Constructive Mathematics, vol. 97. London Mathematical Society Lecture Notes Series. Cambridge University Press (1987)
8. Feferman, S.: On the strength of some semi-constructive theories. In: Feferman, S., Sieg, W. (eds.) Proof Categories and Computation: Essays in honor of Grigori Mints, pp. 109–129. College Publications, London (2010)
9. Feferman, S.: Logic, mathematics, and conceptual structuralism. In: Rush, P. (ed.) The Metaphysics of Logic, pp. 72–92. Cambridge University Press (2014)
10. Ferreira, F.: A new computation of the Σ-ordinal of KPω. J. Symb. Log. **79**, 306–324 (2014)
11. Ferreira, F., Oliva, P.: Bounded functional interpretation. Ann. Pure Appl. Log. **135**, 73–112 (2005)
12. Friedman, H.: Countable models of set theories. In: Mathias, A.R.D., Rogers Jr., H. (eds.) Cambridge Summer School in Mathematical Logic. Lecture Notes in Mathematics, vol. 337, pp. 539–573. Springer, Berlin (1973)
13. Hindley, J.R., Seldin, J.P.: Introduction to Combinators and λ-Calculus, vol. 1. London Mathematical Society Student Texts. Cambridge University Press (1986)
14. Howard, W.A.: A system of abstract constructive ordinals. J. Symb. Log. **37**(2), 355–374 (1972)
15. Kohlenbach, U.: Applied Proof Theory: Proof Interpretations and their Use in Mathematics. Springer Monographs in Mathematics. Springer, Berlin (2008)
16. Pozsgay, L.: Liberal intuitionism as a basis for set theory. In: Scott, D. (ed.) Axiomatic Set Theory (Part 1), vol. XIII. Proceedings of Symposia in Pure Mathematics, pp. 321–330. American Mathematical Society (1971)
17. Rathjen, M.: Relativized ordinal analysis: the case of Power Kripke-Platek set theory. Ann. Pure Appl. Log. **165**, 316–339 (2014)
18. Sacks, G.: Higher Recursion Theory. Perspectives in Mathematical Logic. Springer, Berlin (1990)
19. Salipante, V.: On the consistency strenght of the strict Π^1_1 reflection principle. Ph.D. thesis, Universität Bern (2005)
20. Simpson, S.G.: Subsystems of Second Order Arithmetic. Perspectives in Mathematical Logic. Springer, Berlin (1999)

Author Biography

Fernando Ferreira is Professor of Mathematics at Universidade de Lisboa. He received his Ph.D. at Pennsylvania State University in 1988 (under Stephen Simpson). He was a Fulbright Scholar at Harvard University (Spring 2004) and Tinker Visiting Professor at Stanford University (Fall 2009). He has written papers in weak systems of arithmetic and analysis, proof theory (especially functional interpretations) and philosophy and foundations of mathematics. He also wrote two papers on the problem of truth in Parmenides and Plato. He is a corresponding member of Academia das Ciências de Lisboa.

Proof Theory of Constructive Systems: Inductive Types and Univalence

Michael Rathjen

Abstract In Feferman's work, explicit mathematics and theories of generalized inductive definitions play a central role. One objective of this article is to describe the connections with Martin–Löf type theory and constructive Zermelo–Fraenkel set theory. Proof theory has contributed to a deeper grasp of the relationship between different frameworks for constructive mathematics. Some of the reductions are known only through ordinal-theoretic characterizations. The paper also addresses the strength of Voevodsky's univalence axiom. A further goal is to investigate the strength of intuitionistic theories of generalized inductive definitions in the framework of intuitionistic explicit mathematics that lie beyond the reach of Martin–Löf type theory.

Keywords Explicit mathematics · Constructive Zermelo–Fraenkel set theory · Martin–Löf type theory · Univalence axiom · Proof-theoretic strength

MSC 03F30 · 03F50 · 03C62

1 Introduction

Intuitionistic systems of inductive definitions have figured prominently in Solomon Feferman's program of reducing classical subsystems of analysis and theories of iterated inductive definitions to constructive theories of various kinds. In the special case of classical theories of finitely as well as transfinitely iterated inductive definitions, where the iteration occurs along a computable well-ordering, the program was mainly completed by Buchholz, Pohlers, and Sieg more than 30 years ago (see [13, 19]). For stronger theories of inductive definitions such as those based on Feferman's

---

M. Rathjen (✉)
Department of Pure Mathematics, University of Leeds, Leeds LS2 9JT, UK
e-mail: rathjen@amsta.leeds.ac.uk

G. Jäger and W. Sieg (eds.), *Feferman on Foundations*, Outstanding Contributions to Logic 13, DOI 10.1007/978-3-319-63334-3_14

intutitionic *Explicit Mathematics*[1] ($\mathbf{T}_0^i$) some answers have been provided in the last 10 years while some questions are still open.

The aim of the first part of this paper is to survey the landscape of some prominent constructive theories that emerged in the 1970s. In addition to Feferman's $\mathbf{T}_0^i$, Myhill's *Constructive Set Theory* (**CST**) and Martin–Löf type theory (**MLTT**) have been proposed with the aim of isolating the principles on which constructive mathematics is founded, notably the notions of constructive function and set in Bishop's mathematics.

Martin–Löf type theory with infinitely many universes and inductive types (W-types) has attracted a great deal of attention recently because of a newly found connection between type theory and topology, called *homotopy type theory* (HoTT), where types are interpreted as spaces, terms as maps and the inhabitants of the iterated identity types on a given type A are viewed as paths, homotopies and higher homotopies of increasing levels, respectively, endowing each type with a weak ω-groupoid structure.

Homotopy type theory, so it appears, has now reached the mathematical mainstream:

> *Voevodsky's Univalent Foundations require not just one inaccessible cardinal but an infinite string of cardinals, each inaccessible from its predecessor.* (M. Harris, *Mathematics without apologies*, 2015).

By Univalent Foundations Harris seems to refer to **MLTT** plus Voevodsky's *Univalence Axiom* (**UA**). To set the stage for the latter axiom, let us recall a bit of history of extensionality and universes in type theory. Simple type theory, as formulated by A. Church in 1940 [16], already provides a natural and elegant alternative to set theory for representing mathematics in a formal way. The stratification of mathematical objects into the types of propositions, individuals and functions between two types is indeed quite natural. In this setup, the axiom of extensionality comes in two forms: the stipulation that two logically equivalent propositions are equal and the stipulation that two pointwise equal functions are equal. Some restrictions of expressiveness encountered in simple type theory are overcome by dependent type theory, yet still unnatural limitations remain in that one cannot express the notion of an arbitrary structure in this framework. For instance one cannot assign a type to an arbitrary field. Type theory (and other frameworks as well) solve this issue by introducing the notion of a *universe type*. Whereas most types come associated with a germane axiom of extensionality inherited from its constituent types following the example of simple type theory, it is by no means clear what kind of extensionality principle should govern universes. A convincing proposal was missing until the work of V.

[1] Feferman introduced the theory of explicit mathematics in [20]. There it was based on intuitionistic logic and notated by T_0. The same notation is used e.g. in [13, 34, 48] but increasingly T_0 came to be identified with its classical version. As a result, we adopt the notation $\mathbf{T}_0^i$ to stress its intuitionistic basis and reserve $\mathbf{T}_0$ for the classical theory.

Voevodsky with its formulation of the extensionality axiom for universes in terms of equivalences. This is the univalence axiom, which generalizes propositional extensionality.

Harris's claim that an infinite sequence of inaccessible cardinals is required to model **MLTT** plus Voevodsky's *Univalence Axiom* is a pretty strong statement. Recent research by Bezem, Huber, and Coquand (see [10]), though, indicates that **MLTT** + **UA** has an interpretation in **MLTT** and therefore is proof-theoretically not stronger than **MLTT**. But what is the strength of **MLTT**? As there doesn't seem to exist much common knowledge among type theorists about the strength of various systems and how they relate to the other constructive frameworks as well as classical theories used as a classification hierarchy in reverse mathematics and set theory, it seems reasonable to devote a section to mapping out the relationships and gathering current knowledge in one place. In this section attention will also be payed to the methods employed in proofs such as interpretations but with a particular eye toward the role of ordinal analysis therein.

Section 8 of this paper will be concerned with extensions of explicit mathematics by principles that allow the construction of inductive classifications that lie way beyond **MLTT**'s reach but still have a constructive flavor. The basic theory here is intuitionistic explicit mathematics $\mathbf{T}_0^i$. In $\mathbf{T}_0^i$ one can freely talk about monotone operations on classifications and assert the existence of least fixed points of such operators. There are two ways in which one can add a principle to $\mathbf{T}_0^i$ postulating the existence of least fixed points. **MID** merely existentially asserts that every monotone operation has a least fixed point whereas **UMID** not only postulates the existence of a least solution, but, by adjoining a new functional constant to the language, ensures that a fixed point is uniformly presentable as a function of the monotone operation.

The question of the strength of systems of explicit mathematics with **MID** and **UMID** was raised by Feferman in [22]; we quote:

> *What is the strength of* $\mathbf{T}_0 + \mathbf{MID}$? *[...] I have tried, but did not succeed, to extend my interpretation of* $\mathbf{T_0}$ *in* $\Sigma^1_2 - AC + BI$ *to include the statement* **MID**. *The theory* $\mathbf{T}_0 +$ **MID** *includes all constructive formulations of iteration of monotone inductive definitions of which I am aware, while* $\mathbf{T_0}$ *(in its* IG *axiom) is based squarely on the general iteration of accessibility inductive definitions. Thus it would be of great interest for the present subject to settle the relationship between these theories.* (p. 88)

As it turned out, the principles **MID** and even more **UMID** encapsulate considerable strength, when considered on the basis of classical $\mathbf{T}_0$. For instance $\mathbf{T}_0 + \mathbf{UMID}$ embodies the strength of Π^1_2-comprehension. The first (significant) models of $\mathbf{T}_0 + \mathbf{MID}$ were found by Takahashi [69]. Research on the precise strength was conducted by Rathjen [56–58] and Glaß, Rathjen, Schlüter [26]. The article [59] provides a survey of the classical case. Tupailo [71] obtained the first result in the intuitionistic setting. This and further results will be the topic of Sect. 3.

2 Some Background on Feferman's $\mathbf{T}_0^i$

The theory of explicit mathematics, here denoted by $\mathbf{T}_0^i$, is a formal framework that has great expressive power. It is suitable for representing Bishop-style constructive mathematics as well as generalized recursion, including direct expression of structural concepts which admit self-application. Feferman was led to the development of his *explicit mathematics* when trying to understand what Errett Bishop had achieved in his groundbreaking constructive redevelopment of analysis in [11]. For a detailed account see [20, 21]. The ontology behind the axioms of $\mathbf{T}_0^i$ is that the universe of mathematical objects is populated by (a) natural numbers, (b) operations (in general partial) and (c) classifications (akin to Bishop's sets) where operations and classifications are to be understood as given intensionally. Operations can be applied to any object including operations and classifications; they are governed by axioms giving them the structure of a *partial combinatory algebra* (also known as *applicative structures* or *Schönfinkel algebras*). There are, for example, operations that act on classifications X, Y to produce their Cartesian product $X \times Y$ and exponential X^Y. The formation of classifications is governed by the *Join*, *Inductive Generation* and *Elementary Comprehension Axiom*.

The language of $\mathbf{T}_0^i$, $\mathfrak{L}(\mathbf{T}_0^i)$, has two sorts of variables. The free and bound variables $(a, b, c, \ldots$ and $x, y, z \ldots)$ are conceived to range over the whole constructive universe which comprises *operations* and *classifications* among other kinds of entities; while upper-case versions of these A, B, C, ... and X, Y, Z, ... are used to represent free and bound classification variables.

$\mathbf{N}$ is a classification constant taken to define the class of natural numbers. $\mathbf{0}$, $\mathbf{s_N}$ and $\mathbf{p_N}$ are operation constants whose intended interpretations are the natural number 0 and the successor and predecessor operations. Additional operation constants are $\mathbf{k}$, $\mathbf{s}$, $\mathbf{d}$, $\mathbf{p}$, $\mathbf{p_l}$ and $\mathbf{p_l}$ for the two basic combinators, definition by cases on $\mathbf{N}$, pairing and the corresponding two projections. Additional classification constants are generated using the axioms and the constants $\mathbf{j}$, $\mathbf{i}$ and $\mathbf{c_n}(\mathbf{n} < \omega)$ for *join*, *induction* and *comprehension*.

There is no arity associated with the various constants. The *terms* of $\mathbf{T}_0^i$ are just the variables and constants of the two sorts. The atomic formulae of $\mathbf{T}_0^i$ are built up using the terms and three primitive relation symbols $=$, App and ε as follows. If q, r, r_1, r_2 are terms, then $q = r$, $\mathsf{App}(q, r_1, r_2)$, and $q \;\varepsilon\; r$ (where r has to be a classification variable or constant) are atomic formulae. $\mathsf{App}(q, r_1, r_2)$ expresses that the operation q applied to r_1 yields the value r_2; $q \;\varepsilon\; r$ asserts[2] that q is in r or that q is classified under r.

We write $t_1 t_2 \simeq t_3$ for $\mathsf{App}(t_1, t_2, t_3)$.

The set of formulae is then obtained from these using the propositional connectives and the two quantifiers of each sort.

[2]It should be pointed out that we use the symbol "ε" instead of "$\in$" deliberately, the latter being reserved for the set–theoretic elementhood relation.

In order to facilitate the formulation of the axioms, the language of $\mathbf{T}_0^i$ is expanded definitionally with the symbol $\simeq$ and the auxiliary notion of an *application term* is introduced. The set of application terms is given by two clauses:

1. all terms of $\mathbf{T}_0^i$ are application terms; and
2. if s and t are application terms, then (st) is an application term.

If s is an application term and u is a bound or free variable we define $s \simeq u$ by induction on the buildup of s:

$$s \simeq u \quad \text{is} \quad \begin{cases} s = u, & \text{if } s \text{ is a variable or a constant ,} \\ \exists x, y[s_1 \simeq x \wedge s_2 \simeq y \wedge \mathsf{App}(x, y, u] & \text{if } s \text{ is an application term}(s_1 s_2) \end{cases}$$

For s and t application terms, we have auxiliary, defined formulae of the form:

$$s \simeq t \;\; := \;\; \forall y(s \simeq y \leftrightarrow t \simeq y).$$

Some abbreviations are $t_1 \dots t_n$ for $((...(t_1 t_2)...)t_n)$; $t \downarrow$ for $\exists y(t \simeq y)$ and $\phi(t)$ for $\exists y(t \simeq y \wedge \phi(y))$.

Gödel numbers for formulae play a key role in the axioms introducing the classification constants $\mathbf{c_n}$. A formula is said to be *elementary* if it contains only free occurrences of classification variables A (i.e., only as *parameters*), and even those free occurrences of A are restricted: A must occur only to the right of ε in atomic formulas. The Gödel number $\mathbf{c_n}$ above is the Gödel number of an elementary formula. We assume that a standard Gödel numbering has been chosen for $\mathfrak{L}(\mathbf{T}_0^i)$; if ϕ is an elementary formula and $a, b_1, \dots, b_m, A_1, \dots, A_n$ is a list of variables which includes all parameters of ϕ, then $\{x : \phi(x, b_1, \dots, b_n, A_1, \dots, A_n)\}$ stands for $\mathbf{c_n}(b_1, \dots, b_n, A_1, \dots, A_n)$; $\mathbf{n}$ is the code of the pair of Gödel numbers $\langle \ulcorner \phi \urcorner$, $\ulcorner (a, b_1, \dots, b_m, A_1, \dots, A_n) \urcorner \rangle$ and is called the 'index' of ϕ and the list of variables.

Some further conventions are useful. Systematic notation for *n-tuples* is introduced as follows: (t) is t, (s, t) is $\mathbf{p}st$, and $(t_1, \dots, t_n)$ is defined by $((t_1, \dots, t_{n\text{-}1}), t_n)$. Finally, t' is written for the term $\mathbf{s_N}t$, and $\bot$ is the elementary formula $\mathbf{0} \simeq \mathbf{0}'$.

$\mathbf{T}_0^i$'s logic is intuitionistic two-sorted predicate logic with identity. Its non-logical axioms are:

I. Basic Axioms

1. $\forall X \exists x(X = x)$
2. $\mathsf{App}(a, b, c_1) \wedge \mathsf{App}(a, b, c_2) \rightarrow c_1 = c_2$

II. App Axioms

1. $(\mathbf{k}ab) \downarrow \wedge\, \mathbf{k}ab \simeq a$,
2. $(\mathbf{s}ab) \downarrow \wedge\, \mathbf{s}abc \simeq ac(bc)$,
3. $(\mathbf{p}a_1a_2) \downarrow \wedge\, (\mathbf{p_1}a) \wedge (\mathbf{p_2}a) \downarrow \wedge\, \mathbf{p_i}(\mathbf{p}a_1a_2) \simeq a_i$ for $i = 0, 1$,
4. $(c_1 = c_2 \vee c_1 \neq c_2) \wedge (\mathbf{d}abc_1c_2) \downarrow \wedge\, (c_1 = c_2 \rightarrow \mathbf{d}abc_1c_2 \simeq a) \wedge (c_1 \neq c_2 \rightarrow \mathbf{d}abc_1c_2 \simeq b)$,

5. $a \,\varepsilon\, \mathbb{N} \wedge b \,\varepsilon\, \mathbb{N} \rightarrow [a' \downarrow \wedge \mathbf{p_0}(a') \simeq a \wedge \neg(a' \simeq 0) \wedge (a' \simeq b' \rightarrow a \simeq b)]$.

III. Classification Axioms

Elementary Comprehension Axiom (ECA)
$\exists X[X \simeq \{x : \psi(x)\} \wedge \forall x(x \,\varepsilon\, X \leftrightarrow \psi(x))]$
for each elementary formula ψa, which may contain additional parameters.

Natural Numbers

(i) $\mathbf{0} \,\varepsilon\, \mathbf{N} \wedge \forall x(x \,\varepsilon\, \mathbf{N} \rightarrow x' \,\varepsilon\, \mathbf{N})$
(ii) $\phi(\mathbf{0}) \wedge \forall x(\phi(x) \rightarrow \phi(x')) \rightarrow (\forall x \,\varepsilon\, \mathbf{N})\phi(x)$ for each formula ϕ of $\mathfrak{L}(\mathbf{T}_0^i)$.

Join (J)
$\forall x \,\varepsilon\, A\, \exists Y fx \simeq Y \rightarrow \exists X[X \simeq \mathbf{j}(A, f) \wedge \forall z(z \,\varepsilon\, X \leftrightarrow \exists x \varepsilon A \exists y(z \simeq (x, y) \wedge y \,\varepsilon\, fx))]$
Inductive Generation (IG)
$\exists X[X \simeq \mathbf{i}(A, B) \wedge \forall x \,\varepsilon\, A[\forall y[(y, x) \,\varepsilon\, B \rightarrow y \,\varepsilon\, X] \rightarrow x \,\varepsilon\, X],$
$\quad \wedge[\forall x \,\varepsilon\, A\, [\forall y\, ((y, x) \,\varepsilon\, B \rightarrow \phi(y)) \rightarrow \phi(x)] \rightarrow \forall x \,\varepsilon\, X\, \phi(x)]]$
where ϕ is an arbitrary formula of $\mathbf{T}_0^i$.

3 Type Theories

The type theory of Martin–Löf from the 1984 book [42] will be notated by $\mathbf{MLTT}^{\mathrm{ext}}$ where the superscript is meant to convey that this is an *extensional* theory. It has all the usual type constructors Π, Σ, $+$, $\mathbf{0}$, $\mathbf{1}$, $\mathbf{2}$, Id, W for dependent products, dependent sums, disjoint unions, empty type, unit type, Booleans, propositional identity types, and W-types, respectively. Moreover, the system comprises a sequence of universe types $\mathcal{U}_0, \mathcal{U}_1, \mathcal{U}_2, \ldots$ externally indexed by the natural numbers. The universe types are closed under the type constructors from the first list and they form a cumulative hierarchy in that $\mathcal{U}_n$ is a type in $\mathcal{U}_{n+1}$ and if A is a type in $\mathcal{U}_n$ then A is also a type in $\mathcal{U}_{n+1}$.

In the version of [42] the identity type was taken to be extensional whereas in the more recent versions, e.g. [45] and the one forming the basis for homotopy type theory (see [33]), it is considered to be intensional. The intensional version will simply be denoted by **MLTT**. For the proof-theoretic strength, though, it turns out that the difference is immaterial. The reasons will be explained below, but perhaps a first good approximation comes from the observation that (exact) lower bounds can be established by interpreting certain set theories in type theory in such a way that the extensional identity type can be dispensed with in these interpretations, although for validating certain forms of the axiom of choice, e.g. the **ΠΣ**W-**AC** axiom to be discussed below, chunks of extensionality are still required. Since we shall be discussing (partial) conservativity results of extensional over intensional type theory below, let's recall the differences.

Definition 3.1 A key feature of Martin–Löf's type theory is the distinction of two notions of identity (or equality). *Judgemental identity* appears in judgements in the two forms $\Gamma \vdash s = t : A$ and $\Gamma \vdash A = B$ type between terms and between types, respectively. The general equality rules (reflexivity, symmetry, transitivity) and substitution rules, simultaneously at the level of terms and types, apply to these judgements as further inference rules.[3] But there is also *propositional identity* which gives rise to types $\mathsf{Id}(A, s, t)$ and allows for internal reasoning about identity.

The rules for the extensional identity type are the following[4]:

$$(\mathsf{Id} - \text{Formation}) \quad \frac{\Gamma \vdash A \text{ type} \qquad \Gamma \vdash a : A \qquad \Gamma \vdash b : A}{\Gamma \vdash \mathsf{Id}(A, a, b) \text{ type}}$$

$$(\mathsf{Id} - \text{Introduction}) \quad \frac{\Gamma \vdash a : A}{\Gamma \vdash \mathsf{refl}(a) : \mathsf{Id}(A, a, a)}$$

$$(\mathsf{Id} - \text{Uniqueness}) \quad \frac{\Gamma \vdash p : \mathsf{Id}(A, a, b)}{\Gamma \vdash p = \mathsf{refl}(a) : \mathsf{Id}(A, a, b)}$$

$$(\mathsf{Id} - \text{Reflection}) \quad \frac{\Gamma \vdash p : \mathsf{Id}(A, a, b)}{\Gamma \vdash a = b : A}.$$

Reflection has the effect of rendering judgemental identity undecidable, i.e., the (type checking) questions whether $\Gamma \vdash a = b : A$ or $\Gamma \vdash a : A$ hold become undecidable. On the other hand, the set-theoretic models and many recursion-theoretic models of type theory (see [6, 8, 48]) validate extensionality, lending it an intuitive appeal.

For the intensional identity type, the foregoing rules of formation and introduction are retained, however, uniqueness and reflection are jettisoned, getting replaced by elimination and equality rules which are motivated by Leibniz's principle of indiscernibility, namely that identical elements are those that satisfy the same properties. Though instead of capturing identity by quantifying (impredicatively) over all properties (as in Principia), the entire family of identity types $(\mathsf{Id}(A, x, y))_{x,y:A}$ is viewed as being inductively generated with sole constructor refl (see [33, 45]). The elimination and equality rules are the following:

---

[3]See [45, Chap. 5] or [33, A.2.2], where they are called structural rules.

[4]The rules are essentially the ones used in [42], except that [42] has a constant r as the sole canonical element of all inhabited types $\mathsf{Id}(A, a, b)$. Here we use $\mathsf{refl}(a)$ to make the comparison with the intensional case more transparent. In [42], Id – Uniqueness and Id – Reflection are called I-equality and I-elimination, respectively.

$$(\mathsf{Id}-\text{Elimination})\quad \frac{\begin{array}{l}\Gamma \vdash a : A \\ \Gamma \vdash b : A \\ \Gamma \vdash c : \mathsf{Id}(A, a, b) \\ \Gamma,\ x : A,\ y : A,\ z : \mathsf{Id}(A, x, y) \vdash C(x, y, z)\ \text{type} \\ \Gamma,\ x : A \vdash d(x) : C(x, x, \mathsf{refl}(x))\end{array}}{\Gamma \vdash \mathsf{J}(c, d) : C(a, b, c)}$$

$$(\mathsf{Id}-\text{Equality})\quad \frac{\begin{array}{l}\Gamma \vdash a : A \\ \Gamma,\ x : A,\ y : A,\ z : \mathsf{Id}(A, x, y) \vdash C(x, y, z)\ \text{type} \\ \Gamma,\ x : A \vdash d(x) : C(x, x, \mathsf{refl}(x))\end{array}}{\Gamma \vdash \mathsf{J}(\mathsf{refl}(a), d) = d(a) : C(a, a, \mathsf{refl}(a))\,.}$$

An immediate consequence of these rules is the indiscernibility of identical elements expressed as follows. For every family $(C(x))_{x:A}$ of types there is a function

$$f : \Pi_{x,y:A}\Pi_{p:\mathsf{Id}(A,x,y)}[C(x) \to C(y)]$$

such that with $1_{C(x)}$ being the function $u \mapsto u$ on $C(x)$ we have $f(x, x, \mathsf{refl}(x)) = 1_{C(x)}$.

Foregoing extensional identity and using the induction principle encapsulated in Id – elimination and Id – equality in its stead, is crucial to the more subtle homotopy interpretations of type theory.

4 Constructive Set Theories

Constructive Set Theory was introduced by Myhill in a seminal paper [44], where a specific axiom system **CST** was introduced. Through developing constructive set theory he wanted to isolate the principles underlying Bishop's conception of what sets and functions are, and he wanted "these principles to be such as to make the process of formalization completely trivial, as it is in the classical case" ([44], p. 347). Myhill's **CST** was subsequently modified by Aczel and the resulting theory was called *Constructive Zermelo–Fraenkel set theory*, **CZF**. A hallmark of this theory is that it possesses a type-theoretic interpretation (cf. [2, 5]). Specifically, **CZF** has a scheme called Subset Collection Axiom (which is a generalization of Myhill's Exponentiation Axiom) whose formalization was directly inspired by the type-theoretic interpretation.

The language of **CZF** is the same first order language as that of classical Zermelo–Fraenkel Set Theory, **ZF** whose only non-logical symbol is $\in$. The logic of **CZF** is intuitionistic first order logic with equality. Among its non-logical axioms are *Extensionality*, *Pairing* and *Union* in their usual forms. **CZF** has additionally axiom schemata which we will now proceed to summarize. Below $\emptyset$ stands for the empty set and $v+1$ denotes $v \cup \{v\}$. A set-theoretic formula is said to be *restricted* or *bounded* or Δ_0 if it is constructed from prime formulae using $\neg, \wedge, \vee, \rightarrow$ and only restricted quantifiers $\forall x{\in}y,\ \exists x{\in}y$.

Infinity:[5]

$$\exists x\,[\forall u\big(u{\in}x \leftrightarrow (\emptyset = u \ \vee\ \exists v{\in}x\ u = v+1)\big) \ \wedge\ \forall z\,(\emptyset \in z \wedge \forall y \in z\ y+1 \in z \rightarrow x \subseteq z)].$$

Set Induction: For all formulae ϕ,

$$\forall x[\forall y \in x\phi(y) \rightarrow \phi(x)] \rightarrow \forall x\phi(x).$$

Restricted or Bounded Separation: For all *restricted* formulae ϕ,

$$\forall a\exists b\forall x[x \in b \leftrightarrow x \in a \wedge \phi(x)].$$

Strong Collection: For all formulae ϕ,

$$\forall a\big[\forall x \in a\exists y\phi(x, y) \ \rightarrow\ \exists b\,[\forall x \in a\,\exists y \in b\,\phi(x, y) \wedge \forall y \in b\,\exists x \in a\,\phi(x, y)]\big].$$

Subset Collection: For all formulae ψ,

$$\forall a\forall b\exists c\forall u\,\big[\forall x \in a\,\exists y \in b\ \psi(x, y, u) \rightarrow \\ \exists d \in c\,[\forall x \in a\,\exists y \in d\ \psi(x, y, u) \wedge \forall y \in d\,\exists x \in a\,\psi(x, y, u)]\big].$$

The Subset Collection schema easily qualifies as the most intricate axiom of **CZF**.

We shall also consider an additional axiom that holds true in the type-theoretic interpretation of Aczel if the type theory is equipped with W-types. To introduce it, we need the notion of a regular set. The formula in the language of **CZF** defining the property of a set A that it is *regular* states that A is transitive, and for every $a \in A$ and set $R \subseteq a \times A$ if $\forall x \in a\,\exists y\,(\langle x, y\rangle \in R)$, then there is a set $b \in A$ such that

$$\forall x \in a\,\exists y \in b\,(\langle x, y\rangle \in R) \ \wedge\ \forall y \in b\,\exists x \in a\,(\langle x, y\rangle \in R).$$

In particular, if $R : a \rightarrow A$ is a function, then the image of R is an an element of A. Let $\mathsf{Reg}(A)$ denote this assertion. With this auxiliary definition we can state the

Regular Extension Axiom **REA**

$$\forall x\exists y[x \subseteq y \wedge \mathsf{Reg}(y)]\,.$$

[5]This axiom asserts the existence of a unique set usually called ω. Note that the second conjunct in $[\ldots]$ entails the usual induction principle for ω with regard to set properties (or equivalently Δ_0 formulae).

4.1 The Axiom of Choice in Constructive Set Theories

Among the axioms of set theory, the axiom of choice is distinguished by the fact that it is the only one that one finds mentioned in workaday mathematics. In the mathematical world of the beginning of the 20th century, discussions about the status of the axiom of choice were important. In 1904 Zermelo proved that every set can be well-ordered by employing the axiom of choice. While Zermelo argued that it was self-evident, it was also criticized as an excessively non-constructive principle by some of the most distinguished analysts of the day, notably Borel, Baire, and Lebesgue. At first blush this reaction against the axiom of choice utilized in Cantor's new theory of sets is surprising as the French analysts had used and continued to use choice principles routinely in their work. However, in the context of 19th century classical analysis only the Axiom of Dependent Choices, **DC**, is invoked and considered to be natural, while the full axiom of choice is unnecessary and even has some counterintuitive consequences.

Unsurprisingly, the axiom of choice does not have a unambiguous status in constructive mathematics either. On the one hand it is said to be an immediate consequence of the constructive interpretation of the quantifiers. Any proof of $\forall x \in A\, \exists y \in B\, \phi(x, y)$ must yield a function $f : A \to B$ such that $\forall x \in A\, \phi(x, f(x))$. This is certainly the case in Martin–Löf's intuitionistic theory of types. On the other hand, it has been observed that the full axiom of choice cannot be added to systems of extensional constructive set theory without yielding constructively unacceptable cases of excluded middle (see [18]). In extensional intuitionistic set theories, a proof of a statement $\forall x \in A\, \exists y \in B\, \phi(x, y)$, in general, provides only a function F, which when fed a proof p witnessing $x \in A$, yields $F(p) \in B$ and $\phi(x, F(p))$. Therefore, in the main, such an F cannot be rendered a function of x alone. Choice will then hold over sets which have a canonical proof function, where a constructive function h is a canonical proof function for A if for each $x \in A$, $h(x)$ is a constructive proof that $x \in A$. Such sets having natural canonical proof functions "built-in" have been called *bases* (cf. [70], p. 841).

Some constructive choice principles In many a text on constructive mathematics, axioms of countable choice and dependent choices are accepted as constructive principles. This is, for instance, the case in Bishop's constructive mathematics (cf. [11]) as well as Brouwer's intuitionistic analysis (cf. [70], Chap. 4, Sect. 2). Myhill also incorporated these axioms in his constructive set theory [44].

The weakest constructive choice principle we shall consider is the *Axiom of Countable Choice*, $\mathbf{AC}_\omega$, i.e. whenever F is a function with domain ω such that $\forall i \in \omega\, \exists y \in F(i)$, then there exists a function f with domain ω such that $\forall i \in \omega\; f(i) \in F(i)$.

A mathematically very useful axiom to have in set theory is the *Dependent Choices Axiom*, **DC**, i.e., for all formulae ψ, whenever

$$(\forall x \in a)\,(\exists y \in a)\,\psi(x, y)$$

and $b_0 \in a$, then there exists a function $f : \omega \to a$ such that $f(0) = b_0$ and

$$(\forall n \in \omega)\,\psi(f(n), f(n+1)).$$

Even more useful is the *Relativized Dependent Choices Axiom*, **RDC**. It asserts that for arbitrary formulae ϕ and ψ, whenever

$$\forall x\big[\phi(x) \rightarrow \exists y\big(\phi(y) \wedge \psi(x, y)\big)\big]$$

and $\phi(b_0)$, then there exists a function f with domain ω such that $f(0) = b_0$ and

$$(\forall n \in \omega)\big[\phi(f(n)) \wedge \psi(f(n), f(n+1))\big].$$

In addition to the "traditional" axioms of choice stated above, the interpretation of set theory in type theory validates several new choice principles which are not well known. To state them we need to introduce various operations on classes.

Remark 4.1 Let $\mathbf{CZF}_{Exp}$ denote the modification of **CZF** with Eponentiation in place of Subset Collection.

In almost all the results of this paper, **CZF** could be replaced by $\mathbf{CZF}_{Exp}$, that is to say, for the purposes of this paper it is enough to assume Exponentiation rather than Subset Collection. However, in what follows we shall not point this out again.

Definition 4.2 (**CZF**) If A is a set and B_x are classes for all $x \in A$, we define a class $\prod_{x\in A} B_x$ by:

$$\prod_{x\in A} B_x := \{f \mid f : A \to \bigcup_{x\in A} B_x \,\wedge\, \forall x \in A(f(x) \in B_x)\}. \tag{1}$$

If A is a class and B_x are classes for all $x \in A$, we define a class $\sum_{x\in A} B_x$ by:

$$\sum_{x\in A} B_x := \{\langle x, y\rangle \mid x \in A \wedge y \in B_x\}. \tag{2}$$

If A is a class and a, b are sets, we define a class $\mathbf{I}(A, a, b)$ by:

$$\mathbf{I}(A, a, b) := \{z \in 1 \mid a = b \,\wedge\, a, b \in A\}. \tag{3}$$

If A is a class and for each $a \in A$, B_a is a set, then

$$\mathbf{W}_{a\in A} B_a$$

is the smallest class Y such that whenever $a \in A$ and $f : B_a \to Y$, then $\langle a, f\rangle \in Y$.

Lemma 4.3 (**CZF**) *If A,B,a,b are sets and B_x is a set for all $x \in A$, then $\prod_{x\in A} B_x$, $\sum_{x\in A} B_x$ and $\mathbf{I}(A, a, b)$ are sets.*

Proof [55, Lemma 2.5]. □

In the following we shall introduce several inductively defined classes, and, moreover, we have to ensure that such classes can be formalized in **CZF**.

We define an *inductive definition* to be a class of ordered pairs. If Φ is an inductive definition and $\langle x, a\rangle \in \Phi$ then we write

$$\frac{x}{a}\ \Phi$$

and call $\frac{x}{a}$ an *(inference) step* of Φ, with set x of *premisses* and *conclusion* a. For any class Y, let

$$\Gamma_\Phi(Y) = \left\{a \mid \exists x\left(x \subseteq Y \ \wedge\ \frac{x}{a}\ \Phi\right)\right\}.$$

The class Y is Φ-*closed* if $\Gamma_\Phi(Y) \subseteq Y$. Note that Γ is monotone; i.e. for classes Y_1, Y_2, whenever $Y_1 \subseteq Y_2$, then $\Gamma(Y_1) \subseteq \Gamma(Y_2)$.

We define the class *inductively defined by* Φ to be the smallest Φ-closed class. The main result about inductively defined classes states that this class, denoted $\mathbf{I}(\Phi)$, always exists.

Lemma 4.4 (**CZF**) (Class Inductive Definition Theorem) *For any inductive definition Φ there is a smallest Φ-closed class $\mathbf{I}(\Phi)$.*

Proof [2], Sect. 4.2 or [4], Theorem 5.1. □

Lemma 4.5 (**CZF + REA**) *If A is a set and B_x is a set for all $x \in A$, then $\mathbf{W}_{a\in A} B_a$ is a set.*

Proof This follows from [3], Corollary 5.3. □

Lemma 4.6 (**CZF**) *There exists a smallest* $\mathbf{\Pi\Sigma}$*-closed class, i.e., a smallest class* $\mathbf{Y}$ *such that the following hold:*
(i) $n \in \mathbf{Y}$ *for all* $n \in \omega$;
(ii) $\omega \in \mathbf{Y}$;
(iii) $\prod_{x\in A} B_x \in \mathbf{Y}$ *and* $\sum_{x\in A} B_x \in \mathbf{Y}$ *whenever* $A \in \mathbf{Y}$ *and* $B_x \in \mathbf{Y}$ *for all* $x \in A$.

Likewise, there exists a smallest $\mathbf{\Pi\Sigma I}$*-closed class, i.e. a smallest class* $\mathbf{Y}^*$*, which, in addition to the closure conditions* (i)–(iii) *above, satisfies:*
(iv) $\mathbf{I}(A, a, b) \in \mathbf{Y}^*$ *whenever* $A \in \mathbf{Y}^*$ *and* $a, b \in A$.

Proof [55, Lemma 2.8]. □

Definition 4.7 The $\mathbf{\Pi\Sigma}$-generated sets are the sets in the smallest $\mathbf{\Pi\Sigma}$-closed class. Similarly one defines the $\mathbf{\Pi\Sigma I}$, $\mathbf{\Pi\Sigma W}$ and $\mathbf{\Pi\Sigma WI}$-generated sets.

A set P is a *base* if for any P-indexed family $(X_a)_{a\in P}$ of inhabited sets X_a, there exists a function f with domain P such that, for all $a \in P$, $f(a) \in X_a$.

$\mathbf{\Pi\Sigma-AC}$ is the statement that every $\mathbf{\Pi\Sigma}$-generated set is a base. Similarly one defines the axioms $\mathbf{\Pi\Sigma I-AC}$, $\mathbf{\Pi\Sigma WI-AC}$, and $\mathbf{\Pi\Sigma W-AC}$.

The *presentation axiom*, **PAx**, states that every set is the surjective image of a base.

Lemma 4.8

(i) (**CZF**) $\mathbf{\Pi\Sigma-AC}$ *and* $\mathbf{\Pi\Sigma I-AC}$ *are equivalent.*
(ii) (**CZF** + **REA**) $\mathbf{\Pi\Sigma W-AC}$ *and* $\mathbf{\Pi\Sigma WI-AC}$ *are equivalent.*

Proof [55, 2.12]. □

4.2 Large Sets in Constructive Set Theory

Large cardinals play a central role in modern set theory. This section deals with large cardinal properties in the context of intuitionistic set theories. Since in intuitionistic set theory $\in$ is not a linear ordering on ordinals the notion of a cardinal does not play a central role. Consequently, one talks about *"large set properties"* instead of *"large cardinal properties"*. When stating these properties one has to proceed rather carefully. Classical equivalences of cardinal notion might no longer prevail in the intuitionistic setting, and one therefore wants to choose a rendering which intuitionistically retains the most strength. On the other hand certain notions have to be avoided so as not to imply excluded third. To give an example, cardinal notions like measurability, supercompactness and hugeness have to be expressed in terms of elementary embeddings rather than ultrafilters.

We shall, however, not concern ourselves with very large cardinals here and rather restrict attention to the very first notions of largeness introduced by Hausdorff and Mahlo, that is, inaccessible and Mahlo sets and the pertaining hierarchies of inaccessible and Mahlo sets.

We have already seen one notion of largeness, namely that of a regular set. In **ZFC**, a regular set which itself is a model of the axioms of **CZF** is of the form V_κ with κ a strongly inaccessible cardinal.[6] In the context of **CZF** this notion is much weaker.

Definition 4.9 If A is a transitive set and ϕ is a formula with parameters in A we denote by ϕ^A the formula which arises from ϕ by replacing all unbounded quantifiers $\forall u$ and $\exists v$ in ϕ by $\forall u \in A$ and $\exists v \in A$, respectively.

[6]Note that **CZF** with classical logic is the same theory as **ZF**.

We can view any transitive set A as a structure equipped with the binary relation $\in_A = \{\langle x, y\rangle \mid x \in y \in A\}$. A set-theoretic sentence whose parameters lie in A, then has a canonical interpretation in $(A, \in_A)$ by interpreting $\in$ as $\in_A$, and $(A, \in_A) \models \phi$ is logically equivalent to ϕ^A. We shall usually write $A \models \phi$ in place of ϕ^A.

A set I is said to be *weakly inaccessible* if I is a regular set such that $I \models \mathbf{CZF}^-$, where $\mathbf{CZF}^-$ denotes the theory **CZF** bereft of the set induction scheme.[7]

The strong regular extension axiom, **sREA**, states that every set is an element of a weakly inaccessible set.

There is a more 'algebraic' way of expressing weak inaccessibility. Stating it requires some definitions.

Definition 4.10 For sets A, B we denote by $\mathbf{mv}(^A B)$ the collection of all full relations from A to B, i.e., of those relations $R \subseteq A \times B$ such that $\forall x \in A\, \exists y \in B\, \langle x, y\rangle \in R$. A set C is said to be *full in* $\mathbf{mv}(^A B)$ if for all $R \in \mathbf{mv}(^A B)$ there exists $R' \in \mathbf{mv}(^A B)$ such that $R' \subseteq R$ and $R' \in C$.

For a set A, define $\bigwedge A$ to be the set $\{x \in 1 \mid \forall u \in A\, x \in u\}$, where $1 = \{\emptyset\}$.

Proposition 4.11 $(\mathbf{CZF}^-)$ *A set I is weakly inaccessible if and only if I is a regular set such that the following are satisfied:*

1. $\omega \in I$,
2. $\forall a \in I\, \bigcup a \in I$,
3. $\forall a \in I\, [a$ *inhabited* $\rightarrow \bigcap a \in I]$,
4. $\forall A, B \in I\, \exists C \in I\;\; C$ *is full in* $\mathbf{mv}(^A B)$.

Proof [5, 10.26].

We will consider two stronger notions.

Definition 4.12 A set I is called *inaccessible* if I is weakly inaccessible and for all $x \in I$ there exists a regular set $y \in I$ such that $x \in y$.

A set M is said to be *Mahlo* if M is inaccessible and for every $R \in \mathbf{mv}(^M M)$ there exists an inaccessible $I \in M$ such that

$$\forall x \in I\, \exists y \in I\, \langle x, y\rangle \in R.$$

4.3 Fragments of Second Order Arithmetic

The proof-theoretic strength of theories is commonly calibrated using standard theories and their canonical fragments. In classical set theory this linear line of consistency strengths is couched in terms of large cardinal axioms while for weaker theories the

[7]Note that if the background set theory validates set induction for Δ_0 formulae then a transitive set will be automatically a model of the full set induction scheme, and thus a regular set I will satisfy $I \models \mathbf{CZF}$.

line of reference systems traditionally consist of subsystems of second order arithmetic. The observation that large chunks of mathematics can already be formalized in fragments of second order arithmetic goes back to Hilbert and Bernays [31], and has led to a systematic research program known as *Reverse Mathematics*. Below we give an account of the syntax of $\mathcal{L}_2$ and frequently considered axiomatic principles.

Definition 4.13 The language $\mathcal{L}_2$ of second-order arithmetic contains number variables $x, y, z, u, \ldots$, set variables $X, Y, Z, U, V, A, B, C, \ldots$ (ranging over subsets of $\mathbb{N}$), the constant 0, function symbols $Suc, +, \cdot$, and relation symbols $=, <, \in$. Suc stands for the successor function. We write $x+1$ for $Suc(x)$. *Terms* are built up as usual. For $n \in \mathbb{N}$, let $\bar{n}$ be the canonical term denoting n. Formulae are built from the prime formulae $s = t$, $s < t$, and $s \in X$ using $\wedge, \vee, \neg, \forall x, \exists x, \forall X$ and $\exists X$ where s, t are terms. Note that equality in $\mathcal{L}_2$ is only a relation on numbers. However, equality of sets will be considered a defined notion, namely $X = Y$ if and only if $\forall x[x \in X \leftrightarrow x \in Y]$. As per usual, number quantifiers are called bounded if they occur in the context $\forall x(x < s \rightarrow \ldots)$ or $\exists x(x < s \wedge \ldots)$ for a term s which does not contain x. The $\mathbf{\Sigma}^0_0$-formulae are those formulae in which all quantifiers are bounded number quantifiers. For $k > 0$, $\mathbf{\Sigma}^0_k$-formulae are formulae of the form $\exists x_1 \forall x_2 \ldots Q x_k \phi$, where ϕ is $\mathbf{\Sigma}^0_0$; $\mathbf{\Pi}^0_k$-formulae are those of the form $\forall x_1 \exists x_2 \ldots Q x_k \phi$. The union of all $\mathbf{\Pi}^0_k$- and $\mathbf{\Sigma}^0_k$-formulae for all $k \in \mathbb{N}$ is the class of *arithmetical* or $\mathbf{\Pi}^0_\infty$*-formulae*. The $\mathbf{\Sigma}^1_k$-formulae ($\mathbf{\Pi}^1_k$-formulae) are the formulae $\exists X_1 \forall X_2 \ldots Q X_k \phi$ (resp. $\forall X_1 \exists X_2 \ldots Q x_k \phi$) for arithmetical ϕ.

The basic axioms in all theories of second-order arithmetic are the defining axioms of $0, 1, +, \cdot, <$ and the *induction axiom*

$$\forall X(0 \in X \wedge \forall x(x \in X \rightarrow x+1 \in X) \rightarrow \forall x(x \in X)),$$

respectively the *scheme of induction*

$$\mathbf{IND} \qquad \phi(0) \wedge \forall x(\phi(x) \rightarrow \phi(x+1)) \rightarrow \forall x \phi(x),$$

where ϕ is an arbitrary $\mathcal{L}_2$-formula. We consider the axiom scheme of $\mathcal{C}$*-comprehension* for formula classes $\mathcal{C}$ which is given by

$$\mathcal{C}\text{-}\mathbf{CA} \qquad \exists X \forall u(u \in X \leftrightarrow \phi(u))$$

for all formulae $\phi \in \mathcal{C}$ (of course, X must not be free in ϕ).

For each axiom scheme $\mathbf{Ax}$ we denote by $(\mathbf{Ax})$ the theory consisting of the basic arithmetical axioms, the scheme $\mathbf{\Pi}^0_\infty$-$\mathbf{CA}$, the scheme of induction and the scheme $\mathbf{Ax}$. If we replace the scheme of induction by the induction axiom, we denote the resulting theory by $(\mathbf{Ax})_0$. An example for these notations is the theory $(\mathbf{\Pi}^1_1\text{-}\mathbf{CA})$ which contains the induction scheme, whereas $(\mathbf{\Pi}^1_1\text{-}\mathbf{CA})_0$ only contains the induction axiom in addition to the comprehension scheme for $\mathbf{\Pi}^1_1$-formulae.

In the basic system one can introduce defined symbols for all primitive recursive functions. Especially, let $\langle , \rangle : \mathbb{N} \times \mathbb{N} \longrightarrow \mathbb{N}$ be a primitive recursive and bijective pairing function. The xth section of U is defined by $U_x := \{y : \langle x, y\rangle \in U\}$. Observe that a set U is uniquely determined by its sections on account of $\langle,\rangle$'s bijectivity. Any set R gives rise to a binary relation $\prec_R$ defined by $y \prec_R x := \langle y, x\rangle \in R$. Using this coding we can formulate the *$\mathcal{C}$-axiom of choice* scheme for formula classes $\mathcal{C}$ which is given by

$$\mathcal{C}\text{-}\mathbf{AC} \qquad \forall x \exists Y\, \psi(x, Y) \to \exists Z\, \forall u\, \psi(x, Z_x),$$

for all formulae $\psi \in \mathcal{C}$ (Z must not be free in ψ).

Another important principle is *Bar induction*:

$$\mathbf{BI} \qquad \forall X\big[\mathbf{WF}(\prec_X) \wedge \forall u\big(\forall v \prec_X u\phi(v) \to \phi(u)\big) \to \forall u\phi(u)\big]$$

for all formulae ϕ, where $\mathbf{WF}(\prec_X)$ expresses that $\prec_X$ is well-founded, i.e., $\mathbf{WF}(\prec_X)$ stands for the formula

$$\forall Y\,\big[\forall u\big[(\forall v \prec_X u\ v \in Y) \to u \in Y\big] \to \forall u\ u \in Y\big].$$

Universes in type theory (with W-types) bear a strong relation to β-models which are models of the language of $\mathcal{L}_2$ or set theory for which the notion well-foundedness is absolute.

Definition 4.14 Any set A of natural numbers gives rise to a set $\mathfrak{X}_A := \{A_i \mid i \in \mathbb{N}\}$ of sets of natural numbers. A is said to be a *β-model* if the $\mathcal{L}_2$-structure

$$\mathfrak{A} := (\mathbb{N}, \mathfrak{X}_A, 0, 1, +, \cdot, \in)$$

is a β-model, i.e., $\mathfrak{A} \models \mathbf{\Pi}^0_\infty$-$\mathbf{CA}$, and whenever $Y \in \mathfrak{X}_A$ and $\mathfrak{A} \models \mathbf{WF}(\prec_Y)$ then $\prec_Y$ is well-founded.

Obviously, the notion, the notion of β-model can be expressed in $\mathcal{L}_2$.

An intuitionistic $\mathcal{L}_2$-theory. There is an interesting version of second order arithmetic, which will be used in theory reductions, that classically has the same strength as full second order arithmetic, (Π^1_∞-**CA**), but when based on intuitionistic logic is of the same strength as $\mathbf{T}^i_0$.

Definition 4.15 **IARI** is a theory in the language of second order arithmetic. The logical rules of **IARI** are those of intuitionistic second order arithmetic. In addition to the usual axioms for intuitionistic second order logic, axioms are (the universal closures of):

1. **Induction**:

$$\phi(0) \wedge \forall n[\phi(n) \to \phi(n+1)] \to \forall n\phi(n)$$

   for all formulae ϕ.
2. **Arithmetic Comprehension Schema**:

$$\exists X \forall n[n \in X \leftrightarrow \psi(x)]$$

for ψ arithmetical (parameters allowed).

3. **Replacement**:

$$\forall X[\forall n \in X \exists ! Y \phi(n, Y) \rightarrow \exists Z \forall n \in X\, \phi(n, Z_n)]$$

for all formulas ϕ. Here $\phi(n, Z_n)$ arises from $\phi(n, Z)$ by replacing each occurrence $t \in Z$ in the formula by $\langle n, t\rangle \in Z$.

4. **Inductive Generation**:

$$\forall U \forall X \exists Y\big[\mathbf{WP}_U(X, Y) \wedge (\forall n[\forall k(k \prec_X n \rightarrow \phi(k)) \rightarrow \phi(n)] \rightarrow \forall m \in Y\, \phi(m))\big],$$

for all formulas ϕ, where $k \prec_X n$ abbreviates $\langle k, n\rangle \in X$ and $\mathbf{WP}_U(X, Y)$ stands for

$$\mathsf{Prog}_U(X, Y) \wedge \forall Z[\mathsf{Prog}_U(X, Z) \rightarrow Y \subseteq Z]$$

with $\mathsf{Prog}_U(X, Y)$ being $\forall n \in U[\forall k(k \prec_X n \rightarrow k \in Y) \rightarrow n \in Y]$.

Remark 4.16 (**IARI**) Note that $\mathbf{WP}_U(X, Y)$ and $\mathbf{WP}_U(X, Y')$ imply $Y = Y'$, i.e. $\forall n(n \in Y \leftrightarrow n \in Y')$. Therefore, if $\mathbf{WP}_U(X, Y)$, then

$$\forall n \in U[\forall k \prec_X n\, \phi(k) \rightarrow \phi(n)] \rightarrow \forall m \in Y\, \phi(m)$$

holds for all formulae ϕ.

The latter principle will be referred to as *"induction over the well–founded part of* $\prec_X$". In the rest of this section we shall write WF(U, X) for the (extensionally) uniquely determined Y which satisfies $\mathbf{WP}_U(X, Y)$.

The main tool for performing the well-ordering proof of [34] in **IARI** is the following principle of transfinite recursion.

Proposition 4.17 *(**IARI**) If* $\mathbf{WP}_U(X, Y)$ *and* $\forall n \in Y \forall W \exists ! V\, \psi(n, W, V)$, *then there exists* Z *such that*

$$\forall n \in Y\, \psi(n, \bigcup\{(Z)_k : k \prec_X n\}, (Z)_n).$$

Proof See [48, 6.4]. □

5 On Relating Theories I

The first result relates intuitionistic explicit mathematics to constructive set theory and a fragment of **MLTT**. Let $\mathbf{MLT}_{1W}V$ be the fragment of **MLTT** with only one universe $\mathcal{U}_0$ where the W-constructor can solely be applied to families of types in $\mathcal{U}_0$ but one can also form the type $\mathsf{V} := \mathsf{W}_{(A:\mathcal{U}_0)}A$ (something that could be called the type of Brouwer ordinals of $\mathcal{U}_0$). We shall also consider the type theory $\mathbf{MLT}_{1W}$ which is the fragment of $\mathbf{MLT}_{1W}V$ without the type V.

A principle of omniscience. Certain basic principles of classical mathematics are taboo for the constructive mathematician. Bishop called them *principles of omniscience*. The limited principle of omniscience, **LPO**, is an instance of the law of excluded middle which usually serves as a line of demarcation, separating "constructive" from "non-constructive" theories. In the case of **CZF**, adding the law of excluded middle even just for atomic statements of the form $a \in b$ results in an enormous increase in proof strength, pushing it up beyond that of Zermelo set theory. However, **LPO** can be added to **CZF** without affecting its proof-theoretic strength. **LPO** has the pleasant side effect that one can carry out elementary analysis pretty much in the same way as in any standard text book.

Definition 5.1 Let $2^{\mathbb{N}}$ be Cantor space, i.e. the set of all functions from the naturals into $\{0, 1\}$. Limited Principle of Omniscience (**LPO**):

$$\forall f \in 2^{\mathbb{N}}\, [\exists n\, f(n) = 1 \quad \vee \quad \forall n\, f(n) = 0].$$

Theorem 5.2 *The following theories have the same proof-theoretic strength and therefore prove (as a minimum) the same Π^0_2 statements of arithmetic:*

(i) *Intuitionistic explicit mathematics,* $\mathbf{T}^i_0$.
(ii) *Constructive Zermelo–Fraenkel set theory with the* Regular Extension Axiom, **CZF + REA**.
(iii) *Constructive Zermelo–Fraenkel set theory augmented by* **RDC** *and the* strong Regular Extension Axiom, **CZF + sREA + RDC**.
(iv) **CZF + REA + ΠΣW-AC + RDC + PAx**.
(v) *The extensional type theory* $\mathbf{MLT}^{ext}_{1W}\mathsf{V}$.
(vi) $\mathbf{MLT}_{1W}\mathsf{V}$.
(vii) *The extensional type theory* $\mathbf{MLT}^{ext}_{1W}$.
(viii) $\mathbf{MLT}_{1W}$.
(ix) *The classical subsystem of second order arithmetic* $(\mathbf{\Sigma}^1_2\text{-}\mathbf{AC}) + \mathbf{BI}$ *(same as* $(\Delta^1_2\text{-}\mathbf{CA}) + \mathbf{BI}$*).*
(x) *The intuitionistic system* **IARI** *of second order arithmetic.*
(xi) *Classical Kripke-Platek set theory* **KP** *(cf. [7]) plus the axiom asserting that every set is contained in an admissible set. (This theory is often denoted by* **KPi***.)*
(xii) *Intuitionistic Kripke-Platek set theory,* **IKP***, plus the axiom asserting that every set is contained in an admissible set. (This theory will be notated by* **IKPi***.)*
(xiii) **CZF + REA + RDC + LPO**.

Proof The equivalence of (i), (ii), (iii), (iv), (v), (vi), (vii), (viii), (ix), (x), and (xi) follows from [48], Theorem 3.9, Proposition 5.3, Theorems 5.13 and 6.13 plus the extra observation that the interpretation of **IRA** in $\mathbf{MLT}^{\text{ext}}_{1W}$ defined in [48, Definition 6.5] and proved to be an interpretation in [48, Theorem 6.9] actually only requires the intensional identity type. It was already observed by Palmgren [46] that the interpretations of theories of iterated, strictly positive inductive definitions in type theory works with the intensional identity, and the same argument applies here.

The equivalence of (ii) and (iii) follows from [52, Theorem 4.7], where the principle **sREA** is denoted by INAC.

The proof-theoretic equivalence of (xi) and (xii) follows since the intuitionistic version is a subtheory of the classical one and the well-ordering proof for initial segments of the ordinal of **KPi** can already be carried out in the intuitionistic theory.

For (xiii) we rely on [61]. That the theory **CZF** + **REA** + **RDC** + **LPO** has a realizability interpretation in ($\mathbf{\Sigma}^1_2$-**AC**) + **BI** follows by an extension of the techniques used in [61, Theorem 6.2]. The proof furnished a realizability model for **CZF** + **RDC** + **LPO** that is based on recursion in the type-2 object $E : (N \to N) \to N$ with $E(f) = n + 1$ if $f(n) = 0$ and $\forall i < n\, f(n) > 0$ and $E(f) = 0$ if $\forall n\, f(n) > 0$. Recursion in E is formalizable in the theory of bar induction, i.e. ($\mathbf{\Pi}^0_\infty$-**CA**) + **BI**, which is known to have the same strength as **CZF** (see [61, Theorem 2.2]). The same recursion theory (or partial combinatory algebra) can be employed in extending the modeling of a type structure given in [61, Sect. 5] to the larger type structure needed for **CZF** + **REA** + **RDC** + **LPO**. This is achieved by basically taking the type structure in [48, 5.8] but changing the underlying partial combinatory algebra to the one obtained from recursion in the type two object E rather than the usual one provided by the partial recursive functions on $\mathbb{N}$.

It is very likely that the interpretation also validates **ΠΣW-AC** and **PAx**, but this hasn't yet been checked.

At any rate, we have shown the proof-theoretic equivalence of all theories. □

The foregoing proof establishes the claimed results, however, we'd like to look at Theorem 5.2 in more detail, especially at its proof(s) and the information one can extract from it.

For starters, what does the phrase "same proof-theoretic strength" mean? At a minimum it means that the theories ought to be finitistically equiconsistent. Here it means that they prove at least the same Π^0_2 statements of the language of first-order arithmetic. But more can be shown. A result we will be working toward is that many of the intuitionistic theories of Theorem 5.2 prove the same arithmetical statements. In particular it will be shown that the extensional and the intensional type theories prove the same arithmetical statements. An arithmetical statement gives rise to a type via the propositions-as-types paradigm, so by conservativity of one type theory over another with respect to arithmetic statements we mean that the same arithmetical types are provably inhabited in both theories.

The question of the relation between intensional and extensional type theories has been addressed before by Hofmann in [32]. The set-up there, though, is somewhat different in that the intensional type theory TT_I of [32] is not a pure intensional type theory. It has two extensional rules called *functional extensionality* and *uniqueness of identity*:

$$\text{(ID-UNI-I)} \quad \frac{\Gamma \vdash s : A \qquad \Gamma \vdash p : \mathsf{Id}(A, s, s)}{\Gamma \vdash IdUni(A, s, p) : \mathsf{Id}(\mathsf{Id}(A, s, s), p, \mathsf{refl}(s))}$$

$$\text{(EXT-FORM)} \quad \frac{\Gamma \vdash f, g : \Pi_{(x:A)} B(x) \qquad \Gamma, x : A \vdash p(x) : \mathsf{Id}(B(x), fx, gx)}{\Gamma \vdash Ext(f, g, p) : \mathsf{Id}(\Pi_{(x:A)} B(x), f, g)\,.}$$

These rules are not provable in the purely intensional context, so as a result, we are pursuing a different question here.

Proposition 5.3 $\mathbf{T}_0^i$ *can be interpreted in* **CZF + REA**. *The interpretation preserves (at least) all arithmetic statements.*

Proof The proof of [48] Theorem 3.9 provides an interpretation of $\mathbf{T}_0^i$ in **CZF + REA** which is essentially a class model of $\mathbf{T}_0^i$ inside **CZF + REA**. Having defined an applicative structure, the classifications are defined inductively along the (intuitionistic) ordinals. This is inspired by Feferman's construction of a model of $\mathbf{T}_0^i$ in [20, Theorem 4.1.1]. Inspection of the translation confirms that arithmetic statements get preserved. □

Proposition 5.4 *(i)* **CZF + REA** *has an interpretation in* $\mathbf{MLT}_{1\mathsf{W}}\mathsf{V}$.
(ii) **CZF + REA + ΠΣW-AC + RDC + PAx** *has an interpretation in* $\mathbf{MLT}^{ext}_{1\mathsf{W}}\mathsf{V}$.

Proof (i) and (ii) follow from [3]. The interpretation uses the type V and two propositional functions

$$\dot{=} : \mathsf{V} \times \mathsf{V} \to \mathcal{U}_0$$
$$\dot{\in} : \mathsf{V} \times \mathsf{V} \to \mathcal{U}_0$$

to interpret $=$ and $\in$. For (i), the identity type does not play any role. For (ii) one needs the extensionality of function types. □

Proposition 5.5 **CZF + REA + ΠΣW-AC + RDC + PAx** *is conservative over* **CZF + REA + FT-AC** *for statements of finite type arithmetic (i.e., of the language of* $\mathbf{HA}^{\omega}$*).*

Proof From [55, Theorem 5.23] it follows that **CZF + REA + ΠΣW-AC + RDC + PAx** and **CZF + REA + ΠΣW-AC** prove the same sentences of finite type arithmetic (and more) since the inner model $\mathsf{H}(\mathsf{Y}^*_{\mathsf{W}})$ satisfies **CZF + REA + ΠΣW-AC + RDC + PAx**, assuming **CZF + REA + ΠΣW-AC** in the background.

By [54, Theorem 4.33], there is an interpretation of **CZF + REA + ΠΣW-AC** in **CZF + REA**. Inspection shows that, in the presence of **FT-AC**, the meanings of statements of finite type arithmetic are preserved under this interpretation. □

Proposition 5.6 *For* θ *a sentence of arithmetic let* $\|\theta\|$ *be the corresponding type term according to the propositions-as-types translation. If*

$$\mathbf{MLT}^{ext}_{1\mathsf{W}}\mathsf{V} \vdash t : \|\theta\|$$

for some term t, then

$$\mathbf{CZF} + \mathbf{REA} + \mathbf{FT\text{-}AC} \vdash \theta_{set}$$

with θ_{set} denoting the standard set-theoretic rendering of θ.

Proof Assume $\mathbf{MLT}^{\mathrm{ext}}_{1\mathsf{W}}\mathsf{V} \vdash t : \|\theta\|$. The interpretation $^\wedge$ of $\mathbf{MLT}^{\mathrm{ext}}_{1\mathsf{W}}\mathsf{V}$ into $\mathbf{CZF} +$ $\mathbf{REA}$ given in [55, Sect. 6] yields $\mathbf{CZF} + \mathbf{REA} \vdash (t : \|\theta\|)^\wedge$. Inspection shows that $(t : \|\theta\|)^\wedge$ is a statement about the finite type structure over ω. One then sees, with the help of **FT-AC**, that θ_{set} holds. This is similar to the proof of [55, Theorem 3.15]. □

Theorem 5.7 $\mathbf{CZF} + \mathbf{REA} + \mathbf{\Pi\Sigma}\mathsf{W}\text{-}\mathbf{AC} + \mathbf{RDC} + \mathbf{PAx}$ *is conservative over*

$$\mathbf{IKP} + \forall x\, \exists y\, [x \in y \ \wedge\ y \texttt{ is an admissible set}]$$

for arithmetical statements.

Proof We shall use the shorthand **IKPi** for the latter theory. By Proposition 5.5 it suffices to show that $\mathbf{CZF} + \mathbf{REA} + \mathbf{FT\text{-}AC}$ is conservative over **IKPi** for arithmetic statements. [48, Theorem 5.11] shows that $\mathbf{MLT}^{\mathrm{ext}}_{1\mathsf{W}}\mathsf{V}$ has an interpretation in the classical theory **KPi** where types are interpreted as subsets of ω and crucially dependent products of types are interpreted as sets of indices of partial recursive functions. This also furnishes an interpretation of $\mathbf{CZF} + \mathbf{REA} + \mathbf{FT\text{-}AC}$ in **KPi** since the former is interpretable in $\mathbf{MLT}^{\mathrm{ext}}_{1\mathsf{W}}\mathsf{V}$. The interpretation also works for **IKPi** as definition by (transfinite) Σ-recursion works in intuitionistic **KP** as well (see [4, Sec. 11] and [5, Sec. 19]). The inductive definition of 5.8 in [48] proceeds along the ordinals and focusses on successor ordinals, seemingly requiring a classical case distinction as to whether an ordinal is a successor or a limit or 0, but this is actually completely irrelevant.

Now, the upshot of this hereditarily recursive interpretation is that every Π^0_2 theorem of $\mathbf{CZF} + \mathbf{REA} + \mathbf{FT\text{-}AC}$ is provable in **IKPi**. To be able to extend this approach to all of arithmetic, one needs a more abstract type structure such that interpretability entails deducibility. The conservativity of $\mathbf{HA}^\omega + \mathbf{FT\text{-}AC}$ over **HA**, due to Goodman [27, 28], provides the template. The two steps of Goodman's second proof have been neatly separated by Beeson [9] to construct a general methodology for showing an intuitionistic theory T to be conservative over another theory S for arithmetic statements. The idea is to combine two interpretations, where the first uses functions that are recursive relative to a generic oracle and the second step is a forcing construction. The same idea has been used by Gordeev [29], and in more recent times by Chen and Rathjen in [14, 15, 62], establishing several conservativity results.

The oracle $\mathcal{O}$ will be a fixed but arbitrary partial function from $\mathbb{N}$ to $\{0, 1\}$. A partial function ϕ is recursive relative to $\mathcal{O}$ if it is given by a Turing machine with access to $\mathcal{O}$. During a computation the oracle may be consulted about the value of $\mathcal{O}(n)$ for several n. If $\mathcal{O}(n)$ is defined it will return that value and the computation will continue, but if $\mathcal{O}(n)$ is not defined no response will be coming forward and the computation will never come to a halt. The idea of the second interpretation step is that on account of $\mathcal{O}$'s arbitrariness it can be interpreted in many ways. Given an arithmetic statement θ, an oracle $\mathcal{O}_\theta$ can be engineered so that in a forcing model realizability of θ with functions computable relative to $\mathcal{O}_\theta$ entails the truth of θ. The final step, then, is achieved by noticing that for arithmetic statements forcibility (where the forcing conditions are finite partial functions on $\mathbb{N}$) and validity coincide. For details we'll have to refer to [14, 15]. □

Definition 5.8 Below we shall speak about arithmetical statements in various theories with differing languages. There is a canonical translation of the language of first and second order arithmetic into the language of set theory. However, it is perhaps less obvious what arithmetical statements mean in the context of type theory.

The terms of the language of **HA** are to be translated in an obvious way, crucially using the type-theoretic recursor for the type $\mathbb{N}$. In this way each term t of **HA** gets assigned a raw term $\hat{t}$ of type theory. For details see [42, pp. 71–75], [8, XI.17] [70, Chap. 11, Sect. 2]. An equation $s = t$ of the language **HA** is translated as a type-expression $\mathsf{Id}(\mathbb{N}, \hat{s}, \hat{t})$. For complex formulas the translation proceeds in the obvious way.

We then say that two type theories TT_1 and TT_2 prove the same arithmetical statements if for all sentences A of **HA**,

$$TT_1 \vdash p : \hat{A} \text{ for some } p \text{ iff } TT_2 \vdash p' : \hat{A} \text{ for some } p',$$

where $\hat{A}$ denotes the type-theoretic translation of A.

Recall that **IKPi** is the theory **IKP** $+ \forall x\, \exists y\, [x \in y \wedge y$ `is an admissible set`].

Theorem 5.9 *The following theories prove the same arithmetical statements, i.e. statements of the language of first order arithmetic (also known as Peano arithmetic).*

(i) $\mathbf{T}_0^i$.
(ii) **CZF** + **REA**.
(iii) **CZF** + **REA** + $\mathbf{\Pi\Sigma}$**W-AC** + **RDC** + **PAx**.
(iv) $\mathbf{MLT}_{1\mathsf{W}}^{ext}\mathsf{V}$.
(v) $\mathbf{MLT}_{1\mathsf{W}}\mathsf{V}$.
(vi) $\mathbf{MLT}_{1\mathsf{W}}^{ext}$.
(vii) $\mathbf{MLT}_{1\mathsf{W}}$.
(viii) **IARI**.
(ix) **IKPi**.

Proof Let θ be an arithmetic sentence. Then we have

$$\begin{aligned} \mathbf{T}_0^i \vdash \theta &\Rightarrow \mathbf{CZF} + \mathbf{REA} \vdash \theta \\ &\Rightarrow \mathbf{CZF} + \mathbf{REA} + \mathbf{\Pi\Sigma}\mathsf{W}\text{-}\mathbf{AC} + \mathbf{RDC} + \mathbf{PAx} \vdash \theta \\ &\Rightarrow \mathbf{IKPi} \vdash \theta \end{aligned}$$

by Proposition 5.3 and Theorem 5.7. Now it follows from Jäger's article [34] and from [36] that every initial segment of the proof-theoretic ordinal of **IKPi** is provably well-founded in $\mathbf{T}_0^i$, and thus, if **IKPi** $\vdash \theta$, then $\mathbf{T}_0^i$ is sufficient to show that there is an infinite intuitionistic cut-free proof of θ. By induction on the length of the proof it then follows that all sequents in the proof are true, yielding that $\mathbf{T}_0^i \vdash \theta$. The upshot is that the theories of (i), (ii) and (iii) prove the same arithmetic statements. Furthermore, if $\mathbf{MLT}^{\text{ext}}_{1\mathsf{W}}\mathsf{V} \vdash t : \|\theta\|$ for some term t, then **CZF** + **REA** $\vdash \theta$ by Proposition 5.6 and hence $\mathbf{T}_0^i \vdash \theta$.

So to finish the proof it would suffice to show that $\mathbf{T}_0^i \vdash \theta$ yields $\mathbf{MLT}_{1\mathsf{W}}\mathsf{V} \vdash s : \|\theta\|$ for some term s. Now [48, Sec. 6] shows that the intuitionistic theory **IARI** has the same proof-theoretic ordinal as **IKPi** and $\mathbf{T}_0^i$. So from $\mathbf{T}_0^i \vdash \theta$ it follows that **IARI** $\vdash \theta$. By [48, Theorem 6.9] we then get $\mathbf{MLT}_{1\mathsf{W}}\mathsf{V} \vdash s : \|\theta\|$ for some term s, completing the circle. □

Remark 5.10 Ordinal analysis played a crucial role in the proofs of Theorems 5.2 and 5.9. Having the same proof-theoretic ordinal allowed us to infer that $\mathbf{T}_0^i$, **IKPi** and **IARI** prove the same arithmetic statements.

For a long time [34] was also the only proof that enabled one to reduce the classical theories $(\Delta^1_2\text{-}\mathsf{CA}) + \mathsf{BI}$ and **KPi** to classical $\mathbf{T}_0$. There is now also a proof by Sato [65] for the reductions in the *classical* case that avoids proof-theoretic ordinals. However, determining the strength of other important fragments of **MLTT** (such as the ones analyzed by Setzer in [66]) still requires the techniques of ordinal analysis.

Remark 5.11 We conjecture that also the theory **CZF** + **sREA** + $\mathbf{\Pi\Sigma}$W-**AC** + **RDC** + **PAx** (or at least **CZF** + **sREA** + $\mathbf{\Pi\Sigma}$W-**AC** + **RDC**) proves the same arithmetical statements as any of the theories featuring in Theorem 5.9. As the latter relies on a substantial number of results from the literature, several of them would have to be revisited and possibly amended to establish this.

6 On Relating Theories II: MLTT and Friends

So far we have only gathered results concerning theories that are of the strength of Martin–Löf type theory with one universe. The earlier quote by Harris speculated on the strength of type theory with infinitely many universes. As it turns out, similar techniques can be applied in this context as well.

To begin with, we shall define versions of explicit mathematics, second order arithmetic and constructive set theory featuring analogues of universes.

6.1 $\mathbf{T}_0^i$ with Universes

Definition 6.1 Systems of explicit mathematics with universes have been defined and studied in several papers (cf. [37–39]) and were probably first introduced by Feferman [23].

By $\mathbf{T}_0^i + \bigcup_n \mathsf{U}_n$ we denote an extension of $\mathbf{T}_0^i$ whose language has infinitely many classification constants $\mathsf{U}_0, \mathsf{U}_1, \ldots$ and the following axioms for each constant U_n.

1. $\mathbf{N} \,\varepsilon\, \mathsf{U}_n$ and $\mathsf{U}_i \,\varepsilon\, \mathsf{U}_n$ for $i < n$.
2. $\forall x \,\varepsilon\, \mathsf{U}_n \,\exists X\, x = X$ (i.e. every element of U_n is a classification).
3. For every elementary formula $\psi(x, \vec{v}, X_1, \ldots, X_r)$ with all classification variables exhibited and which does not contain constants U_i with $i \geq n$,
$$\forall X_1, \ldots, X_r \,\varepsilon\, \mathsf{U}_n \,\exists Y\, [Y \,\varepsilon\, \mathsf{U}_n \wedge Y \simeq \{x : \psi(x, \vec{v}, X_1, \ldots, X_r)\}] .$$
4. $\forall X \,\varepsilon\, \mathsf{U}_n\, [\forall x \,\varepsilon\, X\, \exists Y \,\varepsilon\, \mathsf{U}_n\, fx \simeq Y \rightarrow \exists Z[Z \in \mathsf{U}_n \wedge Z \simeq \mathbf{j}(X, f)]]$.
5. $\forall X, Y \,\varepsilon\, \mathsf{U}_n\, \exists Z\, [Z \in \mathsf{U}_n \wedge Z \simeq \mathbf{i}(X, Y)]$.

In other words, a classification U_n is a universe containing $\mathbf{N}, \mathsf{U}_0, \ldots, \mathsf{U}_{n-1}$ closed under elementary comprehension, join and inductive generation.

By $\mathbf{T}_0^i + \bigcup_{i<n} \mathsf{U}_i$ we denote the theory with just the universes $\mathsf{U}_0, \ldots, \mathsf{U}_{n-1}$ and their pertaining axioms.

6.2 Universes in Intuitionistic Second Order Arithmetic

It is also useful to have a many universes version of **IARI** to obtain an intuitionistic theory of second order arithmetic which can be easily interpreted in **MLTT**. One idea would be to adopt the notion of β-model from Definition 4.14 to serve as a notion of universe. However, a β-model comes with an explicit countable enumeration of its sets and therefore it would be difficult if not impossible to model such structures in **MLTT**. Instead, an option is to add set predicates $\mathfrak{U}_0, \mathfrak{U}_1, \ldots$ to the language $\mathcal{L}_2$ that are intended to apply to sets of natural numbers with the aim of singling out collections of sets that have universe-like properties.

Definition 6.2 The theory **IARI**+$\bigcup_n \mathfrak{U}_n$ has additional predicates $\mathfrak{U}_0, \mathfrak{U}_1, \ldots$ for creating new atomic formulas $\mathfrak{U}_n(X)$ ($n \in \mathbb{N}$), where X is a second order variable. We use abbreviations like $\forall X \in \mathfrak{U}_n\, \varphi$ and $\exists X \in \mathfrak{U}_n\, \varphi$ for $\forall X\, (\mathfrak{U}_n(X) \rightarrow \varphi)$ and $\exists X\, (\mathfrak{U}_n(X) \wedge \varphi)$, respectively. If ψ is any formula of this language, then $\psi^{\mathfrak{U}_n}$ arises from ψ by relativizing all second order quantifiers to $\mathfrak{U}_n$, i.e., replacing all quantifiers QX in ψ by $QX \in \mathfrak{U}_n$.

In addition to the axioms of **IARI** there are the following pertaining to the new predicates.

1. The predicates $\mathfrak{U}_n$ are cumulative, i.e. $\forall X\, [\mathfrak{U}_i(X) \rightarrow \mathfrak{U}_j(X)]$ whenever $i \leq j$.
2. **Induction**:
$$\phi(0) \wedge \forall u[\phi(u) \rightarrow \phi(u+1)] \rightarrow \forall u \phi(u)$$
for all formulae ϕ.

3. **Arithmetic Comprehension Schema for $\mathfrak{U}_n$:**

$$Y_1, \ldots, Y_r \in \mathfrak{U}_n \rightarrow \exists X \in \mathfrak{U}_n \forall u[u \in X \leftrightarrow \psi(u, Y_1, \ldots, Y_r)]$$

if $\psi(u, Y_1, \ldots, Y_r)$ is a formula with all free second order variables exhibited, in which all second order quantifiers are of the form $QX \in \mathfrak{U}_i$ for some $i < n$, and moreover, no predicates $\mathfrak{U}_j$ for $j \geq n$ occur in it.

4. **Replacement**:

$$\forall X \in \mathfrak{U}_n[\forall u \in X \exists !Y \in \mathfrak{U}_n \phi(u, Y) \rightarrow \exists Z \in \mathfrak{U}_n \forall u \in X\, \phi(u, Z_u)]$$

for all formulas ϕ. Here $\phi(u, Z_u)$ arises from $\phi(u, Z)$ by replacing each occurrence $t \in Z$ in the formula by $\langle u, t\rangle \in Z$.

5. **Inductive Generation**:

$$\forall U \in \mathfrak{U}_n \forall X \in \mathfrak{U}_n \exists Y \in \mathfrak{U}_n \left[\mathbf{WP}_U(X, Y) \wedge (\forall u[\forall v(v \prec_X u \rightarrow \phi(v)) \rightarrow \phi(u)] \rightarrow \forall x \in Y\, \phi(x))\right],$$

for all formulas ϕ, where $v \prec_X u$ abbreviates $\langle v, u\rangle \in X$ and $\mathbf{WP}_U(X, Y)$ stands for

$$\mathsf{Prog}_U(X, Y) \wedge \forall Z[\mathsf{Prog}_U(X, Z) \rightarrow Y \subseteq Z]$$

with $\mathsf{Prog}_U(X, Y)$ being $\forall y \in U[\forall z(z \prec_X y \rightarrow z \in Y) \rightarrow y \in Y]$.

By $\mathbf{IARI}+\bigcup_{i<m}\mathfrak{U}_i$ we denote the theory with only the additional predicates $\mathfrak{U}_0, \ldots, \mathfrak{U}_{m-1}$ and their pertaining axioms.

Definition 6.3 Recall the notion of inaccessible set defined in 4.7. For $n > 0$, $\mathbf{Inacc}(n)$ stands for the set-theoretic statement that there are n-many inaccessible sets $I_0 \in \ldots \in I_{n-1}$. Let $\mathbf{Inacc}(0)$ stand for $0 = 0$.

β-models were introduced in Definition 4.14. By $\mathbf{Beta}(n)$ we denote the statement of second order arithmetic asserting that there are n many sets $A_0, \ldots, A_{n-1}$ which are β-models of $\mathbf{\Sigma}^1_2$-$\mathbf{AC}$ such that $A_0 \in \ldots \in A_{n-1}$, where for sets X, Y of natural numbers $X \in Y$ is defined by $\exists u\ X = Y_u$.

For $n > 0$, let $\mathbf{MLT}_{n\mathsf{W}}V$ be the fragment of $\mathbf{MLTT}$ with n-many universes $\mathcal{U}_0, \ldots, \mathcal{U}_{n-1}$, where the W-constructor can solely be applied to families of types in $\mathcal{U}_0, \ldots, \mathcal{U}_{n-1}$ but one can also form the type $\mathsf{V} := \mathsf{W}_{(A:\mathcal{U}_{n-1})}A$, i.e. a W-type over the largest universe $\mathcal{U}_{n-1}$. We shall also consider the type theory $\mathbf{MLT}_{n\mathsf{W}}$ which is the fragment of $\mathbf{MLT}_{n\mathsf{W}}V$ without the type V. Below we assume that $n > 0$.

Theorem 6.4 *(i)* $\mathbf{T}^i_0 + \bigcup_{i<n}\mathsf{U}_i$ *has an interpretation in* $\mathbf{CZF} + \mathbf{REA} + \mathbf{Inacc}(n)$. *The interpretation preserves (at least) all arithmetic statements.*

(ii) $\mathbf{CZF} + \mathbf{REA} + \mathbf{Inacc}(n-1)$ *has an interpretation in* $\mathbf{MLT}_{n\mathsf{W}}\mathsf{V}$.

(iii) $\mathbf{CZF} + \mathbf{REA} + \mathbf{Inacc}(n-1) + \mathbf{\Pi\Sigma}\mathsf{W}\text{-}\mathbf{AC} + \mathbf{RDC} + \mathbf{PAx}$ *has an interpretation in* $\mathbf{MLT}^{ext}_{n\mathsf{W}}\mathsf{V}$.

(iv) $\mathbf{MLT}^{ext}_{n\mathsf{W}}\mathsf{V}$ *has an interpretation in the classical set theory* **KPi** *plus an axiom asserting that there exist* $n-1$*-many recursively inaccessible ordinals.*

(v) $(\mathbf{\Sigma}_2^1\text{-}\mathbf{AC}) + \mathbf{BI} + \mathbf{Beta}(n)$ *has an interpretation in* **KPi** *plus the existence of n-many recursively inaccessible ordinals.*

(vi) **KPi** *plus the existence of n-many recursively inaccessible ordinals has a sets-as-trees interpretation in* $(\mathbf{\Sigma}_2^1\text{-}\mathbf{AC}) + \mathbf{BI} + \mathbf{Beta}(n)$.

(vii) *The intuitionistic system* $\mathbf{IRA} + \bigcup_{i<n-1} \mathfrak{U}_i$ *of second order arithmetic can be interpreted in* $\mathbf{MLT}_n\mathsf{W}$.

(viii) $\mathbf{CZF} + \mathbf{REA} + \mathbf{RDC} + \mathbf{Inacc}(n)$ *has a realizability interpretation in* **KPi** *plus the existence of n-many recursively inaccessible ordinals.*

(ix) *All the above theories have the same proof-theoretic strength and prove (at least) the same* Π_2^0*-statements of arithmetic.*

Proof The interpretations are extensions of those discussed in the previous section, taking more universes into account. We can only indicate the steps. The interpretation of $\mathbf{T}_0^i$ in $\mathbf{CZF} + \mathbf{REA}$ can be lifted to an interpretation of $\mathbf{T}_0^i + \bigcup_{i<n} \mathsf{U}_i$ into $\mathbf{CZF} + \mathbf{REA} + \mathbf{Inacc}(n)$. The latter theory possesses a sets-as-types interpretation in intensional Martin–Löf type theory with $n + 1$ universes.

$\mathbf{CZF} + \mathbf{REA} + \mathbf{\Pi\Sigma W\text{-}AC} + \mathbf{RDC} + \mathbf{PAx} + \mathbf{Inacc}(n - 1)$ possesses a sets-as-types interpretation in $\mathbf{MLT}^{\text{ext}}_{n\mathsf{W}}\mathsf{V}$. In turn, $\mathbf{MLT}^{\text{ext}}_{n\mathsf{W}}\mathsf{V}$ can be interpreted in classical Kripke-Platek set theory **KPi** plus an axiom asserting that there are at least $n -$ 1-many recursively inaccessible ordinals, following the Ansatz of [48, Theorem 5.11]. $(\mathbf{\Sigma}_2^1\text{-}\mathbf{AC}) + \mathbf{BI} + \mathbf{Beta}(n)$ can be easily interpreted in **KPi** plus n-recursively inaccessible ordinals.

The proof-theoretic equivalence ensues from an ordinal analysis of the 'top theory', **KPi** plus the existence of n-many recursively inaccessible ordinals, together with proofs that any ordinal below the proof-theoretic ordinal of that theory is provably well-founded in $\mathbf{T}_0^i + \bigcup_{i<n} \mathsf{U}_i$ as well as $\mathbf{IARI}+ \bigcup_{i<n} \mathfrak{U}_i$. Neither the ordinal analysis nor the well-ordering proofs are available from the published literature. The ordinal analysis of **KPi** plus the existence of n-many recursively inaccessible ordinals, though, can be obtained in a straightforward way by extending the one given for **KPi** in [36] or rather its modern version in [12]. It also follows from the ordinal analysis of the much stronger theory **KPM** given in [47] by restricting the treatment therein to the pertaining small fragments. For the well-ordering proof substantially more work is required; details will be published in [63]. □

Theorem 6.5 *The following theories have the same proof-theoretic strength and prove the same* Π_2^0*-statements of arithmetic:*

(i) $\mathbf{T}_0^i + \bigcup_n \mathsf{U}_n$.

(ii) **CZF** *plus* $\mathbf{Inacc}(n)$ *for all* $n > 0$.

(iii) $\mathbf{CZF} + \mathbf{\Pi\Sigma W\text{-}AC} + \mathbf{RDC} + \mathbf{PAx}$ *plus* $\mathbf{Inacc}(n)$ *for all* $n > 0$.

(iv) *The extensional type theory* $\mathbf{MLTT}^{ext}$.

(v) **MLTT**.

(vi) *The classical subsystem of second order arithmetic* $(\mathbf{\Sigma}_2^1\text{-}\mathbf{AC}) + \mathbf{BI}$ *plus* $\mathbf{Beta}(n)$ *for all* $n > 0$.

(vii) *Classical Kripke-Platek set theory* **KP** *plus for every* $n > 0$ *an axiom asserting that there are at least n-many recursively inaccessible ordinals.*

(viii) **IARI**+ $\bigcup_n \mathfrak{U}_n$.
(ix) **CZF** + **RDC** + **LPO** *plus the axioms* **Inacc**(n) *for all* $n > 0$.

Proof This follows directly from the previous theorem. □

The latter theorem also shows that the strength of **MLTT** is dwarfed by that of (Π^1_2-**CA**). It corresponds to a tiny fragment of second order arithmetic which itself is a tiny fragment of **ZF**, so there are aeons between **MLTT** and classical set theory with inaccessible cardinals.

Theorem 6.6 *The following theories prove the same arithmetical statements:*

(i) $\mathbf{T}^i_0 + \bigcup_n \mathsf{U}_n$.
(ii) **MLTT**.
(iii) *The extensional type theory* $\mathbf{MLTT}^{ext}$.
(iv) **CZF** *plus* **Inacc**(n) *for every* $n > 0$.
(v) **CZF** + $\bigcup_n$ **Inacc**(n) + $\bigcup_n$ **ΠΣW-AC** + **RDC** + **PAx**.
(vi) **IARI**+ $\bigcup_n \mathfrak{U}_n$.

Proof The methods for proving this were described in the proof of 5.9. Details will appear in [63]. □

Finally, it should be mentioned that Martin–Löf type theory with stronger universes (e.g. Mahlo universes) has been studied by Setzer (cf. [67]).

6.3 Adding the Univalence Axiom

The quote (1) from Harris' book [30] claimed that modeling Voevodsky's univalence axiom (**UA**) requires infinitely many inaccessible cardinals (for a definition of **UA** see [33, Sec. 2.10]). While the simplicial model of type theory with univalence developed in the paper [41] by Kalpulkin, Lumsdaine and Voevodsky is indeed carried out in a background set theory with inaccessible cardinals, it is by no means clear that the existence or proof-theoretic strength of these objects is required for finding a model of type theory with **UA**. In actuality, Bezem, Coquand and Huber in their article [10] provided a cubical model of type theory that also validates **UA**. Crucially, their modeling can be carried out in a constructive background theory such as **CZF** + $\bigcup_n$ **Inacc**(n) + $\bigcup_n$ **ΠΣW-AC** + **RDC** + **PAx**. Thus it follows that adding **UA** does not increase the strength of type theory and that no inaccessible cardinals are required. Hence in view of Theorem 6.5 we have the following result.

Corollary 6.7 **MLTT** *has the same proof-theoretic strength as* **MLTT** + **UA**. *Thus* **MLTT** + **UA** *shares the same proof-theoretic strength with all theories listed in Theorem 6.5, in particular with classical Kripke-Platek set theory* **KP** *augmented by axioms asserting that there are at least n-many recursively inaccessible ordinals for every* $n > 0$.

7 On Relating Theories III: Omitting W

The proof-theoretic strength of type theories crucially depends on the availability of inductive types and to a much lesser extent on its universes. Relinquishing the W-type brings about an enormous collapse of proof power (cf. [49–51]). Letting **MLTT**$^-$ be **MLTT** bereft of the W-type constructor, we arrive at a theory no stronger than the system $\mathbf{ATR}_0$ of reverse mathematics (see [68, I.11]), having the famous ordinal Γ_0 as its proof-theoretic ordinal. According to Feferman's analysis (see [24, 25]), Γ_0 delineates the limit of a notion of predicativity that only accepts the natural numbers as a completed infinity (which was first adumbrated in Hermann Weyl's book *"Das Kontinuum"* from 1918 [72]). Peter Hancock conjectured in the 1970s the ordinal of **MLTT**$^-$ to be Γ_0. Feferman [23] and independently Aczel (see also [1]) proved *Hancock's Conjecture*. There is also a version of **CZF** with inaccessible sets of strength Γ_0, due to Crosilla and Rathjen [17], which does not have set induction. Thus the set-theoretic analogue to eschewing W-types consists in leaving out the principle of set induction. In the next theorem we denote by $\mathbf{ATR}_0^i$ the intuitionistic version of $\mathbf{ATR}_0$ (see [49, Definition 4.10] for details). By **CZF**$^-$ we denote Constructive Zermelo–Fraenkel set theory without set induction but with the Infinity axiom strengthened as follows:

$$0 \in \omega \wedge \forall y[\, y \in \omega \rightarrow y+1 \in \omega] \tag{4}$$

$$\forall x\,[0 \in x \wedge \forall y(y \in x \rightarrow y+1 \in x) \rightarrow \omega \subseteq x] \tag{5}$$

(for details see [17, Definition 2.2]). Likewise we denote by **KP**$^-$ the theory without the set induction scheme but with the infinity axioms (4) and (5).

The notion of weak inaccessibility used below is the one from Definition 4.9. For $n > 0$ let $\mathbf{wInacc}(n)$ be the statement that there exist weakly inaccessible sets $x_0, \ldots, x_{n-1}$ such that $x_0 \in \ldots \in x_{n-1}$.

A restricted form of **RDC** is Δ_0-**RDC**: For all Δ_0-formulae θ and ψ, whenever

$$(\forall x \in a)\big[\theta(x) \rightarrow (\exists y \in a)\big(\theta(y) \wedge \psi(x, y)\big)\big]$$

and $b_0 \in a \wedge \phi(b_0)$, then there exists a function $f : \omega \rightarrow a$ such that $f(0) = b_0$ and

$$(\forall n \in \omega)\big[\theta(f(n)) \wedge \psi(f(n), f(n+1))\big].$$

Theorem 7.1 *The following theories share the same proof-theoretic strength and ordinal Γ_0, and prove the same Π^0_2-sentences of arithmetic:*

(i) **MLTT**$^-$.
(ii) *The extensional version of* **MLTT**$^-$.
(iii) $\mathbf{ATR}_0$.
(iv) $\mathbf{ATR}_0^i$.
(v) **CZF**$^-$ $+ \forall x\, \exists y\, [\, x \in y \wedge y$ *is weakly inaccessible* $] + \Delta_0$-**RDC**.
(vi) **CZF**$^-$ $+ \{\mathbf{wInacc}(n) \mid n > 0\} +$ **RDC**.

(vii) $\mathbf{KP}^- + \forall x\, \exists y\, [x \in y \wedge y \textit{ is admissible }]$.

Proof We only have to establish that all theories have proof-theoretic ordinal Γ_0. For extensional $\mathbf{MLTT}^-$ this follows from [23]. The lower bound part, namely that $\mathbf{MLTT}^-$ has at least the strength Γ_0 is due to Jervell [40]. So we are done with (i) and (ii). That $\mathbf{ATR}_0$ has ordinal Γ_0 is well known. For $\mathbf{ATR}_0^i$ this follows from the observation in [49, Lemma 4.11] that the well ordering proof for any ordinal notation below Γ_0 uses only intuitionistic logic. The determination of the ordinal for the system in (v) and (vi) is due to Crosilla and Rathjen [17, Corollary 9.14] with the validation of Δ_0-**RDC** and **RDC** coming from [52, Theorem 4.17] and [52, Theorem 4.16], respectively. The proof-theoretic analysis of the system in (vii) is due to Jäger [35]. □

We also conjecture that all of the intuitionistic theories from the above list, i.e., $\mathbf{MLTT}^-$, the extensional version of $\mathbf{MLTT}^-$, $\mathbf{ATR}_0^i$, and $\mathbf{CZF}^- + \forall x \exists y\, [x \in y \wedge \text{y is weakly inaccessible}]$ prove the same arithmetic statements using the usual techniques. But we have not yet checked that. What is known is that $\mathbf{ATR}_0^i$ embeds in all of these theories (see [49]).

A final question concerns the status of the univalence axiom. Do we get more strength when we add **UA** to $\mathbf{MLTT}^-$? It turns out that we just have to check whether the cubical model construction from [10] can be carried out in one of the theories from the list. Inspection of [10] reveals that

$$\mathbf{CZF}^- + \forall x \exists y\, [x \in y \wedge y \text{ is weakly inaccessible}] + \Delta_0\text{-}\mathbf{RDC}$$

suffices as a background theory for all the constructions, except W-types.

Corollary 7.2 *The univalent type theory* $\mathbf{MLTT}^- + \mathbf{UA}$ *is of the same strength as* $\mathbf{MLTT}^-$ *and* $\mathbf{ATR}_0$ *and all the other systems from Theorem 7.1. Therefore its proof-theoretic ordinal is* Γ_0.

8 Monotone Fixed Point Principles in Intuitionistic Explicit Mathematics

Martin–Löf type theory appears to capture the abstract notion of an inductively defined type very well via its W-type. There are, however, intuitionistic theories of inductive definitions that at first glance appear to be just slight extensions of Feferman's explicit mathematics (see Feferman's quote from Sect.1) but have turned out to be much stronger than anything considered in Martin–Löf type theory. They are obtained from $\mathbf{T}_0^i$ by the augmentation of a monotone fixed point principle which asserts that every monotone operation on classifications (Feferman's notion of set) possesses a least fixed point. To be more precise, there are two versions of this principle. **MID** merely postulates the existence of a least solution, whereas **UMID** provides a uniform version of this axiom by adjoining a new functional constant to the language, ensuring that a fixed point is uniformly presentable as a function of the monotone operation.

Definition 8.1 For *extensional equality* of classifications we use the shorthand "$=_{\text{ext}}$", i.e.

$$X =_{\text{ext}} Y = \forall v(v\varepsilon X \leftrightarrow v\varepsilon Y).$$

Further, let $X \subseteq Y$ be a shorthand for $\forall v(v\varepsilon X \to v\varepsilon Y)$. To state the monotone fixed point principle for subclassifications of a given classification A we introduce the following shorthands:

$$\begin{array}{ll} \mathbf{Clop}(f, A) & \text{if } \forall X \subseteq A\ \exists Y \subseteq A\ fX \simeq Y \\ \mathbf{Ext}(f, A) & \text{if } \forall X \subseteq A\ \forall Y \subseteq A\ [X =_{\text{ext}} Y \to fX =_{\text{ext}} fY] \\ \mathbf{Mon}(f, A) & \text{if } \forall X \subseteq A\ \forall Y \subseteq A\,[X \subseteq Y \to fX \subseteq fY]. \\ \mathbf{Lfp}(Y, f, A) & \text{if } fY \subseteq Y \wedge Y \subseteq A \wedge \forall X \subseteq A\left[fX \subseteq X \to Y \subseteq X\right] \end{array}$$

When f satisfies $\mathbf{Clop}(f, A)$, we call f a *classification operation on* A. When f satisfies $\mathbf{Clop}(f, A)$ and $\mathbf{Ext}(f, A)$, we call f *extensional* or an *extensional operation on* A. When f satisfies $\mathbf{Clop}(f, A)$ and $\mathbf{Mon}(f, A)$, we say that f is a *monotone operation on* A. Since monotonicity entails extensionality, a monotone operation is always extensional.

Now we state $\mathbf{UMID}_A$.

$\mathbf{MID}_A$ (Monotone Inductive Definition on A)

$$\forall f\,[\mathbf{Clop}(f, A) \wedge \mathbf{Mon}(f, A) \to \exists Y\,\mathbf{Lfp}(Y, f, A)].$$

$\mathbf{UMID}_A$ (Uniform Monotone Inductive Definition on A)

$$\forall f\,[\mathbf{Clop}(f, A) \wedge \mathbf{Mon}(f, A) \to \mathbf{Lfp}(\mathbf{lfp}(f), f, A)].$$

$\mathbf{UMID}_A$ states that if f is monotone on subclassifications of A, then $\mathbf{lfp}(f)$ is a least fixed point of f.

Let V be the universe, i.e. $V := \{x : x = x\}$. By **MID** and **UMID** we denote the principles $\mathbf{MID}_V$ and $\mathbf{UMID}_V$, respectively.[8]

The strength of the various classical versions was determined as a result of several papers [26, 56–58]. The **MID** case is dealt with in [26, 59]. [59] provides a survey of all known results in the classical case. $\mathbf{UMID}_{\mathbb{N}}$ was shown to be related to subsystems of second order arithmetic based on Π^1_2 comprehension.

To relate the state of the art in these matters we shall need some terminology. Below we shall distinguish between the classical and the intuitionistic version of a theory by appending the superscript c and i, respectively. For a system S of explicit mathematics we denote by $S\restriction$ the version wherein the induction principles for the

[8] The acronym for the principle **MID** in Feferman's paper [21], Sect. 7 was MIG $\restriction$.

natural numbers and for inductive generation are restricted to sets. $\mathbf{IND}_{\mathbb{N}}$ stands for the schema of induction on natural numbers for arbitrary formulas of the language of explicit mathematics. $(\Pi^1_2 - \mathbf{CA})_0$ denotes the subsystem of second order arithmetic (based on classical logic) with $\mathbf{\Pi}^1_2$-comprehension but with induction restricted to sets, whereas $(\Pi^1_2 - \mathbf{CA})$ also contains the full schema of induction on $\mathbb{N}$.

[57, 58] yielded the following results:

Theorem 8.2 *(i)* $(\Pi^1_2 - \mathbf{CA})_0$ *and* $\mathbf{T}^c_0\restriction + \mathbf{UMID}_{\mathbb{N}}$ *have the same proof-theoretic strength.*

(ii) $(\Pi^1_2 - \mathbf{CA})$ *and* $\mathbf{T}^c_0\restriction + \mathbf{IND}_{\mathbb{N}} + \mathbf{UMID}_{\mathbb{N}}$ *have the same proof-theoretic strength.*

The first result about $\mathbf{UMID}_{\mathbb{N}}$ on the basis of intuitionistic explicit mathematics was obtained by Tupailo in [71].

Theorem 8.3 $(\Pi^1_2 - \mathbf{CA})_0$ *and* $\mathbf{T}^i_0\restriction + \mathbf{UMID}_{\mathbb{N}}$ *have the same proof-theoretic strength.*

[71] uses a characterization of $(\Pi^1_2 - \mathbf{CA})_0$ via a classical μ-calculus (a theory which extends the concept of an inductive definition), dubbed $\mathbf{ACA}_0(\mathcal{L}^\mu)$, given by Möllerfeld [43] and then proceeds to show that $\mathbf{ACA}_0(\mathcal{L}^\mu)$ can be interpreted in its intuitionistic version, $\mathbf{ACA}^i_0(\mathcal{L}^\mu)$, by means of a double negation translation. Finally, as the latter theory is readily interpretable in $\mathbf{T}^i_0\restriction + \mathbf{UMID}_{\mathbb{N}}$, the proof-theoretic equivalence stated in Theorem 8.3 follows in view of Theorem 8.2.

The proof of [71], however, does not generalize to $\mathbf{T}^i_0\restriction + \mathbf{IND}_{\mathbb{N}} + \mathbf{UMID}_{\mathbb{N}}$ and extensions by further induction principles. The main reason for this is that adding induction principles such as induction on natural numbers for all formulas to $\mathbf{ACA}_0(\mathcal{L}^\mu)$ only slightly increases the strength of the theory and by no means reaches the strength of $(\Pi^1_2 - \mathbf{CA})$. In order to arrive at a μ-calculus of the strength of $(\Pi^1_2 - \mathbf{CA})$ one would have to allow for transfinite nestings of the μ-operator of length α for any ordinal $\alpha < \varepsilon_0$. As it seems to be already a considerable task to get a clean syntactic formalization of transfinite μ-calculi (let alone furnishing double negation translation thereof), this paper will proceed along a different path. In actuality, much of the work was already accomplished in [57], where it was shown that $(\Pi^1_2 - \mathbf{CA})_0$ and $(\Pi^1_2 - \mathbf{CA})$ can be reduced to operator theories $\mathbf{T}^{\mathbf{OP}}_{<\omega}$ and $\mathbf{T}^{\mathbf{OP}}_{<\varepsilon_0}$, respectively. A careful axiomatization of the foregoing theories in conjunction with results from [56] showed that they lend themselves to double negation translations and thus can be translated into their intuitionistic counterparts. As the intuitionistic theories can be easily viewed as subtheories of $\mathbf{T}^i_0\restriction + \mathbf{UMID}_{\mathbb{N}}$ and $\mathbf{T}^i_0\restriction + \mathbf{IND}_{\mathbb{N}} + \mathbf{UMID}_{\mathbb{N}}$, respectively, one can conclude the following result.

Theorem 8.4 *(i)* $(\Pi^1_2 - \mathbf{CA})_0$ *and* $\mathbf{T}^i_0\restriction + \mathbf{UMID}_{\mathbb{N}}$ *have the same proof-theoretic strength.*

(ii) $(\Pi^1_2 - \mathbf{CA})$ *and* $\mathbf{T}^i_0\restriction + \mathbf{IND}_{\mathbb{N}} + \mathbf{UMID}_{\mathbb{N}}$ *have the same proof-theoretic strength.*

Proof See [64]. □

Through Theorem 8.4 one also gets a different proof of Theorem 8.3 which does not hinge upon [43].

Remark 8.5 Virtually nothing is currently known about the strength of $\mathbf{T}_0^i + \mathbf{MID}$ and variants. In the classical case there is a close relationship with parameter-free Π_2^1-comprehension. It would be very interesting to investigate whether the strength of **MID** diminishes in the intuitionistic setting.

The strength of explicit mathematics with principle like $\mathbf{UMID}_{\mathbb{N}}$ and even **MID** considerably exceeds that of Martin–Löf type theory. This has a bearing on foundational questions such as the limit of constructivity or the limits of different concepts of constructivity. In [53, 60] an attempt is made to delineate the form of constructivism underlying Martin–Löf type theory, suggesting that $\mathbf{T}_0^i + \mathbf{UMID}_{\mathbb{N}}$ lies beyond its scope.

Acknowledgements Part of the material is based upon research supported by the EPSRC of the UK through grant No. EP/K023128/1. This research was also supported by a Leverhulme Research Fellowship and a Marie Curie International Research Staff Exchange Scheme Fellowship within the 7th European Community Framework Programme. This publication was also made possible through the support of a grant from the John Templeton Foundation.
Thanks are also due for the invitation to speak at the American Annual Meeting of the Association for Symbolic Logic (University of Connecticut, Storrs, 23 May, 2016) where the material of the first seven sections was presented.

References

1. Aczel, P.: The strength of Martin–Löf's intuitionistic type theory with one universe. In: Miettinen, S., Väänänen, S. (eds.) Proceedings of Symposia in Mathematical Logic, Oulu, 1974, and Helsinki, 1975, Report No. 2, University of Helsinki, Department of Philosophy, pp. 1–32 (1977)
2. Aczel, P.: The type theoretic interpretation of constructive set theory: choice principles. In: Troelstra, A.S., van Dalen, D (eds.) The L.E.J. Brouwer Centenary Symposium, pp. 1–40. North Holland, Amsterdam (1982)
3. Aczel, P.: The type theoretic interpretation of constructive set theory: inductive definitions. In: Marcus, R.B., et al. (eds.) Logic, Methodology and Philosophy of Science VII, pp. 17–49. North Holland, Amsterdam (1986)
4. Aczel, P., Rathjen, M.: Notes on constructive set theory, Technical report 40, Institut Mittag-Leffler, The Royal Swedish Academy of Sciences (2001). http://www.mittag-leffler.se/preprints/0001/, Preprint No. 40
5. Aczel, P., Rathjen, M.: Notes on constructive set theory, 243 p, Preprint (2010). http://www1.maths.leeds.ac.uk/~rathjen/book.pdf
6. Aczel, P.: On relating type theories and set theories. In: Altenkirch, T., Naraschewski, W., Reus, B. (eds.) Types '98. Lecture Notes in Computer Science 1257. Springer, Berlin (2000)
7. Barwise, J.: Admissible Sets and Structures. Springer, Berlin (1975)
8. Beeson, M.: Foundations of Constructive Mathematics. Springer, Berlin (1985)
9. Beeson, M.: Goodman's theorem and beyond. Pac. J. Math. **84** (1979)
10. Bezem, M., Coquand, T., Huber, S.: A model of type theory in cubical sets. In: Matthes, R., Schubert, A. (eds.) 19th International Conference on Types for Proofs and Programs (TYPES 2013), pp. 107–128 (2013)
11. Bishop, E.: Foundations of Constructive Analysis. McGraw-Hill, New York (1967)
12. Buchholz, W.: A simplified version of local predicativity. In: Aczel, P., Simmons, H., Wainer, S. (eds.) Leeds Proof Theory 90, pp. 115–147. Cambridge University Press, Cambridge (1993)

13. Buchholz, W., Feferman, S., Pohlers, W., Sieg, W.: Iterated inductive definitions and subsystems of analysis. In: LNM 897. Springer (1981)
14. R.-M. Chen.: Independence and conservativity results for intuitionistic set theory, Ph.D. thesis, University of Leeds, Leeds (2010)
15. R.-M. Chen., M. Rathjen.: Conservativity à la Goodman for intuitionistic set theories, in preparation
16. Church, A.: A formulation of the simple theory of types. J. Symb. Log. **5**, 56–68 (1940)
17. Crosilla, L., Rathjen, M.: Inaccessible set axioms may have little consistency strength. Ann. Pure Appl. Log. **115**, 33–70 (2002)
18. Diaconescu, R.: Axiom of choice and complementation. Proc. Amer. Math. Soc. **51**, 176–178 (1975)
19. Feferman, S.: Proof theory of classical and constructive inductive definitions. A forty year saga, pp. 7–30, 1968–2008
20. Feferman, S.: A language and axioms for explicit mathematics. Algebra Log. Lect. Not. Math. **450**, 87–139 (1975)
21. Feferman, S.: Constructive theories of functions and classes. In: Crossley, J.N. (ed.) Logic Colloquium '78, pp. 159–224 (1979)
22. Feferman, S.: Monotone inductive definitions. In: Troelstra, A.S., van Dallen, D. (eds.) The L.E.J. Brouwer Centenary Symposium, pp. 77–89. North-Holland (1982)
23. Feferman, S.: Iterated inductive fixed-point theories: application to Hancock's conjecture. In: Patras Logic Symposion, pp. 171–196. North-Holland, Amsterdam (1982)
24. Feferman, S.: Systems of predicative analysis. J. Symb. Log. **29**, 1–30 (1964)
25. Feferman, S.: Systems of predicative analysis II. Representations of ordinals. J. Symb. Log. **33**, 193–220 (1968)
26. Glaß, T., Rathjen, M., Schlüter, A.: The strength of monotone inductive definitions in explicit mathematics. Ann. Pure Appl. Log. **85**, 1–46 (1997)
27. Goodman, N.: The theory of the Gödel functionals. J. Symb. Log. **41**, 574–583 (1976)
28. Goodman, N.: Relativized realizability in intuitionistic arithmetic of all finite types. J. Symb. Log. **43** (1978)
29. Gordeev, L.: Proof-theoretic analysis: Weak systems of functions and classes. Ann. Pure Appl. Log. **38**, 1–121 (1988)
30. Harris, M.: Mathematics without Apologies (2015)
31. Hilbert, D., Bernays, P.: Grundlagen der Mathematik II. Springer, Berlin (1938)
32. Hofmann, M.: Conservativity of equality reflection over intensional type theory. In: Berardi, S., Coppo, M. (eds.) Types for Proofs and Programs, International Workshop TYPES'95. Lecture Notes in Computer Science 1158, pp. 153–164. Springer (1995)
33. Homotopy Type Theory: Univalent Foundations of Mathematics. The Univalent Foundations Program (2013)
34. Jäger, G.: A well-ordering proof for Feferman's theory T_0. Archiv f. Math. Logik **23**, 65–77 (1983)
35. Jäger, G.: Theories for admissible sets: a unifying approach to proof theory. Bibliopolis, Naples (1986)
36. Jäger, G., Pohlers, W.: Eine beweistheoretische Untersuchung von $\Delta^1_2 - \mathrm{CA} + \mathrm{BI}$ und verwandter Systeme. Sitzungsberichte der Bayerischen Akademie der Wissenschaften, Mathematisch-Naturwissenschaftliche Klasse (1982)
37. Jäger, G., Strahm, T.: Upper bounds for metapredicative Mahlo in explicit mathematics and admissible set theory. J. Symb. Log. **66**, 935–958 (2001)
38. Jäger, G., Studer, T.: Extending the system T_0 of explicit mathematics: the limit and Mahlo axioms. Ann. Pure Appl. Log. **114**, 79–101 (2002)
39. Jäger, G., Kahle, R., Studer, T.: Universes in explicit mathematics. Ann. Pure Appl. Log. **109**, 141–162 (2001)
40. Jervell, H.R.: Constructive universes I. In: Müller G.H., Scott D.S. (eds.) Higher Set Theory. Lecture notes in Mathematics No. 669, pp. 73–98. Springer, Berlin (1978)

41. Kalpulkin, C., Lummsdaine, P.L., Voevodsky, V.: The simplicial model of univalent foundations. Preprint (2012). arXiv:1211.2851)
42. Martin-Löf, P.: Intuitionistic Type Theory. Bibliopolis, Naples (1984)
43. Möllerfeld, M.: Generalized Inductive Definitions. The μ-calculus and Π^1_2-comprehension. Ph.D. dissertation, Universität Münster (2002)
44. Myhill, J.: Constructive set theory. J. Symb. Log. **40**, 347–382 (1975)
45. Nordström, B., Petersson, K., Smith, J.M.: Programming in Martin-Löf's Type Theory. Clarendon Press, Oxford (1990)
46. Palmgren, E.: Type-theoretic interpretations of iterated, strictly positive inductive definitions. Arch. Math. Logic **32**, 75–99 (1993)
47. Rathjen, M.: Proof-Theoretic Analysis of KPM. Arch. Math. Log. **30**, 377–403 (1991)
48. Rathjen, M.: The strength of some Martin-Löf type theories, p. 39. Department of Mathematics, Ohio State University, Preprint (1993)
49. Rathjen, M.: The strength of Martin-Löf type theory with a superuniverse. Part I. Arch. Math. Log. **39**, 1–39 (2000)
50. Rathjen, M.: The strength of Martin-Löf type theory with a superuniverse. Part II. Arch. Math. Log. **40**, 207–233 (2001)
51. Rathjen, M., Palmgren, E.: Inaccessibility in constructive set theory and type theory. Ann. Pure Appl. Log. **94**, 181–200 (1998)
52. Rathjen, M.: The anti-foundation axiom in constructive set theories. In: Mints, G., Muskens, R. (eds.) Games, Logic, and Constructive Sets, pp. 87–108. CSLI Publications, Stanford (2003)
53. Rathjen, M.: The constructive Hilbert programme and the limits of Martin-Löf type theory. Synthese **147**, 81–120 (2005)
54. Rathjen, M.: The formulae-as-classes interpretation of constructive set theory. In: Schwichtenberg, H., Spies, K. (eds.) Proof Technology and Computation, pp. 279–322. IOS Press, Amsterdam (2006)
55. Rathjen, M., Tupailo, S.: Characterizing the interpretation of set theory in Martin-Löf type theory. Ann. Pure Appl. Log. **141**, 442–471 (2006)
56. Rathjen, M.: Monotone inductive definitions in explicit mathematics. J. Symb. Log. **61**, 125–146 (1996)
57. Rathjen, M.: Explicit Mathematics with the monotone fixed point principle. J. Symb. Log. **63**, 509–542 (1998)
58. Rathjen, M.: Explicit Mathematics with the monotone fixed point principle. II: Models. J. Symb. Log. **64**, 517–550 (1999)
59. Rathjen, M.: Explicit Mathematics with monotone inductive definitions: a survey. In: Sieg, W. et al. (eds.) Reflections on the Foundations of Mathematics: Essays in Honor of Solomon Feferman. Lecture Notes in Logic 15, pp. 329–346 (2002)
60. Rathjen, M.: The constructive Hilbert programme and the limits of Martin–Löf type theory. In: Lindström, S., Palmgren, E., Segerberg, K., Stoltenberg-Hansen, V. (eds.) Logicism, Intuitionism, and Formalism. What has Become of Them? Synthese Library, vol. 341, pp. 397–433. Springer (2009)
61. Rathjen, M.: Constructive Zermelo-Fraenkel set theory and the limited principle of omniscience. Ann. Pure Appl. Log. **165**, 563–572 (2014)
62. Rathjen, M.: Turing's 'Oracle' in proof theory. In: Cooper, S.B., van Leeuwen, J. (eds.) Alan Turing: His Work and Impact, pp. 198–202. Elsevier, Amsterdam (2013)
63. Rathjen, M.: Well-ordering proofs in intuitionistic theories and conservativity. In preparation
64. Rathjen, M., Tupailo, S.: On the strength of UMID in intuitionistic explicit mathematics. (2004) Unpublished
65. Sato, K.: A new model construction by making a detour via intuitionistic theories II: interpretability lower bound of Feferman's explicit mathematics T_0. Ann. Pure Appl. Log. **166**, 800–835 (2015)
66. Setzer, A.: A well-ordering proof for the proof theoretical strength of Martin-Löf type theory. Ann. Pure Appl. Log. **92**, 113–159 (1998)

67. Setzer, A.: Extending Martin-Löf type theory by one Mahlo-universe. Arch. Math. Log. **39**, 155–181 (2000)
68. Simpson, S.G.: Subsystems of Second Order Arithmetic, 2nd edn. Cambridge University Press, New York (2009)
69. Takahashi, S.: Monotone inductive definitions in a constructive theory of functions and classes. Ann. Pure Appl. Logic **42**, 255–279 (1989)
70. Troelstra, A., van Dalen, D.: Constructivism in Mathematics, vol. II. North Holland, Amsterdam (1988)
71. Tupailo, S.: On the intuitionistic strength of monotone inductive definitions. J. Symb. Log. **69**, 790–798 (2004)
72. Weyl, H.: Das Kontinuum, kritische Untersuchungen über die Grundlagen der Analysis. Verlag von Veit und Comp, Leipzig (1918)

Author Biography

Michael Rathjen is currently Professor of Pure Mathematics at the University of Leeds. He received his Ph.D. in 1988 and Habilitation in 1992 from the University of Münster. After teaching at the Ohio State University in Columbus (1991–1993), he became a Heisenberg Fellow taking him to mathematical logic research centers at Ohio State, UCLA, Stanford, and Uppsala. In 1996 he joined the Mathematics Department of the University of Leeds. From 2002 till 2006 he was a Professor of Mathematics at the Ohio State University. His research interests include proof theory (especially ordinal analysis of impredicative theories), explicit mathematics, type theory, higher type computability, constructive and intuitionistic set theories, alternative set theories, and the philosophy of mathematics and logic.

Part IV
From Logic to Philosophy

Predicativity and Feferman

Laura Crosilla

Dedicated to Professor Solomon Feferman

Abstract Predicativity is a notable example of fruitful interaction between philosophy and mathematical logic. It originated at the beginning of the 20th century from methodological and philosophical reflections on a changing concept of set. A clarification of this notion has prompted the development of fundamental new technical instruments, from Russell's type theory to an important chapter in proof theory, which saw the decisive involvement of Kreisel, Feferman and Schütte. The technical outcomes of predicativity have since taken a life of their own, but have also produced a deeper understanding of the notion of predicativity, therefore witnessing the "light logic throws on problems in the foundations of mathematics." [30, p. vii] Predicativity has been at the center of a considerable part of Feferman's work: over the years he has explored alternative ways of explicating and analyzing this notion and has shown that predicative mathematics extends much further than expected within ordinary mathematics. The aim of this note is to outline the principal features of predicativity, from its original motivations at the start of the past century to its logical analysis in the 1950–1960s. The hope is to convey why predicativity is a fascinating subject, which has attracted Feferman's attention over the years.

I thank Andrea Cantini for helpful comments on a draft of this article. I also profited from remarks by Gerhard Jäger and Wilfried Sieg, as well as two anonymous referees. I gratefully acknowledge funding from the School of Philosophy, Religion and History of Science of the University of Leeds and from the European Research Council under the European Union's Seventh Framework Programme (FP/2007–2013)/ERC Grant Agreement n. 312938.

L. Crosilla (✉)
School of Philosophy, Religion and History of Science, University of Leeds,
LS2 9JT, Leeds, UK
e-mail: Laura.Crosilla@gmail.com

G. Jäger and W. Sieg (eds.), *Feferman on Foundations*, Outstanding Contributions to Logic 13, DOI 10.1007/978-3-319-63334-3_15

Keywords Impredicative definitions · Predicativity given the natural numbers Vicious circle principle · Invariance · Γ_0

2010 Mathematics Subject Classification 00A30 · 03A05.

1 Introduction

The distinction between predicative and impredicative definitions has its origins in the writings of Poincaré and Russell and was instigated by the discovery of the set-theoretic paradoxes.[1] According to one characterization of (im)predicativity, a definition is impredicative if it defines an entity by reference to a totality to which the entity itself belongs, and it is predicative otherwise. Adherence to predicativity was proposed as a way of avoiding vicious circularity in definitions and resulted in the creation, by Russell, of ramified type theory; it also motivated a first development of a predicative form of analysis by Weyl.[2] A new phase for predicativity began in the 1950s, with a logical analysis of predicativity which employed state-of-the-art logical machinery for its analysis. That work culminated with an important chapter in proof theory whose principal outcome was the determination of the limit of predicativity by means of ordinal analysis.[3] In addition, Feferman's work and the so-called "Reverse Mathematics programme" have since clarified that large portions of everyday mathematics can be already carried out in predicative settings.[4]

Feferman's engagement with predicativity extends well beyond his celebrated contributions to the determination of the proof-theoretic limit of predicativity, as over the years he has explored alternative ways of explicating and analyzing this notion, as well as assessing the reach of predicative mathematics. This had two principal purposes: to offer further support for the original logical analysis of predicativity and to highlight the significance of predicative mathematics, both within mathematical logic and ordinary mathematics, with a particular attention to scientifically applicable mathematics. Feferman has also offered unrivalled expositions of predicativity.[5]

One of the difficulties in writing on predicativity is what might be called a lack of consensus on this notion. The early writings on predicativity by Poincaré, Russell and Weyl at the turn of the 20th century are rich of stimulating ideas, and deserve further scrutiny, however, to a contemporary technically-trained eye they often appear as insufficiently clear, opening up the way for a number of possible interpretations of

---

[1] See, for example, [69–73, 80–82].

[2] See [79, 82, 97, 99].

[3] See [16, 54, 85, 86].

[4] See e.g. [26, 33, 36, 88, 89].

[5] See, for example, [34], and [16, 24–31, 33].

predicativity.[6] The subsequent logical analysis of predicativity of the 1950-60's shed light on important aspects of predicativity, employing an array of precise logical instruments that were unavailable at the beginning of the past century. Notwithstanding that fundamental work, predicativity still raises complex questions from historical, philosophical and technical perspectives.[7] In addition, further difficulties are induced by the emergence over the years of a *plurality* of forms *of predicativity*, some within a classical and some within an intuitionistic context.[8]

The aim of this note is to offer an outlook of predicativity, sketching the most important (and hopefully less controversial) features of this notion, the original motivations, as well as the logical work that was inspired by the desire to clarify it. In particular, I shall focus on the *classical* form of predicativity that goes under the name of *predicativity given the natural numbers*; this has been extensively studied mathematically and is at the heart of Feferman's work.[9] I hope to convey why predicativity is a fascinating subject, which has attracted Feferman's attention over the years, and why it is an area of research offering the potential to substantially enrich today's philosophy of mathematics. Like Feferman, I also think that an account of predicativity ought to begin from the early discussions on predicativity, which clarify how we arrived at the notion of predicativity given the natural numbers and its logical analysis.

2 Predicativity: The Origins

The early debates on predicativity were prompted by the discovery of the set theoretic paradoxes, which gained particular attention after Russell's famous letter to Frege in 1902. The general context of the early discussions on predicativity are renowned reflections by prominent mathematicians of the time on new concepts and methods

---

[6]Feferman [34] writes: "Though early discussions are often muddy on the concepts and their employment, in a number of important respects they set the stage for the further developments, and so I shall give them special attention."

[7]In his introduction to a chapter on the ordinal analysis of predicativity, Pohlers [68, p. 134] writes: "The notion of predicativity is still controversial. Therefore we define and discuss here predicativity in a pure mathematical – and perhaps oversimplified – setting." See [22, 50, 55, 56, 96] for discussions pertaining to the logical analysis of predicativity. See also the discussion on "metapredicativity" in [52]. As to the philosophical and historical aspects of predicativity, see, e.g., [32, 37, 48, 59, 65–67].

[8]Predicativity-related themes have appeared in different forms over the years, both in classical and constructive settings. In fact, predicativity is gaining renewed prominence today especially in the constructive context. I shall postpone to another occasion a discussion of other forms of predicativity, as the constructive predicativity which characterizes Martin-Löf type theory see, e.g., [60–62] and forms of "strict predicativity" [64, 65, 67]. See also [11, 12].

[9]In the following, I shall also write "predicativity" to denote predicativity given the natural numbers. See Sect. 3.4 for some remarks on the notion of predicativity given the natural numbers.

of proof, which had emerged in mathematics from the nineteenth century.[10] These debates are well-known as they gave rise to influential foundational programs as logicism, formalism and intuitionism. In the case of Poincaré and Weyl's writings on predicativity, one finds severe criticism of the new methodology, and, especially, of the concept of arbitrary set which emerged from Cantorian set theory.[11] Adherence to predicativity offered a way of securing a safe concept of set, one that is not prey to the set-theoretic paradoxes and also avoids the arbitrariness implicit in the new concept of set.

The term predicativity itself emerged in an animated discussion between Poincaré and Russell which spanned from 1905 to 1912. Notwithstanding the remarkably different views of Poincaré and Russell, for instance, on the role of formalization within mathematics, they both converged on holding impredicative definitions the cause of the onset of the paradoxes, and attempted to clarify a notion of predicativity, adherence to which would avoid inconsistencies. Through Russell and Poincaré's confrontation a number of ways of capturing impredicativity and explaining its perceived problematic character emerged.

A first observation was that paradoxes as, for example, that of *Russell's class* of all those classes that are not members of themselves, typically display a form of *vicious circularity*.[12] In modern terminology we may define

$$R = \{x \mid x \notin x\}$$

by application of the *Naive Comprehension schema*: given any formula φ in the language of set theory, we form the class of all the x's that satisfy φ, that is, $\{x \mid \varphi(x)\}$. Then we have that $R \in R$ if and only if $R \notin R$. A circularity arises here as R's definition *refers* to the class of all classes, to which R itself is supposed to belong.

Observations along similar lines gave rise to a characterization of impredicativity as follows: *a definition is impredicative if it defines an entity by reference to a totality to which the entity itself belongs*.[13] In particular, a definition is impredicative if it defines an entity by quantifying over a totality which includes the entity to be defined. A definition is then *predicative* if it is not impredicative. Given this notion of impredicative definition, one may call an entity (e.g. a class) impredicative if it

---

[10]See e.g. [92, 95].

[11]See, in particular, [73, 97]. See also [2, 42, 58] for a discussion of arbitrary sets and [66] for an analysis of Weyl's conception of predicative set and its reception by Weyl's contemporaries.

[12]See [69–71, 81, 82]. Note that the term "class" is used here as in Russell and Poincaré's texts, that is, to refer to a generic collection. Hence it should be carefully distinguished from the notion of proper class that is found in contemporary set theory. In the original literature one frequently finds also the word "totality". In this section I shall try to avoid the use of the term "set", since the latter has in the meantime acquired additional connotations (as set in e.g. ZFC) that should not be presupposed in this discussion.

[13]See [69–71, 80, 81]. See e.g. [47, p. 455] for discussion. Note also that today the distinction between predicative and impredicative definitions is typically framed as relating to sets. However, Russell and Poincaré's discussions are concerned with definitions of different kinds of entities, including propositions, properties, etc.

can *only* be defined by an impredicative definition.[14] Russell famously introduced his *"Vicious–Circle Principle" (VCP)* to ban impredicative definitions. This had a number of formulations, like, for example: *"no totality can contain members defined in terms of itself"* [82, p. 237]. Another is to be found in [83, p. 198]:

> [...] whatever in any way concerns *all* or *any* or *some* of a class must not be itself one of the members of a class.

The latter formulation forbids definitions which quantify over a totality to which the definiendum belongs, and is at the heart of Russell's implementation of the VCP in his type theory (see Sect. 2.1).

Two examples may better clarify the notion of impredicative definition and its perceived difficulties; the first one is given by the logicist definition of natural number, and the second by the Liar paradox. The first example is significant not only because it is of central importance for the logicist project pursued by Russell, but as it clarifies that the discussion on impredicativity, which originated from an analysis of paradoxes of various kind, extended quickly beyond the case of the paradoxes. Let

$$N(n) := \forall F[F(0) \wedge \forall x(F(x) \rightarrow F(Suc(x))) \rightarrow F(n)].$$

According to this definition, the concept of natural number is defined by reference to all properties F of the natural numbers. A circularity arises here as the property N itself is within the range of the first quantifier. As a consequence, N is defined by reference to itself. The difficulty with this definition is typically explained as follows[15]: suppose we wish to determine whether the predicate N holds for a specific natural number, say 3. It would seem that we need to check for *every* property of the natural numbers, F, whether F holds of 3, that is, whether:
$\forall F[F(0) \wedge \forall x(F(x) \rightarrow F(Suc(x))) \rightarrow F(3)]$.

However, the property "to be a natural number", which is expressed by the predicate N, is one of the properties of the natural numbers. That is, to find out whether $N(3)$ holds, we need to be able to clarify whether the following holds:

$$N(0) \wedge \forall x(N(x) \rightarrow N(Suc(x))) \rightarrow N(3).$$

Therefore it would seem that we need to determine whether $N(3)$ holds prior to determining whether $N(3)$ holds.[16]

---

[14] The issue of how we establish whether an entity is impredicative (and in which context) is more complex than this coarse characterization of impredicativity may suggest. This complexity was further addressed by the development of Russell's type theory, Weyl's [97] and the logical analysis of predicativity to be discussed below.

[15] Here I shall follow [7, p 48].

[16] Carnap [7, p. 48] concludes that this definition of natural number is "circular and useless". It is worth recalling that Carnap in [7] also hints at a form of platonism, attributed to Ramsey (but not endorsed by Carnap), which finds no fault with impredicative definitions. See also [47, 73] for further discussion.

Let us consider another example of impredicative definition that is discussed by Russell [82, p. 224]: the Liar paradox. In this case it is instructive to see how Russell himself analyzed the paradox. Russell first of all observes that the sentence "I'm lying" is the same as: "There is a proposition which I am affirming and which is false." He also notices that this, in turn, can be rephrased as: "It is not true for all propositions p that if I affirm p, p is true." He then concludes that "[t]he paradox results from regarding this statement as affirming a proposition, which must therefore come within the scope of the statement." Russell's conclusion is that the notion of *all* propositions is *illegitimate*, "for otherwise, there must be propositions (such as the above) which are about all propositions, and yet can not, without contradiction, be included among the propositions they are about." In fact, Russell further claims that "[w]hatever we suppose to be the totality of propositions, statements about this totality generate new propositions which, on pain of contradiction, must lie outside the totality." The worry here is that an impredicative definition of an entity (e.g. a proposition) would seem to *generate* a new element of the very class that was employed to define it. As a consequence, "there must be no totality of propositions", and statements such as "all propositions" must be *meaningless*.

A second characterization of predicativity emerged from Poincaré's renewed analysis in [72, 73]. Central to this characterization is the thought that an impredicative definition seems to *generate* new elements of a class which is used (e.g. as a domain of quantification) within that definition. Here Poincaré's criticism of impredicativity is deeply interrelated with a reflection on infinity and the role of definitions in mathematics. For the French mathematician a *definition* is a *classification*: it separates the objects which satisfy, from those which do not satisfy that definition, and it arranges them in two distinct classes. Poincaré also highlights a sort of incompleteness of infinite classes: they are open-ended and unfinished, so that definitions which refer to their totality might become problematic. For example, in the case of the definition of Russell's class, R, above, it would seem that we need first of all to *fix* the class of all classes, say C, *prior* to defining R by reference to C. But then the definition of R would seem to extend C by a new class, R itself. And this process may be repeated at will.[17]

Poincaré's discussion hints towards a distinction between predicative and impredicative classes that appeals to a form of "invariance" of predicative definitions: *a predicative classification is one that can not be "disordered" by the introduction of new elements.* This gives rise to a new characterization of predicativity which does not directly appeal to circularity, and can be so expressed in modern terminology: *a definition is predicative if the class it defines is invariant under extension.*[18]

[17] Poincaré's texts make use of other examples, more directly drawn from the mathematical practice. See also [13, 14] for a similar reading of Russell's paradox. Cantini [4] proposes a detailed analysis of Poincaré's ideas.

[18] See [55] for discussion of this characterization from a modern logical perspective. See also [18, 39].

Poincaré [72, p. 463] writes:

> Hence a distinction between two species of classifications, which are applicable to the elements of infinite collections; the *predicative* classifications, which can not be disordered by the introduction of new elements; the *non predicative* classifications which are forced to remain without end by the introduction of new elements.[19]

For Poincaré impredicative definitions are problematic as they treat as completed (French "arrêté") infinite classes which are instead "in fieri", open-ended or incomplete by their very nature. Predicative definitions, instead, guarantee that the classes so defined are stable or invariant. Poincaré does not spell out this notion of invariance in any detail, being very critical of formal endeavors; however, he indicates that the relations between the elements of the class and the class itself should not admit of change as we progress introducing new elements through our definitions. He also points towards a kind of genetic construction of predicative classes, which are built up from some initial elements step by step: we construct new elements of a predicative class by defining them in terms of the initial elements, we then define further new elements from the latter, and so on. In the case of infinite classes, this process is without end. A related but more precise account of a predicative conception of set is to be found in [97], as further discussed in Sect. 2.2.

2.1 Russell's Way Out

The analysis of the paradoxes and their relation with impredicativity turned out to be extremely fruitful for the development of mathematical logic, starting from Russell's own implementation of the vicious circle principle through his type theory.[20] Russell's way out from the paradoxes is well-known, as it introduced a regimentation of classes through a hierarchy of types and orders.[21] For Russell the paradoxes were due to the assumption that any propositional function gives rise to a class: the class of all the objects that satisfy it.[22] As discussed above, of particular concern were classes defined impredicatively. To avoid impredicativity, in setting up ramified type theory Russell [82] made two distinguishable moves as follows. The first move amounts to associating a *range of significance* to each propositional function, that is, a collection of all arguments to which the propositional function can be meaningfully applied. In Russell's terms: "within this range of arguments, the function is true or false; outside this range, it is nonsense." [82, p. 247] The ranges of significance then form types,

[19] My translation; italics by Poincaré. The word "disordered" translates the French "bouleverseé".

[20] See [5] for a rich discussion of the impact of the paradoxes on mathematical logic.

[21] Russell's ideas on type theory appeared first in an appendix to [79], and were further developed (with ramification) in [82] and then in [99].

[22] In the present context we may follow [34], and identify the notion of propositional function with that of open formula, i.e. a formula with a free variable, say $\varphi(x)$. Note, however, that the interpretation of the notions of proposition and propositional function in Russell is complex. See e.g. [57].

and these are arranged in **levels**: first we have a type of individuals, and then types which are ranges of significance of propositional functions defined on the individuals, and so on. The crucial point is that as a consequence of this regimentation of classes, expressions such as $x \in x$ and $x \notin x$ are simply ill–formed, since in $z \in w$, w must be of the next-higher level than z. Accordingly, Russell's paradox (and other set-theoretic paradoxes) do not carry through.

It was subsequently realized by Chwistek and Ramsey that if one implements only this restriction, then one obtains a formalism that is interesting in its own right.[23] Today this goes under the name of simple type theory and its formulation was subsequently simplified by Church [8]. Simple type theory seems sufficient to block all set theoretic paradoxes; however, it does not eliminate all impredicativity. The second move, ramification, has the effect of eliminating all impredicativity.[24] As discussed above, for Russell one of the lessons of the paradoxes was that impredicative totalities, as, for example, the totality of all propositions, are illegitimate; hence quantification over them makes no sense. He therefore introduced, alongside a notion of *level* for ranges of significance of propositional functions, a notion of **order** for propositional functions, and required that *a propositional function can only quantify over propositional functions of lower order than its own*. Thus in ramified type theory, one has *first order* propositional functions, *second order* ones, etc.; in addition, the second order propositional functions can quantify on the first order ones, but not on propositional functions of order higher than one, and so on.

In this way one apparently blocks not only the set theoretic paradoxes, but semantic paradoxes as the Liar, too. This is analyzed as follows by Russell [82, p. 238]:

> if Epimenides asserts "all first-order propositions affirmed by me are false", then he asserts a second order proposition; he may assert this truly, without asserting truly any first order proposition, and thus no contradiction arises.

While ramified type theory fully complies with predicativity, it also turns out to make the development of mathematics awkward. This may be seen by considering again the definition of natural number discussed above, which requires a universal quantifier over properties of the natural numbers. When appropriately re-formulated in a ramified context, this definition gives rise to only partial renderings of the notion of natural number, one for each order of propositional functions, and therefore it does not offer a general definition of the concept of natural number. As a consequence, many proofs by induction do not carry through in their usual form, as they would require the full generality of universally quantified statements; for example, in ramified type theory we can not prove in full generality that if m and n are finite numbers, so is $m + n$.[25] These difficulties prompted Russell [82] to introduce the axiom of

---

[23] See [9, 74].

[24] See [49] for a discussion of the reasons that might support Russell's (and Whitehead's) choice of a ramified type theory over a simple type theory.

[25] See [82]. See also [63]. See [10, 31] for introductory expositions of the ideas underlying ramified type theory and the difficulties it encounters.

reducibility, which, however, has the effect of reinstating impredicativity. Reducibility is so presented in [82, p. 242-3]: "every propositional function is equivalent, for all its values, to some predicative function", where a function φ of one argument x is predicative if it is "of the order next above x".[26] This axiom was strenuously criticized for being introduced for purely pragmatic reasons and for being ad hoc.[27] For example, Russell [98, p. 50] wrote:

> Russell, in order to extricate himself from the affair, causes reason to commit harikari, by postulating the above assertion [the axiom of reducibility] in spite of its lack of support by any evidence.

.

2.2 *Das Kontinuum*

With Weyl's "Das Kontinuum" [97] we have another approach to predicativity which also played a significant role for the subsequent logical analysis of predicativity, and especially Feferman's work. Weyl's aim was to develop a predicative form of analysis, founded on a concept of set which is immune from paradoxes and vicious circularity. Weyl's concern was that impredicativity affected not only set theory in general, but it was to be found already at the heart of analysis, as the Least Upper Bound principle (LUB) requires impredicative reasoning.[28] One of Weyl's fundamental achievements was to show how to circumvent this difficulty without resorting to ramification or reducibility.[29] The result is a predicative (in fact, arithmetical) treatment of large portions of 19th century analysis.

Weyl [97] expounds in detail a concept of *predicative set*: a set is the *extensional counterpart of a property* and may be seen as if it were constructed step by step from some primitive domain of objects by application of elementary operations over it. The "production" of sets from an initial domain is expressed first in full generality, and then specialized to the particular case of the natural numbers as starting

---

[26]Russell [82, p. 243] also writes: "Thus a predicative function of an individual is a first-order function; and for higher types of arguments, predicative functions take the place that first-order functions take in respect of individuals. We assume, then, that every function is equivalent, for all its values, to some predicative function of the same argument.".

[27]Wilfried Sieg has informed me about perceptive discussions by Hilbert and Bernays on predicativity and Russell's logicism, including the axiom of reducibility. See e.g. Hilbert's lecture notes from 1917/18 entitled "Prinzipien der Mathematik" and those from 1920 entitled "Probleme der mathematischen Logik" published in [15]. See also [87] for discussion. Sieg [87] also draws important correlations between [97] and Hilbert and Bernays' work around 1920. As suggested by Sieg, the relations between Hilbert and Bernays' writings and Weyl's [97] deserve more thorough investigations.

[28]The (LUB) states that every bounded, non–empty subset M of the real numbers has a least upper bound. See [16, 26] for discussion.

[29]Weyl, in particular, made use of sequential rather than Dedekind completeness, the first being amenable to predicative treatment. See also [26].

domain, which is of relevance for the development of analysis. One begins with an initial domain (or basic category) of objects, and "certain individual, immediately exhibited 'primitive' properties" which apply to the objects of this domain [97, p. 28].[30] One then considers derived properties which arise from the primitive ones (as clarified below) and takes sets to be the extensional counterparts of primitive and derived properties. Weyl [97, p. 20] writes: "to every primitive or derived property P there corresponds a set (P)", the set of all the objects which have the property P. Crucially sets are identified extensionally, that is, "the same set corresponds to two such properties P ad P' if and only if every object (of our category) which has the property P also has the property P'." [97, p. 20] The step from primitive properties to derived ones is discussed in the first section of "Das Kontinuum", where Weyl describes the formation of judgments.[31] The starting point is once more a given basic domain of objects and some primitive properties which apply to the objects of that domain. One then forms simple (i.e. atomic) judgments affirming that the primitive properties hold of the objects of the basic domain. The next step is given by taking combinations of these judgments by means of the ordinary logical operations, but with the crucial constraint that *quantifiers are only allowed to range over the basic domain.*[32] In this way one essentially obtains first-order definable properties of the objects of the initial domain; sets then arise as extensions of such properties (modulo extensionality). Weyl calls *"mathematical process"* [97, p. 22] the formation described above of a new "system" of sets from a basic initial domain and certain primitive properties of its objects.

A particularly important application of the mathematical process arises when the initial domain is the natural numbers. Here, from the contemporary logician's perspective, Weyl's concept of set gives rise to subsets of the natural numbers obtainable by application of the comprehension schema restricted to *arithmetical formulas*, that is, to those formulas that do not quantify over sets (but may quantify over natural numbers). This restriction to number quantifiers in the comprehension principle aims at preventing vicious–circular definitions of subsets of the natural numbers.[33]

An aspect of particular foundational interest is that Weyl, as Poincaré before him, takes the natural numbers with *full mathematical induction* as starting point,

[30] In addition to properties, Weyl [97] also considers relations, here omitted for simplicity.

[31] The notion of "judgment" is so clarified by Weyl [97, p. 5]: a "*judgment* affirms a *state of affairs*".

[32] Weyl also considers a principle of substitution [97, p. 10]. In addition, in the paradigmatic case of the natural numbers as basic domain, one also applies a principle of iteration, as further discussed below.

[33] As remarked by Feferman [26] see also [31], it is not completely clear how strong is the system Weyl sketches in [97]. Feferman has, however, verified that system W of [26], which is inspired by [97], suffices to carry out all of Weyl's constructions in "Das Kontinuum". System W is a conservative extension of Peano Arithmetic, PA [38]. As clarified in Sect. 3.3, Feferman has also shown that W allows for the development of a more extensive portion of contemporary analysis, compared with [97].

as intuitively given.[34] The comparison with Russell is instructive, as Russell aimed at a *definition* of the natural number concept, to witness its logical nature. Instead both Poincaré and Weyl criticized any attempt at founding the concept of natural number (and in particular the principle of mathematical induction) on logic or on the concept of set, given the fundamental role the natural numbers play within all of mathematics. Weyl [97, p. 48] wrote: "*the idea of iteration, i.e., of the sequence of the natural numbers, is an ultimate foundation of mathematical thought, which can not be further reduced*". The natural numbers, for Weyl, are *"individuals"*, in the sense that they can be characterized uniquely by means of their properties: starting from an initial element, the iteration of the successor operation allows us to *characterize uniquely each natural number in elementary terms, and by exclusive appeal to its predecessors.* Weyl [97, p. 15] also writes that "it is impossible for a number to be given otherwise than through its position in the number sequence, i.e. by indicating its characteristic property." This also justifies Weyl's adoption of bivalence for statements on the natural numbers and their assumption as paradigmatic initial domain for the mathematical process. The latter now gives rise to a system of sets as extensions of arithmetical properties of the natural numbers.

It is important to clarify why Weyl takes sets as extensions of first-order definable properties. Weyl [97, p. 20] writes:

> Finite sets can be described in two ways: either in *individual* terms, by exhibiting each of their elements, or in *general* terms, on the basis of a rule, i.e., by indicating properties which apply to the elements of the set and to no other objects. In the case of infinite sets, the first way is impossible (and this is the very essence of the infinite).

He also writes [97, p. 23]:

> No one can describe an infinite set other than by indicating poperties which are characteristic of the elements of the set. And no one can establish a correspondence among infinitely many things without indicating a *rule*, i.e. a relation, which connects the corresponding objects with one another.

Weyl's "arithmetical" sets are the extensional counterparts of arithmetical rules or laws, and may be seen as if they were obtained through application of a fixed set of elementary operations starting from the natural numbers (with iteration). This is contrasted by Weyl with the concept of *arbitrary* set which had recently emerged within set theory, and which is characterized by the absence of any requirement of law or rule of formation. In his criticism of the concept of arbitrary set Weyl is once more in agreement with Poincaré, who drew a direct connection between the debate on impredicativity and the lack of explicit definability of impredicatively defined sets [73]. An arbitrary set for Weyl is "a 'gathering' brought together by infinitely many individual arbitrary acts of selection, assembled and then surveyed as a whole

---

[34] Poincaré see, e.g., [71] states that mathematical induction is synthetic a priori. Note also that Weyl expresses mathematical induction by appeal to a principle of iteration. See [30, p. 264-5] for discussion. Poincaré and Weyl fully realized the significance of the assumption of unrestricted mathematical induction. This is further clarified by a comparison with approaches to predicativity which instead introduce restrictions on induction [64, 65].

by consciousness" and, as such, it is "nonsensical" [97, p. 23]. Weyl instead shows how predicative sets may be "produced" step by step from the safety of the natural numbers by application of a rule or a uniform condition.

Weyl's remarkable achievement in the second part of [97] was to show that his arithmetical concept of set suffices to develop a fundamental portion of 19th century analysis. "Das Kontinuum" is also particularly interesting from a philosophical perspective, as it clearly puts forth a predicativist position: Weyl is adamant that what can not be predicatively accounted for, needs to be relinquished.

3 The Logical Analysis of Predicativity Given the Natural Numbers

Interest in predicativity declined after [97] for a number of reasons, like, for example, the realization that simple type theory was apparently sufficient to block the set-theoretic paradoxes.[35] In addition, the rapid accreditation of impredicative set theory as standard foundation, especially in the form of the Zermelo–Fraenkel system with choice, ZFC, played a crucial role in the downfall of predicativity.[36]

The technical results obtained in [82, 97, 99], however, paved the way for subsequent work in mathematical logic which eventually gave rise to a new phase for predicativity starting in the 1950s: a *logical analysis of predicativity*. A crucial point to note is that the motivation prompting the new discussions on predicativity differed profoundly from the ones which had given rise to the first debates on predicativity outlined above. In this respect, already Gödel proposed a shift of attitude in his influential appraisal of Russell's contribution to mathematical logic in [47]. There Gödel clearly expressed the view that predicativity is a fruitful concept which can give rise to mathematical progress, but that it should be pursued "independently of the question whether impredicative definitions are admissible."[37] Gödel's observations mark the beginning of a study of predicativity which, although of relevance for the philosophical debates on the foundations of mathematics, is carried out independently of predicativism; its principal aims are no more to secure the ultimate justification of (a portion of) mathematics, but to draw a clearer demarcation of the *boundary* between predicative and impredicative mathematics. We may distinguish two main objectives: (1) the determination of a theoretical *limit* of predicativity; and (2) the clarification of the *extent* of predicative mathematics.

It is important to recall the notion of predicativity that has been so investigated. This takes inspiration from Poincaré and, especially, Weyl's writings, and is characterized by the assumption, at the start, of the natural numbers with full mathematical

[35]See [9, 74].

[36]See also [34] for additional thoughts on what "pushed predicativity to the sidelines.".

[37]Gödel [47] also mentions a prominent example of the fruitfulness of predicativity: the constructible hierarchy [45, 46], inspired by Russell's idea of ramification.

induction.[38] For this reason it has been termed *"predicativity given the natural numbers"*, and, as in Weyl, it uses classical logic.[39] A difference with Weyl's approach is that the new logical analysis of predicativity also aims at exploring *how far* can we extend beyond the natural numbers in a predicatively justified way; it therefore focuses on a notion of predicativity given the natural numbers that stretches beyond Weyl's arithmetical predicativity. In fact, Russell's original idea of ramification and a distinctive use of ordinals (and ordinal notations) played a crucial role in setting out this form of predicativity.

3.1 *The Limit of Predicativity*

The literature from the 1950s and 1960s witnesses the complexity of the task of clarifying the limit of predicativity, which saw the involvement of a number of prominent logicians, as Feferman, Gandy, Kleene, Kreisel, Lorenzen, Myhill, Schütte, Spector and Wang. The first attempts at a logical analysis of predicativity focused on issues of definability of sets of natural numbers and highlighted a connection between predicativity and the recently developed concept of the hyperarithmetical hierarchy.[40] This consists of a hierarchy of sets of natural numbers which can be equivalently characterized in a number of ways. The simplest characterization is in terms of definability, and sees the hyperarithmetical sets as those sets of natural numbers that can be defined equivalently by a Σ^1_1 and by a Π^1_1 formulas, also called the Δ^1_1 sets.[41] Given this characterization of the hyperarithmetical sets, the relation between the hyperarithmetical hierarchy and predicativity might at first seem problematic, as a hyperarithmetical set is defined by formulas with unrestricted set quantifiers. However, a more constructive rendering of the hyperarithmetical sets was given by Kleene in terms of iteration of the so-called Turing jump through the recursive ordinals.[42]

---

[38]See also Sect. 3.4 for more on the notion of predicativity given the natural numbers.

[39]The use of classical logic marks a crucial difference with the form of predicativity that is to be found in e.g. Martin-Löf type theory [60].

[40]The hyperarithmetical hierarchy has a central place in the development of mathematical logic because of its prominence within a number of fundamental areas in mathematical logic: definability theory, recursion theory and admissible set theory. This witnesses the centrality within logic of themes that pertain to the predicativity debate, and further explains the interest of this notion form a logical point of view.

[41]In the language of second order arithmetic, a Σ^1_1 formula is one of the form: $\exists X\, \varphi(X)$, with φ an arithmetical formula, that is, a formula that does not quantify over sets (but may quantify over natural numbers). Note that here the upper case letter X denotes a second order variable, standing for a set of natural numbers. A Π^1_1 formula is one of the form $\forall X\, \psi(X)$, with ψ an arithmetical formula.

[42]See [53], see also [84]. See [16, 55] for further clarification of why this may be seen as offering a predicative justification for this kind of second order quantification. Kreisel [55] offers additional considerations that directly relate to Poincaré's notion of invariance discussed above.

Another way of bringing the relation with predicativity to light is by drawing a correlation between the hyperarithmetical hierarchy and (a fragment of) the ramified analytic hierarchy. The latter essentially represents an implementation of Russell's idea of ramification to the particular case of second order arithmetic, now, however, with orders extending into the transfinite. The idea is to define a hierarchy analogous to Gödel's constructible hierarchy [45, 46], but at successor steps to collect definable *subsets of the natural numbers*. More precisely, let $Def^2(X)$ be the set of all those $A \subseteq N$ such that A is definable over X in second order arithmetic, that is, there is a formula $\varphi(x)$ of second order arithmetic such that for all n, $n \in A \iff (\varphi(n))^X$. Here the notation $(\varphi(n))^X$ indicates that all second-order quantifiers in φ range over X. Then we let $R_0 := \emptyset$, and $R_{\alpha+1} := Def^2(R_\alpha)$; at limit ordinal λ, we take $R_\lambda := \bigcup_{\xi<\lambda} R_\xi$. It is clear that the step from each level of the ramified analytic hierarchy to its successor is predicatively justified, as all second order quantifiers range over previous levels of the hierarchy. However, the ramified analytic hierarchy *as a whole* is problematic from a predicative perspective, since it presupposes the notion of "arbitrary" ordinal, i.e. "arbitrary" well-ordering relation, which, from a predicative perspective, is "as meaningless as the notion of 'arbitrary' set" [16, p. 9]. To overcome this difficulty a first thought was to introduce a "proviso of autonomy" on the ordinals used as indexes of the hierarchy: each ordinal used is to be determined by a well-ordering relation that, considered as a set of ordered pairs, is already admitted as predicative [16, p. 9]. We may call the resulting (suitably specified) ordinals "predicatively definable ordinals" [34]; then it turns out that by crucial results by Spector [91] and Kleene [53] the predicatively definable ordinals do not go beyond the recursive ordinals. In particular, Kleene [53] showed that $R_{\omega_1^{CK}} = HYP$, where ω_1^{CK} is the first non-recursive ordinal and HYP is the collection of hyperarithmetical sets. These results brought Kreisel [55] to tentatively identify the predicatively definable sets with those definable within $R_{\omega_1^{CK}}$ and thus also with the hyperarithmetical sets.[43]

The proposed identification of the realm of predicativity with the hyperarithmetical hierarchy, however, turned out to rely on the assumption of the countable ordinals up to the first non-recursive ordinal, ω_1^{CK}, along which to iterate the construction of the ramified analytic hierarchy. Feferman [34] writes:

> Though the considerations leading to the identification of the predicative ordinals, resp. sets of natural numbers, with the recursive ordinals, resp. hyperarithmetical sets, have a certain plausibility, they ignored one crucial point if predicativity is only to take the natural numbers for granted as a completed totality, namely that they involve in an essential way [...] the impredicative notion of being a well-ordering relation.

For these reasons a new phase in the logical analysis of predicativity began, which was prompted by another suggestion by Kreisel [54]. Kreisel [54] put forth a hierarchy of *formal systems* that would canonically represent predicative reasoning and called for the determination of its limit. A remarkable consequence of this new course of inquiry is that it shifted the focus of research from *definability* issues to

[43]See also [95].

provability issues.[44] The celebrated upshot of that research was the determination of the limit of predicativity by Feferman and Schütte (independently) [16, 85, 86] by means of proof-theoretic techniques. Russell's original idea of ramification had once more a crucial role, as a transfinite progression of systems of ramified second order arithmetic indexed by ordinals was introduced as a tool for determining a precise limit of predicativity by appeal to ordinal analysis.[45] The subsystems of second order arithmetic that make up the levels of the hierarchy, RA_α, are characterized by principles of ramified comprehension which express closure under the appropriate ramified definitions and essentially give rise to a formal version of the conditions we saw for the ramified analytic hierarchy. Each level of the hierarchy, therefore, can be seen as predicatively justified, since quantification is suitably restricted to previous levels. Once more, a fundamental issue turned out to be how to specify the iteration that justifies the ascent to higher levels of the hierarchy. Here one introduces a suitable "boot-strapping" condition. A crucial difference with the previous attempts is in that the ordinals indexing the hierarchy are not only those that can be *defined* by well-ordering relations within the hierarchy, but those which can also be *proved* to be such relations at *previous stages* of the hierarchy. That is, one carefully introduces a notion of *predicatively provable ordinal*, which has the purpose of guaranteeing that one progresses up along the hierarchy to a stage α only if α has already been *recognized* as predicative, i.e. if at a previous stage of the hierarchy we have a proof that it is an ordinal. The fundamental contribution of Feferman and Schütte was to determine that the least non-predicatively provable ordinal is an ordinal known as Γ_0.[46] Therefore, in proof theory $\boldsymbol{\Gamma_0}$ is often referred to as **the limit of predicativity**.

It is important to note that the limit of predicativity so determined is an "external limit". As clearly acknowledged by Feferman (see, e.g., [16]), one takes an impredicative stance and attempts to clarify the limit of predicativity from "the outside". The convinced predicativist will not recognize the limit Γ_0, as it lies beyond his reach, its very definition being impredicative. Gandy [44] writes: "The role played by Γ_0 for predicative systems is closely analogous to that played by ϵ_0 for finitist systems. Γ_0 is not a predicatively definable ordinal, but he who understands Γ_0 understands

[44] See also the review by Gandy [44].

[45] Here ordinals are not to be considered set-theoretically, rather as notations from a suitable ordinal notation system. See [68] for details on ordinal notation systems.

[46] See e.g. [68, Ch. 1] for details. In the branch of proof theory known as ordinal analysis, suitable (countable) ordinals, termed "proof-theoretic ordinals", are assigned to theories as a way of measuring their consistency strength and computational power. The "proof-theoretic strength" of a theory is then expressed in terms of such ordinals. The countable ordinal Γ_0 is the proof-theoretic ordinal assigned to the progression of ramified systems mentioned above. It is relatively small in proof-theoretic terms. As a way of comparison, it is well above the ordinal ϵ_0 which encapsulates the proof-theoretic strength of Peano Arithmetic, but it is much smaller than the ordinal assigned to a well–known theory, called ID_1, of one inductive definition. The latter ordinal is known in the literature as the Bachmann–Howard ordinal [3]. The strength of ID_1 is well below that of second order arithmetic, which is in turn much weaker than full set theory. For surveys on proof theory and ordinal analysis see, for example, [76–78].

the consistency, the potentialities and the limitations of predicative proof." This once more clarifies the deep change in attitude between the early discussions on predicativity and its logical analysis, as the latter is an attempt at understanding predicativity rather than arguing for it as a foundational stance.

3.2 Proof-Theoretic Reducibility

Feferman has explored a number of alternative perspectives over the years, as a way of corroborating the analysis of predicativity. In particular, one of the aims was to avoid direct appeal to the idea of ramification, which is particularly artificial and distant from mathematical practice. For example, already in the fundamental article [16], Feferman introduced a progression of formal systems which are based on a Hyperarithmetic Comprehension principle, therefore exploiting the previous observation of an important connection between predicative subsets of the natural numbers and hyperarithmetical sets. In that paper Feferman also introduced a single formal system IR which does not make direct reference to provability or definability.[47] Feferman [16] then established that also these two approaches give rise to Γ_0 as limit.[48] The fact that a number of distinct approaches to predicativity converged to the ordinal Γ_0 was then seen as confirmation of the thesis that Γ_0 marks the limit of predicativity, in a similar way as the convergence of different characterizations of computability are usually taken to support Church's thesis. The study of alternative routes to predicativity was also suggested by the desire to clarify which parts of mathematics can be given predicative form. As ramified systems are cumbersome to work in, one needs a way of assessing the predicativity of other systems which are better suited to the practical needs of a codification of ordinary mathematics. The notion of *proof–theoretic reducibility* was therefore appealed to for this purpose.[49] In order to assess the predicativity of a formal system T it suffices to appropriately "translate" it in (that is, proof-theoretically reduce it to) one of the ramified systems. The latter, thus, act as *canonical systems of reference* in terms of which the predicativity of other systems can be assessed. The outcome is a notion of *predicative justification*: a formal system is considered predicatively justified if it is proof–theoretically reducible to a system of ramified second order arithmetic indexed by an ordinal less than Γ_0. In addition, a notion of *locally predicative justification* was introduced, which applies to the case in which a system T is proof-theoretically reducible to the union of all the RA_α. In this case each theorem in T may be considered predicative, although the system T in its whole is not predicatively justified. A well-known locally pred-

---

[47] See [50, p. 283] for discussion.

[48] Further approaches to predicativity were explored, for example, in [17, 19, 20, 22, 23]. See also [21]. More recently, Feferman has developed the notion of "unfolding"; see [40, 41] and [93]. See also [6] for relations between Feferman's work on predicativity and theories of truth.

[49] See e.g. [28] for discussion of this notion, and [34] for an informal account of its application to an analysis of predicativity.

icatively justifiable system is Friedman's system ATR_0, which has been extensively studied in the Reverse Mathematics programme [43, 89].

3.3 *Predicativity and ordinary mathematics: the extent of predicativity*

Weyl's aim in "Das Kontinuum" was to clarify how far can we proceed in developing analysis from the bare assumption of the natural numbers with full induction and by iterating elementary properties and relations over them. His work gives a first, partial answer, to the important question of the relation between predicative mathematics with ordinary, or everyday, mathematics.[50]

A number of mathematical logicians in the 1950s felt that the early debates on impredicativity had left unresolved the question of which role impredicativity plays within ordinary mathematics. Wang [95, p. 244] clearly expressed this concern, when he observed that the use of uncountable (or indenumerable) and impredicative sets "remains a mystery which has shed little light on any problems of ordinary mathematics. There is no clear reason why mathematics could not dispense with impredicative or absolutely indenumerable sets."

In his fundamental article, Feferman [16, p. 3–4] writes:

> It is well known that a number of algebraic and analytic arguments can be systematically recast into a form which can be subsumed under elementary (first order) number theory. [...] It is thus not at first sight inconceivable that predicative mathematics is already (formally) sufficient to obtain the full range of arithmetical consequences realized by impredicative mathematics.

As Feferman quickly clarifies, not every elementary statement can be so obtained. The logical analysis of predicativity in [16] readily provides us with a counterexample: the very arithmetical statement expressing the consistency of predicative analysis. However, Feferman suggests that one could argue that "*all mathematically interesting statements about the natural numbers, as well as many analytic statements, which have so far been obtained by impredicative methods can already be obtained by predicative ones*".

The fundamental question of whether predicative mathematics is "already (formally) sufficient to obtain the full range of arithmetical consequences realized by impredicative mathematics" has been addressed by combining an appeal to the notion of proof-theoretic reducibility (that enables us to work in syntactically convenient systems) with a careful *case by case* logical analysis of ordinary mathematics. Here Weyl's pioneering work in "Das Kontinuum" has been a fundamental reference,

---

[50]The expression "ordinary mathematics" refers to mainstream mathematics, and has been so characterized, for example, by Simpson [89, p. 1]: "that body of mathematics which is prior to or independent of the introduction of abstract set-theoretic concepts". That is: "geometry, number theory, calculus, differential equations, real and complex analysis, countable algebra, the topology of complete separable metric spaces, mathematical logic and computability theory".

especially for Feferman's investigations [26, 34]. More precisely, Feferman [26] has carefully analysed Weyl's text and proposed a system, W, which codifies in modern terms Weyl's system in "Das Kontinuum". System W is particularly weak proof-theoretically, as it is as strong as Peano Arithmetic; as a consequence, it lies well within predicative mathematics. Feferman has verified that large portions of contemporary analysis can be carried out on its basis, in fact "most of classical analysis and substantial portions of modern analysis" [36]; therefore he has significantly improved on Weyl's [97].

Another source of insight is the research carried out within Friedman and Simpson's program of Reverse Mathematics [89], which has analyzed large portions of ordinary mathematics from a logical point of view. The principal outcome of these studies is a further confirmation that large parts of ordinary mathematics can be framed within predicative systems.[51] More surprisingly, it typically turns out that if a theorem can be established predicatively, it can already be carried out within a system not stronger than Peano Arithmetic. In fact, a finitary system suffices for most cases.[52]

The outcome of this research is that, once analyzed in detail, the prima facie necessity of abstract features of ordinary mathematics turns out to be avoidable in many cases. As a consequence, a substantial portion of ordinary impredicative mathematics is eliminable in favor of predicative mathematics. This is a striking result, highly unexpected from the perspective of Weyl's contemporaries. In fact, as suggested by Feferman, these insights have the potential of enriching the philosophy of mathematics in a number of ways. For example, they may have an impact on current discussions on indispensability of mathematics to science. Feferman [28] has argued that the case can be made that all scientifically applicable mathematics can be codified by predicative theories (in fact, by system W). The above mentioned work has brought Feferman to formulate the following "working hypothesis":

All of scientifically applicable analysis can be developed predicatively.

If, indeed, weak predicative systems turned out to be formally sufficient to develop all of scientifically applicable mathematics, this would imply the dispensability, at least from a formal point of view, of impredicative mathematics - when we restrict consideration to the mathematics that is required by our best scientific theories. This could then imply that an appeal to indispensability arguments to support the belief in the existence of mathematical entities would only grant, in the most favorable case, a rather limited ontology. As a consequence the above research might contribute to a more careful assessment of the possible outcomes of indispensability arguments, and the kind of platonism they might support, if they were to succeed.[53]

---

[51] See [89] for details and [90] for independence results.

[52] See [26, 34, 88] for informal discussions and further references.

[53] Note that Feferman draws different conclusions on the impact of the logical research on indispensability arguments in the philosophy of mathematics [28].

3.4 Predicativity Given the Natural Numbers

I conclude with some remarks on the notion of predicativity given the natural numbers that is the focus of the logical analysis of predicativity. At first the logical analysis of predicativity aimed at clarifying the notion *predicatively definable from* the natural numbers. In Sect. 3.1 I have emphasized that a shift of focus occurred at the end of the 1950s, so that predicativity given the natural numbers aimed at explicating a notion of *predicatively provable presupposing* the natural numbers [44]. From a philosophical perspective, it is natural to wonder how we should read the presupposition of the natural numbers. As it turns out, the relevant literature provides us with a number of different answers to this question. As reviewed above, Weyl, for example, suggested that the natural numbers, and in particular the idea of iteration, are "an ultimate foundation of mathematical thought", in fact, a "pure" intuition [97, p. 48]. In particular, the natural numbers are "individuals", classical logic applies to them and they can act as domain of quantification; therefore they can be employed for building predicative sets step-by-step by repeated application of elementary operations over them.

In Feferman's writings on predicativity one sometimes reads that the natural numbers (with full mathematical induction) can be regarded as "given" in some sense.[54] Sets are instead considered as no more than definitions, *façons de parler*, or convenient idealizations; as such they need to undergo appropriate constraints to avoid vicious circularity in definitions [16, p. 1–2].[55]

Sometimes the distinction between predicative and impredicative definitions or entities is presented in epistemological terms, and predicativity is seen as an instrument for clarifying what is implicit in our understanding of the natural numbers.[56] Feferman [24, p. 449] writes:

> That there is a fundamental difference between our understanding of the concept of natural numbers and our understanding of the set concept, even for sets of natural numbers, seems to me undeniable. The study of predicativity, as what is implicit in accepting the structure of natural numbers, is thus of special foundational significance. This is *not* to say that only what is predicative is 'justified'. What we are dealing with here are questions of relative conceptual clarity and foundational status [...].

The latter point is significant, and demonstrates, once more, the crucial difference between the attitude of the logicians that studied predicativity from the 1950s and that of the mathematicians that forged this notion at the turn of the 20th century,

---

[54] In the following I shall often omit explicit reference to the unrestricted principle of mathematical induction, and simply write "the natural numbers"; however, I shall presuppose that in the case of predicativity given the natural numbers full induction is also assumed.

[55] Feferman [16, 34] also describes the predicativist position as one that takes the natural numbers as a "completed totality", and views the rest in potentialist terms. However, I could find no further elucidation of the notion of complete totality, beyond the claim that we can use classical logic to reason about it. In [32, 35], Feferman proposes to read the "giveness" of the natural numbers in terms of realism in truth value (restricted to the natural numbers). A fundamental theme that emerges within Feferman's discussions on predicativity is an opposition, analogous to Weyl's, to arbitrary sets, and in particular to the powerset of an infinite set (see e.g., [33]).

[56] See e.g. [29, 54].

in particular Weyl [97]. Perhaps it also explains why there is insufficient clarity in the logical discussion on the philosophical aspects: the aim of the logician is not a defence of predicativism but a clarification of predicativity. In fact, the logician is primarily interested in clarifying the consequences of given assumptions. A crucial question is: which mathematical constructions and which portions of mathematics can we develop from the assumption of certain mathematical entities and operations over them? A conceptual clarification of the mathematical facts is then seen as prior to a clarification of the underlying philosophical stances that determine the choice of certain assumptions; Feferman [34] writes: "[t]he potential value for philosophy then is to be able to say in sharper terms what arguments may be mounted for or against taking such a stance."

In fact, a number of authors, including Feferman, have suggested that the notion of predicativity is more profitably understood as a relative rather than an absolute notion: we analyze what is predicative given some prior assumption, as, for example, the natural numbers. One might take, however, different starting points. From this perspective Gödel's constructible hierarchy may also be framed as an example of predicativity, one which may be seen as reducing all kinds of impredicativity to one special kind: "the existence of certain large ordinal numbers (or well-ordered sets) and the validity of recursive reasoning for them" [47, p. 464]. A weaker form of predicativity, compared with predicativity given the natural numbers, is instead obtained if one considers restrictions to the induction principle as in [64, 65, 67].

Predicativity may now become a tool for an analysis of mathematics, helping us distinguish different portions of the mathematical landscape, distinct for the assumptions and the methodology they require. In other terms, the logical analysis of (forms of) predicativity becomes an instrument for a finer understanding of contemporary mathematics, which addresses the question of which concepts and methods are necessary for the development of given portions of contemporary mathematics.[57]

4 Conclusion

The history of predicativity is witness to a remarkable example of cross-fertilization between philosophy of mathematics and mathematical logic. A critical reflection on the new abstract concepts and methods that were introduced in mathematics in the 19th century gave rise to proposals for the development of mathematics on predicative grounds. Adherence to predicativity was proposed as a way of avoiding vicious circularity in definitions and resulted in Russell's ramified type theory and Weyl's predicative analysis. A clarification of the notion of predicativity and its mathematical implications stimulated further technical advances, and saw the involvement of

[57] Wilfried Sieg has drawn my attention to a passage in Hilbert's 1920s lectures [15, p. 363-4] which suggests that this view of predicativity is in agreement with a Hilbertian perspective. It is also worth observing that with the shift of the logical analysis of predicativity to proof-theoretic considerations this enterprise gained a clear Hilbertian character.

prominent logicians, especially in the 1950-1960s. In particular, Feferman has contributed to the determination of the limit of a notion of predicativity given the natural numbers, and has offered, over the years, new ways of explicating this notion of predicativity.

Beyond the purely logical interest of predicativity, this notion may play a role in the philosophy of mathematics. Compliance with predicativity requirements enables us to carve a restricted concept of set; in particular, in the case of predicativity given the natural numbers we have a concept of set that is deeply rooted in the natural numbers. This, in turn, may be used to assess which mathematical concepts and theories can be developed purely on the basis of this more constrained concept of set, and which ones instead require an essential appeal to more abstract and complex notions. The logical analysis of predicativity has particularly highlighted the crucial role for predicativity of two components: some initial entity, e.g. the natural number set with full mathematical induction, and the iteration of elementary operations over it. This opens up the way for a number of notions of predicativity, which may be employed to help us clarify the difference between distinguishable conceptual spheres of mathematical activity.

References

1. Benacerraf, P., Putnam, H.: Philosophy of Mathematics: Selected Readings. Cambridge University Press, Cambridge (1983)
2. Bernays, P.: Sur the platonisme dans les mathématiques. L'Enseignement mathématique **34**, 52–69 (1935). Translated in [1] with the title: On Platonism in Mathematics
3. Buchholz, W., Feferman, S., Pohlers, W., Sieg, W.: Iterated Inductive Definitions and Subsystems of Analysis. Springer, Berlin (1981)
4. Cantini, A.: Una teoria della predicatività secondo Poincaré. Rivista di Filosofia **72**, 32–50 (1981)
5. Cantini, A.: Paradoxes, self-reference and truth in the 20th century. In: Gabbay, D. (ed.) The Handbook of the History of Logic, pp. 5–875. Elsevier (2009)
6. Cantini, A., Fujimoto, K., Halbach, F.: Feferman and the Truth, this volume (2017)
7. Carnap, R.: Die Logizistische Grundlegung der Mathematik. Erkenntnis **2**(1), 91–105 (1931). Translated in [1]. (Page references are to the translation)
8. Church, A.: A formulation of the simple theory of types. J. Symb. Logic **5**, 56–68 (1940)
9. Chwistek, L.: Über die Antinomien der Prinzipien der Mathematik. Mathematische Zeitschrift **14**, 236–43 (1922)
10. Coquand, T.: Type theory. In: Zalta, E.N. (ed.) The Stanford Encyclopedia of Philosophy (2015)
11. Crosilla, L.: Constructive and intuitionistic ZF. In: Zalta, E.N. (ed.) Stanford Encyclopedia of Philosophy (2014). http://plato.stanford.edu/entries/set-theory-constructive/
12. Crosilla, L.: Error and predicativity. In: Beckmann, A., Mitrana, V., Soskova, M. (eds.) Evolving Computability. Lecture Notes in Computer Science, vol. 9136, pp. 13–22. Springer International Publishing (2015)
13. Dummett, M.: Frege: Philosophy of Mathematics. Harvard University Press, Cambridge (1991)
14. Dummett, M.: What is mathematics about? In: George, A. (ed.) The Seas of Language, pp. 429–445. Oxford University Press (1993). Reprinted in [51], pp. 19–30
15. Ewald, W., Sieg, W.: David Hilbert's Lectures on the Foundations of Arithmetic and Logic, 1917–1933. Springer, Heidelberg (2013)
16. Feferman, S.: Systems of predicative analysis. J. Symb. Logic **29**, 1–30 (1964)

17. Feferman, S.: Predicative provability in set theory. Bull. Am. Math. Soc. **72**, 486–489 (1966)
18. Feferman, S.: Persistent and invariant formulas for outer extensions. Compositio Math. **20**, 29–52 (1968)
19. Feferman, S.: Systems of predicative analysis. II. Representations of ordinals. J. Symb. Logic **33**, 193–220 (1968)
20. Feferman, S.: Predicatively reducible systems of set theory. In: Axiomatic Set Theory (Proceedings of Symposium Pure Mathematics, Vol. XIII, Part II, University California, Los Angeles, California, 1967), pp. 11–32. American Mathematical Society Providence, R.I (1974)
21. Feferman, S.: Impredicativity of the existence of the largest divisible subgroup of an abelian p-group. In: Model Theory and Algebra (A Memorial Tribute to Abraham Robinson). Lecture Notes in Mathematics, vol. 498, pp. 117–130. Springer, Berlin (1975)
22. Feferman, S.: A more perspicuous formal system for predicativity. In: Boffa, M., van Dalen, D., McAloon, K. (eds.) Logic Colloquium '78'. North Holland, Amsterdam (1979)
23. Feferman, S.: Iterated inductive fixed-point theories: application to Hancock's conjecture. In: Patras Logic Symposion (Patras, 1980). Studies in Logic Foundations Mathematics, vol. 109, pp. 171–196. North-Holland, Amsterdam-New York (1982)
24. Feferman, S.: Proof theory: a personal report. In: Appendix to the Second edition of [94], pp. 447–485 (1987)
25. Feferman, S.: Hilbert's program relativized: proof-theoretical and foundational reductions. J. Symb. Logic **53**(2), 364–384 (1988)
26. Feferman, S.: Weyl vindicated: Das Kontinuum seventy years later. In: Cellucci, C., Sambin, G. (eds.) Temi e prospettive della logica e della filosofia della scienza contemporanee, pp. 59–93. CLUEB, Bologna (1988)
27. Feferman, S.: What rests on what? The proof-theoretic analysis of mathematics. In: Czermak, J. (ed.) Proceedings of the 15th International Wittgenstein Symposium Philosophy of Mathematics, Part I. Verlag Hölder–Pichler– Tempsky, Vienna (1993)
28. Feferman, S.: Why a little bit goes a long way. Logical foundations of scientifically applicable mathematics. In: PSA 1992, Philosophy of Science Association (East Lansing), vol. 2, pp. 442–455 (1993). Reprinted in [30]
29. Feferman, S.: Kreisel's unwinding program. In: Odifreddi, P. (ed.) Kreiseliana, pp. 247–273. About and Around Georg Kreisel, A K Peters (1996)
30. Feferman, S.: In the Light of Logic. Oxford University Press, New York (1998)
31. Feferman, S.: The significance of Hermann Weyl's Das Kontinuum. In: Hendricks, V., Pedersen, S.A., Jørgense, K.F. (eds.) Proof Theory. Kluwer, Dordrecht (2000)
32. Feferman, S.: Comments on "Predicativity as a philosophical position" by G. Hellman, Rev. Int. Philos. **229**(3), 313–323 (2004)
33. Feferman, S.: The development of programs for the foundations of mathematics in the first third of the 20th century. In: Petruccioli, S. (ed.) Storia della scienza. Istituto della Enciclopedia Italiana, vol. 8, pp. 112–121. Translated as Le scuole di filosofia della matematica (2004)
34. Feferman, S.: Predicativity. In: Shapiro, S. (ed.) Handbook of the Philosophy of Mathematics and Logic. Oxford University Press, Oxford (2005)
35. Feferman, S.: Conceptions of the continuum. Intellectica **51**, 169–189 (2009)
36. Feferman, S. 2013 , Why a little bit goes a long way: predicative foundations of analysis. Unpublished notes dating from 1977-1981, with a new introduction. Retrieved from the address: https://math.stanford.edu/~feferman/papers.html
37. Feferman, S., Hellman, G.: Predicative foundations of arithmetic. J. Philos. Logic **22**, 1–17 (1995)
38. Feferman, S., Jäger, G.: Systems of explicit mathematics with non-constructive μ-operator, Part I. Ann. Pure Appl. Logic **65**(3), 243–263 (1993)
39. Feferman, S., Kreisel, G.: Persistent and invariant formulas relative to theories of higher order. Bull. Am. Math. Soc. **72**, 480–485 (1966)
40. Feferman, S., Strahm, T.: The unfolding of non-finitist arithmetic. Ann. Pure Appl. Logic **104**(1–3), 75–96 (2000)
41. Feferman, S., Strahm, T.: Unfolding finitist arithmetic. Rev. Symb. Logic **3**(4), 665–689 (2010)

42. Ferreirós, J.: On arbitrary sets and ZFC. Bull. Symb. Logic (3) 2011
43. Friedman, H.: Systems of second order arithmetic with restricted induction, I, II (abstracts). J. Symb. Logic **41**, 557–559 (1976)
44. Gandy, R.O.: Review of [16]. Math. Rev. (1967)
45. Gödel, K.: The consistency of the axiom of choice and of the generalized continuum-hypothesis. Proc. Natl. Acad. Sci. U. S. A. **24**, 556–557 (1938)
46. Gödel, K.: The Consistency of the Axiom of Choice and of the Generalized Continuum-Hypothesis with the Axioms of Set Theory, vol. 17. Princeton University Press (1940)
47. Gödel, K.: Russell's mathematical logic. In: Schlipp, P.A. (ed.) The philosophy of Bertrand Russell, pp. 123–153. Northwestern University, Evanston and Chicago (1944). Reprinted in [1]. Page references are to the reprinting
48. Hellman, G.: Predicativism as a philosophical position. Revue Internationale de Philosophie **3**, 295–312 (2004)
49. Hodes, H.T.: Why ramify? Notre Dame J. Formal Logic **56**(2), 379–415 (2015)
50. Howard, W.A.: Some Proof Theory in the 1960's. In: Odifreddi, P. (ed.) 'Kreiseliana. About and Around Georg Kreisel', pp. 274–288. A K Peters (1996)
51. Jacquette, D. (ed.): Philosophy of Mathematics: An Anthology. Wiley-Blackwell (2001)
52. Jäger, G.: Metapredicative and explicit Mahlo: a proof-theoretic perspective. In: R. Cori et al. (ed.) Proceedings of Logic Colloquium 2000. Association of Symbolic Logic Lecture Notes in Logic, vol. 19, pp. 272–293. AK Peters (2005)
53. Kleene, S.C.: Quantification of number-theoretic functions. Compositio Mathematica **14**, 23–40 (1959)
54. Kreisel, G.: Ordinal logics and the characterization of informal concepts of proof. In: Proceedings of the International Congress of Mathematicians (August 1958), pp. 289–299. Gauthier–Villars, Paris (1958)
55. Kreisel, G.: La prédicativité. Bulletin de la Societé Mathématique de France **88**, 371–391 (1960)
56. Kreisel, G.: Principles of proof and ordinals implicit in given concepts. In: Kino, R.E.V.A., Myhill, J. (eds.) Intuitionism and Proof Theory, pp. 489–516. North-Holland, Amsterdam (1970)
57. Linsky, B.: Propositional functions and universals in principia mathematica. Austr. J. Philos. **66**(4), 447–460 (1988)
58. Maddy, P.: Naturalism in Mathematics. Oxford University Press, Oxford (1997)
59. Mancosu, P.: From Brouwer to Hilbert. The Debate on the Foundations of Mathematics in the 1920s. Oxford University Press, Oxford (1998)
60. Martin-Löf, P.: An intuitionistic theory of types: predicative part. In: Rose, H.E., Shepherdson, J.C. (eds.), Logic Colloquium 1973. North–Holland, Amsterdam (1975)
61. Martin-Löf, P.: Intuitionistic Type Theory. Bibliopolis, Naples (1984)
62. Martin-Löf, P.: The Hilbert–Brouwer controversy resolved? In: van Atten, E.A. (ed.) One Hundred Years of Intuitionism (1907 – 2007), Publications des Archives Henri Poincaré, pp. 243–256 (2008)
63. Myhill, J.: The undefinability of the set of natural numbers in the ramified principia. In: Nakhnikian, G. (ed.) Bertrand Russell's Philosophy, pp. 19–27. Duckworth, London (1974)
64. Nelson, E.: Predicative Arithmetic. Princeton University Press, Princeton (1986)
65. Parsons, C.: The impredicativity of induction. In: Detlefsen, M. (ed.) Proof, Logic, and Formalization, pp. 139–161. Routledge, London (1992)
66. Parsons, C.: Realism and the debate on impredicativity, 1917–1944. Association for Symbolic Logic (2002)
67. Parsons, C.: Mathematical Thought and Its Objects. Cambridge University Press, Cambridge (2008)
68. Pohlers, W.: Proof Theory: The First Step into Impredicativity, Universitext. Springer, Berlin (2009)
69. Poincaré, H.: Les mathématiques et la logique. Revue de Métaphysique et Morale **13**, 815–835 (1905)

70. Poincaré, H.: Les mathématiques et la logique. Revue de Métaphysique et de Morale **14**, 17–34 (1906)
71. Poincaré, H.: Les mathématiques et la logique. Revue de Métaphysique et de Morale **14**, 294–317 (1906)
72. Poincaré, H.: La logique de l'infini. Revue de Métaphysique et Morale **17**, 461–482 (1909)
73. Poincaré, H.: La logique de l'infini. Scientia **12**, 1–11 (1912)
74. Ramsey, F.P.: Foundations of mathematics. Proc. Lond. Math. Soc.**25** (1926). Reprinted in [75]
75. Ramsey, F.P.: Foundations of Mathematics and Other Logical Essays. Routledge and Kegan Paul (1931)
76. Rathjen, M.: The higher infinite in proof theory. In: Makowsky, J.A., Ravve, E.V. (eds.) Logic Colloquium '95'. Springer Lecture Notes in Logic, vol. 11. Springer, New York, Berlin (1998)
77. Rathjen, M.: The realm of ordinal analysis. In: Sets and proofs (Leeds, 1997). London Mathematical Society Lecture Note Series, vol. 258, pp. 219–279. Cambridge Univ. Press, Cambridge (1999)
78. Rathjen, M.: Theories and ordinals in proof theory. Synthese **148**(3), 719–743 (2006)
79. Russell, B.: The Principles of Mathematics. Routledge (1903)
80. Russell, B.: 'Les paradoxes de la logique'. Revue de métaphysique et de morale **14**, 627–650 (1906)
81. Russell, B.: On some difficulties in the theory of transfinite numbers and order types. Proc. Lond. Math. Soc. **4**, 29–53 (1906)
82. Russell, B.: Mathematical logic as based on the theory of types. Am. J. Math. **30**, 222–262 (1908)
83. Russell, B.: Essays in Analysis. In: Lackey, D. (ed.). George Braziller, New York (1973)
84. Sacks, G.E.: Perspectives in Mathematical Logic. Higher recursion theory. Springer, Berlin (1990)
85. Schütte, K.: Eine Grenze für die Beweisbarkeit der Transfiniten Induktion in der verzweigten Typenlogik. Archiv für mathematische Logik und Grundlagenforschung **7**, 45–60 (1965)
86. Schütte, K.: Predicative well–orderings. In: Crossley, J., Dummett, M. (eds.) Formal Systems and Recursive Functions. North–Holland, Amsterdam (1965)
87. Sieg, W.: Hilbert's programs: 1917–1922. Bull. Symb. Logic **5**(1), 1–44 (1999)
88. Simpson, S.G.: Partial realizations of Hilbert's program. J. Symb. Logic **53**(2), 349–363 (1988)
89. Simpson, S.G.: Subsystems of Second Order Arithmetic. Perspectives in Mathematical Logic. Springer, Berlin (1999)
90. Simpson, S.G.: Predicativity: the outer limits. In: Reflections on the foundations of mathematics (Stanford, CA, 1998). Lecture Notes Logic **15**, Assoc. Symbol. Logic, Urbana, IL, 130–136 (2002)
91. Spector, C.: Recursive well-orderings. J. Symb. Logic **20**, 151–163 (1955)
92. Stein, H.: Logos, logic and logistiké. In: Asprey, W., Kitcher, P. (eds.) History and Philosophy of Modern Mathematics, pp. 238–59. University of Minnesota, Minneapolis (1988)
93. Strahm, T.: Unfolding schematic systems, this volume (2017)
94. Takeuti, G.: Proof Theory, 2nd edn. North Holland, Amsterdam (1987)
95. Wang, H.: The formalization of mathematics. J.Symb. Logic **19**(4), 241–266 (1954)
96. Weaver, N.: Predicativity beyond Γ_0 (2005). Preprint submitted to the arXiv repository: arXiv:math/0509244
97. Weyl, H.: Das Kontinuum: Kritische Untersuchungen über die Grundlagen der Analysis, Veit, Leipzig. Translated in English, Dover Books on Mathematics, 2003. (Page references are to the translation) (1918)
98. Weyl, H.: Philosophy of Mathematics and Natural Science. Princeton University Press (1949). An expanded English version of Philosophie der Mathematik und Naturwissenschaft, München, Leibniz Verlag (1927)
99. Whitehead, A.N., Russell, B.: Principia Mathematica, 3 Vols. Cambridge University Press, Cambridge (1910, 1912, 1913); 2nd edn (1925) (Vol 1), 1927 (Vols 2, 3); abridged as Principia Mathematica to *56. Cambridge University Press, Cambridge (1962)

Author Biography

Laura Crosilla graduated in philosophy from the University of Florence and obtained her Ph.D. in mathematical logic from the University of Leeds. She has held postdoctoral positions at the Ludwig-Maximilian University Munich, the University of Florence and the University of Leeds. Her research interests include constructive set theory, proof theory and constructive type theory. Her main research interest at present is the philosophy of mathematics, with particular focus on constructive mathematics and predicativity.

Sameness

Dag Westerståhl

Abstract I attempt an explication of what it means for an operation across domains to be the *same* on all domains, an issue that (Feferman, S.: Logic, logics and logicism. Notre Dame J. Form. Log. **40**, 31–54 (1999)) took to be central for a successful delimitation of the *logical* operations. Some properties that seem strongly related to sameness are examined, notably isomorphism invariance, and sameness under extensions of the domain. The conclusion is that although no precise criterion can satisfy all intuitions about sameness, combining the two properties just mentioned yields a reasonably robust and useful explication of sameness across domains.

Keywords Logical constants · Isomorphism invariance · Extension
Generalized quantifiers

2010 Mathematics Subject Classification 03B65 · 03C80 · 91F20

1 Introduction

In three fairly recent papers, Sol Feferman discussed the question of *logicality*, of what makes some symbols logical and others not.[1] Driven by a "personal feeling that the logical operations do not go beyond those represented in FOL" [5], he explored different ways of characterizing logicality, starting with his analysis and critique

[1] [4–6].

*Sol Feferman was an inspiration to me for most of my adult life and, for almost as long, a friend. I'm delighted to have been asked to contribute to this volume in his honor. As to the paper, thanks to Lauri Hella and Jouko Väänänen for helpful conversations on the model-theoretic parts, and to Stanley Peters who saved me from a rather embarrassing mistake (remaining ones are mine, of course). I also want to thank the two extremely encouraging and patient editors of this volume.

D. Westerståhl (✉)
Department of Philosophy, Stockholm University, Stockholm, Sweden
e-mail: dag.westerstahl@philosophy.su.se

G. Jäger and W. Sieg (eds.), *Feferman on Foundations*, Outstanding Contributions to Logic 13, DOI 10.1007/978-3-319-63334-3_16

of the Tarski–Sher thesis. Tarski's invariance criterion, from [16], was that logical notions are those which are *permutation invariant*; a property I will call PERM. [15] strengthened the requirement to *isomorphism invariance* (ISOM).[2] While granting that ISOM is a necessary feature of logical constants, Feferman argued that it is far from sufficient, and suggested several stronger criteria that essentially narrow down the range of logical constants to those definable in FOL.

In this note I focus on just *one* part of his critique of the Tarski–Sher thesis. In [4], he thought of this as "perhaps the strongest reason for rejecting the Tarski– Sher thesis", and formulated it as follows: "No natural explanation is given by it of what constitutes the *same* logical operation over arbitrary basic domains".

Let us call this criterion SAME. For a basic domain M, let $\langle M \rangle$ be a typed universe over M, relative to some suitable type system. We are considering *operations across domains*, i.e. operations O that with each M associate an object $O_M \in \langle M \rangle$ of a given type.

(SAME) O satisfies SAME if O_M is the *same* operation on each M.

The question, of course, is what "same operation" should mean, and indeed if it is possible to give it a "natural explanation". The notion might seem inherently vague. There are, however, a number of fairly strong intuitions about SAME, so let us start with these.

2 Some Intuitions

Feferman used an example from McGee of an operation across domains that clearly does *not* satisfy SAME: "a logical connective which acts like disjunction when the size of the domain is an even successor cardinal, like conjunction when the size of the domain is an odd successor cardinal, and like a biconditional at limits" [9]:577. There are plenty of other, and simpler, examples in the same vein, for example, a quantifier which acts like $\forall$ on universes with at least 100 elements, and as $\exists$ on smaller universes. Thus:

(I1) There are numerous examples of operations across universes that behave differently on different universes, and therefore do not satisfy SAME.

From the examples illustrating (I1) it seems uncontroversial that ISOM does *not* imply SAME. Perhaps a slightly more debatable question is if the implication in the other direction holds. But if h is an bijection from M to M', and hence lifts to a bijection from each M_τ to M'_τ (see Sect. 3), it seems to me like a very reasonable

[2]Thereby avoiding to treat as logical, for example, an operator which behaves as the universal quantifier on universes containing the number 0, and as the existential quantifier on other universes. Also, Sher restricted attention to operations of type level at most 2, and Feferman in general follows this restriction.

requirement of sameness that an operation O_M should map along h to the corresponding operation on M'_τ, that is, that $h(O_M) = O_{h(M)}$, which is precisely what ISOM says. So I think the following intuition is quite strong too.

(I2) Operations satisfying SAME also satisfy ISOM.

Note that (I2) does not derive from the standard thought that ISOM captures *topic neutrality*, and therefore is a necessary feature of logicality. Rather, we are simply using the idea that isomorphic models have, in the strictest sense, *the same structure*.

Our next intuition about sameness is something practically everyone would agree to:

(I3) The (interpretations of the) usual first-order symbols $\forall, \exists, \neg, \wedge, \vee, \rightarrow$ satisfy SAME.

There is a simple but important lesson to be drawn from (I1) and (I3):

(1) The class of SAME operations is not closed under definability (even first-order definability).

Now one might think this already disqualifies the SAME notion from playing any significant part in characterizing logicality. Indeed, the role of definability for the notion of logicality is itself an interesting and intricate issue, discussed in [2, 4, 5, 9], and recently, from a general model-theoretic perspective, in [3].

But that issue is not the topic of the present note. My aim is just to see what sense can be made of SAME. I conclude from (1) that *if* there is a precise version of this idea, then its role for the notion of logicality would at most be as a requirement on *primitive* logical constants, not defined ones.

Before moving on, we should note that [4] proposed to implement SAME (at least partly) by strengthening ISOM to HOM: invariance under *homomorphisms*. He proved that the operations definable in first-order logic *without equality* are precisely those definable from *monadic* operations across domains satisfying HOM. Granted that HOM is accepted, this of course contradicts (1).

Later, however, Feferman abandoned HOM as a criterion of logicality, not (at least not explicitly) for reasons related to SAME, but rather because (a) equality does not satisfy HOM, and (b) the restriction to monadic types seems rather arbitrary, and once you admit operations of non-monadic types, a host of non-first-order operations become definable; indeed, as [2] showed, the HOM operations of arbitrary type are exactly those definable in $L_{\infty\infty}$ without equality (which, as Feferman points out, for a language with finitely many predicate symbols, has essentially the same expressive power as full $L_{\infty\infty}$).

It also seems to me that HOM has a serious problem with SAME. It requires invariance even when you shrink the universe (homomorphically) to a single point. But, just to take an example, couldn't it be that the action of an operation across domains, say, on some set arguments, essentially depends on whether these are *disjoint* or not? Such dependencies will get lost if the domain is shrunk enough. It would at least require some independent argument, I think, to disqualify such operations from the start, which is what HOM does. For these reasons, I will ignore HOM in what follows, and maintain that (1) is a feature of the notion of sameness we are after.

3 Types and Quantifiers

I restrict attention, in much of what follows, to the operations on domains usually called (generalized or Lindström) *quantifiers*. This is because intuitions, and facts, are simpler for quantifiers, or more generally for operations of type level ≤ 2. Eventually, of course, one would like to be able to deal with operations of all types.

The type system used in [4], let us call it TFT, is (with slight notational differences) as follows: e, t are basic *type symbols*, complex types symbols have the form $\langle\sigma_1 \ldots \sigma_n, \tau\rangle$. Given a domain M, $M_e = M$, $M_t = \{0, 1\}$, and $M_{\langle\sigma_1 \ldots \sigma_n, \tau\rangle} = M_\tau^{M_{\sigma_1} \times \cdots \times M_{\sigma_n}}$. Finally, $\langle M\rangle = \bigcup_\tau M_\tau$, where τ varies over the set of type symbols.

For many purposes, *relational* types are somewhat simpler to deal with. Every type symbol τ in TFT can be uniquely written as

$$\tau = \langle\sigma_{11} \ldots \sigma_{1k_1}, \ldots, \langle\sigma_{n1} \ldots \sigma_{nk_n}, \tau_0\rangle \ldots\rangle$$

where τ_0 is either e or t. Following [21], call τ *individual* if $\tau_0 = e$ and *Boolean* if $\tau_0 = t$. Boolean types are relational but their arguments need not be; cf. $\langle\langle e, e\rangle, t\rangle$. Define the *strictly relational* types of TFT inductively as follows:

> τ is *strictly relational* iff τ is either basic or of the form $\langle\sigma_1 \ldots \sigma_n, t\rangle$, where each σ_i is strictly relational. (e is included for convenience here.)

For these types, a simpler *relational type system*, call it RT, is the following: the only basic type symbol is e, and complex type symbols have the form $(\sigma_1, \ldots, \sigma_n)$. Given M, $M_e = M$, and $M_{(\sigma_1,\ldots,\sigma_n)} = \mathcal{P}(M_{\sigma_1} \times \cdots \times M_{\sigma_n})$.[3] Then it is easy to verify that RT is isomorphic to the strictly relational part of TFT, and that operations across domains in this part of TFT transfer in an obvious way to operations across domains in RT, and vice versa.

Finally, for (generalized) quantifiers, there is the widely used type notation from [8]; we can extend it slightly to include 0-ary relations (i.e. truth values) as arguments, so that propositional connectives are quantifiers too. Thus, a *quantifier of type* $\langle k_1, \ldots, k_n\rangle$ (where each $k_i \geq 0$) is an operation Q across domains such that, for each M, Q_M is an n-ary relation between relations over M, of arities $k_1, \ldots, k_n$, respectively.[4] For example, if Q is of type $\langle 1, 0, 3\rangle$, then Q_M relates a subset of M, a truth value, and a subset of M^3. A binary propositional connective like $\wedge$ has type

---

[3]Stipulating that the cartesian product of the empty sequence () of sets is $\{\emptyset\}$, we have $M_{()} = \mathcal{P}(\{\emptyset\}) = \{\emptyset\{\emptyset\}\} = \{0, 1\}$, so () corresponds to the type t in TFT.

[4]If π_m is the TFT type symbol $\langle e^m, t\rangle$ of m-ary relations between individuals (and $\pi_0 = t$), type $\langle k_1, \ldots, k_n\rangle$ corresponds to $\langle\pi_{k_1} \ldots \pi_{k_n}, t\rangle$. Similarly in RT, it corresponds to $((e^{k_1}), \ldots, (e^{k_n}))$.

$\langle 0, 0\rangle$; here we can think of $\wedge_M$ as the usual truth function (which is independent of M).[5] A quantifier of type $\langle k_1, \ldots, k_n\rangle$ is *(pure) monadic*, if each k_i is ≤ 1 (is $= 1$).

In familiar notation, $L(Q, Q', \ldots)$ is first-order logic (FOL) with added quantifiers $Q, Q', \ldots$: for each quantifier one adds a corresponding variable-binding operator of the same type, and extends the syntax and the definition of satisfaction accordingly; details can be found in any textbook or survey paper dealing with generalized quantifiers. For logics L and L', $L \leq L'$ means that every L-sentence is equivalent to (has the same models as) some L'-sentence, $L \equiv L'$ means that $L \leq L'$ and $L' \leq L$, and $L < L'$ means that $L \leq L'$ and $L' \nleq L$.

In logic, one generally requires $Q, Q', \ldots$ to satisfy ISOM, which has the effect that $L(Q, Q', \ldots)$-sentences preserve their truth value across isomorphic models. This is of course consonant with intuition (I2).

4 Extension and a Clash of Intuitions

A serious candidate for a precise notion of sameness is the property of *Extension* (EXT), introduced by van Benthem.[6] For quantifiers it is the following condition:

> A quantifier Q of type $\langle k_1, \ldots, k_n\rangle$ satisfies EXT if, whenever $M \subseteq M'$ and $R_i \subseteq M^{k_i}$ for $1 \leq i \leq n$, $Q_M(R_1, \ldots, R_n) \Leftrightarrow Q_{M'}(R_1, \ldots, R_n)$.

Indeed, this generalizes to all types in RT:

(EXT) An operator O across domains of type $(\sigma_1, \ldots, \sigma_n)$ is EXT if $M \subseteq M'$ implies that $O_M = O_{M'} \restriction M$ (where $O_{M'} \restriction M = O_{M'} \cap (M_{\sigma_1} \times \cdots \times M_{\sigma_n})$).

The reason this works is that types in RT are *monotone* in the following sense:

(2) If τ is an RT-type and $M \subseteq M'$, then $\mathcal{M}_\tau \subseteq \mathcal{M}'_\tau$.

(2) is in stark contrast to TFT types[7]:

[5] I am using the same notation for the conjunction symbol and its interpretation (its corresponding quantifier), and similarly later on for other quantifier symbols; this should not generate confusion.

[6] [19] made EXT part of the property of *conservativity* for type $\langle 1, 1\rangle$ quantifiers. In [20] it was isolated as a separate property, applicable to all types, and given the natural label 'extension'. [21] calls it 'context insensitivity'.

[7] Call a TFT-type symbol τ *truth-functional* if it does not contain e, and *extended truth-functional* if it is either truth-functional or of the form

$$\langle \sigma_{11} \ldots \sigma_{1k_1}, \ldots, \langle \sigma_{n1} \ldots \sigma_{nk_n}, e\rangle \ldots\rangle$$

where each σ_{ij} is truth-functional. [24] shows that the monotone types in TFT are exactly the extended truth-functional ones, and that all other TFT-types τ are *disjoint* in the sense that $M \neq M'$ implies $\mathcal{M}_\tau \cap \mathcal{M}'_\tau = \emptyset$.

One way to make TFT-types monotone is to allow *partial* functions rather than just total ones. But this raises other issues for a formulation of EXT; the situation is further discussed in [24].

(3) For most TFT-types τ, if M is a proper subset of M', then $\mathcal{M}_\tau \cap \mathcal{M}'_\tau = \emptyset$.

This is the main reason it is more problematic to define a natural version of EXT for arbitrary TFT-types.

As far as I know, van Benthem did not explicitly connect EXT to the SAME idea; that connection is discussed extensively in [11], Ch. 3.4–5. But surely EXT is in itself a very strong notion of sameness across domains. It says that only the elements of M belonging to some tuple in some R_i matter for the truth value of the statement $Q_M(R_1, \ldots, R_n)$. Extending the universe beyond that has no effect: the operation remains the 'same'. More precisely, for $R \subseteq M^k$, let $\mathit{field}(R) = \{a \in M : \exists (a_1, \ldots, a_k) \in R\ \exists j\ a = a_j\}$, and let $\mathit{field}(R_1, \ldots, R_n) = \bigcup_{i=1}^n \mathit{field}(R_i)$. Then EXT for Q is equivalent to:

If $\mathit{field}(R_1, \ldots, R_n) \subseteq M, M'$, then $Q_M(R_1, \ldots, R_n) \Leftrightarrow Q_{M'}(R_1, \ldots, R_n)$.

Here is a different formulation of essentially the same notion. The quantifiers Q of type $\langle k_1, \ldots, k_n \rangle$ we are interested in are usually *definable* in standard set theories, like ZF. That is, Q can be introduced by a ZF-formula $\psi(M, \overline{R})$ with at most the free variables $M, \overline{R}$, a *definition* of Q, such that

$$\forall M \neq \emptyset \forall R_1 \subseteq M^{k_1} \ldots \forall R_n \subseteq M^{k_n} (Q_M(R_1, \ldots, R_n) \leftrightarrow \psi(M, R_1, \ldots, R_n))$$

Then:

(4) A definable quantifier Q is EXT iff it has a definition in which the variable M does not occur.

Proof If Q has such a definition, it is immediate that EXT holds. Conversely, suppose Q is EXT, and defined by $\psi(M, \overline{R})$. We have

$$Q_M(\overline{R}) \Leftrightarrow Q_{\mathit{field}(\overline{R})}(\overline{R})$$

Thus, $Q_M(\overline{R}) \Leftrightarrow \psi(\mathit{field}(\overline{R}), \overline{R})$, and the right-hand side can be written as a ZF-formula where at most the variables $\overline{R}$ occur. □

Thus, the following is a strong intuition about sameness:

(I4) EXT implies SAME.

But now we have a clash of intuitions: (I2) and (I4) are incompatible! For if both were true, it would follow that EXT implies ISOM, which is manifestly not the case. Many EXT quantifiers, such as the so-called Montagovian individuals $(I_a)_M(B) \Leftrightarrow a \in B$, and quantified noun phrase denotations like $(\mathit{most}^{\mathit{student}})_M(B) \Leftrightarrow |\mathit{student} \cap B| > |\mathit{student} - B|$, where a is a *fixed* individual and *student* a *fixed* set, are EXT but not ISOM.

It is not surprising — in fact, I think it is inevitable — that some intuitions about sameness clash. In the case of (I2) and (I4), we have already seen that they are based on quite different ideas about sameness. I will come back to the question of what we should conclude from this, but for now, in this and the next section, I take a closer look at the status of EXT in the literature. To begin, here are some sample facts:

(5) a. The propositional connectives are EXT.
b. The quantifiers $\exists_{\geq n}$ are EXT (where $(\exists_{\geq n})_M(B) \Leftrightarrow |B| \geq n$), and more generally, the quantifiers $\exists_{\geq \kappa}$ for any cardinal κ.
c. The universal quantifier $\forall$ is not EXT.
d. The trivial quantifiers $\mathbf{0}$ and $\mathbf{1}$ (in each type) are EXT.[8]
e. The class of EXT quantifiers (in any type) is closed under conjunction, disjunction, and (outer) negation.
f. The *Rescher quantifier* Q^R and the *Chang quantifier* Q^C are not EXT (where we have $(Q^R)_M(B) \Leftrightarrow |B| > |M-B|$, and $(Q^C)_M(B) \Leftrightarrow |B| = |M|$).[9]
g. Under EXT, PERM is equivalent to ISOM, for all operations across domains of relational type.

As to (5-a), note that we are talking about propositional negation. *Predicate negation* is of type $\langle\langle e,t\rangle, \langle e,t\rangle\rangle$, which is not relational. Together with (I3); (5-c) and (5-f) indicate that the converse of (I4) fails, at least if these quantifiers are taken to satisfy SAME. Regarding (5-e), it seems a rather reasonable assumption that the class of SAME quantifiers (in any type) is also closed under boolean operations. But note that certain types of quantifiers also have a natural notion of *inner* negation. For example, in type $\langle 1, 1\rangle$, we define $(Q\neg)_M(A, B) \Leftrightarrow Q_M(A, M - B)$. Clearly, EXT is *not* preserved under inner negation. In general, the class of EXT quantifiers is not closed under (first-order) definability, since e.g. $\forall$ is definable from $\exists$ and $\neg$. (5-g), finally, requires a small (but straightforward) argument, not given here.

I discuss EXT in connection with other quantifiers from mathematical logic in the next section, but let me end this section with some remarks on the role of EXT in formal semantics for natural language. *Noun phrases*, like *Mary, the tallest man on earth, most students, every professor*, can be taken to denote type $\langle 1\rangle$ quantifiers, and *determiners*, like *every, some, most, at most ten, infinitely many*, to denote type $\langle 1, 1\rangle$ quantifiers. For example,

$$most_M(A, B) \Leftrightarrow |A \cap B| > |A - B|$$

and thus, for all M,
[[most]] $= most$
[[most students]]$_M(B) \Leftrightarrow most_M$([[student]], B)
Now it seems to be a linguistic fact that *all* determiner denotations, i.e. *all* type $\langle 1, 1\rangle$ quantifiers that interpret natural language determiners, are EXT. An explanation

[8]E.g. in type $\langle 1, 2\rangle$, $\neg\mathbf{0}_M(A, R)$ for all $A \subseteq M$ and $R \subseteq M^2$, and $\mathbf{1}_M(A, R)$ for all $A \subseteq M$ and $R \subseteq M^2$.

[9]In the literature, the label 'Rescher quantifier' is often used for a stronger quantifier of type $\langle 1, 1\rangle$, let us call it *more*, where $more_M(A, B) \Leftrightarrow |A| > |B|$ (which, incidentally, is EXT). (For example, the survey article [10] uses 'Q^R' for *more*.) My usage here is historically more accurate, since precisely the type $\langle 1\rangle$ quantifier Q^R was introduced in the abstract [13]. Rescher also mentioned the relativization of Q^R, which is the quantifier *most*, defined by $most_M(A, B) \Leftrightarrow |A \cap B| > |A - B|$; *most* is strictly stronger than Q^R, but strictly weaker than *more*; see, for example, [11], Ch. 14.3.

for this is that determiners serve to *restrict the domain of quantification* to the first (noun) argument, and this enforces EXT.[10]

In fact, the EXT property is ubiquitous in natural language. The non-monadic quantifiers that feature in the interpretation of linguistic constructions such as reciprocals, branching, cumulative quantification, possessives, etc., are all EXT; see [11] for examples and discussion. Likewise, almost all type $\langle 1 \rangle$ noun phrase denotations satisfy EXT; in particular, all denotations of *restricted* noun phrases of the form [determiner + noun], when the noun has a fixed interpretation (*most students, every professor*).

Indeed, the only exceptions seem to be certain noun phrases containing a word, such as the English *thing*, which can be taken to denote the universe. Thus, the denotations of *everything, most things, all but five things*, are not EXT.[11] Since *every, most, all but five*, are all EXT, it seems that the 'blame' for this state of affairs must be placed on words like *thing*; more on this in Sect. 6 below.

5 Mathematical Quantifiers and EXT

A quick survey of quantifiers discussed in model theory — for example, as presented in [10] — reveals that many of them are EXT, but some are not. We shall look at just a few examples. The interest of these quantifiers lies in the model-theoretic properties of the corresponding logics, such as varying degrees of compactness, Löwenheim-Skolem properties, etc. But here the task is just to assess the status of EXT in a model-theoretic context.

We have already encountered Q^R and Q^C as non-EXT quantifiers. Cardinality quantifiers ($\exists_{\geq \aleph_\alpha}$ and variants), and cardinality comparison quantifiers like the *Härtig quantifier* ($I_M(A, B) \Leftrightarrow |A| = |B|$) and $more_M(A, B) \Leftrightarrow |A| > |B|$ (see note 9, are EXT.)

The *Magidor-Malitz quantifiers* Q^n_κ, defined, for $R \subseteq M^n$, by

$$(Q^n_\kappa)_M(R) \Leftrightarrow \exists X \subseteq M(|X| = \kappa \;\&\; X^n \subseteq R),$$

are EXT, and so are the related *Ramsey quantifiers*, a typical variant of which is defined as follows: for Q of type $\langle 1, 1 \rangle$, $A \subseteq M$, and $R \subseteq M^n$, let

$$(Ram^n(Q))_M(A, R) \Leftrightarrow \exists X \subseteq A(Q_M(A, X) \;\&\; X^n - Id^n_X \subseteq R)$$

---

[10]More precisely, it means that determiner denotations are *relativizations* of type $\langle 1 \rangle$ quantifiers (see Sect. 5), which, for type $\langle 1, 1 \rangle$ quantifiers, can easily be shown to be equivalent to saying that they satisfy EXT and *conservativity*:

(CONSERV) For all M and all $A, B \subseteq M$, $Q_M(A, B) \Leftrightarrow Q_M(A, A \cap B)$.

[11]These denotations are $\forall$, Q^R, and $(\exists_{=5})\neg$, respectively. On the other hand, *something*, or *at least three things*, also contain *thing*, but have EXT denotations.

where $Id_X^n = \{(a, \ldots, a) \colon (a, \ldots, a) \in X^n\}$. That is, if Q is EXT, so is $Ram^n(Q)$.

Now consider *partially ordered* quantifiers, i.e. quantifiers that express non-linearly ordered prefixes of $\forall x$ and $\exists y$. For most purposes, it suffices to consider the *Henkin prefix*

$$\left\{\begin{matrix} \forall x\ \exists x' \\ \forall y\ \exists y' \end{matrix}\right\} \psi(x, x', y, y')$$

This expresses the type $\langle 4\rangle$ *Henkin quantifier* Q^H, defined by

$$(Q^H)_M(R) \Leftrightarrow \exists f, g \colon M \to M\ \forall a, b \in M\ R(a, f(a), b, g(b))$$

Q^H does *not* satisfy EXT: indeed, $(Q^H)_M(R)$ and $M \subsetneq M'$ implies $\neg(Q^H)_{M'}(R)$. But this is 'made up for' by the fact that Q^H *relativizes*.

To explain this, recall that for any quantifier Q of type $\langle k_1, \ldots, k_n\rangle$, the *relativization* Q^{rel} of Q is the type $\langle 1, k_1, \ldots, k_n\rangle$ quantifier defined by

$$(Q^{rel})_M(A, R_1, \ldots, R_n) \Leftrightarrow Q_A(R_1 \restriction A, \ldots, R_n \restriction A)$$

A quantifier is *relativized* if it is equal to Q^{rel} for some Q, and we say that Q *relativizes* if Q^{rel} is definable in $L(Q)$. (Note that it is always the case that Q is definable in $L(Q^{rel}$.) Now, Q^H relativizes, since it is easy to check that $(Q^H)^{rel}$ is expressible by the following $L(Q^H)$-sentence:

$$Q^H xx'yy'(Px \wedge Py \ \to\ Px' \wedge Py' \wedge Rxx'yy')$$

We will get back to what this has to do with EXT, but first we consider another group of familiar quantifiers.

Order quantifiers express properties of a (usually linear) order relation R, for example, that R is a well-order, or is dense and contains a countable dense subset, or has cofinality ω, or is isomorphic to some given ordered set. Other order quantifiers compare two order relations, R and S, saying, for example, that they are isomorphic.

Two things are striking (from our perspective) about the discussion of these quantifiers in the literature. *First,* R is sometimes taken to be an ordering of the *universe*, and sometimes just an ordering of its *field*. In the former case, the quantifier is not EXT; in the latter, it usually is. *Second,* not much attention is paid to this difference, apparently since it usually doesn't matter for the relevant model-theoretic facts which version is chosen. Let us see what the general situation is.

Given any quantifier Q of type $\langle k_1, \ldots, k_n\rangle$, we obtain a 'universal version' Q^{uni} of the same type:

$$(Q^{uni})_M(R_1, \ldots, R_n) \Leftrightarrow Q_M(R_1, \ldots, R_n)\ \&\ \mathit{field}(R_1, \ldots, R_n) = M$$

Q^{uni} is definable from Q, since *field*($\overline{R}$) is defined by the first-order formula $\varphi_{\overline{R}}(x)$:

(6) $\varphi_{\overline{R}}(x) := \bigvee_{i=1}^{n} \exists x_1 \ldots \exists x_{k_i}(P_i x_1 \ldots x_{k_i} \wedge \bigvee_{j=1}^{k_i} x = x_j)$

Now, Q^{uni} is practically *never* EXT:

(7) If Q^{uni} is not the trivial quantifier **0**, then Q^{uni} is not EXT.

Proof If $Q^{uni} \neq \mathbf{0}$, there is $\mathcal{M} = (M, \overline{R})$ such that $Q_M(\overline{R})$ and $\mathit{field}(\overline{R}) = M$. Let $M' = M \cup \{a\}$, where $a \notin M$. Then we have that $(Q^{uni})_M(\overline{R})$ holds but not $(Q^{uni})_{M'}(\overline{R})$, so Q^{uni} does not satisfy EXT. □

In the cases mentioned above, the (non-EXT) order quantifiers saying *of the universe* that R is a well-order of it, or an order of a given order type, or dense with a countable dense subset, or has cofinality ω, all have the form Q^{uni}, where Q is EXT and says the same thing about the field of R. Similarly for the isomorphism quantifier.

The following general facts hold about EXT and the expressive power of Q, Q^{rel}, and Q^{uni}.

Proposition 1 *Let L be a logic and Q a quantifier of type* $\langle k_1, \ldots, k_n \rangle$.[12]

(a) Relativized quantifiers satisfy EXT.
(b) EXT *quantifiers relativize. Hence, if Q is* EXT, *then Q is definable in L iff* Q^{rel} *is definable in L.*
(c) If Q is EXT, *then* $L((Q^{uni})^{rel}) \equiv L(Q)$. *Moreover, if L relativizes,*[13] *then Q is definable in L iff* Q^{uni} *is definable in L.*

Proof (a): This is immediate from the definition of relativization.
(b): If Q is EXT, then, with $\overline{R} = R_1, \ldots, R_n$: $(Q^{rel})_M(A, \overline{R}) \Leftrightarrow Q_A(\overline{R}\restriction A) \Leftrightarrow Q_M(\overline{R}\restriction A)$ (by EXT). This condition is expressed in $L(Q)$ by the sentence (where $\overline{x_i} = x_{i1} \ldots x_{ik_i}$, A interprets P, and R_i interprets P_i, $1 \leq i \leq n$):

$$Q\overline{x_1}, \ldots, \overline{x_n}((P_1\overline{x_1} \wedge \bigwedge\nolimits_{j=1}^{k_1} Px_{1j}), \ldots, (P_n\overline{x_n} \wedge \bigwedge\nolimits_{j=1}^{k_n} Px_{nj}))$$

(c): We first observe that, for $A \subseteq M$,

$$\begin{aligned}(Q^{uni})^{rel}_M(A, \overline{R}) &\Leftrightarrow (Q^{uni})_A(\overline{R}\restriction A)\\ &\Leftrightarrow Q_A(\overline{R}\restriction A)\ \&\ \mathit{field}(\overline{R}\restriction A) = A\\ &\Leftrightarrow Q_M(\overline{R}\restriction A)\ \&\ \mathit{field}(\overline{R}\restriction A) = A \qquad \text{(by EXT)}\end{aligned}$$

This shows that $L((Q^{uni})^{rel}) \leq L(Q)$. Moreover, for the special case $A = \mathit{field}(\overline{R})$, we obtain

$$Q_M(\overline{R}) \Leftrightarrow (Q^{uni})^{rel}_M(\mathit{field}(\overline{R}), \overline{R})$$

which shows that $L(Q) \leq L((Q^{uni})^{rel})$.

[12] We assume that L extends FOL and is closed under substitution of formulas for predicate letters.
[13] If $L = L(Q_1, Q_2, \ldots)$, then L relativizes if each Q_i^{rel} is definable in L. Similarly for other logics; for example, FOL, $L_{\omega_1\omega}$, and $L_{\omega_1\omega_1}$ relativize, but not e.g. $L_{\omega\omega_1}$.

The second claim in (c) follows from the first: The left-to-right direction holds trivially because Q^{uni} is definable in $L(Q)$, and the other direction follows using this easy folklore fact:

(8) If L relativizes, then Q is definable in L iff Q^{rel} is definable in L.

□

By (a) of the proposition, for quantifiers like Q^H that are not EXT but still relativize, we can, if required, 'pretend' they are EXT, by using their relativizations instead, which do satisfy EXT and have the same expressive power.[14]

This doesn't work for order quantifiers of the form Q^{uni}, since these quantifiers *usually do not relativize*, even though Q is EXT and hence relativizes.[15] But by (c) of the proposition, Q and Q^{uni} still do not differ with respect to definability in many common logics. For example, the quantifier 'R is a well-order of the universe' is definable in $L_{\omega_1\omega_1}$, by conjoining the order axioms with the sentence

$$\neg\exists x_0 \ldots x_n \ldots (\bigwedge\nolimits_{n<\omega} Px_{n+1}x_n)$$

Since $L_{\omega_1\omega_1}$ relativizes, it follows that the stronger quantifier 'R is a well-order of its field' is also definable in $L_{\omega_1\omega_1}$.[16]

Finally, let us get back to the (pure) monadic quantifiers. If Q is of type $\langle 1, \ldots, 1\rangle$, then

$$(Q^{uni})_M(A_1, \ldots, A_n) \Leftrightarrow Q_M(A_1, \ldots, A_n) \ \&\ A_1 \cup \ldots \cup A_n = M$$

Although Proposition 1 still applies, part (c) has no interest, since in general, Q^{uni} has no interesting connection to Q. For example, $\forall^{uni} = \exists^{uni} = (Q^R)^{uni} = \forall$, $(\neg\exists)^{uni} = (\neg Q^R)^{uni} = \mathbf{0}$. Similarly, $(all^{uni})_M(A, B) \Leftrightarrow B = M$. The latter quantifier is neither EXT nor CONSERV (note 10), and in a natural language (or any other?) context there seems to be no reason to consider it.

Summing up then, it seems fair to say that failure of EXT usually plays no significant role for generalized quantifiers studied in mathematical logic. Many of these quantifiers already satisfy EXT. Among the ones that don't, some, like the Henkin quantifier, relativize, and thus it doesn't matter if one uses them or their relativized versions which do satisfy EXT, since they have the same expressive power. As to the group of order quantifiers, the (non-EXT) version describing properties of orders

[14] Q^H essentially involves the whole universe, which is reflected in the fact that $(Q^H)^{uni} = Q^H$.

[15] I have not attempted a general statement and proof of this fact here, but intuitively it should be fairly clear. Consider, for example, WO^{uni}, where $WO_M(R)$ says that R is a well-ordering of its field. Let η be the order type of the rationals, and let $\mathcal{M} = (M, A, R)$ be a model where R is a linear order of M with the order type $\eta + \omega + \eta$ and A is the subset corresponding to ω, and let $\mathcal{M}' = (M', A', R')$ be a similar model, but where R' has the order type $\eta + \omega + \omega^* + \omega + \eta$, and A' corresponds to $\omega + \omega^* + \omega$. Then $(WO^{uni})^{rel}_M(A, R)$ holds and $(WO^{uni})^{rel}_{M'}(A', R')$ fails, but $\mathcal{M}$ and $\mathcal{M}'$ cannot be distinguished in $L(WO^{uni})$. Indeed, over $\mathcal{M}'$, $L(WO^{uni})$ is equivalent to FOL, since there are no definable well-orders of the universe in $\mathcal{M}'$.

[16] Of course, in this case, this is immediate from a slight adjustment of the order axioms; the point here was to bring out the general situation.

of the whole universe is usually strictly weaker than the (EXT) version where only the field of the relation is ordered. Here we have seen that, as regards definability in logics that themselves allow relativization, the difference between the two versions disappears. Of course, this does not show that this difference is irrelevant to other model-theoretic properties of these quantifiers that have been studied. Still, my claim would be that *usually*, when the weaker version is used, this is because it slightly simplifies certain arguments, but these arguments can easily be adapted to the stronger version as well.

Remark: I am not aware of any study of this issue in the literature. Restricting attention to the property of compactness, here are three examples: (1) The well-order quantifier *WO* (note 15) is such that neither $L(WO)$ nor $L(WO^{uni})$ is compact. (2) The cofinality quantifier Q^{cf}_{ω}, says of a linear order that it has cofinality ω in its field. [14] showed that $L(Q^{cf}_{\omega})$ is fully compact, but, for example, the proof in [17] uses $(Q^{cf}_{\omega})^{uni}$ instead. But the result about $L((Q^{cf}_{\omega})^{uni})$ is only apparently weaker, since the proof is easily adjusted to deal with the stronger quantifier. (3) Let $(Q^D)_M(R) \Leftrightarrow R$ is a dense linear ordering of its field with a countable dense subset. Then $L(Q^D)$ is countably compact, and hence, so is $L((Q^D)^{uni})$. Indeed, the same proof of compactness can be given in both cases, by translation into stationary logic $L(aa)$, which is countably compact; see [10].[17] *End of remark.*

The situation is similar for monadic quantifiers, even though the version which requires the quantifier to 'act' on the whole universe is usually totally uninteresting. Here, the only non-EXT monadic quantifiers in the model-theoretic literature that I know of are of type $\langle 1 \rangle$, notably Q^R, Q^C, and of course $\forall$ (and variants such as $(\exists_{\geq n})\neg$, etc.). $\forall$ relativizes (as does its variants), so one can just as well use $\forall^{rel} = all$. But Q^R and Q^C do not. Further, neither $L(Q^R)$ nor $L(Q^C)$ is compact; in each there is a sentence with arbitrarily large finite models but no infinite model. Both logics satisfy an upward Löwenheim-Skolem theorem, hence none of them can characterize $(\omega, <)$, but it is not hard to show that this structure can be characterized in both $L((Q^R)^{rel}) = L(most)$ and $L((Q^C)^{rel})$.[18]

---

[17] However, Jouko Väänänen pointed out an exception (*p.c.*): the quantifier (not mentioned in [10]) $(Q^{card})_M(R) \Leftrightarrow R$ is a $|field(R)|$-like linear ordering of its field, that is, each proper initial segment has cardinality $< |field(R)|$. Then the sentence saying that R is discrete with a least element, and that for all a, $(Q^{card})_M(R \restriction \{b\colon bRa\})$, characterizes $(\omega, <)$, so $L(Q^{card})$ is not compact. On the other hand, it can be shown with techniques that go back to Vaught and Fuhrken that $L((Q^{card})^{uni})$ is (countably) compact. So my claim is indeed tentative and does not hold across the board; it may be still of some interest to find sufficient conditions for when it does.

[18] In other respects, $L(Q^R)$ and $L(Q^C)$ are quite different: $L(Q^C)$ reduces to FOL on finite models, whereas [1] showed that FOL $< L(Q^R) < L(most)$ even on finite models (illustrating the need for type $\langle 1, 1 \rangle$ quantifiers in natural language semantics).

6 'Thing'

I said that the only thing that seems to lead to violations of EXT in a natural language context are words like the English 'thing'. This word arguably denotes an operation across domains, which we may call a *unary predicate*, of relational type (e), defined by

$$thing_M = M \text{ for all } M$$

We note that *thing* itself is PERM and EXT (and hence ISOM, by (5-g)). Indeed, let the unary predicate *empty* be defined by $empty_M = \emptyset$ for all M. Then we have:

(9) The only unary predicates satisfying PERM and EXT are *thing* and *empty*.

This is a special case of a more general fact.

Proposition 2 ([22]) *The operations O across domains of relational type (e^n) that satisfy* PERM *and* EXT *are exactly the ones that are first-order definable without parameters by Boolean combinations of formulas of the form $x_i = x_j$, $1 \leq i, j \leq n$.*

A typical example of type (e^3) would be

$$O_M(R) = \{(a, b, c) \in M^3 : a = b \text{ or } b \neq c\}$$

which is defined in pure first-order logic with identity by the formula

$$x_1 = x_2 \vee \neg x_2 = x_3$$

Getting back to the unary case, if we count *empty* as trivial, we can say that *thing* is the only non-trivial *logical* unary predicate.[19] Even though it 'refers to the universe' in an obvious sense, it satisfies EXT and therefore, at least according to (I4), SAME.

In natural language semantics, it appears that all instances of non-EXT operations across domains arise from replacing a fixed predicate, such as *students* in *most students*, with *thing*: *most things* = Q^R. Curiously, *thing* itself is EXT, as is of course *most*. This kind of replacement doesn't always destroy EXT: for example, from *at least ten students* we get *at least ten things* = $\exists_{\geq 10}$. It can be shown that preservation of EXT holds precisely when the determiner denotation in question is *symmetric*, in the sense that $Q_M(A, B)$ implies $Q_M(B, A)$ for all M and all $A, B \subseteq M$.[20]

More generally, if we in a relativized quantifier of type $\langle 1, k_1, \ldots, k_n \rangle$ (which is always EXT) replace the first argument by *thing*, we obtain the original quantifier (which may or may not be EXT).

---

[19] [1] uses *thing* as a logical constant in its formal language.

[20] Provided Q is CONSERV and EXT (see [11], Ch. 6.1).

7 Taking Stock

After these observations about EXT, we can get back to SAME. As we saw in Sect. 2, the strongest intuitive cases of non-SAME quantifiers are those where Q 'behaves differently' on different M. Here are some more examples. Define (for all $M \neq \emptyset$)

$$\forall even_M = \begin{cases} \forall_M & \text{if } |M| \text{ is even} \\ \exists_M & \text{otherwise} \end{cases}$$

In other words (for all $A \subseteq M$),

$$\forall even_M(A) \Leftrightarrow \begin{cases} A = M & \text{if } |M| \text{ is even} \\ A \neq \emptyset & \text{otherwise} \end{cases}$$

This seems like a clear case: $\forall even$ does *not* satisfy SAME. But what about the following two test cases (restricting attention to finite universes)?

$$test1_M(A) \Leftrightarrow \begin{cases} |A| \geq (|M| + 2)/2 & \text{if } |M| \text{ is even} \\ |A| \geq (|M| + 1)/2 & \text{if } |M| \text{ is odd} \end{cases}$$

$$test2_M(A) \Leftrightarrow \begin{cases} |A| \geq (|M| + 2)/2 & \text{if } |M| \text{ is even} \\ |A| \geq (|M| - 1)/2 & \text{if } |M| \text{ is odd} \end{cases}$$

There is a strong inclination to treat *test1* and *test2* on a par: either both satisfy SAME or both don't; the difference between them seems minimal. And guided by the clear case of $\forall even$, we may think the latter option is most reasonable.

But now we have another clash of intuitions. We have so far assumed that Q^R satisfies SAME, but, in fact,

$$test1 = Q^R$$

(on finite universes)! So what shall we do: say that *test1* but not *test2* satisfies SAME, or that both do (but $\forall even$ does not), or that none of them does?

We will see below that, for quantifiers *of this particular kind*, there is in fact a principled way to count *test1* and $\forall$, but not *test2* or $\forall even$, among the quantifiers that behave 'the same' on all universes. However, for arbitrary quantifiers and more generally for operations across universes of arbitrary types, I think the example just reinforces the conclusion we have already drawn: there simply is no precise demarcation of the quantifiers, let alone the general operations across domains, that accords with *all* of our intuitions about sameness.

One might object that there could a be a *vague* criterion that fits *most* of our intuitions: there would then be clear cases of operations that satisfy the criterion, and clear cases that don't, but there would also be *borderline cases* where we don't know exactly what to say. This may well be the case. But, first, so far there is no specification of such a criterion that goes beyond the basic intuitions we have discussed in this note. Second, and more importantly, our goal was precisely to *replace* this vague notion by an exact criterion. My contention, then, is that it is not possible, in this way, to be faithful to all of those intuitions.

Certainly, this is not something I have demonstrated; presumably, it is not something that *could* be demonstrated. But cases analogous to *test1* and *test2* can be constructed *ad libitum*. More generally, along the lines above one can generate sequences $Q_1, \ldots, Q_n$ of quantifiers such that the difference between each Q_i and Q_{i+1} seems so small that they should not differ with respect to sameness, but where Q_1 is a familiar quantifier normally thought of as being the same across different domains, whereas Q_n is an 'artifical' quantifier without that property. Similarly, we saw that the intuitions behind (I2) concerning ISOM and the intuitions behind (I4) concerning EXT, if taken to hold unrestrictedly, are at odds with each other.

8 A Proposal

The conclusion just reached may seem disappointing. But we can end on a more positive note. The situation is precisely of a kind, I think, where a Carnapian *explication* is called for. In the present case, here is a proposal:

- Define: SAME = ISOM + EXT

In other words, stipulate that the conjunction of one necessary — on *one* intuition — and one sufficient — on *another* intuition — condition for sameness together constitute a necessary *and* sufficient condition.[21] ISOM says that the operation across domains is preserved under structure-preserving maps. EXT says that the part of the domain which is outside the arguments of the operation never matters. Let's require both.

This means, in effect, that we keep (I2) but reject (I4). It also means that, for example, $\forall$ is does not satisfy SAME. As I have argued, effects of this kind cannot be avoided.

A proposal like this is not right or wrong, but more or less *fruitful*. To evaluate it, one needs to look at its consequences, which is essentially what I have done in this note. First, it entails that sameness is not closed under (first-order) definability, but we already accepted that. It is closed, however, under the usual Boolean operations.

---

[21] As we noted, the proposal that SAME = PERM + EXT is equivalent. We can also formulate the proposal more succinctly: say that an n-ary operation O across domains of relational type is *closed under injections* if for all M, all relations R_i over M of suitable type, and all injections π from M to M',

(INJ) $O_M(R_1, \ldots, R_n) \Leftrightarrow O_{M'}(\pi(R_1), \ldots, \pi(R_n))$

It is easy to check that INJ $\equiv$ ISOM + EXT. However, the ideas behind ISOM and EXT are quite different, so they might as well be kept separate.

Second, many operations, in particular quantifiers, that seem to be defined in the same way across different universes will not satisfy our precise version of SAME. But we found reasons to believe that this is inevitable. What matters instead is whether the proposal has unacceptable consequences for familiar operations. Since sameness is not closed under definability anyway, we may be satisfied if these familar operations are *definable* (in some suitable sense) from SAME operations. This seems indeed to hold in most cases. In particular, as regards the quantifiers in mathematical logic we looked at, all of them are ISOM and many are EXT, and most of the non-EXT ones are either definable from EXT quantifiers, or can be replaced without loss by their EXT versions (their relativizations).

This is even clearer for natural language quantifiers: the only non-EXT ones that occur seem to be obtained from EXT quantifiers (such as determiner denotations) and the unary (and ISOM and EXT) predicate *thing*. On the other hand, many noun phrase denotations and some determiner denotations are not ISOM. Examples of latter are possessives like *Mary's* or *most students'*. But it seems correct to say in these cases that, as determiner denotations, these are indeed *not* SAME. The (standard) truth conditions for *Q students' A are B* are

$$Poss(Q, [\![\text{student}]\!], R)_M(A, B) \Leftrightarrow$$

$$Q_M(\{a \in [\![\text{student}]\!] \colon \exists b \in A\, R(a, b)\}, \{a \colon R_a \subseteq B\})$$

where R is a possessive relation and $R_a = \{b \colon R(a, b)\}$.[22] If we, as is reasonable, take $Q' = Poss(Q, [\![\text{student}]\!], R)$ to be the denotation of *Q students'*, then Q' is EXT (provided Q is CONSERV and EXT), but not ISOM (even if Q is). As long as $[\![\text{student}]\!]$ and R are fixed, $Poss(Q, [\![\text{student}]\!], R)$ is not the same across different domains. But if they are taken as variable arguments, we obtain an ISOM and EXT type $\langle 1, 2, 1, 1\rangle$ quantifier Q'' defined by $Q''(C, R, A, B) \Leftrightarrow Poss(Q, C, R)(A, B)$.

It could perhaps be argued that the type $\langle 1\rangle$ Rescher quantifier Q^R and the Chang quantifier Q^C, which are ISOM but not EXT and do not relativize, are particularly problematic cases for our proposal, since the intuition that these are the same on each domain may seem quite strong. These quantifiers have somewhat different logical properties than their relativizations, not with respect to compactness but with respect to being able to characterize the natural number ordering. So we cannot always use the relativizations instead.

Note that Q^R and Q^C are *monotone increasing* type $\langle 1\rangle$ quantifiers, in the sense that $(Q^R)_M(A)$ and $A \subseteq A' \subseteq M$ implies $(Q^R)_M(A')$, and similarly for Q^C. For these quantifiers, we can, if we want, (under a few extra assumptions) replace EXT by a weaker requirement of sameness.

[22] This is a special case of the more general truth conditions for possessives discussed at length in [12].

8.1 Smooth Quantifiers

Consider ISOM and monotone increasing type ⟨1⟩ quantifiers over *finite* universes. (This reduces Q^C to the universal quantifier.) For these, we know exactly which ones relativize:

Proposition 3 ([7, 23]) *A quantifier of the above kind relativizes if and only if it is first-order definable.*

Also, we know exactly which ones are EXT. The following is easy to show with standard methods from e.g. [19] or [11].

Proposition 4 *A quantifier of the above kind is* EXT *if and only if it is either the trivial quantifier* $\mathbf{0}$ *or* $\exists_{\geq k}$ *for some* $k \geq 0$.

There are uncountably many non-first-order definable, and hence non-EXT, quantifiers of this kind. Is there a reasonable Carnapian explication of which of these nonetheless 'behave the same' on all (finite) universes?

Such a property has in fact been discussed in the literature. The idea is that when the universe is extended with *just one* new element, the behavior must not change too much. More precisely, since Q is monotone increasing and ISOM, for each M there is a *minimum size* m, depending only on $|M|$, such that $Q_M(A) \Leftrightarrow |A| \geq m$. Thus Q is determined by a function $f\colon N \to N$ such that $f(n) \leq n+1$; we let $f(|M|) = n+1$ when $Q_M = \emptyset$. Now, the requirement on Q, i.e. on f, is that for all n, $f(n) \leq f(n+1) \leq f(n)+1$. In other words, when one element is added to M, the minimum size either stays the same or increases with 1.

This property was called *smoothness* in [18], who showed that smooth quantifiers are well-behaved in various logical ways, but it was already introduced (under different names) in [19, 20], where one idea was that smooth quantifiers have a particularly uniform behavior that might be a reason to classify them as *logical*. My point here is that smoothness seems a quite reasonable requirement of *sameness* across different universes. When a new element is added, the quantifier is not required to stay *exactly* the same — that only holds, among the quantifiers discussed here, for $\mathbf{0}$ and $\exists_{\geq k}$ — but it is allowed to change 'as little as possible' (as much as the universes changes).

We can now check that $Q^R = test1$ from Sect. 7 is indeed smooth, as is $\forall$, whereas *test2* and $\forall even$ are not: the minimum level of *test2* jumps *two* steps when we go, for example, from a universe with $2k-1$ elements to one with $2k$ elements. Similarly, the level for $\forall even$ jumps two steps from a universe with $2k$ elements to one with $2k+1$ elements.[23]

Thus, if desired, we may use a different stipulation of the SAME property for this particular class of quantifiers: ISOM + SMOOTH, instead of the much more restrictive ISOM + EXT. However, there seems to be no way to extend this idea to quantifiers of arbitrary type.

---

[23] We may note that, although *test2* and $\forall even$ appear to be very different quantifiers, and intuitively the latter seems clearly non-SAME, they have essentially the same 'jumping behavior'.

9 Conclusion

The main conclusion of our discussion is not really surprising: the idea of being the same operation on different domains is indeed vague, even when we grant — as is motivated by fairly clear examples — that this property is not closed under definability. However, there are strong intuitions about sameness, which can be cashed out in precise constraints. I have focused on two: isomorphism invariance (ISOM) and 'extension invariance' (EXT). The seemingly natural idea that the former is necessary and the latter sufficient for sameness cannot be upheld, but I suggested that the conjunction of the two is a fairly successful *explication* of sameness. It is a stipulation that allows almost all quantifiers used in practice, both in model theory and in natural language semantics, to be *definable* from SAME operations. The stipulation does not conform to every intuition about sameness, but this goal cannot be achieved anyway. I also suggested that, for the special case of monotone type $\langle 1 \rangle$ quantifiers over finite domains, the weaker constraint ISOM + SMOOTH does a better job of accounting for sameness intuitions.

The observations in this note were motivated by the identification in [4] of sameness across domains as a crucial ingredience in the notion of *logical operations*. The delineation of logicality is the more important, and more difficult, task. But I do think that the notion of sameness has some independent interest. Also, since (at least in my view) logicality *is* closed under (suitably specified) definability, we cannot expect sameness to be necessary for logicality. But it seems perfectly feasible to require logical operations of relational type to be definable from ISOM and EXT operations. This is manifestly true for generalized quantifiers: Each ISOM quantifier is definable from an ISOM and EXT quantifier, namely, its relativization.

References

1. Barwise, J., Cooper, R.: Generalized quantifiers and natural language. Linguist. Philos. **4**, 159–219 (1981)
2. Bonnay, D.: Logicality and invariance. Bull. Symb. Log. **14**(1), 29–68 (2008)
3. Bonnay, D., Engström, F.: Invariance and definability, with and without equality. Notre Dame J. Form. Log. **58** (2016)
4. Feferman, S.: Logic, logics and logicism. Notre Dame J. Form. Log. **40**, 31–54 (1999)
5. Feferman, S.: Set-theoretical invariance criteria for logicality. Notre Dame J. Form. Log. **51**, 3–20 (2010)
6. Feferman, S.: Which quantifiers are logical? a combined semantical and inferential criterion. In: Torza, A. (ed.) Quantifiers, Quantifiers and Quantifiers, pp. 19–30. Springer, Berlin (2015)
7. Kolaitis, P., Väänänen, J.: Generalized quantifiers and pebble games on finite structures. Ann. Pure Appl. Log. **74**, 23–75 (1995)
8. Lindström, P.: First order predicate logic with generalized quantifiers. Theoria **32**, 175–71 (1966)
9. McGee, V.: Logical operations. J. Philos. Log. **25**, 567–80 (1996)
10. Mundici, D.: Other quantifiers: An overview. In: Barwise, J., Feferman, S. (eds.) Model-Theoretic Logics, pp. 211–233. Springer-Verlag, Berlin (1985)

11. Peters, S., Westerståhl, D.: Quantifiers in Language and Logic. Oxford University Press, Oxford (2006)
12. Peters, S., Westerståhl, D.: The semantics of possessives. Language **89**(4), 713–759 (2013)
13. Rescher, N.: Plurality-quantification (abstract). J. Symb. Log. **27**, 373–374 (1962)
14. Shelah, S.: Generalized quantifiers and compact logic. Trans. Am. Math. Soc. **204**, 342–364 (1975)
15. Sher, G.: The Bounds of Logic. The MIT Press, Cambridge (1991)
16. Tarski, A.: What are logical notions? Hist. Philos. Log. **7**, 145–154 (1986)
17. Väänänen, J.: Models and Games. Cambridge University Press, Cambridge (2011)
18. Väänänen, J., Westerståhl, D.: On the expressive power of monotone natural language quantifiers over finite models. J. Philos. Log. **31**, 327–358 (2002)
19. van Benthem, J.: Questions about quantifiers. J. Symb. Log. **49**, 443–466 (1984)
20. van Benthem, J.: Essays in Logical Semantics. Kluwer, Dordrecht (1986)
21. van Benthem, J.: Logical constants across varying types. Notre Dame J. Form. Log. 315–342 (1989)
22. Westerståhl, D.: Logical constants in quantifier languages. Linguist. Philos. **8**, 387–413 (1985)
23. Westerståhl, D.: Relativization of quantifiers in finite models. In: van der Does, J., van Eijck, J. (eds.) Generalized Quantifier Theory and Applications, pages 187–205. ILLC, Amsterdam. (1991) Also in Quantifiers: Logic, Models and Computation (same editors), pp. 375–383, CSLI Publications, Stanford (1996)
24. Westerståhl, D.: Constant operators: Partial quantifiers. In: Larsson, S., Borin, L. (eds.) From Quantification to Conversation, pp. 11–35. College Publications, London (2012)

Author Biography

Dag Westerståhl got his Ph.D. with Per Lindström at the University of Gothenburg in 1977, and is currently Professor of Theoretical Philosophy and Logic at Stockholm University. His main research interests are model theory, in particular generalized quantifiers, and formal semantics of natural languages.

Gödel, Nagel, Minds, and Machines

Solomon Feferman

Abstract This is the author's slightly revised Nagel Lecture as it was published in the Journal of Philosophy CVI, 4 (April 2009), pp. 201–219. The permission of the Journal to republish the article is gratefully acknowledged.

Keywords Incompleteness theorems · Mechanism · Anti-mechanism · Turing machines · Undecidability

Some fifty years ago (1957 to be exact), Ernest Nagel and Kurt Gödel became involved in a contentious exchange about the possible inclusion of Gödel's original work on incompleteness in the book, *Gödel s Proof,* then being written by Nagel with James R. Newman.[1] What led to the conflict were some unprecedented demands that Gödel made over the use of his material and his involvement in the contents of the book—demands that resulted in an explosive reaction on Nagel's part. In the end the proposal came to naught. But the story is of interest because of what was basically at issue, namely their provocative related but contrasting views on the possible significance of Gödel's theorems for minds versus machines in the development of mathematics. That is our point of departure for the attempts by Gödel, and later J.R. Lucas and Roger Penrose, to establish definitive consequences of those theorems, attempts which—as we shall see—depend on highly idealized and problematic assumptions about minds, machines, and mathematics. In particular, I shall argue that there is a fundamental equivocation involved in those assumptions that needs to be reexamined. In conclusion, that will lead us to a new way of looking at how the mind

[1] Nagel and Newman, *Gödel's Proof* (New York: University Press, 1958); revised edition edited by Douglas R. Hofstadter (New York: University Press, 2001).

S. Feferman—Revised text of the Ernest Nagel Lecture given at Columbia University on September 27, 2007. I wish to thank Hannes Leitgeb and Carol Rovane for their helpful comments on an earlier version.

S. Feferman (✉)
Stanford, CA, USA

G. Jäger and W. Sieg (eds.), *Feferman on Foundations*, Outstanding Contributions to Logic 13, DOI 10.1007/978-3-319-63334-3_17

may work in deriving mathematics which straddles the mechanist and anti-mechanist viewpoints.

The story of the conflict between Gödel and Nagel has been told in full in the introductory note by Charles Parsons and Wilfried Sieg to the correspondence between them in Volume V of the Gödel *Collected Works*,[2] so I shall confine myself to the high points.

The first popular exposition of Gödel's incompleteness theorems was published by Nagel and Newman in 1956 in an article entitled "Goedel's Proof" for the *Scientific American*.[3] The article was reprinted soon after in the four volume anthology edited by Newman, *The World of Mathematics: A Small Library of the Literature of Mathematics from A'h-mosé the Scribe to Albert Einstein, Presented with Commentaries and Notes.*[4] That was an instant best-seller, and has since been reprinted many times. Newman had been trained as a mathematician but then became a lawyer and was in government service during World War II. Endlessly fascinated with mathematics, he became a member of the editorial board of the *Scientific American* a few years after the war.

Nagel had long been recognized as one of the leading philosophers of science in the United States, along with Rudolf Carnap, Carl Hempel, and Hans Reichenbach. Like them, he was an immigrant to the U.S.A., but unlike them he had come much earlier, in 1911 at the age of ten. Later, while teaching in the public schools, Nagel received his bachelor's degree at City College of New York in 1923 and his Ph.D. in philosophy at Columbia University in 1930. Except for one year at Rockefeller University, his entire academic career was spent at Columbia, where he became the John Dewey Professor of Philosophy and was eventually appointed to the prestigious rank of University Professor in 1967. In his philosophical work, Nagel combined the viewpoints of logical positivism and pragmatic naturalism. His teacher at City College had been Morris R. Cohen, and with Cohen in 1934 he published *An Introduction to Logic and Scientific Method;*[5] one of the first and most successful textbooks in those subjects.

Soon after the appearance of Nagel and Newman's article on Gödel's theorems in *The World of Mathematics,* they undertook to expand it to a short book to be published by New York University Press. Moreover, they had the idea to add an appendix which would include a translation of Gödel's 1931 paper on undecidable propositions together with the notes for lectures on that work that he had given during his first visit to Princeton in 1934.[6] Early in 1957 their editor at the Press, Allan Angoff, approached Gödel for permission to use that material. Though Gödel

---

[2]Gödel, *Collected Works, Volume V: Correspondence H-Z,*, Solomon Feferman et al., eds., (New York: Oxford, 2003), p. 135ff.

[3]*Scientific American,* cxciv (June 1956): 71–86.

[4]In *The World of Mathematics: A Small Library of the Literature of Mathematics from A' h-mosé the Scribe to Albert Einstein, Presented with Commentaries and Notes,* Volume 3 (New York: Simon and Schuster, 1956), pp. 1668–1695.

[5]Nagel and Cohen, *An Introduction to Logic and Scientific Method* (New York: Harcourt Brace, 1934).

[6]Gödel, "Über formal unentscheidbare Sätze der Principia Mathematica und verwandter Systeme I," *Monatshefte fur Mathematik und Physik,* xxxviii (1931): 173–198; reprinted with facing English

said he liked the Nagel–Newman article very much as a nontechnical introduction to his work, he said that he was concerned with some troublesome mistakes[7] that had been made in it and even more so with the interpretation of his results, so he was reluctant about agreeing to the proposal. To encourage him to accept, Nagel paid a lengthy personal visit to him at the Institute in March of 1957. A week later, Gödel wrote Angoff with three conditions. First,

> I would have to write an introduction to the appendix on the one hand in order to mention advances that have been made after the publication of my papers, [and] on the other hand in order to supplement the considerations given in the book, about the philosophical implications of my results. I am not very well pleased with the treatment of these questions that came out in the Scientific American and in the "World of Mathematics". This, for the most part is not the fault of the authors, because almost nothing has been published on this subject, while I have been thinking about it in the past few years.[8]

A second condition, somewhat of a surprise, was that he share in some way in the royalties of the book. But the most contentious condition was the one he put at the beginning of his letter to Angoff: "In view of the fact that giving my consent to this plan implies, in some sense, an approval of the book on my part, I would have to see the manuscript and the proof sheets of the book, including the appendix." This condition was made even more explicit in a follow-up letter:

> Of course I shall have to see the manuscript of the book before I sign the contract, so that I can be sure that I am in agreement with its content, or that passages with which I don't agree can be eliminated, or that I may express my view about the questions concerned in the introduction *(ibid.,* p. 4).

Gödel also wrote Nagel saying that he had made to Angoff "the same suggestions I mentioned to you in our conversation in Princeton." But when Nagel was shown the correspondence with Angoff, he exploded. He thought the proposal to share royalties was "grasping and unreasonable" yet was prepared to accept that. But what really ticked him off, as he wrote Angoff, was Gödel's number one condition:

> I could scarcely believe my eyes when I read his ultimatum that he is not only to see the manuscript of our essay *before* signing the contract with you, but that he is to have the right to eliminate anything in the essay of which he disapproves. In short, he stipulates as a condition of signing the contract the right of censorship.
>
> This seems to me just insulting, and I decline to be a party to any such agreement with Gödel .... If [his] conditions were granted, Jim and I would be compelled to make any alterations Gödel might dictate, and we would be at the mercy of his tastes and procrastination for a

---

translation in Gödel, *Collected Works, Volume I: Publications 1929–1936,* Feferman et al., eds. (New York: Oxford, 1986), pp. 144–195.

[7]See Hilary Putnam, "Review of Nagel and Newman (1958)," *Philosophy of Science* xxvii, 2 (1960): 205–207, for a review of Nagel and Newman's *Gödel's Proof* in which several errors are identified, the most egregious being the misstatement of Gödel's first incompleteness theorem and Rosser's improvement thereof on p. 91.

[8]Gödel, *Collected Works, Volume IV: Correspondence A—G,* Feferman et al., eds. (New York: Oxford, 2003), p. 1.

> period without foreseeable end .... Gödel is of course a great man, but I decline to be his slave.[9]

Nagel's fears about how things might go if they agreed to this condition were indeed well founded, but he left it in Angoff's hands to communicate his displeasure with it. In the event, Angoff tried to avoid direct conflict, and wrote Gödel in a seemingly accomodating but ambiguous way. It was not until August 1957 that Nagel wrote Gödel himself making clear his refusal to accept it:

> ... I must say, quite frankly, that your ... stipulation was a shocking surprise to me, since you were ostensibly asking for the *right to censor* anything of which you disapproved in our essay. Neither Mr. Newman nor I felt we could concur in such a demand without a complete loss of selfrespect. I made all this plain to Mr. Angoff when I wrote him last spring, though it seems he never stated our case to you. I regret now that I did not write you myself, for I believe you would have immediately recognized the justice of our demurrer. *(ibid.*, pp. 152–153).

This was passed over in silence in Gödel's response to Nagel. In the end, the proposal to include anything by him in the book by Nagel and Newman came to naught, and even the specific technical errors which Gödel could have brought to their attention remained uncorrected. Since this was to be the first popular exposition that would reach a wide audience, Gödel had good reason to be concerned, but by putting the conditions in the way that he did he passed up the chance to be a constructive critic.

As an aside, it is questionable whether Gödel indeed recognized the "justice of [their] demurrer." He made a similar request seven years later, when Paul Benacerraf and Hilary Putnam approached him about including his papers on Russell's mathematical logic and Cantor's continuum problem in their forthcoming collection of articles on the philosophy of mathematics. Gödel feared that Benacerraf and Putnam would use their introduction to mount an attack on his platonistic views, and he demanded what amounted to editorial control of it as a condition for inclusion. In that case, the editors refused straight off, but gave Gödel sufficient reassurance about the nature of its contents for him to grant permission to reprint as requested.[10]

1 Gödel's Concerns, Part I: The Formulation of the Incompleteness Theorems

Recall that from his first letter to Angoff, one of the conditions that Gödel put for the proposed appendix to the Nagel and Newman book was that he write an introduction to it "in order to mention advances that have been made after the publication of my papers" as well as "to supplement the considerations, given in the book, about the philosophical implications of my results," with which he was "not very well pleased."

[9]Gödel, *Collected Works, Volume V: Correspondence H—Z,* pp. 138–139.

[10]Cf. Gödel, *Collected Works, Volume II: Publications 1938–1974,* Feferman et al., eds. (New York: Oxford, 1990), p. 166.

As to the first of these, Gödel would have wanted to use the opportunity to put on record what he considered the strongest formulation of his incompleteness theorems. This was signaled in a letter that he composed to Nagel early in 1957 but that was apparently never sent. (A number of letters found in Gödel's *Nachlass* were marked *nicht abgeschickt.*) He there writes that

> [c]onsiderable advances have been made ... since 1934. ...it was only by Turing's work that it became completely clear, that my proof is applicable to *every* formal system containing arithmetic. I think the reader has a right to be informed about the present state of affairs.[11]

What he is referring to, of course, is Alan Turing's analysis in 1937 of the concept of effective computation procedure by means of what we now call Turing machines; Gödel had readily embraced Turing's explication after having rejected earlier proposals by Alonzo Church and Jacques Herbrand. But it was not until eight years after the debacle with Nagel and Newman that Gödel spelled out the connection with formal systems and his own work. That was in a postscript he added to the 1965 reprinting of his Princeton lectures in the volume, *The Undecidable,*[12] edited by Martin Davis:

> In consequence of later advances, in particular of the fact that, due to A.M. Turing's work, a precise and unquestionably adequate definition of the general concept of formal system can now be given, the existence of undecidable arithmetical propositions and the non-demonstrability of the consistency of a system in the same system can now be proved rigorously for *every* consistent formal system containing a certain amount of finitary number theory.
> Turing's work gives an analysis of the concept of "mechanical procedure" (alias "algorithm" or "computation procedure" or "finite combinatorial procedure"). This concept is shown to be equivalent with that of a "Turing machine". *A formal system can simply be defined to be any mechanical procedure for producing formulas, called provable formulas.* For any formal system in this sense there exists one in the [usual] sense ... that has the same provable formulas (and likewise vice versa) ... [Italics mine]

As we will see, there is much more to this identification than meets the eye; in fact, in my view it is the source of a crucial misdirection in the minds versus machines disputes that we shall take up below. By comparison, Nagel and Newman distinguish the consequences of the incompleteness theorems for axiomatic systems and for "calculating machines" in their concluding reflections as follows:

> [Gödel's theorems] show that there is an endless number of true arithmetical statements which cannot be formally deduced from any specified set of axioms .... It follows, therefore, that an axiomatic approach to number theory ... cannot exhaust the domain of arithmetical truth ....[13]
>
> Gödel's conclusions also have a bearing on the question whether calculating machines can be constructed which would be substitutes for a living mathematical intelligence. Such machines, as currently constructed and planned, operate in obedience to a fixed set of directives built in, and they involve mechanisms which proceed in a step-by-step manner. But in

[11]Gödel, *Collected Works, Volume V: Correspondence H—Z,* p. 147.

[12]In Martin Davis, ed., *The Undecidable: Basic Papers on Undecidable Propositions, Unsolvable Problems and Computable Functions* (Hewlett, NY: Raven, 1965); reproduced in Gödel, *Collected Works, Volume I: Publications 1929–1936,* p. 369.

[13]Nagel and Newman, "Goedel's Proof," *The World of Mathematics,* p. 1694.

> the light of Gödel's incompleteness theorem, there is an endless set of problems in elementary number theory for which such machines are inherently incapable of supplying answers, however complex their built-in mechanisms may be and however rapid their operations *(ibid.,* p. 1695).

Of course, Gödel could charitably read their informal idea of calculating machines simply to be explicated by the notion of Turing machines, but as we saw above, he would have gone beyond that to stress that there is no essential difference between these results.

The first thing that Gödel was surely reacting to on the philosophical rather than the technical side was the statement by Nagel and Newman that

> [the incompleteness theorems] seem to show that the hope of finding an absolute proof of consistency for any deductive system in which the whole of arithmetic is expressible cannot be realized, if such a proof must satisfy the finitistic requirements of Hilbert's original program *(ibid.,* p. 1694).

What must have annoyed Gödel was that this, together with their further reflections on the significance of the incompleteness theorems concerned matters to which he had given a good deal of thought over the years, in some respects with overlapping conclusions, in others contrary ones, but in all cases in much greater depth and with much greater care and precision. The trouble was that almost none of his thought on these questions had been published. With respect to the significance of the incompleteness theorems for Hilbert's finitist consistency program this was something that he had made only one brief cautious comment about at the end of his 1931 paper on undecidable propositions. But it was a matter he kept coming back to all through his life, as we found when we unearthed unpublished lectures and seminar presentations in his *Nachlass.* The first time his further views on this would begin to come out in print would be in 1958, a year after the imbroglio with Nagel and Newman.[14] Since my main concern in the remainder of this article is with the minds versus machines debate as it relates to the incompleteness theorems, I shall leave that matter at that.

[14]That was in the article, "Über eine bisher noch nicht benützte Erweiterung des finiten Standpunktes," *Dialectica,* xii (1958): 280–287; reprinted with facing English translation in Gödel, *Collected Works, Volume II: Publications 1938–1974,* pp. 240–251. As a sign of his concern with the issues involved, Gödel worked on a revision of that until late in his life, 1972, "On an Extension of Finitary Mathematics Which Has Not Yet Been Used," also in Gödel, *Collected Works, Volume II,* pp. 271–280. For the full story, see my forthcoming piece, "Lieber Herr Bernays! Lieber Herr Gödel! Gödel on Finitism, Constructivity and Hilbert's Program," in *Horizons of Truth* (Gödel centenary conference, Vienna, April 27–29, 2006).

2 Gödel's Concerns, Part II: Significance of the Incompleteness Theorems for the Minds Versus Machines Debate

The above quotations from Nagel and Newman's 1956 version of "Goedel's Proof," all come from their final three paragraphs on the "far reaching import" of the incompleteness theorems. (My guess is that Nagel was largely responsible for their formulation, but I have no specific evidence for that.) These, with some rewording but no essential change in content, were to form the entire last chapter, entitled "Concluding Reflections," of the 1958 version of the book *Gödel's Proof,* it is only the 1956 version that Gödel would have seen, however, and to which he would have reacted and so it is from there that I go on to draw the quotations. We now continue with the remaining parts of these paragraphs that concern the potentialities of human thought versus the potentialities of computing machines, already signaled above in their discussion of "calculating machines":

> It may very well be the case that the human brain is itself a "machine" with built-in limitations of its own, and that there are mathematical problems which it is incapable of solving. Even so, the human brain appears to embody a structure of rules of operation which is far more powerful than the structure of currently conceived artificial machines.[15]

And (from the third paragraph)

> The discovery that there are formally indemonstrable arithmetic truths does not mean that there are truths which are forever incapable of becoming known, or that a mystic intuition must replace cogent proof. It does mean that the resources of the human intellect have not been, and cannot be, fully formalized, and that new principles of demonstration forever await invention and discovery .... Nor do the inherent limitations of calculating machines constitute a basis for valid inferences concerning the impossibility of physico-chemical explanations of living matter and human reason. The possibility of such explanations is neither precluded nor affirmed by Gödel's incompleteness theorem. The theorem does indicate that in structure and power the human brain is far more complex and subtle than any nonliving machine yet envisaged *(ibid.).*

And finally,

> Gödel's work is a remarkable example of such complexity and subtlety. It is an occasion not for dejection because of the limitations of formal deduction but for a renewed appreciation of the powers of human reason (ibid.).

Here again Gödel would have reacted with a "been there, done that" annoyance, since he had already laid out his thoughts in this direction fifteen years earlier in what is usually referred to as his Gibbs lecture, "Some Basic Theorems on the Foundations of Mathematics and Their Implications."[16] But once more this is something he had

---

[15] Nagel and Newman, "Gödel's Proof," p. 1695.

[16] Gödel's lecture was the twenty-fifth in a distinguished series set up by the American Mathematical Society to honor the nineteenth century American mathematician, Josiah Willard Gibbs, famous for his contributions to both pure and applied mathematics. It was delivered to a meeting of the AMS held at Brown University on December 26, 1951. See Gödel, *Collected Works, Volume III: Unpublished Essays and Lectures*, Feferman et al., eds. (New York: Oxford, 1995), pp. 304–323.

never published, though he wrote of his intention to do so soon after delivering the lecture; in fact it never appeared in his lifetime. After Gödel died, the text languished with a number of other important essays and lectures in his *Nachlass* until it was retrieved by our editorial group for publication in Volume III of the Gödel *Collected Works.*

There are essentially two parts to the Gibbs lecture, both drawing conclusions from the incompleteness theorems. The first part concerns the potentialities of mind versus machines for the discovery of mathematical truths, and it is that part that we should compare with Nagel and Newman's reflections. The second part is an argument aimed to "disprove the view that mathematics is only our own creation," and thus to support some version of platonic realism in mathematics. George Boolos wrote a very useful introductory note to both parts of the Gibbs lecture in Volume III of the Gödel *Works (ibid.,* pp. 290–304); more recently I have published an extensive critical analysis of the first part, under the title "Are There Absolutely Unsolvable Problems? Gödel's Dichotomy,"[17] and I shall be drawing on that in the following.

What I call Gödel's dichotomy is the following statement that he highlighted in the first part of the Gibbs lecture:

> *Either...the human mind (even within the realm of pure mathematics) infinitely surpasses the powers of any finite machine, or else there exist absolutely unsolvable diophantine problems... (op. cit.,* p. 310, italics Gödel's).

By a *diophantine problem* is meant a proposition of elementary number theory (a.k.a. first order arithmetic) of a relatively simple arithmetical form whose truth or falsity is to be determined; its exact description is not important to us. Gödel showed that the consistency of a formal system is equivalent to a diophantine problem, to begin with by expressing it in the form that no number codes a proof of a contradiction.[18] According to Gödel, his dichotomy is a "mathematically established fact" which is a consequence of the incompleteness theorem. All that he says by way of an argument for it is the following, however:

> [I]f the human mind were equivalent to a finite machine, then objective mathematics not only would be incompletable in the sense of not being contained in any well-defined axiomatic system, but moreover there would exist *absolutely* unsolvable problems ..., where the epithet "absolutely" means that they would be undecidable, not just within some particular axiomatic system, but by *any* mathematical proof the mind can conceive *(ibid.,* italics Gödel's).

By a *finite machine* here Gödel means a Turing machine, and by a *well-defined axiomatic system* he means an effectively specified formal system; as explained above, he takes these to be equivalent in the sense that the set of theorems provable in such a system is the same as the set of theorems that can be effectively enumerated by such a machine. Thus, to say that the human mind is equivalent to a finite machine "even within the realm of pure mathematics" is another way of saying that what the human mind can *in principle* demonstrate in mathematics is the same as the set of

[17] *Philosophia Mathematica,* Series III, xiv (2006): 134–152.

[18] In modern terms, consistency statements belong to the class $\prod_1$, that is, are of the form $\forall x\, R(x)$ with R primitive recursive.

theorems of some formal system. By *objective mathematics* Gödel means the totality of true statements of mathematics, which includes the totality of true statements of first-order arithmetic. The assertion that objective mathematics is incompletable is at first sight simply a consequence of the second incompleteness theorem in the form that for any consistent formal system S containing a certain basic system S_0 of (true) arithmetic, the number-theoretical statement Con(S) that expresses the consistency of S is true but not provable in S.

Examined more closely, Gödel's argument is that if the human mind were equivalent to a finite machine, or—what comes to the same thing—an effectively presented formal system S, then there would be *a true statement that could never be humanly proved,* namely Con(S). So that statement would be *absolutely undecidable* by the human mind, and moreover it would be equivalent to a diophantine statement. *Note however, the tacit assumption that the human mind is consistent,* otherwise, it is equivalent to a formal system in a trivial way, namely one that proves all statements. Actually, Gödel apparently accepts a much stronger assumption, namely that we prove *only* true statements, but for his argument, only the weaker assumption is necessary (together of course with the assumption that the basic system of arithmetic S_0 has been humanly accepted). Also, Gödel's sketch to establish his dichotomy should be modified slightly, as follows: either humanly demonstrable mathematics is *contained* in some consistent formal system S, or not. If it is, then Con(S) is an absolutely undecidable (diophantine) statement. If not, then for each consistent formal system S, there is a humanly provable statement that is not provable in S, that is, the "human mind ... infinitely surpasses the powers of any finite machine."

Note well that Gödel's dichotomy is not a strict one as it stands; Gödel himself asserts that "the case that both terms of the disjunction are true is not excluded, so that there are, strictly speaking, three alternatives." This could happen, for example, if the human mind infinitely surpasses finite machines with respect to certain diophantine problems, but not with respect to *all* of them. And if we drop the word 'diophantine' from its statement, it might be that the human mind could settle all arithmetical problems, but not all problems of higher mathematics. In fact, some logicians conjecture that Cantor's Continuum Problem is an absolutely undecidable problem of that type.

Be that as it may, how does Gödel's conclusion differ from that of Nagel and Newman? They speak of calculating machines "as currently constructed and planned, [that] operate in obedience to a fixed set of directives built in, and [that] involve mechanisms which proceed in a step-by-step manner," where Gödel speaks more precisely of Turing machines. Still, to be charitable, I think that is a reasonable interpretation of what Nagel and Newman had in mind. Second, they do not make the connection between formal systems and calculating machines, where Gödel sees these as amounting to the same thing.[19] But the essential difference is that they seem to come down in favor of the first, anti-mechanist, disjunct of the dichotomy when they say that "[by] Gödel's incompleteness theorem, there is an endless set of problems

[19]Calculating machines are assimilated more closely to axiomatic systems in the concluding reflections of Nagel and Newman, *Gödel's Proof,* p. 100.

in elementary number theory for which such machines are inherently incapable of supplying answers, however complex their built-in mechanisms may be and however rapid their operations" and "the human brain appears to embody a structure of rules of operation which is far more powerful than the structure of currently conceived artificial machines"; furthermore, in the third paragraph they say that "the resources of the human intellect have not been, and cannot be, fully formalized, and ... new principles of demonstration forever await invention and discovery."

There is a lot of evidence outside of the Gibbs lecture that Gödel was also convinced of the anti-mechanist position as expressed in the first disjunct of his dichotomy. That is supplied, for example, in his informal communication of various ideas about minds and machines to Hao Wang, initially in the book, *From Mathematics to Philosophy*,[20] and then at greater length in *A Logical Journey: From Gödel to Philosophy*.[21] So why didn't Gödel state that outright in the Gibbs lecture instead of the more cautious disjunction in the dichotomy? The reason was simply that he did not have an unassailable proof of the falsity of the mechanist position. Indeed, he there says:

> [It is possible that] the human mind (in the realm of pure mathematics) *is* equivalent to a finite machine that, however, is unable to understand completely its own functioning (*op. cit,* p. 309).

And in a related footnote, despite his views concerning the "impossibility of physico-chemical explanations of ... human reason" he says that

> [I]t is conceivable ... that brain physiology would advance so far that it would be known with empirical certainty
>
> 1. that the brain suffices for the explanation of all mental phenomena and is a machine in the sense of Turing;
> 2. that such and such is the precise anatomical structure and physiological functioning of the part of the brain which performs mathematical thinking *(ibid.)*.

And in the next footnote, he says:

> [T]he physical working of the thinking mechanism could very well be completely understandable; the insight, however, that this particular mechanism must always lead to correct (or only consistent) results would surpass the powers of human reason (*ibid.,* p. 310).

Some twenty years later, Georg Kreisel made a similar point in terms of formal systems rather than Turing machines:

> [I]t has been clear since Gödel's discovery of the incompleteness of formal systems that we could not have *mathematical* evidence for the adequacy of any formal system; but this does not refute the possibility that some quite specific system ... encompasses all possibilities of (correct) mathematical reasoning ...

---

[20]Wang, *From Mathematics to Philiosophy* (New York: Routledge and Kegan Paul, 1974), pp. 324–326.

[21]Wang, *A Logical Journey: From Gödel to Philosophy* (Cambridge: MIT, 1996), especially Chap. 6.

> *In fact the possibility is to be considered that we have some kind of nonmathematical evidence for the adequacy of such [a system].*[22]

3 Critiquing the Minds Versus Machines Debate

I shall call the genuine possibility entertained by Gödel and Kreisel, *the mechanist's empirical defense* (or *escape hatch)* against claims to have *proved* that mind exceeds mechanism on the basis of the incompleteness theorems. The first outright such claim was made by J.R. Lucas in his 1961 article, "Minds, Machines and Gödel."[23] Both Benacerraf and Putnam soon objected to his argument on the basis that Lucas was assuming it is known that one's mind is consistent. Lucas, in response, has tried to shift the burden to the mechanist: "The consistency of the machine is established not by the mathematical ability of the mind, but on the word of the mechanist," a burden that the mechanist can refuse to shoulder by simply citing his empirical defense.[24]

Roger Penrose is the other noted defender of the Gödelian basis for anti-mechanism, most notably in his two books, *The Emperor s New Mind*,[25] and *Shadows of the Mind.*[26] Sensitive to the objections to Lucas, he claimed in the latter only to have proved something more modest (and in accord with experience) from the incompleteness theorems: "Human mathematicians are not using a knowably sound algorithm in order to ascertain mathematical truth" (*ibid.,* p. 76). But later in that work after a somewhat involved discussion, he came up with a new argument purported to show that the human mathematician cannot even consistently *believe* that his mathematical thought is circumscribed by a mechanical algorithm (*ibid.,* Sects. 3.16 and 3.23). Extensive critiques have been made of Penrose's original and new arguments in an issue of the journal *PSYCHE,* to which he responded in the same issue.[27] And more recently, Stewart Shapiro[28] and Per Lindström[29] have carefully analyzed and then undermined his "new argument." But Penrose has continued to defend it, as he did in his public lecture for the Gödel Centenary Conference held in Vienna in April 2006.

---

[22]Kreisel, "Which Number-theoretic Problems Can Be Solved in Recursive Progressions on Π^1_1 Paths through O?" *Journal of Symbolic Logic,* xxxvii (1972): 311–334, see p. 322; italics added.

[23]Lucas, "Minds, Machines, and Gödel," *Philosophy,* xxxvi (1961): 112–137.

[24]Lucas, "Minds, Machines, and Gödel: A Retrospect," in P.J.R. Millican and A. Clark, eds., *Machines and Thought: The Legacy of Alan Turing,* Volume 1 (New York: Oxford, 1996), pp. 103–124.

[25]Penrose, *The Emperor's New Mind* (New York: Oxford, 1989).

[26]Penrose, *Shadows of the Mind* (New York: Oxford, 1994).

[27]Penrose, "Beyond the Doubting of a Shadow," *Psyche,* II, 1 (1996): 89–129; also at http://psyche.cs.monash.edu.au/v2/psyche-2-23-penrose.html.

[28]Shapiro, "Mechanism, Truth, and Penrose's New Argument," *Journal of Philosophical Logic,* xxxii (2003): 19–42.

[29]Lindström, "Penrose's New Argument," *Journal of Philosophical Logic,* xxx (2001): 241–250, and "Remarks on Penrose's 'New Argument'," *Journal of Philosophical Logic,* xxxv (2006): 231–237.

Historically, there are many examples of mathematical proofs of what cannot be done in mathematics by specific procedures, for example, the squaring of the circle, or the solution by radicals of the quintic, or the solvability of the halting problem. But it is hubris to think that by mathematics alone we can determine what the human mind can or cannot do in general. The claims by Gödel, Lucas, and Penrose to do just that from the incompleteness theorems depend on making highly idealized assumptions both about the nature of mind and the nature of machines. A very useful critical examination of these claims and the underlying assumptions has been made by Shapiro in his article, "Incompleteness, Mechanism, and Optimism,"[30] among which are the following. First of all, how are we to understand the mathematizing capacity of the human mind, since what is at issue is the producibility of an infinite set of propositions? No one mathematician, whose life is finitely limited, can produce such a list, so either what one is talking about is what the individual mathematician *could do in principle,* or we are talking in some sense about the potentialities of the pooled efforts of the community of mathematicians now or ever to exist. But even that must be regarded as a matter of what can be done *in principle,* since it is most likely that the human race will eventually be wiped out either by natural causes or through its own self-destructive tendencies by the time the sun ceases to support life on earth.

What about the assumption that the human mind is consistent? In practice, mathematicians certainly make errors and thence arrive at false conclusions that in some cases go long undetected. Penrose, among others, has pointed out that when errors are detected, mathematicians seek out their source and correct them,[31] and so he has argued that it is reasonable to ascribe self-correctability and hence consistency to our idealized mathematician. But even if such a one can correct all his errors, can he know with mathematical certitude, as required for Gödel's claim, that he is consistent?

As Shapiro points out, the relation of both of these idealizations to practice is analogous to the competence/performance distinction in linguistics.

There are two further points of idealization to be added to those considered by Shapiro. The first of these is the assumption that the notions and statements of mathematics are fully and faithfully expressible in a formal language, so that what can be humanly proved can be compared with what can be the output of a machine. In this respect it is usually pointed out that the only part of the assumption that needs be made is that the notions and statements of elementary number theory are fully and faithfully represented in the language of first-order arithmetic, and that among those only diophantine statements of the form $\mathrm{Con}(S)$ for S an arbitrary effectively presented formal system need be considered. But even this idealization requires that statements of unlimited size must be accessible to human comprehension.

Finally to be questioned is the identification of the notion of finite machine with that of Turing machine. Turing's widely accepted explication of the informal concept

[30] Shapiro, "Incompleteness, Mechanism, and Optimism," *Bulletin of Symbolic Logic,* iv (1998): 273–302.

[31] Cf. Penrose, "Beyond the Doubting of a Shadow," p. 137ff.

of effective computability puts no restriction on time or space that might be required to carry out computations. But the point of that idealization was to give the strongest *negative* results, to show that certain kinds of problems cannot be decided by a computing machine, no matter how much time and space we allow. And so if we carry the Turing analysis over to the potentiality of mind in its mathematizing capacity, to say that mind infinitely surpasses any finite machine is to say something even stronger. It would be truly impressive if that could be definitively established, but none of the arguments that have been offered are resistant to the mechanist's empirical defense. Moreover, suppose that the mechanist is right, and that in some reasonable sense mind *is* equivalent to a finite machine: Is it appropriate to formulate that in terms of the identification of what is humanly provable with what can be enumerated by a Turing machine? Isn't the mechanist aiming at something stronger in the opposite direction, namely an explanation of the mechanisms that govern the production of human proofs?

Here is where I think something new has to be said, something that I already drew attention to in my article on Gödel's dichotomy,[32] but that needs to be amplified. Namely, there is an *equivocation* involved that lies in identifying *how* the mathematical mind works with the totality of *what* it can prove. Again, the difference is analogous to what is met in the study of natural language, where we are concerned with the *way* in which linguistically correct utterances are generated and *not* with the potential totality of *all* such utterances. That would seem to suggest that if one is to consider *any* idealized formulation of the mechanist's position at all, it ought to be of the mind as one *constrained* by the axioms and rules of some effectively presented formal system. Since in following those axioms and rules one has *choices* to be made at each step, *at best* that identifies the mathematizing mind with *the program for a nondeterministic Turing machine,* and *not* with the set of its enumerable statements (even though that can equally well be supplied by a deterministic Turing machine).

One could no more disprove this modified version of the idealized mechanist's thesis than the version considered by Gödel and the others, simply by applying the mechanist's empiricist argument. Nevertheless, it is difficult to conceive of any formal system of the sort with which we are familiar, from Peano Arithmetic (PA) up to Zermelo–Fraenkel Set Theory (ZF) and beyond, actually underlying mathematical thought as it is experienced. And the experience of the mathematical practitioner certainly supports the conclusion drawn by Nagel and Newman that "mathematical proof does not coincide with the exploitation of a formalized axiomatic method," even if that cannot be demonstrated unassailably as a consequence of Gödel's incompleteness theorems.

[32]Feferman, "Are There Absolutely Unsolvable Problems? Gödel's Dichotomy".

4 One Way to Straddle the Mechanist and Anti-mechanist Positions

As I see it, the reason for the implausibility of this modified version of the mechanist's thesis lies in the concept of a formal system S that is currently taken for granted in logical work. An essential part of that concept is that the language of S is fixed once and for all. For example, the language of PA is determined (in one version) by taking the basic symbols to be those for equality, zero, successor, addition, and multiplication ($=, 0, ', +, \cdot$), and that of ZF is fixed by taking its basic symbols to be those for equality and membership ($=, \in$). This forces axiom schemata that may be used in such systems, such as induction in arithmetic and separation in set theory, to be infinite bundles of all possible substitution instances by formulas from that language; this makes metamathematical but not mathematical sense. Besides that, the restriction of mathematical discourse to a language fixed in advance, even if only implicitly, is completely foreign to mathematical practice.

In recent years I have undertaken the development of a modified conception of formal system that does justice to the openness of practice and yet gives it an underlying rule-governed logical-axiomatic structure; it thus suggests a way, admittedly rather speculative, of straddling the Gödelian dichotomy. This is in terms of a notion of *open-ended schematic axiomatic system,* that is, one whose schemata are finitely specified by means of propositional and predicate variables (thus putting the 'form' back into 'formal systems') while the language of such a system is considered to be *open-ended*, in the sense that its basic vocabulary may be expanded to any wider conceptual context in which its notions and axioms may be appropriately applied. In other words, on this approach, *implicit in the acceptance of given schemata is the acceptance of any meaningful substitution instances that one may come to meet,* but which those instances are is not determined by restriction to a specific language fixed in advance.[33]

The idea is familiar from standard presentations of propositional and predicate logic, where we have such axioms as

$$P \wedge Q \rightarrow P \qquad \text{and} \qquad (\forall x)P(x) \rightarrow P(a),$$

and rules of inference such as

$$P, P \rightarrow Q \vdash Q \qquad \text{and} \qquad P \rightarrow Q(x) \vdash P \rightarrow (\forall x)Q(x) (\text{for } x \text{ not free in } P).$$

We do not conceive of logic as applying to a single subject matter fixed once and for all, but rather to any subject in which we take it that we are dealing with well-defined propositions and predicates; logic is there applied by substitution for the proposition

[33] Cf. Feferman, "Gödel's Program for New Axioms: Why, Where, How and What?" in *Gödel '96,* P. Hajek, ed., *Lecture Notes in Logic,* vi(1996): 3–22, and Feferman, "Open-ended Schematic Axiom Systems (abstract)," *Bulletin of Symbolic Logic,* xii(2006): 145, and Feferman and Thomas Strahm, "The Unfolding of Non-finitist Arithmetic," *Annals of Pure and Applied Logic,* civ (2000): 75–96.

and predicate letters in its axioms and rules of inference. Similarly to allow systems like those for arithmetic and set theory to be applicable no matter what subject matter we happen to deal with, we formulate their basic principles in schematic form such as the *Induction Axiom Scheme*

$$P(0) \wedge (\forall x)[P(x) \rightarrow P(x')] \rightarrow (\forall x)P(x)$$

in the case of arithmetic, and the *Separation Scheme*

$$(Va)(\exists b)(\forall x)[x \in b \leftrightarrow x \in a \wedge P(x)]$$

in the case of set theory. (The idea for the latter really goes back to Zermelo's conception of the *Aussonderungsaxiom* as applicable to any "definite" predicate.)

But why, it may be (and is) asked, do I insist on the vague idea of an open-ended language for mathematics? Aren't all mathematical concepts defined in the language of set theory? It is indeed the case that the current concepts of working ("pure") mathematicians are with few exceptions expressible in set theory. But there are genuine outliers. For example a natural and to all appearances coherent mathematical notion whose full use is not set-theoretically definable is that of a category; only so-called "small" categories can be directly treated in that way.[34] Other outliers are to be found on the constructive fringe of mathematics in the schools of Brouwerian intuitionism and Bishop's constructivism[35] whose basic notions and principles are not directly accounted for in set theory with its essential use of classical logic. And it may be argued that there are informal mathematical concepts like those of knots, or infinitesimal displacements on a smooth surface, or of random variables, to name just a few, which may be the subject of convincing mathematical reasoning but that are accounted for in set theory only by some substitute notions that share the main expected properties but are not explications in the ordinary sense of the word. Moreover, the idea that set-theoretical concepts and questions like Cantor's continuum problem have determinate mathematical meaning has been challenged on philosophical grounds.[36] Finally, there is a theoretical argument for openness, even if one accepts the language L of set theory as a determinately meaningful one. Namely, by Tarski's theorem, the notion of truth T_L for L is not definable in L; and then the notion of truth for the language obtained by adjoining T_L to L is not definable in *that* language, and so on (even into the transfinite).

---

[34]Cf. Saunders Mac Lane, *Categories for the Working Mathematician* (Berlin: Springer, 1971), and Feferman, "Categorical Foundations and Foundations of Category Theory," in R.E. Butts and J. Hintikka, eds., *Logic, Foundations of Mathematics and Computability Theory,* Volume I (Dordrecht: Reidel, 1977), pp. 149–165, and Feferman, "Enriched Stratified Systems for the Foundations of Category Theory," in G. Sica, ed., *What Is Category Theory?* (Monza, Italy: Polimetrica, 2006), pp. 185–203.

[35]Cf. Michael Beeson, *Foundations of Constructive Mathematics* (Berlin: Springer, 1985).

[36]Feferman, "Does Mathematics Need New Axioms?" *Bulletin of Symbolic Logic,* vi (2000): 401–413.

Another argument that may be made against the restriction of mathematics to a language fixed in advance is historical. Sustained mathematical reasoning had its origins in ancient Greece some 2500 years ago, and mathematical concepts of number and space have undergone considerable evolution since then. Yet even the earliest results have permanent validity, though not necessarily as originally conceived. While Euclidean geometry is no longer considered to be the geometry of actual space, its consequences as an axiomatic development (suitably refined through the work of Pasch, Hilbert, and others), such as the angle sum theorem and Pythagoras's theorem, are as valid now within that context as in Euclid's time. On the other hand, the origins of number theory as represented in Euclid's *Elements,* including the existence of infinitely prime numbers and the fundamental theorem of arithmetic, retain their direct interpretation and validity. But the language of mathematical practice (often mixed with physical concepts) then and through the many following centuries up to the present obviously cannot be identified with the language of set theory. One thing that would instead account for the continuity of mathematical thought throughout its history is the employment of certain underlying formal patterns, such as those indicated above, that could be instantiated by the evolving concepts that have come to fill out mathematics in the process of its development. And there is no reason to believe that this evolutionary process has come to an end; it would be foolish to believe that only that which can be expressed, say, in the language of set theory, will count as mathematics henceforth.

On this picture, in order to straddle the mechanist/anti-mechanist divide at the level considered here, one will have to identify *finitely many basic forms of mathematical reasoning* which work in tandem to fully constrain and distinguish it. These would constitute the mechanist side of the picture, while the openness as to what counts as a mathematical concept would constitute the anti-mechanist side. The evidence for such would have to be empirical, by showing how typical yet substantial portions of the mathematical corpus are accounted for in those terms while giving special attention to challenging cases. I have suggested steps in that direction,[37] but the program is ambitious and I have only made a start; spelling that out is planned for a future publication. In the meantime, the program itself should be treated as highly speculative, yet—I hope—worthy of serious consideration.

Considered more broadly and apart from the tendentious terms of the mechanism/anti-mechanism debate, I think the goal should be to give an informative, systematic account at a theoretical level of how the mathematical mind works that squares with experience. Characterizing the logical structure of mathematics—what constitutes a proof—is just one aspect of that, important as that may be. Other aspects—the ones that are crucial in making the difficult choices that are necessary for the mathematician to obtain proofs of difficult theorems—such as the role of heuristics, analogies, metaphors, physical and geometric intuition, visualization, and so on, have also been taken up and are being pursued in a more or less systematic way by mathematicians,

[37] As indicated in "Are There Absolutely Unsolvable Problems?".

philosophers, and cognitive scientists.[38] Without abandoning a basic naturalist stance, I see no usable reductive account of the mathematical experience anywhere on the horizon at the neuro-physiological level let alone more basic physico-chemical levels of the sort contemplated by Nagel and Newman and currently sought by Penrose, among others.

Solomen Fefermen

Stanford University

[38]Cf., for example, Efraim Fischbein, *Intuition in Science and Mathematics* (Dordrecht: Reidel, 1987); George Lakoff and Rafael E. Núñez, *Where Mathematics Comes From* (New York: Basic, 2000); Paolo Mancosu, Klaus Frovin Jørgensen, Stig Andur Pedersen, eds., *Visualization, Explanation and Reasoning Styles in Mathematics* (Dordrecht: Springer, 2005); and George Polya, *Mathematics and Plausible Reasoning,* Volumes 1 and 2 (Princeton: University Press, 1968, 2nd edition) among others. In addition, there is a massive amount of anecdotal evidence for the nonmechanical essentially creative nature of mathematical research; for a small sample with further references, cf., for example, Philip J. Davis, and Reuben Hersh, *The Mathematical Experience* (Boston: Birkhäuser, 1981), and David Ruelle, *The Mathematician's Brain* (Princeton: University Press, 2007), and Benjamin H. Yandell, *The Honors Class: Hilbert's Problems and Their Solvers* (Natick, MA: A.K. Peters, 2002).

A Brief Note on *Gödel, Nagel, Minds, and Machines*

Wilfried Sieg

Abstract This note is a brief comment on Feferman's *Gödel, Nagel, Minds, and Machines*. It emphasizes the need to expand proof theory and use its formal tools for the analysis of the informal proofs of mathematical practice. Natural formalization is seen as one important step toward providing what Feferman called for, namely, "an informative, systematic account at a theoretical level of how the mathematical mind works that squares with experience".

Keywords Mechanist thesis · Mechanism and mind · Computability · Natural formalization · Theory of proofs

Feferman's paper *Gödel, Nagel, Minds, and Machines* is a revision of the Nagel Lecture he presented on 27 September 2007 at Columbia University in New York City. The lecture and this paper were intended for a broad audience, steeped neither in mathematical details of Gödel's proofs, nor in methodological problems concerning a precise mathematical definition of "computability", nor in philosophical controversies surrounding the mind and machine issue. Feferman presents these important and intertwined matters in a lucid way. He discusses in a self-contained way also the contentious interaction between Gödel and Nagel[1] and the neglected, but insightful remarks on minds and machines in (Nagel and Newman, [5, 6]). Thus, there is no need to describe the context for this paper in order to support a reader who is eager to engage in this discussion at the intersection of philosophy, mathematics and theoretical computer science – with a significant link to cognitive science.

[1]Feferman's remarks are based on the correspondence between Gödel and Nagel – beginning with a letter from Gödel on 25 February 1957 and ending with a short note again from Gödel on 29 August 1957 – and the Introductory Note to that correspondence by Parsons and Sieg; all of this can be found in volume V of *Gödel's Collected Works*, pp.135–154.

W. Sieg (✉)
Department of Philosophy, Carnegie Mellon University,
Baker Hall 161 5000 Forbes Avenue, Pittsburgh, PA 15213, USA
e-mail: sieg@cmu.edu

G. Jäger and W. Sieg (eds.), *Feferman on Foundations*, Outstanding Contributions to Logic 13, DOI 10.1007/978-3-319-63334-3_18

The clear exposition is informed by Feferman's special expertise as a mathematical logician and as a historian of mathematical ideas. As a logician, he has worked on the "arithmetization of metamathematics" underlying Gödel's proofs and on "progressions of theories" that partially overcome the limitations of formal theories revealed by the incompleteness theorems. As a historian, he has deep knowledge of the evolution of Gödel's thought through his work over more than two decades as editor-in-chief of Gödel's *Collected Works*. In the last part of this essay, Feferman proposes a new way of "straddling" the mechanist and anti-mechanist divide on the mind and machine issue. He views his proposal as "highly speculative", but based on a perspective obtained by investigating how the mind may work when deriving mathematics.

The "mechanist thesis" is the claim that the mathematical mind is, in some sense, equivalent to a Turing machine. Famously, Lucas [4] and Penrose [7] attempted to refute this thesis by an appeal to Gödel's incompleteness theorems. Gödel pursued the same goal, with more subtle considerations, in his Gibbs Lecture (Gödel [2]). Feferman asserts, however, that there is no "unassailable proof of the falsity of the mechanist position" and attributes that fact *in part* to what he calls "*the mechanist's empirical defense* (or *escape hatch*)". It may very well be that, as Kreisel has claimed, "some specific system ... encompasses all possibilities of (correct) mathematical thinking". (p. 212.) The *distinctive part* of Feferman's critical position, reacting to the escape hatch, is this: the very formulation of the mechanist thesis involves an *equivocation* "that lies in identifying *how* the mathematical mind works with the totality of *what* it can prove". (p. 215) He asks a central question: Isn't the mechanist really aiming at "an explanation of the mechanisms that govern the production of human proofs"?

Feferman thinks that the familiar formal systems like Peano arithmetic PA or Zermelo Fraenkel set theory ZF can't serve that purpose, as they do not underlie "mathematical thought as it is experienced". (p. 215) He continues:

> And the experience of the mathematical practitioner certainly supports the conclusion drawn by Nagel and Newman that "mathematical proof does not coincide with the exploitation of a formalized axiomatic method," even if that cannot be demonstrated unassailably as a consequence of Gödel's incompleteness theorems. (pp. 215–216)

These observations make Feferman consider as implausible even the modified mechanist thesis concerning the production of human proofs. It should be noted, however, that the implausibility rests on a very narrow notion of formalization: it is to be carried out in the basic language of a theory – for ZF in just the language of = and $\in$. Feferman's major logical suggestion addressing that problem is a different conception of formal systems. That conception is not tied to a formal language fixed in advance, but rather (1) it specifies the form of axiom schemata by means of propositional and predicate variables, and (2) it allows an open-ended, expanding development of its language. In particular, axiom schemata like induction in the case of PA and separation in the case of ZF may then be appropriately instantiated with formulae of the expanded language.

> In other words, on this approach, implicit in the acceptance of given schemata is the acceptance of any meaningful substitution instances that one may come to meet, but which those instances are is not determined by restriction to a specific language fixed in advance. (p. 216)

It is in this way that Feferman sees a way of straddling the mechanist and anti-mechanist divide, as (1) would reflect the mechanist position, whereas (2) would do some justice to the anti-mechanist perspective.[2]

Feferman counsels that – ultimately and independently of the mind & machine issue – we should aim for "an informative, systematic account at a theoretical level of how the mathematical mind works that squares with experience". To achieve such a more adequate account, he suggests considering "the role of heuristics, analogies, metaphors, physical and geometric intuition, visualization, and so on". I agree that those are important aspects of mathematical experience and that they should be investigated. However, characterizing the "logical structure of mathematics–what constitutes a proof" is a crucial task that has not been fully addressed. We have investigated formal proofs for metamathematical purposes, but how are they related to ordinary mathematical proofs? How can the structure of formal proofs reflect that of ordinary proofs? How, finally, can formal proofs be used as a tool for the analysis of informal proofs? Answering these questions should be a substantial task for *proof theory*. When describing the formal system of elementary number theory in his [1], Gentzen remarked: "The objects of proof theory shall be the *proofs* carried out in mathematics proper." (p. 499)

Feferman is completely right when asserting that the formal development in the systems PA or ZF *in their respective fixed language* does not underlie mathematical experience. A more realistic, yet nevertheless formal development certainly involves (definitional) extensions and thus takes into account the conceptual organization of parts of mathematics in a crucial way. There are exciting developments in interactive theorem proving, but for the purpose at hand the *natural formalization* of parts of mathematics and the step towards the *automated search for humanly intelligible proofs* offers the greatest opportunities for exploring a new chapter of proof theory.[3]

References

1. Gentzen, G.: Die Widerspruchsfreiheit der reinen Zahlentheorie. Mathematische Annalen **112**, pp. 493–565 (1936)
2. Gödel, K.: Some basic theorems on the foundations of mathematics and their implications; delivered as the 25th Gibbs Lecture on 26 December 1951; published in volume III of Gödel's Collected Works, pp. 304–323. Oxford University Press (1995)

---

[2]In a quite different way, Gödel and Turing tried to straddle this divide as far as the extending development of mathematics is concerned; that is discussed in my (2013).

[3]The beginnings of such investigations go back to the 1970s when human- and machine-oriented approaches to theorem proving were contrasted. More recently, Gowers and collaborators have pursued automatic theorem proving in a radically human-oriented way; see (Gowers [3]). In [8] I have discussed my own perspective on the automated, but heuristically informed search for conceptually structured proofs.

3. Gowers, T.: Interview. Not. Am. Math. Soc. **63**(9), pp. 1026–1028 (2016)
4. Lukas, J.R.: 1959 Minds, machines and Gödel; given in 1959 to the Oxford Philosophical Society; published in Philosophy XXXVI, pp. 112–127 (1961)
5. Nagel, E., Newman, J.R.: Gödel's Proof. Scientific American CXCIV, pp. 71–86 (1956)
6. Nagel, E., Newman, J.R.: Gödel's Proof. New York University Press, New York (1958)
7. Penrose, R.: The Emperor's New Mind: Concerning Computers, Minds, and the Laws of Physics. Oxford University Press, Oxford (1989)
8. Sieg, W.: Searching for proofs (and uncovering capacities of the mathematical mind). In: Feferman, S., Sieg, W. (eds.) Proofs, Categories and Computations – Essays in Honor of Grigori Mints, pp. 189–215. College publication (2010)
9. Sieg, W.: Gödel's philosophical challenge (to Turing). In: Copeland, B.J., Posy, C.J., Shagrir, O. (eds.) Computability – Turing, Gödel, Church, and Beyond, pp. 183–202. MIT Press (2013)

Author Biography

Wilfried Sieg is Patrick Suppes Professor of Philosophy at Carnegie Mellon University and Fellow of the American Academy of Arts and Sciences. He is known for mathematical work in proof and computation theory, historical work on modern logic and mathematics, and philosophical essays on the nature of mathematics. Papers joining these aspects of his research were published in the collection *Hilbert's Programs and Beyond*, Oxford University Press, 2013.

Feferman on Set Theory: Infinity up on Trial

Peter Koellner

"Look into infinity, all you see is trouble."
— Bob Dylan, *"Trouble"*

Abstract In this paper I examine Feferman's reasons for maintaining that while the statements of first-order number theory are "completely clear" and "completely definite," many of the statements of analysis and set theory are "inherently vague" and "indefinite." I critique his five main arguments and argue that in the end the entire case rests on the brute intuition that the concept of subsets of natural numbers—along with the richer concepts of set theory—is not "clear enough to secure definiteness." My response to this final, remaining point will be that the concept of "being clear enough to secure definiteness" is about as clear a case of an inherently vague and indefinite concept as one might find, and as such it can bear little weight in making a case against the definiteness of analysis and set theory.

Keywords Predicativity · Set theory · The continuum hypothesis
Definiteness · Semi-constructive systems

2010 Mathematics Subject Classification 00A30 · 03F35 · 03E60 · 03E50

It is an honor to be able to contribute a paper for a volume celebrating the work of Feferman. I first encountered his work when I was an undergraduate. I read "Systems of Predicative Analysis" (1964) and was simply blown away. Here one found a perfectly clear analysis of the concept of predicativity given the natural numbers, along with several theorems to support it. I thought: "This is how conceptual analysis

P. Koellner (✉)
Department of Philosophy, Harvard University, Cambridge, MA 02138, USA
e-mail: koellner@fas.harvard.edu

G. Jäger and W. Sieg (eds.), *Feferman on Foundations*, Outstanding Contributions to Logic 13, DOI 10.1007/978-3-319-63334-3_19

ought to be done!" That paper was one of the main reasons I decided to concentrate my efforts on the intersection of philosophy and mathematics.

I read everything by Feferman that I could get my hands on. I found myself in agreement with pretty much everything that he wrote. In the beginning this extended to set theory. Like Feferman I too was skeptical of set theory. But as I learned more set theory that skepticism began to wane. More and more I became convinced by the arguments in favor of new axioms. This one point, concerning set theory, emerged as the only substantial point where I found myself in disagreement with Feferman.

In this paper I want to focus on that one point of disagreement, not because of a desire to be contrarian, but because it just doesn't sit well with me—how can we disagree on this point when we agree on so much? One of us must be wrong, and I really hope it isn't me.

The paper is really the continuation of a conversation that we have been having for many years. The focus will be on Feferman's reasons for maintaining that statements like the continuum hypothesis (CH) are not definite. But to set the stage I will first discuss the realm of mathematics where he thinks that the statements *are* definite. This part of the paper is really a tribute to his celebrated work on predicativity, the work that first sparked my interest.[1] I will then turn, in the remaining part of the paper, to his reasons for thinking that the distinctive statements of analysis and set theory are not definite.[2] I've combed through his work and have found five arguments:

(1) Feferman maintains that CH has effectively ceased to be regarded as definite by the mathematical community and that this fact provides "considerable circumstantial evidence to support the view that CH is not definite."[3]
(2) Feferman thinks that the concept of arbitrary subset (of an infinite domain) is inherently vague in the sense that (a) it is vague and (b) it cannot be sharpened without violating what it is supposed to be about.
(3) Feferman argues that given the alleged lack of clarity of the powerset operation that the only recourse to establishing the definiteness of statements of set theory (or even analysis) is an untenable form of platonism, one which faces certain insurmountable philosophical problems.
(4) Feferman's reasons for thinking that CH is indefinite are partly based on the metamathematical results in set theory; in particular the results showing that "CH is independent of all remotely plausible axioms extending ZFC, including all large cardinal axioms that have been proposed so far."[4]

---

[1] A more thorough account will appear in [28].

[2] Throughout this paper I will use 'analysis' and 'second-order number theory' interchangeably.

[3] [17], p. 1.

[4] [15], p. 127.

(5) Feferman takes the formal results on indefiniteness—in particular the result of Rathjen showing that CH is indefinite relative to the semi-constructive system SCS^+—as providing evidence that CH is indefinite.

I will argue that (1) has no force and that when the dust settles (3), (4), and (5) all reduce to (2) and that in the end the entire case rests on the brute intuition that the concept of subsets of natural numbers—along with the richer concepts of set theory—is not "clear enough to secure definiteness." My response to this final, remaining point will be that the concept of "being clear enough to secure definiteness" is about as clear a case of an inherently vague and indefinite concept as one might find, and as such it can bear little weight in making a case against the definiteness of analysis and set theory.[5,6]

Sol and I never got the opportunity to discuss this paper. Shortly after it was completed I received the shocking news that he had died. Although it may seem rather personal for an academic paper, I would like to include a letter that I wrote to my sister when I learned that he had died:

> Dear Kel,
>
> I got some sad news today. Sol died yesterday.
>
> I was going to visit him in a couple of weeks.
>
> I'm really going to miss him. I had always been inspired by his work. We agreed on a great many things but we strongly disagreed on set theory. Much of our conversation was about set theory. We would argue and argue.
>
> The last time I saw him was in New York. We were there for a conference in honor of our friend Charles Parsons. Sol gave a talk on set theory. My talk was supposed to be a critical response to his talk. When we arrived we were all sitting around a table together, sort of like being in the back stage before a concert. The person doing the introductions asked us for our titles. I was a bit nervous, since I didn't have a title. Sol said his title was "Parsons and I: Sympathies and Differences". I was asked next. I immediately said (imitating his rhythm): "Sol and I: *Differences*!" Everybody laughed, especially Sol. He reached over, shook my hand, and said "Nicely played."
>
> We were like that. We strongly disagreed, but we are playful about it. And we really liked each other. It was so nice when I visited him at his place in January. We just hung out. We watched part of Ken Burns' documentary "The Civil War" on a little TV screen in the

[5]The title involves a reference to Feferman's "Infinity in Mathematics: Is Cantor Necessary?", which opens with what is perhaps the coolest epigraph ever used in a paper:

"Infinity is up on trial ..."
— Bob Dylan, *Visions of Johanna*.

[6]I would like to thank Douglas Blue, J.T. Chipman, Solomon Feferman, Gabriel Goldberg, Charles Parsons, Michael Rathjen, Wilfried Sieg, Thomas Strahm, and Hugh Woodin for helpful comments and discussion. I would also like to thank the The Journal of Philosophy for giving me permission to repeat passages that occur [28], which was a high-level summary of the present paper, one that I delivered as a reply to Sol in the conference described below.

kitchen. I loved being with him. I slept like a baby in his house. To think, that I could be so comfortable, that I could just hang out with this man who was two generations my senior.

The very last time I saw him was after the dinner after the conference. We were in a taxi. Wilfried Sieg was sitting in between us. In my reply to his talk I was very harsh with him. He hadn't had a real opportunity to reply. In the car he let me have it. He said "You can't seriously believe $AD^{L(\mathbb{R})}$!" I replied: "Yes, I certainly do!" And then we got into it. Wilfried was calmly staring straight ahead while we argued back and forth. When we arrived at the hotel I popped out of the car quickly, hurried around the back, and met him as he was getting out. I grabbed his arms firmly and lifted him out, and then steadied him as we walked into the hotel. While doing this I was conscious of how brittle he felt. We walked together, slowly. He said: "Peter, you are my favorite person to argue with." I told him I would like to come and visit him again. His eyes were a bit watery. He said he would like that. I didn't go with them in the elevator, I told them I wanted to go for a walk. We said goodbye. I watched the elevator doors close.

Love,
Peter

1 The Realm of the Definite

It will be useful to begin by saying something about Feferman's general philosophical framework.

Feferman is an avowed anti-platonist. In place of platonism he espouses what he calls *conceptual structuralism*, an "ontologically non-realist philosophy of mathematics"[7] according to which "the basic objects of mathematical thought exist only as socially shared mental conceptions."[8] I will say a bit more about this view below, but for now I would just like to point out that it is part of this view that "there are differences in clarity or definiteness between basic conceptions."[9] This is the component of the view that allows Feferman to maintain that number theory is "completely clear" and "completely definite", while analysis and set theory are not.

Indeed he maintains that our conception of the natural numbers is so clear that (on the epistemic dimension) the standard axioms of PA (with open-ended induction) are "evident on our conception"[10] and that (on the semantical, or metaphysical dimension) it is "completely definite"[11] in the sense that each statement of number theory has a definite truth value regardless of whether or not we can determine that truth value on the basis of evident principles. It is for these reasons that Feferman "grant[s] the natural numbers a position of primacy in our mathematical thought."[12]

---

[7] [19], p. 74.

[8] [15], pp. 2–3.

[9] [15], pp. 2–3.

[10] [4], p. 70.

[11] [15], p. 7.

[12] [13], p. 314.

It is thus of interest to ask: *How much mathematics can one secure on the basis of the concept of natural numbers?* There are several aspects to this question. For our purposes we will be interested in which additional notions and statements should be deemed *completely clear and definite* on the basis of our conception of the natural numbers.

Feferman maintains that our conception of the natural numbers as a completed totality over which arbitrary arithmetical statements are definite enables us to get a completely clear grasp on *arithmetically definable* sets of natural numbers. Statements involving quantification over the arithmetically definable sets of natural numbers are thus completely clear and definite. With the resources of an expanded language—where we allow second-order quantification over the arithmetically definable sets of natural numbers—we can define sets of natural numbers that were previously undefinable. This new domain is also completely clear and definite on the basis of our concept of natural number. This process can be iterated into the transfinite. This is the *predicative* conception of sets of natural numbers, in contrast to the *impredicative* conception.

The trouble that Feferman (following Poincaré and Weyl) has with impredicative definitions is that they involve treating the collection of arbitrary subsets of natural numbers as a clear and definite, completed totality and yet, according to Feferman, "[o]n the face of it, such definitions are justified only by a thorough-going platonistic philosophy of mathematics."[13] It is "by way of rejection of the set-theoretic platonistic ontology, and more specifically of that part of it which warrants reference to the supposed totality of arbitrary subsets of any infinite set, that one is led to an alternative *definitionistic* view of sets as the extensions of properties successively seen to be defined in a non-circular way."[14] On the predicativist conception "only the natural numbers can be regarded as 'given' to us" and sets are regarded as "created by man to act as convenient abstractions (*façons de parler*) from particular conditions or definitions."[15]

The stages of the predicative hierarchy can be described as follows: For a collection $\mathcal{C}$ of sets of natural numbers, let $\mathrm{Def}(\mathcal{C})$ be the *definable powerset* relative to $\mathcal{C}$. Now iterate the process as follows:

$$\begin{aligned} \mathcal{R}_0 &= \mathbb{N} \\ \mathcal{R}_{\alpha+1} &= \mathrm{Def}(\mathcal{R}_\alpha), \text{ and} \\ \mathcal{R}_\lambda &= \textstyle\bigcup_{\alpha<\lambda} \mathcal{R}_\alpha \text{ for limit } \lambda. \end{aligned}$$

This hierarchy is known as the *ramified analytic hierarchy.*

The question then becomes: *How far are we justified in iterating this process, on the basis of our conception of natural number?* It would take us too far afield to discuss the details of Feferman's elegant answer to this question. Suffice it to say that

[13] [13], p. 314.
[14] [13], p. 316.
[15] [3], p. 1.

it involves passing from the domains $\mathcal{R}_\alpha$ to the second-order ramified theories RA_α that axiomatize them, and imposing an *autonomy constraint* to the effect that (roughly speaking) one only accepts those RA_α for which there exists an earlier $\beta < \alpha$ such that RA_β has already been accepted and RA_β proves that "α is acceptable." The least ordinal which cannot be reached in this bootstrapping fashion was determined independently by Feferman and Schütte. It is known as Γ_0. The proposal, then, is that $\mathrm{RA}_{<\Gamma_0}$ captures the informal notion of *predicativity given the natural numbers*.[16]

The next step was to provide alternative, independent characterizations of predicativity given the natural numbers, so as to bolster the thesis that $\mathrm{RA}_{<\Gamma_0}$ does indeed capture the informal notion of predicativity given the natural numbers. In a series of papers Feferman gave alternative characterizations, first in terms of the notion of the definite powerset,[17] second in terms that avoided the notion of a well-ordering,[18] third in terms of reflective closure,[19] and finally in terms of the notion of unfolding.[20] In each case he showed that the system in question was proof-theoretically equivalent to $\mathrm{RA}_{<\Gamma_0}$. The situation is thus parallel to the situation with other instances of conceptual analysis, most notably, the analysis of the notion of computability, where various independent characterizations were subsequently shown to be equivalent.

I think that this is a remarkable achievement. Indeed it is hard to think of a more impressive and successful instance of *philosophical* analysis. For our purposes the important point is that granting that our conception of the natural numbers is completely clear and definite, the above analysis tells us that we can also accept (locally at least) the conception of predicative analysis also as completely clear and definite.[21]

This is not to say that $\mathrm{RA}_{<\Gamma_0}$ encompasses *all* of the mathematics that Feferman regards as clear and definite. In fact, he has made it clear that he is "not now nor never has been a predicativist," though he has been a "sympathizer."[22] For him the above analysis only tells us how far we can go on the basis of the concept of natural numbers alone. But there are other conceptions that he accepts as clear and definite, although it is to be expected that as we climb further up the hierarchy of conceptions their clarity and definiteness will gradually diminish.

Feferman has made it clear in personal correspondence that he accepts constructive systems at the level of Π^1_1-CA and even $(\Delta^1_2\text{-CA}) + \mathrm{BI}$.[23] At the upper limit he has expressed skepticism about the possibility of any "evidently constructive" system at the level of $(\Pi^1_2\text{-CA}) + \mathrm{BI}$.[24] And I suspect that he would agree with the more recent

---

[16]See [3].

[17][3].

[18][4].

[19][6].

[20][8, 21].

[21]The qualification 'locally at least' is necessary since the totality $\mathrm{RA}_{<\Gamma_0}$ is not itself predicatively characterizable.

[22][13], pp. 313–314.

[23]Letter of Jan. 4, 2016.

[24][11], Sect. 6.

arguments of Rathjen and Martin-Löf to the effect that there cannot be an evidently constructive system at this level.[25]

But even though he accepts *constructive* systems at these levels it is important to note that he does not accept the corresponding classical systems. For example, he maintains that

> Π^1_1-CA (or dually Σ^1_1-CA) is justified only on the assumption of the meaningfulness of impredicative definitions, and that in turn implicitly assumes that there is a well-determined totality of subsets of the natural numbers which exists independently of any means of (human) definition or construction.[26]

I think that he would have to say the same of Π^1_1-CA_0 since the issue is the comprehension principle. It follows that he cannot accept any statement which (by reverse mathematics) is provably equivalent to Π^1_1-CA_0. For example, the Cantor-Bendixson theorem is such a statement. So I think he has to maintain that the Cantor-Bendixson theorem is indefinite. In any case, whether or not this is the level of analysis where indefiniteness kicks in for Feferman, he is certainly explicit in maintaining that there are statements of analysis that are indefinite. For example, he denies that there is "a fact of the matter whether all projective sets are Lebesgue measurable or have the Baire property, and so on."[27]

The point I wish to make is that although Feferman concentrates on CH (a statement of third-order arithmetic), his position is much more radical—he holds that there are statements of second-order arithmetic (quite likely as low down as the Cantor-Bendixson theorem) that are inherently unclear and indefinite.

Let us now turn to Feferman's arguments concerning inherent unclarity and indefiniteness. There are five main arguments. I will discuss each in turn.

2 Circumstantial Evidence

> *Feferman maintains that* CH *has effectively ceased to be regarded as definite by the mathematical community and that this fact provides "considerable circumstantial evidence to support the view that* CH *is not definite."* [28]

2.1 Feferman's Case

The argument turns on a thought experiment involving the Millenium Prize Problems. Here is the background: On May 24, 2000 in Paris, the Clay Mathematics Institute announced seven Millenium Prize Problems, each with a reward of one million

[25]See [32, 35].

[26][7], pp. 200–201.

[27][12], p. 405.

[28][17], p. 1.

dollars. One of the problems on the list—the Riemann Hypothesis—also appeared on the famous list of problems that Hilbert presented in Paris on August 9, 1900. But, as Feferman notes, CH, which was the first problem on Hilbert's list, is not on the Millenium Prize list.

Feferman then imagines a discussion between the scientific advisory board and a group of set theorists to determine whether CH is suitable for inclusion on the list. The set theorists explain some of the advances in the program for large cardinal axioms—in particular, the work of Martin, Steel, and Woodin—which demonstrate that large cardinal axioms settle the classic undecided statements of second-order number theory and statements further up in the complexity hierarchy. They then give an account of Woodin's first approach to settling CH, the approach based on maximality. The board ends up concluding that CH is not suitable for inclusion.

This is taken to show that (a) the scientific board (and the mathematical community more generally) thinks that CH is not a definite mathematical problem, and this in turn is taken to provide "considerable circumstantial evidence" that (b) CH is not definite.[29]

2.2 *Response*

I will be rather brief with this argument since it is weaker than the other arguments that Feferman provides.

(1) Even if (a) were true, the conclusion (b) would only follow by an appeal to authority. But arguments that appeal to authority have little force. Moreover, in this specific case, it is surprising to find Feferman appealing to the authority of the mathematical community concerning matters of definiteness, since he also thinks that this very community disagrees with him on matters of definiteness. For example, he writes, "[t]here is no doubt that the mathematical community as a whole takes the concept of the set of *arbitrary* real numbers as a definite, robust concept, and thus, indirectly, the concept of the set of *arbitrary* subsets of the natural numbers."[30]

(2) Moreover, I think that Feferman is mistaken in maintaining that the scientific board's behaviour is evidence of (a). The whole issue is tied up with an ambiguity involved in the word 'problem.' One must draw a distinction between (i) a problem in the sense of a *statement* (a statement that is an open problem) and (ii) a problem in the sense of a *task* (the task of settling an open problem). Feferman wants to argue that the *statement* CH is indefinite. But the scientific advisory board's conjectured behaviour show at most that they regard the *task* of settling CH as insufficiently definite to warrant placing one million dollars on the implementation of the task. This is entirely reasonable. I think that every set theorist would agree that the task of settling CH is not sufficiently definite to warrant inclusion on the list since it is known that any resolution of CH is going to involve subtle issues concerning the

[29][17], p. 1.
[30][20], p. 7.

justification of new axioms. But this has no bearing on the question of whether the scientific advisory board (or the set theorists who are in agreement) thinks that CH is not a definite *statement*.[31]

To answer the question of whether the *statement* CH is not definite, one is going to have to dig deeper. So let us turn to Feferman's other, more direct, arguments.

3 Conceptual Clarity

Feferman thinks that the concept of arbitrary subset (of an infinite domain) is inherently vague in the sense that (a) it is vague and (b) it cannot be sharpened without violating what it is supposed to be about.

3.1 Feferman's Case

The claim that our conception of the subsets of natural numbers lacks the clarity and definiteness of our conception of the natural numbers, is one that occurs fairly early in Feferman's writings. In a paper of 1979 he says that what distinguishes our conception of the natural numbers from our conception of the subsets of natural numbers is that "*we have a complete and clear mental survey of all of the objects being considered, together with the basic interrelationships between them*" and that "the logical operation of quantification over the natural numbers has a *definite character in our understanding* which quantification over functions, etc. lacks."[32] It is of interest to note that already in this early passage he employs both the notions of *clarity* and *definiteness*, saying that our conception of the natural numbers is completely clear and definite, while our conception of the subsets of natural numbers is not.

In later passages, he adds a new element. For in addition to saying that the notions of set theory are not clear, he says that they are *inherently* unclear (or inherently vague):

> I, for one, am a pessimist or, better, anti-platonist about the Continuum Hypothesis: I think that the problem is an inherently vague or indefinite one, as are the propositions of higher set theory more generally.[33]

Soon thereafter he tells us more about what he means by the 'inherent' in 'inherent vagueness':

---

[31]To underscore this point consider a hypothetical scenario involving an arithmetical statement. Suppose that the scientific board were presented with a number-theoretic statement that was equivalent to the consistency of ZFC + "There is a supercompact cardinal." They could quite reasonably refrain from putting it on the list because they could quite reasonably think that any positive resolution would require subtle issues surrounding the justification of new axioms. But this would not provide "considerable circumstantial evidence" that the number-theoretic statement is indefinite.

[32][4], p. 70.

[33][9], p. 7, Introduction.

> Rather, it's a conception we have of the totality of "arbitrary" subsets of the set of natural numbers, a conception that is clear enough for us to ascribe many evident properties to that supposed object (such as the impredicative comprehension axiom scheme) but which cannot be sharpened in any way to determine or fix that object itself.[34]

But is it true that it cannot be sharpened? For could we not sharpen it, say, by restricting to the *constructible* subsets of natural numbers, that is, those subsets of natural numbers that appear in L? For, after all, it would seem that L is perfectly clear. In later writings Feferman guards against this move by revising his claim, now saying that the concept of subsets of natural numbers cannot be sharpened *without violating what that notion is supposed to be about*:

> Moreover, I would argue that it is *inherently vague,* in the sense that there is no reasonable way the notion can be sharpened without violating what the notion is supposed to be about. For example, the assumption that all subsets of the reals are in L or even $L(\mathbb{R})$ would be such a sharpening, since that violates the idea of "arbitrariness." In the other direction, it is hard to see how there could be any non-circular sharpening of the form that there [are] as many such sets as possible. It is from such considerations that I have been led to the view that the statement CH is inherently vague and that it is meaningless to speak of its truth value; the fact that no remotely plausible axioms of higher set theory serve to settle CH only bolsters my conviction.[35]

This point is repeated in his most recent writings on the topic[36] and we are told there that this is the "*main reason* that has led [him] to the view that CH is not definite."[37]

To summarize: The concept of subsets of natural numbers is not merely unclear, it is *inherently* unclear in the sense that it cannot be clarified without violating the essential feature of "arbitrariness" and, moreover, this essential feature of "arbitrariness" is not something that can be clarified in a non-circular way. Furthermore, this is the main reason that Feferman thinks that CH is not definite.

3.2 *Response*

(1) I agree that "[w]hat we are dealing with here are *questions of relative conceptual clarity and foundational status*"[38] and I am willing to grant that our conception of the natural numbers is clearer than our conception of the subsets of natural numbers.

(2) I do not strictly speaking agree with the claim that there can be no way to sharpen our concept of subsets of natural numbers without *violating what it is supposed to be about*, since that makes it look as though there is something implicit in that conception that *implies* that V cannot be any fine-structural inner model. But I do accept the related claim that there is no way to sharpen our concept of subsets of

---

[34] [12], p. 405.

[35] [12], pp. 410–411.

[36] See, for example, [15], p. 130 and [17], p. 2 and p. 21.

[37] [17], p. 2, my emphasis.

[38] [14], p. 619.

natural numbers that merely unfolds that concept. More precisely, I think that it isn't *analytic* of our concept of subsets of the natural numbers that all such subsets are in L or any other fine-structural inner model. The reason is that there is nothing on the face of it in our concept of subsets of natural numbers that involves mention of these models (the definitions of which are quite technical) and it would be far-fetched to maintain that a "deeper analysis" of the concept would imply, say, that all subsets of natural numbers are in L or any of the other fine-structural inner models.

(3) I also agree that the key component of "arbitrariness" cannot be clarified in more fundamental terms.

(4) But although I agree with (2) and (3), I don't see either as raising a problem. I see them rather as a sign that we are dealing here with a *primitive* concept. The same point about there being "no non-circular sharpening" applies to all primitive concepts. For example, it applies to our conception of the natural numbers. One cannot clarify or explain the concept of natural numbers in more fundamental terms. All attempts to give a more primitive explication lead back to the same concept in a different guise. For example, one might try to explain the domain of natural numbers as the domain obtained by starting with 0 and applying the successor operation a finite number of times. But here the conception of natural numbers appears, hidden, in a different guise, in the reference to *finite number of times*. The hallmark of a primitive concept—indeed the defining characteristic of being a primitive concept—is that such a concept cannot be defined or explained in more fundamental terms. So in *this* respect our conception of subsets of natural numbers is on a par with our conception of natural numbers.

The key difference, then, between our conception of the natural numbers and our conception of subsets of natural numbers is that the former is *clearer than* the latter. That is something with which I am happy to agree. But the key question at hand is whether the former, but not the latter, is *clear enough to secure definiteness*. This is where I think the case falters. Feferman is making essential use of the concept of *being sufficiently clear to secure definiteness*. I would like to say that the concept of *being sufficiently clear to secure definiteness* is not sufficiently clear to secure definiteness. It is about as good an example of an inherently unclear and indefinite concept as one might find. In this regard it bears more kinship to the concept of a "feasible number" than to the concept (taken at face value) of subsets of natural numbers, and as such I don't think it can bear much weight in a case against the definiteness of the latter concept. To find an argument to that effect we will have to dig even deeper.

4 Metaphysical

Feferman argues that given the alleged lack of clarity of the powerset operation that the only recourse to establishing the definiteness of statements of set theory (or even analysis) is an untenable form of platonism, one which faces certain insurmountable philosophical problems.

4.1 Feferman's Case

The argument involves the topic of realism in mathematics. This is a very difficult topic with a long history. Nevertheless, I think it is possible to isolate the key steps in Feferman's argument and critique it in a general way without entering the more delicate issues and without mounting a defense of a tenable brand of realism. In short, my criticism will be internal.

In outline the argument is as follows: (1) There are two opposing views on the nature of mathematics: platonism and conceptual structuralism. (2) The definiteness of set theory requires platonism. (3) Platonism is untenable because it faces an insurmountable epistemological problem, namely, the access problem raised by Benacerraf. (4) Conceptual structuralism, in contrast, does not face the access problem.

Let us now spell this out in detail.

4.1.1 Two Opposing Extremes

In the spectrum of views concerning the nature of mathematics there are two opposing extremes that have animated much of the discussion in the literature. At one extreme we have the view that the mathematical realm is completely "independent" of our practices and that its objects lie "outside" of the space-time manifold in an "eternal," "immutable" realm, known as the "third realm." This view is often called *platonism* (or *realism*). At the other extreme we have the view that the mathematical realm is really just a "projection of our practice." This view, in its general form, might be called *projectionism.* (As we shall see later the particular form of projectionism that Feferman holds is the view mentioned earlier, namely, *conceptual structuralism.*)

These two extremes have complementary strengths and weaknesses. Platonism has no problem in explaining the apparent objectivity of mathematics (by reference to the third realm) but it has difficulty explaining what by its lights would appear to be a miracle, namely, how our practice of proving things "down here" can manage to track the nature of things "up there." In other words, while this view solves the *objectivity problem*, it runs into difficulty when it comes to the *access problem.* In contrast, projectionism fares well when it comes to the access problem (since, on this view, the objects are not independent items "up there" that we have to track "down here," but are rather constituted by what we do "down here") but it has trouble in overcoming the objectivity problem, since on this view mathematics would seem to be like fiction, where we have complete sovereignty, and that is something that does not appear to be the case in mathematics.

Something like these two extremes (in the special setting of set theory) is at play in the opening section of Feferman's classic 1964 paper on predicativity. He begins by considering "two extremes of what sets are conceived to be", namely, the *Platonistic* conception and the *predicative* conception:

> From one point of view, often identified as the *Platonistic* or *Cantorian* conception, sets have an existence which is independent of human definitions and constructions. The words "arbitrary set" are often used to emphasize this independence. Various statements about sets are readily recognized to be correct under this conception, for example the axioms of comprehension and of choice. Other statements, such as the continuum hypothesis and its generalizations remain undecided on this conception. (However, the inability of humans to decide such questions can no more be charged as a defect of this conception than can their inability to decide certain number-theoretical statements on the basis of the usual conceptions of the natural numbers.)
>
> In contrast, the other extreme is what we shall refer to as the *predicative* conception. According to this, only the natural numbers can be regarded as "given" to us (and, in the even more severe nominalist point of view, not even these abstract objects are available to us). In contrast, sets are created by man to act as convenient abstractions (*façons de parler*) from particular conditions or definitions.[39]

These are the two opposing extremes concerning the nature of mathematics. But are they really the only options?

4.1.2 Set Theory Requires Platonism

Feferman argues that they are. He argues that if one rejects the predicativist view of sets (the one extreme) then one *must* embrace a brand of platonism of the problematic kind (the other extreme). For example, concerning impredicative definitions he writes:

> On the face of it, such definitions are justified only by a thorough-going platonistic philosophy of mathematics.[40]

And concerning set theory more generally he maintains that to accept $P(\mathbb{N})$ as a definite totality "is to accept the problematic realist ontology of set theory."[41]

In other words, for Feferman, when it comes to analysis and set theory, the two opposing extremes concerning the nature of mathematics seem to be the only options.

4.1.3 Problems with Platonism

He then makes it clear that he is not a platonist. In fact, he finds the view "philosophically preposterous":

> It will soon be clear to the reader that I am a convinced antiplatonist in mathematics. Briefly, according to the platonist philosophy, the objects of mathematics such as numbers, sets, functions, and spaces are supposed to exist independently of human thoughts and constructions, and statements concerning these abstract entities are supposed to have a truth value

---

39 [3], pp. 1–2.

40 [13], p. 314.

41 [19], p. 81.

> independent of our ability to determine them. Though this accords with the mental practice of the working mathematician, I find the viewpoint philosophically preposterous.[42]

The trouble with platonism, according to Feferman, is that it "it faces well-known difficulties [that] have left it with few if any adherents."[43] These difficulties are the *access problem* that we alluded to earlier. Feferman asks: "If [the continuum] has only *Platonic* existence, how can we access its properties?"[44] And, in a later paper he is more explicit about the problem:

> The set-theoretical view of $P(A)$ is justified by a thorough-going platonism, as accepted by Gödel. According to this view, sets in general have an existence independent of human thoughts and constructions, and in particular, for any set A, $P(A)$ is the definite totality of arbitrary subsets of A. A major problem with this view is the classic one of epistemic access raised most famously by Paul Benacerraf (1965).[45]

In contrast, his own view of conceptual structuralism does not face the access problem:

> The open-ended view, on the other hand, rests on conceptual structuralism, for which epistemic access is no problem, but the determinateness of various statements then comes into question.[46]

To see how his conceptual structuralism evades the access problem let us say a few more words about it.

4.1.4 Conceptual Structuralism

Feferman's version of conceptual structuralism is "an ontologically non-realist philosophy of mathematics."[47] He espouses non-realism *about objects* across the board. However, it is important to note that Feferman does not espouse non-realism *about truth* across the board. Indeed the question of realism about truth is equivalent to the question of definiteness, and, as we have seen, he embraces definiteness (and hence realism about truth) with regard to number theory.

According to conceptual structuralism "the basic objects of mathematical thought exist only as socially shared mental conceptions" and "[t]he objectivity of mathematics is a special case of intersubjective objectivity that is ubiquitous in social reality."[48] It is through closely tethering mathematics to mathematical practice that one overcomes the access problem.

---

[42] [9], Introduction, p. ix.

[43] [17], p. 9.

[44] [10], p. 107.

[45] [17], p. 18. In this quotation I have used '$P(A)$' in place of Feferman's '$S(A)$'. As Charles Parsons pointed out to me, the reference should be to Benacerraf's 1973 paper.

[46] [17], p. 18.

[47] [19], p. 74.

[48] [20], pp. 2–3.

The view allows for "differences in clarity or definiteness between basic conceptions" and this turns out to be key when it comes to the question of definiteness. For, even though the view tethers mathematics to human practice "[o]ne may speak of what is true in a given conception". But "that notion of truth may only be partial". "Truth in full is applicable only to completely clear conceptions."[49] In other words,

> this view of mathematics does not require total realism about truth values. That is, it may simply be undecided under a given conception whether a given statement in the language of that conception has a determinate truth value, just as, for example, our conception of the government of the United States is underdetermined as to the presidential line of succession past a certain point.[50]

4.1.5 Summary

To repeat, in outline the argument is as follows: (1) There are two opposing views on the nature of mathematics: platonism and conceptual structuralism. (2) The definiteness of set theory requires platonism. (3) Platonism is untenable because it faces an insurmountable epistemological problem, namely, the access problem. (4) Conceptual structuralism, in contrast, does not face the access problem. All the more reason to embrace conceptual structuralism.

4.2 Response

Before getting to the main point of my response I want to make a few preliminary points about each of the steps in the above argument.

Re. (1). There is a tendency in much philosophical literature to frame issues in dramatic terms by maintaining that one is faced with an iron dichotomy. But more often than not the dichotomy turns out to be a false one. I believe that the present case is such a case. The form of platonism that Feferman has in mind—and which he thinks is "philosophically preposterous"—bears much in common with what Tait calls "superrealism." But there are other, more tenable forms, such as the "default realism" that Tait defends.[51] Between the two extremes there are many options. The truth surely lies between.

Re. (2). I do not see why set theory requires a form of superrealism that is "philosophically preposterous." For example, Feferman's conceptual structuralism is a flexible framework. There is nothing in it per se that does not enable one to endorse conceptual structuralism with regard to set theory. After all, Feferman is able to avoid what he regards as a preposterous form of platonism with regard to *number theory* by adopting conceptual structuralism in that setting, and, in doing so, he secures realism

---

[49] [17], p. 11.

[50] [17], p. 13.

[51] See [38, 39], Chap. 4.

about truth values, which is the main point at issue. Why can't we do the same in the case of *set theory*? I will return to this matter below, since it will lead to my main point.

Re. (3). Even though I am no sympathizer with the kind of superrealism that Feferman seems to have in mind, I think we should be wary of the kind of epistemological argument that he employs. More generally, for a domain X that on the face of it is a secure and legitimate discipline, I think we should be hesitant to embrace arguments of the form: "Domain X faces insurmountable epistemological difficulties; therefore we have to give up our claims to knowledge regarding X." The reason we should be wary of such arguments is that epistemology doesn't exactly have the greatest track record. Epistemologists have still not solved the problem of induction! But that shouldn't lead us to reject common sense claims or basic claims of physics. It should lead us to acknowledge that epistemology is a very difficult subject. In my view epistemology doesn't occupy an intellectual high ground from which we can undermine common sense claims, let alone claims of the queen of the sciences. It's just one of many disciplines, one which itself is not on particularly sure footing.

Re. (4). It is true that conceptual structuralism doesn't face the access problem. But that doesn't mean that it is without problems. For it faces the objectivity problem. Recall that according to conceptual structuralism "[t]he objectivity of mathematics is a special case of intersubjective objectivity that is ubiquitous in social reality."[52] I think that the assimilation of mathematical objectivity with intersubjective objectivity in social reality is a mistake. To see why let us examine counterfactuals in each domain[53]: Compare (a) "if there were no countries, then PK would not be a Canadian citizen" with (b) "if there were no humans (and hence no human concepts), then there would not be infinitely many prime numbers." In the case of (b) there are two readings. On the first reading we imagine how things look in the counterfactual situation. In that situation there are no humans, and hence no human concepts, and so, evaluating the counterfactual in the counterfactual situation we find that it comes out true. On the second reading we employ the conceptual apparatus that we have here, and *then* we consider the counterfactual situation, and, evaluating the counterfactual *from our present conceptual standpoint* we find that it comes out false—it is still true that there are infinitely many primes in the counterfactual situation envisaged, a situation where by design there are no humans around to recognize or articulate that mathematical fact. In short, on the second reading the fact—the infinitude of the prime numbers—is not "injured" by the absence of people. The situation with (a) is, however, entirely different. In this case the second reading cannot get a foothold. There is no way of thinking of me being a Canadian citizen (or there being any Canadian citizens at all) in the counterfactual situation. Citizenship is too intimately tied to the political structures. Take away the structures and you take away citizenship. In short, there is an asymmetry between the case of mathematical objectivity and social objectivity, one that is not tracked by Feferman's assimilation.

---

[52][17], p. 12.

[53]For a very clear discussion of this distinction see [2].

Let me now come to the main point, which I alluded to above: The point is that Feferman applies conceptual structuralism to number theory and, in conjunction with doing this, he embraces definiteness and realism in truth values. Why then does he not make the same move with set theory?

There is a hint of an answer to this question in his remarks on the role of mathematical models in physics. For in addition to thinking that no mythical third realm (the platonic realm) can secure the definiteness of set theory, he also thinks that the physical world cannot secure the definiteness of set theory. His reason for thinking this (when focusing, for example, on the case of the continuum) is that "one must ask whether the *mathematical structure* of the real number system can be identified with the *physical structure*, or whether it is instead simply an *idealized mathematical model* of the latter, much as the laws of physics formulated in mathematical terms are highly idealized models of aspects of physical reality."[54] Here I completely agree that the so-called "physical world" is not able to step in and secure the definiteness of analysis and set theory. Physicalists maintain that it must, and often maintain that it can. But in claiming that it can, they make questionable idealizing assumptions, for example, that space and time are either infinite in extent or infinitely divisible. But for all we know "the physical world" is finite in extent and granular at the level of the Planck length, thus making it truly finite. The mistaken move in most of these arguments is to confuse a *mathematical model of the physical world* with *the physical world "in and of itself"* (whatever *that* might be). So here I agree with Feferman.

In summary, I agree with both the claim that the definiteness of analysis and set theory cannot be secured by appeal to a mythical third realm (of the sort characterized by Feferman) and with the claim that it cannot be secured by appeal to the physical world. But notice that Feferman thinks the same thing about number theory, that is, he thinks that in the case of number theory the definiteness of the statements is not secured by a mythical third realm or by the physical world. He thinks, rather, that it is secured by the conception alone. But then why can't the conception alone also save the day in the case of analysis and set theory? The answer can only be: The conception alone can save the day in the case of number theory since in that case the conception is completely clear. But the conception alone cannot save the day in the case of analysis or set theory since in these cases the conception is not completely clear.

But then we are back to clarity! We are not getting a new argument to the effect that the concept of subsets of natural numbers is not completely clear and definite. Rather, the entire case *rests* on the claim that the concept of subsets of natural numbers is not completely clear.

---

[54][10], p. 107.

It should be noted that to a certain extent Feferman *does* think that the concept of subsets of natural numbers is clear. This come out in his discussion of the question of consistency.

> I, for one, have absolutely no doubt that PA and even PA_2 are consistent, and no genuine doubt that ZF is consistent, and there seems to be hardly anyone who seriously entertains such doubts. Some may defend a belief in the consistency of these systems by simply pointing to the fact that no obvious inconsistencies are forthcoming in them, or that these systems have been used heavily for a long time without leading to an inconsistency. To an extent, those kinds of arguments apply to NF, which has been studied and worked on by a number of people. My own reason for believing in the consistency of these systems is quite different. Namely, in the case of PA, we have an absolutely clear intuitive model in the natural numbers, which in the case of PA_2 is expanded through the notion of arbitrary subset of the natural numbers. Finally, ZF has an intuitive model in the transfinite iteration of the power set operation taken cumulatively. This has nothing to do with a belief in a platonic reality whose members include the natural numbers and arbitrary sets of natural numbers, and so on. On the contrary, I disbelieve in such entities. But I have as good a conception of what arbitrary subsets of natural numbers are *supposed* to be like as I do of the basic notions of Euclidean geometry, where I am invited to conceive of points, lines and planes as being utterly fine, utterly straight, and utterly flat, resp.[55]

In other words, Feferman's reasons for believing the consistency of PA_2 and ZFC are not inductive, but rather rest on having a clear conception of what the corresponding structures are *supposed* to look like.[56]

I am not sure that I understand the wedge between (a) being clear enough to give the structure and secure definiteness and (b) being clear enough to give a clear sense of what the structure is *supposed* to be and to thereby secure consistency. What is the extra element required to bridge the gap? It can't be the third realm. It can't be the physical world. But then what is it? It must be couched in the conception alone.

I asked him about this in personal correspondence. Here's his response:

> A picture is just a picture, [and] so doesn't qualify as a model. But it is a picture of something that is supposed to be an ω-model of PA_2. So in that picture, all arithmetical consequences of PA_2 are correct.
>
> The question then is, whether that picture is clear enough for one to accept PA_2 on its own basis. That's what I resist, because of my difficulty with the idea that there actually is a totality of arbitrary subsets of $\mathbb{N}$. But considered purely axiomatically, PA_2 tells us (at least part of) what we should accept about what holds in a picture of "arbitrary" sets of natural numbers.
>
> By way of comparison, I can picture the Euclidean plane ("perfectly flat") of points ("perfectly fine") and lines ("perfectly straight"). In that picture I recognize that Euclid's axioms hold, and more (such as the "missing" axioms about existence of points of intersection). But I don't accept Euclid's axioms straight out on its own basis. Of course, I can consider Euclidean geometry as a nice subject for axiomatic and metamathematical study.[57]

The comparison with Euclidean geometry, and the talk of whether there "actually is a totality of arbitrary subsets of $\mathbb{N}$," makes it look as though *in addition* to the conception

---

55 [11], p. 72.

56 See also [17], p. 19, [12], p. 411, and [19], p. 79.

57 Personal communication: Letter of March 14, 2016.

there needs to be something "external" to which that conception corresponds. But in the case of the concept of natural numbers Feferman embraces realism about truth values and yet he does not think that *in addition* to the conception there needs to be something "external" to which that conception corresponds, whether it be something in a mythical third realm (say, the "true" natural number structure) or something in physical reality (say, a physical instantiation of the natural number structure). So I think that this interpretation is a mistake and that what he is really questioning is simply whether there is an actual totality in the sense of *completed* totality of subsets of natural numbers, as opposed to a *potential* totality, and *this* question ultimately comes down to the question of whether the concept of subsets of natural numbers is completely clear.

Earlier we were told that the concept of natural numbers is "clear enough to secure definiteness" and that the concept of subsets of natural numbers is "inherently unclear" and as such "not clear enough to secure definiteness." Now we are told that the concept of subsets of natural numbers is, in fact, clear—it is clear enough to give a picture of what the subsets of the natural numbers are *supposed* be, and, in doing so, it is clear enough to give us confidence in the consistency of PA_2. It is just not clear *enough* to secure definiteness. So, it's clear, just not clear enough.

This further weakens my grasp of the concept of clarity at play here. I think that the concept of being clear enough to secure consistency (and what the structure is *supposed* to be like) but not clear enough to secure definiteness is itself inherently unclear. In any case, if we are to find an additional argument—one beyond the brute claim that our concept of subsets of natural numbers is not clear enough to secure definiteness—then we are going to have to move on.

5 Metamathematical

Feferman's reasons for thinking that CH *is indefinite are partly based on the metamathematical results in set theory, in particular the results showing that "* CH *is independent of all remotely plausible axioms extending* ZFC, *including all large cardinal axioms that have been proposed so far."* [58]

5.1 Feferman's Case

The background is this: Gödel proposed large cardinal axioms as a means of settling CH. It was subsequently shown that (in a sense that can be made precise) large cardinal axioms settle every question of complexity strictly below that of CH but that they cannot settle CH. Since then there have been two main approaches to making a case for axioms that settle CH: first, the approach based on forcing axioms (of which Woodin's case for the axiom $(*)$ is the most sophisticated) and second,

58[15], p. 127.

the approach based on inner model theory, more precisely, Woodin's work on the possibility of an "Ultimate-L."

Feferman has not discussed these two main approaches in detail. Instead he has focused on the results of Levy-Solovay and others, which show that the standard large cardinal axioms are invariant under small forcing and hence cannot settle CH.[59]

5.2 Response

A great deal of work has been done toward making a case for axioms that settle CH, but since Feferman does not engage with the details of this material I will set it aside.

Feferman's claim is that we should be led by the fact that large cardinal axioms (and, more generally, all remotely plausible axioms) do not settle CH to the conclusion that CH is an indefinite statement. I don't see how the former can lead us to the latter. For the question at hand is the question of determinateness and this concerns the conception, not our epistemic situation. Indeed, Feferman makes exactly this point, through a comparison with the situation in number theory.

> However, the inability of humans to decide such questions can no more be charged as a defect of this conception than can their inability to decide certain number-theoretical statements on the basis of the usual conceptions of the natural numbers.[60]

Feferman is a realist about the truth-values of statements in number theory "independently of whether we can establish them one way or the other"[61] precisely because this is a matter that concerns the conception and not our epistemic access. In summary, the semantical or ontological matter of definiteness is independent of the epistemic matter of whether we can establish the statements one way or the other, something that Feferman is aware of, and in fact appeals to in underscoring the nature of his realism about truth values when it comes to number theory.

Moreover, Feferman is in a similar situation when it comes to number theoretic statements. For example, let φ be the statement asserting the consistency of ZFC + "There is a supercompact cardinal." Feferman would maintain that φ has "not been settled by any remotely plausible assumption." But this would not lead him to conclude that this statement "is an inherently indefinite problem which will never be 'solved'," for this statement is number theoretic and hence definite by his lights. More generally, it seems clear that, given his limitative framework, there are true Π^0_1 statements of the form "S is consistent" that will *never* be solved by any new axiom that he regards as remotely plausible. For example, if the statement φ is true, then it is independent of any large cardinal axioms that Feferman deems remotely plausible. Yet this alone would not lead him to conclude that these statements are indefinite. It

---

[59] [5], pp. 72–73, [12], pp. 404–405, and [15].

[60] [3], p. 1.

[61] [17], p. 15.

follows that *not-(currently)-being-settled-by-remotely-plausible-assumptions* is not (even by Feferman's lights) sufficient for indefiniteness.[62]

5.3 Further

Feferman doesn't discuss other meta-mathematical arguments but such arguments are common in the literature and folklore. This seems as good a place as any to discuss them. I want to stress, however, that in doing so I make no claim that Feferman endorses these arguments—in fact, in some of the cases I know that he *doesn't* endorse the arguments. My goal in supplementing the above discussion is to make the point that from a meta-mathematical point of view the situation in set theory is much closer to the situation in number theory than is commonly thought.

5.3.1 Categoricity

In the case of number theory we have a categoricity theorem in the second-order setting. More precisely, PA_2 is categorical. In the case of set theory there is also a categoricity theorem in the second-order setting, only in this case one only gets *quasi*-categoricity, unless one adds a hypothesis specifying the height of the universe. More precisely, ZFC_2 is *quasi-categorical* in the sense that given two models one is isomorphic to an initial segment of the other; and if one restricts to certain fragments of this theory—like PA_3—or one adds a negative hypothesis—such as that there are no inaccessible cardinals—then one gets full categoricity. So, as far as statements like CH are concerned, number theory and set theory are on a par with regard to categoricity results.

These results are theorems (concerning second-order systems) and as such there can be no dispute about them. What has been disputed is the interpretation of such results and the question of their foundational significance. But here again I believe that there is a parallel: I agree with Feferman in thinking that in *neither* case do the theorems *secure* the definiteness of the concepts in question, since in each case one begs the question in the meta-language by appealing to higher-order quantification.[63]

It is worth mentioning that there is another, weaker version of categoricity, one that does not employ full second-order logic, namely, what has been called *schematic* or

[62]It is often pointed out that statements like φ are not "natural." That may be true but it is not relevant to our present discussion, which concerns the question of definiteness. Feferman is maintaining that (Footnote 62 continued)
all of the statements of number theory have a determinate truth value, not just the "natural" ones. And he is providing arguments which apply to statements regardless of whether they are "natural" or not. (It would be far-fetched to say "not-(currently)-being-settled-by-remotely-plausible-assumptions secures indefiniteness ..." and then add "but only for "natural" statements."

[63]See the last section of [24] for further discussion.

internal categoricity. The idea dates back a paper of Parsons.[64] The basic idea is that in being committed to the natural numbers we are committed to accepting induction for any predicate that we come to accept as definite. Now, suppose two people have their respective number systems, $\langle N, 0, S, \ldots\rangle$ and $\langle N', 0', S', \ldots\rangle$. If each admits as definite the other's number predicate—N, respectively, N'—and allows it to figure in the range of their induction scheme, then together, through communication, they can show that the natural mapping f, sending 0 to $0'$ and $S(n)$ to $S'(f(n))$, is an isomorphism.

But once again there is a parallel in the case of set theory, as has been pointed out by Martin.[65] And, once again, I think that Feferman and I are in agreement in thinking that the schematic categoricity theorems do not *secure* the definiteness of the concepts in question but rather merely enable one to *articulate* one's views concerning the definiteness of the concepts in question, views that must be secured by independent arguments. So again, there is a parallel between the case of number theory and the case of set theory.

5.3.2 Non-standard Models

There is an approach to finding an asymmetry between the two cases that rests on non-standard models. In the case of number theory when you build a non-standard model its non-standardness is immediately revealed since the standard part is a proper substructure of the entire structure. Some have thought that the case is quite different in set theory.

But the case in set theory is surprisingly similar, as far as recognizable-non-standardness is concerned, a point that has been made by Martin.[66] Take a case like Cohen's method of forcing. There are two standard model-theoretic ways of formalizing forcing. In the first approach one starts with a *countable* transitive model M of the theory in question, say, ZFC, and one adds a generic object. The point is that all of this is done *within* set theory and it is revealed in the first step that one is dealing with a non-standard model since it is revealed in the first step that one is dealing with a countable object. In the second approach one builds class-size models—and so here the countability criticism is no longer applicable—but here one builds a *Boolean-valued* model $V^{\mathbb{B}}$ and so, once again, it is immediately recognized that it is non-standard. So, once again, number theory and set theory are parallel in this regard.

64[33]. See also [34].

65[30, 31].

66[30].

5.3.3 Facility in Constructing Models

Some have cited the facility with which we can manipulate models of set theory as evidence that certain statements of set theory, like CH, are not definite. For example, given a countable transitive model M of ZFC we can force to obtain an extension $M[G_0]$ that satisfies CH and then force again to obtain an extension $M[G_0][G_1]$ that satisfies $\neg$CH and then force again to obtain an extension $M[G_0][G_1][G_2]$ that satisfies CH, and so on. We can flip CH on and off like a switch. This complete control has led some to think that it indicates that CH is not definite.[67]

But exactly the same thing holds in the setting of first-order number theory, the reason being that there are number theoretic Orey sentences. Let φ be an Orey sentence for PA.[68] Such a sentence has the feature that if we take a model M_0 of PA we can end-extend it to obtain a model M_1 in which φ holds and then end-extend that model to obtain a model M_2 in which $\neg\varphi$ holds, and then end-extend that model to obtain a model M_3 in which φ holds, and so on. If the above argument concerning CH is a good one then this "god-like" control should indicate that φ is not a definite statement. But Feferman (and most people) think that *all* statements of first-order number theory are definite.

5.3.4 Flexible Orey Sentences

One might try to argue that set theory is different in that there are *flexible* Orey sentences like PU and CH.[69] For example, PU has the feature that in addition to being an Orey sentence for ZFC it is also an Orey sentence for ZFC supplemented with certain large cardinals (measurable, strong, etc.) until at a certain point (in the region of Woodin cardinals) it ceases to be an Orey sentence because it is outright settled (in this case positively) by the large cardinal axioms in question. And CH has the feature that in addition to being an Orey sentence for ZFC it is also an Orey sentence for ZFC supplemented with *any* of the presently known large cardinal axioms. So, perhaps this is a metamathematical difference between number theory and set theory.

It turns out that one can construct number theoretic sentences that have the aforementioned metamathematical features of PU and CH. This is something that Woodin and I showed in unpublished work from some time back. Here I would like to briefly describe the results.

---

[67] For example, this argument is put forward by [22].

[68] One can choose theories other than PA and one can arrange that φ is quite simple—it can be Δ^0_2 (provably over PA (or over whichever background theory one is working with)), as is demonstrated in the first appendix.

[69] 'PU' is short hand for the statement that all projective sets have the uniformization property. This is a statement of (schematic) second-order number theory, and so concerns the concept of subsets of natural numbers. CH, in contrast, is a statement of third-order number theory, one that concerns the concept of subsets of the set of subsets of natural numbers.

To begin with, to place ourselves in a purely number theoretic setting we must replace large cardinal axioms with number theoretic analogues. The most natural way to do this is to speak of the Π^0_1 consequences of large cardinal axioms. More precisely, suppose L is a large cardinal axiom. The theory ZFC + L and the theory PA + $\bigcup_{n<\omega} \mathrm{Con}((\mathrm{ZFC}+L)\restriction n)$ are mutually interpretable, which is to say that they have the same Π^0_1 consequences. Thus, the latter, which is purely arithmetical, captures exactly the Π^0_1 consequences of ZFC + L.

Our question then is whether there are flexible number theoretic Orey sentences which are analogues of PU and CH in the following sense: A number theoretic analogue of PU is a number theoretic sentence φ that is an Orey sentence for PA, remains an Orey sentence for PA + $\bigcup_{n<\omega} \mathrm{Con}((\mathrm{ZFC}+L)\restriction n)$ for a certain stretch of the large cardinal axiom L, and then ceases to be an Orey sentence, by being outright settled by theories of the above form when the large cardinal axiom L is strengthened. An number theoretic analogue of CH is an arithmetical sentence φ that is an Orey sentence for PA and remains an Orey sentence for PA + $\bigcup_{n<\omega} \mathrm{Con}((\mathrm{ZFC}+L)\restriction n)$ for *all* presently known large cardinal axioms.

We can now state two theorems, which were proved jointly with Hugh Woodin:

Theorem 1 *Suppose $\langle \psi_i : i < \omega \rangle$ is a recursive sequence of Π^0_1-statements that are of increasing strength; that is, such that for all $i < j$, $\mathrm{PA} \vdash \psi_j \rightarrow \mathrm{Con}(\psi_i)$. There is a Π^0_2-statement φ such that for all $i < \omega$,*

$$\mathrm{PA} \vdash \text{“}\varphi \textit{ is an Orey sentence for } \mathrm{PA} + \psi_i.\text{”}$$

Theorem 2 *For each $i < \omega$, there is a Π^0_{i+2}-sentence φ such that for all Π^0_i-sentences ψ*

$$\mathrm{PA} \vdash \mathrm{Con}(\mathrm{PA}+\psi) \rightarrow \big(\mathrm{Con}(\mathrm{PA}+\psi+\varphi) \wedge \mathrm{Con}(\mathrm{PA}+\psi+\neg\varphi)\big).$$

It follows from these theorems that there are flexible number theoretic Orey sentences which are analogues of PU and CH in the sense described above. Thus, even in this case we have a parallel between number theory and set theory.

In summary: As far as Feferman's metamathematical arguments are concerned, and as far as the additional metamathematical arguments that I have considered are concerned, there is a parallel between the case of number theory and set theory and we still do not have a case—independent of the brute claim that the concept of subsets of natural numbers is not sufficiently clear to secure definiteness—that analysis and set theory, in contrast to number theory, are indefinite.

6 Formal Results on Indefiniteness

Feferman takes the formal results on indefiniteness—in particular the result of Rathjen showing that CH *is indefinite relative to the semi-constructive system* SCS^+*—as providing evidence that* CH *is indefinite.*

6.1 Feferman's Case

In the spirit of getting exact about informal claims, Feferman "proposed [a] logical framework for distinguishing definite from indefinite concepts."[70] The basic idea is that classical logic is the logic appropriate for definite concepts, while intuitionistic logic is the logic appropriate for indefinite concepts. Thus, for example, if one wishes to articulate and investigate the view of someone who, like Feferman, maintains that the concept of natural numbers is definite, while the concept of subsets of natural numbers is not definite, then an appropriate system would be a system of *semi-constructive* set theory involving two logics, classical logic for the number theoretic component and intuitionistic logic for the remainder. A statement φ is then defined to be (formally) *definite* with regard to such a system S if $S \vdash \varphi \vee \neg\varphi$. So, in the case under consideration, it will be immediate that statements of number theory are definite (this having been built in from the start). The question then arises: *Which other statements can be shown to be definite relative to such a system S?*

Feferman introduced two semi-constructive systems of set theory, which will here be labeled 'SCS' and 'SCS^+'. The first system is the one briefly described above. The second system aims to capture the view of a descriptive set theorist who maintains that the concept of subsets of natural numbers is definite, but makes no explicit claims about the definiteness of the concept of subsets of the set of subsets of natural numbers, or concepts involving further iterations of the powerset operation. It involves classical logic for second-order number theory and intuitionistic logic beyond.[71]

Feferman conjectured and Rathjen proved that CH is indefinite relative to SCS^+.[72] Feferman takes this result "as further evidence in support" of the claim that CH is indefinite.[73]

Our question is: *Does this result provide evidence that* CH *is indefinite?*

---

[70] [17], p. 2. See [16] for the technical work on these systems. See [17–19], for an informal, philosophical discussion.

[71] It would take us too far afield, and it would not be especially illuminating, to pause to lay out the axioms of these two systems. For a precise account of these systems the interested reader can consult [36]. (I am using 'SCS' for the system Rathjen also calls 'SCS' and I am using 'SCS^+' for the system he calls T, that is, the system SCS + "$\mathbb{R}$ is a set".).

[72] [36].

[73] [18], p. 23.

6.2 *Response*

It seems antecedently unsurprising that if one only assumes classical logic for a limited domain—say first-order number theory (or second-order number theory)—then one will not be able to *prove* that classical logic holds for distinctive statements of a richer domain—such as second-order number theory (or third-order number theory). In particular, it is unsurprising that if one just assumes that classical logic holds for second-order number theory then one won't be able to show that the law of excluded middle (LEM) holds for a distinctive Σ_1^2 statement like CH.

Moreover, not only is it unsurprising, it is hard to see how this fact is relevant to the question of whether CH is definite. The people who maintain that CH is definite do not maintain that it can be shown to be definite by assuming only that the concept of subsets of natural numbers is definite. Generally, they maintain that the concept of subsets of the set of subsets of natural numbers is definite. But Feferman's system SCS^+, on the face of it, starts with the assumption that this concept is *indefinite* and asks whether one can prove that despite that prima facie assumption one can prove that it is definite.

Perhaps a comparison is helpful: Feferman maintains that statements of number theory are definite. He maintains this because he maintains that the concept of natural numbers is definite. Imagine now that we ask him to *prove* the definiteness of statements of number theory by working in a system which does not have LEM for statements of number theory but which assumes classical logic only for a much weaker fragment. I believe he would be unmoved by this demand. And I think he would be right to be unmoved. Similarly, I do not see the point of asking someone to *prove* the definiteness of CH by working in a system which does not have LEM for statements of third-order number theory but which assumes classical logic only for a much weaker fragment.

There is one further main point that I would like to make, and for this it will be necessary to investigate Feferman's two systems—SCS and SCS^+—in more detail.

6.2.1 Regarding SCS^+

Let's start with the second system, SCS^+, since this is the system that figures prominently in Feferman's discussion of CH. This system is proposed as a system that explicates the working perspective of the descriptive set theorist. Descriptive set theory is the study of "definable" sets of reals. Here the notion of being a "definable" set of reals is really a hierarchy of notions, ranging from being a Borel set of reals, to being a projective set of reals, to being a set of reals existing in $L(\mathbb{R})$, to being a set of reals existing in $\text{HOD}(\mathbb{R})$. It tends to be the case that statements which are difficult—and in some cases, impossible—to settle in ZFC when one considers "arbitrary" sets of reals, become tractable when one considers "definable" sets of reals. For example, as is well known, Cantor was unsuccessful in his attempts to settle CH, and, as we now know, it was not through any lack of ingenuity, but rather because it is *in principle impossible* to settle CH within ZFC. Nevertheless, he was able to prove that CH holds for certain "definable" sets of reals, in particular the closed sets of reals—in the sense that every infinite closed set of reals is either countable or contains a perfect set and so has size the continuum—and it was subsequently shown that it also holds (in this sense) for much richer classes of definable sets.

In using SCS^+ to explicate the views of the descriptive set theorists it would seem that the underlying idea is that the descriptive set theorists think that statements about "definable" sets of reals are definite, while they are agnostic or uncertain about the definiteness of "arbitrary" sets of reals. It is then of interest to ask how far one can get (in terms of establishing definiteness) within this framework.

Let's examine definiteness in SCS^+ and see how well it articulates the views of the descriptive set theorists. At the start we have that all statements of second-order number theory are definite. So far so good. This is in accord with the views of the descriptive set theorists. Let us now march our way up the hierarchy of definability. It turns out that statements about Borel sets come out as definite; for example, the statements that all Borel sets are Lebesgue measurable, have the perfect set property, and are determined, all come out as definite in SCS^+. It also turns out that statements about projective sets come out as definite; for example, the statements that all projective sets are Lebesgue measurable, have the perfect set property, are determined, etc. come out as definite in SCS^+. Thus far we are doing well. All of this is in accord with the views of the descriptive set theorists.

But things change when we get to $L(\mathbb{R})$. Let $\mathrm{AC}^{L(\mathbb{R})}$ be the statement that "there is a well-ordering of $\mathbb{R}$ in $L(\mathbb{R})$". It turns out that $\mathrm{AC}^{L(\mathbb{R})}$ is indefinite from the point of view of SCS^+, assuming the consistency of large cardinals at the level of $\mathrm{AD}^{L(\mathbb{R})}$.[74]

Theorem 3 *Assume* Con(ZFC + *"There are ω-many Woodin cardinals"*). *Then* $\mathrm{SCS}^+ \nvdash \mathrm{AC}^{L(\mathbb{R})} \vee \neg\mathrm{AC}^{L(\mathbb{R})}$.

Likewise $\mathrm{AD}^{L(\mathbb{R})}$ comes out as indefinite in SCS^+. So, in this regard, $\mathrm{AD}^{L(\mathbb{R})}$ is just like CH. But in contrast to CH it does not concern arbitrary subsets of reals, it only concerns definable ones, and it is a statement that the descriptive set theorists, at least modern descriptive set theorists, regard as definite.

So the system SCS^+ gives a mixed verdict on statements of descriptive set theory, when measured with regard to the community of modern descriptive set theorists. In some cases it is in alignment with that community and in other cases it is out of alignment with that community. For example, it classifies statements of second-order number theory (and even "schematic" statements of second-order number theory, like PD) as definite, and in this respect it is in alignment with the views of modern descriptive set theorists. But it classifies other statements (like $\mathrm{AC}^{L(\mathbb{R})}$ and $\mathrm{AD}^{L(\mathbb{R})}$) as indefinite, and in this respect it is *out* of alignment with the views of modern descriptive set theorists.

---

[74]This result and the remaining results in this section were proved jointly with Hugh Woodin. It seems likely that the hypothesis can be weakened to just ZFC but the technical hurdles are difficult and it doesn't affect the point I wish to make. I am confident that the hypothesis is true, and hence that the conclusion follows. If one has an issue with the hypothesis we have a whole other debate ... In any case, the skeptical reader can take the theorem as it stands.

The lack of alignment becomes even more dramatic when we look not just at what the community of modern descriptive set theorists deem *definite* but at what they deem *true*. The modern descriptive set theorists who have discussed the more philosophical aspects of set theory—such as the search for new axioms and the question of the definiteness of CH—have made a case for axioms of definable determinacy, including *both* PD and $\mathrm{AD}^{L(\mathbb{R})}$ (and hence that $\mathrm{AC}^{L(\mathbb{R})}$ is false). In fact, it is hard to see how one might maintain PD and *not* maintain $\mathrm{AD}^{L(\mathbb{R})}$. Certainly everyone who maintains the former maintains the latter, and for exactly the same reasons. One of the reasons is that these principles follow from large cardinal axioms and the large cardinal axioms required to prove the latter are only minimally stronger than those required to prove the former. Indeed I doubt that there is even a coherent view in which one can make a case for the former large cardinal axioms without simultaneously making a case for the larger large cardinal axioms.[75] Another reason is the phenomenon whereby PD and $\mathrm{AD}^{L(\mathbb{R})}$ follow from every known "natural" theory of sufficiently strong consistency strength and, in the case of the latter statement, the consistency strength required is only slightly greater than in the case of the former statement. Indeed it is hard to see how one could draw a principled line between the two statements in terms of the consistency hierarchy, maintaining that some of those below the line were plausible while none of those above the line were plausible. In any case, I do not know of a single descriptive set theorist who maintains that PD is definite (or true) and maintains that $\mathrm{AD}^{L(\mathbb{R})}$ is indefinite (or false). So in this regard SCS^{+} does not capture the views of the modern descriptive set theorists.

It should be mentioned that there are other frameworks that have been proposed which fare better in capturing the views of modern descriptive set theorists. In these frameworks $\mathrm{AD}^{L(\mathbb{R})}$ comes out as definite.[76] I have spent some time investigating such frameworks, with the aim of seeing how one might make a solid case for the claim that CH is either absolutely undecidable or indefinite.[77] But in each case there has always been an interesting twist—there has always been an escape hatch in which one might break the asymmetry and make a case for one side or the other.

But let us return to Feferman's views. Feferman believes much more than that CH is indefinite—he believes that $\mathrm{AD}^{L(\mathbb{R})}$ is indefinite. And his system SCS^{+} implies much more than that CH is (formally) indefinite—it implies that $\mathrm{AD}^{L(\mathbb{R})}$ is (formally) indefinite. So the focus of this discussion should really be $\mathrm{AD}^{L(\mathbb{R})}$. This is precisely the point where modern descriptive set theorists have argued that there is a convincing case for new axioms.

---

[75]The former follows from the existence of infinitely many Woodin cardinals. The latter follows from the existence of infinitely many Woodin cardinals with a measurable above them all. The assumption of Woodin cardinals is much stronger than the assumption of measurable cardinals.

[76]See for example [29, 37, 40].

[77]See [23, 29].

6.2.2 Regarding SCS

But when it comes to Feferman's view even SCS^+ itself is a bit of a red herring. For recall that Feferman maintains something much stronger than that CH is indefinite. He maintains that the concept of the subsets of natural numbers is indefinite. In particular, he maintains not only that $\mathrm{AD}^{L(\mathbb{R})}$ is indefinite, he also maintains that PD and much more is indefinite. For recall that he denies

> that it is a fact of the matter whether all projective sets are Lebesgue measurable or have the Baire property, and so on.[78]

So, of the two systems that he has introduced, SCS is much closer than SCS^+ to articulating his own position with regard to matters of definiteness. Let us then investigate this system and see what comes out as indefinite within it.

Let R be the statement that "there is a non-constructible real", that is, the statement that there is a real which is not in L.

Theorem 4 $\mathrm{SCS} \nvdash R \vee \neg R$.

In other words, the statement that there is a non-constructible real is not definite from the point of view of SCS.

This creates a bit of tension for Feferman, for it would seem that he would have to maintain that R is *definite*, since it would seem that he would have to maintain that R is *false*. For, although he has made it clear that he is not a predicativist, both his positive views about what is justified and his negative views about what is not justified all lead him to constructions that do not break out of L. For example, Feferman certainly accepts L_{Γ_0} since this covers the realm of predicativity and he accepts Γ_0 as completely clear and definite. But he probably accepts much larger fragments of L. He might even accept $L_{\omega_1^{\mathrm{CK}}}$, although I am not sure of this. In any case, it seems clear that everything he would accept—in terms of iterating the definable powerset—will fall far short of $L_{(\omega_1)^L}$, for after all, he does not accept the uncountable infinite. In short, he will never break the $V = L$ barrier. He will never even come close. And this seems to be something that is clear upon reflection on his considered views. So, it would seem that he must maintain that R is false and hence definite.

There are also statements much closer to home which come out as indefinite. In fact, any statement which is equivalent to ATR_0 over RCA_0 comes out as indefinite:

Theorem 5 $\mathrm{SCS} \nvdash \varphi \vee \neg\varphi$ *where* φ *is any of the following statements*:

(a) *Σ_1-separation: Every pair of disjoint analytic sets can be separated by a Borel set.*
(b) *Perfect Set Theorem: Every uncountable closed (or analytic) set has a perfect subset.*
(c) *Comparability of Countable Well-orderings.*

---

[78][12], p. 405.

(d) *The Ulm theory for countable reduced Abelian p-groups.*
(e) Δ_0 *Determinacy.*

In fact, the result for the statement in (e) is much stronger. For using a result of [1] it turns out that there is a recursively enumerable game G such that the statement "G is determined" is indefinite with respect to SCS.

Nevertheless, Feferman has made it clear to me in conversation that he thinks that open determinacy is in fact definite.[79] Feferman has told me that to fully capture his view concerning the extent of what is definite one would have to supplement SCS by adding structure beyond that of the natural numbers, with the aim of ensuring that open determinacy and its kin come out as definite. I think that this is the right course for him to take. It is certainly the right course for *us* to take, since I don't think there can be reasonable doubt that open determinacy is definite.

6.2.3 Conclusion

In the previous two sections we considered two case studies—one concerning SCS^+ and one concerning SCS—and we showed that in each case the systems proposed fail to capture the views on indefiniteness of the target audience that they are designed to articulate: The first fails to capture the views of the descriptive set theorists since it has PD coming out as definite while it has $AD^{L(\mathbb{R})}$ coming out as indefinite; the second fails to capture Feferman's views and those who agree with him on matters of indefiniteness since it has the Riemann hypothesis coming out as definite while it has open determinacy coming out as indefinite.

The natural course to take is to supplement each system by adding additional structure, so as to bring it into alignment with the views of the target audience. But when we start doing this we see that the enterprise is rather ad hoc. The fact that we are inclined to make these modifications underscores the point that what is really guiding the enterprise is the pre-theoretic views on definiteness. The systems are *designed* to track those views. If we are not willing to adjust our views in light of the results of the proposed system, but instead are inclined to hold our views fixed and modify the systems, then how can we maintain that the systems are giving us insight into what is definite and what is not?

In summary: (a) The systems proposed thus far fail to articulate the views of their target audience. (b) We are inclined to hold the pre-theoretic views fixed and adjust the systems to bring them in step with those pre-theoretic views. (c) This is evidence that the systems are not giving us insight into what's definite and what's not. (d) In any case, we do not yet have positive evidence that this approach is going to give us insight into what's definite and what's not, and, given the nature of the enterprise, we have reason to believe that it will not.

[79] It is unclear to me that he should say this. For open determinacy is *logically equivalent* over RCA_0 (a system Feferman accepts) to the statement that all countable well-orderings are comparable, and it does not seem that Feferman can say that the latter is definite since it involves what he regards as a bankrupt notion, namely, the notion of a well-ordering.

Once again we have not been given an independent argument for the claim that the concept of subsets of natural numbers is inherently unclear and indefinite, but rather the entire enterprise is being guided by brute, pre-theoretic intuitions as to what is completely clear and definite and what is inherently unclear and indefinite. So, once again, we are back to clarity.

7 Conclusion

I have argued that in the end the entire case ultimately rests on the brute intuition that the concept of subsets of natural numbers—along with the richer concepts of set theory—is not "clear enough to secure definiteness." And my response to this is that the concept of "being clear enough to secure definiteness" is about as clear a case of an inherently unclear and indefinite concept as one might find. It is a concept that does not enjoy the sort of intersubjective objectivity that is a hallmark of Feferman's conceptual structuralism. For these reasons it cannot carry much weight in a foundational enterprise, especially one aimed at arguing that the concept of subsets of natural numbers—a conception that on the face of it *does* enjoy the sort of intersubjective objectivity that Feferman celebrates—is not a definite concept.

In this paper my aim has been *negative* in that I have concentrated on rebutting Feferman's arguments to the effect that the concept of subsets of natural number—along with the richer concepts of set theory—are indefinite. But I have not advanced any *positive* arguments to the effect that this concept (or these richer concepts) *are* definite.

I wish I could continue and say something positive on behalf of my defendant. In short, I agree with Feferman when he writes:

> [T]he objectivity of mathematics lies in its stability and coherence under repeated communication, critical scrutiny and expansion by many individuals often working independently of each other.[80]

It is precisely this—the stability and coherence under repeated communication, critical scrutiny and expansion by many individuals often working independently of each other—that has led me to believe that the case for new axioms in set theory is a solid one.[81] But that's another story.

References

1. Blass, A.: Complexity of winning strategies. Discret. Math. **3**(4), 295–300 (1972)
2. Einheuser, I.: Counterconventional conditionals. Philos. Stud. **127**, 459–482 (2006)

[80] [20], p. 3.

[81] For some of the elements that go into making a positive case see the expository articles [25–27], and the references therein.

3. Feferman, S.: Systems of predicative analysis. J. Symb. Log. **29**, 1–30 (1964)
4. Feferman, S.: A more perspicuous formal system for predicativity. In: Lorenz, K. (ed.) Konstruktionen versus Positionen, vol. I, pp. 69–93. Walter de Gruyter, Berlin (1979)
5. Feferman, S.: Infinity in mathematics: is Cantor necessary? In: Toraldo di Francia, G. (ed.) L'infinito nella scienzia, pp. 151–209. Istituto della Enciclopedia Italiana, Rome (1987). Reprinted in [9]
6. Feferman, S.: Reflecting on incompleteness. J. Symb. Log. **56**, 1–49 (1991)
7. Feferman, S.: What rests on what? Proof-theoretic analysis of mathematics. Proceedings of the 15th international Wittgenstein symposium. In: Czermak, J. (ed.), Philosophy of Mathematics, Part 1. Verlag Hölder-Pichler-Tempsky, Vienna, pp. 141–171. Reprinted in [9], Chap. 10, pp. 187–208 (1993)
8. Feferman, S.: Gödel's program for new axioms: why, where, how and what? In: Hajek, P. (ed.), Gödel 96: Logical Foundations of Mathematics, Computer Science, and Physics, vol. 6. Lecture Notes in Logic. AK Peters Ltd., pp. 3–22 (1996)
9. Feferman, S.: In the Light of Logic, Logic and Computation in Philosophy. Oxford University Press, Oxford (1998)
10. Feferman, S.: Does mathematics need new axioms? Am. Math. Mon. **106**, 99–111 (1999)
11. Feferman, S.: Does reductive proof theory have a viable rationale? Erkenntnis **53**(1/2), 63–96 (2000)
12. Feferman, S.: Why the programs for new axioms need to be questioned. Bull. Symb. Log. **6**(4), 401–413 (2000)
13. Feferman, S.: Comments on, "Predicativty as a philosophical position", by G. Hellman, Revue Internationale de Philosophie **3**, 313–323 (2004)
14. Feferman, S.: Predicativity. In: Shapiro, S. (ed.), The Oxford Handbook of Philosophy of Mathematics and Logic, pp. 590–624. Oxford University Press (2005)
15. Feferman, S.: The 5 questions. In: Hendricks, V.F., Leitgeb, H. (eds.), Philosophy of Mathematics: 5 Questions, pp. 115–135. Automatic Press/VIP (2007)
16. Feferman, S.: On the strength of some semi-constructive theories. In: Feferman, S., Sieg, W. (eds.) Proof, Categories and Computation: Essays in honor of Grigori Mints, pp. 109–129. College Publications, London (2010)
17. Feferman, S.: Is the continuum hypothesis a definite mathematical problem? for the Exploring the Frontiers of Incompleteness (EFI) Project at Harvard Universitty, 2010–2012 (2011)
18. Feferman, S.: The continuum hypothesis is neither a definite mathematical problem nor a definite logical problem. In: Koellner, P. (ed.) Exploring the Frontiers of Incompleteness. Harvard University Press (2014). Forthcoming
19. Feferman, S.: Logic, mathematics, and conceptual structuralism. In: Rush, P. (ed.) The Metaphysics of Logic. Cambridge University Press (2014). Forthcoming
20. Feferman, S.: Parsons and I: Sympathies and differences. Talk given in honor of Charles Parsons at Columbia University (2016)
21. Feferman, S., Strahm, T.: The unfolding of non-finitist arithmetic. Ann. Pure Appl. Log. **104**, 75–96 (2000)
22. Hamkins, J.: The set-theoretic multiverse. Rev. Symb. Log. **5**(3), 416–449 (2012)
23. Koellner, P.: Truth in mathematics: the question of pluralism. In: Bueno, O., Linnebo, Ø. (eds.) New Waves in Philosophy of Mathematics, pp. 80–116. Palgrave Macmillan, New Waves in Philosophy (2009)
24. Koellner, P.: Strong logics of first and second order. Bull. Symb. Log. **16**(1), 1–36 (2010)
25. Koellner, P.: Independence and large cardinals. In: Zalta, E.N. (ed.) Stanford Encyclopedia of Philosophy. Stanford University (2011). ⟨http://plato.stanford.edu/entries/independence-large-cardinals⟩
26. Koellner, P.: The continuum hypothesis. In: Zalta, E.N. (ed.) Stanford Encyclopedia of Philosophy. Stanford University (2013). ⟨http://plato.stanford.edu/entries/continuum-hypothesis⟩
27. Koellner, P.: Large cardinals and determinacy. In: Zalta, E.N. (ed.) Stanford Encyclopedia of Philosophy. Stanford University (2013). ⟨http://plato.stanford.edu/entries/large-cardinals-determinacy⟩

28. Koellner, P.: Feferman on predicativity (In preparation) (2016)
29. Koellner, P., Woodin, W.H.: Incompatible Ω-complete theories. J. Symb. Log. **74**(4) (2009)
30. Martin, D.A.: Multiple universes of sets and indeterminate truth values. Topoi **20**(1), 5–16 (2001)
31. Martin, D.A.: Completeness or incompleteness of basic mathematical concepts (2015). ⟨http://www.math.ucla.edu/~dam/booketc/efi.pdf⟩
32. Martin-Löf, P.: The Hilbert-Brouwer controversy resolved? In: Atten, M.V., Boldini, P., Bourdeau, M., Heinzmann, G. (eds.) One Hundred Years of Intuitionism (1907–2007), pp. 243–256. Publications des Archives Henri Poincaré, Birkhauser Verlag AG, Basel (2008)
33. Parsons, C.: The uniqueness of the natural numbers. Iyyun **39**, 13–44 (1990)
34. Parsons, C.: Mathematical Thought and its Objects. Cambridge University Press (2008)
35. Rathjen, M.: The constructive Hilbert program and the limits of Martin-Löf type theory. Synthese **147**, 81–120 (2005)
36. Rathjen, M.: Indefiniteness in semi-intuitionistic set theories: On a conjecture of Feferman. J. Symb. Log. **81**(2), 742–754 (2016)
37. Steel, J.: Gödel's program. In: Kennedy, J. (ed.) Interpreting Gödel: Critical Essays. Cambridge University Press, pp. 153–179 (2014)
38. Tait, W.W.: Truth and proof: the platonism of mathematics. Synthese, **69**, 341–370 (1986). Reprinted in [39]
39. Tait, W.W.: The Provenance of Pure Reason: Essays in the Philosophy of Mathematics and Its History, Logic and Computation in Philosophy. Oxford University Press (2005)
40. Woodin, W.H.: The Continuum Hypothesis, the generic-multiverse of sets, and the Ω-conjecture. In: Kennedy, J., Kossak, R. (eds.) Set Theory, Arithmetic, and Foundations of Mathematics: Theorems, Philosophies, vol. 36. Lecture Notes in Logic. Cambridge University Press (2011)

Author Biography

Peter Koellner is Professor of Philosophy at Harvard University. He received his Ph.D from MIT in 2003. His main areas of research are mathematical logic (specifically set theory, large cardinals, and determinacy) and the philosophy of mathematics (specifically, the question of the justification of new axioms).

Feferman's Skepticism About Set Theory

Charles Parsons

Abstract Solomon Feferman has expressed skepticism or reserve about set theory, especially higher set theory, in many writings, and in his mathematical work he has largely stayed away from set theory. The paper undertakes to describe and diagnose Feferman's attitude toward set theory, especially higher set theory. Section 1 discusses his opposition to Platonism in relation to some understandings of what Platonism is. Section 2 discusses his interest in predicativity, his analysis of predicative provability, and his sympathy for "predicativism," although he denies being an adherent and has used impredicative methods in his own metamathematical work. It also notes his reconstruction of Hermann Weyl's attempt to construct the elements of analysis on a predicative basis. Section 3 concerns his attitude toward proof theory. Section 4 turns more to philosophy. His anti-platonism is compared with the well known platonism of Gödel. Unlike Gödel, Feferman views concepts as human creations. He notes that basic mathematical concepts can differ in clarity and argues that less clear concepts, in particular those of set theory, can give rise to questions that do not have definite answers. It is questioned whether Feferman's conceptual structuralism gives mathematics the degree of objectivity that its application in science requires. In Sect. 5 remarks are made about Hilary Putnam's criticism of Feferman's claim that a predicative system conservative over PA is adequate for the mathematics applied in science. The difference is seen to turn on Putnam's scientific realism.

Keywords Logic · Set theory · Predicativity · Proof theory · Skepticism

---

This paper was in virtually final form before Sol Feferman's unfortunate death in July 2016. That event implies that this volume will not contain the reply by him that was originally envisaged. However, *Feferman* [17], presented in a symposium at Columbia University in April 2016, contains ideas that he would probably have expressed in his reply.

Thanks to Wilfried Sieg for comments on earlier versions and to him and Peter Koellner for discussions of Feferman's views over an extended period.

C. Parsons (✉)
Department of Philosophy, Harvard University, Emerson Hall 25 Quincy St,
Cambridge, MA 02138, USA
e-mail: parsons2@fas.harvard.edu

G. Jäger and W. Sieg (eds.), *Feferman on Foundations*, Outstanding
Contributions to Logic 13, DOI 10.1007/978-3-319-63334-3_20

Subject Classification Logic and foundations

Solomon Feferman was a mathematician and philosopher of very wide cultivation and interests, who has produced over a long career a formidable body of research in mathematical logic and foundations of mathematics, as well as a lot of expository writing that articulates a philosophical point of view. That point of view would not be easy to characterize, and to do so in a comprehensive way is a larger task than I will undertake in this essay. One characteristic of his attitude is a certain distance from developed set theory, particularly those parts of the subject that depend on large cardinals or seek to give a definitive answer to questions undecidable in ZFC.[1] Cantor's continuum problem is known to be undecidable in ZFC even with the help of the known large cardinal axioms, so that it is natural that it should be a particular target of his doubts.

1 Feferman's Anti-platonism: Preliminaries

When he gives brief expositions of his point of view, he emphasizes opposition to "platonism." His characterizations of his "antiplatonism" tend to be brief. The term "platonism" is rather ambiguous in discussions of the philosophy of mathematics, but Feferman does say enough to place him at least roughly on the map. In much American philosophical discourse, platonism is contrasted with nominalism, so that any point of view that admits abstract entities at all (or at least classes) counts as platonist. By this standard, Feferman is certainly a platonist. The term is also treated as more or less synonymous with "realism," as that term is used with reference to mathematics. With respect to much of mathematics (but probably not higher set theory), Feferman is a realist by an undemanding standard proposed by William Tait:

> I want to save this term ["realism"] for the view that we can truthfully assert the existence of numbers and the like without explaining the assertion away as something else. Realism in this sense is the default position: when one believes mathematics is meaningful and has, as one inevitably must, finally become convinced that mathematical propositions cannot be reduced to propositions about something else or about nothing at all, then one is a realist. ([43], p. 91.)

The default realism that Tait states here and advocates in his writings can be held about some parts of mathematics and not others.[2] The same is true of "platonism" as defined in the classic essay Bernays [20], which amounts to acceptance of bivalence for mathematical propositions. Platonism according to Bernays is a little more demanding than default realism according to Tait, because the latter is a view that a constructivist who rejects classical logic can take of the mathematics he accepts.

[1]Consideration of some large cardinal notions seems to have been forced on him by developments in proof theory.

[2]For further discussion of Tait on realism, see Sect. 3 of Parsons [15].

Both Tait's default realism and Bernays's "platonism" are at bottom methodological stances. There is no doubt that Feferman would agree with Tait's default realism about the mathematics he is comfortable with. It is not clear where the limit of that mathematics lies. I don't think Feferman has undertaken to describe and explain a sharp limit.[3] It is clear that any such limit would be short of the higher reaches of set theory.

A more delicate question concerns platonism as defined by Bernays. Feferman has worked in proof theory for a major part of his career and has collaborated actively with other proof theorists. But it is a general methodological principle of proof theory that results should be proved constructively. It would follow that the proof-theoretic work that Feferman has done (and that of others that he has relied on) is not platonist in Bernays's sense.

I think Feferman's view would be that the failure of an argument in proof theory to be constructive would not imply that we do not have reason to take the result to be true but rather that it would not accomplish a certain desirable kind of reduction. Many results are proved by model-theoretic methods and only subsequently proved by constructive proof-theoretic methods. I don't think Feferman has any quarrel with that way of proceeding, but if what is desired is a kind of reduction to a constructive theory, only the constructive proof-theoretic proof accomplishes this.

In rejecting what he calls platonism, I think Feferman understands it in a more metaphysical way than would fit the understandings canvassed above. I will postpone until later further articulation and examination of what his anti-platonism consists in and what philosophical arguments he gives for it. More mathematical considerations are central to his buttressing his point of view, and I think it likely that it derives to an important degree from his mathematical experience.

That mathematical considerations should be at the root of Feferman's views should not surprise us, since he is of course first of all a mathematician. But I think that in order to clarify the roots of his attitude toward set theory, it is helpful and perhaps necessary to examine some of his mathematical work. He only began to write expository and philosophical essays in the mid-1970s. By then he already had a substantial amount of mathematical work under his belt.[4]

Some of Feferman's earliest work was in proof theory in a general sense, although he had not yet worked in the modified Hilbert program that was being brought to the United States at the time by Georg Kreisel. His graduate work was in Berkeley, dominated by Tarski, at the time of the birth of modern model theory. Although he developed a taste for that subject and made some contributions to it, it was never his central focus.

---

[3]It is likely that he would see a difficulty of principle in establishing such a limit with mathematical precision; cf. the remarks in Sect. 2 below on his analysis of predicativity.

[4]The earliest in date of publication of any of the essays in *In the Light of Logic* (Feferman [10]) is 1979.

At Stanford he was a colleague of Paul Cohen, and during the latter's work on his epoch-making independence results, Feferman gave him comments and advice.[5] He did not really join the group of logicians who produced a large number of independence results in set theory in the wake of Cohen's work. He did, however, apply Cohen's method in the paper [2], published when the method was new.[6] The paper did contain mainstream set-theoretic results, that the Boolean Prime Ideal Theorem is independent of ZF, and that the existence of a definable well-ordering of the continuum is independent of ZFC + GCH. But what is more interesting and revealing is his application of forcing to first- and second-order arithmetic. At the outset Feferman remarks, "The notion of forcing is syntactic and the notion of generic (sequence of) sets is obtained directly from it. However, the motivations behind the introduction of these notions is essentially model-theoretic" (p. 325). Feferman then uses forcing to prove results about hyperarithmetic sets of natural numbers. He has so far as I know not returned to the method of forcing in later work.

2 Predicativity and Predicativism

Feferman has been called a predicativist by some.[7] However, he has firmly denied this, although the qualification shows that he is less distant from that position than most writers on the foundations of mathematics, even constructivists:

> Though I am not and never have been a predicativist, I have to admit to being a sympathizer since I am an avowed anti-platonist, at least insofar as set theory is concerned, and I grant the natural numbers a position of primacy in our mathematical thought. ([14], pp. 313–14.)

Feferman might assent to the famous dictum of Leopold Konecker, "Die natürlichen Zahlen hat der liebe Gott gemacht; alles andere ist Menschenwerk,"[8] although he would not have drawn the radical consequences Kronecker drew from it or even accepted the predicativist interpretation proposed by Paul Bernays ([19], p. 41 n. of reprint).

---

[5]See his introductory note to Gödel's letters to Cohen in Gödel [25]. Cohen did not allow his letters to Gödel to be published, although some information about their content can be gleaned from Feferman's introduction.

[6]Feferman's own papers will be cited only by the number in the list of references.

[7]One such person is Hilary Putnam in *Putnam* [37], p. 137. In fairness to Putnam, it should be noted that he has primarily in mind those essays in which Feferman argues that the mathematics applied in natural science can be developed in theories that are conservative extensions of PA and predicative by his own lights. (Cf. the remarks on Putnam at the end of Sect. 5 below.)

[8]Usually translated as, "God created the natural numbers; everything else is the work of man." As we shall see in Sect. 4, Feferman could accept the theological framework only as metaphor.

Early in his career Feferman obtained one of his most widely known results, an analysis of predicative provability in analysis.[9] In the period around 1960 several logicians, led by G. Kreisel, worked on the analysis of predicative definability and advanced the hypothesis that any predicatively definable set of natural numbers is hyperarithmetic. The converse posed conceptual problems, and Kreisel proposed changing focus to predicative provability. In his classic paper "Systems of predicative analysis" ([1]), Feferman considered the question what are the limits of predicative provability in analysis (second-order arithmetic), and by considering both ramified and unramified systems gave a convincing argument as to what the limits are. According to this analysis, usual formulations of ramified analysis are predicative if and only if the levels are indexed by recursive ordinals less than the first strongly critical ordinal Γ_0. He constructed an unramified system IR which encodes all of predicative analysis according to his analysis and verified that its proof-theoretic ordinal is also Γ_0.[10] Later work showed that the analysis is robust. For example, *Feferman* [3] describes a system of set theory that is a conservative extension of IR for formulae of second-order arithmetic and so has the same provable ordinals.

One criterion of predicativism is that the predicativist should regard all mathematics that is not predicative as illegitimate, although he might accept as legitimate mathematics that is predicatively reducible by some criterion.[11] It is hard to see how such a predicativist should think the analysis of predicativity in Feferman's work to be a sensible or even intelligible enterprise. Feferman's analysis used means that are impredicative by his own criterion. About his system of predicative set theory PS he writes:

> Necessarily (just as with IR), the system PS does not have predicative justification as a whole; if it did we could extend it by reflection. What is claimed is that the ideal predicativist can recognize as correct each particular axiom and application of a rule of inference.[12]

I have not found a place in Feferman's writings on this subject where he apologizes for stepping beyond predicativity in his analysis of it. I would not expect one, since research programs that he worked on for years involved methods that are impredicative according to his own analysis. One such program is the modified Hilbert program, pursued by many logicians from Gerhard Gentzen to members of a younger generation than Feferman himself. In general, the analysis of what a given foundational standpoint allows typically involves the use of methods going beyond that standpoint (cf. note 10 above).

---

[9]Throughout it is predicativity relative to the natural numbers that is under consideration. That is usually called predicativity *simpliciter*. Clearly that is what is at issue in the quotation just given.

[10]The results concerning ramified systems were obtained independently by Kurt Schütte, who also brought out the significance of the ordinal Γ_0. See *Schütte* [38, 39]. I myself was fascinated by Feferman's and Schütte's work. In 1970 I obtained a Gödel functional interpretation (in Spector's extensional version) of IR and some related systems. See the abstract *Parsons* [31].

[11]I don't believe that in the earlier history of foundations predicativism was a very well-defined position. Cf. the comments on this issue in *Parsons* [34], pp. 63–65.

[12]Reference [3], p. 486. Some years later, William Tait said virtually the same thing about his own analysis of fintism. See *Tait* [42], pp. 527, 533 of original.

Although he disclaims being a predicativist, Feferman describes himself as a "sympathizer." That partly arises from a taste for using the minimal assumptions needed to arrive at desired mathematical results, as is reflected in his remark, "A little bit goes a long way," embodied in the title of [8]. I believe there is a more philosophical reason, which will be considered in Sect. 4.

Feferman [6] does the substantial work of reconstructing Hermann Weyl's attempt to rebuild classical analysis on a predicative basis. With some modification of the development in Weyl [44], he established that the basic analysis of continuous functions can be done in a second-order system with only arithmetic comprehension, thus in a rather weak subsystem of full predicative analysis (e.g. IR). Feferman does not enter far into Weyl's polemical arguments, either to criticize them or to defend them, but concentrates on the development of predicative analysis, first through analysis of Weyl's text and then through reconstructing Weyl's work in more modern terms but avoiding stronger assumptions. I will leave until Sect. 4 the question how Feferman might view the polemical arguments.

3 Constructivism and Proof Theory

As is well known, the enterprise of proof theory survived the blow to it resulting from Gödel's incompleteness theorems by abandoning the requirement that its method should be "finitary" and allowing constructive methods more liberally understood. That is the way the most significant early result of post-Gödel proof theory, Gentzen's proof of the consistency of PA, was viewed at least retrospectively. Although the dominant constructive conception at the time was intuitionism, researchers hesitated to use the full intuitionist arsenal developed by Brouwer and Heyting.[13] Gödel in [22] criticized this possibility briefly but sharply. When proof theory revived after the war, more explicit methods were used.

A second change, which Feferman has emphasized, was in the goals of proof theory. It became much less focused on proving consistency and conceived the proof-theoretical analysis of a formal theory more broadly. The pioneer of this change was Georg Kreisel, whose early writings stressed the additional information that could be obtained from proof-theoretic consistency proofs. Central to this enterprise was the characterization of the strength of theories by ordinals. Very often proof-theoretic results imply or contain conservative extension results. In addition to ordinal analysis, this is emphasized in Feferman's later surveys of proof-theoretic work.

Feferman's interest in this subject would suggest a preference for constructive methods over non-constructive. But although he has been interested in the formalization of constructive mathematics, he doesn't show a preference for constructive methods in mathematics generally. So the preference for constructivity is specific to

[13] One could ask whether this is still true of work in the 1980s and later, where work on stronger classical systems aimed at proof-theoretic reduction to Per Martin-Löf's intuitionistic theory of types, a much more powerful theory than had been envisaged in traditional intuitionism.

proof theory. It is stretched rather far in the most ambitious proof-theoretic project led by him, the volume [4], but it is stretched further in later work by others that he has followed closely and on which he has commented.

In two survey articles [5, 7] Feferman has organized the subject so that a proof-theoretical analysis reduces a theory of one type to one of another type, where the latter theory may claim to be constructive (but possibly still stretching the concept of constructivity), but the reduction itself is carried out by much weaker means, generally reducible to PRA. In important cases the interest of the reduction may depend on the constructive character of the theory to which reduction is carried out, but it is not necessary for the mathematical soundness of the proof of reduction. He is more doubtful of the value of proving consistency by constructive methods than other leading proof theorists (see [12], Sect. 4). In this context he remarks that the notions of finitist and constructive proof are vague, in particular because there are many varieties of constructivism (ibid., p. 70).

A program that Feferman originated and has pursued for many years is that of developing and analyzing formal systems of what he calls explicit mathematics. The theories he and his collaborators developed and investigated have both classical versions and constructive versions (with intuitionist logic), and the full theories are substantially stronger than predicative theories (by Feferman's own lights). However, the initial inspiration seems to have come from Feferman's study of Errett Bishop's *Foundations of Constructive Analysis* [21]. An initial aim was to develop a theory that would formalize analysis as developed by Bishop in a way faithful to his distinctive approach. The systems T_0 and T_1 that Feferman devised at the outset turned out to be proof-theoretically quite strong, as noted above. But he asserts that Bishop's actual work can be formalized in a subsystem that is conservative over PA.

Feferman has had an enduring interest in constructive methods and theories. However, although he was evidently impressed by the achievement of Bishop, he has not been especially interested in a program of reconstructing mathematics (analysis in particular) on a constructive basis. In fact he has been more interested in predicative reconstruction, although the reason for that may be that it has been pursued less by others.[14] I do not know whether he advocates what I elsewhere call a critical view of logic, according to which the application of ordinary logic, in particular the law of the excluded middle, is open to question where quantifiers range over an infinite domain.[15] He has been all but silent on the subject. However, at the very least he agrees that there are alternatives, embodied in constructive developments. As we shall see shortly, however, he is confident of the consistency of the main theories that apply classical logic over infinite domains, short of set theory involving very large cardinals.

[14] Feferman indicates that this is a reason in the Preface to [10], p. ix.

[15] See Parsons [35]. I identify as advocates of the view Brouwer, Weyl, and Hilbert.

4 Philosophical Considerations

Two theses of a philosophical character are repeated in a number of Feferman's writings. One is what he generally calls "anti-platonism." We have already discussed that in a preliminary way in Sect. 1. It was already clear at that point that we had not got to the essential point, since the forms of platonism or realism that we canvassed were ones that he either assents to or embodies in his actual practice. We will try shortly to see what the more metaphysical way in which he understands "platonism" is, and we will consider what rejection of such a view implies about one's attitude toward higher set theory.

The other is more a methodological matter and is widely scattered through his mathematical work but is emphasized most strongly in the essays in the last part of *In the Light of Logic*. That is a preference for developing branches of mathematics with minimal assumptions, to show that "a little bit goes a long way." Success in developing mathematics in this way vindicates the parsimony of assumptions that he seems to favor. Nonetheless, I will consider criticisms of that point of view.

For the present I will concentrate on anti-platonism as briefly described above. A powerful example of the opposing view can be found in the writings of Gödel, especially in the philosophical essays he wrote between the 1940s and early 1960s, either reprinted or first published in volumes II and III of his *Collected Works* (*Gödel* [23, 24]). It is tempting to conjecture that Feferman formed his anti-platonist views in reaction to Gödel's, in the wake of the study of Gödel's writings occasioned by his role as chief editor of Gödel's works. If that were so, it would be likely that Feferman's mathematical experience would already have disposed him against Gödel's version of platonism. However, although Feferman had no publications expressing his anti-platonist point of view during Gödel's lifetime, he put forward an early version of his "theses of conceptual structuralism" in a talk at Columbia University in December 1977 and embodied them in a paper, "Mathematics as objective subjectivity," which was circulated afterward but never published.[16] Some of these theses will be commented on shortly.

Elsewhere I have singled out four "elements" of Gödel's platonism: (1) Mathematics has a real content, as opposed to being tautologous or a reflection of conventions or the use of language; (2) Mathematics is inexhaustible; no formal system or definite conceptual framework can capture it completely; in particular, the concept of set is, in Michael Dummett's terms, indefinitely extensible; (3) Gödel defended set theory robustly and was not deterred by independence results from maintaining that the axioms of set theory refer to a "well-determined reality" in which statements such as CH not decided by currently available axioms are true or false; (4) concepts are

[16]While working on this paper I was not able to find my copy of that manuscript, although I was present at the talk. But see the comments below on the theses as published in [15].

also objective, as objective as sets, although Gödel did not formulate a view of the logic of concepts that he found convincing.[17]

Feferman's writings don't show that he has a problem with (1), which was anyway directed against views of the Vienna Circle that had largely lost their influence when he began to write on the philosophy of mathematics. He would have to accept a weak version of (2), forced by Gödel's incompleteness theorems. He very likely agrees that in principle set theory can be extended indefinitely by ever stronger axioms of infinity, but even in the case of axioms less strong than those at the present frontier of the subject, he has expressed skepticism about their truth, and I think he is not without doubt about the consistency of stronger axioms. (3) is where he dissents most strongly, but as we shall see his arguments are at least partly mathematical. As to (4), Feferman has not directly taken on Gödel's realism about concepts. But the"conceptual structuralism" that he has advocated means that he would not accept Gödel's view that mathematics involves concepts that are in no way human creations. I don't think he could assent to Gödel's statement in his 1951 Gibbs Lecture that "The truth, I believe, is that these concepts form an objective reality of their own, which we cannot create or change, but only perceive and describe."[18] Gödel went on to describe the "platonistic view" as including the claim that "mathematics describes a non-sensual reality, which exists independently both of the acts and the dispositions of the human mind."[19] It is quite clear that according to Gödel this reality includes concepts. And at least the part of Gödel's statement that I have quoted expresses a view that Feferman rejects in repeated statements.

On this point, Feferman emphasizes the thesis that mathematical concepts are human creations, though influenced by experience. A rationalist like Gödel might concede that there is a legitimate sense of the word 'concept' in which this is true; after all, there is a long tradition, of which Kant is a well-known instance, according to which concepts are representations, one of the tools of the mind in cognition. Consider the following thesis:

1. The basic objects of mathematical thought exist only as mental conceptions, though the source of these conceptions lies in everyday experience in manifold ways (counting, ordering, matching, combining, separating, and locating in space and time).[20]

The rationalist would hold that there is another sense of the word 'concept', according to which concepts are quite independent of the mind for their existence and essential properties and relations, although the human mind does have access to them; thus Gödel talks of "perceiving" concepts, and earlier Frege wrote of "grasping" senses that have the same objective character. Clearly Feferman holds that such a sense of 'concept' belongs to the platonist view that he rejects.

---

[17] See *Parsons* [33], pp. 97–99 of reprint. For some aspects of this view, particularly (1) (which is close to but not quite the same as Tait's default realism), 'realism' is a more appropriate term. In his discussions with Hao Wang Gödel preferred the term 'objectivism', which has not been taken up by others.

[18] *Gödel* [24], p. 320.

[19] Ibid., p. 325.

[20] Reference [15], p. 170. This is the first of the ten "theses of conceptual structuralism."

Crucial to his view of set theory is another thesis in his list:

> 5. Basic conceptions differ in their degree of clarity. One may speak of what is true in a given conception, but that notion of truth may be only partial. Truth in full is applicable only to completely clear conceptions.

With many others, Feferman finds the notion of natural number especially clear; in particular it is sufficiently clear to determine every statement in the language of first-order arithmetic as true or false, "independently of whether we can establish them one way or the other" (ibid., p. 5). But he declines to conclude that the numbers have a mind-independent existence, either as a structure or as a set of definite individuals.[21]

Many have found a difference between the clarity of the conception of natural number and that of real number and again between that and the conception of arbitrary set of real numbers. It is clear that Feferman also discerns a difference of clarity at both boundaries. Of course third-order arithmetic, which allows quantification over sets of real numbers, can state the continuum hypothesis (CH), and Feferman has argued at length that the continuum problem is not a definite problem. We shall comment on his arguments shortly.

Admitting such differences of clarity is quite compatible with a greater degree of realism than Feferman admits even at the level of the natural numbers; after all, one might hold that arbitrary sets have a mind-independent existence even though we do not have a clear enough conception of them to be certain that CH has a determinate truth value, or that it does not.[22]

Some will question whether conceptual structuralism, as outlined in the theses Feferman states, offers an adequate basis for mathematical objectivity, even at the level of the theory of natural numbers. Feferman notes that mathematics is highly intersubjective (see especially thesis 10), not only in a given culture at a particular time but over long history and between cultures, even if there is not complete understanding and agreement across those distances. He writes, "The objectivity of mathematics is a special case of intersubjective objectivity that is ubiquitous in social reality." Citing *Searle* [40], Feferman notes that there many objective facts involving social institutions that are only there by human agreement, involving institutions like property, marriage, money, and government. An example would be, "As of 2015, Charles and Marjorie Parsons own a house in Springfield, New Hampshire."

Anyone would agree that a statement like that is objective by everyday standards. Even someone from a society that has no conception of property could probably be got to understand it, at least if the teaching effort is started early enough in life. A more subtle understanding would include grasping the fact that Springfield is a

---

[21]It follows, I believe, that he would reject Michael Dummett's view that asserting bivalence of statements independently of whether we have any means of deciding their truth is a criterion of realism.

[22]Donald A. Martin is at least inclined to that view; see *Martin* [28]. However, Martin holds that if CH does not have a determinate truth-value, then a structure that would be a standard model of third-order arithmetic does not exist. In his language, the concept would not be instantiated. It is still possible to investigate mathematically what is implied by the concept. See further *Martin* [29].

town, and New Hampshire a state, of the United States. But perhaps less central than understanding is that its validity would be even more widespread.

However, is it as objective as '$7 + 5 = 12$' or the field equations of General Relativity? These statements apply to parts of the universe so remote that inhabitants of them (if they exist) could have had no idea of myself, my wife, the United States and its subdivisions, or home ownership as understood in our species. They also apply to possible evolutions of the cosmos that would not have given rise to our species at all.

However, it is not the purpose of this essay to examine whether Feferman has an adequate account of the objectivity of mathematics. However, it is relevant whether his account supports the difference he himself claims between the arithmetic of natural numbers and set theory, especially that part of set theory that depends on large cardinals. There is clearly a difference in his mind between these two levels of set theory.

It would be a matter of wide agreement that there is a difference in clarity between the conceptions of the natural numbers and that of the universe of sets. There is certainly not as much agreement as to how to understand this difference. We might begin with the weakest assumption that would justify regarding these structures as legitimate objects of mathematical investigation, the consistency of the basic theories that have been around for some time: PA for the natural numbers and ZFC for sets. Feferman does not doubt the consistency of either. He places less weight than some on the "empirical" argument that these theories have been worked with for many years by a great variety of mathematicians, and no inconsistency has been discovered. He writes:

> I, for one, have absolutely no doubt that PA and even PA_2 are consistent, and no genuine doubt that ZF is consistent, and there seems to be hardly anyone who seriously entertains such doubts. ... Namely, in the case of PA, we have an absolutely clear intuitive model in the natural numbers, which in the case of PA_2 is expanded through the notion of arbitrary subset of the natural numbers. Finally, ZF has an intuitive model in the transfinite iteration of the power set operation taken cumulatively. ([12], p. 72)

Although in this case as in others he disclaims belief in a "platonic reality" encompassing numbers and sets, he says, "I have as good a conception of what arbitrary subsets of natural numbers are *supposed* to be like as I do of the basic notions of Euclidean geometry" (ibid.) I think his confidence in the consistency of ZF would extend to extensions of it with small large cardinals, certainly inaccessibles and very likely also Mahlo cardinals.

Nonetheless, although there are brief allusions to other views that set theorists might hold, he pretty consistently associates accepting set theory with accepting a metaphysical platonistic picture, of sets as existing and having the relations they have quite independently of our own knowledge or even thought, as a reality that, in Gödel's words quoted above, "we cannot create or change but only perceive and describe."

It is hard to put one's finger on what this "independence" amounts to. Truth in general is independent of what we believe about the matter at hand or even what individuals can conceive.[23] Furthermore, set theory can be used in the description of a world in which there are no human beings to think anything. But Feferman may well agree that is already true of the natural numbers or the reals of predicative analysis. Given what he has written about the continuum problem, Feferman clearly dissents from a proposition that may not be an integral part of platonist philosophy but is certainly associated with it, namely what Gödel expressed by saying that "if the meanings of the primitive terms of set theory ... are accepted as sound, it follows that the set-theoretical concepts and theorems describe some well-determined reality, in which Cantor's conjecture must be true or false."[24]

Since researchers in set theory have different opinions about the continuum problem, in particular about whether CH is determinately true or false, it would be difficult to argue that the view of Gödel just cited is essential to the practice of set theory. One can hardly doubt that set theory is a going concern. Of course Feferman admits this; he makes it clear in his writings. He notes that it is permeated by mathematical logic, but of course the same is true of his own mathematical enterprise. One might, somewhat uncharitably, regard it as just the study of different models of ZF. Feferman's statement that he is confident that ZF is consistent does not imply more than that the arithmetic models yielded by the completeness theorem exist. But I would guess that he would concede the existence of models with standardness properties, ω-models or β-models, for example, or at least that we have a coherent enough conception of such models to reason about them. One would still have an interesting mathematical subject even if one regarded *all* large cardinal axioms as hypotheses.[25] Actual set theorists can probably be grouped on a scale between those who adopt an attitude close to this one and those who hold, or are inclined to, the Gödelian view that the axioms of set theory (including, at least up to a point, large cardinal axioms) describe a "well determined reality."

Even those who adopt the latter attitude may not hold it in the a priori form suggested by Feferman's characterization of the platonism he opposes. They may simply judge, on the basis of the course of set theoretic research, that the development of the subject tends toward unity. In particular, as regards CH, they see some promise in programs that aim to decide it but regard even the question whether CH has a determinate truth-value as open.

Feferman sometimes expresses the attitude of the "ordinary" mathematician, not a logician, and asks whether certain propositions shown to be independent of usual axiom systems have "mathematical interest" independent of the logical result that they are invented to establish. The issues are discussed at length in Chap. 12 of [10]

---

[23]In the case of institution-dependent facts such as my example of home ownership, it is not independent of what *anyone* thinks. Would it still be true that we own the house in Springfield if it came to be widely believed, in particular by those in authority, that the records in the Registry of Deeds of Sullivan County, New Hampshire, are meaningless pieces of paper or records of a hopelessly distant past? (We bought the house in 1973.)

[24]*Gödel* [23], p. 260, from the 1964 version of "What is Cantor's continuum problem?".

[25]An even more radical view in this direction is expressed in *Hamkins* [26].

(pp. 237–43) and taken up again in [11, 13]. He expresses some skepticism about what is shown by certain results showing independence of mathematical propositions of PA or stronger systems, well into set theory. Reference [11] and the symposium [13] that it gave rise to bear the implicitly anti-Gödelian title "Does mathematics need new axioms?" He goes over a number of these examples and has different views in different cases about whether they are "of mathematical interest." A question he raises is whether they can be proved by ordinary mathematical reasoning or only through metamathematics.

He comments that no previously formulated open problem is known to be independent of standard formal systems, not even PA ([11], p. 108). In a very famous case, Fermat's Last Theorem, I believe that the verdict is still out. *McLarty* [30] argues that it would require substantial further work in number theory to show that the theorem is provable in PA or a conservative extension of it. McLarty observes that the actual proof uses Grothendieck universes, which would amount to assuming inaccessible cardinals. It is apparently not an especially difficult problem to eliminate that assumption, so that the proof is given in ZFC. According to McLarty, universes were used in some other proofs of important results in number theory. He believes that an "unwinding" that would not be especially deep mathematically would suffice to yield a proof that could be formalized in something like 7th or 8th order arithmetic. I think his view is that it is virtually inconceivable that a proof of the theorem could have been *discovered* in our time by someone working with those means, still less with PA, even if it should be shown in the end that the theorem is provable in PA.

If McLarty is right, the work in algebraic geometry and number theory deriving from Grothendieck's ideas would be a much more sustained and powerful instance of something Feferman acknowledges in a field in which he has been directly involved: the proof-theoretical analysis of stronger systems of second-order arithmetic, where uncountable ordinals and even inaccessibles are used as heuristic tools to formulate well-orderings that are in the end recursive. The examples of Fermat's Last Theorem and (apparently) of others in algebraic geometry and number theory seem to show that, in addition to the logician's question of how powerful the means are that are *necessary* for a proof of the theorem, it is natural to ask what means are most conducive to the discovery of the theorem and to the development of the field. Those means may include assumptions that can be eliminated after the fact, but sometimes with non-trivial further work.

There is another issue of "independence" in Gödel's writings that is relevant to Feferman's anti-platonism. Gödel talks of sets and concepts as "existing independently of our definitions and constructions."[26] Gödel has in mind principally Russell's ramified theory of types, which without the axiom of reducibility would avoid commitment to propositional functions (and therefore classes) that are not definable. The demand that sets be definable in a non-circular way is a natural one that was voiced by prominent mathematicians in the early twentieth century. In Weyl's predicative

[26] *Gödel* [23], p. 128 (from "Russell's mathematical logic").

construction of analysis, for example, this condition is easily satisfied, since sets are defined as extensions of arithmetical predicates. However, it is still satisfied as one ascends into ramified analysis, since although sets may be defined in a way that involves quantification over sets, that can always be understood as quantification over sets defined at an earlier stage.[27] Weyl's arguments in [44, 45] that standard arguments in analysis involve a vicious circle presuppose the conception of sets as extensions of predicates or properties that are close shadows of predicates. In *Parsons* [32], Sect. 1, I maintain that if one presupposes this conception, these arguments are defensible.[28]

Above I remarked that Feferman does not engage Weyl's polemical arguments. He expresses sympathy with the view of sets that underlies Weyl's arguments (e.g. [10], p. 54). However, he is evidently not convinced by them, although he may believe that the platonism that he rejects is the only alternative. Perhaps he is also too much of a structuralist to put very much weight on a conception of what sets *are*.[29]

I shall make only a few remarks about the other aspect of Feferman's anti-platonism, his undertaking to develop parts of mathematics with assumptions of minimal strength and his emphasizing the maxim, "A little bit goes a long way." He has argued in Chaps. 12–14 of [10] that all of "scientifically applicable" mathematics can be developed in a system W that is a conservative extension of PA. That all scientifically *applicable* mathematics can be developed in this system is a conjecture, but he does make a case for the claim that this is true for all mathematics actually applied in science, with a small number of exceptions which he considers marginal. This work converges with that in the "reverse mathematics" program originated by Harvey Friedman and pursued energetically by Stephen Simpson and his students.[30] However, we should qualify the meaning of the conjecture along the lines suggested above in connection with McLarty's discussion of Fermat's Last Theorem. What has been shown in the analysis of the actual mathematics used is that it *can* be formalized in W or some system of comparable strength. Much of it is presented in the instruction given to scientists and in the relevant textbooks in a way that does not emphasize axiomatic economy, and that is probably not what the discoverers of the relevant theorems had primarily in view. So the economy is something that is achieved after the mathematics and a lot of relevant, mostly later, mathematical logic have been discovered.

[27] Of course if the ascent is into the transfinite, the issues arise that are addressed by the autonomy condition on ordinal levels in Feferman's own analysis.

[28] This is quite in accord with Gödel's classic analysis of Russell's vicious circle principle.

[29] I am not confident enough in my understanding of Feferman's systems of explicit mathematics to say whether they are naturally interpreted so that all sets are definable. Evidently this could be so only with the help of inductively defined predicates.

[30] See *Simpson* [41].

5 Concluding Remarks

Feferman has written in a number of places that he considers the continuum problem "inherently vague"; he evidently does not expect the efforts being made, in particular by W. Hugh Woodin, to lead to a generally accepted solution. Recently he has written two versions of a paper on the subject, both of which he has made available on his web site. The titles are revealing; the first version, the text for his actual lecture on the subject at Harvard University in 2011, is called "Is the continuum problem a definite mathematical problem?" The revised version, [18], is entitled "The continuum problem is neither a definite mathematical problem nor a definite logical problem." That would suggest that the reaction to his lecture and subsequent critical comments by Peter Koellner have worked more to reinforce his conviction than to undermine it. However, what interests me most is another difference between the two papers. The first goes into more general matters: Feferman's own conceptual structuralism and the reason why the continuum problem is not on the list of Millenium Problems, a list of outstanding unsolved problems put forth by the Clay Mathematics Institute. The revised version concentrates much more on mathematical arguments, reasons for doubting that the results achieved in the existing programs will prove to be definitive rather than still leaving equally persuasive alternatives with different verdicts about the power of the continuum.

I am not qualified to assess these arguments. The point of bringing up the difference between the two versions of Feferman's paper is to remind the reader of something about their author: He is first of all a mathematician. I find persuasive the conjecture that his philosophy has followed his mathematics: In the course of his work as a mathematical logician, always having in view other parts of mathematics, what has been paramount for him is what is mathematically of interest and what leads to definite and fruitful results. Central to his mathematical interests have been proof theory, constructivity, and predicativity and relatively economical extensions of the latter. Although he is informed about set theory, he is not a set theorist or, as we have seen, gripped by set theory as *mathematics*. This may be surprising in a student of Alfred Tarski. In some autobiographical remarks, he has told us how certain accidents enabled him early on to develop mathematical interests that were not especially Tarskian.[31]

The aim of this essay has been to understand Feferman's skepticism about set theory and not to defend or criticize it. But I do want to mention one criticism of his stance, which affects more his quest for minimal axiomatic assumptions than his anti-platonism directly. Hilary Putnam writes,

> Let us turn now to the claim that predicative set theory is adequate to the needs of physics. I would not dare to challenge Solomon Feferman's claim ... to have shown that the theorems that are needed in the "applications" of physical science can all be derived within predicative mathematics. What I am skeptical about is whether the predicative mathematician can answer the difficulty that I have raised as an objection to Wittgenstein's finitism. The statement

[31] Of great importance was Feferman's work with Leon Henkin in 1955–56, when Tarski was away. See his [9, 16], as well as his autobiography in this volume.

> "There is a point corresponding to every triple of real numbers" (alternatively, "There is a sphere the coordinates of whose center are any given triple of real numbers"), is not, as far as I can see, *expressible* without quantifying over *all* triples of real numbers, which is just what predicative analysis forbids! (*Putnam* [36], pp. 200–01.)

In remarks Putnam goes on to make, he gives the impression that the predicative system claimed to be adequate for the mathematics applied in physics will be ramified, so that the object language will talk only of real numbers of one or another order. But of course the system W that Feferman claims to be adequate for the mathematics applied in science is unramified, so that so far as the system goes, one can talk of all real numbers, or all triples of real numbers. However, that reply is rather superficial, since even within predicative constraints one can introduce more real numbers than W allows, and this process can be repeated transfinitely.

I don't think Feferman has yet commented on this objection of Putnam. But we do find the following remark:

> Alternatively, one might reply that the continuum has *physical existence* in space and/or time. But then one must ask whether the mathematical structure of the real number system can be identified with the physical structure, or whether it is instead simply an idealized mathematical model of the latter, much as the laws of physics formulated in mathematical terms are highly idealized models of aspects of physical reality. ([11], p. 107)

I don't have the knowledge of physics necessary to sort out the implicit controversy between Feferman and Putnam. However, I will observe that Putnam has frequently proclaimed that he is a scientific realist. Feferman seems not to stand on that ground, although he is somewhat reticent about issues in the philosophy of science. I think it likely that Feferman's position is not compatible with scientific realism as Putnam understands it. At any rate, I doubt that he and Putnam could easily come to an understanding about these issues.

Solomon Feferman has built up a very impressive legacy of research in the foundations of mathematics, first of all mathematical but also philosophical and historical. His skepticism about higher set theory would not be widely shared in the logical community, but it has probably been extremely fruitful for Feferman himself and his many collaborators, and even for those who disagree with him.

Let me conclude by saying briefly why I do not share Feferman's attitude, in spite of my very great respect for him and my own early background in constructive proof theory. John Steel describes the aim of exploring proposed new axioms for set theory as to "maximize interpretative power."[32] That would be a more precise and modern version of Georg Cantor's claim that the essence of mathematics is its freedom. Set theory explores possibilities that were unimaginable before Cantor, and its higher reaches only came to the consciousness of mathematicians much more recently. Much research in set theory since the 1960s has taken a direction opposite to Feferman's quest for the most axiomatically economical foundations. Rather, it has sought to probe the limits of coherent mathematical thought and to see to what extent unity is maintained when one attempts to scale such heights. Although I can

---

[32] See p. 423 of his contribution to *Feferman et al.* [13].

only be a philosophical spectator of this enterprise, I continue to find that it offers deep insights into the possibilities of mathematical thought, and I don't expect my attitude to be changed even if Feferman should be vindicated in his expectation that no widely accepted solution to the continuum problem will emerge.

There are other reasons for embracing set theory. I have tried to say briefly why it engages me as a philosopher.

References

1. Feferman, S.: Systems of predicative analysis. J. Symb. Log. **29**, 1–30 (1964)
2. Feferman, S.: Some applications of forcing and generic sets. Fundam. Math. **56**, 325–345 (1964a)
3. Feferman, S.: Predicative provability in set theory. Bull. Am. Math. Soc. **72**, 486–489 (1966)
4. Feferman, S.: (With Wilfried Buchholz, Wolfram Pohlers, and Wilfried Sieg.) Iterated Inductive Definitions and Subsysems of Analysis. Lecture Notes in Mathematics, vol. 897. Springer, Berlin (1981)
5. Feferman, S.: Hilbert's program relativized: Proof-theoretical and foundational reductions. J. Symb. Log. **53**, 364–384 (1988)
6. Feferman, S.: Weyl vindicated: Das Kontinuum 70 years later. In Carlo Celluci and Giovanni Sambin (eds.), Temi e prospettive della logica e della filosofia della scienza contemporanee, vol. I, pp. 59–93. Bologna: CLUEB (1998). Reprinted with minor corrections and additions as chapter 13 of [10]
7. Feferman, S.: What rests on what? The proof-theoretic analysis of mathematics. In Proceedings of the 15th International Wittgenstein Symposium: Philosophy of Mathematics, vol. 1, pp. 147–171. Vienna: Hölder-Pichler-Tempsky (1998). Reprinted as chapter 13 of [10]
8. Feferman, S.: Why a little bit goes a long way. Logical foundations of scientifically applicable mathematics. In PSA 1992, vol. 2, pp. 442–455. East Lansing, Michigan: Philosophy of Science Association (1998). Reprinted as chapter 13 of [10]
9. Feferman, S.: My route to arithmetization. Theoria **63**, 168–181 (1997)
10. Feferman, S.: In the Light of Logic. New York and Oxford: Oxford University Press (1998)
11. Feferman, S.: Does mathematics need new axioms? Am. Math. Mon. **106**, 99–111 (1999)
12. Feferman, S.: Does reductive proof theory have a viable rationale? Erkenntnis **53**, 63–96 (2000)
13. Feferman, S.: (With Penelope Maddy, Harvey Friedman, and John R. Steel.) Does mathematics need new axioms? Bull. Symb. Log. **6**, 401–446 (2000a)
14. Feferman, S.: Comments on, Predicativity as a philosophical position, by G. Hellman. Revue internationale de philosophie **58**, 313–323 (2004)
15. Feferman, S.: Conceptions of the continuum. Intellectica **51**, 169–189 (2009)
16. Feferman, S.: A fortuitous year with Leon Henkin. In María Manzano, Ildikó Sain, and Enrique Alonso (eds.), The Life and Work of Leon Henkin: Essays on his Contributions, pp. 35–40. Birkhäuser (2014)
17. Feferman, S.: Parsons and I: Sympathies and differences. J. Philos. **113**, 234–246 (2016)
18. Feferman, S.: The Continuum Hypothesis is neither a definite mathematical problem nor a definite logical problem. Forthcoming in Koellner [27]

Other Writings

19. Bernays, P.: Die Philosophie der Mathematik und die Hilbertsche Beweistheorie. Blätter für deutsche Philosophie **4**, 326–367 (1930). Reprinted with Postscript in Abhandlungen zur.

Philosophie der Mathematik (Darmstadt: Wissenschaftliche Buchgesellschaft, 1976), pp. 17–61
20. Bernays, P.: Sur le platonisme dans les mathématiques. *L'enseignement mathématique* **34**, 52–69 (1935)
21. Bishop, E.: Foundations of Constructive Analysis. McGraw-Hill, New York (1967)
22. Gödel, K.: Vortrag bei Zilsel. Shorthand notes for a lecture to an informal seminar in Vienna (1938). Transcription and translation in Gödel [24]
23. Gödel, K.: In: Feferman, S., Dawson Jr., J.W., Kleene, S.C., Moore, G.H., Solovay, R.M., van Heijenoort, J. (eds.) Collected Works, Volume II: Publications 1938–1974. Oxford University Press, New York (1990)
24. Gödel, K.: In: Feferman, S., Dawson Jr., J.W., Goldfarb, W., Parsons, C., Solovay, R.M. (eds.) Collected Works, Volume III: Unpublished Essays and Lectures. Oxford University Press, New York (1995)
25. Gödel, K.: In: Feferman, S., Dawson Jr., J.W., Goldfarb, W., Parsons, C., Sieg, W. (eds.) Collected Works, Volume IV: Correspondence A-G. Clarendon Press, Oxford (2003)
26. Hamkins, J.D.: The set-theoretic multiverse. Rev. Symb. Log. **5**, 416–449 (2012)
27. Koellner, P. (ed.): Exploring the Frontiers of Incompleteness. Forthcoming (201?)
28. Martin, D.A.: Gödel's conceptual realism. Bull. Symb. Log. **11**, 207–224 (2005)
29. Martin, D.A.: Completeness and incompleteness of basic mathematical concepts. Forthcoming in Koellner [27]
30. McLarty, C.: What does it take to prove Fermat's last theorem? Grothendieck and the logic of number theory. Bull. Symb. Log. **16**, 359–377 (2010)
31. Parsons, C.: Gödel-spector interpretation of predicative analysis and related systems (abstract). J. Symb. Log. **36**, 707 (1971)
32. Parsons, C.: Realism and the debate on impredicativity 1917–1944. In: Sieg, W., Sommer, R., Talcott, C. (eds.) Reflections on the Foundations of Mathematics: Essays in honor of Solomon Feferman. Lecture Notes in Logic, vol. 15, pp. 372–389 (2002). Association for Symbolic Logic and A. K. Peters (Natick, Mass.) Reprinted with Postscript in Parsons [34]
33. Parsons, C.: Gödel, Kurt Friedrich (1906–1978). In: Shook, J.R. (ed.) Dictionary of Modern American Philosophers, vol. 2, pp. 940–946. Thoemmes Press, Bristol (2005). Reprinted with editorial changes as "Kurt Gödel" in Parsons [34]
34. Parsons, C.: Philosophy of Mathematics in the Twentieth Century. Selected Essays. Harvard University Press, Cambridge (2014)
35. Parsons, C.: Infinity and a critical view of logic. Inquiry **58**, 1–19 (2015)
36. Putnam, H.: Indispensability arguments in the philosophy of mathematics. In: De Caro, M., Macarthur, D. (eds.) Philosophy in the Age of Science. Physics, Mathematics, and Skepticism, pp. 181–201. Harvard University Press, Cambridge (2012)
37. Putnam, H.: Reply to Charles Parsons. In: Auxier, R.E., Anderson, D.R., Hahn, L.E. (eds.) The Philosophy of Hilary Putnam. The Library of Living Philosophers, vol. 34, pp. 134–143. Open Court, Chicago (2015)
38. Schütte, K.: Predicative well-orderings. In: Crossley, J.N., Dummett, M.A.E. (eds.) Formal Systems and Recursive Functions, pp. 280–303. North-Holland, Amsterdam (1965)
39. Schütte, K.: Eine Grenze für die Beweisbarkeit der transfiniten Induktion in der verzweigten Typenlogik. Archiv für mathematische Logik und Grundlagenforschung **7**, 45–60 (1965a)
40. Searle, J.R.: The Construction of Social Reality. The Free Press/Macmillan, New York (1995)
41. Simpson, S.G.: Subsystems of Second-Order Arithmetic. Springer, Berlin (1999), 2nd edn., Association for Symbolic Logic and Cambridge University Press, Cambridge (2009)
42. Tait, W.W.: Finitism. J. Philos. **78**, 524–546 (1981). Reprinted in Tait 2005
43. Tait, W.W.: The Provenance of Pure Reason. Essays on the Philosophy of Mathematics and its History. Oxford University Press, New York (2005)
44. Weyl, H.: Das Kontinuum. Kritische Untersuchungen über die Grundlagen der Analysis. Leipzig, Veit (1918)
45. Weyl, H.: Der circulus vitiosus in der heutigen Begründung der Analysis. Jahresbericht der Deutschen Mathematiker-Vereinigung **28**, 85–92 (1919)

Author Biography

Charles Parsons is Edgar Pierce Professor of Philosophy, Emeritus, at Harvard University and was for many years on the faculty of Columbia University. He was an editor, with Solomon Feferman and others, of Kurt Gödel's posthumous works (volumes III–V of the Collected Works). He is author of *Mathematics in Philosophy* (1983) and *Mathematical Thought and its Objects* (2008) and has published articles on proof theory, philosophy of mathematics, and historical figures such as Kant, Hilbert, Bernays, and Gödel.

Name Index

G. Jäger and W. Sieg (eds.), *Feferman on Foundations*, Outstanding Contributions to Logic 13, DOI 10.1007/978-3-319-63334-3

T

V

W